Modeling and Managing Interdependent Complex Systems of Systems

MODELING AND MANAGING INTERDEPENDENT COMPLEX SYSTEMS OF SYSTEMS

Yacov Y. Haimes

WILEY

IEEE PRESS

This edition first published 2019

Registered Office
John Wiley & Sons, Inc., 111 River Street, Hoboken, NJ 07030, USA

Editorial Office
111 River Street, Hoboken, NJ 07030, USA

For details of our global editorial offices, customer services, and more information about Wiley products visit us at www.wiley.com.

Wiley also publishes its books in a variety of electronic formats and by print-on-demand. Some content that appears in standard print versions of this book may not be available in other formats.

Library of Congress Cataloging-in-Publication Data

Names: Haimes, Yacov Y., author.
Title: Modeling and managing interdependent complex systems of systems /
by Yacov Y. Haimes.
Description: Hoboken, NJ : John Wiley & Sons, 2018. | Includes
bibliographical references and index. |
Identifiers: LCCN 2018000550 (print) | LCCN 2018009974 (ebook) | ISBN
9781119173700 (pdf) | ISBN 9781119173694 (epub) | ISBN 9781119173656
(cloth)
Subjects: LCSH: Systems engineering. | System analysis.
Classification: LCC TA168 (ebook) | LCC TA168 .H28 2018 (print) |
DDC 003–dc23
LC record available at https://lccn.loc.gov/2018000550

Cover Design: Wiley
Cover Image: © Digital_Art/Shutterstock

Set in 10.5/13.5pt Times Ten by SPi Global, Pondicherry, India

Printed in the United States of America

V10007110_011519

Contents

Foreword

PHILOSOPHICAL AND HISTORICAL PERSPECTIVES ON UNDERSTANDING COMMONALITIES CHARACTERIZING COMPLEXITY

The growing interest by the systems modeling community in the concept and in the literature on complexity deserves a fresh reflection on its essence and on its evolving definitions and characterizations. For systems modelers, the starting point begins by focusing on what constitutes complexity and how to understand, model, and manage it. The English language fails to provide a succinct definition of the term complexity in one short or long sentence. This is because each of the two words – "modeling and managing" – used in the title of this book has multiple connotations, interpretations, and associations of the term complexity depending on the individuals using the terms and the specific context in which they are used.

We define and model complexity in this book via the interdependencies and interconnectedness (I-I) characterizing complex systems of systems (SoS) (Complex SoS). We further model and quantify the I-I by building on the shared/common states and other essential entities (shared decisions, resources, functions, policies, decision makers, stakeholders, and organizational setups) within and among the subsystems that, in their totality, constitute Complex SoS. Indeed, the above, along with hierarchical decomposition and higher-level coordination, encompass the essence of the modeling, theory, methodology, and practice espoused in this book. We build on the fact that all outputs from a system are functions of the states of that system and thus also of the decisions and all other inputs to the system. This fact is of particular significance to modeling Complex SoS. For example, Chen (2012) offers the following succinct definition of *state variable*: "The state $x(t_0)$ of a system at time t_0 is the information at t_0 that together with the input $u(t)$, for $t \geq t_0$, determines uniquely the output $y(t)$ for all $t \geq t_0$."

Indeed, the states of a system are commonly a multidimensional vector that characterizes the system as a whole and plays a major role in estimating its future behavior for any given input. Thus, (i) the behavior of the states of the system as a function of time enables modelers to determine, under certain conditions, the system's future behavior for any given input, or initiating event – and (ii) the shared states and other essential entities within and among the subsystems and systems constitute the essence of the multifarious attributes of the I-I characterizing Complex SoS.

Thus, in modeling Complex SoS, we exploit the I-I characterizing Complex SoS that are manifested via shared states and other essential entities in multiple ways. The following sample of modeling methodologies beyond Chapter 1 includes (i) decomposition and multilevel-hierarchical coordination (Chapters 2 and 4) with a primer on modeling risk and uncertainty in Part II of Chapter 2; (ii) hierarchical holographic modeling (HHM) (Chapter 3); (iii) multiple conflicting, competing, and noncommensurate goals and objectives and the associated tradeoffs (Chapter 5); (iv) hierarchical coordinated Bayesian modeling of Complex SoS (Chapter 6); (v) hierarchical-multiobjective modeling and decision making of Complex SoS (Chapter 7); (vi) modeling economic interdependencies among Complex SoS (Chapter 8); (vii) guiding principles for modeling and managing Complex SoS (Chapter 9); (viii) modeling cyber–physical Complex SoS – four case studies (Chapter 10); (ix) global supply chain as Complex SoS (Chapter 11); (x) understanding and managing the organizational dimension of Complex SoS (Chapter 12); (xi) software engineering – the driver of cyber–physical Complex SoS (Chapter 13); (xii) infrastructure preparedness for communities as Complex SoS (Chapter 14); and (xiii) modeling safety of highway Complex SoS via fault trees (Chapter 15).

Throughout this book, we introduce the reader, via examples and case studies, to decomposition, hierarchical modeling, multilevel decision making, and optimization and to multiobjective tradeoff analyses. Decomposition is employed to decouple the I-I characterizing Complex SoS. We postulate that decisions made at the subsystem's lower levels of the hierarchy can serve as a pretext that they are "independent." The discrepancies and conflicts, fundamental differences, and the associated tradeoffs are harmonized at the highest levels of the model's hierarchical decision-making process.

Starting in the 1960s, many scholars aimed at identifying the fundamental commonalities that characterize modeling and managing Complex SoS. Most of the theory and methodology that were developed employed decomposition using pseudo-variables at the lower levels of the hierarchical models and were ultimately harmonized at a higher level of the hierarchy. Over the years, we continued to study and improve our modeling perspectives supported by new tools and methodologies that led to a better understanding and more useful modeling of the I-I that constitute Complex SoS. In the past, modeling the I-I was directed at the coupled decisions and decision makers that characterized Complex SoS. This was mostly achieved by the deployment of pseudo-variables, which enabled the reliance on decomposition at lower levels of the hierarchy, and a higher-level hierarchical coordination of tightly interdependent and interconnected systems and subsystems.

Previous methods developed for modeling Complex SoS were aimed at advancing theory and methodology for uncoupling the I-I that characterize them. In this book, we will also study and identify interdependencies and interconnections by seeking a better comprehension of their essence and their dominant contributions to the complexity of SoS. We address this challenge by *identifying* the I-I of Complex SoS manifested via shared states and other essential entities. We also embrace the fact that all outputs from a system are functions of the states of that system and the latter are functions of all decisions and all inputs to the system. This notion is also of particular significance and central to modeling Complex SoS. For example, to determine the reliability and functionality of a car, one must know the *states* of the fuel, oil, tire pressure, and other mechanical and electrical systems. All systems are characterized at any moment by their respective states and the conditions thereof, and these conditions are subject to continuous variation and fluctuation. Similarly, the states of health of a human are multifaceted, including blood composition and pressure, among myriad others, and the I-I that exist among the states of biological systems.

The time frame has always been recognized as a major driver of what we term complexity. This is due to the fact that all systems continue to evolve, emerge, and thus change, while the capability of our modeling tools to keep pace with these changes continues to lag behind. Our inability to model the dynamic changes that characterize Complex SoS remains an impediment that characterizes and impairs our modeling and managing the I-I characterizing Complex SoS. We embrace the fact that complexities cannot, by their essence and definition, be compounded, packaged, understood, or modeled via one "straightjacket" modeling schema. Rather, we have to keep building on what we have learned from past contributions developed by other scholars, researchers, and practitioners, and augment this past knowledge into our current thinking, thereby creating new and improved theories and methodologies. Furthermore, seeking to discover what makes the I-I of Complex SoS so difficult to model will ultimately help us better manage them. This is not a fatalistic view of modeling complexity, rather a sober understanding of the reality characterizing Complex SoS.

COMPLEXITY, INTERDEPENDENCY, INTERCONNECTEDNESS, AND REINVENTION OF FAULT TREES

For decades engineers and scientists have explored the modeling power of fault trees in their quest to study and discover connections between two or among several systems that may lead to catastrophic failure of safety-critical systems. The fundamental difference characterizing the previous use of fault trees and our present reinvention stems from the basic characteristics of the two approaches. In this book we investigate and identify the genesis of the I-I by exploring the *shared/common* states and other essential entities within the systems and subsystems that comprise Complex SoS. By doing so, we also discover and quantify the genesis of potential failure of the entire Complex SoS, whether the interdependencies and interconnections are manifested by connections in series and/or in parallel. In this book we also benefit from decades of experience that engineers and scientists have gained from the intrinsic power of fault trees. Furthermore, to model and improve our understanding of the I-I that characterize Complex SoS, we have reinvented the use of fault trees via an innovative interpretation of the contributions that they offer systems modelers. We further exploit the I-I characterizing Complex SoS by tracing (via fault trees) prospective and inevitable failures due to their inherent specific connections via shared states and/or other shared essential entities. This process enables us to determine early in the modeling cycle "what not to do" during planning, design, and future decision making. By investigating the essence of the I-I characterizing Complex SoS, we can discover future failures that could be avoided. *In the parlance of fault-tree analysis, the shared states and other essential entities are translated into systems connected in series or in parallel, rather than being seen as completely independent.*

There exists an insightful correlation and lesson to be learned from the spread of disease in the human body due to the I-I that are enabled by the continuous flow of blood nourishing every cell (subsystem) of every organ (system) and ultimately of the entire body as Complex SoS. Similarly, all cyber–physical infrastructures are, in their essence, Complex SoS, and their modeling, understanding, and management can be characterized by using their shared states and other essential entities (e.g. communication channels, decisions, decision makers, resources, and organizational setups). Our ability to observe, study, and learn from the behavior of the animal kingdom as Complex SoS and develop knowledge based on lessons learned have been central to the insight from which we benefit today.

Although the above observations, as well as the theoretical discoveries, seem obvious to us now, they do shed light on, and provide insightful understanding of, the genesis of the I-I characterizing both living entities and cyber–physical Complex SoS. This finding constitutes another building block in the repertoire of the theory, methodologies, and tools that enable modelers of Complex SoS to gain invaluable insight into deciphering the genesis of the I-I that characterize Complex SoS.

Consider the nearly two-decade-old perspectives on complexity offered by scholars in the 1999 Special Issue of the journal *Science:*

Goldenfeld and Kadanoff (Science, 1999, p. 87) state:

> "To us, complexity means that we have structure and variations. Thus, a living organism is complex because it has many different working parts, each formed by variations in the working out of the same genetic coding…a complex world is interesting because it is highly structured. A chaotic world is interesting because we do not know what is coming next."

Whitesides and Ismagilov (Science, 1999, p. 89) state:

> "Complexity is a word rich with ambiguity and highly dependent on context (citing Mainzer, 1977). Chemistry has its own understanding of this word. In one

characterization, a complex system is one whose evolution is very sensitive to initial conditions or to small perturbations, one in which the number of independent interacting components is large, or one in which there are multiple pathways by which the system can evolve."

Weng et al. (Science, 1999, p. 92) state:

"Biological signaling pathways interact with one another to form complex networks. Complexity arises from the large number of components, many with isoforms that have partially overlapping functions; from the connections among components; and from the spatial relationship between components. The origins of the complex behavior of signaling networks and analytical approaches to deal with the emergent complexity are discussed here."

Koch and Laurent (Science, 1999, p. 96) state:

"Advances in the neurosciences have revealed the staggering complexity of even 'simple' nervous systems. This is reflected in their function, their evolutionary history, their structure, and the coding schemes they use to represent information. These four viewpoints need all play a role in any future science of 'brain complexity.'"

Parrish and Edelstein-Keshet (Science, 1999, p. 99) state:

"One of the most striking patterns in biology is the formation of animal aggregations. Classically, aggregation has been viewed as an evolutionarily advantageous state, in which members derive the benefits of protection, mate choice, and centralized information, balanced by the costs of limiting resources. Consisting of individual members, aggregations nevertheless function as an integrated whole, displaying a complex set of behaviors not possible at the level of the individual organism. Complexity theory indicates that large populations of [biological] units can self-organize into aggregations that generate pattern, store information, and engage in collective decision making. This begs the question, are all emergent properties of animal aggregations functional or are some simply pattern? Solutions to this dilemma will necessitate a closer marriage of theoretical and modeling studies linked to empirical work addressing the choices, and trajectories, of individuals constrained by membership in the group."

In September 1999, the author of this book organized a three-month-long seminar series and invited nine experts on complexity and complex systems to participate in and contribute to it. The following themes were presented and discussed:

- *Modeling Risk in Infrastructures of Large-Scale Complex Systems*
- Yacov Y. Haimes
- *Adaptive Complexity Theory and the Engineering and Management of Large Systems*
- Andrew P. Sage
- *What is Complexity and What Can Models Tell Us About It?*
- Mitch Waldrop
- *Life Beyond Chaos: Non-linear Dynamics in Ecology*
- Carl Zimmer
- *Autonomous Control of Complex Systems*
- Mohammed Jamshidi
- *Origins of Complexity in Cell Signaling Networks*
- Ravi Iyengar
- *Complexity and Critical Infrastructures*
- Steven Rinaldi
- *Complexity in Optimization*
- Leon Lasdon
- *Understanding and Managing Complex Systems*
- Mihajlo Mesarovic
- *Patterns in Nature: The Epiphenomenology of Aggregation*
- Julia Parrish
- Epilogue

In sum, the multifarious interpretations of complexity by the many scholars, who have studied it during the last several decades, attest to its intricacy and the associated challenges it affords to its modeling. This book is dedicated to the understanding of complexity through the discovery of its specific attributes, thereby enhancing our ability to effectively manage complexity with new analytical models and help us improve our understanding and management of Complex SoS. For pedagogical purpose, concepts, theory and methodologies are introduced throughout the book via case studies, including transportation, cyber-physical infrastructure, bridges, software engineering, electricity, communications, water resources, and others.

REFERENCES

Chen, C. (2012). *Linear System Theory and Design*, 4e. New York: Oxford University Press.

Goldenfeld, N. and Kandanoff, L. (1999). Simple lessons from complexity. *Science* **284** (5411): 87–89.

Koch, C. and Laurent, G. (1999). Complexity and the nervous system. *Science* **284** (5411): 96–98.

Parrish, J. and Edelstein-Keshet, L. (1999). Complexity, pattern, and evolutionary trade-offs in animal aggregation. *Science* **284** (5411): 99–101.

Weng, G., Bhalla, U., and Iyengar, R. (1999). Complexity in biological signaling systems. *Science* **284** (5411): 92–96.

Whitesides, G. and Ismagilov, R. (1999). Complexity in chemistry. *Science* **284** (5411): 89–92.

Acknowledgments

Writing this acknowledgment is probably one of the most rewarding moments in the preparation of this book on complexity, because each of the individuals cited here played some significant role during what might be viewed as the "life cycle" of this project. Even with a careful accounting, there will likely be some individuals who have been inadvertently missed. A great sage once said: "From all my teachers I learned and became an educated person, but my students contributed the most to my knowledge and wisdom." This statement epitomizes the gratitude that I owe to more than 120 of my doctoral and masters students whom I have had the privilege of serving as their theses advisor and from whom I learned the most.

This book on complexity was made possible through the generous support and technical help of many individuals to whom I owe heartfelt gratitude. My long-term professional collaboration with Duan Li, Kenneth Crowther, Zhenyu Guo, Eva Andrijcic, Joost Santos, Vira Chankong, Zhenyu Yan, Joshua Bogdanor, and Bryan Lewis and numerous other graduate students, and the many papers that we published together during the more than four decades, have had a major impact on the scope and contents of this book. I will always cherish their contributions to my professional growth. I also want to acknowledge my current colleagues at the University of Virginia, Jim Lambert and Barry Horowitz, for our daily conversations and association.

The painstaking and judicious technical editorial work of Pat Levine is most appreciated and heartily acknowledged. I would like to thank undergraduate students Madeleine Fleshman, Claire Trevisan, and Tyler Brown who labored long hours converting and retyping the text and modifying the figures and tables. Material from papers published jointly with several of my colleagues and graduate students have been incorporated into this book. These colleagues are Clyde Chittister, Duan Li, Kenneth Crowther, Barry Horowitz, Jim Lambert, Zhenyu Guo, Joost Santos, Kash Barker, Steve Chase, Andy Anderegg, and Keith Hipel. The seminal works by Professor Hipel on conflict resolution associated with complex systems of systems have made their marks in the systems engineering field as well as in this book.

I am especially grateful to Rosemary Shaw, who, in addition to managing the Center for Risk Management of Engineering Systems, has worked tirelessly by my side with an abundance of grace and enthusiasm to bring this book to publication.

I am most appreciative and grateful to my editor Brett Kurzman and Victoria Bradshaw at Wiley, USA for their continued support and encouragement.

Special thanks to the Wiley UK and India production team of Kshitija Iyer and Vishnu Priya for their expert production and tireless dedication. I am also thankful to the publishers who granted me and Wiley and the other publishers' permission to reproduce material from published in their journal articles.

I thank my wife Sonia for her constant encouragement and loving support throughout the demanding time and commitment to bringing this book to fruition.

I dedicate this book to my wife Sonia and to Rosemary, the Center's Manager.

1

Modeling and Managing Interdependent Complex Systems of Systems: Fundamentals, Theory and Methodology

PART I: AN OVERVIEW

1.I.1 INTRODUCTION

What does it mean to label systems as interdependent and interconnected complex systems of systems (Complex SoS)? Do we measure their complexity in terms of their subsystems' multiple attributes and perspectives, their functionalities and resources, the number of shared states and decisions, resources, decision makers, and stakeholders, or in terms of their culture and organizational structure, etc.? Modeling is an amalgamation or symbiosis of the arts and the sciences. As the artist reconstructs images and ideas, scenes, people, and structures, so do the modelers of Complex SoS when they decompose and restructure the subsystems "from the inside out and from the outside in" and relate the components to each other through their natural, physical, organizational, and functional attributes, recreating the interdependent and interconnected entity. Using the building blocks of mathematical models (to be discussed in subsequent sections) and ultimately by exploiting the shared states and other essential entities among the subsystems, the modeler and other users are able to better understand Complex SoS. The term other common/shared essential entities includes shared decisions, decisionmakers, stakeholders, resources, organizational behavior and norms, policies and procedures, management, culture, and others. We adopt the premise that models are built to answer specific questions; they must be as simple as possible and as complex as required. Thus, modeling the natural environment and the constructed environment such as organizations, or a combination thereof, represents a similar challenge. Namely, how many perspectives of a single system must be considered by modelers to achieve close-to-a-holistic model(s) in response to the required needs? And are we able to conceive of and discover all the essential attributes, characteristics, and perspectives of Complex SoS? Such open-ended questions reinforce the notion that the modeling process is a journey of discovery, imagination, and creativity. When we think we have succeeded, we are likely to be proven wrong. This assertion ought to be interpreted constructively and philosophically, but never fatalistically. In other words, the modeling

Modeling and Managing Interdependent Complex Systems of Systems, First Edition. Yacov Y. Haimes.

process is an open-ended continuous journey of learning and exploration that is characterized by successes and failures through which progress is made and, eventually, models are declared representative and valuable.

What does it mean to characterize systems as Complex SoS? Indeed, the emergence of the complexity characterizing Complex SoS requires a reevaluation of their modeling, management, and communication. The evolution of the terms complexity and complex systems, their differing connotations during the last 50 years, and the ways in which they have led us to model and manage complexity are the subject of this book. Current models for Complex SoS are insufficient because too often they rely on the same modeling schema used for single systems. These models commonly fail to incorporate the complexity derived from the networks of interdependencies and interconnectedness (I-I) characterizing Complex SoS.

In their essence, most cyber–physical, organizational, and governmental enterprises, now and in the future, belong to Complex SoS. Understanding their complexity and being able to characterize them can lead us to reevaluate our theory and methodologies as applied to single systems; more specifically being cognizant of and responsive to the emergent nature of Complex SoS, given the *Evolving Base*. The *Evolving Base*, discussed in Chapter 9, is represented by the following dynamic shifting rules and realities for each subsystem and for the entire Complex SoS: (i) goals and objectives; (ii) stakeholders, decision makers, and interest groups; (iii) organizational, political, and budgetary baselines; (iv) reorganization and reallocation of key personnel; (v) emergent technology and its deployment; and (vi) requirements, specifications, delivery, users, and clients (Haimes, 2012b).

In modeling Complex SoS, holism must be equally applied to natural and constructed environments, as well as to human and community activities and behaviors. The challenge is how to model the interface and the interplay among these activities that are not independent; rather, their I-I are one manifestation of Complex SoS.

The above discussion is harmonious with the philosopher Jacob Bronowski's (1978) seminal statements:

> The world is totally connected. Whatever explanation we invent at any moment is a partial connection, and its richness derives from the richness of such connections as we are able to make. (p. 96)
>
> There is no nerve without the muscle and no muscle without the nerve in the total animal. This is the same statement as the one I made about the total connection of the world…. (p. 99)

Of the human senses, Bronowski argues that arts mediated by the sense of light, like sculpture and painting, and arts that mediated by speech and sound, like the novel, drama, and music, dominate our outlook. Most of the time we use vision to give us information about the world and sound to give us information about other people in the world. How do we translate and build on Bronowski's "vision" and "sound" in our modeling of Complex SoS? What kind of "instruments" do we need to model Complex SoS? In modeling, we commonly build on (i) domain knowledge, (ii) human and organizational behavior, (iii) the role of cyber–physical infrastructure in today's quality of life of communities and individuals, (iv) systems engineering theory and methodology, (v) databases, and (vi) modeling experience, among others. What is the role of inference and perception in translating a system and its environment from reality into an abstract vision that is built on Bronowski's and on other philosophers' ideas in support of the fundamentals of state-space theory (Bellman and Dreyfus, 1962, Nise, 2014)? The art and science of modeling is but an interpretation of the common multiple perspectives of Complex SoS used by modelers, namely, natural, physical, structural, organizational, or human behavior.

Fundamentally, this construable process represents a mental translation that implies a subjective cognitive understanding of each of the multiple perspectives of each system and their integration as a Complex SoS. Conceivably, two different modelers would interpret and perceive systems, subsystems, and, ultimately, the integrated Complex SoS, differently, given the amalgamation of the arts and sciences on which the modeling process is built. It is here where state-space theory contributes to harmonizing the modeling process of Complex SoS. In particular, given the large number of states (variables) required to model and represent the multiple subsystems and their multitude of perspectives, as well as the necessity for brevity yet representativeness, modelers from different disciplines, and thus different perspectives, will naturally tend to be influenced by their unique personalities and backgrounds.

Furthermore, the large number of states that might be generated through the iterative, learn-as-you-go modeling process necessitates the selection of a representative subset of shared states and other essential entities. Recall that we define essential entities to connote shared/common decisions, decision makers, stakeholders, resources, organizational setups, and history, among others. *This selection of a minimum number of shared states and other essential entities with which to identify critical I-I is the first step in identifying invaluable precursors to future impending failures.* Note that the I-I within Complex SoS constitute the essence of the sources of risk thereto. This step converts systems that heretofore were marginally connected in parallel to becoming connected in series. This process is pivotal for discovering one of the major sources of risk facing Complex SoS and the most important result of modeling the I-I within and among systems and subsystems. Working together collaboratively, modelers can develop better models by augmenting the ingenuity of other modelers and scholars, as they collectively focus on and interpret the genesis of the I-I characterizing the subsystems and, eventually, the entire Complex SoS. Alternatively, it is possible to envision separate modeling efforts by multiple modelers with a subsequent attempt to integrate the models to yield a better and more representative set of attributes of the overall Complex SoS. We ought to not overlook the modelers' inherent ingenuity, background, talent, experiences, and innovativeness, contributing to the iterative modeling process that is characterized by a trial-and-error and a learn-as-you-go process. In other words, the multipath exploration process that characterizes the modeling effort necessarily implies and even requires the intellectual creativity and energy of modelers of Complex SoS – a process that commonly yields to a better representation of the modeling efforts.

In his book *Ageless Body Timeless Mind*, the physician, philosopher, and author Deepak Chopra (1994) suggests the following three "models" of humans: *physiology*, *mental capacity*, and *spirituality*. No one would negate the notion that the human body is an interdependent and interconnected Complex SoS. Indeed, each organ is by itself a system of systems composed of multiple subsystems. The basic question is, can we model or represent a complete understanding of a person when we ignore one of the above three attributes identified by Chopra? The same principle of completeness/representativeness must apply to the natural and constructed Complex SoS. From several perspectives, Complex SoS are opaque. Our observations, studying and reading documents, consulting with knowledgeable experts, and exploring and exploiting all sources of information relevant to Complex SoS are important and invaluable. Nevertheless, this tedious and essential modeling process does not reduce the inherent intricacy characterizing Complex SoS. Moreover, the above solicited and collected information ought not lead us to the illusion that what we have observed and learned constitute the entire reality. *Rather, we ought to augment our acquired knowledge with an endless learn-as-you-go modeling process*. Thus, our notion in this book is that the modeling of Complex SoS is an intricate amalgamation of the arts, sciences, and engineering, guided by the ingenuity of systems modelers. This amalgamation of the visible and invisible, and the interplay between the arts and the sciences in the modeling of Complex SoS, is in many ways analogous to the architectural design of high-rise buildings and the ultimate translation of the design into the reality of a physical structure. Indeed, architects and systems modelers share some similarity in building on the arts and sciences in their specialties, although each discipline uses different crafts in its work. Modelers of systems and Complex SoS use the building blocks of mathematical and simulation models, among others, while architects use in their crafts of drawing and scale models to reflect weight, force, and balance as well as aesthetics, among other things. Both address the essential sequence of translating their conceptual, analytical, or other models into their ultimate realization.

Furthermore, the fact that all single systems and Complex SoS – natural or the constructed environment – are dynamically changing and evolving necessarily requires consideration of the time frame in modeling, implicitly or explicitly. Modeling such changing systems requires the use of dynamic models. This requirement adds an enormous challenge to modelers, who often revert, when possible, to steady-state models, taking cover under the adage that "models must be as simple as possible, but as complex as required." In other words, since the essence of modeling is an amalgamation of art and science, imagination, judgment, and experience, then "assumptions" become an essential instrument that modelers use to navigate between the grace of static simplicity and the harshness of dynamic complexity, with the required and challenging balance between the two. Of course, the choice between static and dynamic models is only one of the challenges facing modelers of Complex SoS. Not all submodels of subsystems necessarily require the same inherent characterization, e.g. linear vs. nonlinear, static vs. dynamic, deterministic vs. stochastic–probabilistic, lumped parameter vs. distributed parameter, or discrete vs. continuous. Here again, modelers necessarily resort to the essential guidance provided by the arts and sciences, namely, the creativity and imagination that constitute the foundation of the modeling process. This never-ending process of tradeoffs is necessarily resolved with justified assumptions by the modelers and by their ultimate users.

1.1.2 CAPTURING THE ESSENCE OF A SYSTEM VIA MODELING

There is an unfortunate imbalance in the curricula of most undergraduate and graduate programs in systems engineering and in industrial engineering and operations research that is driven by a focus on *system optimization* versus *systems modeling*. Such imbalance in education and subsequent experiences could lead to optimizing a system with a poorly constructed or misrepresentative model. In system optimization, we assume knowledge of the systems model under specific assumptions, where for each set of inputs we can generate, or probabilistically estimate, the outputs. For example, in the context of risk management, no effective risk management policy options can be developed, nor can the associated tradeoffs among all critical costs, benefits, and risks be evaluated, without having constructed a model, or a set of interdependent models, that represents the essence of the system, or of the Complex SoS.

Students and other professionals often ask: "What is systems engineering?" Indeed, systems engineering is distinguished by a practical philosophy that advocates holism in modeling and cognition in decision making. This philosophy is grounded in the arts, natural and behavioral sciences, and engineering. Hence, the systems engineering discipline is supported by a complement of modeling methodologies, tradeoffs among multiple non-commensurate, competing, and conflicting objectives, optimization and simulation techniques, data management procedures, and decision-making approaches. The ultimate purpose of systems engineering is to (i) build an understanding of the nature of systems and Complex SoS, their functional behavior, and interaction with their environment; (ii) improve the decision-making process in planning, design, development, operation, and management; (iii) collect appropriate databases with which to populate the systems models; and (iv) identify, quantify, and evaluate risks, uncertainties, and variability within the decision-making process.

One way to gain a better understanding of systems engineering is to consider the well-publicized ideas of Stephen R. Covey in his best-selling book, *The Seven Habits of Highly Effective People* (Covey, 1989), and to relate these seven habits to various steps that constitute systems thinking, or the systems approach to problem solving. Covey's journey for personal development, as detailed in his book, has much in common with the holistic systems concept that constitutes the foundation of the field of systems engineering. Viewed in parallel, the two philosophies – Covey's and the systems approach – have a lot in common. Analyzing a system cannot be a selective process, subject to the single perspective of the analyst who is responsible for deciphering the maze of disparate databases and knowledge. Rather, a holistic approach is one that encompasses the multiple visions and perspectives at play, supported by vast pools of data and other information. Such a systemic process is imperative to successfully understand and address the natural and the constructed environment, including organizational systems, which at their core are composed of interconnected, interactive, and interdependent Complex SoS – the theme of this book.

Systems engineering cannot be practiced effectively, if at all, without models – analytical, conceptual, or simulation. Models, experiments, and simulations are conceived and built to answer specific questions. A mathematical model is a set of equations that describes and represents the *essence* of the real system. The *Merriam-Webster Dictionary* defines *essence* as "The most significant element, quality, or aspect of a thing" (2017). The equations describe the various aspects of the problem; they identify the functional relationships among all of the system's components, elements, and its environment; they establish measures of effectiveness and constraints and, thus, indicate what data should be collected to deal with the problem quantitatively. These equations could be algebraic, differential, linear, or nonlinear or take other forms depending on the nature of the system being modeled.

In general, models can help us assess the consequences of a course of action, given what we know, or what we think we know, what we need to know, or what and where additional knowledge is needed to build a more effective model for decision making. Furthermore, mathematical models are the imperative mechanisms with which to perform quantitative systems engineering. They are built and used to help systems engineers, managers, and decision makers better understand and manage a system using its relevant and/or critical interdependent and interconnected subsystems; namely, a Complex SoS. In the medical sciences, for example, there are mathematical models that use various states of the patient, e.g. temperature and blood pressure, to help in a diagnosis. Such models are important to correctly understand the human body as a Complex SoS.

Modeling has a strong element of art because successful models must build on the artistic traits of experimentation, imagination, creativity, independent thinking, vision, and entrepreneurship. Systems modelers must possess and merge values and traits offered by both the arts and the sciences. However, in contrast to scientific knowledge, whose validity can and must be proven, mathematical models cannot always be successfully subjected to such metrics. In fact, the more complex the system to be modeled, the lower the modeler's ability to verify or validate emerging models. Some scholars even argue that no complex model can be verified or validated, which is due, in part, to the dynamic and probabilistic characteristics of all natural and constructed Complex SoS. Heisenberg's (1930) uncertainty principle is at work here as well, namely, once the system's model is deployed, the essence of the system will change.

Models can help answer limited questions about the behavior of systems under both steady-state conditions and dynamic forced changes. The multiple perspectives that characterize each system, and the entire Complex SoS, require developing models that represent the essence of the multiplicity of perspectives, attributes, functions, and dimensions of the system. Physical, chemical, biological, and other natural laws serve as the first principles and the foundation for such models. Although mostly necessary, these natural laws are not sufficient for model construction because of the intricacy of systems and of Complex SoS. Furthermore, the influence of organizational and other emergent forced changes (EFCs) from within or outside the Complex SoS affects it positively or negatively. The term EFCs connotes internal or external changes that may positively or negatively affect one system, or the entire Complex SoS. The multiple perspectives of any system, whether it is the human body, the environment, a bridge, a building, or an airplane, cannot be adequately modeled using a single model – a fact that presents a challenge to modelers. Thus, what is needed is a mechanism or a systemic framework, with which to augment natural and physical laws with human and organizational behavior, imagination, inventions, innovation, entrepreneurship, out-of-the-box thinking, and boundless experimentation.

Natural and constructed environments and organizational systems are, at their core, composed of interdependent and interconnected Complex SoS. A wildlife refuge in Alaska and a large suspension bridge over the Bosporus in Istanbul connecting Europe with Asia are two examples of such Complex SoS. The wildlife refuge may support ample species with diverse life cycles; their interdependency constitutes a complex ecosystem. The bridge over the Bosporus may be perceived merely as an infrastructure that is constructed of steel, asphalt, cement, and light fixtures. Nothing could be further from the truth. The suspended cables in the bridge are 60 cm in diameter and contain countless numbers of bundled steel wires. Similarly, the hanging cables, the towers, the bridge itself, and the myriad supporting invisible physical infrastructures all constitute an interdependent and interconnected Complex SoS. Clearly, there is a need to understand the science and engineering that ultimately determine the reliability, sustainability, and safety of Complex SoS, including bridges. Such understanding may employ expertise in civil and structural engineering and systems engineering, the arts and sciences, and organizational and behavioral sciences, among other fields of study. Systems engineers are commonly the integrators of contributions made by experts in these diverse disciplines. Indeed, systems integration, where Humpty Dumpty is put together so that the system can function as intended, cannot be successfully performed in earnest without a heavy reliance on systems modeling.

1.I.3 A BRIEF HISTORY OF MODERN SYSTEMS ENGINEERING

Systems engineering has many parents. During his distinguished career, Albert Einstein attempted to develop a unified theory that embraced all forces of nature as a system. Feynman et al. (1963) described a hierarchy or continuum of physical laws as distinct systems or disciplines that are cooperating and interdependent. Modern systems foundations are attributed to various scholars. Among them is Norbert Wiener, who in 1948 published his seminal book *Cybernetics*. Wiener's work was an outgrowth or response to the development of computer technology, information theory, self-regulating machines, and feedback control. In the second edition of *Cybernetics* (Wiener, 1961), Wiener commented on the work of Leibniz:

> At this point there enters an element which occurs repeatedly in the history of cybernetics – the influence of mathematical logic. If I were to choose a patron saint for cybernetics out of the history of science, I should have to choose Leibniz. The philosophy of Leibniz centers about two closely related concepts – that of a universal symbolism and that of a calculus of reasoning. From these are descended the mathematical notation and the symbolic logic of the present day.

Ludwig von Bertalanffy (1968) coined the term *general systems theory* around 1950, which is documented in his seminal book *General System Theory: Foundations, Development, Applications* (Bertalanffy, 1976). Of particular interest (pp. 9–11):

> In the last two decades we have witnessed the emergence of the "system" as a key concept in scientific research. Systems, of course, have been studied for centuries, but something new has been added.... The tendency to study systems as an entity rather than as a conglomeration of parts is consistent with the tendency in contemporary science no longer to isolate phenomena in narrowly confined contexts, but rather to open interactions for examination and to examine larger and larger slices of nature. Under the banner of systems research (and its many synonyms) we have witnessed a convergence of many more specialized contemporary scientific developments. So far as can be ascertained, the idea of a "general system theory" was first introduced by the present author prior to cybernetics, systems engineering and the emergence of related fields .... Although the term "system" itself was not emphasized, the history of this concept includes many illustrious names. As "natural philosophy", we may trace it back to Leibniz; to Nicholas of Cusa with his coincidence of opposites; to the mystic medicine of Paracelsus; to Vico's and ibn-Kaldun's vision of history as a sequence of cultural entities or "systems"; to the dialectic of Marx and Hegel, to mention but a few names from a rich panoply of thinkers.

Kenneth Boulding, an economist, published his 1953 work *General Empirical Theory* (Boulding, 1953) and claimed that it was the same as the general systems theory advocated by Bertalanffy. The Society for General Systems Research was organized in 1954 by the American Association for the Advancement of Science. The society's mission was to develop theoretical systems applicable to more than one traditional department of knowledge.

Several modeling philosophies and methods have been developed over the years to address the complexity of modeling large-scale systems and to offer various modeling schema. In his book *Methodology for Large-Scale Systems*, Sage (1977) addressed the need for value systems that are structurally repeatable and capable of articulation across interdisciplinary fields that can be used to model the multiple dimensions of societal problems. Blauberg et al. (1977) pointed out that, for the understanding and analysis of a large-scale system, the fundamental principles of *wholeness* (representing the integrity of the system) and *hierarchy* (representing the internal structure of the system) must be supplemented by the principle of "the multiplicity of description for any system." To capture the multiple dimensions and perspectives of a system, Haimes (1981) introduced Hierarchical Holographic Modeling (HHM), which is the subject of Chapter 3 and is applied throughout this book. Recognizing that a system may be subject to a multiplicity of management, control, and design objectives. Zeigler (1984) addressed such modeling complexity in his book *Multifaceted Modelling and Discrete Event Simulation*. Zeigler introduced the term *multifaceted* "to denote an approach to modeling that recognizes the existence of multiplicities of objectives and models as a fact of life" (p. 8). In his book *Synectics, the Development of Creative Capacity*, Gordon (1968) introduced an approach that uses metaphoric thinking as a means to solve complex problems. Hall (1989) developed a theoretical framework, which he termed *metasystems methodology*, to capture the multiple dimensions and perspectives of a system. Other early seminal works in this area include *Social Systems – Planning and Complexity* on societal systems and complexity by Warfield (1976) and *Systems Engineering* (Sage, 1992). Sage identified several phases of the systems engineering life cycle. His analyses embraced multiple perspectives including the structural definition, the functional definition, and the purposeful definition. The multiple volumes of the *Systems and Control Encyclopedia: Theory, Technology, Applications* (Singh, 1987) offer a plethora of theories and methodologies for modeling large-scale and complex systems. Thus, multifaceted modeling, meta-systems, HHM, and other contributions in the field of large-scale systems constitute the fundamental philosophy upon which systems engineering is built.

Indeed, several modeling philosophies and methods have been developed over the last seven decades to address the complexity of modeling large-scale systems and to offer various modeling schema. They are included in the following volumes: *Views on General Systems Theory* (Mesarović, 1964), *General Systems Theory* (Macko, 1967), *Systems Theory and Biology* (Mesarović, 1968), *Advances in Control Systems* (Leondes, 1969), *Theory of Hierarchical, Multilevel Systems* (Mesarović et al., 1970), *Methodology for Large-Scale Systems* (Sage, 1977), *Systems Theory: Philosophical and Methodological Problems* (Blauberg et al., 1977), *Hierarchical Analyses of Water Resources Systems: Modeling and Optimization of Large-Scale Systems* (Haimes, 1977), and *Multifaceted Modelling and Discrete Event Simulation* (Zeigler, 1984). Haimes (1981) developed Hierarchical Holographic Modeling (HHM) for Complex SoS; Gheorghe (1982) presented the philosophy of systems engineering as it is applied to real-world systems. Haimes and Macko (1973), Hall (1989), Macko and Haimes (1978), Haimes et al. (1990), and Haimes (2007, 2008, 2012a) developed a theoretical framework to capture the multiple dimensions and perspectives of a system and (Lasdon, 1991) published a seminal book on optimization theory for large systems; indeed, Lasdon is among the pioneers who contributed to decomposition and hierarchical coordination of large-scale systems. Other works include those by Sage (1977, 1992, 1995), Shenhar (1994), and Sage and Rouse (1999). Eisner (1993), Maier (1998), and Sage and Cuppan (2001) together provide valuable insights into SoS and definitions of emergent behavior of complex systems in the context of SoS.

Most of the works on systems of systems have been devoted to their organizational, functional, and structural nature. There has been comparatively little inquiry into the problem of modeling Complex SoS, and most of the contributions within the last two decades have focused on their description, classification, and characterization. For example, Ottino (2003) reviewed three major tools for quantitative modeling and studying complex systems: nonlinear dynamics, agent-based models, and network theory. Shalizi (2006) reviewed the main methods and techniques of complex systems, which include tools for analyzing data, constructing and evaluating models, and measuring complexity. Chang and Harrington (2005) provided a comprehensive description of agent-based models of organizations. Amaral and Ottino (2004) described network theory and its importance in augmenting the framework for the quantitative study of complex systems. Lloyd and Lloyd (2003) presented a general method for modeling complex systems in terms of flows of information. Page (1999) discussed robust computational models. In an analysis of the challenges associated with complex systems engineering, Johnson (2006) provided a comprehensive review of emergent properties and how they affect the engineering of complex systems. Bar-Yam (2003a) reviewed past lessons learned from problems with systems engineering historically and suggested adopting an evolutionary paradigm for complex systems engineering. Within the application of complex systems theory to a multiscale analysis of military littoral warfare, Bar-Yam (2003b) suggested the necessity of considering the specific organizational and technological requirements needed to perform effectively in a highly complex environment. In health care, Funderburk (2004) presented a brief survey of several formal dynamic and/or network-based models that are relevant for health-care policy development and evaluation. Tivnan et al. (2007) described the formulation, successful replication, and critical analysis of Levinthal's model of emergent order for economic firms. Jamshidi (2009a, b) edited two volumes on systems of systems engineering. In the preface to the first volume (2009a), he wrote: "The SoS [Systems of Systems] concept presents a high-level viewpoint and explains the interactions between each of the independent systems. However, when it comes to engineering and engineering tools of SoS, we have a long way to go. This is the main goal of this volume" (p. *ix*). Indeed, Jamshidi confirmed the need for concerted efforts in modeling Complex SoS.

Sage and Biemer (2007) argued that no universally accepted definition of SoS is currently available. Sage and Cuppan (2001) built their analyses on five properties of SoS suggested by Maier (1998). During the past decade, several disciplines have recognized the importance of addressing the management, and thus the modeling, of their SoS including finance, health care, defense, and physical–cyber infrastructure systems. De Laurentis (2008), Lewe et al. (2004), Parker (2010), and Dahmann and Baldwin (2008) all suggest that SoS problems require a new modeling paradigm that can account for the multiplicity of stakeholders, objectives, interdependencies, and emergent outcomes. Fisher (2006) argued that emergent behavior is inherent in SoS and traditional software and systems engineering methods are inadequate for interpretation of SoS. De Laurentis and Callaway (2004) discussed the need to focus the modeling effort on SoS interdependencies, and they suggested that the evaluation of an individual entity at its own level is of less importance than how it affects the higher levels of the organization of which it is a member. Similarly, Thissen and Herder (2003) claimed that efforts to increase understanding at the overall SoS level are much needed. Aktan and Faust (2003) called for the need for SoS modeling approaches for civil engineering researchers and practitioners. They maintain that an integrated modeling of large-scale infrastructure SoS encompassing engineered, human, and natural elements has been unsuccessful thus far.

The emerging roles of systems engineering in the design, implementation, and management of Complex SoS have resulted in increased interest in engineering systems as SoS and as an emerging multidiscipline. Sousa-Poza et al. (2009) and Keating (2005) articulated several of the critical research challenges that SoS must address and identified a preliminary set of critical research areas for a more integrated research agenda. Maier and Rechtin (2009) indicated that SoS pose specific challenges for design and development, which are distinct from those of conventional systems. The principal challenges include designing for social and technical equilibrium, promoting sequential decision making for technology, and creating system roadmaps with large uncertainty.

Gorod, Sauser, and Boardman (2008) identified distinguishing characteristics as a foundation on which to build an effective SoS management framework. Dahmann et al. (2011) proposed a time-sequenced, incremental development "wave" modeling approach using an implementers' view of systems engineering for SoS. In order to achieve a common purpose, an SoS approach is essential in resolving issues involving heterogeneous, independently operable systems. Successful operation of SoS requires communication, coordination, and negotiation among appropriate individuals and groups across enterprises using an effective protocol (De Laurentis et al., 2007). Multiple criteria decision analysis and conflict resolution using graph models were discussed extensively in Hipel et al. (1993), Li et al. (2004), and Kilgour and Hipel (2005). An application of Extensible Markup Language (XML) to represent data communicated among systems was proposed by Sahin et al. (2007).

The mathematical methods used in the studies of Complex SoS include nonlinear dynamics, graph and network theories, and agent-based modeling and simulation (ABMS) (Barabási and Albert, 1999; Ottino, 2003; Baldwin et al., 2017). Bifurcation and catastrophe theory (Arnold, 1994) have also been used to describe the behaviors of nonlinear systems. These theories focus on and classify phenomena characterized by sudden shifts in behavior arising from small changes in circumstances, analyzing how the qualitative nature of equation-based solutions depends on the parameters that appear in the equation. Various methods and techniques used in complex systems science can also be found in Shalizi (2006), Lloyd and Lloyd (2003), and Page (1999).

The essential characteristics of Complex SoS present serious difficulties for traditional hazard analysis techniques (Alexander et al., 2004). Bristow et al. (2012) argued that risk analysis of extreme events affecting SoS should address the complex, ambiguous, and uncertain aspects of extreme risk and the strategic interactions among multiple participants. Investigations of several accidents of complex systems, such as the Three Mile Island accident (Perrow, 1999), showed that the causes of complex system failure usually include multiple component failures and their unexpected interactions. Perrow pointed out that the root causes of system accidents reside in the properties of complex systems themselves, rather than in the errors that owners and operators make in running them. It is the system's characteristics that make it inherently vulnerable to such accidents. Eusgeld et al. (2011) also discussed the potential failure propagation among infrastructures leading to cascade failures, and they analyzed two modeling alternatives, comparing integrated with coupled models.

1.I.4 BUILDING BLOCKS OF MATHEMATICAL MODELS AND THE CENTRALITY OF STATE VARIABLES IN SYSTEMS MODELING

The systems modeling process relies on the fundamental building blocks of mathematical models: input, output, state variables, decision (control) variables, exogenous variables, uncertain and random variables, and time frame. These are commonly augmented to yield multiple, noncommensurate, and commonly competing objective functions and constraints. Note that these building blocks are not necessarily distinct and they may overlap. For example, input and output may be random. All good managers desire to change the states of the systems they control in order to support better, more effective, and more efficient attainment of the system objectives. At the same time, these managers demand acceptable tradeoffs among the many competing objectives but within an acceptable time frame and cost structure. The objectives and motivations of the stakeholders and decision makers are to determine the desired levels of the states of the system within an acceptable time frame and acceptable tradeoffs. As noted earlier, a large number of states, sub-states, and sub-sub-states characterize Complex SoS. For example, the state of blood might be characterized by white cells and red cells; however, the states of each category can be further subdivided. Thus, we use the term *vital states* to connote selected fundamental and indispensable states that are central to the essence of the Complex SoS as a whole and the associated goals, objectives, and major decisions. As another example, to control the production of steel requires an understanding of the states of the steel at any instant – its temperature, viscosity, and other physical and chemical properties that characterize its quality. Similarly, to know when to irrigate and fertilize a field, a farmer must assess the states of soil moisture and the nutrients in the soil. And, to treat a patient, a physician must first know the temperature, blood pressure, and other states of the patient's physical health. Finally, the body and its systems are continuously bombarded by a variety of bacteria, viruses, and other pathogens. A more detailed characterization and discussion of the vulnerability of a system as a manifestation of its states will be introduced in Part II of this chapter in the context of the resilience and vulnerability of Complex SoS.

Consider the following diverse examples:

- Engineers are asked to determine the safety of a municipality's drinking water. Can they perform this task without determining the state of acidity of the water and the states of turbidity, dissolved oxygen, bacteria, and other pathogens?
- Teachers are required to nominate the top five students in their classes for awards based on their performance. Can they make such a selection without having assessed the states of the prospective candidates' knowledge, talent, competence, aptitude, performance, and learning capabilities?
- Bus drivers are commissioned to transport musicians to a tightly scheduled concert and must guarantee that their busses are mechanically

and otherwise reliable to ensure timely arrival at the concert site. Can they ascertain the reliability and functionality of their buses without knowing the states of the latter's fuel, oil, tire pressure, and other mechanical and electrical components?

The above examples, and almost all other real-world problems, share the following systems-based fact – all are characterized at any moment by their respective *essential* state variables. (The term essential states connotes the minimum number of state variables with which to effectively model a system or Complex SoS.) In reality, all states are under continuous change and natural, EFCs – positive and negative. We earlier defined EFCs as *external or internal forces that may negatively or positively affect specific states of a system*. The fact that the states of a system are functions of the time frame and that most, if not all, systems are dynamic and evolve over time, necessarily implies that representative models ought to be dynamic as well. Complying with this premise means ideal models must be dynamic, which could be a more elaborate task for modelers. At the same time, despite the fact that in reality, real systems may vary over time, not every model will require time-dependent state variables. This is where the artistry, creativity, good judgment, and experience of the modeler come into play. Recall that models are built to answer specific questions and to represent the essence of the system under consideration. Thus, if over time, small or insignificant changes in a state of the system have no important effect on the answers sought from the model, then that state variable may be assumed to be static or time independent. The decision as to whether a state variable should be modeled as static (constant) or dynamic (time dependent) depends on the modeler's ability to select the best representative model topology (structure and form).

In fact, the art and science of systems modeling is characterized by a never-ending tradeoff process by modelers faced with various levels of complexity and detail. However, since models should also be as simple as possible yet as complex as required to answer specific questions, tasks may also include (i) selecting a minimum number of *vital* state variables from each building block of the model to adequately represent the essence of the system, (ii) determining the required complexity of the model (e.g. topology and parameters), and (iii) developing the required database (as appropriate) to populate the model in order to provide meaningful and specific answers. All these and many more specific details, which will be further explored in subsequent sections and chapters of this book, represent real challenges in terms of the model's required complexity, cost, and time of completion, its users and stakeholders, its required databases, the needed level of testing, and, not the least, the scope and specificity of the assumptions made in its construction.

1.I.5 THE CENTRALITY OF THE STATES IN MODELING COMPLEX SYSTEMS OF SYSTEMS

Recall Chen's (2012) succinct definition of *state variable*: "The state $x(t_o)$ of a system at time t_o is the information at time t_o that, together with the input $u(t)$, for $t \geq t_o$, determines uniquely the output $y(t)$ for all $t \geq t_o$." The states of a system, commonly a multidimensional vector, characterize the system as a whole and play a major role in estimating its future behavior for any given input. Thus, the behavior of the states of the system, as a function of time, enables modelers to determine, under certain conditions, its future behavior for any given *input* or *initiating event*. In other words, all systems are characterized at any moment by their respective state (variables) and the conditions thereof, and these conditions are subject to a continuous change. In addition a modeler, who is determined to select only those state variables that represent the critical elements of a system (i.e. essential states), must decide whether those state variables should be modeled as static (constant) or dynamic (time dependent), deterministic or stochastic, etc.

Given that all systems, large and small, can be characterized by their states, we must also recognize the inherent hierarchy of states, sub-states, and sub-sub-states – a crucial attribute in systems modeling. For example, a representative water resources system that supplies water to a large community can be characterized by the states of the water distribution (groundwater and surface water) storage, purification, and sewer systems. The data for each of the states can be further presented by sub-states. As another example, the states of the water distribution system may be represented by the status of the main carriers, local pipes, pumps, and storage tanks. With any Complex SoS, the most critical fact to note is the relationships that exist within and among the states of the system, which necessarily overlap the multiple perspectives of the system represented by the multiple models. In other words, a central role of modeling Complex SoS is to understand the essence of the I-I of the shared states and other essential entities (defined earlier) of the Complex SoS under consideration; namely, to comprehend, or "make a whole," of the various attributes that characterize the multiple systems that constitute Complex SoS. This important task cannot be achieved without domain knowledge of the systems and carefully discovering and identifying those states that characterize the most important aspects of each system and of the Complex SoS as a whole. Furthermore, the fact that all state variables are functions of random and uncertain initiating events requires that our modeling efforts take into account both epistemic and aleatory uncertainties (Paté-Cornell, 1990, 1996; Apostolakis, 1999; Haimes, 2016).

Consider the following definitions of the vulnerability and resilience of a system Haimes (2016):

> *Vulnerability is the manifestation of the inherent states of the system (e.g., physical, technical, organizational, and cultural) that can be subjected to a natural hazard or be exploited to adversely affect (cause harm or damage to) that system.* The vulnerability of a system is multidimensional, a *vector* in mathematical terms. (p. 56)

> The resilience of a system is also *a manifestation of the states of the system and it is a vector that is time- and threat (initiating event)-dependent.* More specifically, *resilience* represents *the ability of the system to withstand a major disruption within acceptable degradation parameters and to recover within an acceptable cost and time.* In other words, resilience is a vector state of the system that is neither abstract nor static, nor deterministic. Moreover, resilience is similar to vulnerability in that it cannot simply be measured in a single unit metric; its importance lies in the ultimate multidimensional outputs of the system (the consequences) for any specific inputs (threats). (p. 57)

The question "What is the resilience of the University of Virginia?" is unanswerable without specifying the specific threat, considering the specific likely vulnerable or affected states of the University of Virginia, and of the timing of the threat. Likewise, questions on the vulnerability and resilience of a system can be answered only when the threat (initiating event) scenario (or a set of scenarios) is identified or the vital states of the system and of Complex SoS are specified. Resilience is not merely an abstract concept; it is a state of the system (composed of a vector of substates) that may have different responses to different inputs (threat scenarios).

This discussion of the centrality of states of the system in modeling will be further explored in Part II of this chapter and throughout this book and will be related to (i) intrinsic meta-modeling coordination, (ii) integration of the multiperspective models, and (iii) the necessity of relying on the states of the system. This is in contrast to relying solely on the extrinsic output-to-input model coordination and integration, which does not build explicitly on the shared (common) states and other essential entities, and the overlapping states among the systems and subsystems that constitute Complex SoS.

1.I.6 THE CENTRALITY OF TIME IN MODELING MULTIDIMENSIONAL RISK AND UNCERTAINTY

Time is central to all decisions, whether connected implicitly or explicitly, and thus to systems modeling. For a pilot, the time frame may be measured in mere seconds; for a planner, it may be years or decades. Indeed, all real-world Complex SoS are characterized by dynamic multiple objectives, often noncommensurate, competing, and in conflict with each other. Chapter 5 is devoted in its entirety to the subject of multiple objectives. A Pareto-optimal policy (solution) is such that improving one objective can be achieved only at the expense of degrading another objective. Pareto-optimal policies associated with such systems models are achieved through the manipulation of the appropriate states of the system, and since the latter are functions of time, the time frame becomes critical to systems modeling. Models that are built to answer specific questions must also be constructed to address the question: What is the impact of current decisions on future options, given the inevitable EFCs? (Recall that the term *emergent forced change was* defined earlier in this chapter.) Uncertainty, commonly viewed as *the inability to determine the true states of a system*, can be caused by *incomplete knowledge* and/or by *stochastic variability*. Two major sources of uncertainty in modeling affect risk analysis (Paté-Cornell, 1990, 1996; Apostolakis, 1999; Haimes, 2016). *Knowledge (epistemic) uncertainty* manifests itself in the selection of appropriate model topology (structure) and model parameters, which can be sources of ignorance (e.g. when modelers lack knowledge of important interdependencies within the states of the system and among other systems). *Variability (aleatory) uncertainty* applies to all relevant and important random processes and other random events. Uncertainty dominates most decision-making processes and is the Achilles' heel for all deterministic, and most probabilistic, models. Uncertainty on the part of the modeler often results in the selection of an incorrect model topology (structure) – for example, selecting a linear model for a highly nonlinear system, thereby inaccurately rendering its parameters, data collection, and processing techniques. Model uncertainties will often be introduced through human errors of both commission and omission. Uncertainty analysis becomes even more imperative in risk analysis of the I-I of emergent Complex SoS. Sources of uncertainty and lack of understanding of the complexity associated with the subsystems of Complex SoS would likely result in (i) adherence to unrealistic assumptions, (ii) a lack of awareness of and accountability to the critical I-I of the Complex SoS under consideration, (iii) poorly selected representative model topology and comprehensiveness, (iv) a dated or insufficient database to populate and calibrate the multiple subsystems models, and (v) essential risk scenarios being poorly represented or structured for all interdependent and interconnected systems. In sum, uncertainty analysis associated with Complex SoS is probably one of the most difficult, albeit important, tasks in the broader risk analysis process.

An adverse initiating event is likely to yield multidimensional probabilistic consequences to each system and to the Complex SoS. These consequences are represented by a complex multidimensional risk function, the modeling and quantification of which present considerable challenges. The selection of appropriate models to represent the essence of a system's multiple perspectives also determines the effectiveness of the entire risk assessment, management, and ultimately the communication processes. In particular, the scope and effectiveness of strategic risk management options are implicitly and explicitly dependent on the system perspectives that are included (or excluded) in the ultimate modeling efforts. In particular, a probable initiating event would necessarily affect only sub-states of a subsystem but not necessarily the entire Complex SoS. Thus, one must model the different probability distribution functions of consequences affecting each subsystem resulting from the same initiating event. Each perspective of a system – manifested through its structure, functionality, the services it provides, the customers it supports, and the other systems on which it depends – will experience specific, and likely, unique consequences resulting from the same initiating event.

Recall that the complexity of SoS stems primarily from the I-I within and among its number of systems and subsystems. Consider, for example, the I-I among three common cyber–physical infrastructures: electricity, communications, and water (and of course the communities they serve). A major initiating event that may cause the failure of the electricity system would result in adverse consequences to the other two cyber–physical infrastructures because of their dependency on electricity; neither one can operate without electricity. *Clearly, there is a need to understand and thus model the pathways through which the failure of one system propagates to other interconnected systems.* For example, an effective risk analysis of a Complex SoS requires a clear understanding of its configurations to enable the identification of specific critical failure modes. This enables the development and deployment of effective risk mitigation and other management strategies. Current risk models of systems that do not consider the inherent interdependencies among systems are likely to be inferior to those models that do. We emphasize throughout this book that the I-I are best understood and modeled through the *shared states* and other essential entities of the Complex SoS as a whole. Recall that the states of Complex SoS represent the smallest set of linearly independent system's outputs, such that the values of the members of the set at time t_o, along with known forcing functions, completely determine the value of all system variables for all $t \geq t_o$. Thus, the behavior of the states of the system, as a function of time and other inputs, enables modelers and risk analysts to determine, under certain conditions, its future behavior for any given *input* or *initiating event*. Indeed, in industrial production and management, the feedback control process is predicated on the fact that the smooth operation of the multiple interconnected subsystems is built on the knowledge of the operational states of each subsystem at each instant.

Systems modelers face nontrivial challenges when selecting the minimum number of states to adequately and effectively represent (model) the subsystems under consideration. Consider, for example, the challenges in selecting the states in the risk modeling process for sustained years of drought, including (i) availability and quantity (states) of groundwater and surface water, (ii) quality (states) of groundwater and surface water, and (iii) human, livestock population, and industrial needs for water. The above states of water quality and quantity have direct impact on the (i) rural or urban populations and on livestock, (ii) agriculture and industry, and (iii) future recovery of the states of water quality and quantity for the entire River Basin.

From the perspective of the *reliability of* Complex SoS, shared states and other essential entities within and among the subsystems represent the essence of the I-I that characterize Complex SoS. Initiating events affecting one or more of the interdependent and interconnected subsystems would necessarily increase the probability of failure of the entire SoS (Haberlin and Haimes, 2018; Lewis and Haimes, 2018). This fundamental fact can be viewed in the parlance of fault-tree analysis (NUREG, 1981) as converting subsystems from being "marginally connected," or in parallel, to becoming "directly connected," or in series. The subject of fault trees will be introduced and elaborated upon throughout this book; also, consult the Appendix in this book. (Chapter 13 of the fourth edition of the book *Risk Modeling, Assessment, and Management* (Haimes, 2016) is devoted to fault-tree analysis.) Also, Chapter 10 of this book presents four case studies with a reliance on fault trees. Nuclear reactors, which epitomize Complex SoS, have always relied on fault-tree analysis to ensure their safety. Thus, from that perspective, subsystems that share states and other essential entities *are most likely to* be affected by adverse initiating events. For example, this phenomenon is most evident in the growing use of cloud-computing technology, where numerous hardware–software subsystems are shared among multiple users (Haimes et al., 2015). Similar results have been demonstrated with I-I among cyber–physical infrastructures. An initiating event may not affect all shared states and other essential entities of different subsystems of a Complex SoS in the same way. This fact necessarily implies the following scenarios when analyzing risk to Complex SoS: (i) When each subsystem has different decision makers, then decisions made to control subsystem A may affect positively or negatively subsystem B that shares subsystem A's states and other essential entities. (ii) When decision makers collaborate among themselves and coordinate their decisions, this can have a positive effect on Complex SoS and improves their overall effectiveness and management. (iii) When different initiating events affect one or more subsystems, the level of shared states and other essential entities among them can positively or negatively affect the risk management process of the entire Complex SoS. Therefore, modelers should recognize and exploit the shared states and other essential entities among subsystems. As a natural example of interdependent subsystems, the states of the Earth and its moon relative to each other and to the sun are known at any instant due to shared gravitational forces. In their case, without any initiating events affecting any of them, the states of their celestial coordinates in space and time can be determined.

Guiding principles for modeling Complex SoS are presented in Chapter 9. The following *Evolving Base* is a sample of emergent components of Complex SoS: (i) goals and objectives; (ii) stakeholders, decision makers, and interest groups; (iii) organizational, political, and budgetary baselines; (iv) reorganization and reallocation of key personnel; and (v) requirements, specifications, delivery, users, and clients. Changes to these components are common to most complex systems, but most notably of the interdependent and interconnected Complex SoS, where they have a more dominant impact. In particular, modelers ought not overlook the likely multiple impacts of the *Evolving Base* on the shared/common states and other essential entities within and among the systems that constitute Complex SoS.

The organizational infrastructure at all levels of the subsystems and the systems of systems necessarily affects the corresponding states and substates. This fundamental fact constitutes a major driver in both the modeling and the management of the I-I of Complex SoS. Decision-making processes of most, if not all, organizations are characterized by a state of flux, given the mobility of executives within the organization and of incoming leadership replacing departing personnel at all levels. New leaders and executives to the organization are commonly hired to infuse new ideas and energy into the organization. A by-product of this process can lead to a recalibration of the goals and objectives of the subsystems, if not of the entire organization Complex SoS.

1.I.7 SYSTEMS MODELING AND INTEGRATION

Consider the laptop computer as Complex SoS; it has become an indispensable enabler for students, laypersons, and professionals alike including the writing of this book. The average laptop is assembled from about 2000 components (subsystems), each of which is designed and manufactured to perform certain critical functions on which the reliability and functionality of the laptop depend. The battery alone (as one subsystem) is assembled from multiple components and subcomponents. The nontrivial task of integrating and connecting the laptop's "2000 subsystems" requires understanding not only the functionality and role of each subsystem but also its effect on the performance of other subsystems. Without relying on systems modeling, such systems integration of multiple subsystems of hardware along with its software could neither be successfully accomplished, nor would the overall computer system's performance be realized.

Effective systems integration of Complex SoS requires accounting for all the system's functions, aspects, and components. For example, software-intensive systems not only require the integration of components but also understanding the functionality that emerges from that integration. Indeed, when two or more components are integrated, they often deliver more than the sum of what each was intended to deliver. Invariably, the integration adds synergy and enhances functionality. Also, the process of risk assessment and management is a requirement for successful systems integration; this is especially true for software-intensive systems.

1.I.8 STRUCTURE, STATES, AND FUNCTIONS OF COMPLEX SYSTEMS OF SYSTEMS

Consider the "translation" of the intricate relationships among *structure*, *states*, and *functions* of a system as suggested by Bronowski (1978). The states of a manufacturing system are directly influenced and affected by an intricate mix of machines, robots, materials, humans, organizational structures, and users of the final product, among others. And all of the above affect the system's functionality and thus the products of the manufacturing Complex SoS. Conversely, the conditions of the states of the manufacturing Complex SoS directly affect the integrity of the structure and its functionality. Furthermore, the I-I within and among the different components/subsystems that characterize Complex SoS may take many forms and levels, each of which defines the structure and functionality of the resulting subsystems and ultimately the entire Complex SoS.

The fact that, by their definition, Complex SoS are composed of interdependent and interconnected systems and subsystems implies that their modeling provides a more holistic vision and representation than when modeled as separate subsystems. In many ways, once a subsystem becomes a part of Complex SoS, it is likely to lose some or much of its autonomy and unique attributes and characteristics. Such change may be manifested via subsystem functionally, its role in the organization, or in the decision-making process. In modeling a single system of Complex SoS, each system thereof must be viewed holistically, especially when considering the natural and constructed environment and systems involving humans and community behavior. Holism represents a challenge in how to model the interface and the interplay between the *structure* of systems and the *nature* of systems.

How much perception and an ability to infer do good modelers need in order to build effective models? This includes the need to translate the reality of Complex SoS and their environments into an abstract vision that draws on modeling experience and expertise, and to use and build on the fundamentals of state-space theory and methodology.

In many ways, modeling can be viewed as an interpretation by the modelers of the multiple perspectives of Complex SoS – natural, physical, structural, organizational, and the influence of human behavior.

1.I.9 THE MULTIFARIOUS PERSPECTIVES AND DIMENSIONS OF COMPLEX SYSTEMS OF SYSTEMS

The complexity resulting from the I-I among the systems and subsystems that characterize Complex SoS can neither be well-understood nor modeled by a single individual. To be effective, the learn-as-you-go dynamic modeling process must be performed by a cross-disciplinary team: one that includes systems modelers and other individuals who possess domain knowledge of the historical, technical, and organizational complexity as well as other characteristics and orientations of the Complex SoS under consideration. In the book *The Wisdom of Teams*, Katzenbach and Smith (2015) identify four major attributes of an effective team: (i) a small number of members; (ii) with complementary experiences, perspectives, and skills; (iii) who are committed to a common purpose and performance goals; and (iv) where all members are mutually accountable. There are no specific norms that guide the composition of modeling teams for a given Complex SoS. Rather, the team is built with core resident modelers selected for their expertise and ability to contribute to the modeling process. The team is often augmented with additional expertise as needs arise over time.

Organizational dynamics and perspectives contributing to, and part of, the flow of EFCs, are likely to affect the states of many subsystems or the entire Complex SoS. (*Recall that the term EFC was defined earlier as emergent forces originating from within or outside one system, or from the entire Complex SoS*, that would affect the Complex SoS *positively or negatively*.) Slow or fast emergent internal and external EFCs, to which all systems and entire Complex SoS are commonly subjected, must continuously be accounted for in the modeling process. In particular, the criticality of slow creeping changes that affect the shared and other essential entities characterizing Complex SoS may not receive the serious consideration required by the modelers, stakeholders, and other principals engaged in the process. The I-I characterizing Complex SoS necessarily require appropriate knowledge and awareness of the emergent nature of the Complex SoS reality. *In other words, subsystem A that is affected by a specific EFC would also affect other subsystems that share states and other essential entities with it.* This concept of interdependence and interconnectedness has basic ramifications on the theory and methodologies that we deploy in modeling and managing Complex SoS in this book.

1.I.10 WHAT HAVE WE LEARNED FROM OTHER CONTRIBUTORS

Reflecting on the history of modern systems theory, and its close ties to the Gestalt psychology first introduced in 1912, we can underestimate neither the intellectual power of the multidisciplinary talent required for modeling Complex SoS, nor the holistic philosophy that has sustained it; thus forcing it to transcend the arts, the humanities, and the natural, social, and physical sciences, as well as engineering, medicine, and law. The fact that systems engineering and systems analysis have continued to grow over the years and contribute to other fields of study can be attributed to the fundamental premise that Complex SoS can be understood only if all the I-I among its systems and within its environment are also understood and accounted for. For more than a century, particular mathematical models, upon which systems-based theory and methodologies were developed, have been deployed in myriad large-scale projects in the natural and constructed environments. Moreover, if we were to identify a single concept that has dominated systems thinking and modeling, it would be *state-space theory* and the Gestalt–holistic philosophy. It can be argued that the art and science of systems modeling have served, in many ways, as the medium through which the holistic systems philosophy has informed and guided not only the practice of engineering but of a broad range of other fields. As the discipline of systems engineering continues to develop and expand into new domains, the need has emerged for new organizational and modeling paradigms to represent Complex SoS.

1.I.11 CONCLUSIONS

No single book on complexity and Complex SoS can do justice to, nor adequately represent, the plethora of perspectives and the multifarious nature of Complex SoS. And this book is of no exception. However, not attempting to try would constitute a fatalistic foresight that would be an enigma to scholarship and to the essential quest for discovery and learning. The vision and challenges that inspired the conception and guided the writing of this book stem from the following: *the need to not only define and represent but also to analytically quantify at least one fundamental characteristic of Complex SoS, namely, their interconnectedness and interdependencies.* This quest was enabled and augmented by the vast literature on modeling and optimizing (as well as in the sense of Pareto optimality associated with multiple noncommensurate, competing, and conflicting objectives), to which this author has contributed several books and copious technical articles since the 1960s.

1.I.12 MODELING AND MANAGING INTERDEPENDENT COMPLEX SYSTEMS OF SYSTEMS: BOOK OVERVIEW

As we present various methods for modeling and managing Complex SoS, it is important to map the course through which these theories, methodologies, case studies, and example problems are presented in the 15 chapters of this book, along with the Appendix.

Chapter 1: Modeling and Managing Interdependent Complex Systems of Systems: Fundamentals, Theory, and Methodology

This chapter is of two parts. Part I provides an overview of the entire 15 chapters of the book. Part II provides a comprehensive discussion on the resilience and vulnerability of Complex SoS. The Appendix, which follows Chapter 15, augments the textbook with systems engineering fundamentals that support basic theory and methodology on Complex SoS. [The book begins with a Foreword titled "Philosophical and Historical Perspectives on Understanding Commonalities Characterizing Complexity"]. The following sections present a general overview of the entire book and 15 chapters highlighting each of the succeeding 14 chapters and the Appendix.

The theme of the book advances the notion that current models for Complex SoS are insufficient, because too often they rely on the same modeling schema used for single systems. These models commonly fail to incorporate the complexity of the networks of I-I characterizing Complex SoS, and consequently the risk analysis and management based on such models suffer. For completeness, we redefine I-I to connote interdependencies and interconnectedness. Revised theoretical and methodological foundations for understanding, modeling, and managing risk to accommodate the unique attributes of Complex SoS are provided by research and case studies. The 15 chapters of this book underscore that effective modeling of Complex SoS lies in adequately understanding and modeling the I-I of systems manifested through shared/common states and other *essential entities* within and among the systems that constitute SoS. The term essential entities connotes shared/common, decisions, decision makers, stakeholder, resources, organizational behavior and norms, policies and procedures, management, culture, and others.A history of the discipline of systems engineering and the development of systems of systems with their unique complexities provide the base upon which to build new methods of modeling and managing interdependent and interconnected Complex SoS.

Chapter 2: Modeling, Decomposition, and Multilevel Coordination of Complex Systems of Systems

This chapter is of two parts. In studying Complex SoS including their technological, societal, and environmental aspects, the efforts in their modeling, as well as in their management, are magnified and often overwhelm the analysis. This is due to the (i) high dimensionality (very large number of variables), (ii) complexity (nonlinearity in the coupling and interactions among the variables), and (iii) dynamic changes and emergent behavior of the resulting models. In modeling the I-I of Complex SoS, one of the major impediments faced by modelers from the natural and behavioral sciences, engineering, and other professions stems from the dynamic and evolving nondeterministic processes that govern the interactions among the system's components. Whenever decentralization of a complex system is needed, the system is further decomposed to enable its effective modeling and ultimate coordination among the subsystems, as well as with the corresponding decision makers and the associated stakeholders. Part I of this chapter expands the concept of the hierarchical-multilevel approach, based on the decomposition of Complex SoS, and the subsequent modeling of the subsystems as independent at the lower levels of the hierarchy. This innovative decentralization utilizes the concepts of strata, layers, and echelons to enable systems modelers to analyze and comprehend the behavior of the subsystems at a lower level of the hierarchy and to transmit the information obtained to fewer subsystems at a higher level. Part II provides a primer on the theory and practice of incorporating probability distribution and uncertainty analysis in modeling Complex SoS. We address Bayesian methods for risk and uncertainty analysis.

Chapter 3: Hierarchical Holographic Modeling of Complex Systems of Systems

Hierarchical Holographic Modeling (HHM) is a holistic philosophy and proven methodology aimed at capturing and representing the essence of the inherent diverse characteristics and attributes of a system – its multiple aspects, perspectives, facets, views, dimensions, and hierarchies. The HHM (Haimes, 1981), which forms the basis for this chapter, emerged from a generalization of a Hierarchical Overlapping Coordination method and is capable of representing fundamental attributes of Complex SoS, which have commonly escaped multiperspective modeling representation.

The fundamental attribute of interdependent and interconnected Complex SoS is their inescapably multifarious nature: hierarchical noncommensurable objectives, multiple decision makers, multiple transcending aspects, and elements of risk and uncertainty. In part, this may be a natural consequence of the fact that most Complex SoS respond to a variety of needs that are basically noncommensurable and may under some circumstances openly conflict.

The HHM reflects a difference in kind from previous modeling schemas and contributes to the theory and methodology of modeling Complex SoS. There is a useful analogy between HHM and the capture of images. Conventional photography captures only two-dimensional planar representations of scenes and is analogous to conventional mathematical modeling techniques that yield "planar" models. Three-dimensional cinematography, however, is similar to the multidimensional schema needed to model the multifarious attributes of interdependent and interconnected Complex SoS. This chapter demonstrates the impracticality of representing, within a single model, all the aspects of an interdependent and interconnected Complex SoS, which may be of interest at any given time to its management, government regulators, students, or any other stakeholder.

Chapter 4: Modeling Complex Systems of Systems with Phantom Systems Models

In this chapter we introduce phantom systems modeling (PSM), a modeling paradigm that is congruent with and responsive to the uncertain and ever-evolving world of emergent systems. The PSM methodology/philosophy serves as an adaptive process, a learn-as-you-go modeling laboratory, where different scenarios of need and stages of development for emergent SoS can be explored and tested. These scenarios build on and extend the basic theory and philosophy of HHM by offering operational guidelines and principles on which to model interdependent and interconnected Complex SoS. In PSM, methodology and technology match, and emergent systems are studied through PSM similar to the way other appropriate models are constructed for systems with different characteristics. Equally agile and adaptive, PSM can be continually manipulated and reconfigured in the attempt to answer difficult emergent challenges. Examples of PSM include difference equations and differential equations for dynamic systems, algebraic equations for static systems, and probabilities for systems that are driven by random events and processes.

Chapter 5: Complex Systems of Systems with Multiple Goals and Objectives

For sound and informative decision-making processes, it is imperative that decision makers also be provided with the tradeoff values associated with the respective objectives. An "optimum" solution may exist for a model; however, for a real-life problem such an "optimum" depends on myriad factors, which include the (i) specificity of each subsystem; (ii) extent of shared states and other essential entities within and among the subsystems; (iii) identity, perspectives, and biases of the modelers, decision makers, and stakeholders; (iv) credibility of the database; and (v) time frame. Therefore, a mathematical optimum for a model does not necessarily correspond to the "optimum" for the real subsystems, nor for the Complex SoS, because multiple decision makers, and thus perspectives and needs, with varied authority are associated with each subsystem and with the Complex SoS as a whole. Each subsystem commonly represents different constituencies, preferences, and perspectives; as elected, appointed, or commissioned; or as public servants, professionals, proprietors, or associates; and connected with a specific level of the various hierarchies of objectives within the subsystems and the Complex SoS as a whole. This chapter outlines methods for achieving resolution to the multiobjective decision making associated with Complex SoS. Note that Complex SoS commonly involve multiple decision makers and decisions or "compromised" solutions are often reached through negotiation, either through the use of group techniques of multiple-criteria decision making (MCDM) or on an ad hoc basis.

Chapter 6: Hierarchical Coordinated Bayesian Modeling of Complex Systems of Systems

In this chapter we incorporate multiple decompositions from multiple perspectives, supported and populated with Bayesian data analysis. This modeling theory, philosophy, and methodology integrate all the direct and indirect relevant information from different levels of the hierarchies while placing more emphasis on relevant direct data. Indeed, by modeling the systems and data from different perspectives (such as via HHM), we can fully extract and exploit information from different dimensions via Bayesian modeling. In sum, we coordinate the results from different decompositions and perform quantitative modeling of Complex SoS supported with and enriched by multiple databases.

Chapter 7: Hierarchical-Multiobjective Modeling and Decision Making of Complex Systems of Systems

Many of the world's cyber–physical critical infrastructure systems fall within the category of Complex SoS. They are commonly composed of interdependent and interconnected subsystems, which in their essence constitute Complex SoS with multiple functions, operations, and stakeholders. Emergent, large-scale engineering systems profiled in this book such as aviation, supply chain, the power grid, and cyber–infrastructure systems, all pose significant challenges to risk modeling and management. The complexity of cyber–physical SoS is characterized by the highly interdependent and interconnected physical, economic, and social components that constitute a major source of EFCs to infrastructure systems. This means identifying relationships among different system components and understanding their impact on the systems so that efficient risk management strategies, including preparedness and response planning, can be deployed. Risk assessment, management, and communications are indispensable tools within which to evaluate the states of the system, reduce its vulnerability, and increase its resilience to EFCs. The development and application of risk analysis theories and methodologies for these cyber–physical infrastructure Complex SoS presented in Chapter 10 are key to their effective and efficient management.

Chapter 8: Modeling Economic Interdependencies Among Three Sectors: Supply Chain, Electricity, and Communications

This chapter introduces the inoperability input–output model (IIM) (Haimes and Jiang, 2001; Haimes, 2016), which is based on the input–output model developed by Leontief (1951a, b) for modeling the impact of the disruption of specified sectors of the global economy. For example, supply chain, electricity, and communications are three safety-critical sectors of every country's economy and key to its population's well-being. They literarily constitute the lifeline of every modern community and transcend cultural, societal, and political borders. Together, they comprise interdependent and interconnected emergent Complex SoS. Furthermore, their I-I represent significant universal sources of risk to the population they serve and to the economy of every country in the world. To model the impact of their disruption on the regional and national levels, we select the supply chain, to which Chapter 11 is devoted to in its entirety. Furthermore, we present (via four case studies in Chapter 10) the I-I that exist among the electricity, communications, and water systems. We quantify the consequences resulting from the I-I that characterize these three safety-critical sectors (supply chain, electricity, and communications) with the IIM. Note, however, that in Chapter 10 we quantify, via fault-tree analyses, the I-I among the Complex SoS by building on OR Gates (systems connected in series) and on AND Gates (systems connected in parallel).

Chapter 9: Guiding Principles for Modeling and Managing Complex Systems of Systems

This chapter presents and updates guiding principles for effective risk modeling, assessment, management, and communication associated with interdependent and interconnected Complex SoS. Risk analysis has become a dominant interest and a requisite area of expertise in almost every discipline as well as in government, public, and private organizations. By its nature, risk analysis is an intricate, dynamic process – an amalgamation of the arts and sciences – tailored to the specific sources of risk to Complex SoS. It follows then that for any system, especially for interdependent and interconnected Complex SoS, the balance between quantitative and qualitative risk analysis will be problem-and domain-specific. The 10 principles set forth in this chapter are intended to guide both quantitative- and qualitative-centered risk analyses. Meeting the challenges associated with defining, modeling, and quantifying the multidimensional risk function of a single system, and even more importantly for interdependent and interconnected Complex SoS, will likely be guided by the specific discipline performing such tasks, and influenced by the specific experiences and expertise of its risk modelers and decision makers. While the disciplines of systems engineering and risk analysis share the wide common denominators of philosophy, theory, methodology, and practice, each discipline has historically evolved along separate pathways, and thus has distinctive followers and protocols.

Chapter 10: Modeling Cyber–Physical Complex Systems of Systems: Four Case Studies

This chapter presents four cyber–physical case studies following the theory and methodology developed for Complex SoS in the previous chapters. The research protocol in each case shares the following common denominators, namely, that it builds on and exploits (i) state-space theory and methodology, (ii) modeling interdependencies and interconnectedness (I-I) via shared states and other essential entities (decisions, decision makers, stakeholders, resources, organizational setup, goals, and objectives, among others), (iii) analytically quantifying via fault-tree analysis (without the reliance on reliability) the I-I within and among the subsystems that compose Complex SoS, (iv) outlining the detailed methodological process, and (v) drawing lessons learned.

Chapter 11: Global Supply Chain as Complex Systems of Systems

The supply chain is the backbone of the global economy of every country, and its success is paramount to the success of individual economies as well as individual businesses with which we interact daily. Declaring the supply chain a safety-critical Complex SoS requires an understanding of its myriad basic intertwined systems and subsystems, which permeate through every country's economy, starting with raw and scarce material and heavy metals to abundant processed and manufactured commodities.

Chapter 12: Understanding and Managing the Organizational Dimension of Complex Systems of Systems

This chapter addresses the impact of organizational culture, vision, and the quality of leadership as critical drivers to effective and successful performance of interdependent and interconnected Complex SoS. The organizational structures vary so widely, especially among the private and public sectors, that they may be branded from their modeling perspectives as unbounded sets. A hierarchy of multiple conflicting, noncommensurate, and competing objectives characterizes both the private and public sector organizations. Moreover, the hierarchy of objectives is associated with the decision makers and stakeholders responsible for different levels of the organization's history, mandate, vision, and structure. Each organizational structure has specific characteristics that require customized modeling efforts. This chapter also demonstrates common denominators among them to enable modelers to exploit the essential attributes that most, if not all, organizational groups share.

Chapter 13: Software Engineering: The Driver of Cyber–Physical Complex Systems of Systems

This chapter illustrates the challenges and associated difficulties in understanding and modeling cyber–physical Complex SoS. The historical, cultural, organizational, and cognitive differences associated with the conception, design, development, and integration of cyber (software) and physical (hardware) systems comprise the ultimate cyber–physical Complex SoS. We often neither understand nor appreciate the two distinctive worlds in which both *software* and *hardware* are conceived, developed, and ultimately integrated. When the design, development, and ultimate integration of cyber and physical systems are executed separately, and without a strict adherence to compatibility between the two components of the cyber–physical Complex SoS, the results are likely to contain myriad predictable and invisible sources of risk.

Chapter 14: Community Preparedness for Risk to Infrastructure Complex Systems of Systems

Protection of Complex SoS may include a variety of risk-related countermeasures, such as detection, prevention, hardening, and containment. These all important risk management policy options are aimed at increasing security. To appreciate the limitations of these security measures when they are not balanced with resiliency, it is important to understand the epistemology of infrastructure risks in terms of threats to Complex SoS and of their vulnerability. In this chapter, we understand the composite of humans, organizations, and human–cyber–physical infrastructures to constitute, in their essence, interdependent and interconnected Complex SoS. These include human stakeholders as well as the multiple functions and operations of the physical infrastructures, such as roads and bridges, telecommunications networks, electric power generation, oil and gas pipelines and installations, and water treatment and supply utilities, to cite a few. Each of these individual entities constitutes a *subsystem*, and their integration with the other subsystems makes up interdependent and interconnected Complex SoS. The subsystems may function autonomously despite their I-I; however, each is susceptible to experiencing unique adverse consequences resulting from an initiating event that affects a subset or all of them. In turn, such consequences may propagate and inflict other disastrous results. This chapter also addresses the challenges in establishing a national preparedness system for terrorism and natural disasters.

Chapter 15: Modeling Safety of Highway Complex Systems of Systems via Fault Trees

Transportation is an emergent safety-critical interdependent and interconnected sector of the economy, which in its essence constitutes Complex SoS. Almost uncounted factors and sources of risk characterize and determine the ultimate safety of transportation on the highways. As a starting point, we investigate the modeling of automobile safety as a function of its design (assuming an average uniform level of safety characterizing all drivers). We focus primarily on how automobile accidents occur and address the broader challenge of quantifying and managing the risk inherent in particular automobile designs. The highway safety of the transportation Complex SoS draws on fault trees for assessing the risk thereto. The modeling process of improving highway safety can be used as a prototype for modeling other associated emergent Complex SoS. For example, there were about 222 million registered drivers in the United States in 2016, and there were over 32 000 fatal motor vehicle crashes in the United States in that year. The frequency and severity of motor vehicle accidents is of such consequence that they comprise the sixth leading cause of death in the United States and the number one cause of death due to injury. Current studies reveal that researchers have made an effective use of accident databases, simulations, and crash testing to examine narrowly defined factors that contribute to automobile safety. An evaluation of the contributions of poor driver judgment, vehicle failure, poor weather conditions, and other causative factors can offer insight into steering and other failures. This can guide the development of technologies to mitigate both the likelihood and the consequences of the associated failures.

Epilogue

Complexity and interdependent and interconnected emergent complex systems of systems, the theme of this book, have mesmerized all of us for decades: students and philosophers, practitioners and modelers, systems engineers and risk analysts, and other professionals engaged in this ever-growing and expanding enterprise. Among the many drivers that continuously redefine Complex SoS are the emergence of new technologies, notably the seemingly seamless integration of the physical infrastructure world with the more amorphous cyber world. Indeed, the new world in which we live, and on which we are becoming increasingly dependent, is the genesis of a new era: the era of the Internet of Things, the era of smart cities, and the era of robotics and automation that result in displacing workers and drivers of vehicles.

This book focuses on emergent Complex SoS and on the myriad associated theories, methodologies, and practices that transcend all 15 chapters.

We expect that the readers of this book will not only benefit from its content but also, most importantly, will build and expand on both the theory and methodologies advanced in this book.

PART II: ON THE RESILIENCE AND VULNERABILITY OF COMPLEX SYSTEMS OF SYSTEMS

1.II.1 INTRODUCTION

Why do farmers irrigate their crops in nonrainy seasons? And why do farmers add fertilizer to the soil? The genesis of both questions stems from the condition of the *states* of soil moisture and nutrients.

Our premise is that the vulnerability and the resilience are a manifestation of the states of a system, and that a threat to a vulnerable system would necessarily lead to risk. And the resulting risk is a function of the (i) specific threat, (ii) specific state of vulnerability of the system to the threat and (iii) specific time frame. In other words, while one may speak generically about risk, and define risk in terms of a probable threat and consequences to a vulnerable system, its quantification calls for a more careful consideration that goes back to Lowrance's definition of "Risk as a measure of the probability and severity of adverse effect" (Lowrance, 1976). Furthermore, a specific threat to a system that affects more than one state of that system would likely yield multiple adverse consequences – each is associated with a specific vulnerable subsystem of the Complex SoS. This statement does not exclude additional risks that may result from cascading effects of the multiple vulnerabilities of Complex SoS. Referring to the above farmer's dilemma consider moisture and acidity as two states that represent the soil's condition. Depriving the crops of irrigation when soil moisture is below the required level would impair the growth of the crops in specific ways. Similarly, depriving the soil of lime to neutralize high soil acidity (threat) would damage the growth of the crops. In other words, in each case, a specific threat to a vulnerable system would yield specific probable consequence (risk). (We define risk as "a measure of the probability and severity of adverse effects" (Lowrance, 1976).) In addition, there may be cascading effects resulting from the impact of one threat (either lack of irrigation in non-rainy season, or lack of adding lime) to other states of the soil. Factoring in soil nutrients, which constitute multiple states (e.g. one state for each critical nutrient), would further complicate the quantification of the multiple risks to which the crops might be subjected. In sum, a farm's soil with appropriate levels of moisture and nutrients can be sufficiently resilient and to safely withstand a short period of vulnerability to drought and a delay in adding fertilizers.

Systems engineering fundamentals are the building blocks of mathematical models and, especially, the notion of state variables (other building blocks are inputs and outputs, objective functions, and random, decision, and exogenous variables). The ultimate goal of all decision makers is to make appropriate decisions with which to manipulate the states of the system to achieve (i) specific goals and objectives, (ii) at acceptable tradeoffs (e.g. cost, assurance, quality, etc.), and (iii) within an acceptable time frame. For example, the farmer would also wish to control the states of soil moisture, acidity, and nutrients of the farm at an acceptable cost and within an appropriate time frame.

The terms "system vulnerability" and "system resilience" have become common in the parlance of risk analysis and systems engineering. This evolution has become more prevalent due to acts of terrorism, frequent natural disasters commonly attributed to climate change, and other acts with adverse effects on communities and on the natural and constructed environment. Sadly, a significant number of archival publications continue to "dance around" the essence of the resilience and vulnerability of a system by ignoring the fundamental tenet that both are manifestations of the states of the system. That is, a system's resilience and vulnerability are functions of the state of the system, the time frame, and the initiating event (threat). We offered the following definitions: *Vulnerability* refers to the inherent states of a given system (e.g. physical, technical, organizational, and cultural) that can be exploited by an adversary to adversely affect (cause harm or damage to) that system. *Intent* is the desire or motivation of an adversary to attack a target and to cause adverse effects. *Capability* is the ability and capacity to attack a target and to cause adverse effects. *Threat* denotes the *intent* and *capability* to adversely affect (cause harm or damage to) the system by *adversely changing its states.* A threat with adverse effects to a vulnerable system may lead to risk (Haimes, 2006). Throughout the rest of this book, the term *threat* will connote a threat with adverse effects. *Resilience*, however, has been more characterized than adequately defined in the literature.

Consider, for example, the following parochial "definitions," none of which refers to resilience as a manifestation (or a function) of the states of the system: (i) Resilience is the ability of a system to absorb external stresses, or a system's capability to create foresight, to recognize, to anticipate, and to defend against the changing shape of risk before adverse consequences occur. (ii) Resilience refers to the inherent ability and adaptive responses of systems that enable them to avoid potential losses, and capability of a system to minimize adverse consequences, and (iii) to recover quickly from adverse consequences.

In this book, and throughout our several earlier publications, we have defined resilience of a system as a manifestation of the states of the system. Most critically, it is a state vector that is threat and time dependent. Thus, resilience, in this book, also connotes the ability of the system to withstand a disruption within acceptable degradation parameters and to recover within an acceptable composite cost and time (Haimes, 2008, 2009, 2016). Moreover, neither vulnerability nor resilience can be measured in a single metric unit because the states of the system form a multidimensional vector, and so are the multidimensional consequences (the outputs) of the system for any specific inputs (threats). Note that the consequence (that is considered as part of the risk metric) is in fact the output of the systems model, and that the input to the system's model is commensurate with the concept of a threat. For example, the risk associated with a cyber attack on a cyber–physical Complex SoS depends on both the resilience of the system to the specific cyber attack and its sophistication. This is because resilience, as a function of the states of a system, can be measured only in terms of the specific threat (input), the system's recovery time, and the associated composite consequences. Thus, different attacks would generate different consequence (output) trajectories for the same resilient system.

Consider the immunization of a population against a major strain of flu virus termed Type B. Assume that the population develops resilience for multiple strains of viruses of Type B, except for an evolving strain of Type A. In this case, even though the population might have resilience (immunity) for Type B, the appearance of strain A into this population would likely be infectious. Here again the risk to the population from a threat is dependent on the specific type of threat, the time frame, and the states of the system; namely, the risk from a threat is dependent on the resilience of the system to the specific threat, and the ability of the (states of the) system to withstand that specific threat.

Likewise, consider any physical infrastructure, such as electric power, transportation, or telecommunication. In any such Complex SoS, the question "What is the resilience of infrastructure x?" is unanswerable because the *answer* implicitly depends upon knowing the specific threat, and the states of the system, and whether infrastructure x would recover following any attack y within an acceptable time and composite costs and risks. Thus, the only way such a question can be answerable is when the threat (or a set of threats) is specifically identified. Indeed, the system's resilience is not merely an abstract attribute of the system; rather, it is a state of the system (composed of a vector of sub-states) for which any specific sub-state may respond differently to different inputs (threats). For example, a water distribution system may have redundancy in its electric power subsystem, and thus it may be resilient to a major storm that would shut down one of the power lines to the water distribution system, leaving the other redundant line intact. On the other hand, suppose the water distribution system is dependent on only one main pipe to supply water to its customers, and it is located in a region susceptible to earthquakes. The system is resilient only to the extent that the main pipe is functioning and it can withstand an earthquake up to level 4 on the Richter scale. However, the system would likely fail during an earthquake of level 5 or 6. Here again, measuring the resilience of the water system is actually measuring the responses of the states of the system to the specific threat (input) and in this case the scale of the earthquake. *We will revisit the resilience and vulnerability of interdependent and interconnected Complex SoS in subsequent chapters.*

One may associate a vector of resilience with each subsystem, given no direct interdependencies exist among the subsystems of a Complex SoS with respect to the specific input/threat. Thus, there is a hierarchy of resilience attributes for any natural or the constructed environment. For example, the human body as a Complex SoS is made up of many subsystems (e.g. the digestive, pulmonary, and auditory systems, among others), each with a set of resilient organs and sub-organs, where the latter depends on the states of each organ and on the inputs (e.g. physical or biological threats). The duration of the output (e.g. a temporary or long-term impaired, or a loss of, functionality of specific organs or sub-organs) is a function of the corresponding affected states. This example reinforces the thesis that system resilience can be measured in terms of the outputs/(responses) given inputs/(threats) to the system. (Note that the inputs to the system, the states of the system, and the outputs are commonly time variant and probabilistic, as will be discussed subsequently.) To further appreciate the centrality of the system's input–output relationship to its resilience (states of the system), consider the fact that despite the resilience of the human body to various physical and biological attacks on it, its ultimate resilience depends upon the states of the body at the time, as well as the type and strength of such attacks.

A system may also be characterized by its specific redundancy and robustness. *Redundancy* refers to the ability of certain subsystems of a system (or of a Complex SoS) to assume the functions of failed subsystems without adversely affecting the performance of the Complex SoS itself. Of course, redundancies constitute an integral part of all safety-critical systems. *Robustness* refers to the degree of insensitivity of a system to perturbations, or to errors in the estimates of those parameters affecting the design choice.

1.II.2 RELATING THE CENTRALITY OF STATE VARIABLES TO THE DEFINITIONS OF RISK, VULNERABILITY, AND RESILIENCE

During the last several decades, the terms vulnerability, resilience, and risk have received multiple diverse definitions and interpretations. The fact is that all three are related, interdependent, and interconnected via the (i) states of the system under consideration; (ii) initiating event, e.g. a threat, or what we generally have called in this book *EFCs to connote internal or external forces that affect the states of the system positively or negatively*; and (iii) time frame. The fundamental difference between the impact of EFCs on a single system and their impact on an interdependent and interconnected emergent Complex SoS stems from the intrinsic characteristics of EFCs, introduced in Part I of this chapter.

Current risk analysis of a single system must be fundamentally extended when applied to Complex SoS. As noted in Part I of this chapter, this complexity stems primarily from the interdependencies and interconnectedness (I-I) within and among the systems and subsystems of SoS. Consider, for example, the I-I among three common cyber–physical infrastructures: electricity, communications, and water (and of course the communities they serve). A major initiating event that may cause the failure of the electricity system would result in adverse consequences to the other two cyber–physical infrastructures because of their dependency on electricity; neither one can operate without electricity. In the parlance of risk analysis, the states of each of the three infrastructures – electricity, communications, and water – could be at (i) different *states* of viable performance and security, (ii) *vulnerable* to different EFCs, and (iii) operating in different *time frames* (duration, day/night, season, etc.). Clearly, there is a need to understand and model the pathway through which the failure of one system propagates to other interdependent and interconnected systems of a Complex SoS. Indeed, an effective risk analysis of Complex SoS requires a clear understanding of the configurations of (i) the interdependencies and interconnections of the states of the systems that compose Complex SoS, (ii) the vulnerabilities and resilience of each system and the Complex SoS as a whole to specific (scenarios) of probable EFCs, and (iii) the time frame of the risk scenarios. Such analyses of risk scenarios would enable the identification of specific critical failure modes and the development and deployment of effective risk mitigation and other risk management strategies. Sadly, current risk models of systems do not consider the inherent I-I among the many systems that comprise Complex SoS; thus, the corresponding risk analysis is likely to be inferior to those models that do.

Recall our premise from Part I of this chapter that the above interdependencies and interconnections of Complex SoS are best understood, and thus modeled, via the *shared states* and other essential entities within the Complex SoS. And, that the term "essential entities" connotes shared/common decisions, decision makers, stakeholders, resources, organizational behavior and norms, policies and procedures, management, culture, and others. Also recall that the states of each system, and of the Complex SoS as a whole, play a significant role in estimating the future behavior of the systems for any given input. We requote here Chen's (2012) conceptual definition of state variables that posits that the "state $x(t_o)$ of a system at time t_o is the information at time t_o that, together with the input $u(t)$, for $t \geq t_o$, determines uniquely the output $y(t)$ for all $t \geq t_o$."

The genesis of the I-I can be traced to the shared/common *states* and to other shared *essential entities that characterize SoS*. Similarly, Nise (2014) emphasizes that state variables represent the smallest set of linearly independent system variables such that the values of the members of the set at time t_o along with known forcing functions completely determine the value of all system variables for all $t \geq t_o$. Thus, the behavior of the states of the system, as a function of time and other inputs, enables modelers and risk analysts to determine, under certain conditions, its future behavior for any given *input* or *initiating event*. Indeed, in industrial production and management, the feedback control process (as an integral part of risk management and quality control) is predicated on the fact that the smooth operations of the multiple interconnected subsystems are built on the knowledge of the operational states of each subsystem at each instant. As a simple example, to determine the reliability and functionality of a car, one must know the *states* of the fuel, oil, tire pressure, and the states of other mechanical and electrical components (due to the dynamic nature of all states).

Systems modelers face nontrivial challenges when selecting the minimum number of states to adequately and effectively represent (model) the subsystems under consideration and, thus, the risk thereto. Challenges in selecting the states in the risk modeling process for sustained years of drought include (i) availability and quantity (states) of groundwater and surface water in storage, (ii) quality (states) of groundwater and surface water in storage, and (iii) human and livestock population and industrial needs for water. The above states have a direct impact on rural or urban populations and on livestock, as well as on agriculture and industry. These states also adversely impact future recovery of the states of water quality and quantity for the entire River Basin.

From the perspective of the reliability of Complex SoS, shared states and other essential entities within and among the subsystems represent the essence of the I-I that characterize Complex SoS. *Initiating events affecting one or more of the interdependent and interconnected subsystems would necessarily increase the probability of failure of the entire SoS* (Haimes, 2018). This fundamental fact could be viewed in the parlance of fault-tree analysis (NUREG, 1981) as converting subsystems from being "marginally connected," or in parallel, to becoming "directly connected," or in series. Nuclear reactors, which epitomize Complex SoS, have always relied on fault-tree analysis to ensure their safety. From the perspective of fault-tree analysis, subsystems that share states and other essential entities *are most likely to* be affected by adverse initiating events. For example, this phenomenon is most evident in the emergent extensive use of cloud-computing technology, where numerous hardware–software subsystems are shared among multiple users (Haimes et al., 2015). Similar results have been demonstrated with the I-I among cyber–physical infrastructures. An initiating event may affect shared states of different subsystems differently. This fact necessarily implies the following three scenarios when analyzing risk to Complex SoS: (i) When each subsystem has different decision makers, then decisions made to control subsystem A may affect positively or negatively subsystem B that shares subsystem A's states and other essential entities. (ii) When decision makers collaborate among themselves and coordinate their decisions, this can have a positive effect on Complex SoS and improve their overall effectiveness and management. (iii) When different initiating events affect one or more subsystems, the level of shared states and other essential entities among them can positively or negatively affect the risk management process of the entire Complex SoS. Therefore, modelers and risk analysts should recognize and exploit these attributes of shared states and other essential entities among subsystems. As a natural example of interdependent subsystems, the states of the Earth and its moon relative to each other and to the sun are known at any instant due to shared gravitational forces. In this case, without any initiating events affecting any of them, the states of their celestial coordinates in space and time can be determined.

Uncertainty analysis becomes even more imperative in risk analysis of emergent Complex SoS. Two major sources of uncertainty in modeling affect risk analysis (Pate-Cornell, 1990, 1996; Apostolakis, 1999). *Knowledge* (*epistemic*) *uncertainty* may manifest itself in faulty selection of appropriate model topology (structure) and model parameters, due to ignorance of the system or of the SoS under consideration. *Variability* (*aleatory*) *uncertainty* applies to all relevant and important events and must not be overlooked. Sources of uncertainty dominate most decision-making processes, especially Complex SoS, and are the Achilles' heel for all deterministic and probabilistic models. Sources of uncertainty and lack of understanding of the complexity associated with one subsystem of SoS would likely result in (i) adherence to unrealistic assumptions, (ii) a lack of awareness of and accountability to the critical I-I of the SoS under consideration, (iii) a poorly selected representative model topology and model comprehensiveness, (iv) a dated or insufficient database to populate and calibrate the multiple models, and (v) risk scenarios essential to any risk assessment, management, and communication process being poorly represented or structured for all interdependent and interconnected systems. In addition, model uncertainties are often introduced through human errors of both commission and omission. In sum, uncertainty analysis associated with SoS is probably one of the most difficult, albeit important, tasks in the broader risk analysis process.

1.II.3 SYSTEMS ENGINEERING AND RELATING VULNERABILITY AND RESILIENCE TO THE RISK FUNCTION

Risk analysis and systems engineering/analysis share a common philosophical approach to problem solving, but they differ in their historical evolution and technical maturity. Both aspire to the *Gestalt*–holistic philosophy in their problem-solving methodologies. Systems modeling frameworks build on a plethora of theories, methods, tools, techniques, and practice to provide, to the extent possible, the instruments with which problems are studied, assessed, understood, managed, and solved (Haimes, 1989, 2009). Risk analysis is similar to systems engineering/systems analysis, which is predicated on the centrality of the *state-space theory and practice and of* their role in determining the resulting outputs (consequences) for each input (initiating event). Note that (i) the *performance capabilities* of a system are a function of its state vector; (ii) a system's *vulnerability* and *resilience vectors* are each a function of the input, its time of occurrence, and (the vector of) the states of the system and of the Complex SoS; (iii) the *consequences* are functions of the time of the event, the states vector of, the vulnerability, and the resilience of the system and of the Complex SoS; (iv) the *states* of a system are time dependent and commonly fraught with variability uncertainties and knowledge uncertainties; and (v) *risk* is a measure of the probability and severity of adverse effects (i.e. consequences). These five premises, among others, imply that *risk is a vector of the same units (dimensions) as the consequences and is a function of (i) time; (ii) the probability of the threat (initiating event) and its specificity (input); (iii) the probability of the consequences, given the threat; (iv) the states of the system (including its performance capability, vulnerability, and resilience); and (v) the vector of the resulting consequences.*

Based on the above discussion, it is appropriate to make the time domain explicit in the questions of the risk assessment process (developed by Kaplan and Garrick (1981)), and to the three original questions, namely, "What can go wrong? What is the likelihood? What are the consequences?," we add a fourth question: "Over what time frame?" (Haimes, 1991).

Consider a sample of the multidimensional vector of consequences from hurricanes *Harvey* and *Irma* that hit the southern part of the United States in 2017: loss of lives; displaced population; destruction of major infrastructure Complex SoS, e.g. electrical grid, transportation, and water supply; and major flooding of homes, roads, and myriad facilities. Other consequences are loss of jobs and erosion of confidence in government and technology, among others. If we were to develop risk scenarios for future hurricanes with an unusually high surge of water, a similar vector of risk components would necessarily emerge from the risk assessment process. Since consequences are measured through a natural vector of noncommensurate attributes, the units of each element of the risk vector ought to correspond respectively to the same units of the vector of consequences for each system that constitute a Complex SoS. Identifying and modeling the I-I among the myriad infrastructures, within and among the affected populations and communities, become a challenging risk modeling and management task.

The above discussion on risk analysis implies that significant modeling efforts are required to first evaluate the vector of consequences for each threat scenario (as functions of the threat (initiating event), the vulnerability and resilience of the Complex SoS and their subsystems, and the time of the event). Then each element of this vector must be paired with the (i) probability of the scenario's occurrence or (ii) probability of the severity of the consequences. This fundamentally complex modeling and analysis process cannot be performed correctly and effectively without relying on the states of the system being studied.

The multifaceted composition of risk to Complex SoS includes the levels of uncertainty and intensity of the initiating events or threats, the time frame, and the dynamic, probabilistic, and often nonlinear natures of the states of all natural and constructed environments on which the system's vulnerability and resilience depend. This intricacy cannot be modeled and understood on an ad hoc basis. In other words, *we must understand, model, and define the complexity of risk, vulnerability, and resilience in a systemic way and through a methodical, theoretically based systems approach, where the states of the system constitute the essence of the analysis.*

In sum, by projecting Heisenberg's uncertainty principle and Einstein's advice on the complexity of theories to the field of risk analysis, we assert, by paraphrasing, that (i) to the extent that quantifying the vulnerability to and the resilience and risk analysis of Complex SoS is precise, it is not real and (ii) to the extent that quantifying the vulnerability to and the resilience and risk analysis of Complex SoS is real, it is not precise.

1.II.4 MODELING AND QUANTIFYING THE CONSEQUENCES AND RISKS TO THREATENED COMPLEX SYSTEMS OF SYSTEMS AND THEIR VULNERABILITY AND RESILIENCE

This section builds on the premise introduced earlier that both vulnerability and system resilience are manifestations of the states of the system. In the following modeling effort, we use an existing discrete, linear, time-invariant, dynamic, and normally distributed stochastic model to formulate the dynamics of the vulnerability and resilience of a system (Guo and Haimes, 2017). The intention is to motivate researchers and practitioners to develop causal relationship models with which to relate the vulnerability and resilience of a system, and of Complex SoS, to policies and actions made for reducing their vulnerability and enhancing their resilience for specific threats.

The literature in systems engineering, operations research, system dynamics, decision analysis, process control, and risk analysis, among others, is replete with tools and assumptions to enable analysts and practitioners to model Complex SoS with simplified models. The most relevant example is control theory, or simply process control. The characterization and quantification of the states of a threatened system analyzed through its high vulnerability and low resilience, and the ultimate quantification of the associated risk function – fall into this modeling paradigm. As noted earlier, fundamental tradeoffs exist between model complexity and solution feasibility or simplicity. Indeed, since models ought to be as simple as possible and as complex as required, then the ultimate choice depends on myriad factors. For example, most graduate curricula in systems engineering, electrical engineering, and process control focus primarily on multidimensional, continuous and discrete dynamic, linear – with time-invariant coefficients – and normally distributed stochastic models. The reason is that relatively simple closed-form solutions for such models exist and are widely used. What follows is a representation of the vulnerability and resilience of a system adopted from a simplified modeling approach.

Let the state vectors of the vulnerability and resilience of the system at time k be represented, respectively, by Eq. (1.1):

$$\mathbf{v}(k)=[v_1(k),v_2(k)],\quad \text{and} \quad \mathbf{r}(k)=[r_1(k),r_2(k)] \tag{1.1}$$

We consider that the dynamical changes of $\mathbf{v}(k)$ and $\mathbf{r}(k)$ are independent of each other. This assumption will enable us to use a linear dynamic model. (It is also possible to define and add to the system of state equations a new state variable that relates the associated interdependency.) We represent the dynamics of the vulnerability and resilience of the system in the following discrete linear, time-invariant, dynamic, and normally distributed stochastic model.

For time (stage) $k=0, \ldots, T-1$, system's vulnerability and resilience are expressed with the set of Eq. (1.2):

$$\begin{aligned} v_1(k+1)&=a_1v_1(k)+b_{11}u_1(k)+b_{12}u_2(k)+w_1(k)\\ v_2(k+1)&=a_2v_2(k)+b_{21}u_1(k)+b_{22}u_2(k)+w_2(k)\\ r_1(k+1)&=a_3r_1(k)+b_{31}u_1(k)+b_{32}u_2(k)+w_3(k)\\ r_2(k+1)&=a_4r_2(k)+b_{41}u_1(k)+b_{42}u_2(k)+w_4(k)\end{aligned} \tag{1.2}$$

where $u_1(k)$ represents the threat (adverse disturbance), $u_2(k)$ represents risk management actions, and $w_i(k)$ for $I=1, 2, 3, 4$ are independent and normally distributed random variables that represent random variability in the states introduced into the model. The initial conditions are represented in Eq. (1.3):

$$\mathbf{v}(0)=\mathbf{v}_0,\mathbf{r}(0)=\mathbf{r}_0, \tag{1.3}$$

Model coefficients a_j and b_{ij} $(i=1, 2, 3, 4; j=1, 2)$ describe the contributions from the previous system states and system inputs to the current system states. These coefficients can be derived from historical data or system simulation. The time horizon is T stages $(k=0, \ldots, T-1)$.

Let the vector of consequences $\mathbf{y}(k)$ of the assumed threat to the system at time k be represented by $\mathbf{y}(k)=[y_1(k), y_2(k)]$; it can be described by the following set of equations:

$$\begin{aligned} y_1(k)&=c_1v_1(k)+d_1r_1(k)+\upsilon_1(k)\\ y_2(k)&=c_2v_2(k)+d_2r_2(k)+\upsilon_2(k)\end{aligned} \tag{1.4}$$

The coefficient vectors are defined as $\mathbf{c}=[c_1, c_2]^T$ and $\mathbf{d}=[d_1, d_2]^T$, where $\mathbf{c}$ and $\mathbf{d}$ are model coefficients and the variables $\upsilon_1(k)$ and $\upsilon_2(k)$ are independent and normally distributed random variables that represent the element of randomness introduced into the model. The coefficients a_i and b_{ij} $(i=1,2,3,4; j=1,2)$ describe the "contributions" from the previous system states and system inputs to the current system states. These coefficients can be derived from historical data or from system simulation. In short, Eqs. (1.1)–(1.4) represent vectors of vulnerability, resilience, and consequences.

1.II.5 ON THE RELATIONSHIP AMONG VULNERABILITY, RESILIENCE, AND PREPAREDNESS

Vulnerability, resilience, and preparedness are considered integral to addressing risks associated with Complex SoS. We noted how a natural and/or human (e.g. a terrorist) threat at a specific time to a given system (i) can adversely affect one or more states of a system, or a Complex SoS; (ii) an initiating event is commonly multidimensional with its probability and time frame; and (iii) where the states of a system, or Complex SoS, are commonly represented by vulnerability and resilience and the probabilistic multidimensional consequences. Moreover, the level and magnitude of the vulnerability and resilience of each system and of the Complex SoS as a whole and consequently the level and magnitude of the associated consequences all are functions of the states of the systems and of the Complex SoS.

The mathematics and engineering literatures on what is commonly termed *state-space theory* quantify the states of a system as functions of time, inputs, and random decision and exogenous variables, and where the outputs (consequences) are functions of the states of the system. Given that both the vulnerability and resilience of a system are manifestations of the respective states of the system, it is logical to quantify the consequences resulting from a threat through the states of the system.

Determining the impacts of current decisions on future options requires a continuous quantification of the dynamically evolving risk function. Intelligence collection and analysis associated with the tracking of EFCs to a targeted physical infrastructure, or cyber–physical infrastructure Complex SoS, constitute an ongoing process of a commonly adaptive, incremental risk modeling, assessment, management, and communication. Bayesian analysis (discussed extensively in Chapters 2 and 6) constitutes a critically important mechanism with which to update the probabilities of specific threats with newly gathered intelligence. Through the theory of scenario structuring, a large number of conceivable scenarios are developed and ultimately reduced to a group of significant and critical ones (Haimes, 2016). Without this last step, the resources required to invest in preparedness for a large number of scenarios would be prohibitive. The often incoherent and inconclusive sources of information and other intelligence on a tracked scenario require continued reassessing and reevaluating their evolution (along with all other tracked scenarios) at each stage of the analysis. As more intelligence and information become available, incremental investment in risk management can be an effective policy through which to minimize potential disasters within the budgeted resources.

In his classic book *Normal Accidents*, Perrow (1999) presents a comprehensive discussion of accidents. Appreciating the interplay among terrorism, natural hazards, and accidents is fundamental to understanding and benefiting from the synergistic results derived from investing for either purpose. A well-planned and well-executed preparedness plan can make threatened systems more resilient against both types of events. Also, it is imperative to understand the difference in the public perception and the psychological response to economic impacts and other devastation resulting from *acts of terrorism* versus *natural hazards and accidents*. Thus, it is constructive to distinguish between the two from sociopolitical perspectives. Indeed, although the consequences from the two events might be similar, the nature of the initiating events in the case of terrorism is critical in terms of public perception and acceptance, economic impacts (e.g. demand reduction), and impact on public policy and overall national security. In other words, it can be viewed as an unacceptable risk to be unprepared for certain types of terrorist attacks, as compared with certain types of natural disasters, because the sentiment of public acceptance can be entirely different depending on the event.

One approach to measuring the resilience of an infrastructure is to predict the trajectory of recovery time following a catastrophic event. Namely, how long would it take to achieve recovery from 10% to 90% of full capability, and at what level of resources? In some sense, cost and recovery time become synonymous with the resilience of Complex SoS and their interdependent and interconnected systems (infrastructures). Consider, for example, the possibility of developing a nationally shared, very secure information infrastructure dedicated to supporting critical Complex SoS. Such a system could add resilience to the country's critical infrastructures, particularly utilities and financial institutions that rely heavily on secure cyberspace to conduct their business. Furthermore, it could potentially be a cost-effective vehicle for reducing risks to critical interdependent infrastructures when compared with the alternative of hardening each of the individual infrastructures separately. Some of the ways that such a system could be used to enhance resilience are to support automation, distributed decision making, information sharing, remote human monitoring and control, automated sensing and control, machine-to-machine communication, and real-time network reconfiguration.

In the next section, we address the prevailing policies of executives of privately owned critical infrastructure, which lead to the tragedy of the commons syndrome.

1.II.6 INFRASTRUCTURE INTERDEPENDENCIES AND THE TRAGEDY OF THE COMMONS

The I-I among infrastructure Complex SoS, such as energy, telecommunications, banking and finance, transportation, and human services, have been widely acknowledged in the literature. Often recognized, but not proactively acted upon, is the central role that the resilience of one infrastructure plays in determining the resilience of other interdependent infrastructures as Complex SoS. This notion, akin to the tragedy of the commons, highlights the role of I-I in the resilience, and ultimate security, of Complex SoS.

Consider, for example, the importance of a resilient water supply system. To varying degrees, the failure of a water supply system (similar to any other interconnected and interdependent system) would affect the performance of other infrastructures. In particular, the operation of wastewater facilities as a Complex SoS may be hampered due to a shortage of finished (fresh)water, emergency services may be strained, and the generation and distribution of electrical power may be disrupted. Furthermore, this Complex SoS is managed by multiple government agencies at the federal, state, regional, and local levels with multiple stake holders, decision makers, and conflicting and often competing objectives. Also, these agencies have different missions, resources, agendas, and timetables. Finally, organizational and human errors and failures are common and may result in dire consequences. Thus, making a water supply infrastructure more resilient would affect the performance of other interdependent systems. Here is where the tragedy of the commons must be understood in the broad infrastructure resilience context.

1.II.7 EPILOGUE

Part II of this chapter highlights the centrality of the relationships among the vulnerability and resilience of, and the risk to, Complex SoS. Note that all three terms/attributes – vulnerability, resilience, and risk, while representing different characteristics of the Complex SoS under consideration, share the following common denominator: All are functions of the (i) states of the Complex SoS, (ii) initiating event/threat, and (iii) time frame.

REFERENCES

Aktan, A.E. and Faust, D. (2003). A holistic integrated systems approach to assure the mobility, efficiency, safety and integrity of highway transportation. *Invited Keynote Paper, First International Conference on Structural Health Monitoring and Intelligence Infrastructure (SHMII-1'2003)*, Tokyo, Japan (13–15 November 2003).

Alexander, R., Hall-May, M., and Kelly, T. (2004). Characterization of systems of systems failures. *Proceedings of the 22nd International System Safety Conference*, Providence, Rhode Island (2–6 August 2004).

Amaral, L. and Ottino, J.M. (2004). Complex networks: augmenting the framework for the study of complex systems. *The European Physical Journal B* **38**(2): 147–162.

Apostolakis, G. (1999). The distinction between aleatory and epistemic uncertainties is important: an example from the inclusion of aging effects into probabilistic safety assessment. *Proceedings of PSA'99* (22–25 August 1999). Washington, DC: American Nuclear Society.

Arnold, V. ed. (1994). *Dynamical Systems V: Bifurcation Theory and Catastrophe*, Encyclopedia of Mathematical Sciences, vol. **5**. Berlin: Springer-Verlag.

Baldwin, W.C., Sauser, B.J., and Boardman, J. (2017). Revisiting "the meaning of of" as a theory for collaborative system of systems. *IEEE Systems Journal* **11**(4): 2215–2226. doi: 10.1109/JSYST.2015.2430755.

Barabasi, A. and Albert, R. (1999). Emergence of scaling in random networks. *Science* **286**: 509–512.

Bar-Yam, Y. (2003a). When systems engineering fails: toward complex systems engineering. *International Conference on Systems, Man and Cybernetics* **2**: 2021–2028.

Bar-Yam, Y. (2003b). Complexity of military conflict: multiscale complex systems analysis of littoral warfare. *Chief of Naval Operations, Report for Contract: F30602-02-C-0158 Multiscale representations phase II: Task 2: Multiscale analysis of littoral warfare.*

Bellman, R.E. and Dreyfus, S.E. (1962). *Applied Dynamic Programming*. Princeton, NJ: Princeton University Press.

Bertalanffy, L.V. (1976). *General Systems Theory: Foundations, Development, Applications*. New York: George Braziller, Inc.

Blauberg, I.V., Sadovsky, V.N., and Yudin, E.G. (1977). *Systems Theory: Philosophical and Methodological Problems*. New York: Progress Publishers.

Boulding, K.E. (1953). *The Organizational Revolution*. New York: Harper and Row.

Bristow, M., Fang, L., and Hipel, K.W. (2012). System of systems engineering and risk management of extreme events: concepts and case study. *Risk Analysis* **32**(11): 1935–1955.

Bronowski, J. (1978). *The Origins of Knowledge and Imagination*. New Haven, CT: Yale University Press.

Chang, M. and Harrington, J. (2005). Agent-based models of organizations. In: *Handbook of Computational Economics II: Agent-Based Computational Economics*, vol. **2**(ed. K. Judd and L. Tesfatsion), 1–66. Amsterdam: North Holland.

Chen, C. (2012). *Linear System Theory and Design*, 4e. New York: Oxford University Press.

Chopra, D. (1994). *Ageless Body, Timeless Mind: The Quantum Alternative to Growing Old*. New York: Harmony, Random House.

Covey, S.R. (1989). *The Seven Habits of Highly Effective People*. New York: Simon and Schuster.

Dahmann, J. and Baldwin, K. (2008). Understanding the current state of US defense systems of systems and the implications for systems engineering. *Systems Conference, 2008 2nd Annual IEEE*, Montreal, Canada (7–10 April 2008), 1–7.

Dahmann, J., Rebovich, G., Lane, J. et al. (2011). An implementers' view of systems engineering for systems of systems. *Proceedings of the IEEE Systems*, Montreal, Quebec (4–7 April 2011).

DeLaurentis, D. (2008). Appropriate modeling and analysis for systems of systems: case study synopses using a taxonomy. *IEEE International Conference on System of Systems Engineering*, Monterey, California (2–5 June 2008).

DeLaurentis, D. and Callaway, R. (2004). A system-of-systems perspective for public policy decisions. *Review of Policy Research* **21**: 829–837. doi: 10.1111/j.1541-1338.2004.00111.x.

DeLaurentis, D., Dickerson, C., DiMario, M. et al. (2007). A case for an international consortium on system-of-systems engineering. *IEEE Systems Journal* **1**(1): 68–71.

Eisner, H. (1993). RCASSE: rapid computer-aided systems of systems engineering. *Proceedings of the 3rd International Symposium of the National Council of System Engineering, INCOSE* **1**: 267–273.

Eusgeld, I., Nan, C., and Dietz, S. (2011). "System-of-systems" approach for interdependent critical infrastructures. *Reliability Engineering and System Safety* **96**(6): 679–686.

Feynman, R.P., Leighton, R.B., and Sands, M. (1963). *The Feynman Lectures on Physics*. Reading, MA: Addison-Wesley.

Fisher, D. (2006). An emergent perspective on interoperation in systems of systems. *Technical Report CMU/SEI-2006-TR-003*.

Funderburk, F. (2004). Organizational culture from a complex dynamic systems perspective: moving from metaphor to action in healthcare. *System Models of Organizational Behavior*. www.necsi.edu (accessed 8 January 2018)

Gheorghe, A.V. (1982). *Applied Systems Engineering*. New York: Wiley.

Gordon, W.J.J. (1968). *Synectics: The Development of Creative Capacity*. New York: Collier Books.

Gorod, A., Sauser, B., and Boardman, J. (2008). System-of-systems engineering management: a review of modern history and a path forward. *IEEE Systems Journal* **2**(4): 484–499.

Guo, Z. and Haimes, Y.Y. (2017). Exploring systemic risks in systems-of-systems within a multi-objective decision framework. *IEEE-SMC Systems* **47**(6): 906–915.

Haberlin, R. and Haimes, Y.Y. (2018). Regional infrastructures as complex systems of systems: a shared state model for regional resilience. *ASCE Journal of Infrastructure Systems* **24**(3). https://doi.org/10.1061/(ASCE) IS. 1943-555X.0000421

Haimes, Y.Y. (1977). *Hierarchical Analyses of Water Resources Systems: Modeling and Optimization of Large-Scale Systems*. New York: McGraw-Hill.

Haimes, Y.Y. (1981). Hierarchical holographic modeling. *IEEE Transactions on Systems, Man, and Cybernetics* **11**(9): 606–617.

Haimes, Y.Y. (1989). Toward a holistic approach to risk assessment and management. *Risk Analysis* **9**(2): 147–149.

Haimes, Y.Y. (1991). Total risk management. *Risk Analysis* **11**(2): 169–171.

Haimes, Y.Y. (2007). Phantom system models for emergent multiscale systems. *Journal of Infrastructure System* **13** (2): 81–87.

Haimes, Y.Y. (2008). Models for risk management of systems of systems. *International Journal of Systems of Systems Engineering* **1**(1/2): 222–236.

Haimes, Y.Y. (2009). On the complex definition of risk: a systems-based approach. *Risk Analysis* **29**(12): 1647–1654.

Haimes, Y.Y. (2012a). Modeling complex systems of systems with phantom system models. *Systems Engineering* **15**(3): 333–346.

Haimes, Y.Y. (2012b). Systems-based guiding principles for risk modeling, planning, assessment, management and communication. *Risk Analysis* **32**(9): 1451–1467.

Haimes, Y.Y. (2016). *Risk Modeling, Assessment, and Management*, 4e. Hoboken, NJ: Wiley.

Haimes, Y.Y. (2018). Risk modeling of interdependent complex systems of systems: theory and practice. *Risk Analysis* **38**: 84–98. doi: 10.1111/risa.12804.

Haimes, Y.Y. and Jiang, P. (2001). Leontief-based model of risk in complex interconnected infrastructures. *ASCE Journal of Infrastructure Systems* **7**(1): 1–12.

Haimes, Y.Y. and Macko, D. (1973). Hierarchical structures in water resources systems management. *IEEE Transactions on Systems, Man, and Cybernetics, SMC* **3**(4): 396–402.

Haimes, Y.Y., Tarvainen, K., Shima, T., and Thadathil, J. (1990). *Hierarchical Multiobjective Analysis of Large Scale Systems*. New York: Hemisphere Publishing.

Haimes, Y.Y., Horowitz, B., Guo, Z. et al. (2015). Assessing systemic risk to cloud-computing technology as complex interconnected systems of systems. *Systems Engineering* **18**(3): 284–299.

Hall, A.D. (1989). *Metasystems Methodology: A New Synthesis and Unification*. Elmsford, New York: Pergamon Press.

Heisenberg, W. (1930). *Die Physikalischen Prinzipien der Quantenmechanik*. Leipzig: Hirzel English translation The Physical Principles of Quantum Theory, Chicago: University of Chicago Press.

Hipel, K.W., Radford, K.J., and Fang, L. (1993). Multiple participant multiple criteria decision making. *IEEE Transactions on Systems, Man, and Cybernetics, Vol. SMC* **23**(4): 1184–1189.

Jamshidi, M. ed. (2009a). *System of Systems Engineering: Innovations for the 21st Century*. Hoboken, NJ: Wiley.

Jamshidi, M. ed. (2009b). *System of Systems Engineering: Principle and Applications*. New York: CRC Press, Taylor & Francis Group.

Johnson, C. (2006). What are emergent properties and how do they affect the engineering of complex systems? *Reliability Engineering and System Safety* **91**(12): 1475–1481.

Kaplan, S. and Garrick, B.J. (1981). On the quantitative definition of risk. *Risk Analysis* **1**(1): 11–27.

Katzenbach, J. and Smith, D. (2015). *The Wisdom of Teams*. Boston: Harvard University Business Press.

Keating, C.B. (2005). Research foundations for system of systems engineering. *IEEE International Conference on Systems, Man and Cybernetics*, Waikoloa, Hawaii (10–12 October 2005), 2720–2725.

Kilgour, D.M. and Hipel, K.W. (2005). The graph model for conflict resolution: past, present, and future. *Group Decision and Negotiation* **14**(6): 441–460.

Lasdon, L.S. (2011). *Optimization Theory for Large Systems*. New York: Dover.

Leondes, C.T. ed. (1969). *Advances in Control Systems*, vol. **6**. New York: Academic Press.

Leontief, W.W. (1951a). Input/output economics. *Scientific American* **185**(4): 15–21.

Leontief, W.W. (1951b). *The Structure of the American Economy, 1919–1939*, 2e. New York: Oxford University Press.

Lewe, J., DeLaurentis, D., and Mavris, D. (2004). Foundation for study of future transportation systems through agent-based simulation. *Proceedings of the 24th ICAS Congress*, Yokohama, Japan (August 2004).

Lewis, B. and Haimes Y.Y. (2018). Risk modeling of communications, navigation, and surveillance complex systems of systems for future aviation. *Systems Engineering* **2018**: 1–10. doi: 10.1002/sys.21423.

Li, K.W., Hipel, K.W., and Kilgour, D.M. (2004). Preference uncertainty in the graph model for conflict resolution. *IEEE Transactions on Systems, Man, and Cybernetics, Part A* **34**(4): 507–520.

Lloyd, S. and Lloyd, T. (2003). *Bits and Bucks: Modeling Complex Systems by Information Flow*, Working Paper Series ESD-WP-2003-01.17. Cambridge, MA: MIT Engineering Systems Division.

Lowrance, W. (1976). *Of Acceptable Risk*. Los Altos, CA: William Kaufmann.

Macko, D. (1967). General System Theory Approach to Multilevel Systems. *Report SRC 106-A-67-44*, Systems Research Center: Case Western Reserve University.

Macko, D. and Haimes, Y.Y. (1978). Overlapping coordination of hierarchical structures. *IEEE Transactions on Systems, Man, and Cybernetics* **8**(10): 745–751.

Maier, M.W. (1998). Architecting principle for systems-of-systems. *Systems Engineering* **1**(4): 267–284.

Maier, M. and Rechtin, E. (2009). *The Art of Systems Architecting*, 3e. Boca Raton, FL: CRC Press.

Merriam-Webster (2017). Essence. www.merriam-webster.com (accessed 24 November 2017).

Mesarović, M.D. (1964). *Views on General Systems Theory*. New York: Wiley.

Mesarović, M.D. ed. (1968). *Systems Theory and Biology*. New York: Springer-Verlag.

Mesarović, M.D., Macko, D., and Takahara, Y. (1970). *Theory of Hierarchical, Multilevel Systems*. New York: Academic Press.

Nise, N. (2014). *Control Systems Engineering*. Pomona, CA: California State Polytechnic University.

Ottino, J.M. (2003). Complex systems. *AIChE Journal* **49**(2): 292–299.

Page, S.E. (1999). Computational models from a to Z. *Complexity* **5**(1): 35–41.

Parker, J. (2010). Applying a system of systems approach for improved transportation. *S.A.P.I.E.N.S.* **3**(2).

Paté-Cornell, M.E. (1990). Organizational aspects of engineering system safety: the case of offshore platforms. *Science* **250**: 1210–1217.

Paté-Cornell, M.E. (1996). Uncertainties in risk analysis: six levels of treatment. *Reliability Engineering and System Safety* **54**(2/3): 95–111.

Perrow, C. (1999). *Normal Accidents: Living with High-Risk Technologies*. Princeton, NJ: Princeton University Press.

Sage, A.P. (1977). *Methodology for Large Scale Systems*. New York: McGraw-Hill.

Sage, A.P. (1992). *Systems Engineering*. New York: Wiley.

Sage, A.P. (1995). *Systems Management for Information Technology and Software Engineering*. New York: Wiley.

Sage, A.P. and Biemer, S.M. (2007). Processes for system family architecting, design and integration. *IEEE Systems Journal* **1**(1): 5–16.

Sage, A.P. and Cuppan, C.D. (2001). On the systems engineering and management of systems of systems and federation of systems. *Information, Knowledge, Systems Management* **2**(4): 325–345.

Sage, A.P. and Rouse, W.B. ed. (1999). *Handbook on Systems Engineering and Management*, 2e. New York: Wiley.

Sahin, F., Jamshidi, M., and Sridhar, P. (2007). A discrete event XML based simulation framework for system of system architectures. *IEEE International Conference on System of Systems Engineering* (4 April 2007), 1–7.

Shalizi, C. (2006). Methods and techniques of complex systems science: an overview. In: *Complex Systems Science in Biomedicine* (ed. T. Deisboeck and J. Kresh). New York: Springer, Inc.

Shenhar, A. (1994). A new systems engineering taxonomy. *Proceedings of the 4th International Symposium of the National Council of System Engineering, INCOSE* **2**: 261–276.

Singh, M.G. ed. (1987). *Systems and Control Encyclopedia: Theory, Technology, Applications*. New York: Pergamon Press.

Sousa-Poza, A., Kovacic, S., and Keating, C.B. (2009). System of systems engineering: an emerging multidiscipline. *International Journal of System of Systems Engineering* **1**(1/2): 1–17.

Thissen, W.A. and Herder, P.M. (2003). *Critical Infrastructures: State of the Art in Research and Application*. New York: Kluwer Academic Publishers.

Tivnan, B., Henderson, S.G., Biller, B. et al. (2007). Modeling organizational adaptation: a replication of Levinthal's model of Emergent order. *Proceedings of the 2007 Winter Simulation Conference*, Washington, DC (9–12 December 2007).

U.S. Nuclear Regulatory Commission (1981). *Fault Tree Handbook*, NUREG-0492. Washington, DC: U.S. Government Printing Office.

Warfield, J.N. (1976). *Societal Systems: Planning, Policy, and Complexity*. New York: Wiley Interscience.

Wiener, N. (1961). *Cybernetics*, 2e. Cambridge, MA: MIT Press.

Zeigler, B.P. (1984). *Multifaceted Modeling and Discrete Simulation*. New York: Academic Press.

2

Modeling, Decomposition, and Multilevel Coordination of Complex Systems of Systems

PART I: DECOMPOSITION AND HIERARCHICAL MODELING

2.I.1 INTRODUCTION

Complex interdependent and interconnected systems of systems (Complex SoS) are commonly composed of multiple subsystems, including technological, societal, environmental, and organizational systems. Thus their modeling is more elaborate than modeling a single system. The high dimensionality of Complex SoS (very large number of systems and subsystems) and thus the intricacy is due to shared states and other *essential entities* among the systems and subsystems, which can overwhelm the analysts and thus the analysis. The term *essential entities* includes shared/common decisions, decision makers, stakeholders, resources, organizational behavior and norms, policies and procedures, management, culture, and others. (Note: The role of the states of a system in modeling was widely explained in Chapter 1.)

The hierarchical-multilevel approach to modeling is based on the (i) decomposition (at the lower level of the hierarchy) of the interdependencies and interconnectedness (I-I) among the subsystems and systems that comprise Complex SoS, (ii) subsequent modeling of the subsystems as "independent" at the lower level of the hierarchy, and (iii) ultimate coordination of the subsystems at the higher level of the hierarchy. This decentralized approach uses strata, layers, and echelons to aid modelers in analyzing and understanding the behavior of the subsystems at a simpler, lower level of the hierarchy and to transmit that information to subsystems at higher levels. There was a concerted effort four to five decades ago to develop theory and methodology for hierarchical and multilevel coordination of complex systems (Bauman, 1966; Lasdon, 1970; Mesarović et al., 1970; Haimes, 1977, 1981, 2012, 2016; Haimes et al., 1990).

Whenever more decentralization of Complex SoS is needed, the system can be further decomposed to enable effective modeling and coordination among the subsystems, as well as coordination among the various decision makers and associated stakeholders. One common approach for achieving an effective decomposition is by introducing new variables into the system, called pseudo-variables, to enable modelers and high-level coordination to treat them as "parameters" at the lower level of the hierarchy and

Modeling and Managing Interdependent Complex Systems of Systems, First Edition. Yacov Y. Haimes.

subsequently to be determined at a higher level in the multilevel hierarchy. This is often accomplished using the Lagrangian function and Lagrange multipliers. (Note that the Lagrangian function and the Lagrange multipliers are introduced and discussed in the Appendix of this book.) For a two-level coordination (or decision making), it is common to refer to the first level as the "primal problem" and to the second level as the "dual problem." The necessary and sufficient condition for a convergence of the first and second levels of the multilevel hierarchy is the union of the primal problem with the second-level dual system (or problem), assuming there is no "duality gap." Furthermore, by minimizing the lower level (as the primal problem) with respect to the model's decision variables and maximizing the dual problem with respect to the Lagrange multipliers, it is possible to accomplish such coordination, assuming there is no duality gap. Convexity ensures convergence and ensures that there is no duality gap between the first and second levels of the Lagrangian function. (The reader who is unfamiliar with duality and the Kuhn–Tucker conditions for optimality may want to consult the Appendix in this book.) Restating the above differently, each subsystem is separately and independently modeled and coordinated with reliance on different coordination methods, based on the nature of the subsystem models as well as on the objectives and constraints of the subsystems. Results from the first level are used at the second-level higher level coordination. The subsystems are joined by the coupling/shared variables that are manipulated at the second, or at a higher level, in order to arrive at an optimal solution for the entire system. Restating the above, one way to achieve subsystem "independence" is by first relaxing, at the lower level of the hierarchical system, one or more of the necessary conditions for optimality, and then satisfying these conditions at the second or higher level. In Chapter 5 we will introduce multiobjective modeling, decision making, and the associated tradeoffs among the various objectives, as well as the importance of the time frame in modeling and in decision making.

To illustrate the hierarchical-multilevel philosophy/approach to modeling and managing Complex SoS, consider a company with N departments, each of which has a manager who reports to the president of the company. In this model, the sole objective of the ith manager is to maximize the performance of his/her department, and that of the president is to maximize the performance of the entire company by imposing internal shadow prices (i.e. the Lagrange multipliers) for the transactions among the departments' products or services. The company's overall optimal policy is thus achieved through an iterative procedure. The separate decision making of each department does not necessarily imply the overall maximization of the company's performance, unless the performance of each department is coordinated at a higher level, i.e. by the president. Most, if not all, decisions are made within multiple objectives that are often in conflict and competition with each other. (See Chapter 5.)

2.1.2 ATTRIBUTES OF DECOMPOSITION AND MULTILEVEL MODELING OF COMPLEX SYSTEMS OF SYSTEMS

In addressing Complex SoS, decomposition, multilevel modeling, and decision making have several significant advantages over conventional modeling methods. The following is a brief summary of these attributes that will be further explored in the broader context of modeling Complex SoS:

1. *Conceptual simplification of complex systems.* Decomposing Complex SoS into several subsubsystems can achieve a conceptual simplification of their modeling process. This is especially important for highly coupled Complex SoS, where multiple outputs of one subsystem are the inputs to others, namely, the subsystems' shared states and other essential entities.
2. *Reduction in dimensionality and in modeling and managing complexity.*
   By identifying the shared states and other essential entities within and among the Complex SoS, modelers using pseudo-variables are able to decompose the latter into "independent" subsystems at a lower level of the hierarchy and coordinate the shared states and other essential entities at higher levels. Often this reduction in variables enables modeling and decision making not otherwise achievable.

3. *More realistic system models.*
   A significant advantage of the hierarchical-multilevel approach is that none of the functions of the system's model needs to be linear; thus more flexible mathematical models can be constructed to represent the real system. This addresses the imbalance that often exists between the complexity of SoS models and their effective modeling and analysis and, when needed, the management and decision making tied to them. These tradeoffs are commonly reflected in the vast number of linearized models in the literature that take advantage of the simplicity and tractability of linear models. By applying decomposition and multilevel decision-making methods, there is no costly sacrifice of realism in modeling since more representative and sophisticated nonlinear mathematical models can be constructed.
4. *Interactions among subsystems can be more effectively modeled and analyzed.*
   Acknowledging the nature of the I-I among the subsystems that make up Complex SoS constitutes a major positive step in their modeling and management. The I-I reflected in the modeling and management processes are critical since all real-world systems are highly coupled and the interactions among their various components can be directly addressed. Water resources and cyber-physical systems can serve as an excellent example. The benefits resulting from the construction of a dam cannot be properly assessed unless the subsequent impact on the water quality, land use, sheet erosion, flood control, and socioeconomic effects, among others, are considered. These couplings, manifested via shared states and other essential entities, are well handled by the hierarchical-multilevel approach.
5. *Applicability to both static and dynamic systems.*
   Decomposition and multilevel modeling of Complex SoS are applicable to both static and dynamic systems. The time frame, which plays an important role in such modeling, need not be imbedded or ignored in the analyses (as is the case in static models). Complex SoS can be modeled by both static algebraic equations and dynamic differential equations. As a result, the tradeoff between simplifying the complexity of modeling Complex SoS and their corresponding management can be more effectively realized.
6. *Different modeling and decision-making methods for each subsystem.*
   Each subsystem in the Complex SoS hierarchy can be handled with different analyses, decision making, or management procedures depending on the nature of the Complex SoS subsystem. This attribute provides important flexibility in modeling and in decision making that enables modelers and analysts to utilize state-of-the-art techniques in the modeling and analysis and supports policy formulation and decision making.
7. *Economic interpretation of the variables.*
   Duality theory of nonlinear programming is central to the computational procedure in hierarchical-multilevel modeling, analyses, and decision making. In particular, the interpretation of the Lagrange multipliers as shadow prices often sheds light on understanding the system's behavior. (See Appendix.)
8. *Applicable to multiobjective analyses.*
   In a hierarchical structure, different objective functions may be associated with each subsystem. Coordination of the resulting overall system becomes a multiobjective analysis problem. In particular, since most objective functions are noncommensurable, the surrogate worth tradeoff (SWT) method can be successfully applied (see Chapter 5 on multiobjective analyses).
9. *Enabling the identification of shared states and other essential entities among the subsystems.*
   Central to modeling and understanding the I-I that characterize Complex SoS is the identification of shared states and other essential entities within and among the subsystems. The multiple perspectives that can be attributed to Complex SoS, and thus to their decomposition, commonly yield invaluable insights into and knowledge of the subsystems.

When modeling Complex SoS, more than one mathematical or conceptual model is likely to emerge; each of these models may focus on a specific aspect of the system, yet all may be regarded as acceptable representations of the system (Macko and Haimes, 1978). Consequently, decomposing Complex SoS often presents a dilemma over the choice of subsystems. For example, an economic system may be decomposed into geographic regions or activity sectors. An electric power management system may be decomposed according to the various functions of the system (e.g. power generation units, energy storage units, transmission units) or along geographic or political boundaries. Another decomposition might be a time-wise decomposition into planning periods. If several aspects of the system are to be dealt with, such as the geographic regions and activity sectors of an economic system, it could be advantageous to consider several decompositions. For example, four major decomposition structures may be identified for water resources systems on the basis of political or geographical, hydrological, temporal, and functional considerations.

2.I.3 GENERAL HIERARCHICAL STRUCTURES

Hierarchical systems or structures are families of subsystems arranged in a hierarchical fashion. Each subsystem is commonly concerned with a specific aspect of the overall Complex SoS and occupies a specific level in the hierarchy. The operation of a subsystem on any given level of the hierarchy is directly and explicitly influenced by higher-level subsystems. Policies and decisions of higher-level subsystems are commonly binding on the lower-level subsystems and reflect a priority of importance in the actions and goals of the higher-level subsystems. Note that fulfillment of the higher-level objectives depends on the actions and performance of the lower-level subsystems. Three kinds of hierarchical-multilevel structures are presented here for dealing with Complex SoS.

2.I.3.1 Multistrata Hierarchy

The multistrata hierarchy is highly effective for modeling Complex SoS. The tradeoffs are between (i) simplicity, a requisite for understanding the I-I that characterize Complex SoS and consequently in the analysis, decision making, and management strategy applied and (ii) comprehensiveness, to account for the unpredictable behavior of Complex SoS. These descriptive levels of a hierarchy are commonly referred to as strata. Lower strata involve more detailed and specialized descriptions of the system than the higher strata. Each stratum has its own concepts and principles and may deal with different aspects of Complex SoS.

2.I.3.2 Multilayer Decision-Making Hierarchy

A multilayer hierarchy is used in resolving complex decision-making situations associated with Complex SoS. Two trivial but important conditions of almost all real-life decision situations are to act without delay, and thereby avoid decisions by default, and to take time to better understand the situation. Layers are essentially levels of decision-making complexity. Parameters in lower-layer problems are fixed by "solutions" of higher-layer problems, and a solution or understanding of the original problem is obtained when all subproblems in the sequence are understood or "solved." A complex decision-making hierarchy naturally emerges in reference to several essential phases of the decision-making process, e.g. gathering information, reducing uncertainties, and selecting a preferred course of action.

2.I.3.3 Multiechelon Hierarchy

The third kind of hierarchy, a multiechelon hierarchy, arises when Complex SoS are explicitly viewed as consisting of many interacting subsystems. Such hierarchies must deal with the mutual relationships among the various subsystems comprising the Complex SoS. Each subsystem is considered as goal seeking, that is, satisfying specific objectives and aspiration levels. In this multiechelon hierarchy, conflicts within subsystems on one echelon are resolved by higher echelon subsystems. The coordination and the resolution of conflicts are commonly accomplished by intervention, which can be achieved by including certain variables in the subsystem's objectives as parameters to be manipulated by higher echelon subsystems. For example, intervention may be of the following kinds: (i) goal intervention that affects goal-related factors, (ii) information gathering that affects outcome expectations, and (iii) constraint intervention that affects available alternative actions.

All three groups of hierarchies may emerge simultaneously in dealing with Complex SoS. The different hierarchies serve different purposes and may be imbedded within one another; the multistrata hierarchies are for modeling purposes; the multilayer hierarchies are for resolving the complexities of decision making; and the multiechelon hierarchies are for organization purposes. The following are features common to all three groups of hierarchies: (i) Higher-level subsystems portray, or are concerned with, a larger portion or broader aspects of the overall Complex SoS, (ii) higher-level subsystems have longer decision horizons and are concerned with longer-range behavior, and (iii) higher-level subsystems have a priority of action over lower-level subsystems, within the Complex SoS.

2.1.4 DECOMPOSITION AND COORDINATION OF COMPLEX SYSTEMS OF SYSTEMS

Several methodologies have been established in the literature for handling large modeling and optimization problems via decomposition and hierarchical coordination schemes (Lasdon, 1964, 1970; Bauman, 1966; Mesarović, 1968; Mesarović et al., 1970; Haimes, 1971, 1973a, b, 1975, 1977; Wismer, 1971; Haimes and Macko, 1973; Leondes, 1973; Lefkowitz, 1966; and others). Common to these methodologies is the ability to (i) restructure the system model such that the overall problem becomes additively separable in the subsystems' variables (at the lower levels of the hierarchy); (ii) form subsystem problems, which are modeled independently of each other; and (iii) achieve an overall effective iterative coordination among the subsystems that make up Complex SoS.

2.1.4.1 General Problem Formulation

It is convenient to define a system modeling problem for a region R and use it as a vehicle in explaining the various decomposition and coordination methodologies. Several applications of the methodologies will be to water resources systems. Although all Complex SoS are characterized by multiple goals and objectives associated with most subsystems, for pedagogical reasons we will address a single objective, leaving the multiple-objective analysis of Complex SoS to Chapter 5, which is devoted in its entirety to multiple-objective tradeoff analysis.

Let the overall optimization problem for a given region R be represented by

$$\max_m f(y,u,m,\alpha), \text{ subject to } g(y,m,u,\alpha) \le 0, \quad y = H(u,m,\alpha) \tag{2.1}$$

where

y is the vector of regional outputs

u is the vector of regional inputs (not manipulable)

m is the vector of manipulable variables, i.e. decision variables

α is the vector of model parameters

G is the vector representing the regional operating constraints

It is an elementary extension to also include equality constraints in the development. Substituting the output relations gives the overall system optimization problem as

$$\max_m f(u,m,\alpha), \text{ subject to } g(m,u,\alpha) \le 0 \tag{2.2}$$

where $f(u,m,\alpha)$ is a scalar-valued function whose value is an indication of the overall system performance. The performance of most water resources systems is characterized by a vector of goals and objectives, which will be addressed in Chapter 5. Introducing multiobjective functions at this stage will only complicate our discussion on hierarchical-multilevel structures.

We will assume hereafter in this section that region R is decomposed into N subregions. Generally, the ith subregion R_i will have an objective function $F_i(x_i, u_i, m_i, \alpha_i, \sigma)$ and operating constraints $G_i(x_i, u_i, m_i, \alpha_i, \sigma) \le 0$. The vector σ is a vector of coordination (pseudo) variables enabling the decomposition. More will be said about σ in the ensuing sections. The vectors u_i, m_i, and α_i are subvectors of u, m, and α, respectively, and x_i is a vector of inputs to subregion R_i from other subregions. In general, the subregions are coupled via their inputs and outputs:

$$x_i = \sum_{j=1}^{N} C_{ij} y_j, \quad i = 1,\dots,N \qquad y_i = H_i(x_i,u_i,m_i,\alpha_i), \quad i = 1,\dots,N \tag{2.3}$$

where y_i is a vector of outputs of subregion R_i and the C_{ij} are coupling matrices representing the I-I.

In general, the ith subregional optimization problems assume the form

$$\max f_i(x_i, u_i, m_i, \alpha_i, \sigma), \text{ subject to } g_i(x_i, u_i, m_i, \alpha_i, \sigma) \le 0 \tag{2.4}$$

The specifics of the subregional optimization problems depend on the specific decomposition and coordination methods used.

Each of the decomposition and coordination methodologies discussed here requires that the overall regional decision-making model be expressed in a form, which is separable in the subregions' variables. Therefore, let us assume that the overall regional problem can be rewritten as

$$\max \sum_{i=1}^{N} f_i(x_i, u_i, m_i, \alpha_i), \text{ subject to } g_i(x_i, u_i, m_i, \alpha_i) \le 0,$$

$$y_i = H_i(x_i, u_i, m_i, \alpha_i), \quad x_i = \sum_{j=1}^{N} C_{ij} y_j \tag{2.5}$$

where $i = 1, \ldots, N$.

2.1.4.2 Nonfeasible or Interaction Balance Methods

Nonfeasible or interaction balance methods form independent subsystem optimization problems by decoupling the subregions and attaching prices to the coupling variables (Lasdon, 1964; Pearson, 1965; Lasdon and Schoeffler, 1966; Macko, 1967; Mesarović et al., 1970; Haimes, 1971; Haimes, 1972; Haimes et al., 1972a, 1972b; and others). The vector of coordination variables is a vector of prices entering the subregions' objective functions. In particular, the ith subsystem optimization problem is then

$$\max_{m_i, x_i} f_i(x_i, u_i, m_i, \alpha_i) + \left(\sum_{j=1}^{N} \sigma_j C_{ji} \right) y_i - \sigma_i x_i,$$

$$\text{subject to } g_i(x_i, u_i, m_i, \alpha_i, \sigma) \le 0, \quad y_i = H_i(x_i, u_i, m_i, \alpha_i) \tag{2.6}$$

The inputs x_i entering the subregion from other subregions are treated as if they could be freely manipulated.

The subregions are coordinated via the prices (presented by Lagrange multipliers) entering their objective functions. The coordinating prices admit optimal subregional solutions that satisfy the coupling equations (Eq. (2.3)), i.e. the interaction balance. If the optimal subregional solutions satisfy the coupling equations, they will be optimal for the overall region. Therefore, the coordinator's objective is to drive the coupling errors to zero.

Barring duality gaps, the coordinating prices in fact minimize the dual overall objective function, which is the sum of the optimal values of the subregion's objective in Eq. (2.6). This function is generally not available to the coordinator. However, it is always convex with respect to the prices, and the optimal subregional solutions readily provide its subgradients with respect to the prices. In particular, the errors in the coupling equations (Eq. (2.3)) are its subgradients. Consequently, the coordinator can use the coupling errors to obtain directions in which to adjust the prices. On the other hand, the coordinator may use the coupling errors to build up a tangential approximation (outer linearization) of the dual overall objective function. It can then obtain new prices by minimizing the approximate function.

2.1.4.3 Dantzig–Wolfe Extensions

Extensions of the Dantzig–Wolfe decompositions provide a different approach to coordinating the subsystems (Dantzig and Wolfe, 1960, 1961; Dantzig, 1963; Lasdon, 1970). The subsystem optimization problems remain precisely the same as they are for the nonfeasible or interaction balance methods. The coordinator, however, obtains new prices by solving a Dantzig–Wolfe master program. This approach requires the overall regional optimization problem to have a concave objective function with convex inequality constraints and linear equality constraints.

The Dantzig–Wolfe master program is an inner linearization of the overall regional optimization problem. The coordinator uses previous subsystem solutions to build up the linear approximation. The solution of the Dantzig–Wolfe master program is a convex combination of previous subregional solutions, and the prices supplied to the subregions are the corresponding optimal multipliers. Although the subregional solutions generally do not satisfy the coupling equations, their convex combinations given by the Dantzig–Wolfe master program do satisfy the coupling equations. Moreover, these convex combinations give increasing values of the overall regional objective function as more and more subregional solutions become available.

2.1.4.4 Feasible Methods

The feasible methods form independent subsystem optimization problems by fixing the subsystem coupling variables. The vector σ of coordination variables is a vector of outputs transferred between the subsystems. In particular, the ith subsystem optimization problem becomes

$$\max_{m_i} f_i(x_i,u_i,m_i,\alpha_i), \text{ subject to } g_i(x_i,u_i,m_i,\alpha) \le 0,$$

$$\sigma_i - H_i(x_i,u_i,m_i,\alpha_i) = 0, \quad \sum_{j=1}^{N} C_{ij}\sigma_j - x_i = 0 \tag{2.7}$$

The subregions are coordinated via the outputs transferred between the subregions.

The coordinating values of σ are feasible outputs that maximize the total of the subregions' optimal returns, namely, the sum of the optimal values of the subregions' objective function in Eq. (2.7). This function is generally not available to the coordinator. However, necessary conditions for σ to be coordinating are that the specified outputs be feasible and the subregions' marginal returns all be equal. Care must be exercised so that the specified output transfers are feasible. This requires the subregions to provide the coordinator with sufficient information to ensure that this will be the case.

2.1.4.4.1 Interaction Prediction Models

The interaction prediction methods supply the subregions with "predicted" values of their inputs and prices on their outputs (Macko, 1967; Mesarović et al., 1970). The subregions assume these predictions are correct, i.e. the outcomes will be as predicted. The vector σ of coordination variables consists of two subvectors: A subvector σ_x of predicted values of the subregional inputs and a subvector σ_y of prices on the subregional outputs. In particular, the ith subregional optimization problem is

$$\max_{m_i} f_i(x_i,u_i,m_i,\alpha_i) + \sigma_{iy}y_i, \text{ subject to } g_i(x_i,u_i,m_i,\alpha) \le 0,$$

$$\sigma_{ix} - x_i = 0, \quad y_i = H_i(x_i,u_i,m_i,\alpha_i) \tag{2.8}$$

The subsystems are coordinated via the predicted values of the subregional inputs and prices on the subsystem outputs. The conditions that σ must satisfy in order to coordinate the subregions are based on the necessary conditions for the overall subsystem optimum. In particular, the predictions must be correct. That is, the optimal solutions must satisfy the coupling equations (Eq. (2.9)), and the prices assigned to the outputs together with the subsystem marginal returns with respect to the predicted input values must satisfy the dual coupling constraints

$$\sigma_{iy} = -\sum_{j=1}^{N} \lambda_j C_{ji} \tag{2.9}$$

where λ_j is the marginal return of the jth subsystem with respect to the predicted input value σ_{ix}. Errors in the subregional coupling equations (Eq. (2.3)) and the dual coupling equations (Eq. (2.9)) are used by the coordinator to update its predictions.

2.1.5 HIERARCHICAL STRUCTURES IN WATER RESOURCES COMPLEX SYSTEMS OF SYSTEMS

The modeling of water resources Complex SoS, such as a River Basin, involves the consideration of and the accounting for a wide spectrum of factors. Economic, technical, environmental, political, legal, and other important considerations necessarily yield to a complex model of high dimensions. The purpose of this section is to analyze this complexity and suggest some remedies in the form of different decomposition and coordination methodologies. A given system description is subject to how the system is viewed. Four major descriptions are readily identified in water resources systems. A brief discussion of each follows:

1. *Temporal description*: A planning time horizon for water supply projects often spans 30–50 years. On the other hand, the planning horizon for hydroelectric power generation plants spans a period of 10–15 years. Planning horizons of water resources for crop and related land use are of the order of 1–2 years. Finally, once a crop has been selected and the water for its seasonal growth has been allocated, horizons of decisions with respect to periodical irrigation within the season are of the order of weeks or days.

2. *Physical–hydrological description*: A River Basin Complex SoS is by definition a hydrological self-contained region separated from adjacent basins by a ridge or other topographical divide. Often water resources management covers a subsystem/region consisting of a complex of several River Basins. The subsystem/region's network of sources may include streams, rivers, lakes, wells, dams, reservoirs, and aqueducts. Clearly, the hydrology characterizing such a region spans local political and economic boundaries. Also, groundwater aquifers usually cross surface River Basins. Thus, conjunctive management of ground water and surface water resources becomes essential.
3. *Political–geographical description*: A regional water resources system often includes different local governments such as metropolitan areas, counties, and states. For example, the North Atlantic region includes 13 states and several River Basins. Modeling for water resources management may consider the geographical–political aspect as a criterion for decomposing the regional area into subregions. In the coupling between this geographical–political aspects, the other three aspects are strong and should not be overlooked.
4. *Goal description*: Most water resources systems have been traditionally analyzed with respect to their goals and economic considerations. Ample models following this pattern are available in the literature. Some models are demand and supply models, models for hydroelectric power generation, and models for water use in irrigation, industrial, and municipal purposes, as well as recreation, navigation, and fishing.

2.I.6 OVERLAPPING DECOMPOSITIONS

The decomposition of a regional area into subregions depends on the viewpoint and aims of the analyst. One decomposition may be with respect to the region's hydrology. The region would be decomposed into subregions, such as River Basins and subbasins, having topographical divisions as their boundaries. A second decomposition may be with respect to political boundaries. The regional area would be decomposed into political subregions such as townships, municipalities, and counties. A third decomposition may be with respect to regional goals and functions. A fourth decomposition would form subregions having distinct goals or purposes related to functions such as irrigation, navigation, hydroelectric power generation, or recreation. In regional water resources management, the major aspects of the regional area cannot be divorced from each other.

The decompositions just cited overlap one another. Hydrological subregions can easily overlap or span political boundaries. Hydroelectric generating stations may be dispersed through a region and not be confined to any one political or hydrological subregion. Indeed, the subregional boundaries in one decomposition generally do not coincide with the subregional boundaries in another decomposition. The problem here is how overlapping decompositions of a regional area can be coordinated.

Consider a regional area that has been decomposed from two different viewpoints such that the subregions of one decomposition overlap the subregions of the other decomposition. To make our discussion more concrete, we assume one decomposition is from the hydrological viewpoint, and the other is from a goal or functional viewpoint. Each subregion of the latter decomposition is concerned with a particular function such as irrigation, navigation, or recreation, among others.

Suppose the overall regional area R has been decomposed into N hydrological subregions indexed by subscripts as previously. The overall regional optimization problem expressed in the subregional variables of this decomposition is given in Eq. (2.5). In the same way, suppose a different decomposition has been obtained from a goal or functional viewpoint and consists of M subregions indexed by superscripts. Suppose the overall regional optimization problem expressed in the subregional variables of the second decomposition is

$$\max \sum_{k=1}^{M} f^k(x^k,u^k,m^k,\alpha^k), \text{ subject to } g^k(x^k,u^k,m^k,\alpha^k) \le 0,$$

$$y^k = H^k(x^k,u^k,m^k,\alpha^k), \quad x^k = \sum_{j=1}^{M} C^{kj} y^j \tag{2.10}$$

where $k=1, \ldots, m$. The variables in Eq. (2.10) have precisely the same interpretation as their counterparts in Eq. (2.5).

We make the following assumptions concerning the relation between the two decompositions. Interactions between the hydrological and goal-functional aspects of the region enter the subregional models via the vectors u_i and u^k as given by Eq. (2.11). In other words, we assume the vector u_i contains the goal function inputs pertinent to the ith hydrological subregion and make a similar assumption for u^k. In particular, we assume the coupling between the two decompositions is given by

$$\begin{aligned} u_i &= \sum_{k=1}^{M} B^{ik} y^k, \; i=1,\ldots,N \\ u^k &= \sum_{i=1}^{N} B_{ki} y_i, \; k=1,\ldots,M \end{aligned} \tag{2.11}$$

The matrices B^{ik} and B_{ki} are coupling matrices indicating which subregional outputs of one decomposition affect the subregions of the other decomposition.

1. *Coordinating overlapping decompositions*: It should be apparent that the coordination methodologies discussed previously may be used to coordinate the two overlapping decompositions. We indicate here the forms of the subregional objective functions and constraints required to implement the coordination methodologies.

   The nonfeasible or interaction balance methods require the subregional objective functions to be augmented as follows:

$$\begin{aligned} & f_i(x_i,u_i,m_i,\alpha_i,\sigma) + \left(\sum_{k=1}^{M} \lambda^k B_{ki}\right) y_i - \lambda_i u_i \\ & f^k\left(x^k,u^k,m^k,\alpha^k,\sigma\right) + \left(\sum_{j=1}^{M} \lambda_i B^{ik}\right) y^k - \lambda^k u^k \end{aligned} \tag{2.12}$$

   No additional constraints are imposed on the subregions. The subregions treat the inputs u_i and u^k as free variables. Coordination of the overlapping decompositions is via the "prices" λ_i and λ^k entering the subregions' objective functions. The augmentations given here are in addition to those required to coordinate the subregions of a given decomposition. The feasible methods leave the subregional objective functions unchanged, but they impose the additional constraints

$$\begin{aligned} \lambda_i - H_i(x_i,u_i,m_i,\alpha_i) &= 0 \\ u_i - \sum_{k=1}^{M} B^{ik} \lambda^k &= 0 \\ \lambda^k - H^k\left(x^k,u^k,m^k,\alpha^k\right) &= 0 \\ u^k - \sum_{i=1}^{N} B_{ki} \lambda_i &= 0 \end{aligned} \tag{2.13}$$

   Coordination of the overlapping decompositions is via the specified values λ_i and λ^k of the subregional outputs y_i and y^k. The interaction prediction methods require the subregional objective functions to be augmented as follows:

$$\begin{aligned} & F_i(x_i,u_i,m_i,\alpha_i,\sigma) + \lambda_{iy} y_i \\ & F^k\left(x^k,u^k,m^k,\alpha^k,\sigma\right) + \lambda^{ky} y^k \end{aligned} \tag{2.14}$$

   In addition, the constraints

$$u_i = \lambda_{iu} \text{ and } u^k = \lambda^{ku} \tag{2.15}$$

   are imposed on the subregions. Coordination of the overlapping decompositions is via "predicted" values of the subregional inputs u_i and u^k and prices on the subregional outputs y_i and y^k.

   In general, different coordination methodologies may be used to coordinate the subregions of a given decomposition and coordinate the overlapping decompositions. For example, feasible methods might be used to coordinate one decomposition, while the other is coordinated via the nonfeasible or interaction balance methods. The overlapping decomposition might then be coordinated via the interaction prediction methods.

2. *Third-level coordination:* The two decompositions form two distinct hierarchical-multilevel structures. The subregions of each decomposition with their objectives and constraints form the first level of the hierarchies. The coordinator of the subregions within a given decomposition occupies the second level of the hierarchy. It coordinates the subregions via the coordination variable σ. On top of the subregional

coordinators is another coordinator, which coordinates the two overlapping decompositions via the coordination variable λ. It joins the two overlapping hierarchies and so occupies the third level.

Our discussion here was limited to only two overlapping decompositions because of notation. Extensions to several overlapping decompositions are elementary.

2.I.7 CASE STUDIES

Several examples of regional planning and management of water resources systems via a hierarchical-multilevel approach are discussed in this section. They are chosen to represent different classes of decomposition and coordination schemes.

2.I.7.1 California Water Project Complex Systems of Systems

The California Water Project (Hall and Shephard, 1967; LeMay, 1969; Hall and Dracup, 1970) is a typical example of a large and complex water resources system. The supply system consists of four rivers (the Trinity, Sacramento, Feather, and American Rivers), their associated dams, 10 reservoirs, power plants, and tunnels. The distribution system consists of six main aqueducts, each of which may have submains, laterals, sublaterals, etc. to reach appropriate delivery points. The system inputs are the hydrology of the four rivers, and the system outputs are hydroelectric power and water for users. The system constraints are evaporation, minimum water release required for fish and navigation control, and maximum release constraint for flood control. The control variables are quantities and time of water release at each reservoir (through the power plant if possible). The planning time horizon considered is 10 years. The objective function for the overall system is to select quantities of firm power, dump power, firm water, and dump water to contract for maximum revenues from the system over the 10-year planning horizon.

The critical hydrological period, which controls the maximum level of firm water and power that can be contracted for, is utilized. In this study the critical period is defined as the period of time in which the historical hydrological record is most critical to the demands of the system. A 10-year critical hydrological record was selected for the study. This was divided into 120 equal time periods (months). Thus the water/power commitments were determined to maximize the revenue for each of these 120 time periods. The overall system was decomposed on a hydrological basis to five reservoir–river subsystems. These subsystems are the Trinity, Shasta, Oroville, Folsom, and San Luis subsystems.

A planning agent has been assumed to exist who is in control of the San Luis reservoir and who assumes the outputs from each of the reservoir subsystems for each of the 120 time periods. Using a linear programming model and these assumed outputs, the planning agent determines the maximum revenues from the entire system for each of the 120 time periods. The dual of the linear programming problem gives him the shadow prices for each output (Note that duality theory and shadow prices are presented in the Appendix.). A Dantzig–Wolfe decomposition has been utilized for this purpose. These shadow prices form a new price for water. The individual reservoir operators use dynamic programming to optimize their returns based on these prices and report to the master planning agent (higher level in the hierarchy) the corresponding quantities of each output they provide during each period. Then the master planning agent uses these new outputs and repeats the cycle. The iteration continues until the improvement of the solution in the last cycle is negligible.

As was mentioned in Section 2.I.4.3, the subsystem optimization problems in the Dantzig–Wolfe approach remain precisely the same as they are for the nonfeasible or interaction balance methods. Figure 2.1 gives a schematic representation of the two-level hierarchy. Note that in the hierarchical system model, the time aspect was imbedded in the hydrological decomposition. The California study thus represents an example of hydrological decomposition where the Dantzig–Wolfe or the interaction balance methods are used for subsystem coordination.

2.I.7.2 Multistage Flash Distillation Process

The problem of designing and operating a standard multistage flash distillation plant naturally lends itself

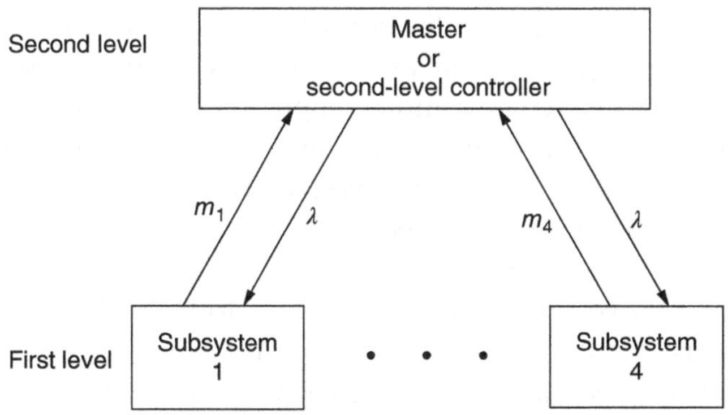

Figure 2.1 Two-level structure.

to a sequential decision-making process with feedback loops. A two-level hierarchical model has been developed for this purpose. The distillation plant is decomposed into five different types of subsystems and a total of $N+1$ subsystems. The $N+1$ subsystems correspond to the number of stages in the distillation column:

1. The last stage in the heat rejection section – Nth stage.
2. The first stage in the heat rejection section – Jth stage.
3. The last stage in the heat recovery section – $(J-1)$th stage.
4. Any other stage in the heat recovery or rejection section – stages 1 through $(J-2)$ and $(J+1)$ through $(N-1)$ (there are $N-3$ such subsystems).
5. The heater considered as stage 0.

Figures 2.2 and 2.3 represent the characteristics of each type of subsystem, including the different types of inputs, outputs, and controls. For each subsystem, constraints corresponding to heat, mass balance, and other technical requirements can be developed. The objective function could be stated as minimizing the cost of producing a given amount of fresh water or

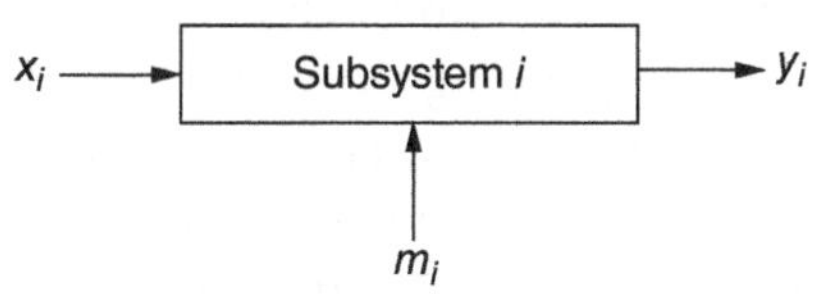

Figure 2.2 Subsystem i.

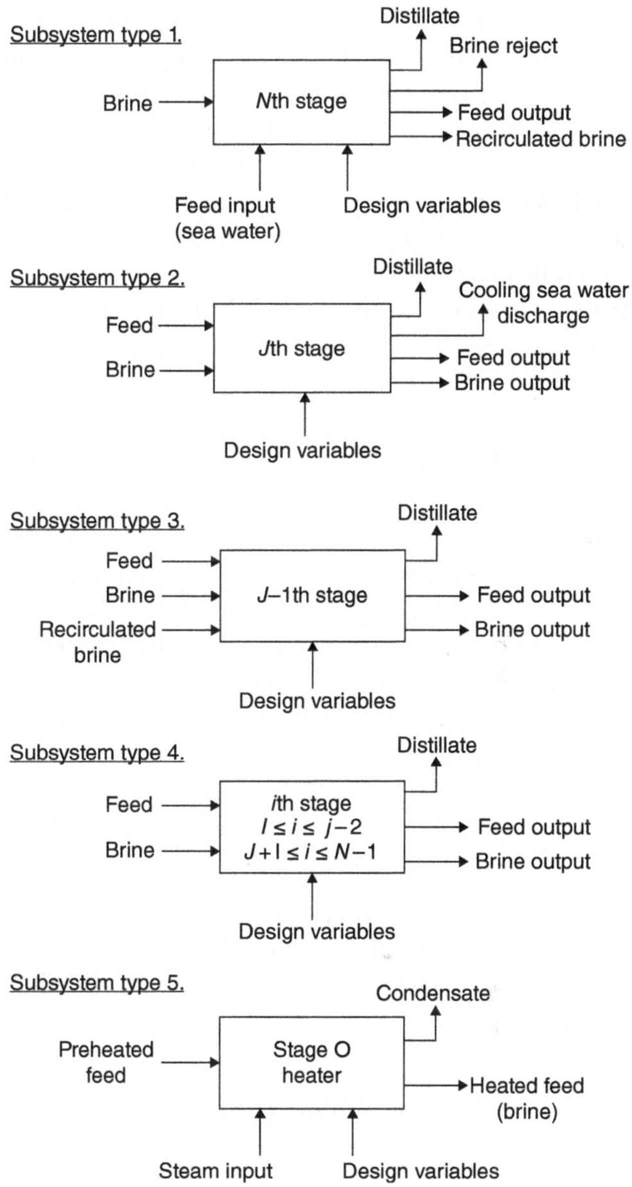

Figure 2.3 Types of subsystems.

maximizing the net benefits obtained through the construction and operation of such a plant in a given region.

The decomposition of this complex system is achieved by introducing pseudo-variables into the system. Each decomposed and independent subsystem is separately and independently optimized at the first level. The subsystems are tied together through coordinator parameters and Lagrange multipliers, which are responsible for the optimization of the whole system at the second level of the hierarchy.

Figure 2.2 represents the ith subsystem, where m_i is the control vector of the ith subsystem (e.g. steam input to heater, feed input, condenser area, vessel size, recirculated brine, etc.), x_i is the input vector into the ith subsystem coming from other subsystems (e.g. preheated feed input, brine input, etc.), and y_i is the vector of the ith subsystem outputs going to other subsystems (e.g. preheated feed output, brine output, etc.) and leaving the system (e.g. distilled water, brine reject, cooling water discharge, etc.). The subsystem couplings are presented in Eq. (2.16):

$$x_i = \sum_{j=1}^{N} C_{ij} y_j \quad y_i = H_i(x_i, m_i), \quad i = 1, \ldots, N \tag{2.16}$$

where the C_{ij} are coupling matrices. Let $f_i(x_i, m_i)$ be the ith subsystem cost function, and let $g_i(x_i, m_i) \leq 0$ be the constraints imposed on the ith subsystem. These constraints may be ignored here for simplicity. They do not contribute to understanding the model, but the model can easily handle them. Forming the Lagrangian in Eq. (2.17),

$$L = \sum_{i=1}^{N} f_i(x_i, m_i) + \sum_{i=1}^{N} \sigma_i \left[\sum\nolimits_{j=1}^{N} C_{ij} y_j - x_i\right] \tag{2.17}$$

The σ_i are Lagrange multipliers assumed to be known in the first level and determined in the second level. L is readily decomposed into N independent subsystems:

$$L = \sum_{i=1}^{N} F_i = \sum_{i=1}^{N} \left[f_i(x_i, m_i) + \left(\sum_{i=1}^{N} \sigma_j C_{ji} \right) y_i - \sigma_i x_i \right] \tag{2.18}$$

At the first level each F_i is minimized subject to $y_i = H_i(x_i, m_i)$, assuming the σ_i are known. At the second level the σ_i are determined by minimizing the error functions $E_i(\sigma)$ with respect to σ:

$$E(\sigma_i) = x_i(\sigma) - \sum_{j=1}^{N} C_{ij} y_i(\sigma). \tag{2.19}$$

This is an iterative process that terminates when convergence is achieved.

2.1.7.3 Water Quality Control and Management

The hierarchical approach for the control and management of water quality has been widely applied. In particular, the utilization of a system of effluent charges for pollution abatement has been demonstrated. One computational approach to the solution of the multilevel model permits the representation of the second level as a regional authority with complete knowledge of the polluter's treatment cost functions, i.e. centralized planning. Conversely, an alternative computational scheme for coordinating first-level decisions operates to satisfy quality conditions. This may be interpreted as the economic law of supply and demand: Raise the tax rate associated with a violated water quality constraint and lower rates associated with a satisfied constraint until equilibrium results.

The interpretation of the second level is of major importance. Should the regional authority levy effluent charges on any waste discharger, the charge may be considered to be a fee for the use of the natural resource of waste assimilation. The amount of the fee will be based on the amount and quality of the effluent. The benefits that would accrue include the following: (i) There is a source of revenue for operating the region; (ii) the cost of pollution is borne by the waste discharger; and (iii) there is an incentive for pollution reduction because the polluter may reduce his charge by improving his treatment.

Furthermore, the regional authority is effective because it can apply various controls over a wide area considering all conditions of flow. Through regional treatment facilities, the regional authority may increase its efficiency by deriving those economies of scale that result from larger plants. Equity is served because the regional authority may charge each polluter proportionally to his load in terms of quantity and quality.

A third computational scheme takes advantage of both of the preceding ones by proposing the construction of a regional treatment plant. The central authority determines the optimal treatment configuration for meeting water quality standards along a River Basin and simultaneously determines *optimal pollution charges* to realize this configuration. The optimal regional plant capacity and regional treatment level are also found, depending on which polluters decide to use the regional facilities. The central authority does not need to know the local treatment cost functions.

The central authority, the third level in the multilevel hierarchy, proposes a taxing structure to the regional plant and the individual polluters. The regional plant, the second level, assumes an optimal plant size and determines a regional treatment level to minimize the regional treatment cost. The individual polluters at the first level determine the optimal local treatment level and decide whether or not to use the regional facilities. According to the first-level decisions, the second level adjusts the regional plant size so that there is no excess capacity. The third level checks water quality constraints and adjusts effluent charges so that all constraints are satisfied at the least cost to society.

The use of hierarchical decomposition was demonstrated with data for the Miami River Basin. A considerable savings over the cost of local treatment is realized when one regional plant is employed. The model has been extended to allow for several regional plants, thus realizing greater savings. This extension of the approach involves a "transportation problem," that of determining which regional plant should be used by each polluter. The possibility of several polluters combining to build a co-op treatment plant is also allowed.

2.I.7.4 Summary

Why hierarchical-multilevel structures? We have presented the basic concepts of hierarchical-multilevel structures and the hierarchical-multilevel approach in planning and management of a Complex SoS, in particular, water resources systems. The essential advantage of the approach is that it breaks down Complex SoS into their component subsystems. It allows these subsystems to be handled on the first level, independent of each other. It might be argued that decomposition is fairly easy; the real problem is resolving the conflicts and interactions among the subsystems and ensuring that specified overall system objectives and constraints are met. The general coordination methodologies presented here take into account the inherent limitations of any real-life decision makers. It distributes the total planning and management task among the component subsystems by localizing the tasks. Localization can have many benefits. For example, disturbances and perturbations can be treated locally and be prevented from propagating throughout the Complex SoS.

The following factors must be taken into account in real applications of the hierarchical-multilevel approach to Complex SoS:

1. Models used for Complex SoS in decision making are usually simplified approximations of their complexity. Because of this and the inevitable presence of uncertainties, operations of Complex SoS are commonly an elaborate process involving multiple decision makers and stakeholders.
2. Real systems are always evolving in uncertain environments, namely, they are not static, but dynamic. Thus, higher-level subsystems of Complex SoS, modeled via hierarchical-multilevel structure, must be tuned to disturbances that affect the subsystems and/or the entire Complex SoS.
3. Subsystems are never actually decoupled. When the nonfeasible or interaction balance methods are used to coordinate the subsystems, implementation of the subsystems' actions must wait until coordination is achieved unless it is possible to combine the subsystems' proposals into a feasible operating plan. Implementing subsystem actions before coordination is achieved introduces disturbances in the system causing suboptimality and, perhaps, violation of constraints.
4. Information is commonly not a free commodity. Data acquisition, processing, and transfer within the Complex SoS require effort and can be costly. Excessive information handling can overburden the system. In some cases information concerning the operation of a subsystem might not be accurate.

2.I.8 SOLVED EXAMPLE PROBLEMS

In this section we solve select problems by building upon the general formulation of the feasible decomposition and the nonfeasible decomposition, also known as the interaction balance method. The two decompositions can be viewed as a duality. In the feasible decomposition, the iterations focus on the state variables where the constraints are always satisfied. In the nonfeasible decomposition, the iterations are focused on the Lagrange multipliers where the constraints are not satisfied until convergence is achieved.

For pedagogical purposes, a simple unconstrained optimization problem will be used as a vehicle to explain the mechanics of the two decomposition methods. Extension of the methodologies to constrained optimization problems will follow. Furthermore, several second-level coordination methods will be introduced and be related to the decomposed system.

Consider the following problem:

$$\min_{x_1,x_2} \left\{ f(x_1,x_2) = (x_1-2)^2 + x_1x_2 + (x_2-1)^2 \right\} \quad (2.20)$$

The simple optimization problem in Eq. (2.20) has two decision variables, x_1 and x_2, that are coupled through the term x_1x_2. For example, these variables x_1 and x_2 may be vectors of N and M dimensions, respectively, each associated with a different department in a company. Without the coupling term, x_1x_2, the problem would be separable, and thus no decomposition would be needed.

In order to uncouple the system, a pseudo-variable, σ, will be introduced:

$$x_1 = \sigma \quad (2.21)$$

The variable x_1 will be replaced by σ whenever it appears in a coupling term. The unconstrained problem in Eq. (2.20) can be replaced by the following:

$$\begin{aligned} &\min_{x_1,x_2} \left\{ f(x_1,x_2,\sigma) = (x_1-2)^2 + \sigma x_2 + (x_2-1)^2 \right\} \\ &\text{subject to} \\ &x_1 = \sigma \end{aligned} \quad (2.22)$$

The system Eq. (2.22) is a constrained optimization problem that can be solved by forming the Lagrangian function $L(\cdot)$ as follows:

$$L(x_1,x_2,\sigma,\lambda) = (x_1-2)^2 + \sigma x_2 + (x_2-1)^2 + \lambda(x_1-\sigma) \quad (2.23)$$

Both the feasible and nonfeasible decomposition will be applied to solve the system Eq. (2.23). For pedagogical purposes, the problem posed by Eq. (2.20) was constructed so that the conditions for the existence of a saddle point to the Lagrangian equation (2.23) are not satisfied. The solution to the original problem in Eq. (2.20) can be readily obtained via calculus to be

$$x_1^* = 2, \ \ x_2^* = 0, \ \ f(x_1^*, x_2^*) = 1$$

2.1.8.1 Feasible Decomposition

In the feasible decomposition, the pseudo-variable σ is determined at the second level of the two-level optimization problem. The Lagrangian presented in Eq. (2.24) is decomposed into two sub-Lagrangians, L_1 and L_2, as follows:

$$L(x_1,x_2,\sigma,\lambda) = L_1(x_1,\lambda,\sigma) + L_2(x_2;\sigma) \quad (2.24)$$

where

$$L_1(x_1,\lambda,\sigma) = (x_1-2)^2 + \lambda(x_1-\sigma) \quad (2.25)$$

$$L_2(x_2;\sigma) = (x_2-1)^2 + \sigma x_2 \quad (2.26)$$

The pseudo-variable, σ, as an argument in the sub-Lagrangians, is preceded by (;) to indicate that it should be viewed as a known parameter at the first-level optimization. Figure 2.4 depicts the transfer of information between the first and second levels.

2.1.8.1.1 First-Level Optimization

There are two subsystems in the first level of the hierarchical structure. In subsystem 1, there seem to be two independent decision variables, x_1 and λ; however, this is not the case, as will be shown subsequently. Necessary conditions for stationary L_1 and L_2 are:

Subsystem 1

$$\frac{\partial L_1}{\partial x_1} = 2(x_1-2) + \lambda = 0 \quad (2.27)$$

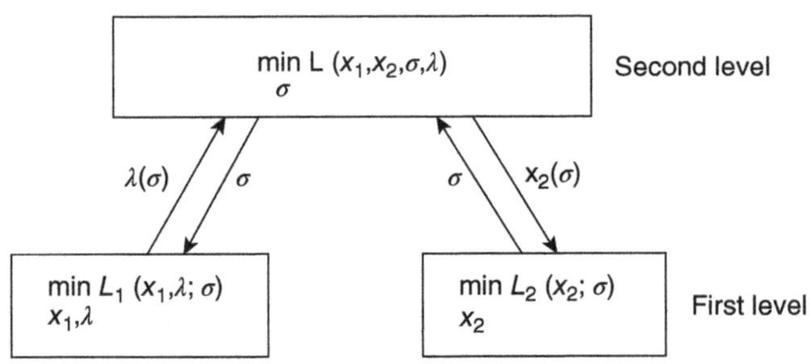

***Figure 2.4** Feasible decomposition.*

or

$$\lambda = 2(2 - x_1) \tag{2.28}$$

$$\frac{\partial L_1}{\partial \lambda} = x_1 - \sigma = 0 \tag{2.29}$$

or

$$x_1 = \sigma \tag{2.30}$$

Combining Eqs. (2.28) and (2.30) yields

$$\lambda(\sigma) = 4 - 2\sigma \tag{2.31}$$

Subsystem 2

$$\frac{\partial L_2}{\partial x_2} = 2(x_2 - 2)1 + \sigma = 0 \tag{2.32}$$

or

$$x_2(\sigma) = 1 - 0.5\sigma \tag{2.33}$$

For any given value of σ determined at the second level, Eqs. (2.31) and (2.33) yield the corresponding optimal decisions at the first level in subsystems 1 and 2. Clearly, sufficiency conditions for a minimum should be checked as well.

2.I.8.1.2 Second-Level Optimization

The total Lagrangian system Eq. (2.23) is optimized at the second level. Thus,

$$\mathrm{d}L = \frac{\partial L}{\partial x_1}\mathrm{d}x_1 + \frac{\partial L}{\partial x_2}\mathrm{d}x_2 + \frac{\partial L}{\partial \lambda}\mathrm{d}\lambda + \frac{\partial L}{\partial \sigma}\mathrm{d}\sigma = 0 \tag{2.34}$$

Since x_1, x_2, and λ are chosen arbitrarily at the first level, then the following relations hold:

$$\mathrm{d}x_1 = \mathrm{d}x_2 = \mathrm{d}\lambda = 0 \tag{2.35}$$

Furthermore, it is assumed that the Lagrangian L possesses minimum; thus, a gradient-type algorithm may be adopted to minimize L with respect to the only decision variable at the second-level optimization, namely, σ:

$$\sigma^{(k+1)} = \sigma^{(k)} - \Delta\frac{\partial L}{\partial \sigma}\left(\sigma^{(k)}\right) \tag{2.36}$$

where $\Delta > 0$.

Substituting for $\partial L/\partial \sigma$ in Eq. (2.36) yields

$$\sigma^{(k+1)} = \sigma^{(k)} - \Delta\left[x_2\left(\sigma^{(k)}\right) - \lambda\left(\sigma^{(k)}\right)\right] \tag{2.37}$$

where the superscript (k) denotes the kth iteration. The results summarized by Eqs. (2.31), (2.33), and (2.37) can be used in solving the two-level optimization problem iteratively.

2.I.8.1.3 Iterative Solution

The iterative procedure may be started with an initial guessed value for σ, e.g. $\sigma^{(1)} = 1$. For convenience, the three equations summarizing the interactions between the first and second levels are listed again:

$$\lambda(\sigma) = 4 - 2\sigma \tag{2.38}$$

$$x_2(\sigma) = 1 - 0.5\sigma \tag{2.39}$$

$$\sigma^{(k+1)} = \sigma^{(k)} - \Delta\left[x_2\left(\sigma^{(k)}\right) - \lambda\left(\sigma^{(k)}\right)\right] \tag{2.40}$$

Substituting $\sigma^{(1)} = 1$ into Eqs. (2.42) and (2.44) yields

$\lambda^{(1)} = 2$, $x_1^{(1)} = 0.5$, $x_1^{(1)} = 1$, $x_2^{(2)} = 0.125$ and $f^{(1)}(1{,}0.5) = 1.75$.

2.I.8.2 Nonfeasible Decomposition

The existence of a saddle point to the Lagrangian function is the heart of the nonfeasible decomposition. The duality concepts are basics in this decomposition, as will be seen subsequently. The pseudo-variables are determined at the first level, where the Lagrange multipliers play the role of the second-level variables. Accordingly, the Lagrangian presented in Eq. (2.35) is decomposed into the following two sub-Lagrangians, L_1 and L_2:

$$L(x_1{,}x_2{,}\lambda{,}\sigma) = L_1(x_1;\lambda) + L_2(x_2;\sigma{,}\lambda) \tag{2.41}$$

where

$$L(x_1;\lambda) = (x_1 - 2)^2 + \lambda x_1 \tag{2.42}$$

$$L_2(x_2{,}\sigma{,}\lambda) = (x_2 - 1)^2 + \sigma x_2 - \lambda\sigma \tag{2.43}$$

Note that the Lagrange multiplier, λ, as an argument in the sub-Lagrangians, is preceded by (;) to indicate that it should be viewed as a known parameter at the first-level optimization. Figure 2.5 depicts the transfer of information between the first and second levels.

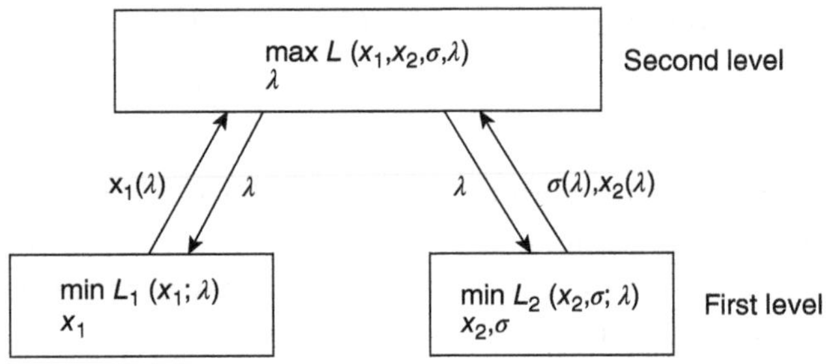

Figure 2.5 *Nonfeasible decomposition.*

2.I.8.2.1 First-Level Optimization

Similar to the feasible decomposition, there are two subsystems at the first level, and the optimization procedure follows the same as well. The decision variable in subsystem 1 is x_1 and in subsystem 2, x_2 and σ. Necessary conditions for stationarity for L_1 and L_2 are:
Subsystem 1

$$\frac{\partial L_1}{\partial x_1} = 2(x_1 - 2) + \lambda = 0 \tag{2.44}$$

$$x_1(\lambda) = 2 - 0.5\lambda \tag{2.45}$$

Subsystem 2

$$\frac{\partial L_2}{\partial x_2} = 2(x_2 - 1) + \sigma = 0 \tag{2.46}$$

$$\sigma = 2(1 - x_2) \tag{2.47}$$

$$\frac{\partial L_2}{\partial \sigma} = x_2 - \lambda = 0 \tag{2.48}$$

$$x_2 - \lambda \tag{2.49}$$

Substituting Eq. (2.46) into Eq. (2.44) yields

$$\sigma(\lambda) = 2(1 - \lambda) \tag{2.50}$$

For any given value of λ determined at the second level, Eqs. (2.42), (2.46), and (2.47) yield the corresponding optimal decisions at the first level in subsystems 1 and 2. Here again, sufficiency conditions for a minimum should be checked.

Several approaches for solving the second-level optimization are discussed in subsequent sections.

2.I.8.2.2 Explicit Second-Level Coordinators

When all first-level decision variables can be expressed as an explicit function of the Lagrange multipliers, as is the case in the example discussed here, the explicit second-level coordinator may be used. Since the only variable at the second level is the Lagrange multiplier, λ, a necessary condition for stationarity to the Lagrangian function is given by Eq. (2.51):

$$\frac{dL}{d\lambda} = x_1 - \sigma = 0 \tag{2.51}$$

Since at the first level both x_1 and σ are given explicitly in terms of λ by Eqs. (2.42) and (2.47), it is straight forward to determine λ directly by substituting into Eq. (2.48):

$$2 - 0.5\lambda = 2(1 - \lambda) \tag{2.52}$$

Solving Eq. (2.49) yields $\lambda = 0$. Substituting the value of $\lambda = 0$ back into the first-level variables yields

$$x_1(\lambda) = 2, \quad x_2(\lambda) = 0, \quad \sigma(\lambda) = 2, \quad \text{and } f(2,0) = 1$$

The above solution is the same one obtained without decomposition and represents the overall optimal solution.

2.I.8.2.3 Dual Second-Level Coordinator

The nonfeasible decomposition provides an excellent opportunity for the use and application of the duality theory. In particular, the first-level optimization can be viewed as solving the primal problem with respect to the original decision and the pseudo-variables. The second-level optimization can be viewed as solving the dual problem with respect to the dual variables, namely, the Lagrange multipliers. The solution of the first-level subsystems, or primal problem, should converge to the solution of the second-level system, the dual problem, if the Lagrangian possesses a saddle point.

The example problem discussed in this section was carefully selected so that the conditions for a saddle point are not satisfied. Therefore, the principle of duality where the primal problem is minimized with respect to the primal variables (x_1, x_2, and σ) and the dual problem is maximized with respect to the dual variable (λ) does not apply in this case, as is shown below.

Assuming that a saddle point exists for the Lagrangian function given by Eq. (2.23), the task of the second-level controller is to maximize the Lagrangian function with respect to λ. Thus, from duality theory the second-level objective is

$$\max_{\lambda \in D} L(x_1^0, x_2^0, \sigma^0, \lambda) \tag{2.53}$$

where x_1^0, x_2^0, and σ^0 are the optimal values obtained from the subsystems' solution (primal problem solution). D is the set of all λs satisfying the Kuhn–Tucker conditions.

The maximization of Eq. (2.53) can be carried out via a gradient type of algorithm. The recursive equation for maximizing the Lagrangian with respect to λ can be written as follows:

$$\lambda^{(k+1)} = \lambda^{(k)} + \Delta' \frac{\partial L}{\partial \lambda} \lambda^{(k)}, \Delta' > 0 \qquad (2.54)$$

where Δ' is a step size. Note, however, that

$$\frac{dL}{d\lambda} = x_1(\lambda) - \sigma(\lambda) = 1.5\lambda \qquad (2.55)$$

Substituting Eq. (2.52) into Eq. (2.51) yields

$$\lambda^{(k+1)} + \lambda^{(k)} + \Delta\lambda^{(k)} \qquad (2.56)$$

where

$$\Delta = 1.5\Delta'$$

It is interesting to note that the recursive Eq. (2.56) will never yield to the optimal value of λ unless the initial value of λ is chosen to be zero. Any negative initial value of $\lambda^{(k)}$ will make all $\lambda^{(k+1)}$ more negative (smaller), where a positive initial value of $\lambda^{(k)}$ will make all $\lambda^{(k+1)}$ more positive (bigger). Therefore, a convergence to $\lambda = 0$ will not be achieved. The reason for the failure of the gradient-type algorithm is that the necessary conditions for the existence of a saddle point are not satisfied, i.e. a minimum does not exist for the Lagrangian with respect to x_1, x_2, and σ. This can be shown by examining the Hessian for sufficient conditions for a minimum.

The Hessian matrix, H, for the Lagrangian is given below:

$$H = \begin{bmatrix} \frac{\partial^2 L}{\partial x_1^2} & \frac{\partial^2 L}{\partial x_1 \partial x_2} & \frac{\partial^2 L}{\partial x_1 \partial \sigma} \\ \frac{\partial^2 L}{\partial x_2 \partial x_1} & \frac{\partial^2 L}{\partial x_2^2} & \frac{\partial^2 L}{\partial x_2 \partial \sigma} \\ \frac{\partial^2 L}{\partial \sigma \partial x_1} & \frac{\partial^2 L}{\partial \sigma \partial x_2} & \frac{\partial^2 L}{\partial \sigma^2} \end{bmatrix} \qquad (2.57)$$

The Lagrangian was decomposed into two subsystems where the decision variable x_1 in subsystem 1 was uncoupled from the decisions x_2 and σ in subsystem 2. Therefore, the cross-terms $\partial^2 L/\partial x_1 \partial x_2$ and $\partial^2 L/\partial x_1 \partial \sigma$ in the Hessian should vanish. The resultant Hessian is thus

$$H = \begin{bmatrix} 2 & 0 & 0 \\ 0 & 2 & 1 \\ 0 & 1 & 0 \end{bmatrix} \qquad (2.58)$$

According to Sylvester's theorem, a sufficient condition for a minimum is that all the major minors of the Hessian be positive definite. Note, however, that the value of the determinant of the Hessian is negative (−2). Therefore, the sufficiency conditions for minimum are not satisfied nor are the necessary conditions for a saddle point to be fulfilled. The economic interpretations of the Lagrange multipliers make this coordinator one of the most promising in the hierarchical-multilevel approach.

2.1.8.3 Example Problem 2.1

Solve the following problem using the nonfeasible decomposition:

$$\min_{x_1 x_2} \{f(x_1, x_2) = x_1 + x_2 + 4x_1^2 - 4x_1x_2 + 2x_2^2 - 5\} \qquad (2.59)$$

subject to the constraint

$$x_1 + x_2 \geq 20 \qquad (2.60)$$

Solution

Let $x_1 = \sigma$. Then

$$\begin{aligned} L &= x_1 + x_2 + 4x_1^2 - 4\sigma x_2 + 2x_2^2 - 5 + \lambda(x_1 - \sigma) \\ &= L_1(x_1;\lambda) + L_2(x_2;\sigma,\lambda) \end{aligned} \qquad (2.61)$$

where

$$L_1(x_1;\lambda) = x_1 + 4x_1^2 - \lambda x_1 \qquad (2.62)$$

$$L_2(x_2,\sigma,\lambda) = x_2 - 2x_2^2 - 5 - 4\sigma x_2 - \lambda\sigma \qquad (2.63)$$

Subsystem 1

$$\min\{L_1 = x_1 + 4x_1^2 + \lambda x_1\} \qquad (2.64)$$

Subsystem 2

$$\min\{L_2 = x_2 + 2x_2^2 - 5 - 4\sigma x_2 + \lambda\sigma\} \qquad (2.65)$$

subject to the constraint

$$\sigma + x_2 \geq 20 \quad (2.66)$$

Necessary conditions for stationarity at the first level are

$$\frac{\partial L_1}{\partial x_1} = 1 + 8x_1 + \lambda = 0 \quad (2.67)$$

or

$$x_1 = -\frac{1}{8}(\lambda + 1) \quad (2.68)$$

$$\hat{L}_2 = L_2 + \mu(20 - \sigma - x_2)$$

$$\hat{L}_2 = x_2 + 2x_2^2 - 5 - 4\sigma x_2 - \lambda x_2 + \mu(20 - \sigma - x_2) \quad (2.69)$$

$$\frac{\partial \hat{L}_2}{\partial x_2} = 1 + 4x_2 - 4\sigma - \mu = 0$$

$$\frac{\partial \hat{L}_2}{\partial x_2} = -4x_2 - \lambda - \mu = 0 \quad (2.70)$$

$$\frac{\partial \hat{L}_2}{\partial x_2} = 20 - \sigma - x_2 \leq 0$$
$$\mu(20 - \sigma - x_2) = 0 \quad (2.71)$$
$$\mu \geq 0$$

Equating Eqs. (2.66) and (2.67)

$$1 + 4x_2 - 4\sigma - \mu = -4x_2 - \lambda - \mu$$

yields

$$4\sigma = 1 + \lambda + 8x_2 \quad (2.72)$$

Solving Eq. (2.68) for $\mu > 0$ (i.e. binding constraint) and substituting into Eq. (2.63) yields

$$4\sigma = 1 + \lambda + 8(20 - \sigma)$$

or

$$\sigma = \frac{1}{12}(161 + \lambda) \quad (2.73)$$

Necessary conditions for stationary at the second level are

$$\frac{\partial L}{\partial \lambda} = x_1 - \sigma = 0$$

Substituting the value of $x_1(\lambda)$ and $\sigma(\lambda)$ from the first-level Eqs. (2.65) and (2.70) yields

$$\frac{1}{12}(161 + \lambda) = -\frac{1}{8}(\lambda + 1)$$

or

$$\lambda = -65 \quad (2.74)$$

Back to the first level:
Substituting the value of $\lambda = -65$ (Eq. (2.65) into Eqs. (2.59) and (2.64)) yields

$$x_1^* = \frac{1}{8}(65 - 1) = 8 \quad (2.75)$$

$$\sigma^* = \frac{1}{12}(161 - 65) = 8 \quad (2.76)$$

Substituting the above values into Eq. (2.63) yields

$$x_2^* = \frac{(32 - 1 + 65)}{8} = 12 \quad (2.77)$$

and the value of the objective function at $x_1^* = 8$, $x_{12}^* = 12$ is

$$f(8,12) = 175$$

2.I.8.4 Summary

There is a growing need for comprehensive modeling schema that are capable of capturing the multifarious aspects (e.g. technological, societal, socioeconomic, political, institutional, managerial) of Complex SoS (e.g. telecommunications, electric power, gas and oil, transportation, water treatment plants, water distribution networks, dams, and levees). In particular, there is a need to understand and model the intricate interconnectedness within a particular infrastructure's components and among the infrastructures. Given the nature of Complex SoS (nonlinear interdependencies and interconnections, probabilistic, and dynamic), a balance between aggregation and reductionism seems imperative.

PART II: INCORPORATING PROBABILITY DISTRIBUTION, AND UNCERTAINTY ANALYSIS IN MODELING COMPLEX SYSTEMS OF SYSTEMS

2.II.1 AN OVERVIEW

Effective modeling of interdependent and interconnected complex systems of systems (Complex SoS) commonly rely on probability and uncertainty analyses for both policy formulation and the subsequent decision-making process.[1] Thus, in Part II we embrace the premise that our knowledge can be measurably enhanced by formally augmenting new evidence or information into the uncertain world within which policies are formulated and decisions are made. More specifically, we will build on Bayes' theorem, which provides a theoretical foundation with which to quantitatively update the credibility of prior information through new evidence. The method of analysis presented in Part II of this chapter recognizes that the construction of posterior probabilities based on intelligence (knowledge), which can be repeated again and again, is harmonious with Bellman's principle of optimality (discussed in the Appendix). In the following discussion we will address tails of probability distributions and extreme events, uncertainty versus variability, the bases for our choices, and when should we use probability-based versus uncertainty-based analyses, or both, for modeling and decision making.

Risk-based decision making and risk-based approaches in decision making, especially for Complex SoS are terms frequently used to indicate that a systemic process dealing with risk and uncertainty is being used to formulate policy options and assess their distributional impacts and ramifications. Today an ever-increasing number of professionals and managers in industry, government, and academia are devoting a large portion of their time and resources to the task of improving their understanding and approach to risk-based decision making.

In this pursuit, they invariably rediscover (often with considerable frustration) the truism that the more you know about a complex subject, the more you realize how much still remains unknown. There are several fundamental reasons for the complexity of this subject. One is that decision making under uncertainty literally encompasses every facet, dimension, and aspect of our lives. It affects us at the personal, corporate, and governmental levels, and it also affects us during the planning, development, design, operation, and management phases of problem solving. Uncertainty colors the decision-making process regardless of whether it (i) involves one or more parties, (ii) is constrained by economic or environmental considerations, (iii) is driven by sociopolitical or geographical forces, (iv) is directed by scientific or technological know-how, or (v) is influenced by various power brokers and stakeholders. First of all, uncertainty is inherent when trying to answer the set of questions posed by William Lowrance: "Who should decide on the acceptability of what risk, for whom, in what terms, and why?" (Lowrance, 1976). The second reason why risk-based decision making is complex is that it is a cross-disciplinary venture with diverse approaches of varying reliability. Some methods, which on occasion produce fallacious results and conclusions, have become entrenched and are hard to eradicate. The third reason is grounded in the need to make tradeoffs among all relevant and important costs, benefits, and risks in a multiobjective decision-making framework, without assigning weights with which to compare risks, costs, and benefits. (Note that Chapter 5 is devoted to multiobjective decision making.)

Consider, for example, the economic impacts on the fairly gross measures of environmental improvements, such as the tons of pollutants removed from the environment. Risk assessment has become a powerful tool for supporting and shaping regulatory decisions, often equal to economic analysis in its usefulness. Risk assessment adds a useful dimension by providing decision makers with estimates of the public health and environmental benefits or the effectiveness associated with regulatory alternatives. In concept, risk assessment, management, and communication provide common measures of what a regulation can be expected to achieve, thus enabling a level playing field on which to critically compare options and to ensure relative quality and consistency within and across programs. For example, the US Environmental Protection Agency commonly makes a concerted effort to hold its regulatory analysis and research to the highest standards possible and practical. *And*, as part of this effort, government agencies commonly issue policies and regulation guidance on risk characterization for risk managers and assessors, along with technology-based information. Such guidance improves risk assessments and enables agency staff to develop a more complete

[1]Part II of this chapter is edited, updated, and drawn from the proceedings of an April 1993 workshop sponsored by the Environmental Protection Agency titled "When and How Can You Specify a Probability Distribution When You Don't Know Much?" organized by the author.

understanding of environmental risks, strive for a greater consistency, compare risk across assessments and across agency programs, and recognize the critical role played by professional judgment in performing and interpreting risk assessment, management, and communication (Haimes, 2016).

2.II.2 RISK OF EXTREME EVENTS

Most analysts and decision theorists recognize a simple yet fundamental philosophical truth. In the face of unforeseen calamities, such as bridges falling and airplanes crashing, we must acknowledge the importance of studying "extreme" events. Modern decision analysts are no longer asking questions about expected risk; instead, they are asking questions about expected catastrophic or unacceptable risk. These analysts and policy makers are focusing their efforts on forming a more robust treatment of extreme events, in both the theoretical and practical sense. Furthermore, managers and decision makers are most concerned with the risk associated with a specific case under consideration, and not with the likelihood of the average adverse outcomes that may result from various risk situations. In this sense, the expected value of risk, which until recently has dominated most risk analysis in the field, not only is inadequate but also can lead to fallacious results and interpretations. In subsequent discussion we will introduce Bayesian analysis (here, in Chapter 6, and elsewhere throughout this book) that incorporates new evidence (knowledge) into our probabilistic- and risk-based analysis. Indeed, people in general are not risk neutral. They are often more concerned with low-probability catastrophic events than with more frequently occurring but less severe accidents. In some cases, a slight increase in the cost of modifying a structure might have a very small effect on the unconditional expected risk (the commonly used business-as-usual measure of risk), but it could make a significant difference to the conditional expected catastrophic risk. Consequently, the conditional expected catastrophic risk adds significant value to many multiobjective risk problems. The reader is referred to the fourth edition of the book *Risk Modeling, Assessment, and Management*, Wiley (Haimes, 2016).

Two difficult questions – how safe is safe enough and what is an acceptable risk – underline the normative, value-judgment perspectives in risk-based decision making. No mathematical or empirical knowledge base today can adequately model the perception of risks in the mind of decision makers. In the study of multiple criteria decision making (MCDM), we clearly distinguish between the quantitative element in the decision-making process, where efficient (Pareto-optimal) solutions and their corresponding tradeoff values are generated, and the normative value-judgment element, where the decision makers use these efficient solutions and tradeoff values to determine their preferred (compromise) solution. In many ways, risk-based decision making can and should be viewed as a type of stochastic MCDM in which some objective functions represent risk functions. This analogy can be most helpful in making use of the extensive knowledge already generated by MCDM (witness the welter of publications and conferences on the subject, e.g. the book of Chankong and Haimes, 1983, 2008).

It is worth noting that there are two modalities to the considerations of risk-based decision making in a multiobjective framework. One is viewing risk (e.g. the risk of dam failure) as an objective function to be traded off with cost and benefit functions. The second modality concerns the treatment of damages of different magnitudes and different probabilities of occurrence as noncommensurate objectives. Probably the most important aspect of considering risk-based decision making within a stochastic MCDM framework is the handling of extreme events.

To dramatize the importance of understanding and adequately quantifying the risk of extreme events, the following statements are adopted from Runyon (1977):

Imagine what life would be like if:

- Our highways were constructed to accommodate the average traffic load of vehicles of average weight.
- Mass transit systems were designed to move only the average number of passengers (i.e. total passengers per day divided by 24 h) during each hour of the day.
- Bridges, homes, and industrial and commercial buildings were constructed to withstand the average wind or the average earthquake.

- Telephone lines and switchboards were sufficient in number to accommodate only the average number of phone calls per hour.
- Your friendly local electric utility calculated the year-round average electrical demand and constructed facilities to provide only this average demand.
- Emergency services provided only the average number of personnel and facilities during all hours of the day and all seasons of the year.
- Our space program provided emergency procedures for only the average type of failure.

Chaos is the word for it. Utter chaos.

Lowrance (1976) makes an important observation on the imperative distinction between the quantification of risk, which is an empirical process, and the determination of safety, which is a normative process. In both of these processes, which are seemingly dichotomous, the influence and imprint of the analyst cannot and should not be overlooked. The essential role of the analyst, sometimes hidden but often explicit, is not unique to risk assessment and management; rather it is indigenous to the process of modeling and decision making.

The analytic challenge facing government agencies and the private sector in risk assessment, management, and communication is to present a full and complete picture of risk, including a statement of *confidence regarding the uncertainties* within the risk assessments, which must be explicitly addressed – one important theme of this chapter. Sometimes, but rarely, risk analysts working on public safety have the luxury of extensive data to describe or model the quantitative relationships among the systems that constitute Complex SoS and their uncertainties. More often, however, data are very limited. The question is, how quantitative should the analysis be, both in estimating risks and representing their uncertainties, when the database is limited? How should probability distribution functions be selected under such circumstances? *Stated differently, when and how can we generate, specify, and use more defensible probability distributions?* Although this question can be addressed generically, special emphasis must be placed on its relevance to the characterization, assessment, and quantification of environmental risk characterizing the interdependencies and interconnectedness (I-I) of Complex SoS. Risk is defined as a measure of the probability and severity of adverse effects. Furthermore, evaluating the central tendency of risk alone would not suffice.

2.II.3 BAYESIAN METHODS AND RISK ANALYSIS

Empirical data are almost always lacking in real-world modeling and managing Complex SoS. While directly relevant data are limited, we have multiple information sources that can provide indirectly relevant data. These are especially useful and essential for modeling Complex SoS, particularly when directly relevant data are scarce.

A Bayesian approach is presented here that is most effective in addressing data limitations. By design, Bayesian methods inherently consider the uncertainty associated with the parameters of a probabilistic model (even if those uncertain parameters are believed to be fixed numbers). Bayesian methodology is often recommended as the proper way to make formal use of subjective information, such as expert opinion and personal judgments or the beliefs of an analyst. Unlike frequentist methods with which they are often contrasted, Bayesian methods have an important advantage – they can always yield a more precise answer, even when few data are available.

There are some limitations to applying Bayesian methods in modeling Complex SoS. A major problem is that sometimes it is difficult to find a precise distribution to be used as prior probability of an appropriate database; also, the posterior probability is generally very sensitive to the prior. Many approaches have been suggested for selecting priors. One, *uninformative* priors, is a uniform prior distribution. Uninformative priors carry no information, such that all the information is from the data embodied in the likelihood function. When data is very limited, the posterior cannot precisely reflect the properties of the underlying system or Complex SoS. Another common method uses the *subjective* prior, that is, the prior distribution is determined by expert subjective opinions. This kind of prior is intended to reflect the analyst's or expert's personal knowledge about the system under investigation. There are many basic questions regarding this method: (i) Who determines the prior?

(ii) Why use this prior? (iii) Based on what evidence? (iv) What does it mean to anyone else? A hierarchical Bayesian model can provide a reasonable prior distribution by integrating data that is indirectly relevant. We will subsequently introduce the Bayesian method for inference in the section on hierarchical Bayesian models.

2.II.3.1 Bayesian Method for Distributions Inference

Given a likelihood $p(x|\theta)$ and prior density $p(\theta)$, with the Bayesian method we can infer the posterior density of parameter θ:

$$p(\theta|x) = \frac{p(x|\theta)p(\theta)}{p(x)} = \frac{p(x|\theta)p(\theta)}{\int_\theta p(x|\theta)p(\theta)d\theta} \tag{2.78}$$

where p denotes probability density (mass for discrete case), θ is a value of the random variable in question (parameters of a system), and x denotes the evidence being considered (observed data generated by the system). $p(\theta)$ is a prior probability that the random variable takes on the value θ. $p(x|\theta)$ is the conditional likelihood function that expresses the probability of the evidence, given a particular value of θ. $p(x)$ is a moralizing factor, which corresponds to the probability of having obtained the observed evidence. We can expand it to be the integral, with respect to θ, of the product of the prior and the likelihood. The moralizing factor is not a function of θ, but has a single value such that the resulting posterior distribution integrates easily to unity. Thus, we can also write the Bayesian rules as

$$p(\theta|x) \propto p(x|\theta)p(\theta) \tag{2.79}$$

where $\propto$ means "proportional to."

The normalizing factor is often difficult to compute analytically, but the use of conjugate pairs (prior and likelihood) can greatly simplify the problem (Gelman et al., 2013). For a nonconjugate pair, we can use computational techniques, such as numerical integration and simulation.

After the posterior density of θ, conditional on the observed data, x, has been obtained by Eq. (2.78), we can predict an unknown observable, $\tilde{x}$. For example, suppose $x = (x_1, \ldots, x_n)$ is the vector of the recorded seconds a runner runs n times for a certain distance and $\theta \sim (\mu, \sigma^2)$ is the unknown expected seconds. We can get the posterior distribution of θ, $p(\theta|x)$ by Eq. (2.78). Let $\tilde{x}$ be the seconds the runner will run in the $(n+1)$st run. The distribution of $\tilde{x}$ is called the posterior predictive distribution, posterior, because it is conditional on the observed x and predictive because it is a prediction for an observable $\tilde{x}$:

$$\begin{aligned} p(\tilde{x}|x) &= \int p(\tilde{x},\theta|x)d\theta = \int p(\tilde{x}|\theta,x)p(\theta|x)d\theta \\ &= \int p(\tilde{x}|\theta)p(\theta|x)d\theta \end{aligned} \tag{2.80}$$

2.II.4 RISK AND UNCERTAINTY

Lowrance (1976) defines risk as a measure of the probability and severity of adverse effects. This definition is harmonious with the mathematical formula used to calculate the expected value of risk to be discussed later:

1. *Risk*. Situations of risk are defined as those in which the potential outcomes (i.e. consequences) can be described in reasonably well-known probability distributions. For example, if it is known that a river will flood to a specific level on the average of once in 20 years, it is a situation of risk rather than uncertainty.
2. *Uncertainty*. In situations of uncertainty, potential outcomes cannot be described in terms of objectively known probability distributions, nor can they be estimated by subjective probabilities.
3. *Imprecision*. In situations of imprecision, the potential outcome cannot be described in terms of objectively known probability distributions, but it can be estimated by subjective probabilities.
4. *Variability*. Variability is a result of inherent fluctuations or differences in the quantity of concern.

Risk and uncertainty arise from measurement errors and from the underlying variability of complex, natural, social, and economic situations. If the analyst is uncertain because of imperfect data or crude analytical tools, the plan is subject to measurement errors. Improved data and refined analytic techniques will obviously help minimize measurement errors. More detailed definitions and a discussion will follow in the subsequent sections and chapters.

Some future demographic, economic, hydrologic, and meteorological events are essentially unpredictable because they are subject to random influences. The question for the analyst is whether the randomness can be described by some probability distribution. If there is a historical database that is applicable to the future, distributions can be described or approximated by objective techniques. If there is no such historical database, the probability distribution of random future events can be described subjectively, based upon the best available insight and judgment.

Risk Assessment Process. The risk assessment process is a set of logical, systemic, and well-defined activities that provide the decision maker with a sound identification, measurement, quantification, and evaluation of the risk associated with certain natural phenomena or man-made activities. The generic term *risk* will connote a multitude of risks. Some authors distinguish between risk assessment and management, and others do not and incorporate risk assessment within the broader risk management label. Although we make a distinction between the two terms in this book, at the same time we recognize that significant overlaps do exist. The following six steps constitute one vision of the entire risk assessment and management process (Haimes, 1991): (i) risk identification; (ii) risk modeling, quantification, and measurement; (iii) risk evaluation; (iv) risk acceptance and avoidance; (v) risk management; and (vi) risk communication.

Indeed, the first three steps relate to the following triplet of *risk assessment* questions posed by Kaplan and Garrick (1981): (i) What can go wrong? (ii) What is the likelihood that it would go wrong? (iii) What are the consequences? Here we add a fourth question, (iv) What is the time frame?

On the other hand, the final two steps – risk acceptance and avoidance and risk management – relate to the following triplet of *risk management* questions posed by Haimes (1991, 2016): (i) What can be done, and what options are available? (ii) What are their associated tradeoffs in terms of all costs, benefits, and risks? (iii) What are the impacts of current management decisions on future options?

Clearly, the risk evaluation step can be associated with both assessment and management activities and is an overlapping step between the two activities. Here, again, we note the importance of the circular-iterative process in systems engineering in general and in risk assessment and management in particular.

2.II.4.1 Risk and Uncertainty Analyses Practiced by Regulatory Agencies

In past years, the focus of regulatory analysis was on economic impacts and on fairly gross measures of environmental improvements, such as the tons of pollutant removed from the environment. Over the last several years, however, the use of quantitative risk assessment within government agencies has increased significantly. Risk assessment has become a powerful tool for supporting and shaping regulatory decisions, often equal to economic analysis in its usefulness. At its best, risk assessment adds a useful dimension by providing the decision maker with estimates of the public health and environmental benefits or effectiveness associated with regulatory alternatives. In concept, risk assessment provides common measures of what a regulation is expected to achieve, thus providing a level playing field to compare options critically and to ensure relative quality and consistency within and across programs.

Indeed, regulatory analysis and research must be afforded the highest standards possible and practical. Government and nongovernment agencies ought to (i) present a full and complete picture of the environmental risks analyzed, (ii) strive for greater consistency and comparability across risk assessments and across agency programs, and (iii) recognize the critical role played by professional judgment in performing and interpreting risk assessments. The above three premises also require that (i) a continuous effort must be made to separate science from policy; (ii) decision makers should be provided with unbiased risk assessment, estimates of uncertainty, and outer ranges of potential risk; (iii) conservative risk management decisions and applications of margins of safety should be made explicitly in the final regulatory decision; and (iv) regulatory decisions that increase net risk by eliminating one risk while increasing a more important risk elsewhere must be carefully avoided.

Indeed, to present a full and complete picture of risk, including a statement of confidence, uncertainties within risk assessments must be explicitly addressed. Environmental risk analysts commonly have the luxury of extensive data to describe the quantitative relationships among sources of risk and their associated uncertainties. More often, however, the available database is very limited. Thus, the challenging composite question facing regulatory agencies is how quantitative should the analysis be, both in estimating risks and representing their uncertainties, when the database is limited? If a quantitative approach is the object, then in which circumstances are Monte Carlo or other analysis methods appropriate? And how should probability distribution functions be selected in these circumstances?

2.II.4.2 Challenges Associated with Translations of Technical Uncertainty to a Decision Maker Who Is Dealing with Other Issues

- **Guiding Questions**. Decision makers maintain that they want the full spectrum of results, but can they handle the complexity? Who is the audience? Should cost be considered in the assessment? Should the risk assessor be concerned only with the technical side, and not with the cost side? How focused should the assessment be? Do political pressures drive the decision more than science?

  The key issue here is, what are the preferences of the decision maker? And, in particular, what is the preference structure (simple risk aversion or aversion to fundamental or epistemic uncertainty)? There is a fundamental principle – risk assessment and risk management should be kept carefully separated.
- **The Ideal Process.** The *ideal* process would involve the decision maker at all stages of the risk assessment process. There should be early discussion of the "burning issues" and development of a plan to study them.

  The decision maker does not have to be involved in the assessment phase. In fact, according to the separation principle, to avoid having the decision maker influence the assessment results by his/her political opinion, the scientific studies and the risk assessment should be performed first. Then the decision maker should be informed and act accordingly.
- **Show a Range of Possibilities.** It is not *necessary* to show the decision maker "the entire distribution." However, the decision maker must at least see the range of possibilities to gain insight about the uncertainties that may or may not be present. The range of possibilities alone is not particularly informative if some values in that range have an extremely low probability.
- **Responsibility.** Communication of technical analysis to any decision maker is the responsibility of the analyst. It does no good to bemoan the "ignorance" of the decision maker. It should be remembered that eventually the public (and the media) may become involved, *so* the decision maker as a single entity does not exist.

  Note the different ways that uncertainty is commonly explained among experts. What is the minimum information that can be given to decision makers that will include all relevant information? What is the assessor's responsibility in making all the tradeoffs (perhaps equity issues) clear to the decision maker? When the decision maker asks for an answer, how much emphasis should be placed on the level of uncertainty (or combination of uncertainties) embedded in the answer?

  Responsibility starts with stating all reasonable hypotheses identifiable at the time of the study, computing the risk under these different hypotheses, and weighting the results by their probabilities as assessed by the relevant experts. Shortcuts in this process can be dangerous.
- **Use of Probability Distributions.** We need to help (or find ways to help) the decision maker "see" probability distribution characterizations as an opportunity to increase flexibility to consider more options and alternatives than might be allowed or feasible if the risk characterization was a "bright line" or single number. The optimal answer at the 90th percentile may be more compatible with respect to some qualitative component not explicitly included in the risk characterization than is the optimal answer at the 95th percentile.
- **Tradeoffs.** When are decisions tough? Simply put, when there are potentially expensive outcomes, in terms of life or dollars, and when there

is uncertainty about the outcomes. These two factors usually combine in complex ways. There is no obvious choice; the decision maker is going to be open for criticism no matter what he or she chooses. One would think, given this situation, the decision maker would like to have as much evidence as possible so that an informed decision could be made; but this is not commonly the case. Additional data potentially confuses the situation and makes the decision harder; therefore, decision makers often like to limit the data. Keeping it simple makes the decision easier and simple heuristics work; there is no need for complicated decision algorithms.

However, in some circumstances, there is a need for "complicated" decision algorithms, for example, when there are fundamental (epistemic) uncertainties associated with different hypotheses or when explicit numerical safety criteria are used. Before throwing away additional data because they could confuse the decision maker, one would want to make sure that they are truly irrelevant (e.g. pertaining to another population than the one of concern).

- **Taxonomy**
  1. Uncertainties and assumptions need to be clearly documented in the assessment, in spite of a decision maker's request to "just give me a number."
  2. Uncertainties should be identified at all levels of the assessment:
     a. Field data limitations (sampling data) as for measurements of chemicals.
     b. Exposure estimates.
     c. Mortality calculations if extrapolated from animals to humans.
     d. Safety margins.
  3. The decision maker may not understand the uncertainties coming from the assessment, and, if they are not understood, they will not be effectively communicated as part of the decision-making process.
  4. The assessor must attempt to outline or describe uncertainties in simple, clear terms.
  5. How well have the outlier (tail of the distribution) subpopulations been examined? Are the effects on end points analyzed radically different for those individuals or events in the tail?

Safety margins are a matter of preferences (i.e. they reflect a desired level of risk aversion). They are a risk management issue that should be kept separated from risk assessment. There is no uncertainty in the choice of safety margins. There is, however, a resulting uncertainty about the actual safety margin that exists, once safety factors and risk assessment results have been combined. But it comes from the uncertainty about the assessment of the risk (data estimations, exposure estimates, and mortality computations) not from the choice of the safety margins.

- **Graphics.** Decision makers are in a difficult position than risk analysts and systems modelers in that they try to balance information of bewildering complexity. One way assessors can help is to develop graphics and figures that convey a welter of information very concisely. This has worked when tried with focus groups of all managers, resulting in the idea that methods for technical data presentation should be developed and tested.

2.II.5 STAKEHOLDERS AND RISK PERCEPTIONS

- **How to Present Data to the Decision Makers.** The risk assessor should present information to the decision maker that will allow the decision maker to address the concerns of those at risk. In this case, it is necessary for the decision maker to define the community at risk and its perception of the risk, whether that perception is accurate or not. The risk assessor needs to keep in mind the outcome and use of his/her assessment to ensure that adequate information is shared with the decision maker for use in clarifying perceptions.
- **How to Separate Risk Measurement and Decision Analysis.** Risk measurement and the decision analysis need to be kept separate. The risk assessment process should be quantitative, exact, and independent of any consideration of perceptions. Perceptions enter into the value-judgment part of the decision analysis but not into the risk analysis. Note that while the quantification needs to be objective, the perception element enters only into the value judgement part of the decision analysis.

- **The Role of Science.** Both the general public and the legal profession appear to view *science* as *certainty,* and in this lies the origin of at least part of the problem in explaining uncertainty to stakeholders. Risk communication researchers have made a notable gain in helping to define and delineate public perception of risk. However, there probably needs to be a more focused effort by individual groups to attempt to understand the risk perceptions of their individual stakeholders, because for each risk assessment group, there are commonly different subgroups of stakeholders.

  What responsibility do we, in the emerging field of risk analysis, have to contribute to changing the public understanding of science, to work with educators to change the beliefs of stakeholders of the future, and to work toward changing stakeholder perceptions of risk?
- **Communication.** One of the issues surrounding the difference in perceptions is that the analyst needs to be sensitive to the fact that the *way* he or she communicates is important. There are excellent well-meaning examples but misguided attempts to impart technical information regarding major risk-inducing cases. This does not mean that the public is ignorant and should be educated. As pointed out in earlier discussion, the level of concern will be moderated by whether the individual feels that the risk is voluntary (i.e. he or she has the option to avoid the risk). Analysts may have minor influence on the opinion of stakeholders, but they should be sensitive to the concerns of those stakeholders, and try to avoid red flag phrases and terminology.
- **Involvement and Communication**. Involvement and communication are key to providing information and removing uncertainty to stakeholders. An early briefing on the problem or issue should provide stakeholders useful information that is understandable (relevant to their lives), let them know that they can be part of the process, and provide the opportunity for further involvement, when appropriate. Regular updates are most important if there are no opportunities for actual participation in the study of the problem or issue. Where there is the potential for environmental inequity, it is even more important for the decision maker to take the initiative and provide a forum for disseminating information.

2.II.6 HOW TO ADDRESS TAILS OF DISTRIBUTIONS

- **Guiding Questions.** What is the definition of tails? Are the populations different in the tail? When do the decisions need to be made? How large is the population? How do you deal with the tail? How do you handle subgroups versus the entire population?
- **Policy Questions.** When should regulatory development be based on tails and when on central tendencies? Should regulations strive for flexibility in making these decisions? How is the impact analysis affected by "tail-hood" (other than in the obvious cases of small populations and unique groups)?
- **Flexibility.** Some inputs are such that we address only the high-risk tail. If it is a subpopulation, an efficient intervention would be much different than if a total population generated the tail. The analyst should be challenged to identify the source of the tail.
- **Cost–Benefit Analysis.** Societies want, as an ideal, zero risk for every individual. We need to convey the cost–benefit tradeoffs of different degrees of attainment of that ideal. Of course, spending all resources on one ideal may not be Pareto-optimal regarding a second problem, or the collection of all problems.
- **Monte Carlo Results.** One must still subject the results of a Monte Carlo analysis to a reality check, particularly at the extremes of the tail. One must examine the output and the combination of factors that lead to these extreme values. Is the combination of factors mutually consistent? Since the tail of the distribution is very sensitive to the assumptions about the shape of the input distributions, one needs to closely examine the sensitivity of the tails to these assumptions.
- **Classification.** The tail of the distribution seems to be somewhat of a misnomer. We are talking either about outliers in a single population or about multiple subpopulations within a total population. The problem of outliers within a single population

depends upon both the cutoff point and the impact on the outliers, including equity issues and social pressures. The multiple population problem is differentiated by whether individuals can be classified into a given population a priori or only after impact (e.g. a genetic propensity for certain types of concerns that can or cannot be determined by a simple test). In the first case, mitigation is possible by addressing the subpopulation; in the second case, mitigation is possible only by addressing the total population. The "tail problem" can be addressed by carefully identifying the *population of interest.*

2.II.7 OVERCONFIDENCE IN MEASUREMENTS INVOLVING UNCERTAINTY AND VARIABILITY

- **Level of Precision.** There can be a tendency for risk assessors and risk managers to read more precision into the analysis than can be justified, given the underlying data. This is a particularly thorny problem when examining the extremes of the tails of the distribution. One has to look closely at the combination of events that lead to these extreme values to ascertain whether they correspond to some realistic, albeit infrequent, situation. Consideration should be given to truncating the upper bound beyond a certain point (e.g. 99.5%).
- **Assumptions.** All modeling begins with some restrictive assumptions about the true state of things. Thus, our models of risk inevitably represent well-defined situations. It is utopian to try to match our confidence or belief (as it appears in the model) to a true state of confidence. The best course of action is to agree on the well-defined situations, which are not real, and let *confidence* refer to our surety about knowledge of situations or scenarios. Overconfidence is part of us, and we cannot eliminate it, or divorce it from our analysis.
- **Confidence.** The topic of "science and certainty" has long been argued by philosophers of science, but in our view, science deals in probabilities. In contrast to that, confidence in one's ideas or work is often the *sine qua non* of professional acceptance, both of an individual and of his or her work. These two items about science and professionalism are in opposition to each other and may lead to overconfidence. The challenge comes in the time frames: In science (theories), an entire complex of ideas may take years to untangle before a community consensus can be reached. In the science (models ~ theories) of risk assessment, the outcome of a particular analysis may be apparent immediately after an action is taken. Analysis of the assessment is also immediate. But how is confidence being judged? Is *one* occurrence of an event that has been assessed as rare with strong confidence a reason to label the assessment immediately as overconfident? Namely, what are the standards for judging confidence in measurements involving uncertainty and variability?

2.II.8 FACILITATE THE TWO-WAY PROCESS BETWEEN ASSESSORS AND DECISION MAKERS

- **Common Understanding.** Both assessors and decision makers need to understand and use the common language of uncertainty and evidence evaluation (i.e. probability in the sense of Laplace and Bayes). Ask clear questions, and acknowledge that we are always uncertain about the answer. Adopt the uniform practice and discipline of "quantifying" or representing uncertainty in the form of probability curves, the specific curves in any instance to be obtained as dictated by evidence. The process of communication must begin immediately between the decision maker and the risk assessor, and it must be meaningful, adequate communication. That is, not only must all parties try to use language that others would understand (i.e. the technicians must adapt their "tech-speak" for the decision maker), but also they must listen to each other. The outcome must be made clear at the beginning of the risk assessment and decision-making process.
- **Standards.** Standards for presenting risk assessments need to be developed to facilitate communication between assessors and managers. There may need to be a training program for managers

in understanding uncertainty. Comfortable routines must be established for a two-way communication about assessments and the associated inherent uncertainty in all assessments. A corollary concern is that such uncertainty needs to be communicated to those who are regulated or affected by the ensuing decisions; this may be the trickiest part of all. In daily life, people expect chance events, but when the word *science* is invoked, the common expectation is surety. The scientific community has not done a proper job of spreading an understanding of the nature of science to those who teach K-12 grades; thus, this challenge of promoting an understanding of uncertainty is a large one indeed. Can we also extend the question to lawyers? How can engineers and scientists work toward helping the legal profession recognize uncertainty and the need for regulatory flexibility based on uncertainty analysis?

- **Objectives.** The need for an early and frequent communication between the risk assessor and the decision maker has been addressed above. Clear objectives should be defined by the decision maker; and the limitations of the assessment should be made clear to the decision maker. This would allow for a refocusing of efforts, whether that involves redirecting the objectives or reevaluating the need for additional research to reduce uncertainty.
- **Relationships**. The communication between the risk assessor and the manager should begin at the start of the assessment and continue until the final risk management decision and beyond. The assessor can judge the direction of the assessment, what data are needed, and what resources to expend, only if the range of possible decisions is known. If the decision maker realizes early on that adequate information is not available and that an interim decision is the best that can be made in the near term, then this would influence the pace of the assessor's work toward a short-term screening assessment while gathering data for a more complete assessment.
- **Subjectivity.** There must be a distinction between the case where a single decision is to be made and the case where a series of decisions are to be made. In the first case, there is only one opportunity, whereas, in the second case, there is a series of opportunities to develop feedback from the decision makers. In the first case, the scope and potential impact of the decision determines how much feedback and iteration may be useful. In the second case, organizational memory must be applied to develop a learning curve for both the assessors and decision makers. Furthermore, all decisions about risk and uncertainty have a subjective element (for example, establishing an acceptable level of confidence for a distribution function is subjective, as is the reliance on poorly validated models).
- **Involving the Public.** The involvement of "the public" (interested and affected parties) in the decision-making process has been increasing. This adds at least an order of magnitude to the difficulty of communicating results to the decision maker (who now, in part, is the public). The public is generally less well equipped to absorb risk assessment results than are government officials and other decision makers.

On dealing with the public, it is clear that when the public's involvement is required, all phases of the analysis must be made explicit and open for review and comment *and* must be presented at a level that is understandable. Also, the public now insists on being involved in the early stages of analysis to make sure that the right alternatives are being considered.

2.II.9 VALUE OF INFORMATION AS A STRATEGY FOR EFFECTIVE USE OF RESOURCES IN DECISION MAKING

- **Scarce Resources.** The value-of-information concept is extremely efficacious in allocating scarce resources when multitudes of problems demand more information. Good analysis of which problems, or parts of a problem, can be most effectively improved by the availability of more data is a powerful approach. For a decision maker, the bottom line is dependent on the questions or objectives of the decision maker. Since the objective may evolve over time (new guidance, legislation, a change in perspective, or lobbying), the "bottom line" may also change. Hence, developing from the start a complete (or more complex) characterization (distributional with uncertainty and variability) may be the best strategy to accommodate a changing target of the decision maker.

- **Data Are Not Necessarily Information**. The old saying "data are not information" seems appropriate here. There are some uncertainties that are more important to reduce than others. This is most useful for setting a research agenda. Just because we can take something down to the *n*th decimal point does not mean it is efficacious. The judgment of the assessor is critical in deciding which information is most valuable given the question that is being asked. Retaining the flexibility to respond to changes is important; however there is a tradeoff between the depth and breadth of the coverage.
- **Communication.** Identification of research needs requires early and frequent communication between the assessor and the decision maker. The decision maker should be aware of the implications of obtaining additional information, which may remove some uncertainty from the process and subsequent decisions, versus the time and cost to obtain such information.

Divergent data may be clarified through additional research (repeat a questionable study); conduct a peer review of a certain consensus about the quality of the data or the interpretation of the data. An independent peer review by objective experts can understandably have different outcomes.

Differences among experts can be resolved through an independent, objective, third party review, but experts do differ, and decisions have to be made based on the best available, least uncertain, most appropriate data.

2.II.9.1 Select the Right Tools and Methods to Fit: Given Data Availability and Knowledge (Type of Decision)

- **The First Step is Decision Analysis.** A necessary first step of any risk assessment should be the problem-formulation stage, where decisions are made as to the population and/or subpopulation that will be the focus, whether one is interested in risk to individuals or to populations, whether acute or chronic effects are the emphasis, and what data sets are appropriate to meet the goals. These up-front decisions must be made in conjunction with the risk manager as well as other analysts who are providing additional components of the analysis. Although one must be cautious of political or other forces exerting undue influence on the risk assessment, the ultimate options available to risk managers must be understood to ensure that the risk assessment meets these needs.
- **"Cookbooks" Do Exist**. They exist for what goes into an assessment for different situations. They scale the intensity of data collection and the effort regarding the scope, importance, and legal or political constraints of the problem. Some of the cookbooks are codified; others resemble orally transmitted traditional wisdom. One wonders what would be learned by collecting and tabulating all the cookbooks. Generally, they fit the problem, but some prescriptions could be greatly improved. Many have not caught up with the dictates of government agencies' risk characterization policy. Furthermore, there is a need for some humility and flexibility. The best way to analyze problems and present their results is not a trivial task. Thus, our "cookbooks" need to have optional recipes; they should encourage experimentation, feedback, and learning, and they should be in loose-leaf binders for a continuous updating. Selecting the right tools and data has a holistic component and requires an iterative approach. Just as Bayesians remind us that we all do have *priors* on some values used in an analysis, we should be explicit about our priors for the analysis. One may start with a mockup, at the very beginning, and update it periodically as the problem is worked on, where we reveal our expectations of what we are dealing with and how much uncertainty could potentially be reduced.

2.II.10 SOURCES OF UNCERTAINTY

- **Qualitative and Quantitative.** When a risk assessment is developed, the sources of uncertainty should be apparent in both qualitative terms and in its quantitative display. All the sources of uncertainty should be a part of the assessment report. There should be a clear delineation of variability inherent in the phenomenon being assessed and a differentiation between variability and uncertainty, when these can be clearly distinguished.

- **Analyze the Impact**. The sources of uncertainty are often listed at the end of risk assessments, but seldom are these uncertainties quantified and their impact analyzed. From the standpoint of the value of the sources of uncertainty to decision making, it is necessary to include these additional components. The question then becomes, can they be quantified and their potential impact determined? That is the task set before the risk assessment community, especially those who work in the area of exposure assessment.
- **To What End?** Is there such a thing as "certain uncertainty?" Can the source of uncertainty be totally discovered? How would knowing the source of uncertainty affect the decision? How certain are we that a coin when flipped will land heads? (A US buffalo nickel is much more likely to land tails when spun.) How about utility assessment? How do we determine the utility of a population? What does it mean? Even for an individual, a utility function cannot be completely defined.

2.II.11 DISINCENTIVES FOR DECISION MAKERS IN MAKING DECISIONS

- **The Role of Conservatism**. There is a spectrum of opinion on the scale of conservatism. To some people, uncertainty means one must be overestimating the true risk. The fact is that *we do not know*! There are many examples in history where decision makers were not conservative enough (e.g. Chernobyl, thalidomide, the Exxon *Valdez*, HIV in the nation's blood supply). Often, it takes a disaster to provide an example of bad decision making. Hindsight is 20/20.
- **Decision Makers' Responsibility and Credibility.** The best leaders in society seem to be those individuals who accept the responsibility for making difficult decisions and who use a process of careful information collection and contemplation. Perhaps an incentive system, rather than disincentive system, for making decisions needs to be developed. Unfortunately, major mistakes seem to get more attention than successes, and if a mistake yields decreased credibility for the decision maker, then much is at risk in decision making. It takes many successes to build credibility and only one mistake to destroy it.

2.II.12 ORGANIZATIONAL PROBLEMS IN USING UNCERTAINTY AND RISK ANALYSIS

- **Assumptions.** Risk managers often require a risk assessor to make simplifying assumptions. The assumptions commonly involve a value judgment. Most risk assessors tend to lean toward the conservative (low risk) direction and often without knowing how conservative is each assumption. These embedded assumptions are hidden and cumulative. Thus, the single number presented to the risk manager is usually very conservative to an unknown and unstated degree. The risk manager may accept this number as "actual" or "best estimate" and add further factors of safety. The above process usually distorts the priorities in resource allocation.
- **Decision Makers' Needs**. There is usually a misunderstanding between decision makers and assessors. Decision makers want information relevant to making a decision. It then depends upon whether the decision maker is a good "manager." Managers are faced with uncertainty in most activities and are paid to make risk decisions. Managers do not want point estimates, but they want their information encapsulated for their decisions.
- **"Decision Truth."** "Biological truth" and "physical truth" are the ultimate objectives. Decisions need to be made, and not making them is also a decision. Therefore we need another "truth," which we might call "decision truth." The truth of decision making is that there is always uncertainty. If we adopt the practice and tradition in an organization of always telling the truth about our uncertainty, then "miracles can happen" in dissipating organizational problems; in fostering

communication and creating consensus, harmony, and teamwork within the organization; and in imbuing credibility and trust among the organization and its external clients, customers, and suppliers.

2.II.13 DEALING WITH DIVERGENT INFORMATION

- **Selecting Experts.** A decision maker can be presented with a separate risk characterization corresponding to each separate expert. There can be several such experts. Do decision makers want to see these separate risk characterizations? Do decision makers want to know or need to know the name of each expert? If a group of experts is intended to be representative of a community (e.g. the scientific community of "knowledgeable" experts in a particular area), then the group of experts should be assembled/sampled/selected in such a way as to be "representative." If the group is intended to "cover the range of experts," then they should be selected to represent such "range." A group that is selected to cover a "range" may not be a "representative" group.
- **Weight the Choices**. When dealing with multiple studies and multiple sets of expert opinions, one needs to establish some "weight of evidence" framework for selecting the data or opinions that will be the basis for the assessment. Often that may come down to a policy formulation (e.g. to be conservative or risk averse). Whatever the choice, the risk assessor must be clear as to what criteria were applied.
- **Do Not Weight.** A weighting approach to commensurate experts' views distorts valuable and useful information. A third party should present the range of results from divergent experts, explaining unambiguously why different underlying assumptions, models, etc. may lead to different results. Similarly, one might explain the disparity between test data and model prediction at the highest (output) level of analysis. Fusion of conflicting information seems to be arbitrary and unsatisfactory for coping with model (premise) sources of uncertainty.
- **Beware of Bias.** Who is charged with being open minded? Can anyone in the decision-making process be truly impartial? Experts are rarely immune to external influence. Consultants have clients; academics have external sources of funds; regulators have precedents. Few people are selfless protectors of the common good, seekers of the truth.
- **Separate the Evidence from Their Interpretations.** We ought to be able to agree on what the evidence and experience are (i.e. on what happened). In this way we may get a "consensus body of evidence." We also need to convert, or express, this consensus (body of evidence) into a probability curve. This process is interpretive, but it is not "person dependent." It must be governed by Bayes' theorem. Thus, we can arrive at a "consensus probability curve" to feed into the decision-making process.

2.II.14 COMPOUNDING MARGINS OF SAFETY

- **Make Margins Visible.** Technical personnel doing part of risk analysis often handle uncertainty by adding margins of safety into estimates or by using standardized methods with built-in margins of safety. In many cases, these margins of safety are not explicitly identified. When the various parts of risk analysis are integrated, the margins of safety are often compound multiplicatively, leading to very high hidden margins of safety, thereby, overstating the risk by a large factor. These margins of safety should be made visible, their compounding effects determined, and provided to the decision maker. The costs associated with different margins of safety should be made explicit. Costs can be direct or indirect; margins of safety should be used, and their visibility and cost should be ascertained.
- **Combining Percentiles**. Multiple 95th percentiles do not combine to give a 95th percentile. The true "percentile" or meaning of such a combination can only be determined by finding the distribution of the function of all variables. For example, a Monte Carlo evaluation of a function, $f(x,y)$, can identify the percentile corresponding to the function evaluated with $X = 95$th percentile of

X and with Y = 95th percentile of Y. Monte Carlo or other distributional characterizations that attempt to consider all combinations of factors (and their likelihoods) are the best ways to determine the implications of combining multiple margins of safety.

2.II.15 RELEVANCY OF BRINGING TYPES OF UNCERTAINTY INTO DECISIONS

- **Different Models.** It is important to identify sources of variation and random error in measurement, variation in population, etc., but there should also be a concern about another form of uncertainty, namely, the challenge of model choice and when dealing with Ames tests, animal bioassay, and epidemiological data. Each of these models has its strengths, given the underlying assumptions that make the results of these models reasonably appropriate, which also depend on the quality of the data and "matching" the type of data to the decision.
- **Use All the Information.** Characterizing risks in terms of distributions provides more opportunity and capability to differentiate among situations, chemicals, or sites than do single-number descriptions of risk. Two single-number risk characterizations can be the same, but the extent of probability distributions on risk can be quite different. For example, two upper bounds can be the same, but the corresponding two means (or medians, 90th, 95th, 99th percentiles) can be very different. The levels of uncertainty in two different situations can be different and need to be highlighted.
- **It Depends on the Decision Maker.** Risk assessors and decision makers should communicate to establish the level of accuracy the decision maker requires. A purely technical decision maker may have options for the documentation of uncertainty. A political decision maker may have little or no need or use for a technical discussion on uncertainties. However, it may be best to develop a concise discussion and summary of uncertainties at each stage of the analysis for consideration by the decision maker, as well as by the assessors.

Discussion of uncertainties needs to be cogent and reflect how they may affect the ultimate quantitative estimate of risk, i.e. whether they expand or decrease the range of uncertainty by X. There is a tradeoff between being honest (clearly articulating all relevant sources of uncertainty) and losing the patience and confidence of the decision maker.

In making a decision, one needs to know all classes of uncertainty. This is totally relevant because decisions cannot be rational without them. How is this done? This is done using the language of probability curves to express the uncertainty. Such probability curves can be derived from the evidence and Bayes' theorem. If possible, the uncertainty and its use in decision making needs to be portrayed in terms that decision makers and laypeople can understand. The "black box" aura needs to be removed. Uncertainty should be presented, and its consequences explained in a user-friendly fashion.

2.II.16 UNCERTAINTY AND VARIABILITY IN ASSURING THE QUALITY OF UNCERTAINTY ANALYSIS

2.II.16.1 Overview of Uncertainty and Variability

- **The Differences Between Uncertainty and Variability.** One can distinguish between two types of uncertainties: *uncertainty* (fundamental, epistemic) due to lack of fundamental knowledge and *variability* (aleatory, uncertainty, randomness) in a population. The former can be reduced by additional fundamental information; the latter can be reduced in principle by exhaustive study. The two types of uncertainties, however, can be "measured" by the same method (probability). When dealing with a single element in the population, both types of uncertainty become the same (lack of knowledge), and the risk is characterized by one probability (e.g. of failure) that represents both types of uncertainty for decision-making purposes. Uncertainty reflects a lack of knowledge that cannot be described by a *probability* distribution. Variability represents heterogeneity across some dimension (population, time, space, etc.) that is represented by a *frequency* distribution. Conceptually, these are very different. Instead of pretending that variability and uncertainty can both be described by probability distributions, one should argue that they are different.

Uncertainty represents random error, systematic error, irreducible lack of knowledge, or lack of an empirical basis for making an estimate. It can be addressed, but not necessarily reduced, by better measurements. Consider the effect of systematic error on tails of distributions – the systematic error may be revealed by better experimental designs, which could have the effect of increasing uncertainty.

- **Determining the Sample Space.** The problem depends on how one defines "individual" in the population under study. If the pattern of exposure is incorporated into the definition of the individual, then the situation described above goes away. A probabilist would say that it all depends on how you define your *sample space.* The same holds for temporal variability; if it is sufficiently important, then it could be incorporated into the definition of an individual.
- **Using Updated Information.** At times, a probability distribution is useful for representing uncertainty, particularly when there are competing theories. The use of subjective probabilities allows us to quantify the uncertainty and provide better information to the decision maker. The distribution can be updated as more information becomes available.
- **Objectivity Versus Subjectivity.** Variability is commonly a property of the system under study, while uncertainty is a property of the analyst(s) studying the system. In practice, variability is incorporated into a study by subdividing the system into a number of cases, which are then analyzed separately. Uncertainty results in a number of different analysis outcomes being possible for each of the cases that result from variability. If knowledge of some aspects of variability (e.g. temporal, spatial, interindividual) is missing, then this knowledge cannot be used to define variability "cases," and its lack will become part of the uncertainty in the analysis.

2.II.16.2 How Can One Separate Uncertainty and Variability in a Data Set?

- **Given Several Data Sets.** It is much easier to think about this separation question regarding a single parameter, for which a data set exists with uncertainties and variability. If several data sets exist, our ability to separate uncertainty and variability for a particular parameter improves, especially if we have repeated measures at several times or repeated measures on an individual basis. One can use replicate samples (same time, place, and/or individual) to get a grasp on measurement error.
- **Given a Single Data Set.** In any case, the only way to separate uncertainty and variability in a single data set is to make an assumption about one or the other (e.g. each of these numbers comes from a binomial or normal distribution reflecting measurement error) and assess what the effect is on the other dimension, "recalibrating" as you go (iteratively).
- **Account for Measurement Error.** In measuring a variable quantity, the measurement process itself can introduce uncertainty (both random and systematic error) such that

$$M = V \times U$$

where
M = measured value
V = one random realization from the heterogeneous population
U = perturbation coefficient used to model uncertainty for realization

In principle, measurement error should be known (i.e. the method should be validated), making it possible to separate measurement error from the sample of measurements to make an inference about variability.

- **Analysis of a Data Set.** Given a sample from a variable quantity, one could use goodness-of-fit techniques to infer the parameter of an appropriate frequency distribution. The limited sample size will result in some confidence intervals on the parameters of the distributions. These confidence intervals represent the uncertainty about the frequency distribution.

2.II.16.3 When Is a Two-dimensional Separation of the Risk Analysis Results Useful?

- **Why It Is Useful?** On a practical level, a two-dimensional separation allows identification of the contribution of variability versus uncertainty. The more appropriate questions are, how does one communicate the results of a two-dimensional assessment of variability and uncertainty and why do we need to make the two-dimensional distinction? Uncertainty characterization of variability provides insight into key subpopulations when one uncertainty has implications for prioritizing research and measurements. Furthermore, a two-dimensional result provides a more technically correct basis for selecting any point value from an analysis, for example, a fraction of a population at or below a given level with a probability (confidence) of x. A two-dimensional representation of the risk can be useful for three reasons: (i) to decide what type of information should be gathered to reduce overall uncertainty, (ii) when the decision involves the repetition of an experiment, and (iii) when the decision maker is "ambiguity averse" (in addition to being risk averse) given epistemic uncertainty and does not wish to base his or her choice on the mean probability of the outcome.

2.II.16.4 What Different Experimental Designs Are Available for Assessment of Uncertainty and Variability?

- **Sensitivity Analysis.** One could evaluate the range and impact of variability in an assessment by doing a type of sensitivity analysis in which the potential outcome's values are presumed to be known (no uncertainty) but those values have different ranges. Then, holding variability fixed (using only one range to represent variability), one could assess the impact of uncertainty in the same way.
- **Measuring Levels of Dependence.** The answer varies with the level of dependence among the variables affected by both types of uncertainty. *Example:* The weight of people can be *measured* independently from some health effect. If the fundamental hypothesis and the distributed variable are dependent, an exhaustive measure (or sampling of the population) of variable X provides information about both the probability of underlying phenomena E (which can be E_1, E_2, E_3, etc.) and about the variability of X, given the "true value" of E. The separation of the two relies on a value of f_x/E_1 (the conditional distribution of X) for each value of E. If there is an experiment that can be performed to capture f_x/E_1, then all is fine. Otherwise, it is necessary to use expert judgment.
- **Beware of Measurement Error.** It is critically important to understand measurement error and its magnitude relative to the variability in the quantity being measured. Thus, one should have some quantitative basis for characterizing the random and systematic error of the measurement technique (e.g. through measurement of replicator, blanks, etc.) before applying the technique to measure the variable quantity of interest. Only when validation of the measurement technique is infeasible, for a reason such as cost, should the subjective approach to characterizing measurement error be emphasized. However, a subjective approach tends to be required, especially when dealing with *systematic* error, which is often the hardest type of error to quantify.

2.II.16.5 Is There a Point Where Uncertainties of Both Types Are Too Broad to Allow Quality Assessment?

- **Distributions Should Represent the State of Knowledge.** One *can* perform a quality assessment with broad distributions, if these distributions properly represent the state of knowledge about uncertainty and if they properly characterize variability. If uncertainty dominates the assessment, then we need to prioritize estimates on research to develop better measurements. If variability dominates, we may be able to identify more homogeneous subpopulations (i.e. stratify the variability) that are especially important (sensitive) in the context of the end point of the assessment. The "broadness" of these distributions does not reflect on the quality of the assessment, but may influence the types of insights we obtain from that assessment. Is it always better to generate a broad distribution of exposure and/or risk than to choose point estimates of unknown conservativeness?

- **It Depends on How We Define a Quality Assessment.** If one correctly determines that there is not enough information available to select between two alternatives, then one has made a quality assessment. On the other hand, if one defines a quality assessment as that which allows a choice between two alternatives, then there will be situations in which a lack of information prevents a quality assessment. A quality assessment will always incorporate uncertainty as opposed to default values or point estimates. If the range of uncertainty is broad, it is even more important to incorporate the uncertainty in the risk assessment so that the risk manager is properly informed about the risk. Likewise, a broad distribution for variability is preferred to a point estimate. Quantifying uncertainty and variability is important for risk analysis of safety-critical systems, since remedies can also have risks and these risks need to be compared. Essentially, the quality of the assessment depends on how well it represents the current state of information. The fact that the decision maker would prefer fewer uncertainties (of both types) is a separate issue. He or she will have to make a decision no matter what. The goal of risk analysis is to be "precise" about one's state of uncertainty.
- **Beware of Tradeoffs.** Often, decision makers are not trying to "get the right answer," because they are commonly working in a sociopolitical arena. If society makes, or has made, a policy decision that, absent other data, animal carcinogens should be treated as human carcinogens, then the "true" uncertainty may be given up in favor of a narrower probability density function (pdf) that is conditional on the model choice, which may be most useful for decision making. In that sense, a broader pdf may be too broad for the practical business of making decisions. This may not be scientifically valid, but that is not the only criterion. Conservatism plays a role in expediting decisions and balancing social costs of type I and type II errors. In statistical hypothesis testing, a type I error is the rejection of a true null hypothesis (also known as a "false positive" finding), while a type II error is retaining a false null hypothesis (also known as a "false negative" finding).

2.II.16.6 When Can We Tell Whether Short-term Variability Will Be the Same as Long-term Variability?

- **When You Can Extrapolate the Lifetime Behavior.** Levels of variability can be reduced in the long run but cannot be reduced to zero. For example, a naive view would be that, over a lifetime, there will be essentially no variability in the amount of aflatoxin each American ingests (assuming we all ate the same quantity of peanuts each day). If concentration is completely random, over a long enough time, we would all be exposed to exactly the same mix of highly contaminated, slightly contaminated, and uncontaminated food. However, in reality the variability evident in each day probably persists to some extent over a lifetime; some people consistently eat brands or types of food that are higher or lower in aflatoxin than what others eat. (Aflatoxin is a type of mold that is considered a human carcinogen.)
- **When You Can Base the Long Term on the Short Term.** One could superimpose frequency distributions of the temporally variable quantities, all made for a different window of time. If the moments match, and the frequency distributions compare well, we could assume the short-term variability for even longer periods.
- **When You Can Use Multiple Data Sets.** Another way to tell that short- and long-term temporal variability are the same is to compare two data sets, one of a short-averaging time and the other of a long-averaging time. It may be possible to take a data set and make a short-averaging time data set by manipulation if enough detail exists in the measurements behind it. Sometimes calling the agency or group that reported the data reveals a wealth of information behind the reported results.
- **Caveat for Extreme Events.** If we are interested in extreme events, then we would not care about characterizing the mean of the long-term frequency distribution. Instead, we might like to characterize a distribution for the 95th percentile (e.g. in the case of a design to deal with floods). In this case, we may have a much more challenging data-gathering requirement in terms of number of samples and time (i.e. we may do a good enough job characterizing the mean but only

need more data to characterize the upper percentiles with an acceptable confidence interval). Such a distribution would represent uncertainty in estimating the long-term 95th percentile of variability. Chapter 8 of the fourth edition of the textbook on *Risk Modeling, Assessment, and Management* (Haimes, 2016) is devoted to risk of extreme events.

2.II.16.7 What Are the Effects of Correlations Among Parameters on the Risk Analysis Results?

- **Model Development.** The effects of including correlations among model parameters on model results are certainly model dependent. In general, the inclusion of correlation will reduce the variance of the model output in a Monte Carlo calculation, but it is possible that the variance of the output will be increased, depending on model structure. Correlations enter a model structure in two ways: (i) Parameters of the model may be naturally correlated (e.g. priority and permeability), or (ii) correlations may be implicit in the structure of the mathematical model.
- **You *Must* Care.** Ignoring correlations can have a large impact on the outcome of an analysis. In practice, ignoring correlations tends to obscure extreme results. However, one can construct examples in which ignoring correlations has a wide range of effect, from increasing uncertainty spread to decreasing uncertainty spread. Neglecting correlation is most important in cases where the correlated variables have similar and longish relative standard deviations and are reasonably well correlated. Over two decades ago, Smith et al. (1992) wrote about simple sums and products of correlated variables. Sometimes neglecting correlation does not matter at all. Sometimes our information of the correlation coefficient is pretty poor, which can be a significant source of uncertainty. If input variables are positively correlated, but the analyst treats them instead as independent variables, then the results will have a dispersion (or scatter) that is too high. On the other hand, if the input variables are highly negatively correlated, the analysis results will have an incorrectly smaller variation (or scatter) than the true variation. The effect of correlation is that the same underlying variable will affect several others; therefore, the same fundamental hypothesis may have a large effect on the result. This is a case where the explicit Bayesian treatment of these dependencies through conditional distributions (or probabilities) is necessary to keep the story straight.
- **Taxonomy of Correlations.**
  A. Correlation between uncertainties: (i) between individuals (i.e. all individuals face the same uncertainty in concentration of a pollutant in drinking water, if the measurement is taken at a central location, whereas uncertainties in measurement would be independent if taken at each household) and (ii) for each individual.
  B. Correlation between variable quantities (e.g. intake rate and body weight).
  C. Cross-correlations between variable quantities and uncertain quantities (e.g. when the magnitude of measurement error depends on the magnitude of the underlying variability).
- **Is the Population or Individual Being Addressed?** This would depend on the risk descriptor of interest – population risk or individual risk. For example, if characterizing risk only has the expected number of cancer deaths in a population, then there would be no need to distinguish between variability and uncertainty, provided that uncertainties are uncorrelated with variability. When addressing individual risks, variability and uncertainty would have obvious implications for the comparison of risks. Problems would arise in trying to trade off high-variability and low uncertainty risks with the converse, high uncertainty and low-variability risks.
- **It Is a Management Issue.** Large variability when correctly and accurately characterized should not affect a direct comparison of risks. Large uncertainty, however, should be taken into account. How that is accomplished will be case specific. The decision maker has two choices: (i) to combine both types of uncertainties and weighting (generally by a utility function) the possible outcomes or (ii) to use two types of utility functions (preferences) for uncertainty and randomness.

The latter method is much more complicated, and there is still a debate as to whether it allows rational and consistent decisions.

- **Prioritizing Remedies.** If there is significant uncertainty in the risk estimate and if there are multiple remedies to reduce the risk, then the uncertainty should be considered in the choice of remedies, particularly if the remedies also pose risk. If risks are being prioritized, then uncertainty should be considered. Variability should be used when determining the cutoff point.

2.II.16.8 Should One Separate the Treatment of Model Uncertainties and Parameter Uncertainties?

- **Yes, They Are Different Issues.** If decision makers find that uncertainties associated with an outcome are too broad to allow a defensible decision, then the solutions, and which resources should be devoted in order to narrow the uncertainties, are different. But it would not be clear where to go next if one did not separate them. Also, one can evaluate the impact of each through sensitivity analysis if they are kept separate, while it is more difficult if they are merged. Finally, if model uncertainties and parameter uncertainties are kept separate, then risk assessments can be updated more clearly and easily as information about parameters changes or as modeling input changes.
- **It Is Problem Dependent.** As for many of the questions in risk assessment, the problem underlying this question does not have a clear answer. Rather, the answer is dependent on the particular analysis under consideration. If there are only a few alternative models and the effects of these models are viewed as being very important in the decision process, then the analyses should be performed separately for each model (i.e. a *ceteris paribus* experimental design, namely, *all other things remaining constant*). On the other hand, if the effects of the models are not seen as important in the decision process, if limited resources prevent the consideration of many alternative analyses, or if the effect of the different models is not anticipated to be large, then model choice can simply be treated as one more (discrete) variable in an uncertainty analysis.
- **You Can Use Expert Judgment.** Models should be weighted using expert judgment so that the uncertainty can be "quantified." Parameters may also be weighted using expert judgment if there is significant uncertainty or distributions may be defined for the parameters if there is variability. It is advantageous to develop a sensitivity analysis for the various parameters, but the uncertainties should also be combined to facilitate the risk manager's decision.
- **You Can Convert Parameter Uncertainties.** Model uncertainties sometimes can be converted to parameter uncertainties, in which case one could treat the model uncertainties as parameter uncertainties and, furthermore, perform sensitivity analysis to determine if those parameters really mattered in the analysis. Other types of model uncertainties include: model topology, scenarios assumed as a basis for the model, model boundaries (if the model treats things as exogenous that ought to be endogenous due to feedback), simplifications of more complex models, lack of a firm theoretical foundation, and temporal or spatial resolution.

2.II.17 CONCLUSIONS

The importance of incorporating risk and uncertainty in modeling and policy formulation and in the decision-making process has been articulated in Part II of this chapter. The distinction between risk and uncertainty and between epistemic and aleatory uncertainty have been demonstrated with multiple examples. Most importantly, in Part II we demonstrated the efficacious contributions of risk and uncertainty to the intricate decision-making process by incorporating new data through Bayes' theorem and with Bayesian analysis.

REFERENCES

Bauman, E.J. (1966). Multilevel optimization techniques with application to trajectory decomposition. Ph.D. dissertation, University of California.

Chankong, V. and Haimes, Y.Y. (1983). *Multiobjective Decision-Making: Theory and Methodology*. New York: North-Holland.

Chankong, V. and Haimes, Y.Y. (2008). *Multiobjective Decision-Making: Theory and Methodology*. New York: Dover.

Dantzig, G.B. (1963). *Linear Programming and Extensions*. Princeton: Princeton University Press.

Dantzig, G.B. and Wolfe, P. (1960). Decomposition principle for linear programs. *Operations Research* **8**: 101–111.

Dantzig, G.B. and Wolfe, P. (1961). The decomposition algorithm for linear programming. *Econometrica* **9**(4): 767–778.

Gelman, A., Carlin, J., Stern, H., et al. (2013). *Bayesian Data Analysis, 3e*. New York: Chapman & Hall.

Haimes, Y.Y. (1971). Modeling and control of pollution of water resources systems via multilevel approach. *Water Resources Bulletin* **7**: 104–113.

Haimes, Y.Y. (1972). Decomposition and multilevel techniques for water quality control. *Water Resources Bulletin* **8**: 779–784.

Haimes, Y.Y. (1973a). Decomposition and multilevel approaching modeling and management of water resources systems. In: *Decomposition of Large Scale Problems* (ed. D.M. Himmelblau), 348–368. Amsterdam: North Holland.

Haimes, Y.Y. (1973b). Multilevel dynamic programming structure for regional water resource management. In: *Decomposition of Large Scale Problems* (ed. D.M. Himmelblau), 369–378. Amsterdam: North Holland.

Haimes, Y.Y. (1975). Hierarchical modeling of regional total water resources systems. *Automatica* **11**: 25–36.

Haimes, Y.Y. (1977). *Hierarchical Analyses of Water Resources Systems: Modeling and Optimization of Large-Scale Systems*. New York: McGraw-Hill.

Haimes, Y.Y. (1981). Hierarchical holographic modeling. *IEEE Transactions on Systems, Man, and Cybernetics SMC* **11**(9): 606–617.

Haimes, Y.Y. (1991). Total risk management. *Risk Analysis* **11**(2):169–171.

Haimes, Y.Y. (2012). Modeling complex systems of systems with phantom system models. *Systems Engineering* **15**(3): 333–346.

Haimes, Y.Y. (2016). *Risk Modeling, Assessment, and Management*, 4e. New York: Wiley.

Haimes, Y.Y. and Macko, D. (1973). Hierarchical structures in water resources systems management. *IEEE Transactions on Systems, Man, and Cybernetics* **3**(4): 396–402.

Haimes, Y.Y., Kaplan, M., and Husar, M. (1972a). A multilevel approach to determining optimal taxation for the abatement of water pollution. *Water Resources Bulletin* **8**: 851–860.

Haimes, Y.Y., Foley, J., and Yu, W. (1972b). Computational results for water pollution taxation using multilevel approach. *Water Resources Bulletin* **8**: 761.

Haimes, Y.Y., Tarvainen, K., Shima, T., and Thadathil, J. (1990). *Hierarchical Multiobjective Analysis of Large-Scale Systems*. New York: Hemisphere Publishing Corporation.

Haimes, Y.Y., Barry, T., and Lambert, J. (1994). When and how can you specify a probability distribution when you don't know much? *Risk Analysis* **14**(5): 661–706. https://doi.org/10.1111/j.1539-6924.1994.tb00280.x

Hall, W.A. and Dracup, J.A. (1970). *Water Resources Systems Engineering*. New York: McGraw-Hill.

Hall, W.A. and Shephard, R.W. (1967). Optimum operating for planning of a complex water resources system. University of California Water Resource Center, Los Angeles, Contribution 122, 75.

Kaplan, S. and Garrick, B.J. (1981). On the quantitative definition of risk. *Risk Analysis* **1**(1): 11–27.

Lasdon, L. (1964). A Multilevel Technique for Optimization. *Systems Research Center SRC-50-64-19*, Case Western Reserve University, Cleveland, OH.

Lasdon, L. (1970). *Optimization Theory for Large Systems*. London: The Macmillan Company.

Lasdon, L.S. and Schoeffler, J.D. (1966). Decentralized plant control. *ISA Transactions* **5**(2): 175–183.

Lefkowitz, I. (1966). Multilevel approach applied to control system design. *Transactions of the ASME* **88**: 2–10.

LeMay, J.L. (1969). Class Notes, School of Engineering and Applied Sciences. University of California, Los Angeles.

Leondes, C.T. ed. (1973). *Control and Dynamic Systems, Advances in Theory and Applications*, vol. **10**. New York: Academic Press.

Lowrance, W.W. (1976). *Of Acceptable Risk*. Los Altos, CA: William Kaufmann.

Macko, D. (1967). General Systems Theory Approach to Multilevel Systems, *SRC 106-A-67-44,* Systems Research Center, Case Western Reserve University, Cleveland, OH.

Macko, D. and Haimes, Y.Y. (1978). Overlapping coordination of hierarchical structures. *IEEE-SMC* **8**: 745–751.

Mesarović, M.D. ed. (1968). *Systems Theory and Biology*. New York: Springer-Verlag.

Mesarović, M., Macko, D., and Takahara, Y. (1970). *Theory of Hierarchical Multilevel Systems*. New York: Academic Press.

Pearson, J.D. (1965). Multilevel Control Systems. *Proceedings IFAC Symposium Adaptive Control*, Teddington, England (14–17 September 1965).

Runyon, R.P. (1977). *Winning the Statistics*. Reading, MA: Addison-Wesley.

Smith, A., Ryan, P.B., and Evans, J. (1992) The effect of neglecting correlations when propagating uncertainty and estimating the population distribution of risk. *Risk Analysis* **12**(4): 467–474.

3

Hierarchical Holographic Modeling and Multilevel Coordination of Complex Systems of Systems

3.1 INTRODUCTION

This chapter addresses the need for the development of a body of prescriptive theory for interdependent and interconnected complex systems of systems (SoS) (Complex SoS). We begin with building on a comprehensive modeling schema termed Hierarchical Holographic Modeling (HHM) (Haimes, 1981, 2016). HHM is capable of representing fundamental attributes of interdependent and interconnected Complex SoS, which have commonly escaped multiperspective modeling representation. Four decades ago, Blauberg et al. (1977) argued that:

> the multiplicity of descriptions for any system: in order to obtain adequate knowledge about a system one has to construct a certain class of its descriptions, each of which can only cover certain aspects of the wholeness and hierarchy of the system. This principle is, in fact, derived from the fundamental relativity of any description of a system. (p. 132)

During the same period in the 1970s, Haimes and Macko (1973), Haimes (1977), and Macko and Haimes (1978) developed a theory/methodology termed Hierarchical Overlapping Coordination (HOC). This theory/methodology emerged from the study of water resource systems that are characterized by multiple interdependent and interconnected systems, which in their essence constitute Complex SoS. In the two previous chapters, we modeled the interdependencies and interconnectedness (I-I) characterizing Complex SoS via the shared/common states and other essential entities within and among the subsystems and systems that make up Complex SoS. Different decompositions of water resources can represent/address multiple objectives, most of which are noncommensurable, competing, and in conflict with each other. However, their simultaneous functions must be at least compatible. In many ways, HOC provides a mathematical framework for modeling, and thus, representation of such systems, permitting alternative decompositions of a single system and showing how the mutual functioning of the resultant plurimodels, may be coordinated. Extensive work on decomposition and higher-level coordination can be found in Macko (1967), Leondes (1969), and Mesarović et al. (1970) and has already yielded very promising results (e.g. Tarvainen, 1980; Haimes and Tarvainen, 1981).

Modeling and Managing Interdependent Complex Systems of Systems, First Edition. Yacov Y. Haimes.

This philosophical and practical approach to modeling Complex SoS highlighted the need for (i) understanding and analyzing them through the fundamental principles of wholeness (adhering to the Gestalt philosophy and representing the integrity of the subsystem and the Complex SoS as a whole) and (ii) constructing a hierarchy to represent their internal structure. Thus the fundamental attribute of Complex SoS is their inescapably multifarious nature – hierarchical noncommensurable objectives, multiple decision makers, and multiple transcending aspects associated with each subsystem, whether temporal, structural, functional, organizational, or human decision making, among other factors. All are shrouded with intertwined shared states and other essential entities. The term essential entities connotes shared/common decisions, decision makers, stakeholders, organizational set-ups and policies, goals and objectives, among others. In part this may be a natural consequence of the fact that Complex SoS are responsive to multiple demands placed on them and, in the process of their emergence or evolution, are accompanied by the need for increasing required resources and justified through a broadening of the purposes served by them.

Note that it is unrealistic to represent within a single model the multiple characteristics, goals and objectives, configurations, and other specific characteristics of a truly Complex SoS. Furthermore, the specific interest at any given time (to its management, to government regulators, to customers, or to any other stakeholder) is likely to vary over time. Our inability to treat this most basic multifarious attribute of Complex SoS from some relevant vantage point with some degree of commonality constitutes a weakness in our theoretic base for modeling Complex SoS. Given this assumption and the notion that even the integrated models we have cannot adequately cover all of a system's aspects, the concept of HHM constitutes an alternative comprehensive theoretical framework for systems modeling and risk identification of Complex SoS.

HHM, which forms the basis for modeling interdependent and interconnected Complex SoS, has emerged from a generalization of HOC. It reflects a difference in kind from previous modeling schemas and contributes to the theory and methodology of modeling Complex SoS. The name is suggested by holography – the technique of lensless photography. The difference between HHM and conventional photography, which captures only two-dimensional planar representations of scenes, is analogous to the differences we see between conventional mathematical modeling techniques (yielding what might be termed "planar" models) and the multidimensional schema proposed by HHM. In the abstract, a mathematical model may be viewed as a one-sided image of the real system that it portrays. For example, with single-model analysis and interpretation, it is quite impossible to identify and document all critical and essential attributes of the Complex SoS or the sources of risk associated not only with the multiple components of an infrastructure (e.g. transportation or a hydroelectric power structure or food processing plants), but also with its welter of societal aspects (functional, temporal, geographical, economic, political, legal, environmental, and institutional).

3.2 HIERARCHICAL HOLOGRAPHIC MODELING

The fundamental attribute of Complex SoS is their inescapably multifarious nature: hierarchical noncommensurate objectives, multiple decision makers, multiple transcending aspects, and elements of risk and uncertainty. In part, this may be a natural consequence of the fact that most Complex SoS respond to a variety of needs that are basically noncommensurable and may under some circumstances openly conflict.

It is impracticable to represent within a single model all the aspects of a truly Complex SoS that may be of interest at any given time to its management, government regulators, students, or any other stakeholder. HHM, which forms the basis for this chapter and book, has emerged from a generalization of HOC, enables a comprehensive modeling elasticity of interdependent and interconnected Complex SoS due to its multidimensional, holographic modeling capabilities.

3.3 DEFINITION AND LITERATURE REVIEW

HHM is a holistic philosophy and methodology aimed at capturing and representing the essence of the inherent diverse characteristics and attributes of a system – its multiple aspects, perspectives, facets, views, dimensions, and hierarchies.

Several modeling philosophies and methods have been developed over the years to address the complexity of modeling interdependent and interconnected Complex SoS and to offer various modeling schema. Four decades ago, in his book *Methodology for Large Scale Systems*, Sage (1977) addressed the "need for value systems, which are structurally repeatable and capable of articulation across interdisciplinary fields," with which to model the multiple dimensions of societal problems. At the same time, Blauberg et al. (1977) pointed out that for the understanding and analysis of a large-scale system, the fundamental principles of wholeness, representing the integrity of the system, and hierarchy representing the internal structure of the system, must be supplemented by the principle of "the multiplicity of description for any system." Recognizing that a system "may be subject to a multiplicity of management, control and design objectives," Zeigler (1984) addressed such modeling complexity in his book *Multifaceted Modeling and Discrete Event Simulation*. Zeigler introduced the term *multifaceted* "to denote an approach to modeling which recognizes the existence of multiplicities of objectives and models as a fact of life" (p. 8). In his book *Synectics: The Development of Creative Capacity,* Gordon (1968) introduced an approach that uses metaphoric thinking as a means to solve complex problems.

Arthur D. Hall III (1989), whose first book on systems engineering was published in 1962, recognized the contributions of HHM in his seminal book *Metasystems Methodology*: "In this way," he wrote, "history becomes one model needed to give a rounded view of our subject within the philosophy of Hierarchical Holographic Modeling (Haimes, 1981) being used throughout this book, defined as using a family of models at several levels to seek understanding of diverse aspects of a subject, and thus comprehend the whole." Hall's theoretical framework, termed *metasystems methodology,* captured the multiple dimensions and perspectives of a system. Other early seminal works in this area include the book on societal systems and complexity by Warfield (1976) and the book *Systems Engineering* (Sage, 1992). For example, in his book Sage identified several phases of the systems engineering life cycle and noted that embedded in such analyses are multiple perspectives – the structural definition, the functional definition, and the purposeful definition. Finally, the multiple volumes of the *Systems and Control Encyclopedia: Theory, Technology, Applications* (Singh, 1987) offered a plethora of theory and methodology on modeling large-scale and complex systems. In this sense, multifaceted modeling, metasystems, HHM, and other contributions in the field of large-scale systems constitute the fundamental philosophy upon which systems engineering and risk analysis are grounded.

3.3.1 Hierarchical Holographic Modeling: Basic Concepts

In the abstract, a mathematical model may be viewed as a one-sided image of the real system that it portrays. With single-model analysis and interpretation, it is quite impossible to clarify and document the multiple perspectives and characteristics of Complex SoS, nor the myriad sources of risk associated not only with the multiple components, objectives, and constraints of Complex SoS but also with its numerous societal aspects (functional, temporal, geographical, economic, political, legal, environmental, sectoral, and institutional).

Central to the mathematical and systems basis of holographic modeling is the overlapping among various holographic models with respect to the objective functions, constraints, decision variables, time frame, and input–output relationships of a basic system or Complex SoS. In this context, holographic modeling may be viewed as the generalization of HOC in the following way.

As discussed in Section 3.2, in HOC, a system's single model is divided into several decompositions to represent the various aspects of the system, and these decompositions are coordinated to yield an improved modeling schema and, thus, analyses and decision making. Coordinating these dissociated models – that is, associating them via holographic modeling methodologies – can be considered a zero-order or degenerate case of holographic modeling.

The term *holographic* refers to the quest to have a multiview image of a system when identifying vulnerabilities to specific inputs (as opposed to a single view or a flat image of the system). Views of risk can include, but are not limited to, economic, health, technical, political, or social. In addition, risks to Complex SoS can be geography related and time related. In order to capture a holographic outcome, the modeling team that performs the analysis must possess a broad array of experience and knowledge.

The term *hierarchical* refers to the quest to understand and model the multiple levels of the Complex SoS hierarchy. HHM recognizes, that for the modeling process to be comprehensive and complete, one must realize that the macroscopic perspectives of the upper management level of an organization can be very different from the microscopic vision subscribed by the leadership at the lower levels. In a particular situation, in the parlance of risk analysis, a microscopic risk at any level of the hierarchy can become a critical factor in making things go wrong. Thus, to perform a complete HHM analysis, the modeling team must include individuals who bring knowledge and expertise from up and down the hierarchy.

HHM has turned out to be particularly useful in modeling large-scale, hierarchical, interdependent, and interconnected Complex SoS, such as cyber–physical civilian and defense infrastructure Complex SoS. The multiple visions and perspectives enabled by HHM add strength to the intricate modeling process and to risk analysis. For example, it has been extensively and successfully deployed to study the risks for government agencies such as the US President's Commission on Critical Infrastructure Protection (PCCIP), the Federal Bureau of Investigation (FBI), the National Aeronautics and Space Administration (NASA), the Virginia Department of Transportation (VDOT), and the National Ground Intelligence Center (NGIC), among others (Haimes, 2016). The HHM methodology/philosophy is grounded on the premise that in the process of modeling Complex SoS, more than one mathematical or conceptual model is likely to emerge. Each of these models may adopt a specific point of view, yet all may be acceptable representations of the infrastructure system. Through HHM, multiple models can be developed and coordinated to capture the essence of many dimensions, visions, and perspectives of infrastructure systems. One example is the study conducted for the PCCIP on the US water supply system. Sixteen different visions/perspectives (head topics) with an additional 94 subvisions (subtopics) were identified as sources of risk (Haimes, 2016).

Perhaps one of the most valuable and critical aspects of HHM is its ability to facilitate the evaluation of sources of risk to subsystems and their corresponding contributions to risks to the entire Complex SoS. For example, in the planning, design, or operational mode, the ability to model and quantify the risks contributed by each subsystem markedly to the entire Complex SoS facilitates identifying, quantifying, and evaluating the management thereof. In particular, HHM has the ability to model the intricate relationships among the various subsystems and to account for all relevant and important elements of risk and uncertainty. This makes for a more tractable modeling process and results in a more representative and encompassing risk modeling and management process.

To present a holistic modeling and view of the elements of interdependent and interconnected Complex SoS that must be included in the modeling process, the HHM approach involves organizing a team of experts with widely varied experience and knowledge bases (technologists, psychologists, political scientists, criminologists, and others). The broader the base of expertise that goes into identifying all critical systems that compose Complex SoS or all potential risk scenarios, the more comprehensive is the ensuing HHM. For example, the result from the elaborate HHM process is the creation of a very large number of risk scenarios, hierarchically organized into sets and subsets. When done well, the set of risk scenarios at any level of the hierarchy would approach a "complete set." The result of the HHM effort is organized into what is called the candidate *scenario model.*

The distinctive attributes of the HHM analysis are summarized below; this modeling approach:

1. Provides a holographic view of a modeled Complex SoS and thus is capable of (i) modeling and relating the I-I among all relevant and important subsystems of Complex SoS and (ii) identifying most, if not all, major sources of risk and uncertainty to interdependent and interconnected Complex SoS.
2. Adds both robustness and resilience to the modeling process by capturing the essence of the I-I characterizing Complex SoS.

3. Provides a more defined responsiveness in modeling development and exploits all sources of available data to populate different holographic models by making use of different databases.
4. Adds more realism to the entire modeling process by recognizing that the limitations of modeling Complex SoS via a single model (which commonly addresses only specific and limited aspects of the system).
5. Provides more responsiveness to the inherent hierarchies of multiple objectives and subobjectives and multiple decision makers associated with Complex SoS.

The impact of HHM in the planning phase may be most profound in modeling Complex SoS and in the way that risks and uncertainties can be integrated into the analysis. From the planning perspective, two major types of risks and uncertainties can be identified. The first type is concerned with the impact of exogenous events on the proposed plan, such as new legislation. The second is concerned with the impact of endogenous events that affect the execution of the plan, such as hardware, software, organizational, or human failures. Since the basic philosophy of HHM is to build a family of models that address different aspects of the system, this is a natural setting in which the impact of both types of risks and uncertainties can be studied in a unified way.

3.4 RELEVANCE OF HHM TO COMPLEX SYSTEMS OF SYSTEMS

Both HHM and HOC methodologies recognize that in the design or planning of Complex SoS, there are families of feasible structures for the subsystems, each of which is responsive to one or more attributes of the overall Complex SoS. In this context, the planning problem is one way of determining a set of decisions or policies based on the various feasible structures. The roles of HHM and HOC are central to the modeling, planning, and managing of Complex SoS because in many ways they are extensions and expansions of the classical mathematical modeling methodologies that have historically dominated the planning and modeling phase for Complex SoS. The impact of these methodologies in the planning phase may be most profound in the way we model the intricacies of the I-I between cyber and physical systems, or when elements of risk and uncertainties can be integrated into the analysis, or in contrast to being ambivalent about emergent forced changes (EFCs), which are precursors to risk and may have dire consequences to the subsystems and to the entire Complex SoS under consideration. The term EFCs connotes emergent internal or external forces that positively or negatively impact a specific system or the entire Complex SoS.

Multiple sources of uncertainty are associated with all systems, small and large, and uncertainty is commonly viewed as the inability to determine the true state of a system, which can be caused by incomplete knowledge and/or by stochastic variability (Haimes, 1981, 2016; Paté-Cornell, 1990, 1996; Apostolakis, 1999). Two major sources of uncertainty in modeling affect risk analysis: (i) Knowledge (epistemic) uncertainty manifests itself in the selection of model topology (structure) and model parameters, among other sources of ignorance (e.g. lack of knowledge of important interdependencies within the states of the system and among other systems). (ii) Variability (aleatory) uncertainty includes all relevant and important random processes and other random events (e.g. EFCs). Uncertainty dominates most decision-making processes and is the Achilles' heel for all deterministic and most probabilistic models. This uncertainty is commonly introduced through the selection of an incorrect model's topology (structure), e.g. linear for a highly nonlinear system, its parameters, data collection, and the employed processing techniques. Model uncertainties will often be introduced through human errors of both commission and omission.

Epistemic and aleatory uncertainties affect every step of the never-ending, learn-as-you-go modeling process, and these in their totality are EFCs, which can be external or internal forces that may adversely or positively affect specific states of that system. The first type of uncertainty, epistemic uncertainty, is commonly concerned with the lack of or incomplete knowledge of facts, affecting and impacting important topology (structures) and the representativeness of the constructed models, as well as by exogenous events, such as evolving global markets and technology, or new legislation impacting the system or Complex SoS under consideration. The second source of uncertainty, variability (aleatory) uncertainty, is concerned with the impact of endogenous random events, such as software or hardware failures, which affect the execution of the plan. Since the basic philosophy of HHM is to build a family of models that represent important and critical aspects of the system, the two types of uncertainty must be addressed, accounted for, and accommodated in the modeling and analysis of interdependent and interconnected Complex SoS.

Numerous applications of HHM for systems modeling and for risk analysis are discussed in subsequent sections of this chapter and throughout the book. One example of a potential application for holographic modeling can be provided by considering one function of the operation of an electric power system for which a sound theoretic framework has yet to be provided, namely, operations and scheduling the maintenance of major equipment. This involves a wide range of decisions for fuel with different time frames, ranging from weeks to months to years. Decisions, which must be made during acceptable time frames, involve the same physical plant, and hence must be made and carried out in a consistent manner. However, the problem is much too complex to be solved in an undivided manner. The uncertainty ranges from severe for the 1-week time horizon to overwhelming for the 1-year horizon. Multi-timescale holographic modeling could conceivably provide an integrated framework for a pluralistic approach to such problems.

3.5 MATRIX ORGANIZATION ILLUSTRATION

This section considers the decomposition and coordination of a matrix organization as a Complex SoS that has more than one hierarchical overlapping structure. The concept and importance of Hierarchical Overlapping Coordination (HOC) is presented through several example problems. To understand HOC as a concept and a methodology, consider first a very simple example. Figure 3.1 depicts a matrix organization of a Complex SoS of an industrial operation. For illustrative purposes, consider its decomposition into a marketing division and a manufacturing division. We assume two sectors in the marketing division: one concerned with product A and one with product B. Likewise, we assume there are three plants in the manufacturing division that are located in different areas. Each of the two product sectors has a manager, and each of the three plants also has a manager. Let us call the decomposition of this structure into a marketing division the "product decomposition" and call the decomposition into a manufacturing division the "plant decomposition." Clearly, the sectors in the product decomposition overlap those in the plant decomposition. The product managers' decisions also overlap the plant managers' decisions. For example, a decision by the manager of product sector A overlaps the decisions of the three plant managers.

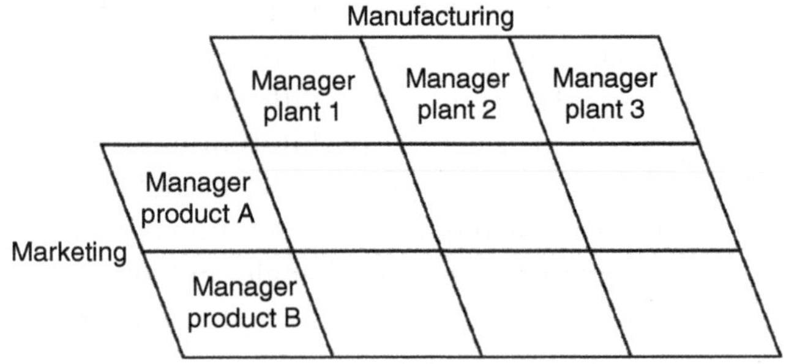

Figure 3.1 *Matrix organization of a production system.*

The hierarchical representation of this overlapping organizational structure is depicted in two ways shown in Figures 3.2 and 3.3. Product managers are concerned with the individual product – its development, marketing, and sales. Plant managers are concerned with the cost and efficiency of the production system. That is, these two different decompositions deal with different aspects of the system. The databases of these two decompositions differ from each other and receive different information from inside and outside the system. It is valuable to consider these different types of decomposition simultaneously.

By considering these different types of decompositions simultaneously, we can expect to obtain a synergistic understanding of the overall Complex SoS and its corresponding sources of risk and uncertainty. The different geographical locations of the three plants, for example, may impose distinctive production constraints due to local environmental regulations. Subsequently, the manufacturing of products A and B at the three plants may be subjected to different risks of cost overrun, time delay in meeting

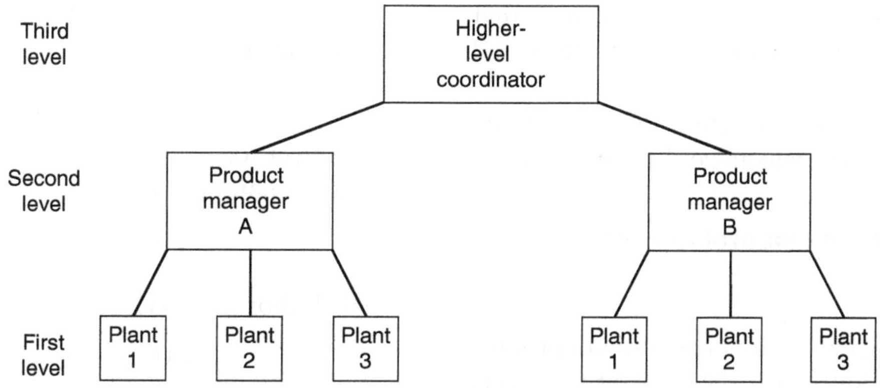

Figure 3.2 *Product–plant decomposition.*

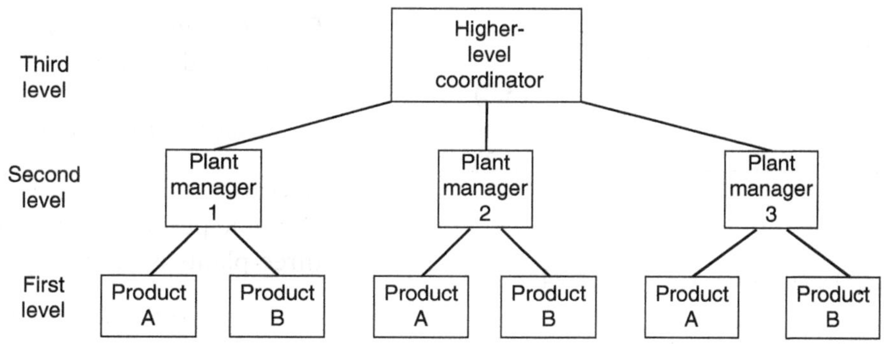

Figure 3.3 *Plant–product decomposition.*

production schedules, or not meeting performance criteria.

Suppose that the overall objective of the matrix production system represented in Figure 3.1 is to maximize a given measure of net profit, with each manager cooperating in order to achieve this goal. Then, a desirable decision-making structure would be one in which (i) each individual manager's decisions are feasible in the overall system and (ii) the information exchange between the product managers and the plant managers leads to a sequence of decisions producing an improved overall benefit that converges to an optimum.

So far, for simplicity, we have been discussing systems with two different hierarchical structures (i.e. decompositions). However, it bears mentioning that Complex SoS sometimes have more than two such structures.

3.5.1 Example Problem

The following example highlights the value of HOC, and thus the importance of HHM in modeling Complex SoS. Consider a furniture company that produces two types of products: Tables ($i = 1$) and chairs ($i = 2$). The company has three manufacturing plants ($j = 1, 2, 3$).

On an average day, the demand for tables is 60 units and the demand for chairs is 120 units. It takes 0.2 h to finish a table and 0.1 h to finish a chair. Assume there are 8 h in a working day for each of the three manufacturing plants (which means a total of 24 working hours per day for all three plants). Also, assume that each plant produces an equal number of chairs and tables.

The profit from one table is \$20 and \$40 from one chair. The objective is to maximize the daily profit:

a. Formulate and solve the problem on a company-wide level.

Solution: Let x_{ij} be the number of units of product $i = 1,2$ to be produced per day at plant $j = 1,2,3.0$.

Maximize daily profit:
$$Z = 20(x_{11} + x_{12} + x_{13}) + 40(x_{21} + x_{22} + x_{23}) \tag{3.1}$$

Subject to:

i. Demand per day:

$$x_{11} + x_{12} + x_{13} \leq 60 \tag{3.2}$$

$$x_{21} + x_{22} + x_{23} \leq 120 \tag{3.3}$$

ii. Labor per day:

$$0.2(x_{11} + x_{12} + x_{13}) + 0.1(x_{21} + x_{22} + x_{23}) \leq 24$$

Result: $Z^* = \$6000$;

$$x_{11} = x_{12} = x_{13} = 20;$$

$$x_{21} = x_{22} = x_{23} = 40$$

b. Formulate and solve the problem from the perspective of each of the two product managers.

Table manager perspective (i = 1)

Maximize daily profit:

$$Z(i=1) = 20(x_{11} + x_{12} + x_{13}) \tag{3.4}$$

Subject to:

i. Demand per day:

$$x_{11} + x_{12} + x_{13} \leq 60 \tag{3.5}$$

ii. Labor per day:

$$0.2(x_{11} + x_{12} + x_{13}) \leq 12 \tag{3.6}$$

Result: $Z^*(i = 1) = \$1200$;

$$x_{11} = x_{12} = x_{13} = 20$$

Table manager perspective (i = 2)

Maximize daily profit:

$$Z(i=2) = 40(x_{21} + x_{22} + x_{23}) \tag{3.7}$$

Subject to:

i. Demand per day:

$$x_{21} + x_{22} + x_{23} \leq 120 \tag{3.8}$$

ii. Labor per day:

$$0.1(x_{21} + x_{22} + x_{23}) \leq 12 \tag{3.9}$$

Result: $Z^*(i = 2) = \$4800$;

$$x_{21} = x_{22} = x_{23} = 40$$

c. Formulate and solve the problem based on the perspective of each of the three plant managers.

Plant 1 manager (j = 1)

Maximize daily profit:

$$Z(j=1) = 20x_{11} + 40x_{21} \tag{3.10}$$

Subject to:

i. Demand per day (assume the demand for the three plants is uniformly distributed):

$$x_{11} \leq 60/3; \quad x_{21} \leq 120/3$$

ii. Labor per day:

$$0.2x_{11} + 0.1x_{21} \leq 8 \tag{3.11}$$

Result: $Z^*(j = 1) = \$2000$;

$$x_{11} = 20; x_{21} = 40;$$

Plant 2 manager (j = 2)

Maximize daily profit:

$$Z(j=2) = 20x_{12} + 40x_{22} \tag{3.12}$$

Subject to:

i. Demand per day (assume the demand for the three plants is uniformly distributed):

$$x_{12} \leq 20; x_{22} \leq 40;$$

ii. Labor per day:

$$0.2x_{12} + 0.1x_{22} \leq 8 \tag{3.13}$$

Result: $Z^*(j = 2) = \$2000$;

$$x_{12} = 20; \quad x_{22} = 40$$

Plant 3 manager ($j = 3$)
Maximize daily profit:

$$Z(j=3) = 20x_{13} + 40x_{23} \quad (3.14)$$

Subject to:

i. Demand per day (assume the demand for the three plants is uniformly distributed):

$$x_{13} \leq 20; \quad x_{23} \leq 40$$

ii. Labor per day:

$$0.2x_{13} + 0.1x_{23} \leq 8 \quad (3.15)$$

Result: $Z^*(j=3) = \$2000$;

$$x_{13} = 20; \qquad x_{23} = 40$$

Summary:
(i) Product decomposition yields a total profit of \$6000 (\$1200 for product $i = 1$ and \$4800 from product $i = 2$). (ii) Plant decomposition yields a total profit of \$6000, equally distributed among all three plants. Both decompositions also yield the same number of tables and chairs produced at each plant. (iii) Although both decompositions yield the same "optimal" solution, each provides a different perspective to the executives of the furniture company.

3.6 THEORETICAL AND PRACTICAL CONTRIBUTIONS OF HHM TO MODELING COMPLEX SYSTEMS OF SYSTEMS

As discussed earlier, in HOC, a Complex SoS model is restructured into several decompositions to represent select critical aspects of the SoS, and these decompositions are coordinated to yield an improved model representation of the entire Complex SoS. Coordinating these dissociated submodels – that is, reassociating them via holographic modeling methodologies – constitutes an important stage in modeling Complex SoS. Given that HHM has been extensively deployed for risk-based scenario structuring (Kaplan et al., 2001; Haimes, 2016), we demonstrate here its practical contributions to risk analysis.

3.7 DECOMPOSITION AND MULTILEVEL COORDINATION OF COMPLEX SYSTEMS OF SYSTEMS

In studying interdependent and interconnected Complex SoS with technological, societal, organizational, and environmental perspectives and dimensions, the efforts in modeling these factors as well as in their ultimate management (deployment of the model) are magnified and often overwhelm the modelers, analysts, and stakeholders. This is due to the models' having (i) high dimensionality (a very large number of variables), (ii) complexity (nonlinearity in the coupling and the interactions among multiple factors, including states, decisions, resources, decision makers, and stakeholders), and (iii) intra- and inter dependencies within and among the many systems and subsystems that constitute Complex SoS.

The decentralized approach presented in Chapter 2 and this chapter, by utilizing the concepts of strata, layers, and echelons, enables modelers and analysts of Complex SoS to analyze and comprehend the behavior of the subsystems at a lower level of the hierarchy as "independent" and to transmit the resulting information to appropriate subsystems at a higher level. Whenever more decentralization is needed, the Complex SoS is further decomposed via the use of pseudo-variables. Then each subsystem is separately and independently modeled with its specific decisions, states, inputs, and other random variables, decision makers and stakeholders, based on the nature of the subsystem models as well as on the objectives and constraints of the subsystems.

The following dilemma is often faced: Developing detailed and comprehensive models, which closely represent the real system, is desired, yet such realistic models are generally too complicated. Moreover, the deployment of decision-making and management strategies and the inherent multiobjective tradeoff analyses are too difficult, if not prohibitive. The above dilemma has often been at least partially circumvented by oversimplifications of the system models. An alternative to the above sacrifice has evolved with the establishment of the theoretical bases and the development of the computational tools and methodologies of decomposition and hierarchical-multilevel coordination (including multiobjective tradeoff analyses).

The concept of hierarchical-multilevel coordination of the subsystems, introduced in Chapter 4 and further expanded upon in other chapters, is based on the decomposition of Complex SoS and subsequent modeling of the subsystems as "independent" at a lower level of the hierarchy through the use of pseudo-variables and other tools. HHM is an additional enabler for modeling and, ultimately, managing interdependent and interconnected Complex SoS. The multiple perspectives associated with the I-I and complexity of SoS are being not only discovered via the HHM process but are also incorporated in the intricate modeling process of Complex SoS. Recall that following this first-level subsystems modeling, the subsystems are then joined by shared states, decisional variables, and so forth, which are manipulated at a second or higher level of the hierarchy in order to arrive at a coordinated interdependency and interconnectedness within and among the entire Complex SoS. Several examples will be introduced in subsequent sections.

3.8 ATTRIBUTES OF DECOMPOSITION AND HIERARCHICAL-MULTILEVEL COORDINATION

Decomposition and hierarchical-multilevel coordination have several significant advantages in modeling Complex SoS. The following brief summary of attributes will be explored throughout this chapter and beyond:

1. *Conceptual simplification of complex systems*: By decomposing the SoS into several subsystems, a conceptual simplification of Complex SoS is achieved. This is especially important for highly coupled Complex SoS, where the subsystems share many states and other essential entities (e.g. shared/common decisions, decision makers, stakeholders, resources, organizational behavior and norms, policies and procedures, management, and culture) and where the outputs of one subsystem are the inputs to others.
2. *Reduction in dimensionality*: The decomposition of Complex SoS yields a reduction in the dimensionality of the modeling efforts at the expense of having to coordinate several submodels of lower dimensions. Often this reduction in dimensionality enables the modeling and management of Complex SoS, which would not otherwise be possible. This attribute has been widely demonstrated in the literature, as will be discussed in subsequent sections.
3. *Effective collection and utilization of multiple databases*: The availability of multiple databases that cover the multiple perspectives of the subsystems comprising Complex SoS constitutes a significant driver in the representativeness of the resulting submodels and, thus, their overall modeling. Clearly, decomposing and modeling each subsystem with its unique representative database, augmenting such a process with other overlapping information, and finally integrating all databases in a coherent and representative fashion enhance the entire data collection and utilization in the modeling and ultimate management of Complex SoS.
4. *More realistic subsystem models*: A significant advantage of decomposition and hierarchical-multilevel coordination of Complex SoS is that each subsystem model may assume a different topology or structure; thus more flexible mathematical models can be constructed to represent the real system. Note that one shortcoming of conventional modeling practices of complex systems or SoS is that they result in an imbalance between the complexity of the system and the necessity of developing oversimplified models; this may be reflected by the large number of linearized models in the literature. By applying the decomposition and multilevel model coordination approach, less costly sacrifice of realism in modeling occurs, as more representative nonlinear and dynamic mathematical models of Complex SoS can be constructed.
5. *The I-I among the subsystems of Complex SoS can be exploited*: This complexity can be handled by the introduction of pseudo-variables. More specifically, assume that subsystems 1 and 2 share with each other one state (or decision) x_1 and x_2, respectively, as a product of the two: $x_1\ x_2$. To decouple the two subsystems, we let $x_1 = \sigma$ for subsystem 2 and ultimately harmonize x_1 and x_2 at the second-level coordination. This is possible since at the lower levels, the subsystems' "independencies" are achieved

via pseudo-variables, as will be discussed in detail in subsequent sections. This is an important attribute of decomposition and higher-level coordination, given that all emergent Complex SoS are, by their essence, highly coupled; thus, the interactions among their various components must be accounted for in the modeling and management process. For example, the I-I that characterize water resources Complex SoS can probably serve as an excellent example in this respect. As another example, the benefits resulting from the construction of a dam cannot be properly assessed unless the impacts of the I-I among water quality, land use, sheet erosion, flood control, socioeconomic effects, and others are considered. These couplings in modeling can be addressed via decomposition and hierarchical-multilevel coordination by resorting to pseudo-variables at the subsystem level of the multilevel hierarchy of the dam Complex SoS.

6. *Applicability to both static and dynamic Complex SoS*: Decomposition and hierarchical-multilevel coordination are applicable to both static and dynamic SoS. The time domain, which plays an important role in most, if not in all SoS, needs not to be imbedded or ignored in the analyses (as is the case in static models). SoS can accordingly be modeled by both static algebraic equations and dynamic differential or difference equations. The tradeoff between the modeling and management of SoS is thus further reduced by the applicability of the approach to both static and dynamic systems.
7. *Use of existing models for SoS*: Available well-documented simulation and mathematical models aimed at investigating specific and selected aspects of one or more subsystems of SoS may be augmented within the decomposition and hierarchical-multilevel modeling process. This feature enables modelers to identify appropriate existing models in the literature and to modify and incorporate them as appropriate. Such a feature is particularly effective for modeling and managing regional SoS.
8. *Economic interpretation*: When the coordination among two or more levels in the hierarchy is accomplished, e.g. via Lagrange multipliers, then the economic interpretation and the relationship of the latter to duality theory can provide additional insight to the modelers. For example, the economic interpretation of Lagrange multipliers as shadow prices, and as dual variables, often sheds more light on the understanding of a system's behavior.
9. *Applicability to multiobjective analyses*: In a hierarchical structure, different objective functions may be associated with each subsystem. The resulting overall system coordination becomes a multiobjective tradeoff analysis problem. In particular, when these objective functions are noncommensurate, then the Epsilon constraint and the surrogate worth tradeoff (SWT) methods can be successfully applied.

3.9 THE ROLE OF POLICY FORMULATION IN THE MANAGEMENT OF COMPLEX SYSTEMS OF SYSTEMS

Policy formulation must be a precursor to decision making for all systems, small and large, and especially for Complex SoS. Given that decisions change the states of the system, it is imperative that analyses be performed prior to making decisions in order to assess their impacts on the essential states of the system and on the system's ability to achieve major goals and objectives. *In other words, policy formulation for Complex SoS constitutes the foundation upon which effective and viable decisions can ultimately be made and can serve as a potent instrument in risk reduction and even prevention.* This can be achieved by developing a representative system's model and by projecting the impacts of current decisions on future options. Sadly, decisions made by governmental and other organizations, which are not supported by proper policy formulation, are made either in haste or without assessing their future impacts and ramifications on other important states of the SoS and the goals and objectives of the enterprise. For example, the ultimate goal of risk analysis is to not only address and manage future critical risk scenarios but to also avoid the creation of unintended consequences, leading to new sources of risk to the Complex SoS. Thus, building on policy formulation prior to decision making constitutes prudent decision-making process, including the management and avoidance of risk.

An effective policy formulation process for Complex SoS can be best accomplished via representative models that can relate cause and effect. For example, using models in conjunction with simulation would enable decision makers to project and learn via event-tree analyses about unfavorable consequences of candidate policies that seem reasonable and effective at the time. This is particularly important in light of the inevitable dynamic changes that affect all Complex SoS as well as single systems. Thus, decision makers who build on well-formulated policies, which are developed with the projection of likely impacts of current decisions on future options, would significantly reduce future unintended consequences and other sources of risk. Simulating various policy options would also provide a window of perspectives to illuminate potential EFCs in evolving states of the subsystems and the SoS as a whole from within and outside the system and the entire SoS.

3.10 THE ROLE OF HIERARCHICAL HOLOGRAPHIC MODELING IN CONFLICT MANAGEMENT OF COMPLEX SYSTEMS OF SYSTEMS

3.10.1 An Overview

A system as complex as individuals as modelers, managers, stakeholders, or constituencies can be viewed from one axis via the many perspectives, attributes, propositions, truths, or outlooks, and the one or the other axis can be selected to form a theoretical image (Haimes and Weiner, 1986). Indeed, individuals have their own truths and each can be as valid as the others. Musil (1933) maintains that partial truths will emerge among groups of decision makers, and their mutual intermeshing will gradually raise truth to higher levels has always proven prejudicial to attribute general validity to a partial truth. Yet, a partial truth could not have been reached without overstating its value. Thus the history of truth is intimately interconnected with the history of emotion, perception, and individual orientation, among others. However, the emotional truth has remained in the "dark."

From the point of view of conflict management, conflicts associated with interdependent and interconnected Complex SoS can be divided into soluble and insoluble, given the associated multiple systems, decision makers, and stakeholders: In *soluble conflicts*, parties, though divided by conflicting interests and differences in tradeoffs, still concur that accommodation may be expected to yield higher payoffs than continuation or conflict. Negotiating accommodation is viewed as a positive-sum game in which all parties will compete for benefits within a framework of shared basic assumptions. Conflict will be limited to specific real issues, and the calculus employed by parties to define their positions will be predominantly rational. Finally – and this represents the most important distinguishing aspect of this category of conflict – consensus will be achievable without requiring changes in images or tradeoffs. *Insoluble conflicts* within Complex SoS are usually associated with a much broader and much less well-defined range of issues, expressed in symbolic and emotion-saturated terms. Parties will view conflict as a zero-sum game in which the purported intransigence of adversaries is assumed to preclude accommodation. Conflicts of this type tend to proliferate conflict issues and to escalate – they are *schismogenetic*.

Conflict management has developed reasonably adequate tools to deal with *soluble conflicts*: A number of versions of hierarchical multiobjective analyses exist, many of which are discussed throughout this book, which can assist parties to a conflict to take what might be called a path of least regret in an "incrementalist" consensus process (see, for example, Haimes, 2016; Chankong and Haimes, 1983, 2008; Bristow et al., 2014; Hipel and Bernath Walker, 2011). Although the analyst employing these tools – in the following discussion referred to as the moderator – will have to feed parties with information and suggestions to facilitate the consensus-seeking process, parties will not have to change their images of the conflict situation and of their adversaries. The need for accommodation will be assumed to be obvious, and negotiation will constitute bargaining on the distribution of burdens and benefits.

In *insoluble conflicts*, the situation is different: It is the very need for consensus and its feasibility that will, above all, be questioned, and initiation of the bargaining process will depend on major restructuring of attitudes, images, and tradeoffs. To achieve such restructuring, a complex learning process will have to be induced to assist parties to move, at least in selected domains, from their habitual affective and symbolic mode of argumentation to a calculus in which real costs and benefits are weighed in rational terms. The image of the conflict will have to be transformed from an intractable zero-sum game to a positive-sum game with real prospects of agreement on an equitable division of benefits.

3.10.2 The Operational Model of Holography

Holography makes full and systematic use of what Gregory Bateson (1980) has called multiple (or, in the border case, double) information. Leaders of Complex SoS in conflict will view their universe of discourse through different and highly distorting lenses. By the analysis of the differences between the adversaries' "maps," the moderator can derive a new category of information on the epistemological process employed by the parties, the distortion that their viewing lenses impose on reality. Through the introduction of this information into the negotiation process, the moderator can induce a learning process that would promote restructuring of images and behavior and thereby facilitate the "priming" of consensus seeking. Being derived from differences between sets of information, the *new type of information will be of a higher logical type*. The moderator will, therefore, find it difficult, if not impossible, to impart this information to parties accustomed to a lower logical type of argument. The new type of information will assist the moderator to draw up a map of issues and of their "affective saturations" and to identify in this map pressing issues with low affective concentrations. Gradual accommodation procedures could be designed for these issues, with a view to resolving the most acute conflict aspects, without having to address issues with high affective concentration. Through the implementation of this partial resolution ("transients" in this chapter), prevailing fixations would be eroded, affective rationality would gradually yield to goal rationality, and a longer-term "systemic" view would be promoted. Parties within the Complex SoS would learn to give proper weight to the responses of the opposite side demonstrated in the negotiation game. Another important aspect of holographic modeling would be the derivation, by comparing information from different phases of the conflict management process, of the influence of the expanding supranational interaction and interdependence on the positions of parties.

The most important conflict management category to which holographic modeling could contribute is *management of internal conflict* that arises in Complex SoS, whether its geneses originate from shared resources, organizational turf, territory, power, or personal conceited power. *This chapter, therefore, will primarily deal with this category of conflict.* Management of Complex SoS conflict will depend upon the prevailing inter- and intrainstitutionalization of relations or *meta-association* as the process will be called in this chapter. *Holographic modeling*, representing the *analytical aspect of meta-association*, will become possible to the extent that parties in conflict see their way to embarking on meta-association. However, once launched, holographic modeling is likely to facilitate and promote more advanced modes of meta-association.

3.10.3 The Need for Hierarchical Holographic Modeling

It has been the past practice in systems modeling to represent the intricacy of Complex SoS through a single (planar) model. This unchallenged notion necessitates that Complex SoS be represented by a set of shared/common states and other essential entities. The term "essential entities," first introduced in Chapter 1, includes *shared decisions*, *resources*, *functions, policies*, *decision makers*, *stakeholders*, *organizational setups*, and other factors within and among the systems and subsystems that constitute Complex SoS. While a single planar model may suffice for describing a relatively simple and small-scale system, it may likely fail to do so for Complex SoS (Haimes, 1981, 2012, 2016).

Recall that what constitutes an I-I Complex SoS encompasses a bewildering variety of factors, e.g. shared components, resources, decisions, constraints, and exogenous variables intertwined in a complex input–output relationship, and they reflect multiple shared and conflicting goals and objectives, often noncommensurable, in response to a hierarchy of constituencies and decision makers.

Also note that common characteristics of Complex SoS include (i) a large number of decision variables, exogenous variables, and state variables; (ii) shared/

common states and other essential entities; (iii) a large number of components (subsystems); (iv) a complex and often nonlinear functional input/output relationship; (v) risks and uncertainties; (vi) a hierarchical organizational structure; (vii) multiple, noncommensurable, competing, and often conflicting objectives; (viii) multiple decision makers; (ix) dynamic changes in the systems' orientation, goals, objectives, constraints, exogenous variables, and input/output relationships; and (x) varied decision makers' preferences.

Central to HHM is a modeling philosophy that is cognizant of these characteristics and responsive to them. More specifically, this philosophy is grounded on the premise that a single model of Complex SoS, no matter how detailed or sophisticated it might be, cannot properly and responsively represent all the above characteristics. The overlapping models in the HHM may have both common and distinct sets of objective function(s), constraints, and input/output causal relationships as well as state, decision, random, and exogenous variables. Often, a certain aspect of a subsystem of a Complex SoS is coupled with a certain constituency and, thus, can be best modeled by means of a separate hierarchical holographic submodel. For example, the flood control planning process of a River Basin Complex SoS is often managed by one agency (the US Army Corps of Engineers), while the agricultural aspect is commonly managed by another agency (the US Department of Agriculture). Corresponding to these two major aspects are distinct databases, decisions, decision makers, stakeholders, organizational setups, missions and mandates, and affected constituencies. Yet, the planning process in this case addresses the same River Basin with some common goals, objectives, and constraints that transcend aspects and constituencies. In this sense, HHM offers a responsive schema to the modeling needs of Complex SoS.

Both the concept and the mathematical foundation of HHM are extensions and generalizations of those of HOC as discussed in Chapter 2 and introduced decades ago (Haimes and Macko 1973). The principal premise of HOC is that a system model may be best decomposed into more than one decomposition, say the α, β, γ, and so on decompositions. For example, the Maumee River Basin in the United States, which runs through parts of Ohio, Michigan, and Indiana, lends itself to five subsystems (planning subareas) in the α-geographical/political decomposition and to eight subsystems (watersheds) in the β-hydrological decomposition (Refer to Figures 4.1 and 4.2 in Chapter 4). Each of the α and β decompositions of the River Basin single model may be related to a different database. For instance, for the α-geographical/political decomposition, census and other commercial data would be the most appropriate database to use. On the other hand, for the β-hydrological decomposition, precipitation, streamflow, and other water quality, data would provide the most appropriate database. It is worth noting that different agencies (and decision-making systems) are functionally and administratively associated with each of these decompositions. Mathematical foundations for HOC with single-objective functions and extensions of the HOC concept to accommodate models with multiple objectives are presented in Chapter 2.

Any progress toward the goal of achieving supplemental ways for improving conflict resolution necessitates, first of all, an understanding of the parties involved. The multiple aspects, perspectives, dimensions, and attributes of conflict resolution are likely to be better understood through the modeling process offered by the HHM philosophy.

Another important role that HHM can play in the context of conflict resolution is in advancing and improving on the negotiation process. One cause of adversarial relationships within and among the systems or subsystems that comprise Complex SoS is a misperception of reality and subsequent acting on false information and facts. One may view this perception process in terms of cognitive images and models generated by each party in the conflict. The genesis of this misconception may be the lack of good communication between the two parties. The HHM philosophy can be instrumental in reducing this misconception and facilitating better systemic communication between two (or more) parties (Haimes and Horowitz, 2004).

3.10.4 Tradeoffs in HHM and in Conflict Resolution

Even a cursory review of the history of conflict shows that, while some conflicts have been smoldering for a long time, inaccessible to resolution efforts and having led to periodic outbreaks, others have vanished overnight. In the contemporary world, most conflicts are perceived to be negative-sum games for all participants, at least in economic terms. The longevity of conflict would therefore suggest that, as Canetti (1960) has pointed out, strong extraeconomic payoffs enter into the calculus of parties in conflict.

Conflict associated with Complex SoS becomes insoluble by the refusal of parties to pay the price involved in reaching consensus, be it power and authority, real or perceived, and economic or extraeconomic. This refusal may be motivated by a traumatic historic experience resulting in the displacement of goal rationality by affective rationality by the existence of fundamental differences in value rationality or by the *need to maintain the conflict as a means of preserving internal stability*.

We can distinguish two categories of efforts to deal with insoluble conflict:

1. Attempts to achieve the impossible and negotiate resolution of the conflict in its totality. Such attempts are doomed to fail, and every failure is liable to increase resistance to future attempts. Even where parties in conflict will, for some reason, feel obliged to conclude some vaguely worded general agreement, implementation is liable to prove abortive.
2. Abandonment of the illusory hope of total resolution and focus instead on what might be called "transients."

A "transient" is an intermediate solution associated with an acute, but less controversial, issue. Issues selected for negotiating transient solutions ought to offer reasonable prospects to be considered negotiable by all parties and should harbor opportunities to produce transformation of their images of the conflict situation. As the word indicates, "transients" are stepping-stones to a more comprehensive solution.

Even the negotiating process by itself, and the modicum of cooperation that it involves, will undoubtedly create spillover effects. In game theory terms, agreement on a transient signifies that parties to an insoluble conflict, while maintaining their refusal to abandon position 3 in general, still see their way to move, in some mutually agreed domains, from position 3 to position 1 and, what may prove to be even more important, to agree on procedures to prevent relapse to position 2. The generalization effects of transients will gradually relax mutual suspicions and promote the growth of goal rationality at the cost of affective rationality; this will prepare the ground for more ambitious rounds of negotiations.

Conflict resolution associated with Complex SoS has, of course, not always taken a gradual path. Some of the most dramatic historical cases of conflict resolution have occurred abruptly, after a traumatic experience, a severe crisis, or a devastating war. Such discontinuities can be conceived to occur on lines represented in *catastrophe theory. The combination of holographic modeling with catastrophe theory might open up new vistas for the management of conflicts approaching the brink of critical instability*.

The theory, methodology, and practice of quantitative tradeoffs of multiple objectives – multiple objectives that may be noncommensurable, in conflict or in competition – have been markedly advanced during the last several decades. The availability of methodologies on multiple objectives and tradeoff analysis discussed in Chapter 5 and throughout this book has made possible consideration of the problems of risk and uncertainty aspects within the context of a multiobjective tradeoff framework (Chankong and Haimes, 1983, 2008; Haimes, 2016). Furthermore, multiobjective tradeoff analysis methods add considerably to the quantitative attributes of the HHM approach. Several higher-level coordination methods for single-objective hierarchical Complex SoS structures are introduced in Chapter 2. Such methods applied to hierarchical structures that have multiple objectives (rather than a single objective) typify the most recent advances in modeling and decision making. These advances are an integral part of the HHM approach and provide the theoretical and/or quantitative basis for it.

Central to conflict resolution associated with Complex SoS is the process of identifying the perceived "negotiable" and "nonnegotiable" issues of concern to the various parties involved in the negotiation process. Recall that the HHM philosophy is grounded on this premise, and, indeed, identifying the objectives that are common ("negotiable") or distinct ("nonnegotiable"), the constraints and the decision variables in each HHM submodel, constitutes an important step in the modeling process. On the other hand, assessing the tradeoffs among the multiple objectives associated with the various policies, organizational turf, benefits, costs, and risks constitutes the heart of the negotiation process. This assessment and evaluation process is commonly conducted separately by each system/subsystem/party, keeping in mind that the objectives and constraints of each "party" are usually not explicitly known to one another. The decision makers' perceptions of the attitude of parties other than their own – the goals and objectives and the corresponding institutional, financial/economic, and political constraints – often dominate the outcome of such negotiations.

3.10.5 Societal Conflict and Meta-Information Generated for Complex SoS by HHM

A conspicuous complicating factor in the analysis of societal phenomena characterizing Complex SoS is the ubiquitous discrepancy between rhetoric and action, between leaders' manifest declaration of intent, and their latent motivation and actual behavior. Chapter 12 examines the issues involving organizations and organizational behavior as Complex SoS. While manifest rhetoric will appear to be the outgrowth of reason and universal values, latent motivation and actual behavior will, on analysis, be shown to be the outgrowth of affective fixations.

A further consideration in conflict management within and among the leaderships of Complex SoS is the need to maintain *stability*. Generally speaking, conflict obviously has a destabilizing effect. However, external conflict has often been maintained to prop up, at least temporarily, internal stability in an essentially unstable Complex SoS. It is notorious that one of the ways to unify a nation plagued by acute internal dissension is to redirect affective energies from internal conflict to external threats, by fostering suspicions about the sinister intentions of real or imaginary competitors and possible adversaries. When a major external conflict is suddenly resolved, withdrawal symptoms and their internal destabilizing effect can be avoided only be reattaching affective energy to a new issue with charismatic appeal.

Due to the difference in logical type, meta-information will not be directly accessible to the parties to the negotiation. To make it accessible, the moderator will have to present, within the negotiating context, the protagonists' descriptions of conflict issues and policies proposed by them as well as the expected reactions of each protagonist to policies as suggested by the others. Exposure to the systemic impact of their policy proposals ought to facilitate access to the meta-information presented by the moderator. Once the breakthrough is achieved, the moderator could suggest that the protagonists turn to negotiating the selected issues.

Because of the great variety of conflict situations that arise within and among the leadership of Complex SoS, it will hardly be possible to enumerate all the types of meta-information that holographic analysis might extract by reviewing multiple descriptions. However, the following listing of categories may suffice to illustrate the types of behaviors. For example, the description of the behavior patterns of a single protagonist in a specific conflict situation within any Complex SoS will yield much less information than the comparison of descriptions of the variety of behavioral patterns exhibited by the same protagonist in alternative "games" in the experimental situations of a negotiation process, after exposure to learning. The former type of description will disclose the protagonists' tradeoffs between objectives in a concrete situation, before learning has taken place; the latter type will disclose the (usually nonlinear) change trends of tradeoffs as a consequence of learning through exposure to a directed negotiation process. At a still higher logical level, information can also be derived concerning the extent to which the change trends of one protagonist will influence the change trends of the others. Consider the following observations:

1. If the description of behavior is considered to be of the first logical type, description of tradeoffs within the subsystems of Complex SoS will be of the second and comparison of tradeoff trends of the third logical type. With the rise of logical type, the quality of information will rise from apperception of phenomena to insights into the dynamics of change, from statements of a descriptive nature to statements about epistemology.
2. The analysis of the changes in tradeoffs of protagonists corresponding to changes within Complex SoS in the experimental situations, and of those attributable to the learning function, will

disclose the balance between affective and goal rationality for issues under negotiation. In addition, the analysis will also disclose the issues in which this balance can be most easily modified by learning. The former comparison will also disclose the areas of high affective charge, which may be assumed to have made the conflict unmanageable in the first place; the latter is the extent to which the negotiation process may be counted on to promote limited consensus coupled with learning (transients).

3. Meta-information can assist in the design of transients, which will, at one and the same time, check escalation and induce a learning process. Absence of learning and adherence to dogmatic argument is typical for an insoluble problem. Only by promoting learning can protagonists be induced to shift to "zones" of communication in which partial consensus is feasible.
4. Discontinuities and anomalies in change trends of tradeoffs will disclose the presence of obstructive institutional routines and of vested interests within and among Complex SoS; they may also show ways of relaxing these constraints and what measures can be taken while these constraints continue to exist.
5. Study of gradients and discontinuities in trends of tradeoffs produced in the experimental context of a bona fide negotiation within Complex SoS would disclose the range in which stability may be expected to persist and the range where instability may be expected to set in. This range will, of course, change by learning. The calculus of "catastrophe theory" could possibly be used in studying the behavior of the system in the instability range.
6. The purpose of the priming transient is not so much to address the central issues of polarization as to create contexts and opportunities in which the protagonists can test the potentialities of relaxation of their ingrained suspicions. Even where consensus within Complex SoS can initially be reached only on some tangential issues, the very fact that a dialogue has been started and that, in some domains, a modicum of cooperative behavior is countenanced or, at the very least, that adversary behavior has been checked, is likely to change the images that protagonists entertain of each other.
7. In addition to their manifest and nominal content, messages delivered in Complex SoS carry also meta-communication signals, "context markers" written, as it were, in "invisible ink" and pointing to the framework within which the message has to be interpreted, to its emotional mood, etc. Sensitivity to meta-communication signals often depends upon past relations between communicating parties within the Complex SoS and will be especially high in conflict situations. Meta-information will assist the moderator to promote harmony between the "context markers" and the content of messages in order to avoid the confusion resulting from discrepancies between them.
8. Finally, in a participative negotiation game using holographic modeling, parties would be encouraged to think in systemic terms. In a dialogue in which protagonists, while pondering over their next move, are confronted with the responses of their adversaries; the latter would escape into emotional rhetoric, which is so popular in internal forums for policy discussion. In response, the protagonists will feel obliged to defend their policy suggestions in rational terms. Since there is no better way to promote thinking than having to expose one's case to a critical public conversation, the negotiation context is likely to promote goal rationality. In time, the negotiating team will also develop team spirit, which will dispose participants to represent the case for consensus in a positive spirit in internal forums.

Thus, HHM can assist parties in deescalating an insoluble conflict that arises in a Complex SoS by encouraging them to pursue the most effective course of consensus seeking. However, it cannot bring parties to the negotiation table as long as their calculus will show that they can get more out of conflict than out of consensus. While the "adrenalin" of open conflict swamps the "bloodstream" of the body politic, little can be done to enhance the influence of goal rationality. Nonetheless, at some point, at least one party will come to realize that the rising cost of conflict escalation is becoming a prohibitive drain on scarce resources. At this point, leaders of the Complex SoS and their adversaries might be induced to abandon the addiction to conflict and participate in a learning process. By that time, the other party (or parties) to the conflict may also have developed second thoughts about the wisdom of continuing (and potentially escalating) the conflict.

When parties to an insoluble conflict within and/or outside Complex SoS do decide to open negotiations, it will not take them long to realize that their problems do not admit a solution that would be deemed equitable by all participants. Even when sitting down to negotiate, parties may still continue to harbor the same adversarial, emotion-saturated images of each other, which have been fueling the conflict in the past, and the discrepancies between the parties' tradeoffs will continue to defy reconciliation. With the meta-information enabled by holographic modeling, the moderator could demonstrate the futility of seeking a final solution and, at the same time, the prospects of proceeding gradually from less sensitive, less central, but acute, issues to the more sensitive, more central, and more long-term issues.

Once the consensus-seeking process is primed and consolidated, more sensitive areas may be addressed which, in the meantime, may have become desensitized through learning. The range of negotiations may be extended to influences of the "relevant external parties" and to the new options produced by the gradual untangling of the conflict.

In all stages of negotiation/bargaining, we have to look out for difficulties created by cultural and other differences. Dialogue employing emotional arguments has a tendency to move rapidly to the level of the lowest common denominator: Complex SoS with a more mature organizational culture will be driven to regress to behavioral patterns characterizing the less mature organizational culture of their adversary but will find it difficult to reconcile such regression with their own value systems.

3.10.6 Conclusions: HHM and Conflict Resolutions

HHM, based on meta-information derived from the analysis of multiple descriptions, will hardly spell the end of insoluble conflicts that arise from within and among Complex SoS. Irrational fixations are too universal and too intractable to admit of such a hope. Our expectations ought to be much more modest and realistic. HHM can still be a useful methodology to assist parties in a seemingly insoluble conflict once they have become preoccupied with the cost and potential consequences of continuation of conflict and to be open to tentative negotiations. The objective of such negotiations would be a partial containment of selected conflict issues and the promotion of a highly focused learning process designed to change images, attitudes, and tradeoffs, as a preparation for later expansion of conflict management. Such dual-purpose negotiations within a Complex SoS could be initiated by organizational leadership or by other well-respected neutral parties. Once the negotiation process has been reopened, the parties can be expected to produce spillover effects in other domains. Indeed, the hope to attain these spillover effects would make it highly advisable to initiate negotiation rounds wherever leverage can be ensured.

As a matter of course, HHM cannot overnight produce fundamental changes in the character of consensus seeking. *What it can do, however, is to assist parties within Complex SoS to contain, in specific issues, the power of emotional and irrational behavioral elements.* In performing this task, HHM modelers ought, at all times, to remain aware of its contributions and limitations.

REFERENCES

Apostolakis, G. (1999). *The distinction between aleatory and epistemic uncertainties is important: an example from the inclusion of aging effects into PSA*. Presented at Proceedings of PSA '99, International Topical Meeting on Probabilistic Safety Assessment, Washington, DC (22–26 August 1999).

Bateson, G. (1980). *Mind and Nature: A Necessary Unity. (Part III)*. Toronto: Bantam Books.

Blauberg, I.V., Sadovsky, V.N., and Yudin, E. (1977). *Systems Theory: Philosophical and Methodological Problems*. New York: Progress Publishers.

Bristow, M., Fang, L., and Hipel, K.W. (2014). From values to ordinal preferences for strategic governance. *IEEE SMC: Systems* **44**(10): 1364–1383.

Canetti, D. (1960). *Crowds and Power*. Harmondsworth: Penguin Books.

Chankong, V. and Haimes, Y.Y. (1983). *Multiobjective Decision Making: Theory and Methodology*. New York: Elsevier.

Chankong, V. and Haimes, Y.Y. (2008). *Multiobjective Decision Making: Theory and Methodology*. New York: Dover.

Gordon, W.J.J. (1968). *Synectics: The Development of Creative Capacity*. New York: Collier Books.

Haimes, Y.Y. (1977). *Hierarchical Analyses of Water Resources Systems: Modeling and Optimization of Large-Scale Systems*. New York: McGraw-Hill.

Haimes, Y.Y. (1981). Hierarchical holographic modeling. *IEEE Transactions on Systems, Man, and Cybernetics-SMC* **11** (9): 606–617.

Haimes, Y.Y. (2012). Modeling complex systems of systems with phantom system models. *Systems Engineering* **15**(3): 333–346.

Haimes, Y.Y. (2016). *Risk Modeling, Assessment, and Management*, 4e. New York: Wiley.

Haimes, Y.Y. and Macko, D. (1973). Hierarchical structures in water resources systems management. *IEEE Transactions on Systems, Man, and Cybernetics* **3**(4): 396–402.

Haimes, Y.Y. and Tarvainen, K. (1981). Hierarchical-multiobjective framework for large scale systems. In: *Multicriteria Analysis in Practice* (ed. P. Nijkamp and J. Spronk), 201–232. London: Gower.

Haimes, Y.Y. and Weiner, A. (1986). Hierarchical holographic modeling for conflict resolution. *Philosophy of Science* **53**(2): 200–222.

Haimes, Y.Y. and Horowitz, B. (2004). Adaptive two-player hierarchical holographic modeling game for counterterrorism intelligence analysis. *Journal of Homeland Security and Emergency Management* **1**(3): 1038.

Hall, A.D. (1989). *Metasystems Methodology: A New Synthesis and Unification*. Elmsford: Pergamon Press.

Hipel, K.W. and Bernath Walker, S. (2011). Conflict analysis in environmental management. *Environmetrics* **22**: 279–293. doi: 10.1002/env.1048.

Kaplan, S., Haimes, Y.Y., and Garrick, B. (2001). Fitting hierarchical holographic modeling into the theory of scenario structuring and a resulting refinement to the quantitative definition of risk. *Risk Analysis* **21**(5): 807–819.

Leondes, C.T. ed. (1969). *Advances in Control Systems*, vol. **6**. New York: Academic Press.

Macko, D. (1967). General Systems Theory Approach to Multilevel Systems. *Report SRC 106-A-67-44*, Systems Research Center, Case Western Reserve University, Cleveland, OH.

Macko, D. and Haimes, Y.Y. (1978). Overlapping coordination of hierarchical structures. *IEEE Transactions, Systems, Man, Cybernetics- SMC* **8**: 745–751.

Mesarović, M., Macko, D., and Takahara, Y. (1970). *Theory of Hierarchical Multilevel Systems*. New York: Academic Press.

Musil, R. (1933). *Der Mann Ohne Eigenschaften*. Hamburg: Rowohlt.

Paté-Cornell, M.E. (1990). Organizational aspects of engineering system safety: the case of offshore platforms. *Science* **250**: 1210–1217.

Paté-Cornell, M.E. (1996). Uncertainties in risk analysis: six levels of treatment. *Reliability Engineering and System Safety* **54**: 95–111.

Sage, A.P. (1977). *Methodology for Large Scale Systems*. New York: McGraw-Hill.

Sage, A.P. (1992). *Systems Engineering*. New York: Wiley.

Singh, M.G. ed. (1987). *Systems and Control Encyclopedia: Theory, Technology, Applications*. New York: Pergamon Press.

Tarvainen, K. (1980). Hierarchical-multiobjective optimization. Ph.D. dissertation. Case Western Reserve University.

Warfield, J.N. (1976). *Social Systems–Planning and Complexity*. New York: Wiley.

Zeigler, B.P. (1984). *Multifaceted Modeling and Discrete Event Simulation*. New York: Academic Press.

4

Modeling Complex Systems of Systems with Phantom System Models

4.1 INTRODUCTION

Modeling simple systems, or complex systems of systems (Complex SoS), necessarily implies determining their properties, constructing the relationships among their inputs and outputs through their state variables and other variables and parameters (e.g. random, decision, and exogenous variables), quantifying intra- and interdependencies within and among their various components and subsystems, and determining the appropriate model topology (structure) and parameters that best represent their essence and functionality. To achieve this, modelers rely extensively on data collection through testing, observation, experimentation, and measurement and through a tedious learning process about the "system," including the use of simulation. In this context, a "system" may connote an existing or newly reconfigured multiple systems that may span physical, organizational, societal, and political entities. (In principle, a similar process may be followed for a newly planned or to be constructed system.)

There is a fundamental difference between the complexity and challenges associated with modeling physical systems – which in their essence are controlled and driven by physical laws – and the challenges that characterize organizational and social systems, as well as combinations of multiple types of systems. For example, to model groundwater systems, modelers build on the basic flow through porous media equations and start with Darcy's law, which characterizes the slow flow of a compressible fluid through porous media. Indeed, a plethora of models of two- and three-dimensional partial differential equations have been over the years successfully developed and deployed worldwide. However, entirely different challenges face modelers who attempt to model Complex SoS that represent a combination of interconnected physical, organizational, social, and political systems. For example, planning and management of regional River Basins or supply chain that transcends international borders, or the Federal Aviation Administration's (FAA) Next Generation termed "NextGen," of restructuring the aviation Complex SoS, whose mission involves the development and integration of multiple technology-based systems and dozens of US and international agencies and organizations. For example, the purpose of the NextGen is to (i) develop technologies and procedures to improve airspace redesign that would enable

Modeling and Managing Interdependent Complex Systems of Systems, First Edition. Yacov Y. Haimes.

more direct routes and more efficient operations, (ii) expand satellite-based surveillance, (iii) improve airport runway access, (iv) increase safety and efficiency on the ground, (v) enhance airspace safety and operations, (vi) use less fuel and reduce emissions and air pollution, and (vii) enable more direct routes, among other goals and objectives (U.S. Federal Aviation Administration, 2017).

Among the many challenges facing modelers of Complex SoS are the need to determine the ways and means with which to enhance their knowledge about each subsystem and the systems of systems as a whole, discover the dynamic behavior, and identify the intra- and interdependencies among the subsystems and their environment – all by adhering to physical and other natural laws, basic principles in economics, and social and organizational behavior, among others. In many ways, modeling is the ultimate trial-and-error interplay between (i) theory and prior knowledge about the system and (ii) experimentation, measurement, and estimation, guided by a learn-as-you-go inquisitive and exploratory process. Models are built to answer specific questions; they must be as simple as possible but as complex as required. This tradeoff is at the heart of model building, given that over-complexity within a model is likely to impair its usefulness. Furthermore, the fact that all systems are ultimately affected by human actions (among others) implies the necessity of recognizing and accounting for human cognition, perception, and behavior.

The nontrivial challenge associated with the modeling process of one system is magnified when modeling Complex SoS. This challenge, which constitutes the theme of this chapter and indeed of the entire book, is addressed by exploring the centrality of the states of a system as the major building blocks of models. This process also represents the mechanism with which to model each subsystem within the Complex SoS and to relate, through metamodeling, the intrinsic interdependencies and interconnectedness (I-I) of common shared and unshared states and other essential entities among the subsystems, which in turn enables a deeper understanding of the entire Complex SoS. The term "essential entities" includes *shared/common decisions*, *decision makers*, *stakeholders*, *resources*, *organizational behavior and norms*, *policies and procedures*, *management*, *culture*, and others.

There is an unfortunate imbalance in the curricula of most universities' undergraduate and graduate programs in systems and industrial engineering and in operations research devoted to system modeling versus system optimization (whether modeling systems with single or multiple objectives). Such imbalance in education and experience, and possibly in knowledge as well, could lead to optimizing a system with a poorly constructed or misrepresentative model. This reality was recognized and gained the interest and contributions of many researchers as early as in the 1960s and 1970s in books, technical reports, and archival papers on *system identification* – often termed as the "inverse problem." For example, see Eykhoff (1974), Graupe (1972), and Haimes (1970, 1977). In system optimization we assume knowledge of the system model, under specific assumptions, where for each set of inputs, we can generate, or probabilistically estimate, the outputs. For example, in the context of decision making or effective risk analysis, no effective risk management policy options can be developed, nor can the associated tradeoffs among all critical costs, benefits, and risks be evaluated, and neither can the impacts of current decisions on future options be assessed, without having constructed a representative model or a set of interdependent models that represent the essence of the system and/or the Complex SoS (Haimes, 2018). The fact that modeling is as much an art as a science and a combination thereof – a tedious investigative trial-and-error, learn-as-you-go process – implies that an equally imaginative approach is necessary to discover through modeling the inner functionality of simple systems or of Complex SoS.

4.2 COMPLEX INTERDEPENDENCIES THAT CHARACTERIZE SYSTEMS OF SYSTEMS

The genesis of the I-I noted earlier can be traced to shared/common *states* and to other shared *essential entities that characterize Complex SoS*, thus serving as an imperative mechanism with which to analytically model the foundational I-I characterizing Complex SoS (Haimes, 2018). Nise (2015) emphasizes that state variables represent the smallest set of linearly independent system variables such that the values of the members of the set at time t_o along with known forcing functions completely determine the value of all system variables for all $t \geq t_o$. Thus, the behavior of the states of the system, as a function of time and other inputs, enables modelers and risk analysts to determine, under certain conditions, its future behavior for any given *inputs* or *initiating events*. The highly interdependent and interconnected cyber–physical and economic infrastructure SoS and other sectors of the economy introduce significant modeling challenges for systems analysts. This fact also makes such SoS more vulnerable to natural and human-caused disruptive events and thus introduces challenges for their protection. Such events disrupt the "business-as-usual" production levels of the affected SoS and lead to a variety of economic losses, such as demand/supply reductions, for one. Interdependency analysis includes data collection, testing, and analysis of ripple effects triggered by various sources of disruption, introduction of new technology; revised and modified goals and objectives, reorganization of technology-based Complex SoS; acts of terrorism, and natural calamities and accidents, among others. In modeling the intricate I-I among the myriad components and subsystems that constitute Complex SoS, it is a challenging task to quantify their causal intra- and interconnected and interdependent relationships. Moreover as noted in Chapter 3 with the introduction of Hierarchical Holographic Modeling (HHM), the I-I span multiple perspectives and dimensions, each of which can be best represented by different databases and models. For example, consider the interdependencies among Complex SoS that comprise critical cyber–physical infrastructures, including telecommunication, electric power, transportation, water supply, and hospitals, among others. Historically, many critical infrastructures around the world were physically and logically separate systems with little interdependence. This situation has rapidly changed, and close relationships among infrastructures can now take many forms. For example, similar to other critical infrastructures, water resources systems heretofore have been designed, built, and operated without a threat to their integrity. Such systems include surface and groundwater sources, water transport, treatment, distribution, storage, and wastewater collection and treatment.

Modeling the multiple perspectives of Complex SoS commonly accounts for:

1. Economic transactions among all sectors of the economy (using the vast rich database on close to 500 sectors of the economy from the US Bureau of Economic Analysis, US Department of Commerce).
2. Physical intra- and interconnections among infrastructures, for example, electric power generation plants, transmission lines, and transformers.
3. Spatial interconnections that result from land use and proximity of independent operations (e.g. pipes and wires imbedded in bridges, tunnels, or buildings and steam pipes underneath roadways).
4. Telecommunication linkages and information sharing the infrastructures.
5. Workforce that enables the operation and product and service delivery upon which all other infrastructures are critically dependent.
6. Organizational, institutional, contractual, and legal interdependencies among the owners or leaders of the infrastructures.
7. Interregional and international interdependencies as a consequence of the supply chain movement and shipment of myriad commodities across the world.

The scope of this chapter does not permit exploring all of the above interdependencies and the many facets of Complex SoS. Several philosophies, methodologies, and approaches for studying and modeling Complex SoS are briefly introduced for continuity including HHM, which is the subject of Chapter 3, and three versions of the Inoperability Input–Output Model (IIM) for modeling interdependent and interconnected Complex SoS, introduced in Chapter 8.

Almost every living entity, all infrastructures, both the natural and constructed environment, and entire households of tools and equipment, all are complex systems, often composed of multiple subsystems, which in their essence constitute Complex SoS. Most subsystems are characterized by a hierarchy of interacting and networked subsystems and are associated with multiple functions, operations, efficiencies, and costs. The subsystems are selected and coordinated according to some existing tradeoffs among multiple objectives and operational perspectives. Clearly, no single model can ever attempt to capture the essence of such Complex SoS – their multiple dimensions and perspectives.

Consider the modeling process of Complex SoS, which relies on the fundamental building blocks of mathematical models: input, output, state variables, decision (control) variables, exogenous variables, uncertain variables, and random variables. Note that these building blocks are not necessarily distinct and may overlap, for example, input and output are likely to be random. All good managers espouse to change the states of the systems they control to gain more effective and efficient attainment of the Complex SoS objectives. System modelers and analysts, in their quest to represent the managers' objectives, would necessarily affect the choice of some important states of the system while emphasizing in the modeling process-specific decisions that control other states of the system. Identifying and quantifying (to the extent possible) the building blocks of a mathematical model of any system constitutes a fundamental step in modeling. Recall that the *states of the system* constitute the most critical building elements of models. Indeed, they are essential for any quantitative modeling process. This is because at any instant the levels of the state variables *are affected* by other building blocks (e.g. decision, exogenous and random variables, and inputs), and these levels determine the outputs of the system and, thus, the associated goals and objectives of each subsystem and of the Complex SoS as a whole. For example, as noted in Chapter 1, to control the production of steel requires an understanding of the states of the steel at any instant – its temperature and other physical and chemical properties. To know when to irrigate and fertilize a farm, a farmer must assess the soil moisture and the level of nutrients in the soil. To treat a patient, a physician first must know the temperature, blood pressure, and other states of the patient's physical health. Consider again the human body and its vulnerability to infectious diseases. Different organs and parts of the body are continuously bombarded by a variety of bacteria, viruses, and other pathogens. However, only a subset of the human body is vulnerable to the threats from yet another subset of the would-be attackers, and due to our immune system, only a smaller subset of the human body would experience adverse effects. Indeed, a system's vulnerability is a manifestation of the inherent *states* of that system, and each of those states is dynamic and changes in response to the inputs and other modeling building blocks.

Moreover, within any *single model*, it is a challenge to quantify the causal relationships among all relevant building blocks of models that represent the Complex SoS, especially the state variables. We address the need to develop a body of prescriptive theory and methodology for modeling Complex SoS. The purpose is to enable analysts to appropriately model and understand the evolving behavior of SoS, due to the continued emergent forced changes (EFCs) imposed on them. We define EFCs as internal or external forces that may affect a system or the entire Complex SoS positively or negatively. Indeed, models, laboratory experiments, and simulations all are designed to answer specific questions, keeping in mind that system models provide responses based on the states of a system under given conditions and model's assumptions. Unprecedented and emerging Complex SoS (e.g. the mission to Mars, the power grid, or the FAA's new air traffic control system) are inherently visionary and at times elusive. They are, by and large phantom entities, grounded on a mix of future needs and available resources, technology, forced developments and changes, and myriad other unforeseen events. This chapter addresses some of these challenges in modeling emergent Complex SoS.

4.3 STUDYING AND MODELING THE MULTIPLE DIMENSIONS OF COMPLEX SYSTEMS OF SYSTEMS

About four decades ago, Haimes and Macko (1973) studied the modeling of large-scale systems and the necessity to consider and account for a wide spectrum of factors, including economic, technical, environmental, political, legal, and other important considerations, which necessarily yield a highly complex model. To analyze the complexity of such models, they "suggested some remedies in the form of different decompositions and coordination methodologies." For example, a River Basin may be modeled through multiple descriptions, including temporal, physical, hydrological, political, and geographical descriptions and goal descriptions (see Figures 4.1 and 4.2). They also pursued the concept of Hierarchical Overlapping Coordination (HOC) (Macko and Haimes, 1978) (see Chapter 2) and HHM (see Chapter 3) for modeling large-scale systems, arguing that: "In the modeling process of large-scale and complex systems, more than one mathematical model is likely to emerge, each of which may focus on a specific aspect of the system; yet all these models may be regarded as acceptable representation[s] of the system." They presented a mathematical formulation for an overlapping coordination between two or more hierarchical structures pertaining to the same system. In the abstract, Blauberg et al. (1977) point out that for understanding and analyzing a large-scale system, the fundamental principles of wholeness (representing the integrity of the system) and *hierarchy* (representing the internal structure of the system) must be supplemented by the multiplicity of descriptions for any system: in order to obtain adequate knowledge about a system, one has to construct a certain class of its descriptions, each of which can only cover certain aspects of the wholeness and hierarchy of the system. This principle is, in fact, derived from the fundamental relativity of any description of a system. Indeed, the fundamental attribute of large-scale systems is their inescapably multifarious nature – hierarchical noncommensurate

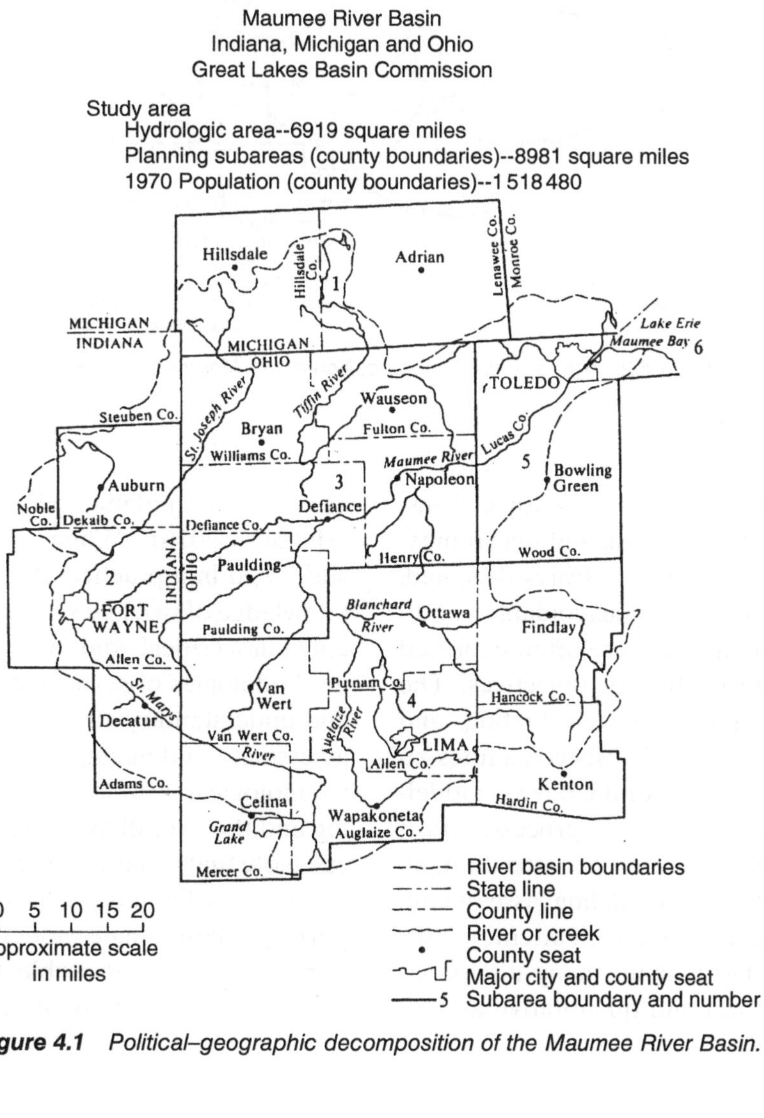

Figure 4.1 *Political–geographic decomposition of the Maumee River Basin.*

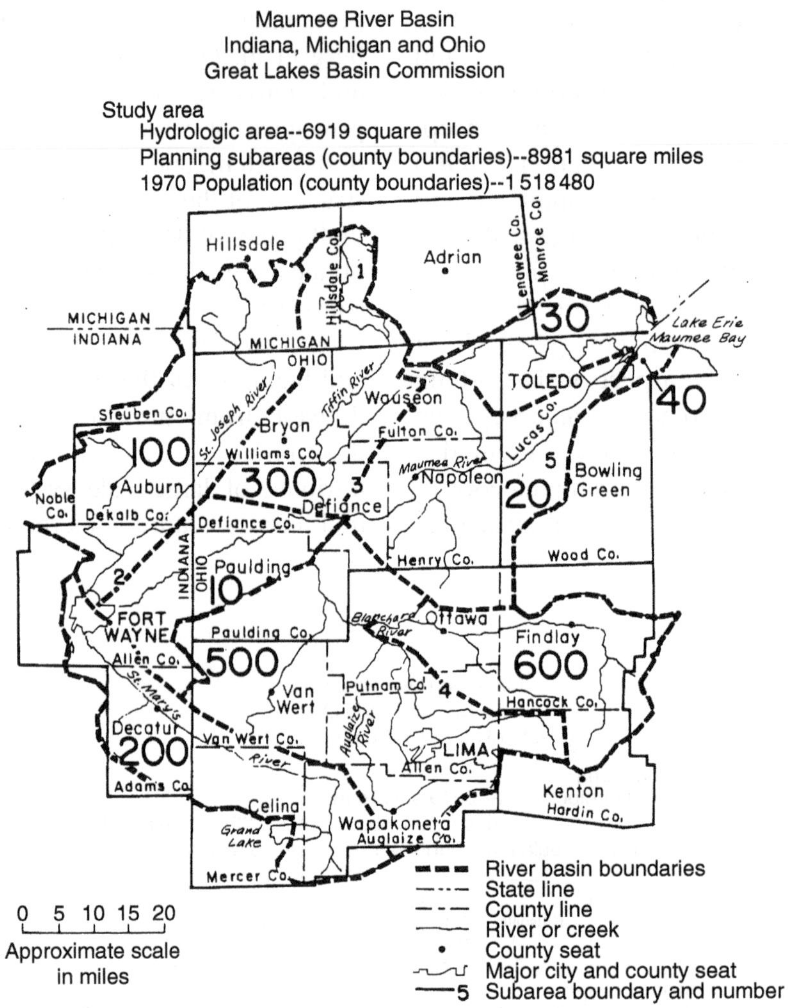

Figure 4.2 *Hydrological decomposition of the Maumee River Basin.*

objectives, multiple decision makers, multiple transcending aspects, and elements of risk and uncertainty.

Engineering systems are almost always designed, constructed, and operated under unavoidable conditions of risk and uncertainty and are often expected to achieve multiple and conflicting objectives. The process of identifying, quantifying, evaluating, and trading off of risks, benefits, and costs constitutes an integral and explicit component of the overall modeling and managerial decision-making process; and it should not be a separate, cosmetic afterthought. The body of knowledge in risk modeling, assessment, and management has gained significant attention during the last three decades; it spans many disciplines and encompasses empirical and quantitative as well as normative, judgmental aspects of decision making (Haimes, 2016). When risk and uncertainty are addressed in a practical decision-making framework, knowledge of risk assessment and management markedly fills a critical void that augments the theories and methodologies of systems engineering. Reflecting on and understanding the nature, role, and place of risk assessment and management in the administration of technological systems constitute the essence of modeling and the overall managerial decision-making process. Note that a mathematical model may be viewed as a one-sided limited image of the real system that it portrays. With single-model analysis and interpretation, it is quite impossible to clarify and document not only the multiple components, objectives, constraints, and EFCs characterizing Complex SoS but also their welter of societal aspects (including functional, temporal, geographical, economic, political, legal, environmental, and institutional). Given this perspective and the notion that even current integrated models cannot adequately cover all relevant and critical aspects of Complex SoS, the theory and philosophy of HHM introduced in Chapter 3 comprise a comprehensive theoretical framework for modeling Complex SoS. As a holistic philosophy/methodology, HHM aimed at capturing and representing the essence of the inherent diverse characteristics and attributes of a system – its multiple aspects, perspectives, facets, views, dimensions, and hierarchies (Haimes, 1981). Arthur D. Hall III, whose first book on systems engineering was published in 1962, recognized the contributions of HHM in his seminal book *Metasystems Methodology* (Hall, 1989, p. 6): "In this way," he wrote,

> history becomes one model needed to give a rounded view of our subject within the philosophy of Holographic modeling (Haimes, 1981) being used throughout this book, defined as using a family of models at several levels to seek understanding of diverse aspects of a subject, and thus comprehend the whole.

Hall developed a theoretical framework, which he termed metasystems methodology, with which to capture the multiple dimensions and perspectives of a system. Other early seminal works in this area include the book on societal systems and complexity by Warfield (1976) and the book *Systems Engineering* (Sage, 1992).

To present a holistic view of the elements that must be included in the model, the HHM approach involves organizing a team of experts with widely varied experience and knowledge bases (e.g. technologists, psychologists, political scientists, criminologists, and others). Note that the broader the base of expertise that goes into identifying elements of the multifarious characteristics of the system or of the SoS, the more comprehensive is the ensuing HHM. The result of the HHM process is the creation of a very large number of visions, perspectives, hierarchies, and attributes of each subsystem and of the Complex SoS as a whole.

4.4 HISTORICAL PERSPECTIVES OF MODELING COMPLEX SYSTEMS OF SYSTEMS

The *systems* concept has a long history. The art and science of systems engineering as a natural philosophy can be traced to Greek philosophers. Although the term system itself was not emphasized in earlier writings, the history of this concept includes many illustrious names, including *Plato* [428–348 BC] and *Aristotle* [384–322 BC]. The writings of *Baron von Leibniz* [1646–1716], a mathematician and philosopher, are directed by holism and systems thinking. He shares with Isaac Newton [1642–1727] the distinction of developing the theory of differential and integral calculus. By quantifying the causal relationships among the interplanetary SoS, Newton represents the epitome of a systems philosopher and modeler. In their seminal book, *Isaac Newton's The Principia*, Cohen and Whitman (1999) write about Newton, the system modeler:

> Newton's discovery of interplanetary forces as a special instance of universal gravity enables us to specify two goals of the Principia: The first is to show the conditions under which Kepler's laws of planetary motion arc exactly or accurately true; the second is to explore how these laws must he modified in the world of observed nature by perturbations on the motions of planets and their moons.

Johann Gottlieb Fichte [1762–1814] introduced the idea of synthesis – one of the fundamental concepts of systems thinking. For example, he argued that "freedom" can never be understood unless one loses it. Thus, the *thesis* is that a man is born free, the loss of freedom is the *antithesis*, and the ability to enjoy freedom and do good works with it is the *synthesis*. In other words, to develop an understanding of a system as a whole (synthesis), one must appreciate and understand the roles and perspectives of its subsystems (thesis and antithesis). *Georg Hegel* [1770–1831], a contemporary of Fichte, was one of the most influential thinkers of his time. Like Aristotle before him, Hegel tried to develop a system of philosophy in which all the contributions of his major predecessors would be integrated. His *Encyclopedia of the Philosophical Sciences* [1817], which contains his comprehensive thoughts in a condensed form, provides important foundations for the concept of holism and the overall systems approach (Hegel, 1952).

Around 1912, *Max Wertheimer*, *Kurt Koffk*, and *Wolfgang Kohler* founded the Gestalt psychology, which emphasizes the study of experience as a *unified whole*. The German word *gestalt* means pattern, form, or shape (*The World Book Encyclopedia*, 1980): Gestalt psychologists believe that pattern or form is the most important part of experience. The whole pattern gives meaning to each individual element of experience. In other words, *the whole is more important than the sum of its parts*. Gestalt psychology greatly influenced the study of human perception, and psychologists used Gestalt ideas in developing several principles – for example, the principle of closure (people tend to see incomplete patterns as complete or unified wholes).

Buckley (1968) states that "Modern Systems Theory, though seemingly springing de nouveau out of the last war effort, can be seen as a culmination of a broad shift in scientific perspective striving for dominance over the last few centuries."

Slovic (2000), among his many seminal works, presents the capabilities of decision makers to understand and make "optimal" decisions in uncertain environments. He concludes that perceptions of risk can be quantified and changed, but often deviate from true measures, even among experts.

During his distinguished career, *Albert Einstein* attempted to develop a unified theory that embraces all forces of nature as a system. Feynman (1963) describes a hierarchy or continuum of physical laws as distinct systems or disciplines that are cooperating and interdependent. Modern systems foundations are attributed to select scholars. Among them is *Norbert Wiener*, who in (1948) published his seminal book *Cybernetics*; Wiener's work was the outgrowth and development of computer technology, information theory, self-regulating machines, and feedback control. In the second edition of *Cybernetics* (1961), Weiner commented on the work of Leibniz:

> At this point there enters an element, which occurs repeatedly in the history of cybernetics-the influence of mathematical logic, If I were to choose a patron saint for cybernetics out of the history of science, I should have to choose Leibniz. The philosophy of Leibniz centers about two closely related concepts-that of a universal symbolism and that of a calculus of reasoning. From these are descended the mathematical notation and the symbolic logic of the present day. (p. 9)

Kenneth Boulding, an economist, published work in 1953 on *general empirical theory* (Boulding, 1953) and claimed that it was the same as the general system theory advocated by Bertalanffy. The Society for General Systems Research was organized in 1954 by the American Association for the Advancement of Science (Bertalanffy, 1968). The society's mission was to develop theoretical systems applicable to more than one traditional department of knowledge.

Several modeling philosophies and methods have been developed over the last four decades to address the complexity of modeling complex large-scale systems and to offer various modeling schema. They are included in the following volumes: *General Systems Theory and Systems Research Contrasting Conceptions of Systems Science* (Mesarovic, 1964), *New Directions in General Theory of Systems* (Mesarović, 1965), *General Systems Theory* (Macko, 1967), *Systems Theory and Biology* (Mesarović, 1968), *Advances in Control Systems* (Leondes, 1969), *Theory of Hierarchical Multilevel Systems* (Mesarović et al., 1970), *Methodology for Large Scale Systems* (Sage, 1977), *Systems Theory: Philosophical and Methodological Problems* (Blauberg et al., 1977), *Hierarchical Analyses of Water Resources Systems: Modeling and Optimization of Large-Scale Systems* (Haimes, 1977), and *Multifaceted Modeling and Discrete Event Simulation* (Zeigler, 1984). *Synectics: The Development of Creative Capacity* (Gordon, 1968) introduced an approach that uses metaphoric thinking as a means to solve complex problems. In the same era, Lowrance (1976) published an influential work considering the science of measuring the likelihood and consequence of uncertain adverse effects that emerge from complex systems. He outlined critical considerations for engineering complex systems that are characterized by uncertainty. Gheorghe (1982) presented the philosophy of systems engineering as it is applied to real world systems. In his book *Metasystems Methodology*, Hall (1989) developed a theoretical framework to capture the multiple dimensions and perspectives of a system. Other works include Sage (1977, 1992, 1995) and Sage and Rouse (1999). Sage and Cuppan (2001) provide a definition of emergent behavior in the context of a SoS. Slovic (2000), among his many far-reaching works, presents the capabilities of decision makers to understand and make "optimal" decisions in uncertain environments. Other books on systems include Fang et al. (1993), Gharajedaghi (2005), Rasmussen et al. (1994), Rouse (1991), Adelman (1991), Zeleny (2005), Blanchard (2003), Kossiakoff and Sweet (2002), Maier and Rechtin (2009), Buede (1999), Blanchard and Fabrycky (2005), Sage and Armstrong (2003), and Hatley et al. (2000). Haimes (2012) and Guo and Haimes (2016, 2017) addressed risk modeling of Complex SoS by building on state-space theory. Eusgeld et al. (2011) discussed the potential failure propagation among infrastructures leading to cascade failures, and they analyzed two modeling alternatives, comparing integrated with coupled models. Haberlin and Haimes (2018) present a case study on the modeling of interconnected and interdependent cyber–physical infrastructures as Complex SoS. Extensive works on SoS have been devoted to their organizational, functional, and structural nature; on the other hand, there has been comparatively less inquiry into the problem of modeling SoS, and much of it has emerged within the last decade. For example, Ottino (2003) reviews three major tools for quantitative modeling and studying complex systems: nonlinear dynamics, agent-based models, and network theory. Shalizi (2006) also reviews the main methods and techniques of complex systems, which include tools for analyzing data, constructing and evaluating models, and measuring complexity. Chang and Harrington (2005) provide a comprehensive description of agent-based models of organizations. Amaral and Ottino (2004) describe network theory and its importance in augmenting the framework for the quantitative study of complex systems. Lloyd and Lloyd (2003) present a general method for modeling complex systems in terms of flows of information. Page (1999) discusses robust computational models. In an analysis of the challenges associated with complex systems engineering, Johnson (2006) provides a comprehensive review of emergent properties and how they affect the engineering of complex systems. Bar-Yam (2003a) reviews past lessons learned from problems with systems engineering over the past and suggests adopting an evolutionary paradigm for complex systems engineering. Within the application of complex system theory, in a multiscale analysis of military littoral warfare, Bar-Yam (2003b) suggests the necessity of considering the specific organizational and technological requirements needed to perform effectively in a high-complexity environment. In health care, Funderburk (2004) presents a brief survey of several formal dynamic and/or network-based models that are relevant for health-care policy development and evaluation. Tivnan et al. (2007) describe the formulation, successful replication, and critical analysis of Levinthal's model of emergent order for economic firms. Jamshidi (2009a, b) edited two volumes on SoS engineering. In the preface of the first volume (2009a), he writes: "The SoS [Systems of Systems] concept presents a high-level viewpoint and explains the interactions between each of the independent systems. However, when it comes to engineering and engineering tools of SoS, we have a long way to go. This is the main goal of this volume." Indeed, Jamshidi confirms the need for concerted efforts in modeling Complex SoS.

Reflecting on the history of modern systems theory, and its close ties to the Gestalt psychology first introduced in 1912, we cannot underestimate the intellectual power of this multidisciplinary field and the holistic philosophy that has sustained it, thus, allowing it to transcend the arts; the humanities; the natural, social, and physical sciences; engineering; medicine; and law. The fact that systems engineering and systems analysis have continued to grow and infiltrate other fields of study over the decades can be attributed to the fundamental premise that a system can be understood only if all the intra- and interdependencies among its parts and its environment are also understood and accounted for – a central guiding premise of this book. For more than a century, particular mathematical models, upon which systems-based theory and methodologies were developed, have been deployed in myriad large-scale projects in the natural and constructed environments. Moreover, if we were to identify a single concept that has dominated systems thinking and modeling, it would be the *state space*. Indeed, the centrality of state variables in this context is so dominant that no meaningful mathematical model of a real system can be built without identifying the states of that system and relating all other building blocks of the model to them (including decision, random, and exogenous variables and inputs and outputs). (More will be discussed on the centrality of state variables in modeling as it relates to the entire theme of this chapter and the book.) In this respect, the art and science of systems modeling has served, in many ways, as the medium through which the holistic systems philosophy has informed the practice not only of engineering but also of a broad range of other fields. As the discipline of systems engineering continues to develop and expand its domains of application, the need for new organizational and modeling paradigms to represent complex systems has emerged and has ultimately led to the study of Complex SoS.

Complex systems are commonly composed of myriad subsystems, which in their essence constitute SoS. Each subsystem of a Complex SoS has multiple functions, operations, efficiencies, costs, and stakeholders. Complex SoS are characterized by a hierarchy of interacting components. Clearly, no single model can ever attempt to capture the essence of such systems – their multiple dimensions and perspectives. Indeed, almost every living entity, all infrastructures, and both the natural and constructed environment, are Complex SoS. For example, different organs and parts of the human body, as an SoS, are continuously bombarded by a variety of bacteria, viruses, and other pathogens; however, only a subset of the (states of the) human body is vulnerable to the threats from yet another subset of the would-be attackers, and due to our immune system, only a smaller subset of the human body would experience adverse effects. Thus composites of low-level, measurable states integrate to define higher-level fundamental state variables that characterize the system. Indeed, both the vulnerability and the resilience of a system are a manifestation of the inherent states of that system, and each state of a system can be dynamic and changed in response to inputs, other random variables, and the building blocks of mathematical models (as discussed in the next section).

The precise definition of SoS, however, is more elusive. In a seminal paper, Sage and Cuppan (2001) directly ask, "What is a system of systems?" They conclude, "Unfortunately, there is no universally accepted definition of these 'super systems.' What distinguishes a system of systems from other systems does not at this point have a definitive answer." In a subsequent paper, Sage and Biemer (2007) provide the following answer to the same question: "No universally accepted definition of an SoS is available at this time." To address this problem, Sage and Cuppan (2001) build on the following properties of SoS suggested by Maier (1998) (p. 271):

> (i) *Operational Independence of the Individual Systems.* A system of systems is composed of systems that are independent and useful in their own right. (ii) *Managerial Independence of the Systems.* The component systems not only can operate independently, they generally are operated independently to achieve an intended purpose. (iii) *Geographic Distribution.* Geographic dispersion of component systems is often large. Often, these systems can readily exchange only information and knowledge with one another, and not substantial quantities of physical mass or energy. (iv) *Emergent Behavior.* The system of systems performs functions and carries out purposes that do not reside in any component system. (v) *Evolutionary Development.* A system of systems is never fully formed or complete. Development of these systems is evolutionary over time and with structure, function and purpose added, removed, and modified as experience with the system grows and evolves over time.

Building on the above five principles, we attempt to improve our understanding of Complex SoS by extending the multiperspective modeling schema (through HHM) (Haimes, 1981, 2016 and Chapter 3) into the Phantom System Models (PSMs).

4.5 RISK MODELING OF INTERDEPENDENT AND INTERCONNECTED COMPLEX SYSTEMS OF SYSTEMS

4.5.1 An Overview

The complexity characterizing our interdependent and interconnected safety-critical cyber–physical infrastructure requires a reevaluation of the way we model, assess, manage, communicate, and analyze the risk thereto. A sample of such interdependent and interconnected cyber–physical Complex SoS introduced in this book include (i) groundwater and surface water carriers and dams, (ii) electrical grid, (iii) telecommunications, (iv) supply chain, (v) multiple modes of transportation, (vi) geographic positioning systems (GPS), (vii) national defense and communication infrastructures, and (viii) myriad others. Current models for risk analysis of the above sample of emergent Complex SoS are insufficient, because too often they rely on the same risk functions and models used for single systems. These models commonly fail to incorporate the complexity derived from the networks of I-I characterizing our cyber–physical infrastructure SoS. Thus, in this chapter and especially in Chapter 3, we reevaluate our currently practiced risk analysis to Complex SoS. Indeed, the key to evaluating the risk to Complex SoS lies in understanding the genesis of characterizing I-I of systems manifested through shared/common *stakeholders*, *decisions*, *decision makers*, *resources*, *functions*, *policies*, and *organizational setups*, among many others.

We noted in Chapter 1 that the terms complexity and complex systems have evolved and assumed different connotations during the last 50 years and thus the way to model and manage complexity. Decision makers and stakeholders by *necessity aspire to a reliable and trustworthy risk analysis, not only for mundane single systems but also especially for interdependent and interconnected Complex SoS.* Indeed, in their essence, most cyber–physical, organizational, and government enterprises, now and in the future, belong to the latter category. Two major fundamentals guide our reevaluation of current practices of risk analysis when addressing the risk to Complex SoS. These are grounded on understanding and evaluating the I-I within and among the Complex SoS, manifested through shared/common states and other *essential entities* within and among the systems and subsystems that make up SoS. We add the *Evolving Base* in Chapter 9, which is represented by the following dynamic shifting rules and realities for each subsystem and for the entire SoS: (i) goals and objectives; (ii) stakeholders, decision makers, and interest groups; (iii) organizational, political, and budgetary baselines; (iv) reorganization and reallocation of key personnel; (v) emergent technology and its deployment; and (vi) requirements, specifications, delivery, users, and clients. *The fundamental departures from performing risk analysis from a single system to Complex SoS constitute the core research needs and methodological challenges, a sample of which is addressed in this chapter.*

4.5.2 Fundamental Differences between Risk Analyses of Single Systems Versus Complex Systems of Systems

Complex SoS are commonly composed of myriad I-I subsystems, each of which can be characterized by a hierarchy of shared or interacting components, each with multiple functions, operations, databases, costs, and stakeholders (Haimes, 2006, 2008, 2009, 2018). All systems, natural and the constructed environment, are commonly subjected to, and thus affected by, multiple adverse initiating events that could be initiated from outside or from within one or more subsystems that constitute Complex SoS. The fundamental differences between the impacts of adverse initiating events on single systems compared with the impacts on Complex SoS stem from several intrinsic characteristics of SoS. Each subsystem can be characterized by a hierarchy of shared or interacting components, each with multiple functions, operations, databases, costs, and stakeholders. Clearly, no single model can ever successfully capture the essence of such systems – their multiple dimensions and perspectives. These inherent natures and characteristics of the systems and subsystems that make up Complex SoS, which differentiate them from single systems, are commonly subjected to fundamentally different sources of risk to each other and thus to the SoS as a whole. Thus, there is the imperative need to review and recalibrate the intricate process of risk modeling, assessment, management, and communication for each of the two categories.

4.5.3 The Role of Shared States among Subsystems in Risk Modeling of Systems of Systems

The evolution of the field of risk analysis, since the founding of the Society for Risk Analysis and its journal Risk Analysis, requires continuous reassessment of its prevailing theory, methodology, and practice. Thus, current risk analysis applied to a single system must be fundamentally extended when applied to Complex SoS, which constitute multiple systems with emergent behaviors. The complexity of these systems results primarily from the I-I within and among the systems and subsystems of SoS. Consider, for example, the I-I among three common cyber–physical infrastructures – electricity, communications, and water (and of course the communities they serve) – introduced in Chapter 10. A major initiating event that may cause the failure of the electricity system would result in adverse consequences to the other two cyber–physical infrastructures, because of their dependency on electricity, and neither one can operate without electricity. *Clearly, there is a need to understand and thus model the pathway through which the failure of one system propagates to other interconnected systems*. An effective risk analysis of Complex SoS requires a clear understanding of the SoS configurations to enable the identification of specific critical failure modes. This allows the development and deployment of effective risk mitigation and other management strategies. Current risk models of systems, which do not consider the inherent interdependencies among other systems, are likely to be inferior to those models that do. This very important issue is being addressed throughout this book. Recall that the premise of this book is that the above I-I characteristics are best understood, and thus modeled, through the *shared states* within the SoS (and other *shared essential entities*) and of the SoS as a whole. (Recall that the term "essential entities" connotes shared decisions, resources, management, policies, decision makers, stakeholders, and organizational setups, among others.) Scientific literature has many definitions of state variables in modeling, because they play a significant role in estimating the future behavior of the systems for any given inputs. Recall Chen's (2012) conceptual definition of state variables posits that the state $x(t_o)$ of a system at time t_o is the information at time t_o that, together with the input $u(t)$, for $t \geq t_o$, determines uniquely the output $y(t)$ for all $t \geq t_o$.

Similarly, Nise emphasizes that state variables represent the smallest set of linearly independent system variables such that the values of the members of the set at time t_o along with known forcing functions completely determine the value of all system variables for all $t \geq t_o$ (2015). Thus, the behavior of the states of the system, as a function of time and other inputs, enables modelers and risk analysts to determine, under certain conditions, its future behavior for any given *inputs* or *initiating events*. Indeed, in industrial production and management, the feedback control process (as an integral part of risk management and quality control) is predicated on the fact that the smooth operations of the multiple interconnected subsystems are built on the knowledge of the operational states of each subsystem at each instant. As a trivial example, to determine the reliability and functionality of a car, one must know the *states* of the fuel, oil, and tire pressure and the states of other mechanical and electrical components (due to the dynamic nature of all states).

Systems modelers face nontrivial challenges when selecting the minimum number of states to adequately and effectively represent (model) the subsystems under consideration and, thus, the risk thereto. Consider, for example, the challenges in selecting the states in the risk modeling process for sustained years of drought including (i) availability and quantity (states) of groundwater and surface water, (ii) quality (states) of groundwater and surface water, and (iii) human and livestock population and industrial needs for water. The above states of water quality and quantity have direct impacts on (a) rural or urban populations and on livestock, (b) agriculture and industry, and (c) future recovery of the states of water quality and quantity for the entire River Basin.

From the perspective of the *reliability of Complex SoS*, shared states and other essential entities within and among the subsystems represent the essence of the I-I that characterizes SoS: *initiating events affecting one or more of the interdependent and interconnected subsystems would necessarily increase the probability of failure of the entire SoS*. This fundamental fact could be viewed in the parlance of fault-tree analysis (NUREG, 1981) as converting subsystems from being "marginally connected," or in "parallel," to becoming "directly connected," or in "series." Nuclear reactors, which epitomize Complex SoS, have always relied on fault-tree analysis to ensure their safety. (See the Appendix for additional information on fault trees.) Thus, from that perspective, subsystems that share states and other essential entities *are most likely to* be affected by adverse initiating events. For example, this phenomenon is most evident in the emergent extensive use of cloud computing technology, where numerous hardware/software subsystems are shared among multiple users (Haimes et al., 2015); also see Chapter 10. Similar results have been demonstrated with the interdependencies and interconnections among cyber–physical infrastructures. An initiating event may not affect shared states of different subsystems in the same way. This fact necessarily implies the following scenarios when analyzing risk to Complex SoS, when (i) each subsystem is likely to have different decision maker(s), then decisions made to control subsystem A may affect positively or negatively subsystem B that shares subsystem A's states and other essential entities; (ii) decision makers collaborate among themselves and coordinate their decisions, this can have a positive effect on Complex SoS and improve their overall effectiveness and management; and (iii) different initiating events affect one or more subsystems, the level of shared states and other essential entities among them can positively or negatively affect the risk management process of the entire SoS. Therefore, modelers and risk analysts should recognize and exploit these attributes of shared states and other essential entities among subsystems. As a natural example of interdependent subsystems, the states of the Earth and its moon relative to each other and to the sun are known at any instant due to shared gravitational forces. In their case, without any initiating events affecting any of them, the states of their celestial coordinates in space and time can be determined.

Uncertainty analysis becomes even more imperative in risk analysis of emergent Complex SoS. Two major sources of uncertainty in modeling affect risk analysis (Paté-Cornell, 1990, 1996; Apostolakis, 1999; Haimes, 2016). *Knowledge* (*epistemic*) *uncertainty* may manifest itself in faulty selection of appropriate model topology (structure) and model parameters, due to ignorance of the system or SoS under consideration. *Variability* (*aleatory*) *uncertainty* applies to all relevant and important events and must not be overlooked. Sources of uncertainty dominate most decision-making processes, especially Complex SoS, and are the Achilles' heel for all deterministic and probabilistic models. Sources of uncertainty and lack of understanding of the complexity associated with one subsystem of Complex SoS would likely result in (i) adherence to unrealistic assumptions, (ii) lack of awareness of and accountability to critical I-I of the Complex SoS under consideration, (iii) poorly selected representative model topology and comprehensiveness, (iv) a dated or insufficient database to populate and calibrate the multiple subsystems models, and (v) essential risk scenarios being poorly represented or structured for all interdependent and interconnected systems. In addition, model uncertainties are often introduced through human errors of both commission and omission. In sum, uncertainty analysis associated with Complex SoS is probably one of the most difficult, albeit important, tasks in the broader risk analysis process.

4.6 RISK AND UNCERTAINTY ANALYSIS OF COMPLEX SYSTEMS OF SYSTEMS

When performing risk analysis, the probability component of the risk function should take *a secondary or tertiary importance to the "not unlikely" extreme and dire consequences*. Planners and stakeholders neither expect miracles nor precise hindsight from modelers and analysts nor are the latter capable of delivering such knowledge or information. Similarly, planners and stakeholders, who are experts in their domain knowledge – with experience in and appreciation of the complexity that characterizes the systems – neither expect miracles from modelers nor would they believe in projected numbers with significant figures of accuracy.

One important contribution of modeling and simulation is enabling the exploration of what might evolve in the future, given the prevalence of EFCs, and by avoiding likely negative or disastrous impacts of current policies on future options. Recall that these EFCs connote internal or external trends in sources of risk to a system that may adversely affect or enhance specific states of that system and consequently affect the entire SoS and may be manifested by the following dynamic shifting rules and realities discussed in more detail in Chapter 9: (i) goals and objectives; (ii) stakeholders, decision makers, and interest groups; (iii) organizational, political, and budgetary baselines; (iv) reorganization and reallocation of key personnel; (v) requirements, specifications, delivery, users, and clients; and (vi) performance and delivery of goods and services from other interdependent infrastructures.

Nonprobabilistic modeling and analyses of high and possible severe consequences should be as equally important as probabilistic models. The key point here is that when consequences to a select system(s) or to the entire Complex SoS would be so *dire* and "not unlikely," then their specific probabilities would not add informative, quantitative value. More specifically, any mathematical model used for the quantification of the risk function by multiplying questionable probabilities with extreme and catastrophic consequences would be impractical and misleading. Indeed, when a catastrophic event is considered "not unlikely" to occur, then assessing the very low probability of that event loses its common contributions to the risk function. (Because multiplying "infinity" representing dire and catastrophic consequences with any probability would be meaningless.)

An important common denominator to long-term planning, risk analysis, and systems modeling is addressing the life cycle of (physical) infrastructure SoS – their planning, requirements, design, construction, operation, maintenance, and replacement. Such a process requires the consideration of (i) multiple current and future needs and demands, (ii) demographic change and new evolving technologies, (iii) sources of financing the infrastructure under consideration in one or in multiple stages, (iv) the time frame during which it is expected to operate and perform its intended functions safely and efficiently, (v) the operating the infrastructure harmoniously with other connected or interdependent infrastructures; and (vi) managing the infrastructure Complex SoS within an acceptable level of risk and uncertainty. Most importantly, considering the above ought to be at acceptable levels or deemed acceptable to the public and the appropriate decision makers and stakeholders. This elaborate, complex, and emergent multiyear process is rarely, if at all, executed by the same entities, given the life cycle spans of years, if not decades.

All decisions (along with all other inputs) directly affect the states of the system. Indeed, specific decisions are made to achieve desired goals and objectives, at acceptable tradeoffs and time frame. Since all goals and objectives are functions of states of the system implies that to achieve desired goals and objectives, one must focus on the appropriate states of the system. Thus, for any given scenario of EFCs, and given the current states of the Complex SoS, the corresponding vulnerability, resilience, and risk to them are all functions of their corresponding affected states at the time of the initiating EFCs. Thus, appropriate risk management options and their associated tradeoffs must be considered to reduce the vulnerability and to enhance the resilience of a system for specific EFCs and thus to reduce the associated risk to an acceptable level (note that the consequences are a multidimensional vector).

4.7 THE MULTIFARIOUS PERSPECTIVES OF THE MAUMEE RIVER BASIN COMPLEX SYSTEMS OF SYSTEMS

In this section we demonstrate, via a case study, the impracticality and ineffectiveness of currently practiced risk analysis to single systems, when applied to a River Basin that transcends hydrological and political boundaries, multiple planning time horizons, multiple conflicting and competing goals and objectives, involving diverse communities, and multiple decision makers and stakeholders. The Maumee River Basin introduced in section 4.3 requires human, agriculture, and environmental protection from flooding and assurance of quality of water supply to population of millions, residing in three states (Indiana, Michigan, and Ohio), that span close to 7000 $mile^2$ of hydrologic area and close to 9000 $mile^2$ of county boundaries. The I-I that characterize the complexity of the River Basin stem from the fact that State and County's political boundaries transcend hydrological boundaries; and State, County, and hydrologic boundaries also transcend each other, as well as all other organizational, political and social divisions. Clearly, more than one analytical or conceptual model is likely to emerge, and each of which may focus on a specific aspect of subsystems or the Complex SoS as a whole; yet all may be regarded as acceptable representations of the River Basin. This phenomenon is particularly common in hierarchical modeling of Complex SoS when focusing on risk and uncertainty, where more than one decomposition approach may be both feasible and desirable, as will be demonstrated subsequently (Haimes, 1981, 2016). Consequently, for risk analysis, decomposing a system or Complex SoS often presents a dilemma over the choice of the perceived critical subsystems. For example, an economic system may be decomposed into geographic regions or activity sectors. An electric power management system may be decomposed according to the various functions of the system (e.g. power generation units, energy storage units, transmission units) or along geographic or political boundaries. Another decomposition might be a timewise decomposition into planning periods. For an effective risk analysis of Complex SoS, several aspects of the system must be addressed, such as the geographic regions and activity sectors of an economic system. Indeed, it is advantageous to consider several decompositions, if each adds value to our understanding and improves our modeling of the systems that make up SoS.

The Maumee River Basin may be modeled from the following four overlapping perspectives: (i) hydrological, (ii) geographical, (iii) organizational, and (iv) political (see Figures 4.1 and 4.2) – as well as multiple objectives and several time planning horizons. The sources of risk to the River Basin as Complex SoS can originate from any of the multiple overlapping perspectives depicted in Figure 4.2, characterizing its systems and subsystems. Thus the intriguing challenges that systems modelers and risk analysts face in deciphering the myriad sources of risk to Complex SoS, in particular, the many sources of risk and their distribution within the multiple towns, municipalities, and states as subsystems, often play a dominant role in the allocation of shared resources. Tradeoffs in risk management must be made among the multiple objectives associated with the affected systems and subsystems. Other examples of the multiple sources of risk to the Maumee River Basin include the risk of drought and floods affecting populations in urban and agricultural areas along with their associated health and community well-being. Also, the basin spans an area of approximately 8000 mile^2 over parts of the states of Ohio, Michigan, and Indiana and has been divided into five planning subareas (PSAs), each one consisting of several counties (political/geographic decomposition).

From risk analysis and flood management perspectives, the basin can also be divided into eight watersheds crossing state and county boundaries (hydrological decomposition), as shown in Figure 4.2. Consider also the following seven major risk management objectives from the functional decomposition perspective: (i) protect agricultural land, (ii) reduce erosion and sedimentation, (iii) enhance water quality, (iv) protect fish and wildlife, (v) enhance outdoor recreational opportunities, (vi) reduce flood damage, and (vii) supply water needs.

Each modeling representation of a specific configuration of the subsystems of the River Basin as Complex SoS represents and uncovers important sources of risk not evident using a single system model. In other words, Complex SoS, such as the Maumee River Basin, must be modeled and the associated sources of risk be assessed and studied from many perspectives and configurations. This intensive modeling process can be achieved and the risk be assessed from the following overlapping shared states and other essential entities: *(i) five PSAs each consisting of several counties (political/geographic risk modeling perspective), (ii) eight watersheds crossing state and county boundaries (hydrological risk modeling perspective), (iii) seven major objectives (functional risk modeling perspective), and (iv) three planning time horizons (temporal risk modeling perspective).*

In sum, we present the multidimensionality of the Maumee River Basin as a representative of Complex SoS, along with the associated large number of decision makers and stakeholders, representing multiple constituencies, while their decisions affect cross-county and cross-State boundaries, laws, and regulations. *Clearly, the above intertwined natural environmental, legal, cultural, organization, and political users and other entities defy and challenge the capability of any single model to represent the essence of the associated risk modeling and analysis process.*

4.8 REFLECTIONS ON RISK TO COMPLEX SYSTEMS OF SYSTEMS

The prevailing theory, methodology, and practice of risk analysis currently applied to single systems must be adapted to respond to the I-I that characterize Complex SoS. For a more effective risk analysis, this evolutionary process requires the participation of multiple disciplines guided by the imperative need to study, understand, and discover the shared states and other essential entities that represent the inherent I-I within and among the systems and subsystems that make up Complex SoS (Haimes, 2018).

In his book *Ageless Body, Timeless Mind*, Deepak Chopra (1994) suggests the following three "models" of human beings: *physiology*, *mental capacity*, and *spirituality*. Here, each aspect is by itself a SoS composed of multiple subsystems. The question is: Can we represent a complete understanding of a person by ignoring one of Chopra's three views? The same principle of representation must apply to the natural and constructed environment, organizations, and other Complex SoS. From several perspectives, *Complex SoS are opaque*; consequently, their risk analysis is markedly more intricate and requires a paradigm shift from the risk analysis currently practiced for a single system. Furthermore, the solicited and collected information should not lead to the illusion that what we have observed and learned constitute the entire reality of the Complex SoS under consideration. Rather, we should augment what we think we know with an endless learn-as-you-go and iterative risk modeling process.

4.9 PHANTOM SYSTEMS MODELS (PSM) AND METAMODELS

4.9.1 The Phantom System Models: Philosophical and Conceptual Foundations

According to *Webster's New International Dictionary*, a phantom is "something that is apparent to the sight or other senses but has no actual substantial existence; something elusive or visionary." The Phantom System Model (PSM) enables research teams to effectively analyze major forced changes in the characteristics and performance of multiscale Complex SoS, such as cyber–physical infrastructures or major socioeconomic SoS. The PSM expands on HHM (see Chapter 3) and on various analytical modeling methods and simulation tools, to present comprehensive views and perspectives on unknowable emergent systems (Haimes, 2007, 2012). Models, experiments, and simulations are conceived and built to answer specific questions. Conventional system models attempt to provide answers based on the responses on the states of a system under given conditions and assumptions. For example, the Leontief Input–Output Economic Model (Leontief 1951a, b, 1986), presented in Chapter 8, enables analysts to ask: What are the interrelationships between production and consumption among the interdependent sectors of the economy? For emergent Complex SoS, analysts may ask an entirely different type of question through the PSM: What kind of a multiscale system and its influencing environment may emerge in the future, and where today's known relationship between production and consumption may or may not hold or be applicable? Answering such questions mandates seeking the "truth" about the unknowable nature of Complex SoS; it requires intellectually bias-free modelers and thinkers, who are empowered to experiment with a multitude of modeling and simulation approaches and to collaborate for appropriate solutions. PSM users will be expected to build on the knowledge generated through the diverse models employed and on the contributions made by analysts of diverse disciplines and expertise.

An artist's first painting is usually not a masterpiece. To achieve this, the artist must usually select themes and explore them to develop relevant knowledge and understanding. The final product can then be carefully designed based upon what is learned through experience. The PSM is a modeling paradigm that is congruent with, and responsive to, the uncertain and ever-evolving world of emergent Complex SoS. In this sense, it serves as an adaptive process, a learn-as-you-go modeling laboratory, where different scenarios of needs and developments for emergent systems can be explored and tested. In other words, to represent and understand the uncertain and imaginary evolution of a future emergent Complex SoS, we need to deploy an appropriate modeling technology that is equally agile and adaptive. One may view the PSM as a matching methodology and technology, where emergent Complex SoS are studied as a panorama of interdependent and interconnected Complex SoS, similar to the way other appropriate models are constructed for systems with different characteristics. Juxtaposition examples are difference equations and differential equations for dynamic systems, algebraic equations for static systems, and probabilities for systems that are driven by random events and processes. The PSM can be continually manipulated and reconfigured in our attempt to answer difficult emergent questions and challenges. For example, the construction/development of different models commonly leads to the integration of databases, models, and simulation tools that have been developed across many disciplines over the last several decades, as an integrated tool can be used to study the impacts of various extreme event scenarios on Complex SoS. At the basic modeling level, there are databases of buildings, businesses, essential facilities, and other basic structural and regional facts that characterize the state of the region under study. These databases are commonly editable to enable exploring agile properties of structures that may change, for example the impact of hurricanes. Scientific models from decades of structural research estimate probabilistic structural damage from wind gusts striking various structural vulnerabilities. Furthermore, PSM can also be viewed as a methodological process for developing tools that will have the flexibility to capture emergent behavior of both regional vulnerabilities and the threats. Outcomes from a PSM may result in a way with which to trace changes in problem definitions, critical variables, critical metrics, available data, among others; and, thus, enable us to measure learning, changing, and improvement in risk management activities over time.

Another example application resulting from PSM might integrate databases of students, training programs, and part-time jobs with probabilistic learning models and simulations of part-time job growth and student success. Such a tool would enable supporting, or rejecting proposals that adolescents could fill a stronger role in skilled labor through vocational training and part-time work during high school while simultaneously preparing for college. PSM can provide a formal framework in which such ideas can be imagined and then realized through modeling and simulation suites that act as large-scale experimental laboratories and, thus, add knowledge of the systems they are discovering. The results of such activities can simultaneously support effective systems modeling and resource allocation for risk management. In other words, *PSM can also serve as the process by which identified emergent risk scenarios can guide the creation of Complex SoS modeling and simulation suites to cost effectively explore and rationalize preparedness against a host of emergent threats that might be unpredictable.*

4.9.2 Philosophical and Conceptual Foundations

Architects, painters, and music composers share similar challenges with analysts who are involved in the art and science of systems modeling. The similarities are manifested in a seemingly endless process of discovery and creativity and in continuous learning through experimentation, measurement, assessment, and trial and error. Creative artists invariably start with a visionary theme through which they deliver one or multiple messages. Through their creative artistic talent and capability and by intuitive inquiries and exploration of a variety of motifs, artists and composers strive to express their visionary themes by answering imaginary or invisible questions (at least to the layperson).

Artists, as the quintessential modelers, represent through their artwork the influence of the culture and social environment within which they live. In an analogous way, systems modelers attempt to represent the multiple perspectives and facets of the system, or Complex SoS under study, so that they may gain a better understanding of the composition of its inherent intra- and I-I, and thus be able to answer specific questions relevant to the Complex SoS under consideration. Thus, both artists and system modelers assume a similar creative, systemic, and challenging task of representation. Finally, not dissimilar to an artistic composition, models ought to be as simple as possible but as complex as required – resulting in a model that offers an acceptable representation of the system and is capable of providing answers and clarifications to the important questions that the model was designed to address. The Human Genome Project may be considered an audacious Complex SoS, fraught with uncertainties and involving participants from multiple disciplines with varied perspectives, experience, skills, and backgrounds.

It is not unrealistic to compare the evolving process of the PSM to the "modeling" experience of children at play. They experiment and explore their uncorrupted imaginative emergent world with Play-Doh® and LEGO® while patiently embracing construction and reconstruction in an endless trial-and-error process with great enjoyment and some successes. The innovation, imagination, and initiatives of modelers experimenting with the PSM on Complex SoS can be instrumental in creating a learning process that can benefit decision makers.

Modeling unprecedented emergent and Complex SoS (e.g. a new national electric power grid SoS, a new and safe generation of cars fueled by hydrogen, or a human space mission to Mars and back), which are inherently elusive and visionary, as well as modeling existing Complex SoS – initially involve phantom entities of multiple perspectives. This modeling effort is driven and constrained by a mix of evolving future needs and available resources, technology, EFCs, and myriad other unforeseen events.

Consider the tradeoffs between the (i) relatively low-intensive modeling of Complex SoS and (ii) inherently invaluable, often unrecognized, and unappreciated efficacy that such modeling generates, including conception, development, construction, and planning for operation of a new generation of physical infrastructures (e.g. water and sewers, electric power grids, transportation systems, communications, public support buildings, etc.). Indeed, the cost associated with bringing to life to each of the above infrastructure Complex SoS could be in the billions of dollars, while the associated modeling cost would be in the millions of dollars. Thus, a ratio of three orders of magnitude ought to encourage and justify the essential investments in modeling.

Models enable us to experiment and test hypotheses, different design options, or to generate responses to, or impacts on, varied policy options. Inversely, by their nature, Complex SoS constitute, in many respects, black holes to modelers that can be penetrated only by acknowledging our inability to directly uncover, understand, or predict their behaviors under different scenarios of disturbances (inputs). We commonly lack sufficient knowledge to assess the causal relationships among the subsystems and to compensate for this shortfall; we revert to multiperspective experimentation aided by the ingenuity, creativity, and domain knowledge of experts, supported by the availability of measurements and databases. There is no assurance that modelers would be able to explain the reasons behind any variability among submodels that comprise Complex SoS; nevertheless, the very inquisitive and learn-as-you-go-process of modeling such variability may highlight limited databases, inconsistent assumptions, unrecognized epistemic and aleatory uncertainties, and a host of other technical or perceptual reasons that ought not to be dismissed. For example, in a closed-loop process control of a system in operation, the automatic controller adjusts the parameters of the system in response to internal or external disturbances or initiating events. In contrast, in response to the initiating events, the adjustment of the parameters in an open-loop process is made by the system's operator or engineer.

In metamodel coordination and integration of the multiple submodels (to be discussed subsequently), the task is exceedingly more complicated, because the modeler assumes the roles of both the closed-loop controller and the open-loop controller. More specifically, the modeler at the metamodeling level makes extensive use of the knowledge generated through lessons learned from the (i) subsystems' coordination, (ii) interdependencies within and among the states of the subsystems, (iii) innovation and creativity in model experimentation, and (iv) intrinsic overlapping and mutual characteristics, functionality, objectives, and states that combine to make all the subsystems a SoS.

4.9.3 Metamodel Coordination and Integration

The essence of metamodel coordination and integration is to build on all relevant direct and indirect sources of information to gain insight into the I-I among the submodels and, on the basis of this insight, to develop representative models of the Complex SoS under consideration. The coordination and integration of the multiple models can be achieved at the metamodeling phase within the PSM, thereby yielding a better understanding of the system as a whole. More specifically, modeling the interdependencies and interconnections within and among the subsystems of Complex SoS requires an understanding of the intricate relationships that characterize the dynamics within and among common/shared states and other essential entities of the subsystems. This very important task is achieved at the metamodeling level of the PSM by observing, estimating, and assessing the outputs for given inputs and by building on the intrinsic shared/common states and other essential entities within and among the subsystems. Note that although the intrinsic common states constitute a key element of the PSM, the extrinsic (input–output) relationships are also very important and support the intrinsic one. Furthermore, for Complex SoS the intrinsic and extrinsic input/output relationships overlap, and the distinction between the two becomes elusive, namely, phantom. Indeed, the selection of the trial inputs to the model and the inquisitive process of making sense of the corresponding outputs are at the heart of system identification and parameter estimation. This is not a one-shot process; rather, it can be best characterized by tireless experimentation, trial and error, parameter estimation and adjustments, and questioning whether the assumed model's topology is representative of the systems and the entire Complex SoS that are being modeled.

The PSM-based intrinsic metamodeling of Complex SoS stems from the basic assumption that some *specific commonalities, interdependencies, interconnectedness, or other relationships must exist between and among any two systems within any system of systems.* (Shared states and other essential entities constitute one such commonality.) More specifically:

1. A single SoS connotes a specific group of subsystems. A subsystem will denote any member system of the Complex SoS. A model of a subsystem will be denoted as a submodel.
2. A metamodel represents the overall coordinated and integrated submodels of the SoS. We define a metamodel as a family of submodels, each representing specific aspects of the subsystem for the purpose of gaining knowledge and understanding of the multiple interdependencies among the submodels and thus allowing us to comprehend the SoS as a whole.
3. *The essence of each subsystem can be represented by a finite number of essential state variables.* (The term essence of a system connotes the quintessence of the system, the heart of the system, that is, everything critical about the system.) Given that a system may have a large number of state variables, the term *essential states (as distinctive from other essential entities) of a system* connotes the minimal number of state variables in a model with which to represent the system in a manner that permits the questions at hand to be effectively answered. Thus, these essential states become fundamental for an acceptable model representation.
4. For a properly defined Complex SoS, any interconnected subsystem will have at least one (typically more) essential state variable(s) and objective(s) and shared states and other essential entities with at least one other subsystem. This requirement constitutes a necessary and sufficient condition for modeling interdependencies among the subsystems (and thus interdependencies across Complex SoS). This ensures an overlapping of states and other essential entities within the subsystems. Of course, the more we can identify and model joint (overlapping) states and other essential entities among the subsystems, the greater will be the representativeness of the submodels and of the metamodel of the SoS.
5. The importance of the availability of multiple, albeit overlapping, databases can be effectively utilized by multiple submodels, each of which is built to answer the specific questions for which it is built. Furthermore, each submodel's characterization, whether modeled separately or in groups, is likely to share common states and other essential entities – a fact that facilitates the ultimate coordination and integration of the multiple submodels at the metamodeling level. Thus, a representative common database that supports the family of SoS must be available.
6. The fusion of multiple submodels via the intrinsic metamodeling coordination and integration enhances our understanding of the inherent behavior and interdependencies of existing and emergent Complex SoS.

4.9.4 PSM-Based Modeling of a Prototype System of Systems: An Example

This section (i) focuses on saltwater intrusion into groundwater systems as consequences of sea-level rise due to climate change and (ii) explores and highlights some concepts associated with modeling a real SoS with PSM. Figure 4.3 depicts the commonly used extrinsic nonreliance on state variables in systems integration (by using inputs from submodels as inputs to others).

In contrast to Figure 4.3, Figure 4.4 depicts the intrinsic reliance on shared and unshared state variables for metamodeling coordination and integration.

Fresh water has been and continues to be a scarce resource in many countries, and groundwater plays a major role in the overall water supply of the United

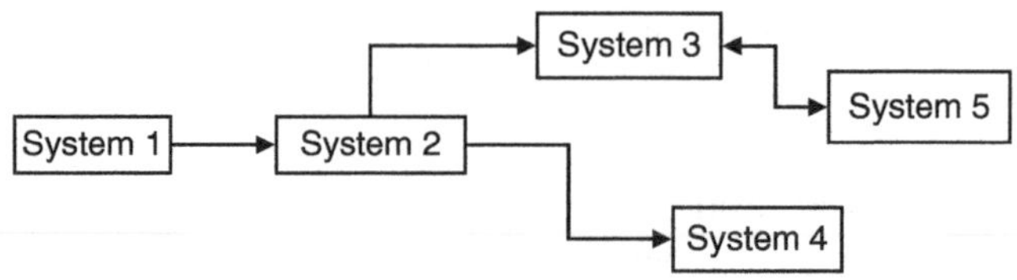

Figure 4.3 *Extrinsic input–output submodel coordination and integration.*

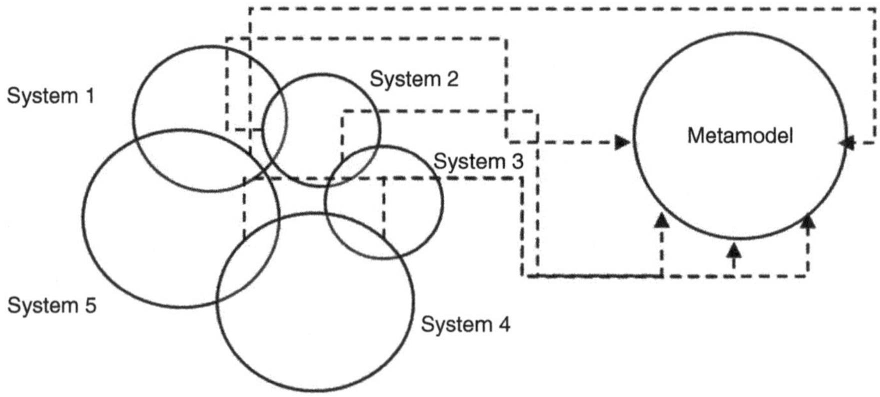

Figure 4.4 *Intrinsic submodel, coordination, and integration via system state variables.*

States and around the world. Many models predict a significant sea-level rise due to climate change, which would cause saltwater intrusion into coastal groundwater aquifer systems. We consider three subsystem models: (i) hydrologic, (ii) agricultural–social, and (iii) regional economic models, where the only inputs are provided from external climatological models. The role of the metamodel, which is composed of the above submodels, is to explore and learn about the I-I among the submodels and to evaluate the information necessary to assess the ultimate multiple impacts of the rise of groundwater salinity (due to the expected sea-level rise resulting from climate change) on crop yield and variety, drinking water quality, farmers' economic well-being, and the regional economy. We envision the following multiple models:

(i) *Hydrologic modeling* effort can focus on a representative set of scenarios of climate change and sea-level rise and can address important questions on the resulting consequences of seawater intrusion into the groundwater system. (ii) *Agricultural–social model* can focus on the impacts of increased groundwater salinity (a) on agriculture, affecting the quality and yield of crops where agriculture is heavily dependent on groundwater and (b) domestic water supply. (iii) *Regional economic model* can focus on the regional economic impacts of the above on the agricultural and domestic use of groundwater. (iv) *Metamodeling* of the groundwater system serves as the coordinator and integrator of the multiple models, building on the shared and unshared state variables.

Let $c(t)$ represent an initiating event of climatological input that impacts seawater level and temperature; let $s_1(t)$ represent seawater level at time t; and let $s_2(t)$ represent the temperature at time t. Note the common and uncommon state variables in the following functional relationships:

1. Groundwater salinity level $s_3(t) = s_3(t, c(t), s_1(t), s_2(t))$.
2. Groundwater yield $s_4(t) = s_4(t, c(t), s_1(t), s_2(t))$.
3. Crop quality and variety $s_5(t) = s_5(t, c(t), s_3(t), s_4(t))$.
4. Income to farmers $s_6(t) = s_6(t, c(t), s_5(t))$.
5. Regional viability of farms $s_7(t) = s_7(t, c(t), s_5(t), s_6(t))$.

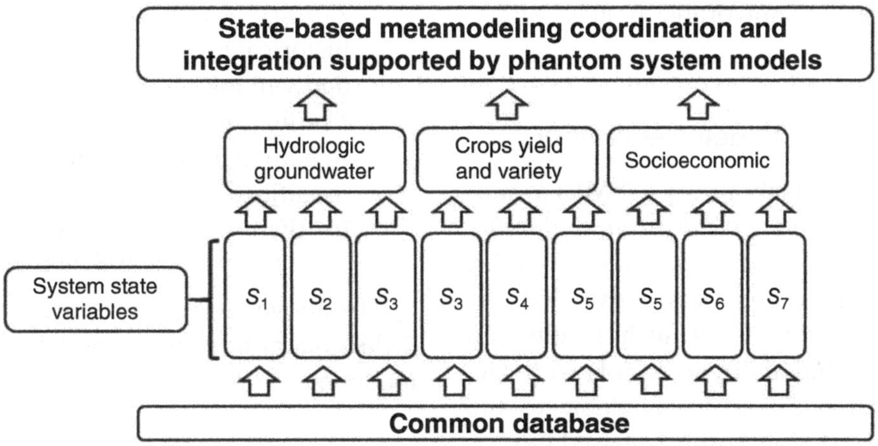

Figure 4.5 *PSM-based metasystem intrinsic coordination via the shared and nonshared state variables of the system.*

Figure 4.5 depicts the PSM-based metasystem intrinsic coordination via the shared and nonshared states and other essential entities. The knowledge and information provided by the state variables (s_1, s_2, …, s_5) enable modelers to learn and better understand the I-I among the different submodels. For instance, the following set of intersections of state variables $s_1 \cap s_2$, $s_2 \cap s_3$, $s_3 \cap s_4$, $s_4 \cap s_4$, $s_5 \cap s_1$ can help modelers to identify causal relationships among the multiple perspectives of the groundwater system.

The effectiveness of the PSM-based metamodel intrinsic coordination and integration is grounded on the:

1. Number of common shared states and other essential entities within and subsystems' models (A minimum of one shared state and one other essential entity is required; otherwise, modelers can reasonably assume that a subsystem is completely independent of the other subsystems).
2. Domain knowledge of each of the subsystems' perspectives to ensure proper and effective modeling of the corresponding subsystems.
3. Appropriate modeling efforts, skills, and expertise invested in modeling each subsystem, including, most importantly, the skill and ability of modelers to learn through infused knowledge, information, and learning generated via the metamodel at higher-level model coordination and knowledge integration.
4. Appropriate modeling methodologies and tools (e.g. analytical, simulation) devoted in modeling each subsystem (which entails the proper selection of model topology/structure and parameters and the incorporation [through the states of the system of systems] of the intra- and interdependencies within and among the subsystems).
5. Availability of proper databases with which to calibrate, test, validate, and verify the model of each subsystem (submodel) under varied conditions.
6. Availability of an appropriate computing laboratory that supports all of the above modeling efforts.

The shared states and other essential entities provide a powerful mechanism for understanding and exploiting the strong interdependencies among the subsystems of Complex SoS. The efficacy of shared states *and other essential entities* among subsystems may be manifested through (i) sharing databases, (ii) realizing that decisions made by the stakeholders of subsystem I can have direct impact on subsystem II, and (iii) encouraging and enticing stakeholders of different subsystems to collaborate on inputs, constraints, and decisions that affect the shared states for win–win outcomes. On the other hand, understanding the potential adverse organizationally induced consequences resulting from unshared states, due to competitiveness among subsystems; and thus, by exploiting unshared states could, for example, (i) defuse potential conflicts among the subsystems and (ii) generate collaborations in the face of joint budgetary constraints or unfavorable policies affecting the subsystems.

4.9.5 Phantom System Models Laboratory

There is a need for a PSM laboratory (PSML) *to support, coordinate, and integrate results from a plurality of computer-based analytical (and simulation models)*, each providing a unique system perspective, with the outlook that the combination of such results can improve our learning and ability to gain knowledge. A PSML configuration can make available to the group of modelers: (i) desired software-based models for a particular analysis, (ii) an array of data sources to support the desired modeling activities, (iii) tools for organizing the components of the modeling system so as to achieve the desired model relationships, and (iv) support for the intrinsic metamodeling coordination and integration. To perform these functions, the PSML ought to provide a structure based upon service-oriented architecture that will enable the user to perform desired modeling efforts. Services may include:

1. *Modeling services* that determine which models will be executed.
2. *Data services* that include organizing the needed data inputs, collecting the desired data outputs, and performing needed data conversions.
3. *Analysis services* that provide postmodeling analysis that compares results from different models and that assesses the sensitivity of results derived from the varying analyses performed by a given model.
4. *Data presentation and visualization services* that include composing the analytical results to aid the modeler and decision makers in interpreting results and presenting the results for different presentation media (print, small/large screen display, etc.).
5. *Workflow services* that determine the sequencing of services that might be executed (e.g. determining which models can be run in parallel and which must be run sequentially) and providing the necessary data to a hardware organizing service to allow proper physical configuration to support the desired modeling effort.
6. *Library services* to provide model descriptions, historical model results, model software designs, and varying model configuration descriptions.
7. *Configuration management and control services* to help manage new model development, model modifications, and existing model integration for the overall PSML system.

The above constitute representative services that a PSML ought to provide modelers in support of their modeling efforts.

4.9.6 Coordinated Hierarchical Bayesian Model (CHBM)

The reliance of direct and indirect information and database is common in system's modeling with sparse database and when empirical data are usually either sparse or lacking, in particular in risk of extreme events (Yan, 2007; Yan and Haimes, 2010). Note that Chapter 6 is devoted in its entirety to the CHBM. Furthermore, with sparse data, important model parameters may not be estimated and tested within an acceptable level of significance. When a large database is available, standard statistical techniques can be applied to estimate the parameters and create a fairly accurate and well-parameterized model. Researchers and practitioners in systems engineering and risk analysis are commonly plagued by the data scarceness problem, which can be prevalent in modeling Complex SoS. On the other hand, it is well known that when estimating the parameters of a model by traditional statistical methods using relatively small datasets, those methods generate "unstable" results with large estimation variance. Consequently, important model parameters cannot be estimated and tested within an acceptable level of significance. In this section we adopt an alternative approach to address this problem at the metamodeling level, through borrowing strength from indirect but relevant data from one subsystem and applying it to another. *Strength-borrowing methods aim to borrow strength from indirect data to compensate for the sparseness of direct data.* Subjective methods include expert evidence solicitation and Bayesian analysis; the latter provides a natural way to combine expert evidence with limited direct data.

We decompose the term "data" into three parts: (i) *Direct data* may represent testing, experimentation, measurements, and observations from a system (or a subsystem) with unknown parameters. (ii) *Indirect data* represent observations from different but related (or similar) subsystems. (iii) *Expert evidence* is information received by soliciting evidence from one or multiple experts. Hierarchical Bayesian models (HBMs), which have been applied in the reliability, risk, and system safety fields, are an objective method suitable for addressing the data sparseness problem (Gelman et al., 2004). Coordinated hierarchical Bayesian models (CHBMs), which borrow strength from indirect data or expert evidence to compensate for the sparseness of direct data (Yan, 2007; Yan and Haimes, 2010), can provide valuable support to the metamodeling process.

The structures of HBM and CHBM are described in Figure 4.6a and b, respectively. In HBM, y_i represents the dataset observed from subsystem i, θ_i represents the parameter for the subsystem i, and η represents the hyperparameter. In CHBM, y_{ij} represents the dataset observed from scenario (i,j), u_i represents the fixed effect of y_{ij} perspective i, α_{ij}

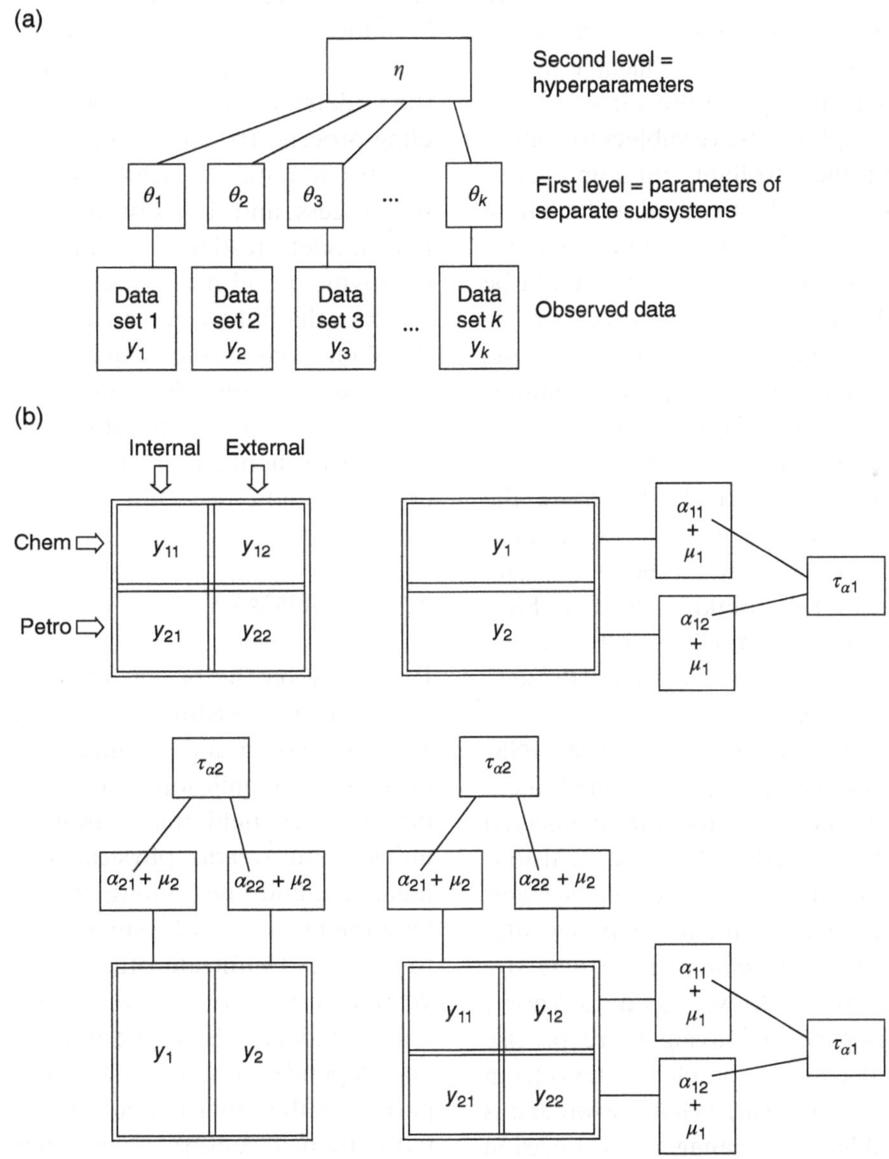

Figure 4.6 *(a) Structure of HBM. (b) Structure of CHBM.*

represents the cross-classified random effects from scenario (i,j), and τai represents the variance of the hyper-distribution of the random effects in τ_i perspective.

As opposed to HBM, where there is only one dimension and a single hierarchy, the CHBM has two cross-hierarchies, each corresponding to one dimension of strength borrowing. Note that the bidimensional model can be easily extended to accommodate multiple dimensional cross-classified random effects as multiple dimensions presenting in a system.

The systems-based approaches presented earlier constitute only a sample of methodologies that support the modeling of Complex SoS through PSM. The challenges associated with modeling SoS necessarily require the reliance on every applicable theory and methodology that can support this effort.

By their nature Complex SoS are subject to continued controllable and uncontrollable random events, and thus they need to be adaptive and agile. These elements of risk and uncertainty necessitate that a systems-based risk assessment and risk management be an integral part of the decision-making process associated with Complex SoS life cycle. Furthermore, from the systems engineering perspective, understanding and effectively responding to and managing these evolving forced changes require an equally agile and flexible multiplicity of models. Furthermore, the PSM philosophy rejects problem solving that relies on a dogmatic modeling approach structured exclusively on a single school of thinking. Rather, PSM attempts to draw on a plural-modeling schema that builds on the multiple perspectives gained through generating multiple models.

The theory and methodology drivers of the PSM enable the construction of appropriate complementary models of interdependent and interconnected Complex SoS, on which to deduce logical conclusions for future actions, whether modeling and systems engineering or in risk management. *Thus, we shift from deciding only what is optimal given what we know to asking*: *(i)* What do we need to know? (ii) What are the impacts of having more precise and updated knowledge about complex systems from a risk-reduction standpoint? (iii) What knowledge is needed for acceptable risk management decision making? Answering these questions mandates seeking the "truth" about the unknowable complex nature of emergent Complex SoS; it requires intellectually bias-free modelers and thinkers who are empowered to experiment with a multitude of modeling and simulation approaches and will collaborate for appropriate solutions.

The PSM has four important functions: (i) identify the shared states and other essential entities within and among the subsystems that make up Complex SoS and that would sufficiently characterize Complex SoS, (ii) enable modelers and analysts to explore cause-and-effect relationships in virtual-to-real laboratory settings, (iii) develop modeling and analysis capability to assess irreversible extreme sources of risks, and (iv) enable modelers, decision makers, stakeholders, and other involved groups who have stakes and interest in Complex SoS to (a) facilitate the understanding of the never-ending intricate modeling process, (b) contribute to and enrich the access to a broad database, which is essential to the modeling process, and (c) assist in and contribute to essential model testing, updating, and its ultimate deployment and usefulness.

Thus, the PSM serves to anticipate and understand the likelihoods and consequences of EFCs around and the associated resulting sources of risks and helps to design and build reconfigured systems that will be sufficiently resilient under EFCs and at acceptable recovery times and costs.

4.10 SUMMARY

Reflecting on the origins of modern systems theory since the introduction of the Gestalt psychology in 1912, we cannot underestimate the intellectual power of the holistic philosophy that has sustained this multidisciplinary field that transcends the arts, humanities, natural and physical sciences, engineering, medicine, and law, among others. The fact that systems engineering and systems analysis have continued to grow and infiltrate other fields of study over the years can be attributed to the fundamental premise that Complex SoS can be understood only if all the interdependencies and interconnections among their parts and the environment are also understood. For more than a century, mathematical models constituted the foundations upon which systems-based theory and methodologies were developed, including their use and deployment on the myriad large-scale projects in the natural and constructed environment. If we were to identify a single concept that has dominated systems thinking and modeling, it would be the state-space theory. Indeed, the centrality of state variables in this context is so dominant that no meaningful mathematical model of a real Complex SoS can be built without identifying the critical states of that system and relating all other building blocks of the model to them (including decision, random and exogenous variables, and inputs and outputs). In this respect, systems modeling has served, in many ways, as the medium with which to infuse and instill the holistic systems philosophy into the practice of engineering and other fields.

In an introduction to the history of European art, William Kloss (2005) writes: "We will place these artists and their masterpieces in the political, religious, and social context of their time, so that we have a profound understanding of both why an artwork was created and how it responded to a particular set of historical circumstances." The creative work of system's modelers is not dissimilar from that of artists. Just as no single model is capable of representing the multiple perspectives of Complex SoS, whether in harmony or in juxtaposition; no symphony by Beethoven could have been composed using one instrument, or one motif, or one theme. The similarities among artists and analysts involved in the art and science of systems modeling are manifested in a seemingly endless process of discovery and creativity and in a continuous process of learning through experimentation and trial and error. Thus, similar to the challenge facing the composer of a symphony and ultimately the conductor of the orchestra performing the piece, the challenge facing modelers is one of amalgamating and assembling the multiple system models to yield the development of a harmonious multimodel representation of the multiperspective Complex SoS.

It is not unrealistic to compare the evolving process of the PSM to the "modeling" experience of children at play. In this sense the PSM is a real-to-virtual laboratory for experimentation, a learn-as-you-go facility, and a process for existing and emergent systems that are not yet completely designed, developed, or understood. The innovation, imagination, and initiatives of modelers experimenting with the PSM on Complex SoS can be instrumental in the creation of a learning process that can ultimately benefit decision makers.

In sum, this chapter advances the following premises:

1. The emergent and dynamic nature of Complex SoS necessarily renders their models to be ephemeral and visionary, building on the intrinsic relationships among the states of the subsystems.
2. This modeling process benefits from a well-designed and well-executed learn-as-you-go process.
3. System models are likely to build on hierarchical and overlapping structures.
4. Since the vulnerability and resilience of a system are manifestations of the states of the system, then the above points have important implications for systems engineering, particularly for identifying sources of risk and understanding system vulnerability and resilience.
5. HBMs and CHBMs, which borrow strength from indirect data or expert evidence to compensate for the sparseness of direct data, can provide valuable support to the metamodeling process.
6. Other systems engineering methods, such as hierarchical decomposition and higher-level coordination, influence diagrams, and others, can be instrumental in the metamodeling process.
7. Building on the intrinsic interplay among the shared and unshared state variables among the subsystems, the philosophy and theory of the PSM provide a modeling paradigm that complements and supplements the commonly used extrinsic (input–output-based) modeling approach.

REFERENCES

Adelman, L. (1991). *Evaluating Decision Support and Expert Systems*. New York: Wiley.

Amaral, L. and Ottino, J.M. (2004). Complex networks: augmenting the framework for the study of complex systems. *The European Physical Journal B* **38**(2): 147–162.

Apostolakis, G. (1999). The distinction between aleatory and epistemic uncertainties is important: an example from the inclusion of aging effects into probabilistic safety assessment. *Proceedings of PSA'99* (22–25 August 1999). Washington, DC: American Nuclear Society.

Bar-Yam, Y. (2003a). When systems engineering fails: toward complex systems engineering. *International Conference on Systems, Man & Cybernetics* **2**: 2021–2028.

Bar-Yam, Y. (2003b). Complexity of military conflict: Multiscale complex systems analysis of littoral warfare. Chief of Naval Operations, *Report for Contract: F30602–02-C-0158* Multiscale representations phase II: Task 2: Multiscale analysis of littoral warfare.

Bertalanffy, L.V. (1968). *General Systems Theory: Foundations, Development, Applications*. New York: George Braziller, Inc.

Blanchard, B.S. (2003). *Systems Engineering Management*, 3e. Hoboken, NJ: Wiley.

Blanchard, B.S. and Fabrycky, W.J. (2005). *Systems Engineering and Analysis*, 4e. Englewood Cliffs, NJ: Prentice-Hall.

Blauberg, I.V., Sadovsky, V.N., and Yudin, E.G. (1977). *Systems Theory: Philosophical and Methodological Problems*. New York: Progress Publishers.

Boulding, K.E. (1953). *The Organizational Revolution*. New York: Harper & Row.

Buckley, W. ed. (1968). *Modern Systems Research for the Behavioral Scientist: A Sourcebook*. Berlin: Walter De Gruyter, Inc.

Buede, D.M. (1999). *The Engineering Design of Systems: Models and Methods*. New York: Wiley.

Chang, M. and Harrington, J. (2005). *Handbook of Computational Economics II: Agent-Based Models of Organizations*. Amsterdam: Elsevier.

Chen, C. (2012). *Linear System Theory and Design*, 4e. New York: Oxford University Press.

Chopra, D. (1994). *Ageless Body, Timeless Mind: The Quantum Alternative to Growing Old*. New York: Random House.

Cohen, B. and Whitman, A. (trans.). (1999). *The Principia: The Authoritative Translation and Guide Mathematical Principles of Natural Philosophy I*. Oakland, CA: University of California Press.

Eusgeld, I., Nan, C., and Dietz, S. (2011). System-of-systems approach for interdependent critical infrastructures. *Reliability Engineering & System Safety* **96**(6): 679–686.

Eykhoff, P. (1974). *System Identification: Parameter and State Estimation*. New York: Wiley.

Fang, L., Hipel, K.W., and Kilgour, D.M. (1993). *Interactive Decision Making: The Graph Model for Conflict Resolution*. New York: Wiley.

Feynman, R.P., Leighton, R.B., and Sands, M. (1963). *The Feynman Lectures on Physics*. Reading, MA: Addison-Wesley.

Funderburk, F. (2004). Organizational culture from a complex dynamic systems perspective: Moving from metaphor to action in healthcare. System Models of Organizational Behavior. www.necsi.edu.

Gelman, A., Carlin, J., Stern, H., and Rubin, D. (2004). *Bayesian Data Analysis*. New York: Chapman & Hall.

Gharajedaghi, J. (2005). *Systems Thinking, 2nd Edition: Managing Chaos and Complexity: A Platform for Designing Business Architecture*. Boston, MA: Butterworth-Heinemann.

Gheorghe, A.V. (1982). *Applied Systems Engineering*. New York: Wiley.

Gordon, W.J.J. (1968). *Synectics: The Development of Creative Capacity*. New York: Collier Books.

Graupe, D. (1972). *Identification of Systems*. New York: Van Nostrand Reinhold Company.

Guo, Z. and Haimes, Y.Y. (2016). Risk assessment of infrastructure system of systems with precursor analysis. *Risk Analysis* **36**(8): 1630–1643. doi: 10.1111/risa.12559.

Guo, Z. and Haimes, Y.Y. (2017). Exploring systemic risks in systems-of-systems within a multiobjective decision framework. *IEEE-SMC Systems* **47**(6): 906–915.

Haberlin, R. and Haimes, Y.Y. (2018). Regional infrastructures as Complex Systems of Systems: A Shared State Model for Regional Resilience. Submitted to *ASCE Journal of Infrastructure Systems* **24**(3).

Haimes, Y.Y. (1970). The Integration of System Identification and System Optimization. *Report UCLA-ENG-7029*, University of California, Los Angeles.

Haimes, Y.Y. (1977). *Hierarchical Analyses of Water Resources Systems: Modeling and Optimization of Large-Scale Systems*. New York: McGraw-Hill.

Haimes, Y.Y. (1981). Hierarchical holographic modeling. *IEEE Transactions on Systems, Man, and Cybernetics* **11**(9): 606–617.

Haimes, Y.Y. (2006). On the definition of vulnerabilities in measuring risk to infrastructures. *Risk Analysis* **26**(2): 293–296.

Haimes, Y.Y. (2007). Phantom system models for emergent multiscale systems. *ASCE Journal of Infrastructure Systems* **13**(2): 81–87.

Haimes, Y.Y. (2008). Models for risk management of systems of systems. *International Journal of Systems of Systems Engineering* **1**(1/2): 222–236.

Haimes, Y.Y. (2009). On the complex definition of risk: a systems-based approach. *Risk Analysis* **29**(12): 1647–1654.

Haimes, Y.Y. (2012). Modeling complex systems of systems with phantom system models. *Systems Engineering* **15**(3): 333–346.

Haimes, Y.Y. (2016). *Risk Modeling, Assessment, and Management*, 4e. Hoboken, NJ: Wiley.

Haimes, Y.Y. (2018). Risk modeling of interdependent complex systems of systems: theory and practice. *Risk Analysis* **38**: 84–98. doi: 10.1111/risa.12804.

Haimes, Y.Y. and Macko, D. (1973). Hierarchical structures in water resources systems management. *IEEE Transactions on Systems, Man, and Cybernetics, SMC* **3**(4): 396–402.

Haimes, Y.Y., Horowitz, B., Guo, Z. et al. (2015). Assessing systemic risk to cloud-computing technology as complex interconnected systems of systems. *Systems Engineering* **18**(3): 284–299.

Hall, A.D. III (1989). *Metasystems Methodology: A New Synthesis and Unification*. Elmsford, NY: Pergamon Press.

Hatley, D.J., Hruschka, P., and Pirbhai, I.A. (2000). *Process for System Architecture and Requirements Engineering*. New York: Dorset House Publishing Company, Inc.

Hegel, G. (1952). *Great Books of the Western World*. Chicago, IL: Encyclopedia Britannica, Inc. (No. 46).

Jamshidi, M. ed. (2009a). *System of Systems Engineering: Innovations for the 21st Century*. Hoboken, NJ: Wiley.

Jamshidi, M. (2009b). *System of Systems Engineering: Principle and Applications*. New York: CRC Press, Taylor & Francis Group.

Johnson, C. (2006). What are emergent properties and how do they affect the engineering of complex systems? *Reliability Engineering and System Safety* **91**(12): 1475–1481.

Kloss, W. (2005). *History of European Art [Lecture Series for the Great Courses]*. Chantilly, VA: The Teaching Company.

Kossiakoff, A. and Sweet, W.N. (2002). *Systems Engineering Principles and Practice*. New York: Wiley.

Leondes, C.T. ed. (1969). *Advances in Control Systems: Vol. 6*. New York: Academic Press.

Leontief, W.W. (1951a). Input/output economics. *Scientific American* **185**(4): 15–21.

Leontief, W.W. (1951b). *The Structure of the American Economy 1919–1939*, 2e. New York: Oxford University Press.

Leontief, W.W. (1986). *Input–Output Economics*, 2e. New York: Oxford University Press.

Lloyd, S. and Lloyd, T. (2003). *Bits and Bucks: Modeling Complex Systems by Information Flow*. Cambridge, MA: MIT Engineering Systems Division Working Paper Series ESD-WP-2003-2001.17.

Lowrance, W.W. (1976). *Of Acceptable Risk*. Los Altos, CA: William Kaufmann.

Macko, D. (1967). General System Theory Approach to Multilevel Systems. *Report SRC 106-A-67-44*, Cleveland, OH: Systems Research Center: Case Western Reserve University.

Macko, D. and Haimes, Y.Y. (1978). Overlapping coordination of hierarchical structures. *IEEE Transactions on Systems, Man, and Cybernetics* **8**(10): 745–751.

Maier, M.W. (1998). Architecting principle for systems-of-systems. *Systems Engineering* **1**(4): 267–284.

Maier, M.W. and Rechtin, E. (2009). *The Art of Systems Architecting*, 3e. Boca Raton, FL: CRC Press.

Mesarović, M.D. (1962). General systems theory and systems research contrasting conceptions of systems science. *Proceedings from the Second System Symposium*, Cleveland, OH (February 1964). Case Institute of Technology.

Mesarović, M.D. ed. (1968). *Systems Theory and Biology*. New York: Springer-Verlag.

Mesarović, M.D., Macko, D., and Takahara, Y. (1970). *Theory of Hierarchical, Multilevel Systems*. New York: Academic Press.

Nise, N. (2015). *Control Systems Engineering*, 7e. Pomona, CA: California State Polytechnic University.

Ottino, J.M. (2003). Complex systems. *AIChE Journal* **49**(2): 292–299.

Page, S.E. (1999). Computational models from a to Z. *Complexity* **5**(1): 35–41.

Paté-Cornell, M.E. (1990). Organizational aspects of engineering system safety: the case of offshore platforms. *Science* **250**: 1210–1217.

Paté-Cornell, M.E. (1996). Uncertainties in risk analysis: six levels of treatment. *Reliability Engineering and System Safety* **54**(2/3): 95–111.

Rasmussen, J., Pejtersen, A.M., and Goodstein, L.P. (1994). *Cognitive Systems Engineering*. New York: Wiley.

Rouse, W.B. (1991). *Design for Success: A Human-Centered Approach to Designing Successful Products and Systems*. New York: Wiley.

Sage, A.P. (1977). *Methodology for Large Scale Systems*. New York: McGraw-Hill.

Sage, A.P. (1992). *Systems Engineering*. New York: Wiley.

Sage, A.P. (1995). *Systems Management for Information Technology and Software Engineering*. New York: Wiley.

Sage, A.P. and Armstrong, J. (2003). *Introduction to Systems Engineering*. Hoboken, NJ: Wiley.

Sage, A.P. and Biemer, S.M. (2007). Processes for system family architecting, design and integration. *IEEE Systems Journal* **1**(1): 5–16.

Sage, A.P. and Cuppan, C.D. (2001). On the systems engineering and management of systems of systems and federation of systems. *Information, Knowledge, Systems Management* **2**(4): 325–345.

Sage, A.P. and Rouse, W.B. ed. (1999). *Handbook on Systems Engineering and Management*, 2e. New York: Wiley.

Shalizi, C. (2006). Methods and Techniques of Complex Systems Science: An Overview 96 pp. arxiv.org/pdf/nlin/0307015.pdf.

Slovic, P. (2000). *The Perception of Risk*. Sterling, VA: Earthscan Publications Ltd.

Tivnan, B., Henderson, S.G., Biller, B., Hsieh, M.H., Shortle, J., Tew, J.D., and Barton, R.R. ed. (2007). Modeling organizational adaptation: A replication of Levinthal's model of emergent order. *Proceedings of the 2007 IEEE Winter Simulation Conference*, Washington, DC (9–12 December 2007).

U.S. Federal Aviation Administration (2017). *NextGen Implementation Plan*. Washington DC: FAA. http://www.faa.gov/nextgen/library/media/NextGen_Implementation_Plan_2014.pdf (accessed 6 January 2018).

U.S. Nuclear Regulatory Commission (1981). *Fault Tree Handbook*, NUREG-81/0492. Washington, DC: U.S. Nuclear Regulatory Commission.

Warfield, J.N. (1976). *Societal Systems: Planning, Policy, and Complexity*. New York: Wiley Interscience.

Wiener, N. (1948). *Cybernetics, or Control and Communication in the Animal and the Machine*. Cambridge, MA: The Technology Press.

Wiener, N. (1961). *Cybernetics*, 2e. Cambridge, MA: MIT Press.

World Book Inc. (1980). *The World Book Encyclopedia*, vol. **8**. Chicago, IL: World Book–Childcraft International, Inc.

Yan, Z. (2007). Risk assessment and management of complex systems with hierarchical analysis methodologies. Ph.D. thesis. University of Virginia, Charlottesville, VA.

Yan, Z. and Haimes, Y.Y. (2010). Cross-classified hierarchical Bayesian models for risk-based analysis of complex systems under sparse data. *Reliability Engineering and System Safety* **95**(7): 764–776.

Zeigler, B.P. (1984). *Multifaceted Modeling and Discrete Simulation*. New York: Academic Press.

Zeleny, M. (2005). *Human Systems Management: Integrating Knowledge, Management and Systems*. Hackensack, NJ: World Scientific Publishing Company.

5

Complex Systems of Systems: Multiple Goals and Objectives

5.1 UNIQUENESS OF MULTIPLE GOALS AND OBJECTIVES TO COMPLEX SYSTEMS OF SYSTEMS

Complex systems of systems (Complex SoS), whether they represent cyber–physical subsystems, organizational setups, or other political, structural, social, or societal entities – all are guided and driven by multiple, often competing and conflicting, goals and objectives. These goals and objectives, which are commonly associated with each subsystem and of the Complex SoS as a whole, constitute a hierarchy of subgoals and sub-objectives. Furthermore, they often continue to evolve over time due to internal and external forces that shape the environment within which they operate. By their nature and organizational setups, these goals and objectives are by and large noncommensurate, competing, conflicting, and ever evolving over time due to changing internal and external circumstances. The nature and organizational structures of Complex SoS require an appropriate representative modeling schema that avoids any precommensuration of natural metrics and other measures associated with each objective and sub-objective. Over a half a century, the field of multiple-criteria decision making (MCDM) has dominated most, if not all, decision-making modeling efforts by avoiding the cost–benefit model. The reason for this evolutionary process is obvious: By their nature and definition, each objective represents a different perspective of the system and thus is measured with different units (e.g. in monetary terms, lives, time frames) that can neither be made commensurable logically nor effectively. Furthermore, measuring them with the same units would violate basic principles in measurement theory. Indeed, the multiobjectives associated with Complex SoS are by their nature hierarchical (e.g. objectives, sub-objectives, sub-sub-objectives) and emerging and evolving over time.

Note that, with some exceptions, an optimum *does not exist in a real sense neither for a single system nor for Complex SoS*. Indeed, an "optimum" solution may at best exist for a model; however, for a real-life problem such an "optimum" depends on myriad assumptions and factors, which include the (i) specificity of each subsystem; (ii) extent of shared states and other essential entities within and among the subsystems (We define shared/common "other essential entities" as shared/common decisions, decision makers (DM), stakeholders, resources, organizational behavior and norms, policies and procedures, management, culture,

Modeling and Managing Interdependent Complex Systems of Systems, First Edition. Yacov Y. Haimes.

and others); (iii) identity, perspectives, and biases of the modelers, DM, and stakeholders; (iv) credibility of the database; and (v) time frame, among others. Therefore, a mathematical optimum for a model does not necessarily correspond to the "optimum" for the real subsystems nor for the Complex SoS as a whole. On the other hand, a solution to a multiobjective decision-making optimization problem is termed noninferior, Pareto-optimal, efficient, or nondominated, if improving one objective function can be achieved only at the expense of degrading at least another one (Pareto, 1896). A formal mathematical definition will be introduced in a subsequent section. For a sound and informative decision-making process, it is imperative that the DM also be provided with the tradeoff values associated with the respective objectives.

In general, multiple DM with varied authorities are associated with each subsystem and for Complex SoS. They may represent different constituencies, preferences, and perspectives; may be elected, appointed, or commissioned; may be public servants, professionals, proprietors, laypersons; and so on. Also, they are often associated, or connected, with a specific level of the various hierarchies of objectives within the subsystems and the Complex SoS as a whole.

Resolutions to a multiobjective decision making associated with Complex SoS commonly involve multiple DM and are often reached through negotiation, either through the use of group techniques of MCDM or on an ad hoc basis. Such solutions are often referred to as compromised solutions. Clearly, "non-win–win" compromised solutions, reached among multiple DM, where one or more lose in the voting or negotiation process (even though the rules of the game have not been violated) are likely to backfire and disrupts an unstable harmony that might otherwise exist. DM in a losing group may be influential enough to sabotage a compromised solution and prevent its implementation. Behind the scenes the "horse-trading" paradigm is a reality that must be accepted as part of human and organizational behavior. If a stalemate arises and a compromised solution is not achievable; e.g. if a consensus rule is followed and one or more DM object to a noninferior solution (to be defined subsequently) that is preferred by all others, then the set of objectives could be enlarged, or the scope of the problem may be broadened. Finally, it is imperative that decisions be made on a timely basis, because a "no-decision" stance could be costly not only to certain subsystems but to the Complex SoS as a whole.

Decision support systems (DSS) are interactive computer-based systems that help DM utilize data, mathematical models, and simulation and optimization methodologies to generate alternative policy options and solve both structured and unstructured problems. This process may assume different variations depending on the specific goals, objectives, and missions of each subsystem and of the SoS as a whole. True DSS must be grounded on the same premises as MCDM; namely, there are always multiple conflicting and competing objectives associated with each subsystem that must be recognized and addressed in the modeling process. From a practical standpoint, DSS and MCDM augment each other. In other words, the goals of MCDM and DSS are the same – to improve the complex decision-making process that is associated with Complex SoS. A similar argument can be made about how MCDM and DSS are related to artificial intelligence (AI), which is the study of ideas that enable computers to be intelligent. The fundamental principle underlying AI is the use of information for learning purposes. Thus, for DM, a DSS will be effective if it incorporates multiobjectives and, at the same time, has the capability of self-learning and model updating.

One should take into account the multiplicity of sources of uncertainty associated with the MCDM process and the Complex SoS. Note the two basic classes of uncertainty – epistemic (knowledge uncertainty) and aleatory (variability uncertainty) – as well as errors associated with the (i) database; (ii) modeling effort; (iii) decision-making process; (iv) DM's perception of their values, needs, and preferences; and (v) rules that guide the decision-making process itself. The complexity of the modeling process of SoS and the diversity of tradeoffs associated with each subsystem of the Complex SoS are likely to complicate and add instability to the preferred solution(s). For example, from the perspective of the modeling process, the values of certain exogenous variables may, in reality, deviate from their assumed nominal values. Constructing and adding one or more new sensitivity functions that are minimized along with the other original multiobjective functions could add some of the needed stability to the resulting preferred solution or selected policy.

5.2 THE SURROGATE WORTH TRADEOFF (SWT) METHOD

5.2.1 Formulation of Multiobjective Optimization Problems

To define a noninferior solution mathematically, consider the following multiobjective function problem, also known as a multiobjective optimization problem (MOP):

$$\text{MOP}: \begin{array}{l} \min\limits_{x \in X} \{f_1(\mathbf{x}), f_2(\mathbf{x}), \ldots, f_n(\mathbf{x})\} \\ X = \{\mathbf{x} | g_i(\mathbf{x}) \leq 0,\ i = 1, 2, \ldots, m\} \end{array} \tag{5.1}$$

where $\mathbf{x}$ is an N-dimensional vector of decision variables, X is the set of all feasible solutions, and $g_i(\mathbf{x})$ is the ith constraint.

Definition A decision $\mathbf{x}^*$ is said to be a noninferior solution to the system posed by the multiobjective optimization problem (MOP) (Eq. (5.1)), if and only if there does not exist another $\bar{\mathbf{x}}$ so that $f_j(\bar{\mathbf{x}}) \leq f_j(\mathbf{x}^*)$, $j = 1, 2, \ldots, n$, with strict inequality holding for at least one objective.

Clearly, the solution to the multiobjective problem posed by Eq. (5.1) is not unique, and some kind of subjective judgment by the DM should be added to the quantitative analysis. Although more than one DM may be involved in the selection of an acceptable and preferred solution, and to avoid complexity in notation, a reference to a DM will also denote multiple DM unless it is specified otherwise.

The various available methodologies for solving Eq. (5.1) differ in three major ways: the (i) procedures used to generate noninferior solutions, (ii) timing and the ways and means used to interact with the DM, and (iii) type of information made available to them in the process (such as tradeoffs). The weighting method, also known as the parametric approach, was the most common method used for solving multiobjective problems, where the MOP (5.1) is essentially converted (in the weighting method) into a scalar optimization $p(w)$ as given below:

$$p(w): \begin{array}{l} \min\limits_{x \in X} \sum_{i=1}^{n} w_i f_i(\mathbf{x}) \\ \sum_{i=1}^{n} w_i = 1,\ w_i \geq 0 \end{array} \tag{5.2}$$

A subjective determination of the levels of the weighting coefficients, w_i, is necessary. Subsequently, this parametric approach may yield meaningful results to the DM only when solved (parametrically) many times for different values of w_i, $i = 1, 2, \ldots, n$. The potential existence of a duality gap is an additional important drawback to this method. Note that the subject of duality is presented in the Appendix.

There exist numerous methods for solving multiobjective problems, such as utility functions; indifference functions; the lexicographic, parametric, and ε-constraint approaches; and the surrogate worth tradeoff (SWT) method (Haimes and Hall, 1974) (Chankong and Haimes, 1983, 2008; Haimes, 2016). Other methods include goal programming, the goal attainment method, the adaptive search approach, interactive approaches, and the ELECTRE method. In subsequent sections we will review the SWT method and its extensions.

5.2.2 Multiobjective Tradeoff Analyses

It would be shortsighted to try to generate Pareto-optimal solutions/policies for objectives that are limited to present aspirations or are not responsive to future needs of the subsystems and of the Complex SoS as a whole. The future impacts of current decisions and policies on future options must be accounted for each subsystem. Therefore, impact analysis for a single system and for Complex SoS should be incorporated into the MCDM process to ensure that (i) the attainment of present goals and objectives can be juxtaposed against potential or perceived future goals objectives (e.g. maximizing present profit vs. maximizing future technological and economic competitiveness through an investment in research and development) and (ii) more flexibility may be added to ensure against adverse future irreversible consequences. For example, evaluating the consequences and future flexibility of two preferred noninferior solutions for any number of subsystems could dictate a distinct choice between two seemingly equivalent options. The value and importance of impact analysis are even more critical for multistage interdependent and interconnected Complex SoS, which are characterized by multiobjectives at each stage of the decision-making process. In other words, a tradeoff between the attainment of present objectives and future flexibility should be incorporated within the MCDM process, whether for a single system or for Complex SoS.

The following basic question commonly arises in the modeling and decision-making process associated with Complex SoS: How many, and which, among multiple perspectives ought to be incorporated in both the modeling process and the decision-making process? Of course, this epistemological set of questions cannot be answered singularly. The art and science of modeling and decision making cannot be packaged with simple dogmas. Furthermore, the modeling and decision-making process, including the negotiated tradeoffs among the multiple conflicting and competing objectives that characterize all systems, and especially Complex SoS, defy shortcuts and oversimplicity. We emphasized earlier that invariably single models cannot adequately capture the multifarious nature of interdependent and interconnected Complex SoS, their bewildering variety of resources and capabilities, their multiple noncommensurable objectives, and their diverse users, constituencies, DM, and stakeholders. When concepts from Hierarchical Holographic Modeling (HHM), discussed in Chapter 3, are incorporated into MCDM, the modeling base is broadened, and an opportunity is provided for a modeling and decision-making framework that is more responsive to users, stakeholders, and DM. Approaches that allow this incorporation seem especially worthwhile for group decision-making situations – a common reality associated with interdependent and interconnected Complex SoS.

5.2.3 The Fine Line between an Inferior and a Noninferior Solution

Modelers and systems analysts of a single system or of Complex SoS place great emphasis on generating only noninferior solutions (i.e. discarding inferior solutions). This emphasis, though justifiable, should be moderately balanced by the fact that a noninferior solution to, for example, a three-objective function could become an inferior solution if one of the three objectives is ignored, or its significance is reduced. Similarly, an inferior solution could become noninferior if the number of objectives is increased, which is prevalent in Complex SoS, while making no changes in the meaning or definition of any objective. This observation is further supported by the fact that the number of objectives that are formally considered within each subsystem and within the SoS as whole is subject to value judgment, the role and mission of each subsystem, and the interdependencies and interconnectedness (I-I) that exist within and among the subsystems, among many other factors. This cautious remark is not unrelated to the overconfidence and reverence that systems analysts place in the "optimality" of a single-objective model.

5.2.4 The ε-Constraint Method

The SWT method recognizes that optimization theory is usually much more concerned with the relative value of additional increments of the various noncommensurable objectives, at a given value of each objective function, than it is with their absolute values. Furthermore, given any current set of objective levels attained, it is much easier to turn to the DM to assess the relative value of the tradeoff of marginal increases and decreases between any two objectives than it is to assess their absolute average values. In addition, procedures for generating Pareto-optimal solutions/policies can be developed to enable us to assess whether one more quantity of one objective is worth more or less than that lost from another at any given level. Ordinal scale can then be used with much less concern for the distortions that relative evaluation introduces into attempts to commensurate the total value of all objectives.

Since the dimension of the decision space N for most real-world problems is generally higher than the dimension of the functional space n (N decisions and n objectives, $N > n$), as a further simplification one should make decisions in the functional space and only later transfer the information to the decision space.

A basic approach to treating noncommensurable objectives is selecting a primary or dominating objective to be optimized while constraining the decisions considered to ensure that some minimum level for all others is attained in the process. If all objectives are equal to or better than this minimum level of attainment with some proposed decision, such a decision can be termed satisfactory. So long as any decision set exists that is satisfactory, it is *unnecessary* to consider any decision that results in a poorer condition in any objective. Hence, this approach will also help reduce the field of decisions to explore.

Let

$$\underline{f}_j = \min_x f_j(\mathbf{x}), \quad \mathbf{x} \in X; \; j = 1, 2, \ldots, n \tag{5.3}$$

The ε-constraint approach replaces $(n-1)$ objective functions by $(n-1)$ constraints as given by $P_k(\varepsilon)$ in Eq. (5.4):

$$\begin{aligned} P_k(\varepsilon): & \min_x f_i(x) \text{ subject to } f_j(\mathbf{x}) \le \varepsilon_j, \\ & j \ne i; \; j = 1, 2, \ldots, n; \; \mathbf{x} \in X \end{aligned} \tag{5.4}$$

where $\varepsilon_j, j \ne i, j = 1, 2, \ldots, n$, are variables $\left(\varepsilon_j = \underline{f}_j + \bar{\varepsilon}_j\right)$, because $\bar{\varepsilon}_j > 0$ are variables.

The levels of satisfactory ε_j can be varied parametrically to evaluate the impact on the single-objective function $f_i(\mathbf{x})$. Of course, the ith objective, $f_i(\mathbf{x})$, can be replaced by the jth objective, $f_j(\mathbf{x})$, and the solution procedure repeated. The equivalence between Eqs. (5.1) and (5.4) is well documented in the literature (Haimes et al., 1971; Chankong and Haimes, 1983, 2008; Haimes 2016). The ε-constraint approach facilitates the generation of noninferior solutions as well as tradeoff functions, as will be discussed later.

By considering one objective function as primary and all others at minimum satisfying levels as constraints, the Lagrange multipliers related to the $(n-1)$ objectives as constraints will be zero or nonzero (Haimes, 2016). If nonzero, that particular constraint does limit the optimum. It will be shown that positive Lagrange multipliers correspond to the noninferior set of solutions. (Also consult the Appendix.) Furthermore, the set of nonzero Lagrange multipliers represents the set of tradeoff ratios between the principal objective and each of the constraining objectives, respectively. Clearly, these Lagrange multipliers are functions of the optimal level attained by the principal objective function, as well as the level of all other objectives satisfied as equality (binding) constraints. Consequently, these Lagrange multipliers form a matrix of tradeoff functions.

The question of the worth ratios still remains after the matrix of tradeoff functions has been computed. The worth ratios are essentially achieved through an interaction with the DM. However, since the worth ratio needs only represent relative worth of the objectives, not the absolute level of worth, any surrogate ratio that varies monotonically with the correct one will suffice.

5.2.5 The Tradeoff Function

The following development shows that the tradeoff functions can be found from the values of the dual variables associated with the constraints in a reformulated problem. Reformulate the system MOP (Eq. (5.1)) with the $P_k(\varepsilon)$ (5.4), where $\varepsilon_j = \underline{f}_j + \bar{\varepsilon}_j, \bar{\varepsilon}_j > 0, j = 2, 3, \ldots, n$, and $\underline{f}_j$ were defined in Eq. (5.3) and $\bar{\varepsilon}_j$ will be varied parametrically in the process of constructing the tradeoff function.

Form the generalized Lagrangian, L, to the system,

$$L = f_1(\mathbf{x}) + \sum_{j=2}^{n} \lambda_{1j} \left[f_j(\mathbf{x}) - \varepsilon_j \right] \tag{5.5}$$

where $\lambda_{1j}, j = 2, 3, \ldots, n$, are generalized Lagrange multipliers. The subscript $1j$ in λ denotes that λ is the Lagrange multiplier associated (in the ε-constraint vector optimization problem) with the jth constraint, where the objective function is $f_1(\mathbf{x})$. Subsequently λ_{1j} will be generalized to associate with the ith objective function and the jth constraint, λ_{ij}. Denote by X the set of all x_i, $i = 1, 2, \ldots, N$, and by Ω the set of all λ_{ij}, $j = 1, 2, \ldots, n$, that satisfy the Kuhn–Tucker condition for Eq. (5.5). The conditions of interest to our analysis are

$$\lambda_{1j} \left[f_j(\mathbf{x}) - \varepsilon_j \right] = 0, \quad \lambda_{1j} \ge 0; \; j = 2, 3, \ldots, n \tag{5.6}$$

Note that if $f_j(\mathbf{x}) < \varepsilon_j$ for any $j = 2, 3, \ldots, n$ (i.e. the constraint is not binding), then the corresponding Lagrange multiplier λ_{1j} equals 0.

The value of $\lambda_{1j}, j = 2, 3, \ldots, n$, corresponding to a binding constraint, is of special interest since it indicates the marginal benefit (cost) of the objective function $f_1(\mathbf{x})$ due to an additional unit of ε_j. From Eq. (5.5), assuming that the solution is global, the following results can be derived:

$$\lambda_{1j}\left(\varepsilon_j\right) = -\frac{\partial L}{\partial \varepsilon_j}, \quad j = 2, 3, \ldots, n \tag{5.7}$$

Note, however, that for $\mathbf{x} \in \hat{X}$, $\lambda_{ij} \in \Omega$ for all j, we obtain

$$f_1(\mathbf{x}) = L \tag{5.8}$$

Thus,

$$\lambda_{1j}\left(\varepsilon_j\right) = -\frac{\partial f_1(\cdot)}{\partial \varepsilon_j}, \quad j = 2, 3, \ldots, n \tag{5.9}$$

In the derivation of the tradeoff functions in the SWT method, only those $\lambda_{ij} > 0$ corresponding to $f_j(\mathbf{x}) = \varepsilon_j$ are of interest (since they correspond to the noninferior solution). Thus, for $f_j(\mathbf{x}) = \varepsilon_j$, Eq. (5.9) can be replaced by Eq. (5.10):

$$\lambda_{1j}(\varepsilon_j) = -\frac{\partial f_1(\cdot)}{\partial f_j(\cdot)}, \quad j=2,3,\ldots,n \tag{5.10}$$

Clearly, Eq. (5.10) can be generalized where the index of performance is the ith objective function of the system (5.1) rather than objective function $f_1(\cdot)$. In this case, the index i should replace the index 1 in λ_{1j}, yielding λ_{ij} Accordingly,

$$\lambda_{ij}(\varepsilon_j) = -\frac{\partial f_i(\cdot)}{\partial f_j(\cdot)}, \quad i \neq j;\ i;\ j=1,2,3,\ldots,n \tag{5.11}$$

For the rest of this section, only $\lambda_{ij}(\varepsilon_j) > 0$ (which correspond to binding constraints) are considered, since there exists a direct correspondence between λ_{ij} associated with the binding constraints and the noninferior set in Eq. (5.1).

The possible existence of a duality gap and its effect on the SWT method are discussed in Chankong and Haimes (1983, 2008) and Haimes, 2016. A duality gap occurs when the minimum of the primal problem is not equal to the maximum of the dual problem. This is the same situation when a saddle point does not exist for the Lagrangian function. Note that if a duality gap does exist, the ε-constraint method still generates all needed noninferior solutions. However, a given value of the tradeoff function λ_{ij} may correspond to more than one noninferior solution. On the other hand, if a duality gap does exist, then not all Pareto-optimal solutions can be generated for the weighting problem $p(w)$ posed in Eq. (5.2).

Definition The *indifference band* is defined to be a subset of the noninferior set where the improvement of one objective function is equivalent (in the mind of the DM) to the degradation of another.

Definition A preferred solution is defined to be any noninferior feasible solution that belongs to the indifference band.

The computational derivation of the tradeoff function λ_{ij} will be demonstrated through the derivation of λ_{ij} as follows.

The system given by Eq. (5.5) is solved for K values of ε_2, say, $\varepsilon_2^1, \varepsilon_2^2, \ldots, \varepsilon_2^k$, where all other ε_j, $j = 3, 4, \ldots, n$, are held fixed at some level ε_j^0. Only those $\lambda_{12}^k > 0$ that correspond to the binding constraints $f_2^k(\mathbf{x}) = \varepsilon_2^k$ $k = 1, 2, \ldots, K$ are of interest, since they belong to the noninferior solution.

Assume that for $\varepsilon_2^1, \lambda_{12}^1 > 0$ with the corresponding solution $\mathbf{x}^1$. Then $f_2(\mathbf{x}^1) = \varepsilon_2^1$. Clearly, not all other λ_{ij}, $j = 3, 4, \ldots, n$, corresponding to this solution $(\mathbf{x}^1)$ are positive. Thus, the following equation is solved:

$$\min_x f_1(\mathbf{x}); \mathbf{x} \in X \text{ so that } f_j(\mathbf{x}) < f_j(\mathbf{x}^1), \quad j=2,3,\ldots,n \tag{5.12}$$

where ε_j^0 were replaced by $f_j(\mathbf{x}^1)$, $j = 3, 4, \ldots, n$. A small variation δ_j may be needed to ensure positive λ_{1j}, $j = 3, 4, \ldots, n$, in the computational procedure. The tradeoff λ_{12} is a function of all ε_j, $j = 2, 3, \ldots, n$ (i.e. $\lambda_{12} = \lambda_{12}(\varepsilon_2, \ldots, \varepsilon_n)$). It will be shown in subsequent discussions that the tradeoff function $\lambda_{ij}(\cdot)$ may be constructed (via multiple regression) in the vicinity of the indifference band.

Similarly, the tradeoff function $\lambda_{13}\sigma$ can be generated, where again the prime objective function is $f_1(\mathbf{x})$, and the system (Eq. (5.5)) is solved for K' different values of $\varepsilon_3^k, k = 1,2,\ldots,K$ with a fixed level of $\varepsilon_2^0, \varepsilon_4^0, \ldots, \varepsilon_n^0$. Similarly, the tradeoff functions f_{1j} can be generated for $j = 4, 5, \ldots, n$. Once all tradeoff functions λ_{1j}, $j = 1, 2, 3, \ldots, n$, have been generated, the prime objective may be changed to the ith, and thus all tradeoff functions λ_{ij}, $i \neq j$; $i, j = 1, 2, 3, \ldots, n$, can be generated. It can be shown, however, that not all λ_{ij} need be generated computationally since the following relationships hold:

$$\lambda_{ij} = \lambda_{ik}\lambda_{kj} \text{ for } \lambda_{ij} > 0; \quad i \neq j; \quad i,j = 1,2,\ldots,n \tag{5.13}$$

In addition, the relationship $\lambda_{ij} = 1/\lambda_{ji}$ for $\lambda_{ji} \neq 0$ can also be used.

5.2.6 The Surrogate Worth Function

The surrogate worth function provides the interface between the DM and the mathematical model. The value of the surrogate worth function W_{ij} is an assessment by the DM as to how much (on an ordinal scale, say, from −10 to +10, with zero signifying equal preference) he or she prefers trading λ_{ij} marginal units of f_i for one marginal unit of f_j, given the values of all the objectives $f_i, \ldots, f_n$ corresponding to λ_{ij}. Note that $W_{ij} > 0$ means the DM does prefer making such a trade, $W_{ij} < 0$ means he or she does not, and $W_{ij} = 0$ implies indifference. A formal definition of W_{ij} is given below:

$$W_{ij} = \begin{cases} >0 & \text{when } \lambda_{ij} \text{ marginal units of } f_i(\mathbf{x}) \text{ are preferred over one marginal unit of } f_j(\mathbf{x}), \text{ given the satisfaction of all objectives at level } \varepsilon_k, \; k=1,2,\ldots,n \\ =0 & \text{when } \lambda_{ij} \text{ marginal units of } f_i(\mathbf{x}) \text{ are equivalent to one marginal unit of } f_j(\mathbf{x}), \text{ given the satisfaction of all objectives at level } \varepsilon_k, \; k=1,2,\ldots,n \\ <0 & \text{when } \lambda_{ij} \text{ marginal units of } f_i(\mathbf{x}) \text{ are not preferred to one marginal unit of } f_j(\mathbf{x}), \text{ given the satisfaction of all objectives at level } \varepsilon_k, \; k=1,2,\ldots,n \end{cases}$$

It is important to note here that the DM is provided with the tradeoff value (via the tradeoff function) of any two objective functions at a given level of attainment of the other objective functions. Furthermore, all tradeoff values generated from the tradeoff function are associated with the noninferior set. Thus, any procedure that can generate a surrogate worth function, which in turn can provide the indifference band of λ_{ij}, $i \neq j$, $i, j = 1, 2, 3, \ldots, n$, will solve the multiobjective problem. In this respect, much of the experience developed and gained in the fields of decision theory and team theory can be utilized in the SWT method.

5.2.7 Transformation to the Decision Space

Once the indifference bands have been determined for λ_{ij}^*, the next and final step in the SWT method is to determine an $\mathbf{x}^*$ that corresponds to all λ_{ij}^*. For each λ_{ij}^* determined from the surrogate worth function via the interaction with the DM, these correspond to $f_j^*(\mathbf{x}), j=1,2,\ldots,n, j \neq i$. These $f_j^*(\mathbf{x})$ are the values of the functions $f_j(\mathbf{x})$ at the equality constraints ε_j so that $f_{ij}^*\left[f_j^* - \varepsilon_j\right] = 0$. Accordingly, the optimal vector of decisions, $\mathbf{x}^*$, can be obtained by simply solving the following optimization problem:

$$\min_{x \in X} f_i(\mathbf{x}) \text{ subject to } f_j(\mathbf{x}) \leq f_j^*(\mathbf{x}), \quad j=1,2,\ldots,n, j \neq i \tag{5.14}$$

Equation (5.14) is a common optimization problem with a single-objective function. The solution of Eq. (5.14) yields the desired $\mathbf{x}^*$ for the total vector optimization problem posed by Eq. (5.1).

The consistency of the DM should not always be assumed. The DM may show nonrational behavior or provide conflicting information at times. The SWT method safeguards against this by cross-checking the resulting λ_{ij}^*. It has been shown elsewhere that one set of $\lambda_{1i}, \ldots, \lambda_{1n}$ will suffice for solving the multiobjective problem posed previously. It is always possible, however, to generate, for example, $\lambda_{12}^*, \ldots, \lambda_{23}^*$ and λ_{13}^* (via an interaction with the DM) and to check that indeed the following relation holds: $\lambda_{13}^* = \lambda_{12}^* \lambda_{23}^*$ (i.e. satisfies the general relationship $\lambda_{ij} = \lambda_{ik} \lambda_{kj}$ for $\lambda_{ij} > 0$; $i, j = 1, 2, \ldots, n$).

In sum, the SWT method can be used to analyze and optimize MOP. The following is a brief summary of this method:

1. It is capable of generating all needed noninferior solutions to a vector optimization problem for Complex SoS.
2. The method generates the tradeoffs between any two objective functions on the basis of duality theory in nonlinear programming. The tradeoff function between the ith and jth objective functions, λ_{ij}, is explicitly evaluated and is equivalent to $-\partial f_i / \partial f_j$.
3. The DM interacts with the systems analyst and the mathematical model at a general and very moderate level. This is done via the generation of the surrogate worth functions, which relate the DM's preferences to the noninferior solutions through the tradeoff functions. These preferences are constructed in the objective function space (more familiar and meaningful to the DM) and only then transferred to the decision space. This is particularly important since the dimensionality of the objective function space is often smaller than that of the decision space. These preferences yield an indifference band where the DM is indifferent to any further tradeoff among the objectives.
4. The SWT method provides for the quantitative and qualitative analysis of noncommensurable objective functions.

5. The method is very well suited to the analysis and optimization of multiobjective functions with multiple DM.
6. The method has an appreciable computational advantage over all other existing methods when the number of objective functions is three or more.

5.3 CHARACTERIZING NONINFERIOR SOLUTIONS

The concept of a proper noninferior solution was first introduced by Kuhn and Tucker (1951), and it was later modified by Geoffrion (1968). A feasible solution, $\mathbf{x}^*$, is a proper noninferior solution if there exists at least a pair of objectives, say, f_i and f_j, for which a finite improvement of one objective is possible only at the expense of some reasonable degradation of the other. More precisely, a proper noninferiority of $\mathbf{x}^*$ implies the existence of a constant $M>0$ such that for each i, $i=1, \ldots, n$, and each $\mathbf{x} \in X$ satisfying $f_i(\mathbf{x})<f_i(\mathbf{x}^*)$, there exists at least one $j \neq i$ with $f_j(\mathbf{x})>f_j(\mathbf{x}^*)$ and $[f_i(\mathbf{x})-f_i(\mathbf{x}^*)][f_i(\mathbf{x}^*)-f_j(\mathbf{x})] \leq M$. Naturally one should only seek, as candidates for the best-compromise solution, proper noninferior solutions. A noninferior solution that is not proper is an *improper noninferior solution.*

Geoffrion (1968) characterizes proper noninferior solutions by showing the following. A sufficient condition for $\mathbf{x}^*$ to be proper and noninferior is that it solves a weighting problem $P(w)$, with w being a vector of strictly positive weights. The condition becomes necessary if convexity for all functions is also assumed. This implies that a necessary and sufficient condition for $\mathbf{x}^*$ to be a proper noninferior solution for a linear MOP is that it solves $P(w)$ with strictly positive weights w.

Chankong (1977) and Chankong and Haimes (1983, 2008; Haimes, 2016) then characterize proper noninferiority by means of the ε-constraint problem discussed in Section 5.4.2. Assuming continuous differentiability of all functions and the regularity of the point $\mathbf{x}^*$ of the binding constraints of $P_k(\varepsilon^k)$, a necessary condition for $\mathbf{x}^*$ to be properly noninferior is that $\mathbf{x}^*$ solves $P_k(\varepsilon^k)$, with all the Kuhn–Tucker multipliers associated with the constraints $f_j(\mathbf{x}) \leq \varepsilon_j$, $j \neq k$, being strictly positive. The condition becomes sufficient if convexity for all functions is further assumed. This condition, as depicted in Figure 5.1, is often easy to verify when the ε-constraint approach is used as a means for generating noninferior solutions. Relationships between improper noninferiority and positivity of the Kuhn–Tucker multipliers can also be established, as displayed in Figure 5.1. Figure 5.2 illustrates a potential use of results depicted in Figure 5.1.

Consider the following vector minimization problem:

$$f_1(\mathbf{x}) = (x_1-1)^2 + (x_2-1)^2$$

$$f_2(\mathbf{x}) = (x_1-6)^2 + (x_2-2)^2$$

$$f_3(\mathbf{x}) = (x_1-2)^2 + (x_2-5)^2$$

and

$$X = \{\mathbf{x} | x \in R^2, x_1 \geq 0, x_2 \geq 0\}$$

It can be shown (Chankong, 1977; Chankong and Haimes, 1983, 2008; Haimes, 2016) through the use of either the weighting problem (note that all objective functions are and must be convex) or the ε-constraint problems that the set of all noninferior solutions consists of all points within and on the

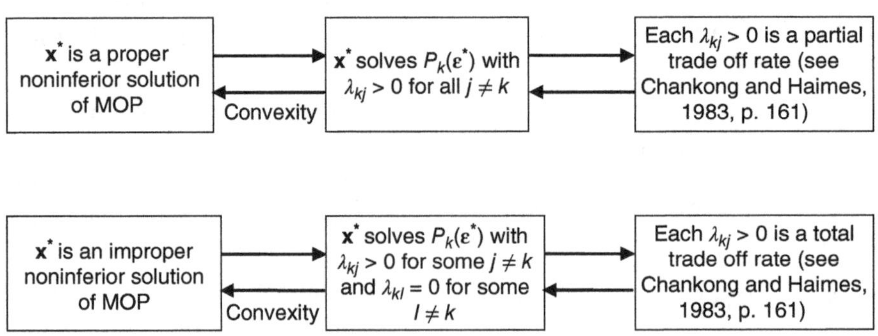

Figure 5.1 *Relationships between proper noninferiority and Kuhn–Tucker multipliers.*

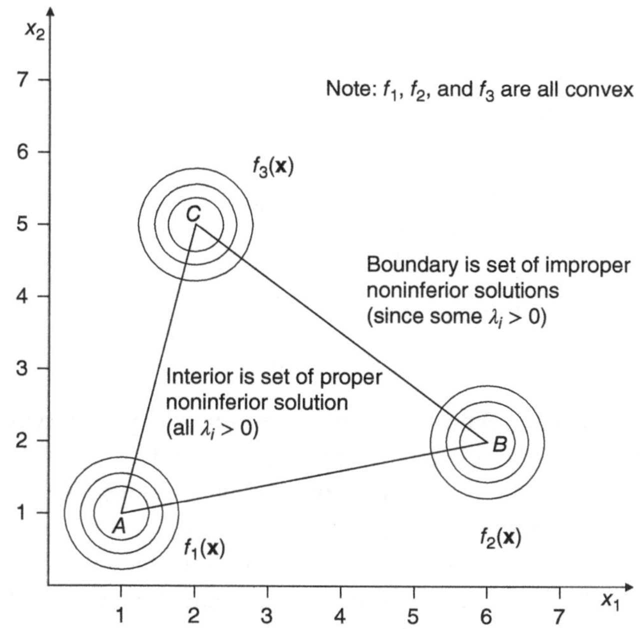

Figure 5.2 *Graphical illustration of relationships between positivity of λ's and proper noninferiority.*

boundary of the triangle ABC in Figure 5.2. If f_1 is taken to be the primary objective in the ε-constraint formulation, then it can be shown that the Kuhn–Tucker multipliers (λ_{12}, λ_{13}), corresponding to each point within the triangle, are strictly positive, while at least one λ_{ij} corresponding to points on the boundary of the triangle is zero. Consequently, each interior point of the triangle is a proper noninferior solution, whereas each boundary point of the triangle is an improper noninferior solution.

5.4 EXAMPLES OF COMPLEX SYSTEMS OF SYSTEMS WITH MULTIOBJECTIVES

5.4.1 Regional Flood Control Versus Hydropower Generation

The planning of water and related land resources in a River Basin (or a region) as Complex SoS is a vital element in the formulation of public policy affecting multiple regions and communities and large population. Such planning is commonly responsive to the inherent multiobjectives and goals and accounts for the tradeoffs among the objectives, including a sample of the following five categories of concern: (i) time horizon (short, intermediate, and long term), (ii) client (various sectors of the public), (iii) nature (aquatic and wildlife habitats), (iv) scope (national, regional, and local needs), and (v) constraints (legal, institutional, environmental, social, political, and economic).

There are many ways and means of identifying and classifying objectives and goals for such a planning effort of I-I Complex SoS. Consider the following four major objectives: (i) national economic development, (ii) regional economic development, (iii) environmental quality, and (iv) social well-being.

The Technical Committee study (Peterson, 1974) identifies nine goals for water resources Complex SoS, which have been divided into two major groups: (i) maintenance of security ((a) environmental security, (b) collective security, and (c) individual security) and (ii) enhancement of opportunity ((d) economic opportunity, (e) recreational opportunity, (f) aesthetic opportunity, (g) cultural and community opportunity, (h) educational opportunity, and (i) individual freedom). In an environmental tradeoff analysis, policies should be established to promote conditions, where human and nature can exist in harmony. Resolution of conflicts should be achieved by balancing the advantages of development against the disadvantages to the environment and the aquatic system. The process is one of balancing the total "benefits," "risks," and "costs" for both people and the environment, where the well-being of future generations is as important as that of present ones. Fundamental to multiobjective analysis is the Pareto-optimal concept.

Consider two major objectives in the operation of reservoir Complex SoS: (i) minimize hydroelectric power generation losses from the reservoir and (ii) minimize flood damages. Obviously, these two objectives are in conflict and competition (see Figures 5.3 and 5.4). Note however that the higher the level of water in the reservoir, the more electric power generation is possible and the less water storage is available for flood control purposes. Clearly, one can identify, within the active storage capacity of that reservoir, a Pareto-optimal region whereby the enhancement of the first objective can be achieved only at the expense or degrading of the second, namely, flood control. Also note that the units of these two objectives are noncommensurable. The first objective, which minimizes the hydropower losses, may be measured in units of energy and not

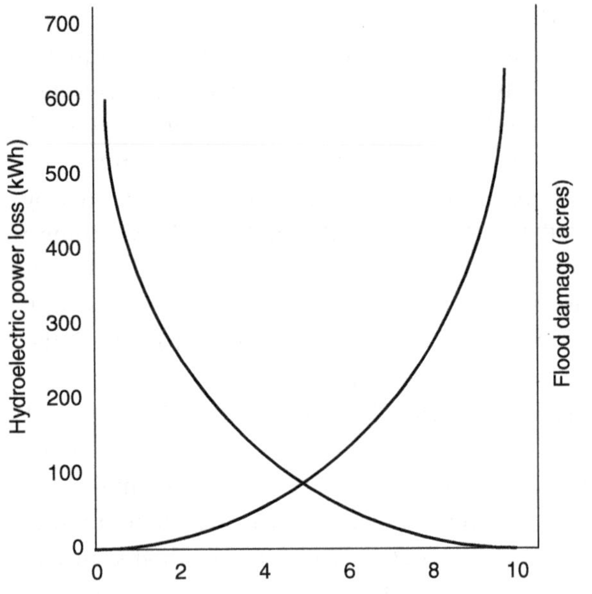

Figure 5.3 Flood damage and hydroelectric power loss in the decision space.

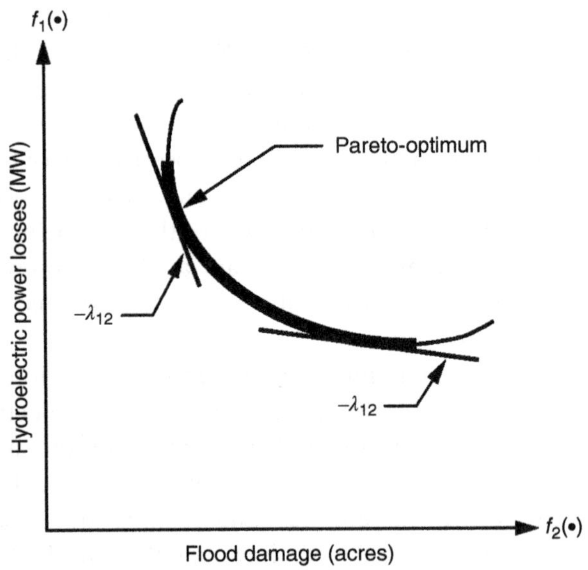

Figure 5.4 Flood damage versus hydroelectric power loss in the functional space.

necessarily in monetary units, where the second flood control objective can be measured in terms of acres of land, livestock, or human life lost.

The objective function $f_1(x)$ represents the hydropower output lost (in kWh), while the objective function $f_2(x)$ represents the expected damage (in acres flooded). The maximum water level possible for the reservoir is 10, where

$$f_1(x) = 1000\,e^{-x}$$
$$f_2(x) = e^{0.65x}$$

and where x denotes the water level at the reservoir.

The MOP is

$$\min_x \begin{cases} f_1(x) = 1000\ e^{-x} \\ f_2(x) = e^{0.65x} \end{cases}$$

subject to the constraint

$$0 < x \le 10$$

Figures 5.3 and 5.4, which are generic graphs typical for these objective functions, show the tradeoffs between flood damage and kilowatt-hours lost in the decision space and functional space, respectively. The water level can be set at a number of levels, all of which are technically Pareto-optimal, because each change in x degrades one objective function while improving the other. Note, however, that at roughly $x = 2$ and $x = 8$, one of the two objective functions stays virtually constant (thus not degraded), while the other objective is improved. Therefore, this range of water levels was chosen for the sample Pareto-optimal solutions shown in Table 5.1.

Table 5.1 presents a set of Pareto-optimal solutions with their associated tradeoff values.

Note that these tradeoffs are calculated using the relationship

$$\lambda_{12} = -\frac{\Delta f_1}{\Delta f_2}.$$

Figure 5.4 is a representation of the tradeoffs in the functional space. Note that $\lambda_{12} > 0$ is a necessary condition for Pareto-optimality, and thus the slope $\Delta f_1/\Delta f_2$ must be negative.

TABLE 5.1 Pareto-Optimal Solutions

Water Reservoir Level (x)	Flood Damage (acres)	Hydropower Loss (kWh)	Tradeoff (slope)
2.0	3.7	135.3	−37.8
2.5	5.1	82.0	−16.6
3.0	7.0	49.8	−7.3
3.5	9.7	30.2	−3.2
4.0	13.5	18.3	−1.4
4.5	18.6	11.1	−0.6
5.0	25.8	6.7	−0.3
5.5	35.7	4.1	−0.1
6.0	49.4	2.5	−0.1

5.4.2 Pedagogical Examples

Two example problems are presented here mainly for pedagogical purposes. There are two objective functions and two decision variables in Example Problem 5.1, and there are three objective functions and two decision variables in Example Problem 5.2. The corresponding solutions are relatively simple; therefore, they do not necessarily demonstrate the actual computational procedures involved in large-scale problems.

Example Problem 5.1 Solve the following multiobjective optimization problem via the SWT method:

$$\min \left\{ \begin{array}{l} f_1(x_1,x_2) = (x_1-2)^2 + (x_2-4)^2 + 5 \\ f_2(x_1,x_2) = (x_1-6)^2 + (x_2-10)^2 + 6 \end{array} \right\} x_1 \geq 0,\ x_2 \geq 0 \tag{5.15}$$

A solution to Eq. (5.15) necessitates the existence of a DM who selects a preferred solution from the noninferior solutions. For simplicity, no constraints are introduced in this example problem.

Solution: The first phase in applying the SWT method is converting Eq. (5.32) into the ε-constraint form presented by Eq. (5.16):

$$\text{Subject to} \quad \left. \begin{array}{l} \min f_1(x_1,x_2) \\ f_2(x_1,x_2) \leq \varepsilon_2 \end{array} \right\} \tag{5.16}$$

Form the Lagrangian function, $L(x_1, x_2, \lambda_{12})$:

$$L(x_1,x_2,\lambda_{12}) = f_1(x_1,x_2) + \lambda_{12}[f_2(x_1,x_2) - \varepsilon_2] \tag{5.17}$$

Substituting Eq. (5.15) into Eq. (5.17) yields

$$L(x_1,x_2,\lambda_{12}) = (x_1-2)^2 + (x_2-4)^2 + 5 + \lambda_{12} \left[(x_1-6)^2 + (x_2-10)^2 + 6 - \varepsilon_2\right] \tag{5.18}$$

Note that the Kuhn–Tucker (1951) necessary conditions for stationarity (see Appendix) are simplified here, since there are no constraints on x_1 and x_2. These conditions are reduced to Eqs. (5.19)–(5.23):

$$\frac{\partial L(\cdot)}{\partial x_1} = 2(x_1-2) + 2\lambda_{12}(x_1-6) = 0 \tag{5.19}$$

$$\frac{\partial L(\cdot)}{\partial x_2} = 2(x_2-4) + 2\lambda_{12}(x_2-10) = 0 \tag{5.20}$$

$$\frac{\partial L(\cdot)}{\partial \lambda_{12}} = \left[(x_1-6)^2 + (x_2-10)^2 + 6 - \varepsilon_2\right] \leq 0 \tag{5.21}$$

$$\lambda_{12}\left[(x_1-6)^2 + (x_2-10)^2 + 6 - \varepsilon_2\right] = 0 \tag{5.22}$$

$$\lambda_{12} \geq 0 \tag{5.23}$$

Equation (5.19) yields

$$\lambda_{12} = \frac{x_1-2}{6-x_1} \tag{5.24}$$

Equation (5.20) yields

$$\lambda_{12} = \frac{x_2-4}{10-x_2} \tag{5.25}$$

Since $\lambda_{12} > 0$ guarantees a noninferior solution, Eqs. (5.21)–(5.23) are reduced to Eqs. (5.26) and (5.27):

$$(x_1-6)^2 + (x_2-10)^2 + 6 - \varepsilon_2 = 0 \tag{5.26}$$

$$\lambda_{12} > 0 \tag{5.27}$$

Note that both Eqs. (5.24) and (5.25) should be satisfied. Therefore, these equations yield Eq. (5.28):

$$\lambda_{12} = \frac{x_1-2}{6-x_1} = \frac{x_2-4}{10-x_2} \tag{5.28}$$

Upper and lower limits on x_1 and x_2 may easily be derived by satisfying Eqs. (5.24), (5.25), and (5.27):

$$2 < x_1 < 6 \tag{5.29}$$

$$4 < x_2 < 10 \tag{5.30}$$

The boundary points 2 and 6 for x_1 and 4 and 10 for x_2 result in either $\lambda_{12} = 0$ or $\lambda_{12} = \infty$.

Solving Eq. (5.28) simplifies the generation of noninferior points as is presented in Table 5.2:

TABLE 5.2 Noninferior Solutions and Tradeoff Values for Example Problem 5.2

x_1	x_2	$f_1(x_1, x_2)$	$f_2(x_1, x_2)$	λ_{12}
2.00	4.00	5.00	58.00	0
2.50	4.75	5.81	45.81	0.14
3.00	5.50	8.25	35.25	0.33
3.50	6.25	12.31	26.31	0.60
4.00	7.00	18.00	19.00	1.00
4.50	7.75	25.31	13.31	1.67
5.00	8.50	34.25	9.25	3.00
5.50	9.25	44.81	6.81	7.00
6.00	10.00	57.00	6.00	∞

$$x_2 = 1.5x_1 + 1 \tag{5.31}$$

The reader should note that noninferior solutions and their corresponding tradeoff values were not generated by varying ε_2, as is suggested by the SWT method, because a closed-form and direct solution was obtained instead. In larger-scale problems with decision variables exceeding even 4 or 5, the above closed form will not be computationally tractable, and noninferior solutions would be generated by varying the ε's. This explanation also applies to Example Problem 5.2, discussed in the next section.

Example Problem 5.2 Solve the following multiobjective optimization problem via the SWT method:

$$\min\left\{\begin{array}{l} f_1(x_1,x_2) = (x_1-2)^2 + (x_2-4)^2 + 5 \\ f_2(x_1,x_2) = (x_1-6)^2 + (x_2-10)^2 + 6 \\ f_3(x_1,x_2) = (x_1-10)^2 + (x_2-15)^2 + 10 \end{array}\right\} \tag{5.32}$$

Solution

Rewrite problem (5.32) into the ε-constraint form:

$$\left.\begin{array}{ll} & \min f_1(x_1,x_2) \\ \text{Subject to constraints} & \\ & f_1(x_1,x_2) \le \varepsilon_2 \\ & f_3(x_1,x_2) \le \varepsilon_3 \end{array}\right\} \tag{5.33}$$

Form the Lagrangian $L_1(\cdot)$ for Eq. (5.33):

$$L(x_1,x_2,\lambda_{12},\lambda_{13}) = f_1(x_1,x_2) + \lambda_{12}[f_2(x_1,x_2) - \varepsilon_2] + \lambda_{13}[f_3(x_1,x_2) - \varepsilon_3] \tag{5.34}$$

Substituting the values of $f_1(\cdot)$, $f_2(\cdot)$, and $f_3(\cdot)$ from Eq. (5.32) into Eq. (5.34) and solving the Kuhn–Tucker necessary conditions (similar to Example Problem 5.1) yields

$$\lambda_{12} = \frac{11x_1 - 8x_2 + 10}{-5x_1 - 4x_2 - 10} \tag{5.35}$$

$$\lambda_{13} = \frac{-6x_1 + 4x_2 - 4}{-5x_1 - 4x_2 - 10} \tag{5.36}$$

Note that there is no requirement for $f_1(x_1, x_2)$ to be the primary objective function with $f_2(x_1, x_2)$ and $f_3(x_1, x_2)$ as constraints. The multiobjective optimization problem (Eq. (5.32)) can be alternatively written in the ε-constraint form as follows:

$$\left.\begin{array}{ll} & \min f_2(x_1,x_2) \\ \text{Subject to constraints} & \\ & f_1(x_1,x_2) \le \varepsilon_1 \\ & f_3(x_1,x_2) \le \varepsilon_3 \end{array}\right\} \tag{5.37}$$

Form the Lagrangian $L_2(\cdot)$ for Eq. (5.37):

$$L_2(x_1,x_2,\lambda_{21},\lambda_{23}) = f_2(x_1,x_2) + \lambda_{21}[f_1(x_1,x_2) - \varepsilon_1] + \lambda_{23}[f_3(x_1,x_2) - \varepsilon_3] \tag{5.38}$$

Again substituting the values of $f_1(\cdot)$, $f_2(\cdot)$, and $f_3(\cdot)$ from Eq. (5.32) into Eq. (5.38) and solving the Kuhn–Tucker necessary conditions yields

$$\lambda_{21} = \frac{-5x_1 - 4x_2 - 10}{11x_1 - 8x_2 + 10} \tag{5.39}$$

$$\lambda_{23} = \frac{-6x_1 + 4x_2 - 4}{11x_1 - 8x_2 + 10} \tag{5.40}$$

Note that Eqs. (5.35), (5.36), (5.39), and (5.40) satisfy Eq. (5.13), which is rewritten here for convenience as

$$\lambda_{ij} = \lambda_{ik}\lambda_{kj} \tag{5.41}$$

for positive λ's and $i \neq j \neq k$ and

$$\lambda_{ij} = \frac{1}{\lambda_{ji}} \tag{5.42}$$

for $i \neq j$, $\lambda_{ji} > 0$.

Similar to Example Problem 5.1, Table 5.3 summarizes several noninferior solutions with the corresponding tradeoff values, Figure 5.5 depicts the noninferior solution in the functional space, and Figure 5.8 depicts the noninferior solution in the decision space x_1, and x_2. Assuming that an interaction with a DM took place and that the values of the tradeoffs λ_{12}^* and λ_{13}^* corresponding to the surrogate worth functions at $W_{12} = 0$ and $W_{13} = 0$, respectively, were obtained, then the preferred solution can be generated by substituting the values of λ_{12}^* and λ_{13}^* into Eqs. (5.35) and (5.36) and solving x_1^* and x_2^*.

TABLE 5.3 Noninferior Solutions and Tradeoff Values for Example Problem 5.2

x_1	x_2	$f_1(x_1, x_2)$	$f_2(x_1, x_2)$	$f_3(x_1, x_2)$	λ_{12}	λ_{13}
4	6.88	17.29	19.73	111.93	0.42	0.19
5	8.25	32.06	10.06	80.56	0.50	0.50
6	9.63	52.70	6.14	54.84	0.70	1.00
7	11.00	79.00	8.00	35.00	1.00	2.00
8	12.38	111.22	15.66	20.86	2.17	5.17

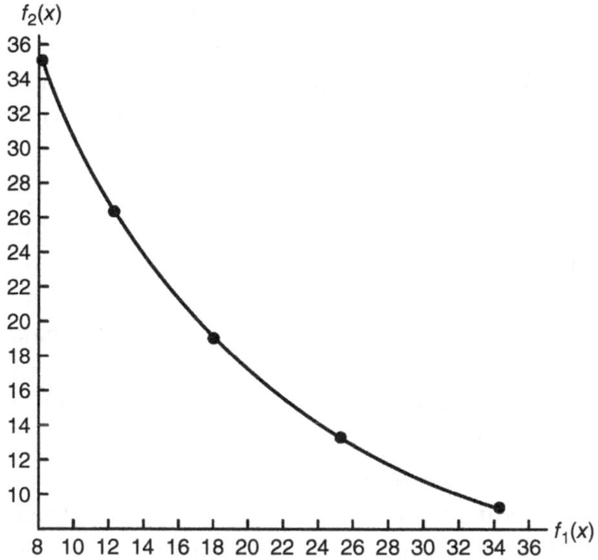

Figure 5.5 *Noninferior solution in the functional space.*

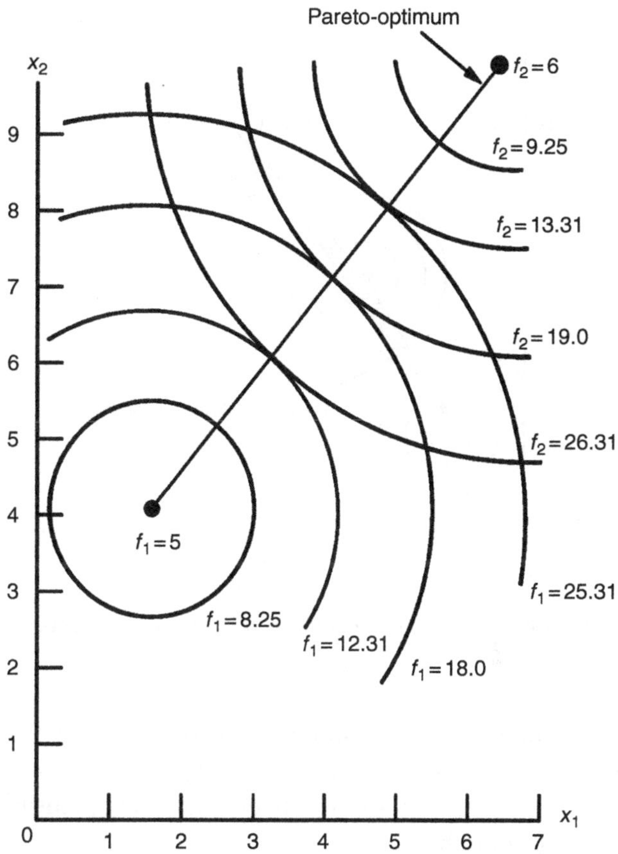

Figure 5.6 *Noninferior solution in the decision space.*

The reader is again reminded that for larger problems a closed-form solution may not be obtained, as is the case in this example, and the generation of noninferior solutions and their corresponding tradeoff values then should be obtained by varying the ε's.

5.4.3 The Limitation of Pareto-Optimal Solutions

Example Problems 5.1 and 5.2 have two common objective functions; however, a third objective function in Example Problem 5.2 has been added to demonstrate an important attribute that characterizes all MOP. *Namely, the set of Pareto-optimal solutions is critically dependent not only on the form but also on the number of objective functions that constitute the system's model.* Note, for example, that the Pareto-optimal set in the decision space for the two-objective optimization problem (Example Problem 5.1) lies on a straight line (see Figure 5.6). On the other hand, Figure 5.7 presents the tradeoff function $\lambda_{12}(f_2)$ versus $f_2(x)$. Yet, by adding a third objective function, the Pareto-optimal set in the decision space now constitutes an entire plane (see the shaded triangle in Figure 5.8). This means that a large number of Pareto-optimal solutions have been added. Conversely, by deleting one or more objective functions, the Pareto-optimal frontier will be reduced markedly.

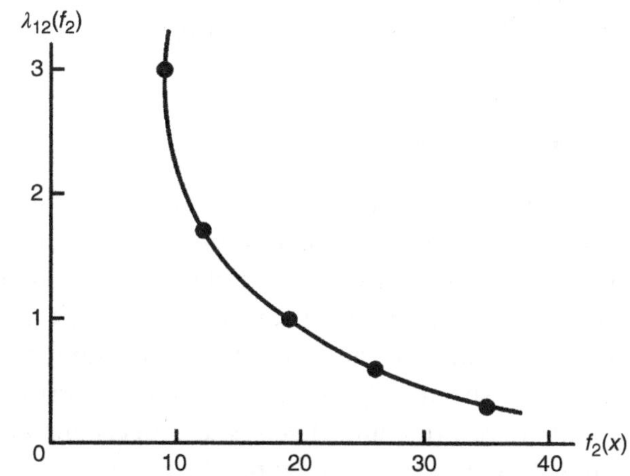

Figure 5.7 *Tradeoff function λ_{12} (f_2) versus $f_2(x)$.*

The direct and sobering conclusion is that a large set of what were previously considered "optimal" solutions (in the Pareto sense) have suddenly become inferior, non-Pareto-optimal solutions. This is indeed a humbling experience for all modelers who consider any Pareto-optimal set to an MOP as a "sacred" and undisputed "optimal set of solutions." In particular, remember that

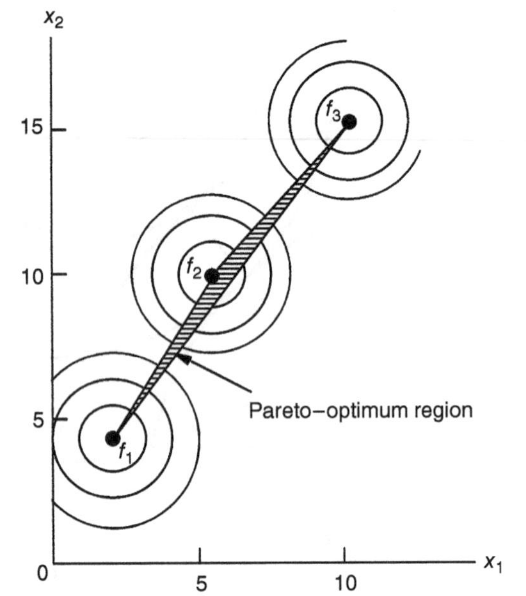

Figure 5.8 *Noninferior solution in the decision space.*

DM commonly have a number of objectives that they desire to optimize, and thus adding or deleting a secondary or a tertiary set of objectives is not only plausible but most probable.

5.5 SEQUENTIAL PARETO-OPTIMAL DECISIONS MADE DURING EMERGENT COMPLEX SYSTEMS OF SYSTEMS: A CASE STUDY

5.5.1 An Overview

This part of Chapter 5 addresses the challenges facing participants, including stakeholders, DM, systems engineers, and risk analysts in modeling and managing the development of ongoing emergent Complex SoS, with a focus on the centrality of state variables and time frame. More specifically, we address here the critical role that shared (common) states and decisions and the time frame play in modeling the I-I among the subsystems that constitute emergent Complex SoS. The theoretical and methodological concepts are harmonized through their relevance to the ongoing emergent Federal Aviation Administration's (FAA) Next Generation (NextGen) Complex SoS project also discussed in Chapter 10. Appropriate parts of the dynamic roadmap for risk modeling, assessment, management, and communication and their companion principles and guidelines are used to guide the analyses. Moreover, decisions made at time t for one subsystem will subsequently change the states of that subsystem as well as affect other interconnected and interdependent subsystems that share states. This phenomenon is profoundly important even when the time frame of the affected states is measured in months or years, as is the case in the application discussed here. The centrality of the time frame in decision making is even more important when the impact of current decisions on future options can affect not only the intended subsystem but can also inadvertently affect other subsystems in the future.

Addressing the third question in risk management – "What are the impacts of current decisions on future options?" – is critically important for emergent Complex SoS, because their conception, the evolution of their requirements and specifications, their design and development, and their ultimate operation can span several years. Furthermore, the sequential decisions made during the development of each subsystem (of these Complex SoS) will most likely affect the development of other subsystems in the future, with the expectation that in their totality, all subsystems are ultimately expected to operate as an integrated harmonious Complex SoS. For example, decisions to achieve specific outcomes made on subsystem A of Complex SoS can change the states of that subsystem as well as affect other interconnected and interdependent subsystems that share states with subsystem A.

The risks and challenges associated with this sequential decision-making process are of paramount importance for Complex SoS. The I-I of these Complex SoS are modeled and represented through shared (common) states and decision variables existing among the subsystems, as advanced by the Phantom System Models (see Chapter 4). To avoid abstraction, and to more clearly communicate the concepts and methodologies introduced here, appropriate examples of the shared states and decisions are linked to the US FAA NextGen to demonstrate their relevance and practicality. In this sense, relating the systems-based methodological approach and its application to the NextGen enterprise becomes one integral theme. The technical presentation is described through the current and ongoing modernization of the National Airspace

System (NAS), one of the most all-encompassing projects ever undertaken by the FAA.

The following premises guide the technical presentation of the case study, given that the time frame for the development of the NextGen spans two decades:

1. Sequential decisions made during the emergence of Complex SoS must be dynamically Pareto-optimal and can be generated using the envelope approach (Li and Haimes, 1987; Haimes, 2016). The criticality of the time frame, within which the ultimate decisions are most likely to accomplish their intended objectives, and the multiple DM and stakeholders, will require agreed-upon policies commensurate with the satisficing principle introduced by Simon (1956).
2. The emergent behavior of Complex SoS arises from the various tradeoffs made by multiple stakeholders who have multiple goals and objectives affecting future decisions and outcomes.
3. The desired emergent behavior of systems of systems, as a whole, represents a collective objective driven by separate subsystem-level decisions made by individual stakeholders based on "local subsystem objectives." Note that the motivation for changes in a subsystem can be twofold: (a) the objectives of the local subsystem and (b) the larger desired emergent behavior for the SoS as a whole. The FAA is generally interested in the desired emergent behavior of the overall enterprise in addition to the local effect. DeLaurentis (2005) was a pioneer in addressing the air transportation as systems of systems.
4. The collaboration among the stakeholders regarding tradeoffs associated with each objective is essential to the optimal sequencing of decisions and to making enterprise strategy-level decisions.

Furthermore, the appropriate elements of the systems-based guiding principles for risk analysis and the related dynamic roadmap discussed in Chapter 9 will be applied to the NextGen enterprise. The challenges associated with the modeling and development of emergent Complex SoS are highlighted, with a focus on the centrality of the state variables and time frame. One theme that transcends the entire discussion in this case study is the critical role that shared (common) states, decisions, and other essential entities and the time frame play in modeling the I-I among the subsystems that constitute emergent Complex SoS.

5.5.2 Methodological Approach

The increasing complexity of shared infrastructure projects that span the public and private sectors has begun to exceed the capabilities of traditional management techniques to plan, construct, and control them. Every new generation of technology and operational practice increases in complexity and ambition. This trend has led to an explosion in the sophistication of government and private industry projects designed to develop and/or acquire technology-based systems. Unpredictable interactions and interdependencies among individual and community users, software–hardware, and organizations commonly characterize these systems. This complexity has imposed a complementary rise in the level of "unpredictable" adverse consequences, in terms of technological risk (not meeting a project's performance criteria, e.g. failing to deliver expected benefits for individual investors or the desired behavior of the emergent system of systems as a whole) and programmatic risk (cost overrun, time delay in project completion.)

In general, there is no one right way to perform the similar complex process of risk modeling, assessment, management, and communication associated with Complex SoS. Often the best approach is driven by the unique characteristics of the systems under consideration. *Regardless, there are certain principles and* guidelines, *introduced in Chapter 9, that apply universally to emergent Complex SoS.*

5.5.2.1 The Complexity of Planned and Emergent Complex Systems of Systems

Emergent Complex SoS are by definition composed of several subsystems and are commonly in a constant state of flux, modification, and readjustment caused by the fluctuations in one or more subsystems. Most notably, these seemingly random driving forces, denoted by the *Evolving Base* are dynamically shifting interdependencies and realities based on changing (see Chapter 9): (i) goals and objectives; (ii) stakeholders,

DM, and interest groups; (iii) organizational, political, and budgetary considerations; (iv) reorganization and reallocation of key personnel; and (v) requirements, specifications, delivery dates, users, and clients.

The literature is replete with definitions of what constitute systems of systems, albeit not how to model them. For example, Maier (1998) identifies five characteristics of systems of systems: (i) operational independence of the individual systems, (ii) managerial independence of the systems, (iii) geographic distribution, (iv) emergent behavior, and (v) evolutionary development–all of which apply to the FAA case study (addressed in Chapter 9).

Complex SoS perform functions and carry out purposes that do not reside in any subsystem. These behaviors are emergent properties of the entire Complex SoS and not necessarily the behavior of any subsystem or a system. One of the major challenges facing professionals in charge of the multiple stages in the development of Complex SoS, such as the FAA's NextGen, is harmonizing the myriad sequential decisions, whether regarding policies and procedures or structural issues, so that decisions made at one point will not conflict with, or adversely affect, future options. By their nature, emergent Complex SoS are composed of multiple intra- and interdependent and interconnected subsystems, each with a unique mission, and human, cultural, organizational, geographical, and technological characteristics that span cyber and physical infrastructures. In addition, a hierarchy of diverse stakeholders and DM, multiple noncommensurate and often conflicting and competing goals and objectives that span multiple time frames, and assorted policies and procedures serving each subsystem–all are features of these Complex SoS. Clearly, it is essential to coordinate the planning, design, specification, and construction of each subsystem, to develop appropriate operational policies and procedures that meet the needs of each subsystem, and to ensure a harmonious integration of the subsystems with each other. This also constitutes a challenging process of risk modeling, assessment, management, and communication. The common denominators that both transcend and link the above activities are the *time frame* and *the shared states and other essential entities* that constrain as well as enable the subsystems to function, to the extent possible, as a harmonious SoS.

One of the challenges facing the developers of emergent Complex SoS is to navigate the sequential temporal decision-making process while being cognizant of the dynamic multiobjective tradeoffs, leading to a satisficing set of policies that are responsive to the broader goals and objectives of emergent systems of systems enterprise. This process is also aimed at harmonizing the myriad decisions associated with each emerging subsystem and within the Complex SoS.

In sum, the development of a sequential multiobjective decision-making process and its deployment on the NextGen SoS case study demonstrate that decisions made, policies formulated, and procedures and regulations promulgated during one period (i) neither would cause future detrimental technical and programmatic risks to the subsystems within NextGen, nor (ii) would adversely affect or impact the multiobjectives and stakeholders associated with the NextGen SoS.

5.5.2.2 The FAA NextGen Complex System of Systems

Applications presented in this chapter are drawn from the FAA NextGen Complex SoS – a multi-year, multi-agency, and multi-billion-dollar project. Decisions for this complex enterprise are made by a combination of individual stakeholders and organizations in order to achieve specific objectives and outcomes. At the same time, the FAA has a desired set of emergent behaviors for the evolution of the NAS, which will be the product of these many separate decisions. NextGen epitomizes interconnected and interdependent Complex SoS.

NextGen is a public–private endeavor involving over 30 major systems and 190 government subsystems, each of which has its own platform and many stakeholders. The private sector consists of hundreds of commercial aircraft operators with hundreds of models of aircraft and thousands of private aircraft operators. These aircraft operate out of over 5000 airports, with local management. All of these components are part of the NextGen emergent system of systems. From the FAA perspective as a regulator and service provider, the decisions for FAA components must serve the diverse population of private sector stakeholders. An aim of NextGen is to create incentives for all participants to invest in capabilities that will contribute to the desired emergent behaviors of safety, flight efficiency, capacity, and the environmental impact of the systems of systems as a whole.

FAA's NextGen Complex SoS is being planned with three major five-year implementation windows, known as Alpha (2010–2015), Bravo (2015–2020), and Charlie (2020–2025). An inherent characteristic of an emergent Complex SoS such as NextGen is the long time frame from its development to its realization. Components such as aircraft and airport modifications will be expected to remain in service *for 20 years or more. Each decision, therefore, will have long-lasting consequences.* Furthermore, the timing of changes must be planned and aligned with changes in other subsystems that share common states. In particular, the coordination of the planned services will require changes in ground systems and synchronizing of aircraft and flight crews.

5.5.2.3 Three Major Sources of Risk that Threaten Emergent Interdependent Systems of Systems

Technical, programmatic, and enterprise performance risks are all concerns of emergent SoS. The first two risks apply to the subsystem level, while enterprise risk is the unexpected consequence in the desired emergent behavior of the Complex SoS as a whole.

More specifically:

1. *Technical risk* occurs when a project fails to meet its performance criteria. This includes hardware and software failures and requirement shortfalls. Technical risk can occur when an emergent forced change (EFC) causes significant changes in the behavior of one or more subsystems, leading to a cascading failure of the Complex SoS. The phrase "emergent forced changes" connotes *trends in external or internal sources of system risk that may adversely affect specific states of that system.*
2. *Programmatic risk* denotes cost overruns and delays in the schedule. This source of risk characterizes many emergent large-scale systems, especially for subsystems with significant software components.
3. *Enterprise performance risk* leads to negative outcomes due to changes in the system. When the performance of an *emergent system of systems* does not meet the collective desired behavior, whether by unanticipated, undetected, misunderstood, or ignored EFCs from within or from outside a system, the state changes may result in a negative emergent behavior. Therefore, it is imperative to be able through scenario structuring, modeling, and risk analysis to envision, discover, and track EFC (Haimes, 1981, 2009, 2016) and to understand the likely consequences for emergent behavior. The interdependency among subsystems of systems of systems is a manifestation of the shared states among them. Consequently, EFC that affect shared states of one subsystem will affect all other subsystems that share the same states. In sum, the essence of interdependencies among subsystems of systems of systems is grounded on the shared states among the subsystems.

Consider the case where a DM acts believing that a subsystem change will create a desired behavior; however, the change is actually insufficient to provide the desired performance outcome. In a hypothetical example from NextGen, consider two capabilities that may each be necessary, yet not sufficient to achieve the desired benefit. The first is an *airborne reroute* path-dependent *capability*, which replaces a series of paper note exchanges and phone calls with electronic exchange of information between multiple parties needed to direct an aircraft to an alternate route, as depicted in Figure 5.9. The second capability, *collaborative trajectory options program*, also shown in Figure 5.9, enables the aircraft operator flight operations center to indicate a preference for airborne trajectory options. Each of these capabilities is necessary to create more flexible routing around weather, but is individually insufficient to do so. It may be that without including both capabilities, the use of airborne reroutes will fail to achieve the desired performance. This could be anticipated by examining the shared states for the two capabilities. Both capabilities involve knowledge of the airborne flight trajectory, procedures for avoiding weather, and operating environments involving weather. The implication is that all capabilities with shared states and other essential entities must be evaluated in collaboration with the independent DM for the necessary and sufficient conditions to create a desired emergent behavior. Such decisions become corequisite.

In other cases, decisions may transfer program or technical risk among DM. This would be the case if

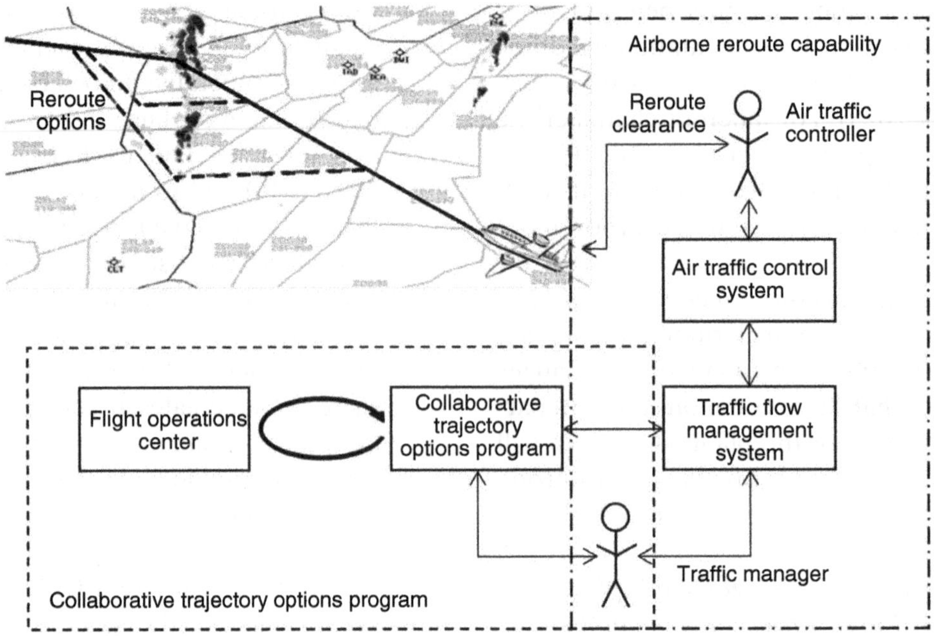

Figure 5.9 *Airborne reroute and airborne trajectory options capabilities.*

one of the capabilities in the example above, either the airborne reroute capability or the airborne trajectory options capability, was implemented with specifications that reduced the risk of one by adding program risk to the other. If coordinated as part of the overall design of a combined capability, then this risk management should lower the overall risk. If this transfer of risk occurs as a result of separate, uncoordinated program designed activities, then this represents a risk transfer from one DM to another, which ultimately can degrade both the desired performance of the Complex SoS as a whole and the level of trust. A key concept in managing the risk of not achieving the desired emergent behavior is that *the critical decisions that warrant additional collaboration among DM can be identified through the analysis of the shared states and can be addressed proactively*. Risk mitigation should include coordination of the sequence and scope of the individual decisions with knowledge of the implications on other decisions.

EFCs are tied to risk management and they cannot be overlooked. Because of their broad consequences to the contractor, the customer, the user, the analyst, and other stakeholders, numerous parties have a stake in the assessment and management of the above sources of risk. To effectively manage these sources of risk, these parties must meet collectively and periodically to ensure that proper risk management is being conducted. No party can be ignored because each brings a different perspective and background knowledge.

5.5.2.4 Criticality of the Time Frame in Sequential Decision Making

One of the characteristics of emergent systems of systems is the long time frame associated with their development and realization. The initiation of a new system starts with defining mission needs and conducting analysis. Such activities can span two decades for completion and would likely be impacted by EFCs that must be projected and addressed. As noted earlier, the planning time frame for the FAA NextGen project is segmented into three major design and development periods of 5 years each, known as Alpha, Bravo, and Charlie. Indeed, decisions made at any time period are "path dependence." Liebowitz and Margolis (2000) eloquently elaborate on the importance of the time frame concept: "Most generally, path dependence means that where we go next depends not only on where we are now, but also upon where we have been." Decisions made during any period can have significant consequences on the state of the system both at the time decisions are made and during subsequent periods; namely, we have path-dependent decisions. By altering the states of the systems of systems, each decision may adversely affect the size, location, functionality, and schedule of future changes. In addition, a large number of major structural and policy decisions exacerbate tracking such forced changes.

Consider, as an example, the following two decisions made in parallel. Decision ***A*** will be an FAA decision to introduce schedule-based merges within a terminal area. Referred to as terminal sequencing and scheduling, this is a flow management capability in which specific times and tolerances are used to smoothly interleave two flows of traffic. Decision ***B*** will be a decision to structure the arrival routes in a way that improves the ability to model and execute flight approaches to tighter tolerances. If decision ***A*** were made first, it would provide insights concerning objectives for decision ***B***. However decision ***A*** may potentially set tolerances that are unachievable by the models or execution methods available at the terminal. If decision ***B*** were made first, it would establish the flows that would be criteria as to whether ***A*** remains a viable option for that terminal. These policy decisions are of a nontransitive property – they must be followed not only at the right time but also in the right sequence. Decision pairs with this nontransitive property may be identified by analysis of decisions that share common states in the system. Note that decisions ***A*** and ***B*** affect the following three shared states: (i) state of regulations, (ii) state of operating environment, and (iii) state of specific technology.

5.5.3 Dynamic Multiobjective Decision Making

Most decisions involve implicit or explicit dynamic tradeoffs among multiple, often conflicting, and noncommensurate goals and objectives. *Indeed, analyzing and understanding the impacts of decisions on multiobjectives and their associated tradeoffs that affect different time frames constitute a foundational maxim in prudent decision making.*

Given the often conflicting, competing, and noncommensurate nature of the multiobjectives associated with NextGen Complex SoS, and because the tradeoffs among them are complex from multiple perspectives, it is impractical to consider assigning them weights in making meaningful decisions. The decision-making process must avoid an overly constrained feasible set for which there is no answer. In this chapter, we adopt the concept of satisficing originally introduced by Herbert Simon (1956), which builds on an even earlier paper by him (Simon 1947).

Any attempt to "optimize" a set of objectives, even within the Pareto-optimal sense, that are limited to present needs and aspirations or that are not responsive to emergent or future forced changes could lead to severe unintended consequences. Informally, *a solution to two-objective functions is termed Pareto-optimal if improving one objective function can be achieved only by degrading the other* (Intriligator, 1970). In other words, given the nature of the evolving, adaptive, incremental, and multistage systems of systems project, present decisions and policies must take into consideration that EFC may produce catastrophic or irreversible consequences on the project in the future. Thus, a systemic approach to addressing the multistage decisions during Alpha, Bravo, and Charlie periods should (i) continuously assess and evaluate precursors to potential EFC (building on the theory of scenario structuring), (ii) balance present multiobjectives with potential or perceived emergent future needs and objectives, and (iii) add more flexibility to present policy formulation to ensure against adverse emergent or unintended catastrophic consequences.

Robustness stems from evaluating the consequences and future flexibility of two preferred Pareto-optimal policies where potential future environments would yield a distinct choice between two options presently perceived as seemingly equivalent. Assessing the temporal impacts on future multiobjective-based options is especially critical for rapidly changing technological systems, such as NextGen Complex SoS. The *envelope approach*, which enables dynamic multiobjective tradeoff analyses (Li and Haimes, 1987), will be introduced below.

Furthermore, a dynamic driver of the NextGen, where the time frame is of paramount importance, introduces another challenge to both modelers and DM. Namely, ignoring or not accounting for the changes in the states of a safety-critical dynamic system as time progresses (e.g. such as the integration of the unmanned aircraft system (UAS) within the NAS) would render static-based decisions misleading, if not disastrous. Recognizing this fact, methods such as event trees, decision trees, and process control can account for the impact of current decisions on future options (Haimes, 2016).

Consider the following two objectives at time k depicted in Figure 5.10: *access*, $f_1(\cdot)$, in the *Y*-axes *representing increased flight capacity resulting from NextGen, and capacity*, $f_2(\cdot)$, *in the X-axes representing improving flight efficiency at time k*. Figure 5.10a presents two Pareto-optimal policies **A** and **B** selected at time $t=k$. When the projections of policies **A** and **B** are plotted for period $t=k+1$ in Figure 5.10b, they reveal that only the "envelope" depicted in the dotted line of the intersection of the two Pareto-optimal frontiers constitutes the new Pareto-optimal frontier (for $t=k+1$). Thus, given the availability of a dynamic model of the two objectives, each policy **A** (associated with UAS integration) and **B** (associated with high performance airspace) at time $t=k$ (see Figure 5.10a) would generate future (new "envelope") Pareto-optimal frontier at $t=k+1$ (see Figure 5.10b). In sum, although at time $t=k$ policies **A** and **B** are equivalent in the sense of Pareto-optimality, their projected consequences at time $k+1$ as depicted in Figure 5.10b in the new Pareto-optimal frontier are not. Consider efficiency, $f_3(\cdot)$, (of the NAS) as a third objective of the NextGen as depicted in Figure 5.11a and b: (i) At $t=k$, the DM desire to maximize access to and capacity of the NAS at time $t=k$ (Figure 5.11a). However, at the next period, $t=k+1$, the DM would want to consider a third objective; namely, (ii) maximize access to and efficiency of the NAS (Figure 5.11b). Similar to Figure 5.10, policy **A** at time $t=k$ would be inferior at $t=k+1$ for Regime I in Figure 5.11b; and policy **B** at time $t=k$ would be inferior at $t=k+1$ for Regime II in Figure 5.11b. In other words, given the *Evolving Base* and the expected changes in circumstances and priorities, DM and policy analysts can use the dynamic multiobjective analysis using the "envelope" approach to project scenarios, given the inevitable *Evolving Base*. More will be discussed in the next section. The envelope approach

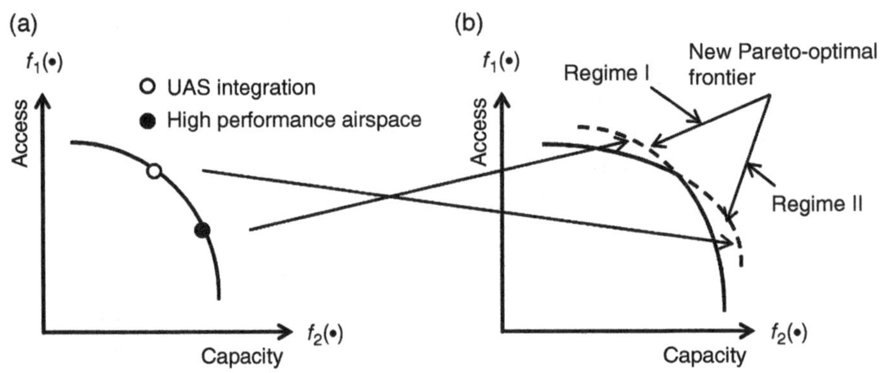

Figure 5.10 *A dynamic Pareto-optimal frontier are related for two objective functions: access versus capacity. (a)* t = k *and (b)* t = k + 1.

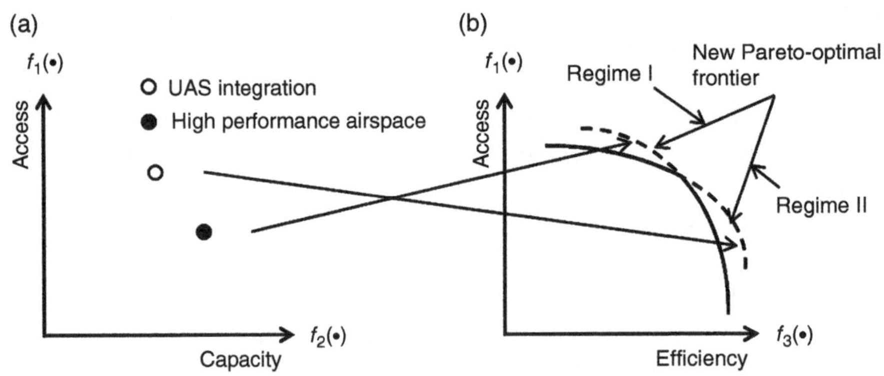

Figure 5.11 *A dynamic Pareto-optimal frontier are related for three objective functions: access versus capacity and access versus efficiency. (a)* t = k *and (b)* t = k + 1.

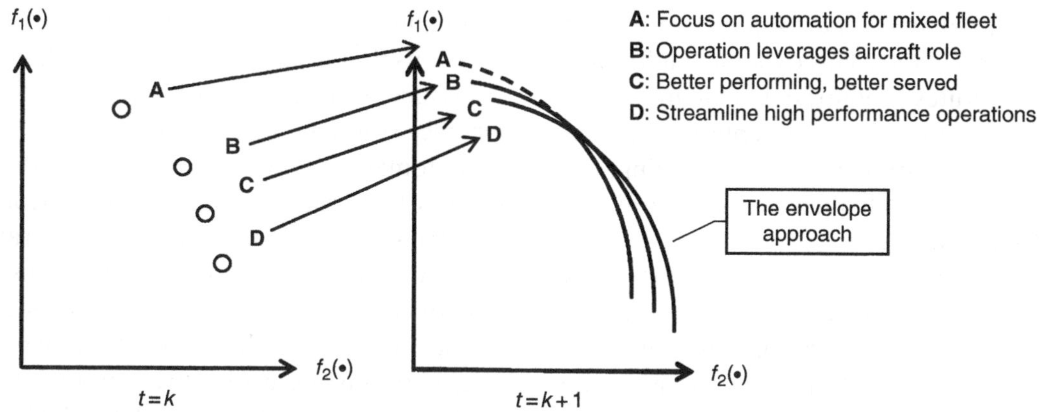

Figure 5.12 *The envelope of the combined Pareto-optimal frontier of policies **A, B, C,** and **D** for* k + *1st period.*

was developed to account for the propagation of Pareto-optimal solutions from one period to another (Li and Haimes, 1987).

Figure 5.12 presents a final example of dynamic Pareto-optimal frontier to n periods through the envelope approach. Note that the solid line's outer curve represents the envelope of the combined Pareto-optimal frontier of policies **A**, **B**, **C**, and **D** for the $k+1$st period. In sum, the value of the dynamic Pareto-optimal frontier using the envelope approach can be explained in the following more generalized example: All four decisions in Figure 5.12 are equivalent in the sense of Pareto-optimality; however, if we project those decisions to time $t=k+1$, each of the policies **A** through **D** would project a different Pareto-optimal frontier, the envelope of which is the only Pareto-optimal frontier for $t=k+1$. More specifically, consider the case where policies **A** through **D** represent during time $t=k$ distinct means to deliver air traffic control services at nontowered airports with (i) different cost effectiveness and (ii) different airport throughput implications. Given future operating environments in period $t=k+1$, the introduction of different classes of UAVs would produce different Pareto-optimal frontiers associated with cost effectiveness and throughput for each option.

The future environment would likely be similarly affected by other changes in period $t=k+1$, such as the use of data communications between pilot and controller, the expansion of surveillance capabilities, changes in the high-altitude airspace flows, and allocation of controller workforce across facilities driven by demand.

5.5.4 Judgment and Values in Complex SoS Decision Making

In every decision process there are factual elements and value elements. Factual elements are those that can be verified scientifically and subjected to scientific machinery and are likely to lead to other verifiable elements. Value elements, on the other hand, defy all forms of scientific verification and treatment. A collection of value elements and their sources constitutes a value system. Judgment, which signifies the act of giving an opinion, is the most common value element in any decision-making process (Chankong and Haimes, 1983, 2008). The decision-making process is often an amalgamation of real or perceived facts and of value judgment.

The development of the emergent NextGen Complex SoS is necessarily driven by both fact-based and judgmental decisions. For the publicly funded and operated NextGen Complex SoS, analytically harmonizing logical inputs with the political decision-making process is often not a smooth process. In such cases, past experiences can skew the sense of where investments are needed to properly assess and manage technological or programmatic risks. Similarly, biases affect the relative value of desired behaviors and outcomes in a multiobjective decision-making environment.

We posit here that the amalgamation of facts and value judgments has an important role to play in the complex dynamic multiobjective decision-making process, where judgment plays an integral role. Modeling and managing NextGen as Complex SoS enable stakeholders and DM to analyze, model, and measure desired emergent behaviors' performance of the systems of systems as a whole. This dynamic multiobjective process constitutes a nontrivial challenge to the community that must decide on acceptable tradeoffs among dynamic, noncommensurate, competing, and conflicting objectives that characterize the overall performance of NextGen.

Similar to the existence of a hierarchy of sub- and sub-sub-states of a system are their sub-objectives and sub-sub-objectives, especially for emergent Complex SoS. The highest level of this structure generally represents the broad overall objectives that are instrumental in initiating the multiobjective decision problem in the first place. The overall objectives of NextGen – safety, efficiency, capacity, and environment – are a case in point. However, these objectives are often vaguely stated and, hence, not operational. For example, the NextGen project must minimize inefficiencies for flight operators while also holding the accident rate to 10^{-6} and simultaneously increasing the national airspace capacity and ensuring stewardship of the environment. On the other hand, farther down the hierarchy, the objectives are more specific and more operational than those at the higher levels, and they are at least perceived as a means to achieve the higher ends represented by the objectives at the higher level. As a result, objectives at the lowest level of the hierarchy are the most specific and the most operational.

For example, the NextGen Complex SoS would simultaneously reduce excess taxi time on the ground, which also reduces the risk of runway incursions representing a win–win – two synergistic outcomes with a positive change. As another example, Figure 5.11 represents a dynamic Pareto-optimal frontier for two objectives: access versus capacity and efficiency. From a capacity perspective, the two capabilities have a clear tradeoff with access, but when considering efficiency, the relative tradeoff may reverse, and then the criteria for the judgment call become important.

Consider the introduction of procedures to integrate UAS within the NAS or alternatively to create performance-based airspace as a means of accommodation of the UAS. The tradeoffs could be quantified, but the acceptable policy on the Pareto-optimal frontier would be a judgment call.

These judgment calls are made more complex by the combination of tradeoffs among shared states highlighted in Table 5.4. For example, three NextGen decisions that fit in such a cluster of shared state variables include a (i) decision to create high performance airspace (segregating certain operations), (ii) decision on automation needs to serve a more diverse fleet in terms of aircraft and navigation performance, and (iii) decision on the aircraft role in separation and collision avoidance. Each of these decisions has a different effect on the three objectives: access, capacity, and efficiency. The criteria or basis for judgment in balancing between these objectives must be clear to the analyst who must study the three decisions as a set. The resulting dynamic multiobjective tradeoffs must be represented in a three-dimensional space.

The US Congress mandated that the FAA develop a plan for the integration of UAS within the NAS within the NextGen enterprise. The known and precursors to unknown multiple sources of risk associated with such integration must be carefully assessed and addressed. Balancing the tradeoffs among the multiple competing goals and objectives aspired by the many stakeholders and DM constitutes a daunting task that can benefit from the dynamic analyses depicted in Figures 5.11 and 5.12. These trades may be informed by fact but are still a judgment call in the end, and they constitute some of the drivers that influence such tradeoffs.

The very nature of emergent systems of systems is that multiple DM direct the evolution of the enterprise. Decisions made by an individual stakeholder reflecting local objectives of one subsystem might be perceived as isolated, but it will likely affect – positively or negatively – other public decisions made to incentivize a desired emergent behavior. Consider NextGen's efforts to increase access to runways in marginal metrological conditions. A capability using improved navigation procedures applicable to all trained aircraft operators would raise the total throughput for the airport by mitigating the loss of parallel runway throughput. An alternative solution uses enhanced visual systems to deliver benefit in the form of greater access as reduced levels of visibility directly to flights that are so equipped and trained. The overall capacity of the airport in reduced visibility is not affected by the second solution. Both decisions impact the local objectives, but they require different investments by aircraft operators and result in different emergent behavior objectives. In the first case all aircraft operators share in the benefit from the investment in training the flight crews regardless of who invests, while in the second case those who invest in the visual system have more exclusive rights to the benefit as only they will receive improved service (a differentiated value for the investor).

5.5.5 Centrality of the States of a System (Revisited)

Recall that the states of a system are the smallest set of independent system variables such that the values of the members of the set at time t_0 along with known inputs, decisions, random, and exogenous variables determine the value of all state variables for all $t > t_0$. The shared states and other essential entities among the subsystems that constitute the NextGen Complex SoS play a major role in modeling, understanding, and ultimately harmonizing (to the extent possible) the myriad time-critical structural and nonstructural policy decisions that are made by the multiple stakeholders engaged in this complex process. Table 5.4 presents shared states among the major NextGen objectives. These shared states play a central role in highlighting and appreciating the complexities in the decision-making process that are caused by the I-I among the subsystems, by the objectives, and by the corresponding stakeholders and DM. Furthermore, the shared states streamline the essential harmonization process associated with reaching acceptable tradeoffs among the dynamic multiobjectives as well as among the principal stakeholders and DM and, ultimately, acting upon those tradeoffs.

The complexity of the quantification of the multidimensional risk function has been modeled by relying on the states of the system. Indeed, the multidimensional probabilistic consequences, resulting from an initiating event, yield a multidimensional risk function whose modeling and quantification can be achieved most effectively through the states of the affected system. Furthermore, both the vulnerability and the resilience of a system to any specific initiating event are manifestation of the states (e.g. *physical, technical, organizational, and cultural*) of the affected system. Thus, the consequences resulting from any specific initiating event, which are necessarily functions of the states and of the specific time frame, are also functions of the vulnerability and resilience of the system to the specific initiating event (Haimes, 2009, 2016). For example, *resilience* represents *the ability of the system to withstand a disruption within acceptable degradation parameters and to recover within acceptable losses and time.* Other characterization of the risk function in terms of robustness, flexibility, and sustainability are discussed in the literature (Jugulum and Frey, 2007; de Neufville and Scholtes, 2011; Cardin et al., 2013; Hu et al., 2013).

The integration of UAS into the NAS, the transition to cloud computing, and the change in pilots' or controllers' roles are often identified as sources of

TABLE 5.4 Shared States Among the Major NextGen Objectives

Efficiency	Safety	Capacity	Environment
1. ATM technology	1. ATM technology	1. ATM technology	
2. Operating environment	**2. Operating environment**	**2. Operating environment**	**2. Operating environment**
3. Personnel and operations	*3. Personnel and operations*	*3. Personnel and operations*	
4. Aircraft responsiveness	4. Aircraft responsiveness	4. Aircraft responsiveness	
5. Information quality	**5. Information quality**		
7. Culture	*7. Culture*	*7. Culture*	
8. Budget		8. Budget	8. Budget
9. Reliability (Six Sigma)	**9. Reliability (Six Sigma)**	**9. Reliability (Six Sigma)**	**9. Reliability (Six Sigma)**
10. Trust	*10. Trust*	*10. Trust*	
11. Standardization	11. Standardization		
12. Congestion	**12. Congestion**		
	13. Product reliability		
21. Runway usage		21. Runway usage	21. Runway usage
23. Fuel usage			**23. Fuel usage**
			24. Noise

vulnerabilities to the NAS. The projected likelihood of these or other vulnerabilities causing specific emergent changes should not be a roadblock to advancing the NAS; instead they should be viewed as keys to shaping the decisions needed to manage the emerging sources of risk to NextGen. Scenario structuring is a useful tool with which to refine and assess the vulnerabilities of NextGen in the *Evolving Base* of changing conditions (to be discussed in subsequent sections and in Chapter 9). For example, in the integration of UAS into the NAS, a decision on the flight rules governing a UAS flight will affect three shared states in the NAS: the (i) controller workload, (ii) vehicle technology requirements, and (iii) pilot qualifications. The EFC to any one of these states will likely alter options available to stakeholders and other DM. Similar to vulnerability, the resilience of a system is also multidimensional, and, in many ways, the vulnerability and resilience of a system are two sides of the same coin.

The current NAS emerged from a series of development efforts that did not explicitly include resilience of the air traffic management capability as a design goal. Individual systems were protected against single point failures, but the collective Complex SoS has no established overall plan for resilience. Thus, the state of resilience is not assured. One of the design objectives of NextGen would be to build in resilience, making individual systems and the entire Complex SoS less vulnerable to EFCs. To avoid re-creating the kind of crystal-like growth that characterizes the present NAS where capabilities were added on as needed, the future must capture a more complete design of services.

5.5.6 Project Life Cycle and Team Risk Management

An often-neglected aspect of project risk management is attention to the entire project life cycle. Manufacturing firms of safety-critical systems, for example, commonly conduct a failure mode and effects analysis (FMEA) and failure mode, effects, and criticality analysis (FMECA) on the product and the assembly line, but they commonly ignore the product development and design processes. Doing so ignores the risks inherent in requirements definition, development, acquisition, and phaseout or upgrade. As early as in 1990s, Sage (1992, 1995) highlighted the different types of risk inherent in the various stages of the life cycle. Ignoring important stages of the life cycle can lead to substantial problems in terms of programmatic risk for both product development at the beginning of the life cycle and product upgrade or replacement at the end. If major sources of risks are not handled sufficiently early, they may magnify their effects later in the project. For example, in information technology acquisitions, errors in the requirements definition phase can lead to costly cascading problems later when the information system fails to meet the customer's needs. As a result, costly modifications may be necessary, causing schedule slips and cost overrun.

Note the NextGen's dynamically shifting operating environment for Complex SoS which frequently alters requirements. For example, an operational tool designed initially to sequence flights to maximize runway occupancy during capability constraints may be consequently required to sequence and space flights for efficiency as fuel prices rise. Versatility and adaptability to uncertain future demands, fleet mix, and general operating environment (the external forced changes) are characteristics of NextGen. A case in point is the need to be cognizant of and responsive to the *Evolving Base* (Haimes, 2012) introduced in Chapter 9, "dynamic shifting rules and realities": (i) goals and objectives; (ii) stakeholders, DM, and interest groups; (iii) organizational, political, and budgetary baselines; (iv) reorganization and reallocation of key personnel; and (v) requirements, specifications, and delivery. For example, requirements necessarily evolve in response to external changes. If they are not adequately managed in adherence to the *Evolving Base*, the resulting consequences can be not only expensive, but they may not serve future needs. Similarly, if the requirements are too fluid, costs and schedules cannot be maintained. Common to all decisions, appropriate tradeoffs must be made.

Managing the risks inherent in any system is contingent upon having sufficient knowledge of the system's structure and operations. Indeed, this knowledge is imperative in order to comprehensively identify the risks associated with project acquisition and management, accurately estimate the probabilities of failure, and correctly predict the consequences of those failures. While the tendency to collect data and information on the project is important, databases are useful only when accompanied by an understanding of the way the system they describe operates. Knowledge of a system provides a means to understand and effectively use information from the system. Obtaining this knowledge is often difficult even for a single system, and the problem is compounded with the development or acquisition of Complex SoS. Knowledge of the many component systems is required, and it is also critical to understand the boundaries where these systems interact and generate new sources of risk. These interactions include project requirements and specifications, design and construction, finance and management, development of new technology, and response to a myriad of changes and conflicting signals from the many participating organizations. Thus, the sheer amount of system knowledge requisite for the risk analysis imposes some difficulties in its collection, dissemination, and integration.

In their book *Working Knowledge*, Davenport and Prusak (1998) suggest that knowledge moves through an organization via markets just as any other scarce resource does. There are buyers, sellers, and brokers of knowledge. Those who possess it will sell their knowledge if properly compensated with money, reciprocity, repute, bonuses, promotions, or other gain. If there is not sufficient compensation for those who sell their knowledge, the transfer will not take place. This market for knowledge has some important implications for risk management. The knowledge necessary to assess the risks to an entire project is spread over many individuals in multiple organizations and at multiple levels in the management hierarchy. For this knowledge to be transferred and collected for the purposes of risk management, an efficient knowledge market must exist. To this end, management and corporate culture are key influences that must facilitate rather than hinder the operation of knowledge markets. First and foremost, trust is required for the exchange of knowledge (Davenport and Prusak, 1998; Covey, 2008).

Team risk management (2009) brings together all of the disparate parties in the risk management effort. Katzenbach and Smith (1999) introduce the following definition: "A team is a small number of people with complementary skills who are committed to a common purpose, performance goals, and approach for which they hold themselves mutually accountable." In other words, when conducting risk management in teams, participants are imbued with a common purpose. Risk management is not externally enforced; rather, it is a process in which everyone participates. When all participants have personal stakes in the process, they are much more likely to share their system knowledge since they can see the potential benefits from doing so. To facilitate this process, organizational boundaries must be overcome. An effective means is to bring people together in face-to-face meetings: individuals from the various participating organizations, from subdivisions within organizations, and from different levels in the management hierarchy.

5.6 SUMMARY

There is no single method to manage the risks associated with the evolution of emergent Complex SoS to achieve a set of desired outcomes. However, the theory, methodology, and principles offered in this chapter indicate that in meeting the goals and objectives set forth for complex organizations as SoS, modelers must sequence and manage their decision-making activities under the following constraining realities: (i) operate within a desired time frame; (ii) address all competing objectives affected by the decisions; (iii) assume linkage to outcomes associated with the alternatives; (iv) operate in the context of decisions made by individual powerbrokers and stakeholders who are striving for state changes beneficial to their objectives and outcomes; (v) address the operating modes explicitly, including degraded modes; (vi) address all questions in the risk assessment, management, and communication processes to identify the EFC that may alter outcomes; and (vii) recognize first-order and indirect impacts on each subsystem, especially those cases that pose a risk of low probability with extreme and dire consequences. Many of the critical decisions were modeled at a high level in terms of six states (service policy, ATM technology, operations and personnel, operating environment, aircraft responsiveness, and information quality) to understand how to apply this reality to the practice of monitoring the *Evolving Base* continuously for such Complex SoS. The insights gained by the FAA show that by focusing on the most significant consequence of each state change, it is possible to cluster decisions and work on mitigation strategies for the sequencing of decisions.

In the following, we recommend a more comprehensive and continuous treatment of NextGen evolution:

1. Develop criteria with which to prioritize decisions made during each period (Alpha, Bravo, and Charlie) and to assess the impact of such decisions on safety, efficiency, environment, and capacity.
2. Focus on decisions where expected impacts are assessed to be high and very high and the associated consequences are deemed to be significant. To accomplish this, apply the risk filtering and ranking method (RFRM) (Haimes et al., 2002; Haimes, 2016) to reduce the "90 decisions" to a manageable high-priority number of decisions. Use the "Consequences" column in the RFRM framework to list the affected (i) states/constraints/objectives, (ii) decisions and inputs, and (iii) systems within NextGen.
3. The specific probabilities required to populate this framework must be built on the collected databases, on expert evidence, and on assumptions made subject to our "learn-as-you-go" principle.

NextGen discussed in here epitomizes the development and evolution of emergent Complex SoS whose completion is expected to span over two decades. Simultaneous with its evolving development, it is subject to a tedious process of modeling, assessing, managing, and communicating the myriad corresponding sources of risk. Consider the following challenges facing the parties involved in this enterprise:

1. How can we correctly estimate the life cycle cost and the overall budget for the NextGen enterprise? Which decisions are the major cost drivers? Are they being taken in proper sequence to drive the most beneficial outcome for the objectives, the outputs, or the states of NextGen?
2. Given that early decisions made during the Alpha time span may determine or constrain the options in ways that would drive the majority of the total cost, how can planners and DM ensure that by eliminating structural or policy options, they are not adversely impacting or eliminating future options? For example, wrong decisions on operational changes even though they are not acquisitions can be very costly because they have direct impacts on workforce or technology cost drivers for the remaining decisions.
3. The FAA is also charged with making a large number of strategic decisions concerning the development of NextGen standards, procedures, and design choices in a short amount of time, many of which will not manifest themselves until several years thereafter. The challenge in making each of these decisions is being able to evaluate its impact not only on the project that the decision pertains to but also on interdependent subsystems of the NAS (Figure 5.13).
4. Since the FAA is also a regulatory government agency, requiring aircraft operators to comply with service policies can represent a transfer of cost to others. For example, solutions that segregate flows to adjacent airports based on performance-based procedures in the flight deck may require operator investment, yet are fundamentally an FAA actions/decisions on its own trade space. The formal rule-making process addresses this type of joint investment decision but may be too onerous to do for each small decision. Packaging multiple decisions into a single dialogue for joint decision making requires understanding the interactions and therefore the sequence of the smaller decisions to construct a fruitful joint investment decision.
5. The developers of the NextGen enterprise must be continuously cognizant of, and responsive to, the *Evolving Base* introduced earlier. These forced changes also have direct impacts on the dynamic multiobjectives and the associated tradeoffs that must be continuously addressed by the concerned DM and stakeholders.
6. The motivation for changes in any subsystem of the NextGen can be twofold: (a) the objectives of the local subsystem and (b) the larger desired emergent behavior for the system of systems as a whole. In particular, the FAA is commonly interested in the desired emergent behavior of the overall enterprise in addition to the local effect. At any time there is a desired set of tolerances for the qualities of service performance. As EFCs

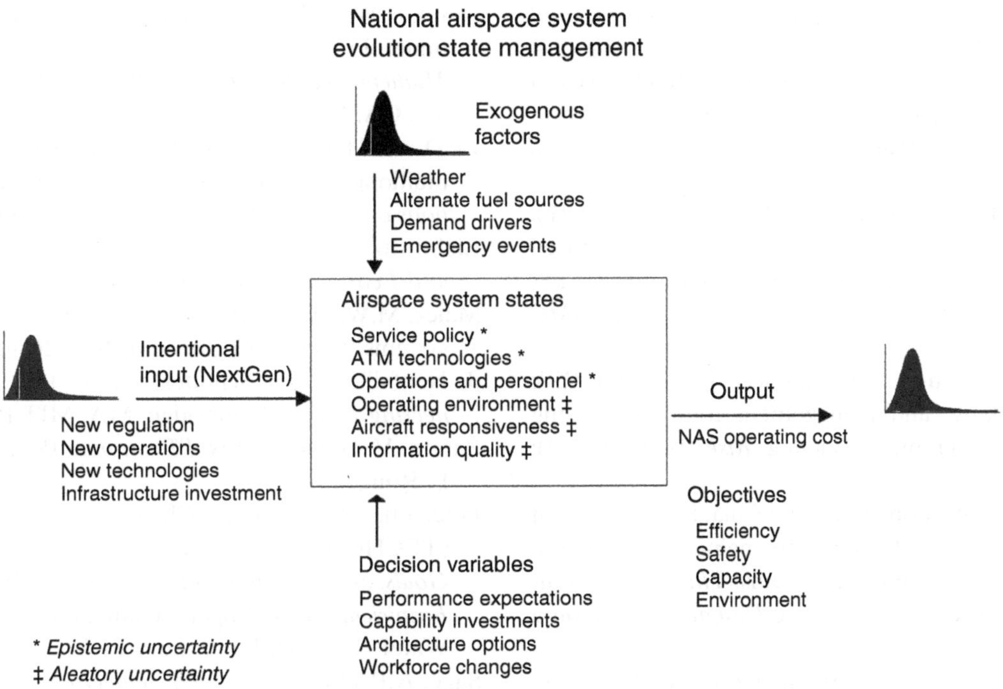

Figure 5.13 *State-space depiction of the National Airspace System.*

influence the operation of the national airspace, the desired emergent behavior may correspondingly change. For example, capacity has long been the primary objective of modernization programs. The new emergent behavior was manifested by higher load factors, fewer flights, and ultimately more pressure for optimal fuel routes. These emergent behaviors can also be geographical with different desired goals – locally and nationally. Thus, the Pareto-optimal frontier is a function of time and of the *Evolving Base*. One way to model this evolution is to observe and react to the changes in the *Evolving Base*, or alternatively, to use scenarios as precursors to forecast the shifts in frontiers and in desired behaviors.

Clearly, the above challenges constitute a daunting systems engineering task.

REFERENCES

Cardin, M.A., Kolfschoten, G.L., Frey, D.D. et al. (2013). Empirical evaluation of procedures to generate flexibility in engineering systems and improve lifecycle performance. *Research in Engineering Design* **24**(3): 277–295.

Chankong, V. (1977). Multiobjective decision making analysis: the interactive surrogate worth trade-off method. Ph.D. dissertation. Case Western Reserve University, Cleveland, OH.

Chankong, V. and Haimes, Y.Y. (1983). Optimization-based methods for multiobjective decision-making: an overview. *Large Scale Systems* **5**: 1–33.

Chankong, V. and Haimes, Y.Y. (2008). *Multiobjective Decision Making: Theory and Methodology*. New York: Dover.

Covey, S.M.R. (2008). *The Speed of Trust: The One Thing that Changes Everything*. New York: Free Press.

Davenport, T.H. and Prusak, L. (1998). *Working Knowledge*. Boston, MA: Harvard Business School Press.

DeLaurentis, D. (2005). Understanding Transportation as a System-of-Systems Design Problem. *43rd AIAA Aerospace Sciences Meeting and Exhibit*, Reno, Nevada (10–13 January 2005).

Geoffrion, A.M. (1968). Proper efficiency and theory of vector maximization. *Journal of Mathematical Analysis and Applications* **22**: 618–630.

Haimes, Y.Y. (1981). Hierarchical holographic modeling. *IEEE Transactions: Systems, Man and Cybernetics* **11** (9): 606–617.

Haimes, Y.Y. (2009). On the complex definition of risk: a systems-based approach. *Risk Analysis* **29**(12): 1647–1654.

Haimes, Y.Y. (2012). Systems-based guiding principles for risk modeling, planning, assessment, management, and communication. *Risk Analysis* **32**(9): 1451–1467.

Haimes, Y.Y. (2016). *Risk Modeling, Assessment, and Management*, 4e. Hoboken, NJ: Wiley.

Haimes, Y.Y. and Hall, W.A. (1974). Multiobjectives in water resources systems analysis: the surrogate worth trade-off method. *Water Resources Research* **10**(4): 615–624.

Haimes, Y.Y., Lasdon, L.S., and Wismer, D.A. (1971). On the bicriterion formulation of the integrated system identification and systems optimization. *IEEE Transactions on Systems, Man, and Cybernetics SMC* **1**: 296–297.

Haimes, Y.Y., Kaplan, S., and Lambert, J.H. (2002). Risk filtering, ranking, and management framework using hierarchical holographic modeling. *Risk Analysis* **22**(2): 383–397.

Hu, J., Cardin, M.-A., Poh, K.-L., and Chia, E.S. (2013). An approach to generate flexibility in engineering design of sustainable waste-to energy systems. *The 19th International Conference on Engineering Design*, Seoul, Korea (19–22 August 2013).

Intriligator, M.D. (1970). *Mathematical Optimization and Economic Theory*. Englewood Cliffs, NJ: Prentice-Hall.

Jugulum, R. and Frey, D.D. (2007). Toward a taxonomy of concept designs for improved robustness. *Journal of Engineering Design* **18**(2): 139–156.

Katzenbach, J.R. and Smith, D.K. (1999). *The Wisdom of Teams*. New York: HarperCollins.

Kuhn, H.W. and Tucker, A.W. (1951). Nonlinear programming. In: *Proceedings, 2nd Berkeley Symposium on Mathematical Statistics and Probability*, 481–492. Berkeley, CA: University of California Press.

Li, D. and Haimes, Y.Y. (1987). The envelope approach for multiobjective optimization problems. *IEEE Transactions on Systems, Man, and Cybernetics* **17**(6): 1026–1038.

Liebowitz, S. and Margolis, S. (2000). *Encyclopedia of Law and Economics*, 981. Cheltenham: Edward Elgar.

Maier, M.W. (1998). Architecting principle for systems-of-systems. *Systems Engineering* **1**(4): 267–284.

de Neufville, R. and Scholtes, S. (2011). *Flexibility in Engineering Design*. Cambridge, MA: MIT Press.

Pareto, V. (1896). *Cours d'Economie Politique*. Lausanne: F. Rouge.

Peterson, D.F. and Chair, Technical Committee (TECHCOM) (1974). *Water Resources Planning, Social Goals, and Indicators: Methodological Development and Empirical Tests*. Logan: Utah State University, Utah Water Research Laboratory PRWG 131–1.

Sage, A.P. (1992). *Systems Engineering*. New York: Wiley.

Sage, A.P. (1995). *Systems Management for Information Technology and Software Engineering*. New York: Wiley.

Simon, H.A. (1947). *Administrative Behavior: A Study of Decisionmaking Processes in Administrative Organization*, 1e. New York: Macmillan.

Simon, H.A. (1956). Rational choice and the structure of the environment. *Psychology Review* **63**(2): 129–138.

6

Hierarchical Coordinated Bayesian Modeling of Complex Systems of Systems

6.1 HIERARCHICAL COORDINATED BAYESIAN MODELING: THEORY AND METHODOLOGY

6.1.1 Introduction

Modeling interdependent and interconnected complex systems of systems (Complex SoS) based only on direct relevant empirical data is deficient in that they lack the encompassing and effective real-world modeling of Complex SoS. The lack of *direct* relevant data can be supplemented by *indirect* relevant data from multiple sources to compensate for the scarceness of direct relevant data. Building on Hierarchical Overlapping Coordination (HOC), introduced in Chapter 3, we develop a new modeling schema for modeling Complex SoS, termed here Hierarchical Coordinated Bayesian Modeling (HCBM), to address this deficiency in appropriate database, with which to populate our modeling of the individual systems that constitute Complex SoS. With HCBM, we decompose the system and the direct and indirect data from different sources and perspectives. For each decomposition, we extract information from various levels of hierarchies of data and deploy Bayesian data analysis. Then, we coordinate the results from different decomposition perspectives. In sum, HCBM can integrate both direct relevant data and indirect relevant data from multiple perspectives and extract information from multiple dimensions, while the emphasis remains on available and direct relevant data (Yan and Haimes, 2010, 2011).

To make rational decisions in modeling, we first need to focus on the needs and objectives of the modeling process. Modeling Complex SoS implies sustained efforts to understand the intricacy of the systems under consideration, supported by necessary assumption, simplification, and abstract encoding. Thus, some information loss is unavoidable, and the resulting mathematical models are generally expected to represent the essence of the real system and, thus, of the real Complex SoS. The role of systems modelers is centered on making adroit choices to iterative transitions from the Complex SoS to their mathematical models and back again while keeping information loss as minimal as possible.

Since we can never develop perfect models of a single system, or of Complex SoS, the mathematical model is built and shaped on the real system by assumption, simplification, and abstract encoding.

Modeling and Managing Interdependent Complex Systems of Systems, First Edition. Yacov Y. Haimes.

On one hand, a critical problem in systems modeling and analysis is the availability of appropriate database to populate our models, which are often scarce. On the other hand, we can obtain data from multiple sources with different structures. We want to integrate all the heterogeneous and limited information to perform holographic systematic modeling and risk analysis. Thus we face the following dilemma: we have information from various sources of different time horizons, but the data that are directly relevant to the Complex SoS under consideration are scarce. For example, suppose we want to estimate the level of radon gas in a home. The measurement process costs money and time, making it difficult to collect sufficient data for a specific household or location. However, we have background data from different hierarchies, e.g. the same county, other counties of the same state, and from other states. Obviously, these data are not as important as the data we collect in the specific location of interest; however, they do provide valuable background information for our modeling efforts of Complex SoS. *The next step is to integrate all the direct and indirect related information while placing more emphasis on the direct relevant data (collected in the specific location).*

Hierarchical modeling and decomposition methodology has been widely used for mathematical modeling of large-scale hierarchical systems. Multiple decomposition methodologies are introduced in Chapters 2 and 4 and throughout the book. A sample of which is highlighted here: Dantzig–Wolfe decomposition (Dantzig and Wolfe, 1960), Lagrangian decomposition (Lasdon and Schoeffler, 1966; Lasdon, 1970; Haimes, 1977), and Bender's partitioning algorithm (Bender, 1962), among others. The basic idea of these methods is to decompose a complex mathematical model (linear or nonlinear) into relatively simpler "independent" submodels, solve them separated, coordinate them with a high-level "coordinator," and thus find "optimal solution" of the complex mathematical model, or Pareto-optimal solutions to multiple objective models (see Chapter 5).

Haimes extended decomposition methodology to systems modeling and proposed HOC (Haimes and Macko, 1978; Haimes et al., 1990) and Hierarchical Holographic Modeling (HHM) (Haimes, 1981; Haimes et al., 1990; 1991). Since alternative decompositions of a single complex system or Complex SoS may be feasible and desirable, HOC decomposes hierarchical Complex SoS from different perspectives and coordinates their multiple decompositions. Other situations may require multiple HOC models with some shared/common states and other essential entities and objectives to be coordinated within HHM. The term "essential entities" (introduced in Chapter 1) includes *shared/common decisions*, *decision makers*, *stakeholders*, *resources*, *organizational behavior and norms*, *policies and procedures*, *management*, *culture*, and others.

In sum, in this chapter we incorporate multiple decompositions from multiple perspectives supported and populated with the Bayesian data analysis. This modeling theory, philosophy and methodology, integrates all the direct and indirect relevant information from different levels of the hierarchies while placing more emphasis on relevant direct data. Indeed, by looking at the systems and data from different perspectives (e.g. via HHM introduced in Chapter 3), we can fully extract and exploit the information from various dimensions. Thus, we coordinate the results from different decompositions and perform quantitative modeling of Complex SoS supported with, and enriched by, multiple databases.

6.1.2 Hierarchical Overlapping Coordination (HOC)

HOC, introduced in Chapter 3, emerged from the study of water resource systems (Haimes and Macko, 1978). These Complex SoS are characterized by distinct coexisting interdependent and interconnected Complex SoS but with different decompositions. The decompositions may address different objectives or other perspectives of Complex SoS; however, their simultaneous functions or perspectives must be compatible. The HOC methodology provides a mathematical framework for representing such systems, permitting alternative decomposition of a single system, and showing how the mutual functioning of the resultant plural models may be coordinated. Applying HOC has yielded very promising results as highlighted in this chapter and elsewhere in this book (Haimes et al., 1990).

In many ways, HHM, the subject of Chapter 3, is a generalization of HOC and constitutes a comprehensive theoretical framework for systems modeling.

In Chapter 3 we listed the following general aspects in decomposing a system with HHM: functional, geographical, economic, political, environmental, institutional, and temporal, among others.

Consider the following advantages of the decomposition of Complex SoS from different perspectives: (i) conceptual simplification of complex systems, (ii) more realistic system models, (iii) interactions among subsystems are permissible, (iv) applicable to both static and dynamic systems, (v) economic interpretation of the variables, (vi) applicable to multiobjective analyses, (vi) reduction in dimensionality, (vii) simple programming and computation procedures, (viii) different computing techniques for subsystem modeling, and (v) use of existing models.

Decomposition can be beneficial for modeling, decision making, and optimization. These advantages are encapsulated in the following three perspectives: (i) Decomposition methods can reflect the internal hierarchical nature of large-scale multiobjective systems. (ii) Tradeoff analysis can be performed among subsystems and the overall system. (iii) Through decomposition, the intricacy of multiobjective Complex SoS can be relaxed by solving several smaller subproblems.

6.1.3 Bayesian Methods and Risk Analysis

Empirical data are almost always lacking in real world for modeling and managing Complex SoS. While directly relevant data are limited, we have multiple information sources, which can provide indirectly relevant data. These are especially useful and essential for modeling Complex SoS, in particular, when directly relevant data are scarce.

A Bayesian approach is presented here, which is most effective in addressing these data limitations. By design, Bayesian methods inherently consider the uncertainty associated with the parameters of a probabilistic model (even if those uncertain parameters are believed to be fixed numbers). Bayesian methodology is often recommended as the proper way to make formal use of subjective information, such as expert opinion and the personal judgments or beliefs of an analyst. Unlike frequentist methods with which they are often contrasted, Bayesian methods have an important advantage: They can always yield a more precise answer, even when few data are available.

There are some limitations to applying Bayesian methods in modeling of Complex SoS. A major problem is that sometimes it is very difficult to come up with a precise distribution to be used as prior probability of a database; also, the posterior probability is generally very sensitive to the prior. Many approaches have been suggested for selecting priors. An *uninformative* prior is a uniform prior distribution. Uninformative priors carry no information, and thus all the information is from the data that is embodied in the likelihood function. When data is very limited, the posterior cannot precisely reflect the properties of the underlying system or Complex SoS. Another common method uses the *subjective* prior, that is, the prior distribution is determined by expert subjective opinions. This kind of prior is intended to reflect the analyst's or expert's personal knowledge about the system under investigation. There are many basic questions with this method: (i) Who determines the prior? (ii) Why use this prior? (iii) Based on what evidence? and (iv) What does it mean to anyone else? A hierarchical Bayesian model can provide a reasonable prior distribution by integrating data that is indirectly relevant. We will subsequently introduce the Bayesian method for inference in the section on hierarchical Bayesian model.

6.1.4 Bayesian Method for Distributions Inference

Given a likelihood $p(x|\theta)$ and prior density $p(\theta)$, with the Bayesian method we can infer the posterior density of parameter θ:

$$p(\theta|x) = \frac{p(x|\theta)p(\theta)}{p(x)} = \frac{p(x|\theta)p(\theta)}{\int_{\theta} p(x|\theta)p(\theta)d\theta} \tag{6.1}$$

where p denotes probability density (mass for discrete case), θ is a value of the random variable in question (parameters of a system), and x denotes the evidence being considered (observed data generated by the system). $p(\theta)$ is a prior probability that the random variable takes on the value θ. $p(x|\theta)$ is the conditional likelihood function that expresses the probability of the evidence given a particular value of θ. $p(x)$ is a moralizing factor, which corresponds to the probability of having obtained the observed evidence. We can expand it to be the integral with respect to θ of the product of the prior and the likelihood. The moralizing factor is not a function of θ but has a single value such that the resulting posterior distribution integrates easily to unity. Thus, we can also write the Bayesian rules as

$$p(\theta|x) \propto p(x|\theta)p(\theta) \tag{6.2}$$

where $\propto$ means "proportional to."

The normalizing factor is often difficult to compute analytically, but the use of conjugate pairs (prior and likelihood) may greatly simplify the problem (Gelman et al., 2013). For nonconjugate pair, we can use computational techniques, such as numerical integration and simulation.

After the posterior density of θ conditional on the observed data x have been obtained by Eq. (6.1), we can predict an unknown observable, $\tilde{x}$. For example, suppose $x = (x_1, \ldots, x_n)$ is the vector of recorded seconds a runner runs n times for a certain distance, and $\theta \sim (\mu, \sigma^2)$ is the unknown expected seconds. We can get the posterior distribution of θ, $p(\theta|x)$ by Eq. (6.1). Let $\tilde{x}$ be the seconds the runner will run in the $(n+1)$th run. The distribution of $\tilde{x}$ is called the posterior predictive distribution, posterior because it is conditional on the observed x and predictive because it is a prediction for an observable $\tilde{x}$:

$$\begin{aligned} p(\tilde{x}|x) &= \int p(\tilde{x},\theta|x)\mathrm{d}\theta = \int p(\tilde{x}|\theta,x)p(\theta|x)\mathrm{d}\theta \\ &= \int p(\tilde{x}|\theta)p(\theta|x)\mathrm{d}\theta \end{aligned} \tag{6.3}$$

Example: *SCADA risk analysis* (Yan and Haimes, 2010, 2011)

Supervisory control and data acquisition (SCADA) is a system that allows operators to monitor and control processes that are distributed among various remote sites for most physical and cyber–physical Complex SoS infrastructures. Many control processes use SCADA systems, including hydroelectric, water distribution and treatment utilities, and natural gas, among others. In the case study discussed here, we assess the risks of a cyber attack on an SCADA serving water supply Complex SoS in city XYZ. Denial of service is the major potential danger for an SCADA system. Denial-of-service attacks intentionally block or severely degrade the communications network. An attacker makes shared resources inoperative by taking up time so that other processes are effectively stopped. This can be accomplished by filling up disk space, modifying network applications, etc. Here, we use the *time to recovery* as a measurement of the consequences after a cyber attack causes an SCADA "denial of service." For example, a disgruntled employee and hackers are the two most likely sources of a cyber attack on an SCADA – operated water supply system.

Suppose the primary concern of the SCADA risk in city XYZ is disgruntled employees, we want to estimate the risk of current SCADA, more specifically, the time to recovery after a cyber attack by disgruntled employees. Since cyber attacks on SCADA that serve Complex SoS are not common events, only four attacks are considered in this case study caused by disgruntled employees and resulting in "denial of service." Table 6.1 estimates the time to recovery of each attack.

We use record 1 as an example. Since the "time to recovery" is nonnegative, lognormal distribution might be reasonable for it. We first take the logarithm of the data to transform them to normal distribution. There are two parameters for a normal distribution, θ and σ^2, which are the mean and variance, respectively. We assume the variance σ^2 is known, which is equal to 2 and we want to estimate the mean θ. Thus, in record 1, the transformed observation $x = \log(z) = 2.433\,613$ is normally distributed with mean θ and variance $\sigma^2 = 2$. We have the likelihood function

$$P(x|\theta) = \frac{1}{\sqrt{4\pi}}\exp\left(-\frac{(x-\theta)^2}{4}\right)$$

TABLE 6.1 Cyber Attack on Current SCADA of City XYZ

Record	City	Attackers	Time to recovery z (h)	$x = \log(z)$
1	XYZ	Disgruntled employee	11.4	2.433 613
2	XYZ	Disgruntled employee	8.2	2.104 134

We assume the prior distribution of θ is a normal distribution with a mean of 1 and unit variance:

$$P(\theta) = \frac{1}{\sqrt{2\pi}}\exp\left(-\frac{(\theta-1)^2}{2}\right)$$

Using Eq. (6.1) we obtain

$$p(\theta|x) \propto p(x|\theta)p(\theta)$$
$$= \frac{1}{\sqrt{4\pi}}\exp\left(-\frac{(x-\theta)^2}{4}\right)\frac{1}{\sqrt{2\pi}}\exp\left(-\frac{(\theta-1)^2}{2}\right)$$
$$\propto \exp\left(-\frac{3\theta^2-2(x+2)\theta}{4}\right)$$

Plugging in the observed value $x = 2.433\,613$, the posterior density of θ is proportional to $\exp\left(-\frac{3\theta^2-8.860\theta}{4}\right)$. After normalization, it is a normal distribution with a mean of 1.4767 and a variance of 0.6667 (the posterior density of θ is still normally distributed because the likelihood (normal) and prior (normal) are conjugate pairs (Gelman et al., 2013)).

The calculation can be repeated for record 2 by using the posterior we got above as the prior for the next application of Bayes' rule. If we have more samples, we can update the distribution recursively. The order in which the sample data arrive generally does not matter. The posterior distribution we eventually obtain is the same, regardless of order of the data. As data accumulates during sequential updates, the initial choice of the first prior has a smaller and smaller influence on the final posterior. Unfortunately, as we have discussed before, empirical data are almost always lacking in real-world risk analyses. With limited samples, the posterior will be very sensitive to the prior. Thus, the selection of prior is an important and even intractable task for Bayesian analysts. The question is: How can a modeler of Complex SoS choose a "good prior" and defend its selection?

In this example, the prior we use is a normal distribution with a mean of 1 and unit variance. Suppose this prior was proposed by modeler 1. Modeler 2 may have a different opinion and suggests a normal prior with a mean of 4 and a unit variance. Figure 6.1 shows the posteriors with 1 sample point (record 1) and 2 sample points (record 1 and 2). To study the convergence, we also tried 10 and 100 sample points. The solid lines are the posteriors with the prior provided by analyst 1 and the dotted lines are

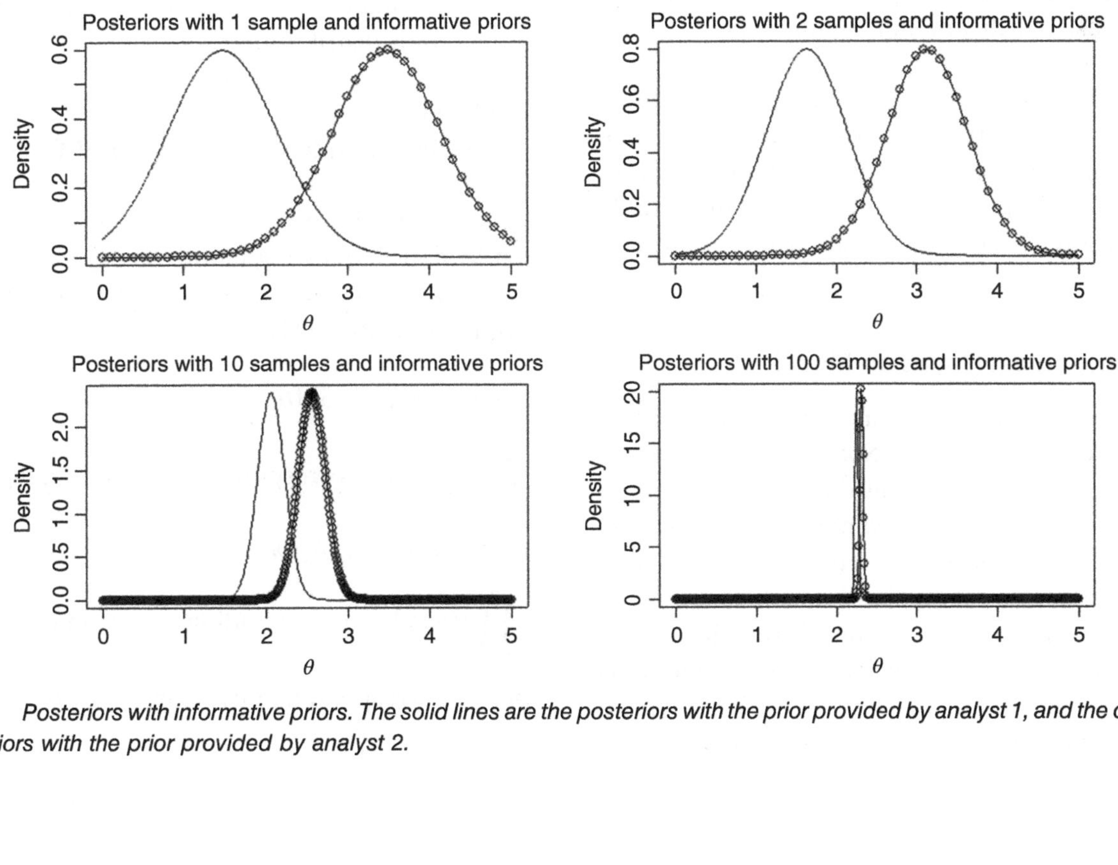

Figure 6.1 Posteriors with informative priors. The solid lines are the posteriors with the prior provided by analyst 1, and the dot lines are the posteriors with the prior provided by analyst 2.

the posteriors with the prior provided by analyst 2. Note that with one or two data points, the two posteriors obviously differ. As more data are collected, their posteriors tend to converge even if the two priors were quite different.

Since the posterior is very sensitive to the prior when data samples are limited, one should be very careful with the selection of prior. An alternative way is to assign a "noninformative" prior to our ignorance or uncertainty about the prior knowledge. For example, assume that analyst 3 gives a normal prior with a mean of 1 and a variance of 1000, and analyst 4 gives a normal prior with a mean of 4 and a variance of 1000. Both the two priors are "noninformative" due to the large variances. The first picture in Figure 6.2 shows the posteriors based on analyst 3 and analyst 4's priors. We can find that the two posteriors are almost overlapped since both of the two prior are noninformative. The second picture in Figure 6.2 compares the posteriors based on noninformative priors with the posteriors based on informative priors. All of them used two data samples from the example. We can find that the posterior distributions with noninformative priors are much flatter than those with informative priors. That is because the noninformative priors do not provide much substantial knowledge. Most of the information comes from the two data samples, which are not enough for posterior inference.

From this example, we can find that both informative and noninformative priors have their problems when the data samples are limited. Thus, selection of prior is a difficult and even intractable task, especially when the data are scarce as in risk analysis.

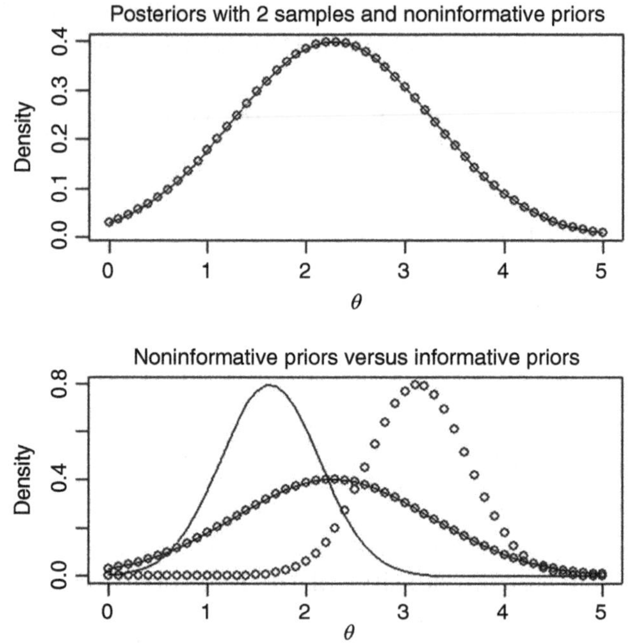

Figure 6.2 *Posteriors with noninformative priors.*

6.1.5 Hierarchical Bayesian Model

An interesting phenomenon in systems engineering and modeling of Complex SoS is that while we may be lacking direct relevant data, we often might be able to obtain indirect relevant data from multiple sources. We want to seek a way to take advantage of both the direct relevant and indirect relevant data by integrating them, while the emphasis remains on the direct relevant data. Hierarchical Bayesian model is very appropriate for this requirement. Hierarchical Bayesian models have been applied in many fields. Nonhierarchical models are usually inappropriate for hierarchical data: with few parameters, they generally cannot fit large data sets accurately, whereas with many parameters, they tend to "overfit" such data. In contrast, hierarchical models can have enough parameters to fit the data well, while using a population distribution to structure some dependence into the parameters (Yan and Haimes, 2010, 2011).

Consider the following pedagogical Complex SoS with hierarchical structure.

Suppose we have k groups of experiments, we can then form the following hierarchy.

A simple hierarchical model is of the following form (Bernardo, 1994):

$$p(x|\theta) = p(x_1,\ldots,x_k|\theta_1,\ldots,\theta_k) = \prod_{i=1}^{k} p(x_i|\theta_i)\ldots\text{corresponding to level 1 in Figure 6.3}$$
$$p(\theta|\phi) = p(\theta_1,\ldots,\theta_k|\phi) = \prod_{i=1}^{n} p(\theta_i|\phi)\ldots\text{corresponding Level 2 in Figure 6.3}$$
$$p(\phi)\ldots\text{corresponding Level 3 in Figure 6.3} \tag{6.4}$$

The basic interpretation is as follows. Data set $x_1, \ldots, x_k$ is available from k different, but related, sources (subsystems of the Complex SoS): for examples, k individuals in a homogeneous population or k clinical trial centers involved in the same study. The first level of the hierarchy specifies parametric model components for each of the k data sets. Because the data sets are under the same decomposition perspective and share some of the same inherent characteristics, the parameters $\theta_1, \ldots, \theta_k$ have the same hyperparameter ϕ, which itself is a random variable. With the Bayesian theorem, we can derive the posterior of θ_i:

$$p(\theta_i|x) = \int p(\theta_i|\phi,x)p(\phi|x)\mathrm{d}\phi \qquad (6.5)$$

where

$$p(\theta_i|\phi,x) \propto p(x|\theta_i)p(\theta_i|\phi)$$
$$p(\phi|x) \propto p(x|\phi)p(\phi)$$
$$p(x|\phi) = \int p(x|\theta)p(\theta|\phi)\mathrm{d}\theta$$

We complete the posterior distribution computation procedure by using normal model as an example.

Consider k independent experiments, with experiment i estimating the parameter θ_i from n_i independent normally distributed data points, $x_{i,j}$, each with known error variance σ^2, that is,

$$x_{i,j} | \theta_i \sim N(\theta_i, \sigma^2), \quad \text{for } i = 1,\ldots,k, \; j = 1,\ldots,n_i \qquad (6.6)$$

Using a standard notation from the analysis of variance, we label the sample mean of each group j as

$$\bar{x}_{i.} = \frac{1}{n_i}\sum_{j=1}^{n_i} x_{ij}$$

With sampling variance $\sigma_i^2 = \frac{\sigma^2}{n_i}$, we can then write the likelihood for each θ_i in terms of the sufficient statistics, $\bar{x}_{i.}$:

$$\bar{x}_{i.} | \theta_i \sim N(\theta_i, \sigma_i^2)$$

For the hierarchical model as discussed above, we assume that the parameters θ_i are drawn from a normal distribution with hyperparameters (μ,τ):

$$P(\theta_1,\ldots,\theta_k|\mu,\tau) = \prod_{i=1}^{k} N(\theta_i|\mu,\tau^2)$$
$$P(\theta_1,\ldots,\theta_k) = \int \prod_{i=1}^{k} [N(\theta_i|\mu,\tau^2)] P(\mu,\tau)\mathrm{d}(\mu,\tau) \qquad (6.7)$$

We assign a noninformative uniform hyperprior distribution to μ, given τ, so that

$$P(\mu,\tau) = P(\mu|\tau)P(\tau)$$

Gelman et al. (2013) gives the joint posterior distribution for all the parameters and hyperparameters:

$$P(\theta,\mu,\tau|x)P(\tau|x)P(\mu|\tau,x)P(\theta|\mu,\tau,x) \qquad (6.8)$$

where

$P(\tau|x)$ is defined as

$$P(\tau|x) \propto \frac{P(\tau)\prod_{i=1}^{k} N(\bar{x}_{i.}|\mu, \sigma_i^2 + \tau^2)}{N(\mu|\mu, V_\mu)}$$
$$\propto P(\tau)V_\mu^{1/2}\prod_{i=1}^{k}(\sigma_i^2 + \tau^2)^{-1/2}\exp\left(-\frac{(\bar{x}_{i.} - \mu)^2}{2(\sigma_i^2 + \tau^2)}\right) \qquad (6.9)$$

with μ and V_μ defined by

$$\mu = \frac{\sum_{i=1}^{k}\frac{1}{\sigma_i^2 + \tau^2}\bar{x}_{i.}}{\sum_{i=1}^{k}\frac{1}{\sigma_i^2 + \tau^2}} \text{ and } V_\mu^{-1} = \sum_{i=1}^{k}\frac{1}{\sigma_j^2 + \tau^2} \qquad (6.10)$$

Thus, $P(\tau|x)$ is a complicated function of τ.

$P(\mu|\tau, x)$ is defined as

$\mu|\tau, x \sim N(\mu, V_\mu)$ with μ and V_μ defined in Eq. (6.10).

$P(\theta|\mu,\tau, x)$ is defined as

$$\theta_i | \mu,\tau,x \sim N(\theta_i, V_i), \quad \text{where}$$
$$\theta_i = \frac{\frac{1}{\sigma_i^2}\bar{x}_{i.} + \frac{1}{\tau^2}\mu}{\frac{1}{\sigma_i^2} + \frac{1}{\tau^2}} \text{ and } V_i = \frac{1}{\frac{1}{\sigma_i^2} + \frac{1}{\tau^2}} \qquad (6.11)$$

It is hard to compute the hierarchical model analytically, even for the conjugate pairs (Gelman et al., 2013). Several computational techniques have been applied, such as numerical integration and the Markov chain simulation.

Hierarchical Bayesian is very appropriate for modeling Complex SoS due to the vast database required for each system and for the entire Complex SoS. We can regard each data set as a subsystem. All the subsystems share some common inherent characteristics, noted in this book as shared states and other essential entities, but each has its own special situation. When the data we can collect for each subsystem are limited, they cannot fully reflect its characteristics. However, we can use that data in other subsystems to extract shared states and other essential entities and other shared characteristics of all the subsystems of the Complex SoS and form the prior knowledge. This can be updated by using the directly relevant data to get the specific estimation of each subsystem. As we have discussed before, lack of directly relevant data is very common in modeling Complex SoS. By using the hierarchical Bayesian model, we can make full use of both data indirectly relevant to the entire Complex SoS (for prior knowledge) and the data directly relevant to the specific subsystem (for updating the prior knowledge) to achieve sufficient analysis. We have shown the disadvantages of both informative and noninformative priors when data are scarce. In hierarchical models, we generally assume noninformative priors for the hyperparameters, and therefore, we cannot get prior information or prior knowledge from prior distribution, which is similar to the nonhierarchical model with noninformative priors. However, unlike nonhierarchical models, the hierarchical structure (e.g. indirect relevant data) provides some useful background information, which can complement the lack of direct relevant data.

In our SCADA Complex SoS example, we have shown that only two data are directly relevant to city XYZ. By studying the data in other cities in the United States, we find some other cities are using the same SCADA systems as city XYZ. The historical data of cyber attack by insider disgruntled employees of these cities are listed in Table 6.2.

TABLE 6.2 Cyber Attack on Current SCADA

City	Attackers	Time to recovery z_i (h)	$x_i = \log(z_i)$
XYZ	Disgruntled employee	11.4	2.43
XYZ	Disgruntled employee	8.2	2.10
City 1	Disgruntled employee	10.60	2.36
City 1	Disgruntled employee	15.70	2.75
City 2	Disgruntled employee	2.90	1.06
City 2	Disgruntled employee	1.97	0.68
City 3	Disgruntled employee	6.80	1.92
City 3	Disgruntled employee	1.20	0.18
City 3	Disgruntled employee	7.90	2.07
City 4	Disgruntled employee	12.53	2.53
City 5	Disgruntled employee	2.70	0.99
City 5	Disgruntled employee	3.02	1.10
City 5	Disgruntled employee	5.40	1.69
City 6	Disgruntled employee	10.70	2.37
City 6	Disgruntled employee	14.22	2.65
City 7	Disgruntled employee	5.20	1.65
City 7	Disgruntled employee	2.10	0.74

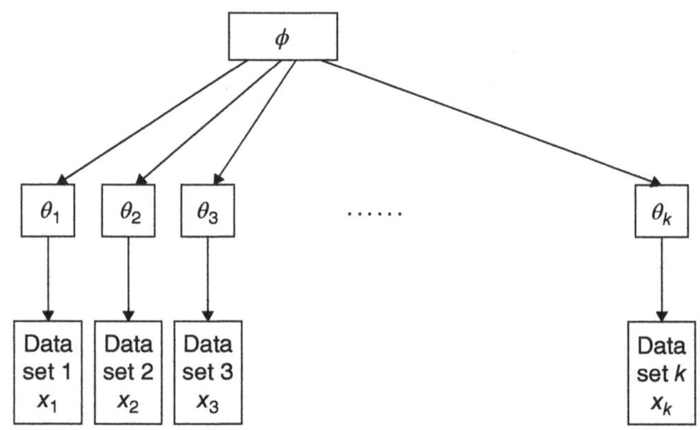

Figure 6.3 *Hierarchical Bayesian model.*

Similar to Figure 6.3, we decompose the data from city perspective and form the hierarchy of Figure 6.4.

For each group (city), the data are very limited. An extreme case is that city 4 has only one sample. Thus, it is hard to estimate the group variance σ_i^2 directly. In this model, we allow the group data variance, σ_i^2, to be unknown. Under the city decomposition, data x_i, $i = 1$, …, 8, are independently normally distributed within each of eight cities, with means θ_i and common variance σ^2. The group means are assumed to follow a normal distribution with unknown mean μ and variance τ^2. Since we do not have much information about the hyperparameters μ and τ, we use noninformative prior information. We express the noninformative prior information by proper distributions with large uncertainties: μ is normally distributed with mean 0 and standard deviation 1000, and τ is uniformly distributed from 0 to 1000.

Thus, we form the following model:

$$x_i \sim N(\theta_i, \sigma^2), \quad i = 0, 1, 2, \ldots, 7$$

$$x_{1,k1} + x_{2,k2} + x_{3,k3} + x_{4,k4} \leq Q$$

$$\mu \sim N(0, 1000^2)$$

$$\tau \sim \text{uniform}(0, 1000)$$

$$\log(\sigma) \sim \text{uniform}(0,1000)$$

To run the MCMC simulation, we used three chains, each with 100 050 iterations (first 49 950 discarded). In Table 6.3 we use the same setup for the attacker decomposition model.

With the posterior distributions of θ_i and common variance σ, we can get the posterior predictive distribution for z_i, that is, the distribution of "time to recovery" after a cyber attack for each city. We show them by histograms in Figures 6.5 and 6.6.

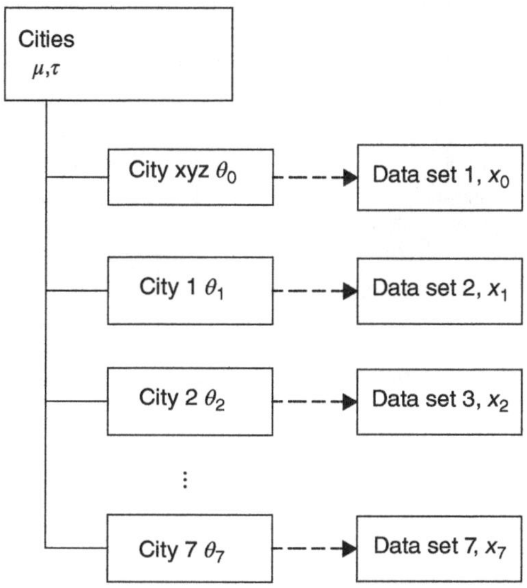

Figure 6.4 City decomposition of the SCADA cyber attack database.

6.2 HIERARCHICAL COORDINATED BAYESIAN MODELING (HCBM) FOR COMPLEX SYSTEMS OF SYSTEMS

6.2.1 Data Decomposition from Multiple Perspectives

In Section 6.1.5, we introduced the hierarchical Bayesian model, by which we decomposed the entire data set of a system into several groups (subsystems) and analyzed them hierarchically by combining information from all the groups without assuming all the θ to be equal. In Complex SoS as we discussed in Section 6.1, we find decomposition from only one perspective to be insufficient and less informative. In the SCADA Complex SoS example, the decomposition perspective was a "city." We decomposed the entire data set from the "city's" perspective and extracted information from the "city's" dimension. However, other decomposition perspectives, or "dimensions," are also possible and desirable. In the SCADA modeling example, another reasonable decomposition perspective might be the "attacker's characteristics." Thus, more than one decompositions on the database is desired. Indeed, each decomposition has its own

TABLE 6.3 Simulation Results for the Posterior Distributions of City Decomposition

	Mean	sd	2.50%	25%	50%	75%	97.50%	R	n_{eff}
θ_0	1.9	0.5	1	1.6	1.8	2.1	2.9	1	1000
θ_1	1.9	0.5	1	1.6	1.9	2.2	3.1	1	1000
θ_2	1.5	0.5	0.5	1.3	1.6	1.9	2.4	1	610
θ_3	1.6	0.4	0.7	1.3	1.6	1.9	2.4	1	630
θ_4	1.8	0.5	0.9	1.5	1.8	2.1	3.1	1	1000
θ_5	1.6	0.4	0.6	1.3	1.6	1.9	2.3	1	620
θ_6	1.9	0.5	1	1.6	1.9	2.2	3.1	1	860
θ_7	1.6	0.4	0.7	1.3	1.6	1.9	2.5	1	1000
μ_θ	1.7	0.3	1.1	1.5	1.7	1.9	2.4	1	1000
σ_θ	0.4	0.4	0	0.2	0.4	0.6	1.4	1	1000
τ_y	0.8	0.1	0.5	0.7	0.9	0.9	1	1	1000
Deviance	43.7	3.4	37.7	41.5	43.3	45.3	51.4	1	1000

pD = 4.7 and DIC = 48.4 (using the rule, pD = var(deviance)/2). For each parameter, n_{eff} is a crude measure of effective sample size, and R is the potential scale reduction factor (at convergence, $R = 1$). DIC is an estimate of expected predictive error (lower deviance is better).

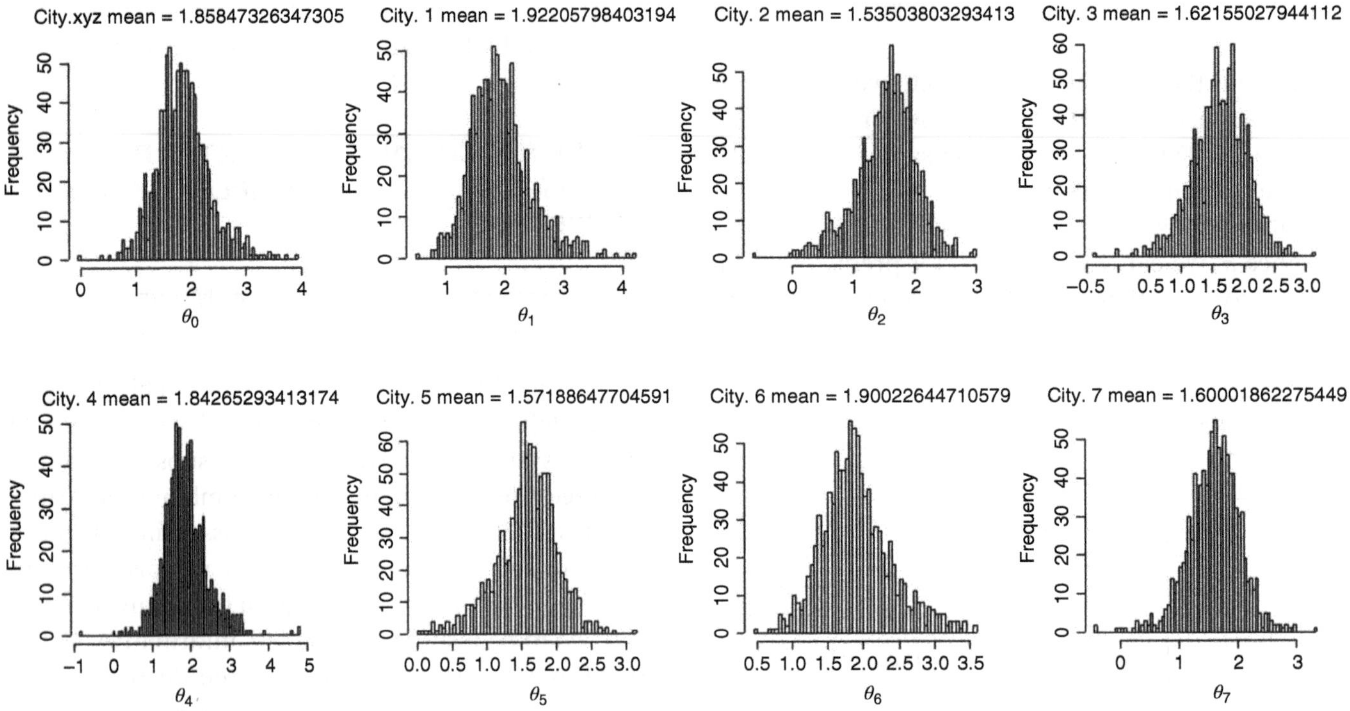

Figure 6.5 *Posterior distribution of* θ_i*, city decomposition.*

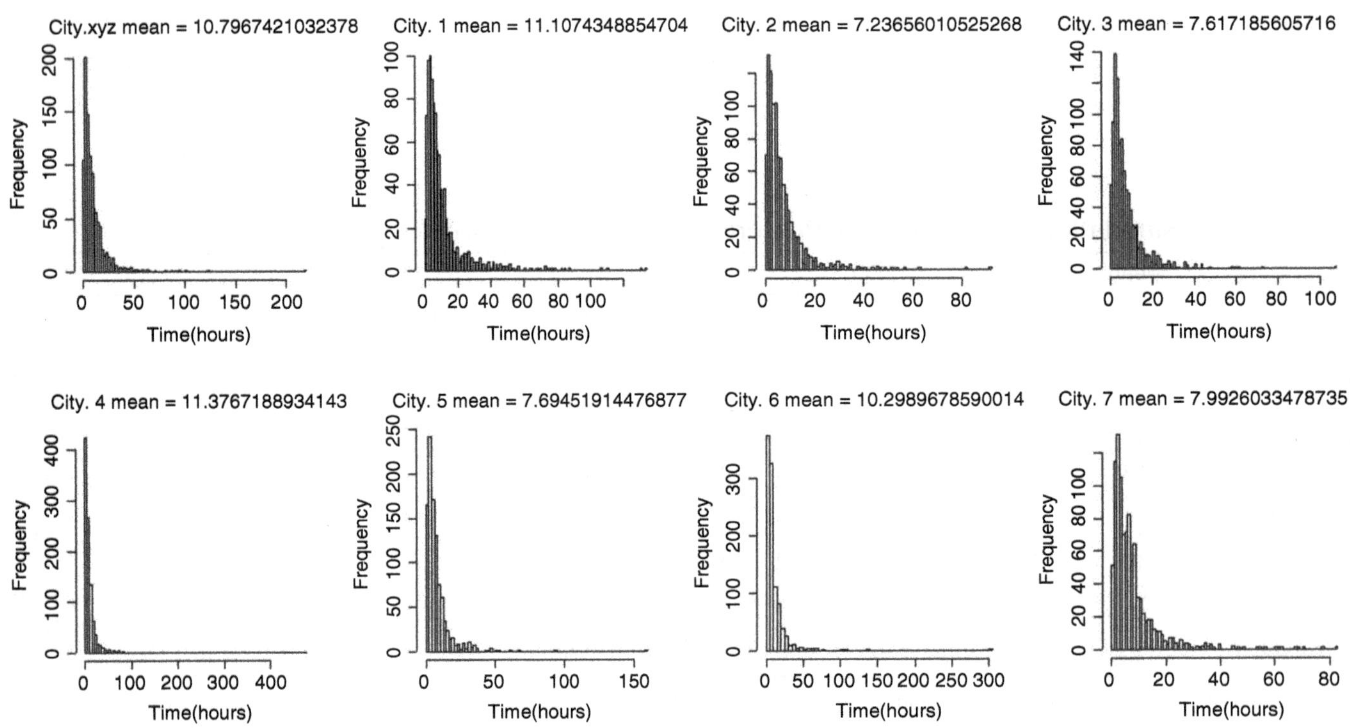

Figure 6.6 *Predictive distribution of "time to recovery" after a cyber attack for each city.*

merits, advantages, and disadvantages, and each represents a very specific database collected and cared for by different agencies and constituencies. Similar to the spirit of HOC, we develop in this chapter "Hierarchical Overlapping Coordination Bayesian Modeling" (HCBM).

In HCBM, we decompose the data into sub-data sets from different perspectives. For each perspective we use hierarchical Bayesian model to do analysis independently. After we get posterior distribution of each perspective, we will seek reasonable methods to coordinate the multiple results from multiple decomposition perspectives. Recall that in Section 6.1 we have decomposed the data set from the "city" perspective. Now we try to study the data from the "attacker type" perspective. Two most likely scenarios of cyber attack to SCADA for water supply system can be characterized by a disgruntled employee and/or a hacker. If the scenario is known, we labeled the attacker type "known." Thus, we have three types: disgruntled employee, hacker, and unknown entity. Table 6.4 presents the complete data set of the eight cities with three attacker types.

From the "city" perspective, we have decomposed the 17 data points with the same attacker type, "disgruntled employee," into 8 groups. If we look at the data from three different attacker types in the same city, XYZ, we can decompose the seven observations into three groups from the "attacker type" perspective. Figure 6.7 shows the two perspective decomposition model.

In Section 6.1, we have modeled and computed the city decomposition. For the "attacker type" decomposition, the model we use is slightly different from the city decomposition. Since for each attacker type, at least two samples are available, we try to estimate the sample variance $\sigma_i'^2$, $i = 1, 2, 3$, rather than assume they are equal and unknown in city decomposition. We still use noninformative priors for the hyperparameters μ' and τ'.

Thus, we form the following model for the attacker decomposition:

$$x_i' \sim N\left(\theta_i', \sigma_i'^2\right), i=1,2,3$$

$$\theta_i' \sim N(\mu', \tau'^2),\ i=1,2,3$$

TABLE 6.4 Complete SCADA Data of Eight Cities with Three Attacker Types

City	Attackers	Time to recovery z_i (h)	$x_i = \log(z_i)$
XYZ	Disgruntled employee	11.4	2.43
XYZ	Disgruntled employee	8.2	2.10
XYZ	Hacker	0.2	−1.61
XYZ	Hacker	6.7	1.90
XYZ	Hacker	9.3	2.23
XYZ	Unknown	2.8	1.03
XYZ	Unknown	7.1	1.96
City 1	Disgruntled employee	10.6	2.36
City 1	Disgruntled employee	15.7	2.75
City 1	Hacker	10.6	2.36
City 1	Hacker	0.5	−0.69
City 2	Disgruntled employee	2.9	1.06
City 2	Disgruntled employee	1.97	0.68
City 2	Hacker	1.1	0.095
City 3	Disgruntled employee	6.8	1.92
City 3	Disgruntled employee	1.2	0.18
City 3	Disgruntled employee	7.9	2.07
City 3	Hacker	0.4	−0.92
City 3	Unknown	1.2	0.18
City 3	Unknown	7.9	2.07
City 4	Disgruntled employee	12.53	2.53
City 4	Hacker	7.6	2.03
City 5	Disgruntled employee	2.7	0.99
City 5	Disgruntled employee	3.02	1.10
City 5	Disgruntled employee	5.4	1.69
City 5	Hacker	4.5	1.50
City 5	Unknown	5.4	1.69
City 6	Disgruntled employee	10.7	2.37
City 6	Disgruntled employee	14.22	2.65
City 6	Unknown	2.8	1.03
City 7	Disgruntled employee	5.2	1.65
City 7	Disgruntled employee	2.1	0.74
City 7	Hacker	2.1	0.74

$$\mu' \sim N(0, 1000^2)$$

$$\tau \sim \text{uniform}(0, 1000)$$

where $i = 1,\ 2,\ 3$ represent disgruntled employee, hacker, and unknown attackers, respectively.

Note that we assume $\sigma_i'^2$ are known by estimating them from the data. Whereas, in the city decomposition, we assume all σ_i^2 are equal and unknown, and thus we gave a noninformative prior distribution for the common σ^2 by log(σ)~uniform(0, 1000).

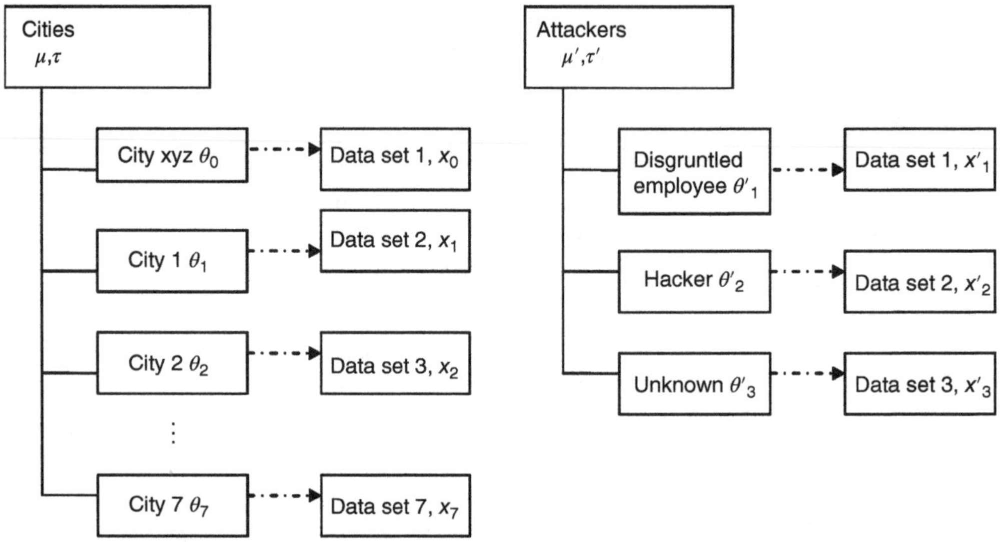

Figure 6.7 *Decomposition from two perspectives.*

TABLE 6.5 Simulation Results for the Posterior Distributions of Attacker Type Decomposition

	Mean	sd	2.50%	25%	50%	75%	97.50%	R	n_{eff}
θ_1	2.2	0.2	1.9	2.1	2.2	2.4	2.6	1	940
θ_2	1.3	1	−0.8	0.7	1.5	2.1	3	1	680
θ_3	1.6	0.5	0.6	1.3	1.6	1.9	2.4	1	1000
μ_θ	1.5	4.4	−4.4	1.2	1.8	2.3	6.8	1	1000
σ_θ	3.3	7.6	0.1	0.5	1.2	2.7	23.5	1	570
Deviance	3.6	2.1	1	2	3.2	4.8	9	1	1000

With MCMC, Table 6.5 shows the posterior distributions of the parameters.

With the posterior distributions of θ_i', we can get the posterior predictive distribution for z_i', that is, the distribution of "time to recovery" after a cyber attack for "attacker type." We show them by histograms in Figures 6.8 and 6.9.

Recall that with hierarchical Bayesian model, we can extract information from different levels of the hierarchies. Furthermore, with HCBM, we can obtain information extraction from different dimensions or perspectives of the hierarchy. In this example, we first used a city decomposition to get the background information from the city dimension, since we believe that the same type of SCADA system in different cities of the United States share some inherently common characteristics. Then we used an "attacker type" decomposition to extract background information from the attacker type dimension. By decomposing the data base from the different perspectives, we can take a full advantage of the limited data, including directly relevant and indirectly relevant data, while the emphasis remains on the direct relevant data. We still need to find a way to coordinate the two or multiple decomposition results.

6.2.2 Coordination of Different Decompositions

By decomposing the entire data set from different perspectives, we fully exploit all the background information of the entire data set from different perspectives and employ them to generate the posterior distribution of each specific subsystem. Now we need to find some ways to coordinate the results from different

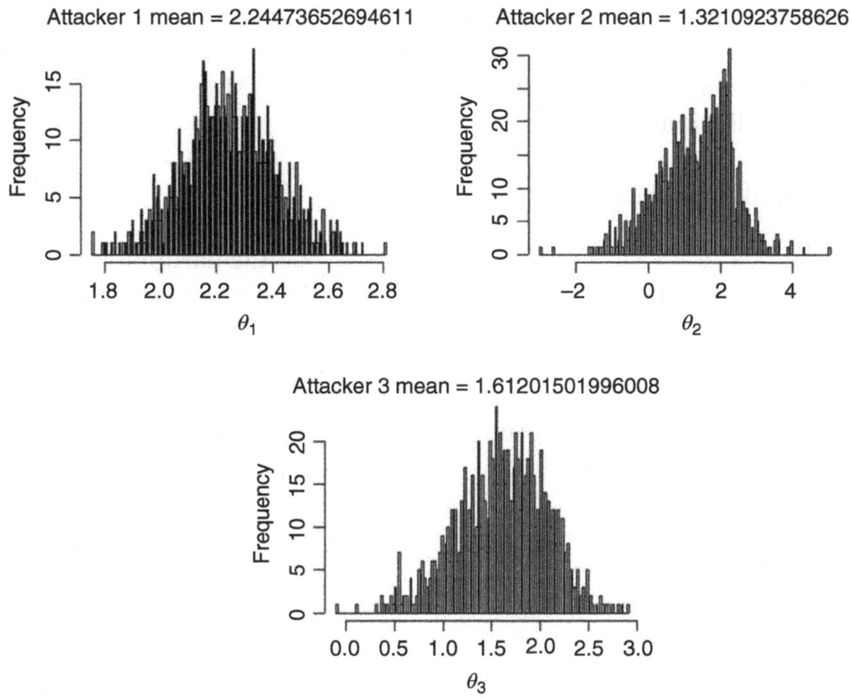

Figure 6.8 *Posterior distribution of θ_i', "attacker type" decomposition.*

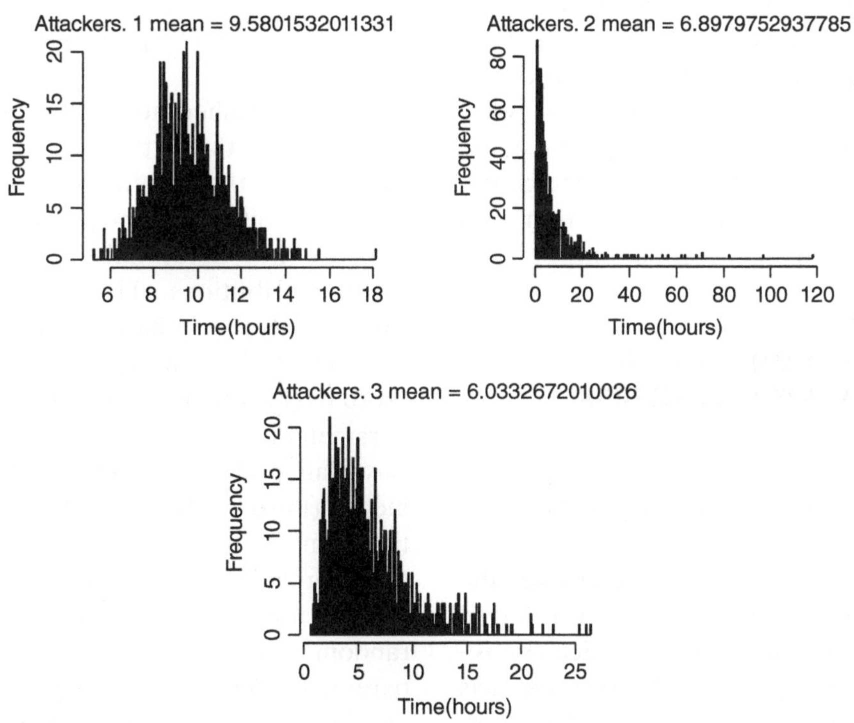

Figure 6.9 *Predictive distribution of "time to recovery" after a cyber attack for each attacker type.*

decompositions, which represent the diverse characteristics of the subsystem of the Complex SoS under investigation. *A mixture distribution* is a direct way to coordinate the predictive distributions from multiple decompositions.

Assume we have m predictive density functions from m decomposition perspectives:

$p_i(y),\ i=1,\ldots,m$, then the mixture density function is

$$p(y)=\sum_{i=1}^{m} a_i p_i(y) \tag{6.12}$$

where

$$\sum_{i=1}^{m} a_i = 1$$

$$p_i(y)=\int p_i(y|\theta)p_i(\theta|x)\mathrm{d}\theta$$

The decision maker will evaluate the importance of each decomposition and assign proper weights on the corresponding predictive density functions.

In the SCADA Complex SoS example, suppose that one decision maker is equally interested in both the "city" and "attacker type" perspectives. Thus $a_1 = a_2 = 0.5$ was set for Eq. (6.12). In practice, since the two predictive posterior distributions were obtained by simulation (city XYZ of Figure 6.6 and attacker Type 1 of Figure 6.9), we just need to generate an equal number of random samples from each distribution to get the mixture distribution.

6.3 INTEGRATING HCBM WITH THE PARTITIONING MULTIOBJECTIVE RISK METHOD (PMRM)

(Note that the Partitioned Multiobjective Risk Method (PMRM) (Haimes, 2016) is added to the Appendix of this book. For more details on the PMRM with solved examples, the reader is referred to Chapter 8 of Haimes (2016).) By data decomposition from multiple perspectives, HCBM extracts information from multiple dimensions and then coordinates the results derived from multiple decompositions. The final result of HCBM can be used as the input for risk analysis. Here we are mainly interested in f_4, which represents the conditional expected value of the upper tail of a distribution.

The PMRM formula for f_4 is

$$f_4 = E(X|X>\beta_2) = \frac{\int_{\beta_2}^{\infty} xp(x)\mathrm{d}x}{\int_{\beta_2}^{\infty} p(x)\mathrm{d}x} \tag{6.13}$$

where β_2 is the partitioning point for high severity and $p(x)$ is the density function.

Using HCBM, we come up with a distribution with two levels:

- *Level 1*: The predictive posterior density function of $\tilde{x}$, conditional on the unknown parameters θ: $p(\tilde{x}|\theta)$. In the SCADA example, $\tilde{x} \sim LN(\theta,\sigma^2)$.
- *Level 2*: The posterior density function of the parameters θ, $p(\theta|x)$. In the SCADA example, $\theta\,|\,x \sim N(\mu,\phi)$.

The predictive posterior density function of $p(\tilde{x})$ (the predictive distribution of the state variable of the system) has been given in Eq. (6.5):

Thus, the formula for f_4 becomes

$$f_4 = \frac{\int_{\beta_2}^{\infty} x \int_{-\infty}^{+\infty} p(\tilde{x}|\theta)p(\theta|x)\mathrm{d}\theta\mathrm{d}\tilde{x}}{\int_{\beta_2}^{\infty}\int_{-\infty}^{+\infty} p(\tilde{x}|\theta)p(\theta|x)\mathrm{d}\theta\mathrm{d}\tilde{x}} \tag{6.14}$$

The established advantage of PMRM is that we can quantify the extreme events of a distribution by studying the conditional expectation. Here we face a new situation: the parameters of the distribution are not deterministic, and they also follow some distributions. This means that (i) the risk (uncertainty) originates from different levels of hierarchies. The ultimate random variable of interest to us (bottom level) is $\tilde{x}$. But $\tilde{x}$ is conditional on its parameters θ (second level). In the hierarchical Bayesian model, θ is conditional on their hyperparameters (third level). (ii) We use a probability distribution to model uncertainties. But often, the parameters of the probability distribution remain nondeterministic. The uncertainty of $\tilde{x}$ may be due to the inherent random nature of the system. The uncertainty of the parameters θ is likely due to our limited information and lack of knowledge about the system. All these uncertainties lead to risk. By integrating HCBM with PMRM, we can account for the system risks as well as the risks from our own lack of knowledge. (iii) The original PMRM model is for systems of extreme events with one random variable (Asbeck and Haimes, 1984) also consult the Appendix in (Haimes, 2016). To apply it to HCBM, we need to extend PMRM to joint distributions.

Since the output of HCBM is simulated distributions, we can get the conditional expected value function, f_4, easily by drawing random samples from the distribution in Figure 6.10. Note that the functions f_4 and f_5 in Table 6.6 connote the conditional expected value of risk and the unconditional (common) expected value of risk, respectively.

By decomposing two different perspectives and coordinating the two Bayesian analysis results of the two decompositions, we can combine and make full use of all the hierarchical and limited information to make rational decisions.

6.3.1 Risk-Based Decision Analysis with HCBM

This section builds on HCBM and the PMRM for risk-based decision analysis of Complex SoS.

In previous sections we have estimated the *time to recovery* after a cyber attack by a disgruntled employee on the water utility's SCADA system by using HCBM. To manage the risk to the subsystems or to the entire Complex SoS, we want to alter parts of the Complex SoS by installing new components. Four alternatives are proposed in Table 6.7:

1. *Alternative 1* consists of three components: (i) outsource the web page hosting; (ii) enforce policy on password sharing, e.g. there is no single system password; and (iii) enforce policy of no personal Internet usage. All Internet usages are logged and randomly checked by administrators. The estimated cost to implement is policy $1000. Compared with the current system, the extra yearly operating cost is $7200.

TABLE 6.7 Cost of SCADA System Risk Management Alternatives

	Discount rate of 7% for 10 years ($)
Alt 1	51 504
Alt 2	72 515
Alt 3	15 904
Alt 4	0

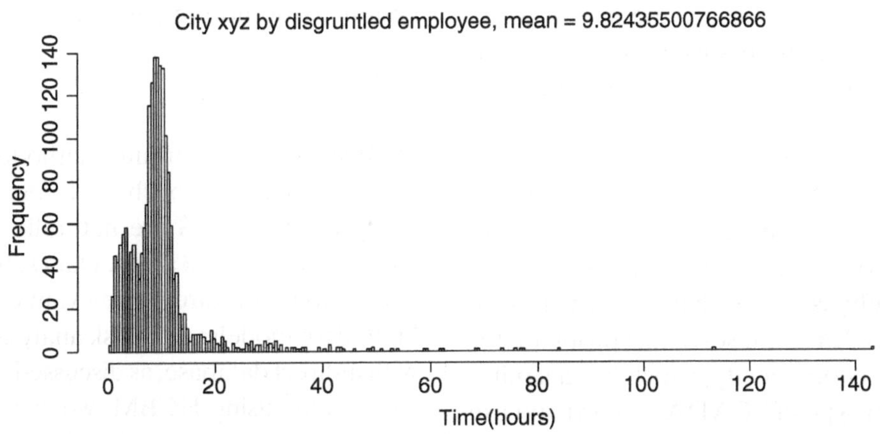

Figure 6.10 *Coordinated distribution of the two decomposition perspectives.*

TABLE 6.6 Expected Value and Conditional Expected Value of the Coordinated Distribution of City XYZ

City	Attacker type	f_5	f_4 $(1-\alpha=0.1)$	f_4 $(1-\alpha=0.05)$	f_4 $(1-\alpha=0.01)$
XYZ	Disgruntled employee	9.824 355	27.98 952	38.22 417	67.66 235

2. *Alternative 2* is the same as 1, but we add an additional component of a token-based authentication server with dialing. The estimated cost to implement is \$32 500 and the extra yearly operating cost is \$6000.
3. *Alternative 3* consists of four components: (i) add to tanks alarm limits that will allow earlier warning, (ii) add alarms to signal unusual periods of excessive pumping, (iii) program the system to enforce password changes, and (iv) separate operations from administrative roles. Designate a specific terminal for administrative access. The estimated cost to implement is \$8000 and the extra yearly operating cost is \$1200.
4. *Alternative 4* is simply the status quo, that is, do nothing and accept the current risk. There is no extra cost for this alternative.

Table 6.7 summarizes the alternatives and net present value at 7% discount and the 10-year cost to maintain each alternative.

We have estimated the state variable (*time to recovery*) of city XYZ's water utility SCADA system under risk management alternative 4 (do nothing) by HCBM. Obviously, we have no data for alternatives 1, 2, and 3, since city XYZ has not yet implemented these. However, as with alternative 4, we have data for other US cities that are using them. We again use the two decomposition perspectives: *city decomposition* and *attacker type* decomposition for alternatives 1, 2, and 3.

For alternative 4, we have some direct relevant data (data from *city XYZ*) for both of the above decompositions, although those direct relevant data are very limited. However, for alternatives 1, 2, and 3, an obvious difficulty is that we have no city decomposition data for city XYZ. To solve this problem, for each alternative we choose a city, which has a similar situation (e.g. size or type of SCADA system) and use the data of that city to represent city XYZ.

We follow the same computation procedure for alternatives 1, 2, and 3. The unconditional expectation f_5 and conditional expectation f_4 of all four alternatives are listed in Tables 6.7 and 6.8 along with the costs of the four alternatives.

Finally, we plot the tradeoffs between the *time to recovery* and the *costs* of the risk management alternatives, using the unconditional and conditional expectations, f_5 and f_4, respectively. Note that Chapter 5 in its entirety is devoted to multiobjectives and the associated tradeoff analyses. Figure 6.11 presents the cost versus time to recovery for the conditional expected value of risk (f_4) and the unconditional expected value of risk (f_5). Recall that the conditional expected value of risk is presented in the Appendix of this book.)

TABLE 6.8 Costs of Risk Management Alternatives Determined by Unconditional and Conditional Expectations

	Cost (discount rate of 7% for 10 years) (\$)	f_5	f_4 $(1-\alpha=0.1)$
Alt 1	51 504	5.56 439	18.9865
Alt 2	72 515	3.23 422	12.2954
Alt 3	15 904	7.19 293	28.3458
Alt 4	0	9.824 355	27.98 952

For the unconditional expectation f_5, all four options are Pareto-optimal. For the conditional expectation f_4, alternative 3 is dominated by alternative 4. For both risk functions f_4 and f_5, alternatives 1, 2, and 4 remain on the Pareto-optimal frontier. We recommend that city XYZ use alternative 1 or 2 to enhance the security of the water supply SCADA system or do nothing and accept the current risk.

6.3.2 Summary

HCBM is an important methodology for modeling Complex SoS and for the risk thereto. This is because many Complex SoS are naturally hierarchical, as discussed in Chapters 3 and 4. Data sources are also hierarchical; however, directly relevant data, which are often limited for modeling and risk analysis, can be augmented with indirect database, as discussed in this chapter. More specifically, using HCBM, we can extract information from different sources of different perspectives of the hierarchy. Subsequently, we can integrate the information from different sources (directly and indirectly relevant data), placing more emphasis on the directly relevant data. For each decomposition perspective, we can use different models, which make our approach very flexible. In the SCADA risk analysis example, for the city decomposition model, we allowed all the group

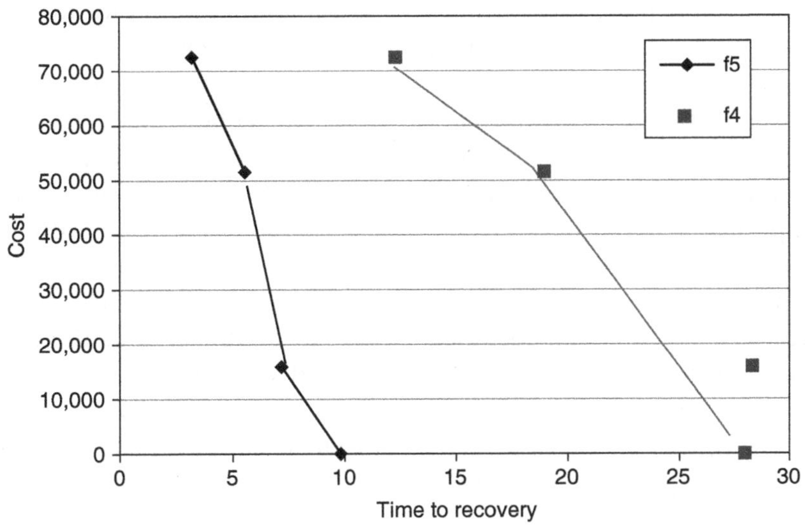

Figure 6.11 *Plot of tradeoff between* time to recovery *and* cost.

variances σ_i^2 to be equal and unknown. And then we assume a noninformative prior distribution for the equal σ^2. For the "attacker type" decomposition model, we estimated the group variances from the data and assume they are known.

Combining the PMRM with HCBM, we were able to quantify the risk to Complex SoS from two sources: uncertainty due to the randomness of the system (predictive posterior function) and uncertainty due to our estimation of the parameters (posterior distribution of the parameters θ).

6.4 MODELING COMPLEX SYSTEMS OF SYSTEMS BY INTEGRATING BAYES' THEOREM WITH DYNAMIC PROGRAMMING

6.4.1 Using Bayes' Theorem for Sequential Decision Making

Bayes' theorem is briefly reviewed in this section to provide further theoretical foundations upon which to quantitatively update the credibility of prior information through new evidence. The method of analysis presented in this chapter is based on recognizing the importance of data in modeling Complex SoS and the importance of constructing posterior probabilities on the basis of new intelligence (knowledge). We then harmonize the seminal contributions of (i) Richard Bellman in recursive modeling of systems via Bellman's principle of optimality in dynamic programming (note that Chapter 11 also builds on Bellman's principle of optimality in dynamic programming, which is added to the Appendix) and (ii) Thomas Bayes for the Bayes' formula used in developing posterior probabilities, discussed in detail in earlier sections and which can be repeated again and again. Furthermore, because of the compatibility between Bayes' formula for posterior probabilities and Bellman's recursive equation in dynamic programming, the two can be embedded into one integrated formulation to improve and (i) populate our models of Complex SoS, testing them, and (ii) optimize sequential decision-making processes that build on different aspects of data and intelligence gathering and analysis. An example problem is developed in this chapter to assist the reader with appreciating the complementarity of the two methods or rather the two philosophies. To assist readers who have not exposed to dynamic programming, the Appendix of the book presents on dynamic programming (with examples).

The use of the Bayesian analysis for tracking positive or negative evolving events through single and multiple classes of scenarios is the subject of this section. We focused earlier on Bayesian analysis, which addresses situations where the intelligence available on multiple scenario classes (e.g. $s_2, \ldots, s_n$)

can be viewed as likelihood functions for updating another scenario class (e.g. s_1). This is pertinent to cases where either a single act of terrorism is planned, or a set of synchronized attacks, such as poisoning food in city A and water in city B. Consider the simple case of tracking two separate scenario classes s_1 and s_2, where $\Pr(s_1)$ is the prior probability of scenario class s_1 being true and $\Pr(s_2 \mid s_1)$ is the probability of scenario class s_2 being true, given s_1, and serving as the likelihood function for scenario s_1. Thus, the posterior probability for scenario s_1 is $\Pr(s_1 \mid s_2)$:

$$\Pr(s_1 \mid s_2) = \frac{\Pr(s_1, s_2)}{\Pr(s_2)} \tag{6.15}$$

This formula can be extended when multiple scenarios are used as likelihood functions to generate a posterior probability for s_1:

$$\Pr(s_1 \mid s_2, s_3, \ldots, s_n) = \frac{\Pr(s_1, s_2, s_3, \ldots, s_n)}{\Pr(s_2, s_3, \ldots, s_n)} \tag{6.16}$$

In addition to the new information supplied by scenario s_2, assume that new evidence e_1, also relevant to scenario s_1, has been gathered by intelligence agencies, with a probability $\Pr(e_1 \mid s_1)$ given scenario s_1 being true. Thus, to calculate the posterior probability of the scenarios s_1, $\Pr(s_1 \mid s_2, e_1)$, we derive the following relationships:

$$\Pr(s_2, e_1, s_1) = \Pr(s_2 \mid e_1, s_1)\Pr(e_1 \mid s_1)\Pr(s_1) \tag{6.17}$$

Similarly,

$$\Pr(s_2, e_1, s_1) = \Pr(s_1 \mid s_2, e_1)\Pr(s_2, e_1) \tag{6.18}$$

Equating Eqs. (6.3) and (6.4) yields the posterior probability for s_1:

$$\Pr(s_1 \mid s_2, e_1) = \Pr(s_2 \mid e_1, s_1)\Pr(e_1 \mid s_1)\Pr(s_1)/\Pr(s_2, e_1) \tag{6.19}$$

Clearly, decentralization or uncoordinated compartmentalization of intelligence sharing and analysis would not bode well for the above Bayesian analysis in cases where there is a coupling between scenarios s_1 and s_2. On the other hand, tracking terrorist scenarios through vigilant intelligence gathering and sharing can serve as a potent mechanism to frustrate such plans and their execution. Paté-Cornell (2002) offers a generalized Bayesian formula for multiple sources of signals. For the purpose of this chapter and following Paté-Cornell's (2002) notation, we augment any new evidence, e, with the scenarios that are being tracked and denote the entire set of scenarios by $\{S\} = (s_1, s_2, s_3, \ldots, s_n)$.

Thus, Eq. (6.16) can be rewritten as

$$\Pr(s_j \mid \{S\, s_j\}) = \Pr(S)/\Pr(\{S\, s_j\}) \tag{6.20}$$

where s_j is the scenario being updated and $\{S\backslash s_j\}$ means the set $\{S\}$ minus the element s_j.

Note that $\Pr(\{S\backslash s_j\}) = \Pr(s_j)\ \Pr(\{S\backslash s_j\} \mid s_j) + \Pr(s_j')\ \Pr(\{S\backslash s_j\} \mid s_j')$, where s_j' denotes the negation of the scenario s_j.

Thus, through the Bayesian analysis, the confidence accumulated in the credibility of a set of scenarios $\{S_1\}$ can be used as likelihood functions for another set of scenarios $\{S_2\}$. This provides an opportunity to build hierarchies of scenarios, which may be formed on the multiple perspectives of concern to the intelligence community.

6.4.2 Dynamic Programming

Dynamic programming can be effectively used for the optimal allocation of resources for scenario tracking, intelligence gathering, and counterterrorism analysis. This is because the Markovian and sequential nature of the required decisions fits nicely into Bellman's principle of optimality on which dynamic programming is based (Bellman and Dreyfus, 1962). Note that Dynamic Programming is introduced with examples in the Section A.1.6.

An optimal policy has the property that whatever the initial state and initial decisions are, the remaining decisions must constitute an optimal policy with regard to the state resulting from the first decision.

This principle will be used in deriving a recursive equation, which relates the state of the system at time k (level of confidence in the corresponding subscenario at stage k) and the decisions to invest in scenario tracking, intelligence gathering, and analysis from stage k to $(k+1)$. The concept of dynamic programming is relatively simple, as will be shown here; however, formulating the recursive equation of real problems can be a challenging task. In particular, tracking multiple main scenarios and their corresponding subscenarios necessarily yields a combinatorial problem for which dynamic programming is well equipped.

The following general sequential resources allocation formulation is introduced to streamline the discussion on the embedded Bayesian analysis and dynamic programming.

First we define the sequential resources allocation problem. Assume we want to allocate the available resources Q to n sequential activities. We define $g_i(x_i)$ to be the return (benefit) for the ith activity from the allocation of x_i resources to it. And we define that the overall objective function $f(\bullet)$in this formulation is the products of the n return functions. Thus, the optimization problem can be formulated as

$$\text{Maximize}\left\{f(q)=\prod_{i=1}^{n} g_i(x_i)\right\} \tag{6.21}$$

Subject to the constraints

$$q=\sum_{i=1}^{n} x_i \le Q, \text{ and } x_i \ge 0 \text{ for } i=1,2,\ldots,n \tag{6.22}$$

where Q represents the total available resources and $q=\sum_{i=1}^{n} x_i$, represents the total resources we will invest in n stages. Note that the objective function in this formulation is defined as the products of the n return functions. In Bayes' calculation, the posterior is proportional to the product of prior and likelihood function. The posterior in one stage can be used as the prior in the next stage. We can update the prior probability by recursive multiplication. Thus, the objective function we defined above is suitable for Bayes' calculation. We will further discuss this issue in subsequent sections.

To formulate the dynamic programming recursive equation, we define

$$\begin{aligned}
&q_i=\sum_{j=1}^{i} x_j, \text{ the cumulative investment at stage } i,\\
&\quad \text{where } q_1=x_1 \text{ and } q_n=q\le Q.\\
&f_1(q_1)=\text{the maximum return from the allocation}\\
&\quad \text{of resource} q_1 \text{ to activity 1}\\
&f_1(q_1)=\text{maximum}\,\{g_1(x_1)\}\\
&\quad 0\le x_1\le q_1\\
&\quad 0\le q_1\le Q\\
&f_n(q_n)=\text{the maximum return from the allocation}\\
&\quad \text{of resource} q_n \text{ to } n \text{ activities}
\end{aligned} \tag{6.23}$$

The n activities also represent stages in the following general recursive equation for the n activities:

$$f_n(q_n)=\text{maximum}\{g_n(x_n)\cdot f_{n-1}(q_{n-1})\} \tag{6.24}$$

Note, however, that $q_{n-1}=q_n-x_n$; thus,

$$\begin{aligned}
&f_n(q_n)=\text{maximum}\{g_n(x_n)\cdot f_{n-1}(q_n-x_n)\}\\
&\quad 0\le x_n\le q_n\\
&\quad 0\le q_n\le Q
\end{aligned} \tag{6.25}$$

The component $f_{n-1}(q_n-x_n)$ in the above recursive equation, which constitutes the posterior probability derived from Bayes' theorem at stage $(n-1)$, will serve as the prior probability at the nth stage.

6.4.3 Bayesian Analysis and Dynamic Programming

Exploiting the synergy resulting from embedding Bayes' theorem with Bellman's principle of optimality seems a natural approach for a class of resource allocation problems. What follows is an embedded formulation of Bayes' theorem and the recursive equation in dynamic programming. Within a given budget, one must determine a feasible path of scenario tracking, intelligence gathering, and analysis, which maximizes the confidence associated with each path, calculated as the posterior probability. Clearly, without dynamic programming the solution to this combinatorial problem could become prohibitive for large networks with many subscenarios. However, for the budget allocation problem depicted here, the scope of available candidates for investment should be sufficiently bounded that dynamic programming integrated with Bayesian

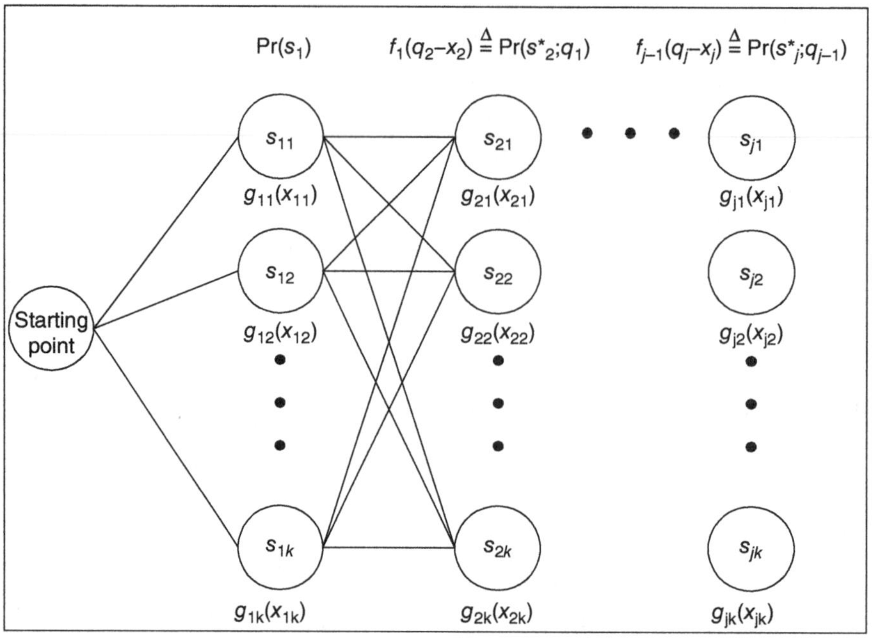

Figure 6.12 *Embedding Bayes' theorem in Bellman's principle of optimality.*

analysis can circumvent the combinatorial calculations and hence effectively reduce the computational effort needed. (To facilitate the presentation of this embedding process, a prototype example presented in the next section is depicted by Figure 6.12.)

The complexity of the problem is reduced by decomposing it into smaller subproblems, which are sequentially coupled to one another through a recursive equation. At the second stage, the following question is asked at each node (subscenario or circle in the network in Figure 6.12: What is the best path of intelligence gathering (originating from stage 1) that maximizes confidence (the posterior probability) and terminates at this node?) Often more than one path may yield the same maximum confidence. In that case, more than one solution can be derived (Yan and Haimes, 2010, 2011).

6.4.4 Model Formulation

In pursuing one main scenario or event s (e.g. food poisoning), we intend to update, with new information, the probabilities of *occurrence* and *nonoccurrence* in multiple subscenarios (stages) j, $j = 1, 2, \ldots, n$. At each stage j, we will consider multiple investment options (q) to pursue new evidence. Let x_{jk} denote the investment at level k, $k = 1, 2, \ldots, K$, made for subscenario (at stage) j, $j = 1, 2, \ldots, n$.

Let $g_j(x_{jk})$ denote the expected level of confidence (measured in probabilities) in the realization of the main scenario (overall event) that is achieved by pursuing intelligence gathering for subscenario, s_j, $j = 1, 2, \ldots, n$ at investment level k, $k = 1, 2, \ldots, K$. Note that the number of subscenarios considered coincides with the number of stages in the dynamic programming. Thus, at stage j, for each subscenario s_j, we will consider K levels of investment in intelligence gathering x_{jk} ($k = 1, 2, \ldots, K$). For a more compact notation, we will refer to $s_j(x_{jk})$ as subscenario s_{jk}.

In the Bayesian model, we will consider n subscenarios (activities). For each subscenario, we will consider K levels of investment. The posterior probability derived from Bayes' theorem at stage $(i - 1)$ will serve as the prior probability at the ith stage. This recursive procedure is reasonable because all the stages (subscenarios) address and relate to the same main event (e.g. food poisoning). We divide the main event (scenario) into different stages (subscenarios) merely to show the main event (scenario) is under different states of information.

In sum, the overall objective of the posterior analysis embedded with the dynamic programming recursive equation is to maximize the expected confidence in the occurrence of the main scenario (expected posterior probability of the main scenario's occurrence) by optimizing the allocation of limited resources into the subscenarios considered under the main scenario. Indeed, all intelligence gathering and analysis activities for the subscenarios will compete for the total limited resources available, Q, where Q is divided into K levels of investments. For example, if four subscenarios are considered, then

$$x_{1,k1} + x_{2,k2} + x_{3,k3} + x_{4,k4} \leq Q \tag{6.26}$$

where $x_{j,\,kj}$ denotes the investment at level k_j in the jth subscenario.

Maximizing the expected confidence in the hypothesis that a terrorist scenario *is underway* could yield a resource allocation policy that ignores intelligence evidence, which expectedly discounts the hypothesis. However, if the risk resulting from a false negative, i.e. an unrecognized scenario, is much greater than the cost of a false positive, then maximizing the posterior probabilities may be acceptable. On the other hand, we also can minimize the posterior probabilities of occurrence (i.e. to minimize the false positive). The two objectives are dual. Both approaches follow the same formulation. In this section we will present our idea by deriving the formulation for the first model (minimizing the false negative). It can be easily generalized to deal with both cases. In practice, the decision makers can set out two teams. One team aims to maximize the posterior probabilities of occurrence (minimize the false negative), and one team aims to minimize the posterior probabilities of occurrence (minimize the false positive). The two resulted posterior probabilities from the two teams will be served as the likelihood's estimated upper bound and lower bound of the risk scenario.

6.4.5 Derivation of the Recursive Equation

We will quantify the objective function to maximize the expected confidence in the occurrence of the main scenario (expected posterior probability of the main scenario's occurrence) and derive the recursive equations for dynamic programming in this section.

In stage 1, we can choose to allocate x_{1k}, $(k = 1, \ldots, K)$ levels of investment in subscenario s_1 for gathering new evidence, e_{1k}, $(k = 1, \ldots, K)$. If we successfully obtain the evidence (the evidence is positive) after the investment, the posterior probability for subscenario s_{1k} is defined using Bayes' formula:

$$\Pr(s_{1k}|e_{1k};x_{1k}) = \frac{\Pr(e_{1k}|s_{1k};x_{1k})}{\Pr(e_{1k};x_{1k})}\Pr(s_{1k}) \tag{6.27}$$

Thus, for the first stage, if we can successfully obtain evidence, e_{1k}, the associated level of confidence gained from the evidence e_{1k} is defined as

$$g_{1k}(e_{1k},x_{1k}) \triangleq \frac{\Pr(e_{1k}|s_{1k};x_{1k})}{\Pr(e_{1k};x_{1k})} \tag{6.28}$$

Here we assume $g_{1k}(e_{1k}, x_{1k}) > 1$ (i.e. $\Pr(e_{1k}|s_{1k}; x_{1k}) > \Pr(e_{1k}; x_{1k})$), given $x_{1k} > 0$. That is, if we successfully obtain the new evidence by investing $x_{1k} > 0$ resources, the expected posterior probability will be increased by the new evidence. Otherwise, we will not consider pursuing this evidence under the objective of minimizing false negative (maximizing expected posterior probability). However, the evidence collection process is a random event. There are generally three major outcomes: (i) We successfully obtain the evidence (the evidence is positive). (ii) The evidence is negative (the hypothesis cannot be substantiated). (iii) We do not know whether the hypothesis is positive or negative even after we have tried to collect the evidence by investing resources. We use e_{1k} to denote the outcome that we finally obtain the evidence (outcome (i)) and use e_{1k}^c to denote outcome (ii) and e_{1k}^0 for outcome (iii).

We have derived formulas for the first outcome, namely, supporting evidence for the main scenario.

On the other hand, should we get the third outcome, namely, we do not know whether the evidence does exist. Thus, no new information will be added even if an investment of resources was made for collecting intelligence. We cannot use Bayes' formula to update the prior probability. Thus, the "posterior" probability will be equal to the prior probability.

We have assumed $g_{1k}(e_{1k}, x_{1k}) > 1$; thus, $g_{1k}(e_{1k}^c, x_{1k}) < 1$. Therefore, if we use the second outcome to update the prior probability, the posterior probability will be less than the prior. Since we want to increase the posterior probability, we will just neglect any outcome that cannot increase the posterior probability, and we define the event $e_{1k}^u = \{e_{1k}^c, e_{1k}^0\}$. Thus, for the second and third outcomes, the posterior probability of subscenario s_{1k} is defined as

$$\Pr\left(s_{1k} \middle| e_{1k}^u; x_{1k}\right) = \Pr(s_{1k}) \tag{6.29}$$

The level of confidence will not be increased after we invest x_{1k} resources for evidence collection. Thus, we have the associated level of confidence gained:

$$g_{1k}\left(e_{1k}^u, x_{1k}\right) \triangleq 1 \tag{6.30}$$

As we have discussed earlier, the objective of maximizing the posterior probability ignores intelligence evidence, which expectedly discounts the hypothesis. For specific scenarios, it is useful to avoid the false negative evidence, especially when the cost of the false negative is much greater than the cost of the false positive, such as the case of weapons of mass destruction (WMD). On the other hand, we can use this objective to get an upper bounder of the likelihood of the risk scenario and use the objective of minimizing the posterior probability to get the lower bounder.

Let the probability that we can obtain the evidence (outcome (i)) be $h(e_{1k}; x_{1k})$ and the probability that we cannot obtain the evidence (outcome (ii) and outcome (iii)) be $h(e_{1k}^u; x_{1k})$. Obviously, we have $h(e_{1k}; x_{1k}) + h(e_{1k}^u; x_{1k}) = 1$. Thus if we invest x_{1k} resources, the *expected* posterior probability for subscenario s_{1k} is defined by

$$\begin{aligned}
\Pr(s_{1k}|x_{1k}) &= \Pr(s_{1k}|e_{1k}; x_{1k}) \times h(e_{1k}; x_{1k}) \\
&\quad + \Pr\left(s_{1k}\middle|e_{1k}^u; x_{1k}\right) \times h\left(e_{1k}^u; x_{1k}\right) \\
&= \frac{\Pr(e_{1k}|s_{1k}; x_{1k})}{\Pr(e_{1k}; x_{1k})} \Pr(s_{1k}) \times h(e_{1k}; x_{1k}) \\
&\quad + \Pr(s_{1k}) \times h\left(e_{1k}^u; x_{1k}\right) \\
&= \Pr(s_{1k}) \\
&\quad \left[\frac{\Pr(e_{1k}|s_{1k}; x_{1k})}{\Pr(e_{1k}; x_{1k})} \times h(e_{1k}; x_{1k}) + h\left(e_{1k}^u; x_{1k}\right)\right]
\end{aligned} \tag{6.31}$$

Thus, for the first stage, the associated *expected* level of confidence gained from the investment x_{1k} is defined as

$$g_{1k}(x_{1k}) \triangleq \frac{\Pr(e_{1k}|s_{1k}; x_{1k})}{\Pr(e_{1k}; x_{1k})} \times h(e_{1k}; x_{1k}) + h\left(e_{1k}^u; x_{1k}\right) \tag{6.32}$$

Note that $\Pr(e_{1k}; x_{1k})$ is not equivalent to $h(e_{1k}; x_{1k})$. The two probabilities differ in the following two aspects:

1. $\Pr(e_{1k}; x_{1k})$ denotes the probability that e_{1k} is positive in a normal situation without conditions. Whereas, $h(e_{1k}; x_{1k})$ is conditional on current situation. The decision makers' estimate of $h(e_{1k}; x_{1k})$ is based on the belief on the current situation of the system and his confidence of obtaining the evidence. Thus, $h(e_{1k}; x_{1k})$ may change from time to time. But $\Pr(e_{1k}; x_{1k})$ is a constant for certain evidence e_{1k}.
2. $h\left(e_{1k}^u; x_{1k}\right)$ is the sum of the probability that the evidence is negative (that is, the evidence does not exist at all) and the probability that we do not know the result (evidence collection mission fails and we do not know if the evidence exists or not).

Given the allocation of x_{1k} resources to obtain new evidence e_{1k} for subscenario s_{1k}, the expected posterior probability for subscenario s_{1k} has been defined in Eq. (6.31). The associated expected level of confidence gained from the evidence e_{1k} has been defined in Eq. (6.32).

Thus from Eqs. (6.31) and (6.32), $f_1(q_1)$ defined in Eq. (6.23) becomes

$$f_1(q_1) = \max_{\substack{0 \le x_{1k} \le q_1 \\ 0 \le q1_1 \le Q}} \left\{ \left(\frac{\Pr(e_{1k}|s_{1k};x_{1k})}{\Pr(e_{1k};x_{1k})} \times h(e_{1k};x_{1k}) + h(e_{1k}^u;x_{1k}) \right) \Pr(s_{1k}) \right\} = \left[\max_{\substack{0 \le x_{1k} \le q_1 \\ 0 \le q_1 \le Q}} \{g_{1k}(x_{1k})\} \right] \Pr(s_1) \tag{6.33}$$

Note that the prior probability at the first stage, $\Pr(s_1)$, in Eq. (6.33) is fixed for all subscenarios s_{1k}; therefore, it is not a part of the optimization with respect to the decision variable levels, x_{11}, x_{12}, ..., x_{1k}. Eq. (6.34) defines the expected posterior probability for stage 1 and, thus, the "optimal" expected prior probability for stage 2:

$$f_1(q_1) = \max_{\substack{0 \le x_{1k} \le q_1 \\ 0 \le q_1 \le Q}} \{\Pr(s_{1k}|x_{1k})\} \triangleq \Pr(s_2^*;q_1) \tag{6.34}$$

where $\Pr(s_2^*;q_1)$ denotes given q_1 resources allocated to stage 1, the "optimal" expected prior probability for stage 2, which is equal to the optimal expected posterior probability for stage 1.

At the second stage, we maximize the expected confidence in the main scenario, s, by proceeding from the optimal level of investment $f_1(q_1)$ in subscenario s_1 for each level of investment $q_1, 0 \le q_1 \le Q$. We achieve this optimum through Bayes' formula and Bellman's recursive equation. Given the optimal expected prior probability, the expected posterior probability for s_{2k} is given by

$$\Pr(s_{2k}|x_{2k}) = \Pr(s_2^*;q_1) \left(\frac{\Pr(e_{2k}|s_{2k};x_{2k})}{\Pr(e_{2k};x_{2k})} \times h(e_{2k};x_{2k}) + h(e_{2k}^u;x_{2k}) \right) \tag{6.35}$$

where $\Pr(s_2^*;q_1)$, which is the "optimal" expected posterior probability generated at stage 1, now constitutes the prior probability at stage 2 and is independent of the collection activity for evidence e_{2k}. (The dependency occurs only through the equation $q_2 = q_1 + x_2$, namely, in the allocation of resource q_2 between the first and second stages.)

The associated level of confidence is given in Eq. (6.36):

$$g_{2k}(x_{2k}) \triangleq \left(\frac{\Pr(e_{2k}|s_{2k};x_{2k})}{\Pr(e_{2k};x_{2k})} \times h(e_{2k};x_{2k}) + h(e_{2k}^u;x_{2k}) \right) \tag{6.36}$$

(Note that $q_1 = q_2 - x_2$.):

Thus, $f_1(q_1) = f_1(q_2 - x_2) \triangleq \Pr(s_2^*;q_1)$

Definition: Let $\Pr\left(s_j^*;q_{j-1}\right)$ represent the prior probability at the stage $j, j = 2, 3, \ldots J$. Therefore, the expected posterior probability of subscenario s_{j-1} is formally defined by Eq. (6.23). Note that $q_{j-1} = q_j - x_j$. Thus,

$$f_{j-1}(q_{j-1}) = f_{j-1}(q_j - x_j) \triangleq \Pr\left(s_j^*;q_{j-1}\right) \quad \text{for } j = 2, \ldots, J \tag{6.37}$$

Given Eqs. (6.35) and (6.36), we can now develop the recursive equation by maximizing the expected posterior probability at the second stage

$$f_2(q_2) = \max_{\substack{0 \le x_{2k} \le q_2 \\ 0 \le q_2 \le Q}} \{\Pr(s_{2k}|x_{2k})\} \tag{6.38}$$

Substituting Eq. (6.35) into Eq. (6.38) yields

$$f_2(q_2) = \max_{\substack{0 \le x_{2k} \le q_2 \\ 0 \le q_2 \le Q}} \left(\left\{ \frac{\Pr(e_{2k}|s_{2k};x_{2k})}{\Pr(e_{2k};x_{2k})} \times h(e_{2k};x_{2k}) + h(e_{2k}^u;x_{2k}) \right) \Pr(s_2^*;q_1) \right\} \tag{6.39}$$

Substituting the definition of $g_{2k}(x_{2k})$ from Eq. (6.36) into Eq. (6.39) yields

$$f_2(q_2) = \max_{\substack{0 \le x_{2k} \le q_2 \\ 0 \le q_2 \le Q}} \{g_{2k}(x_{2k}) \Pr(s_2^*;q_1)\} \tag{6.40}$$

Thus, Bellman's recursive equation for the second stage (second subscenario) is given by Eq. (6.41). Note that now the function, $f_2(q_2)$, constitutes the optimal expected prior probability at the third stage for subscenario s_3.

$$f_2(q_2) = \max_{\substack{0 \le x_{2k} \le q_2 \\ 0 \le q_2 \le Q}} \left\{ \left(\frac{\Pr(e_{2k}|s_{2k};x_{2k})}{\Pr(e_{2k};x_{2k})} \times h(e_{2k};x_{2k}) + h\left(e_{2k}^u;x_{2k}\right) \right) f_1(q_2 - x_{2k}) \right\} \triangleq \Pr\left(s_3^*;q_2\right) \tag{6.41}$$

The general recursive equation for $n-1$ stages is as follows:

$$f_{n-1}(q_{n-1}) = \max_{\substack{0 \le x_{n-1;k} \le q_{n-1} \\ 0 \le q_{n-1} \le Q}} \left\{ \left(\frac{\Pr\left(e_{n-1;k}|s_{n-1;k};x_{n-1;k}\right)}{\Pr\left(e_{n-1;k};x_{n-1;k}\right)} \times h\left(e_{n-1;k};x_{n-1;k}\right) + h\left(e_{n-1;k}^u;x_{n-1;k}\right) \right) f_{n-2}\left(q_{n-1} - x_{n-1;k}\right) \right\} \triangleq \Pr\left(s_n^*;q_{n-1}\right) \tag{6.42}$$

Finally,

$$\begin{aligned} f_n(q_n) &= \max_{\substack{0 \le x_{nk} \le q_n \\ 0 \le q_n \le Q}} \left\{ \left(\frac{\Pr(e_{nk}|s_{nk};x_{nk})}{\Pr(e_{nk};x_{nk})} \times h\left(e_{n;k};x_{n;k}\right) + h\left(e_{n;k}^u;x_{n;k}\right) \right) \Pr\left(s_n^*;q_{n-1}\right) \right\} \\ &= \max_{\substack{0 \le x_{nk} \le q_n \\ 0 \le q_n \le Q}} \left\{ \left(\frac{\Pr(e_{nk}|s_{nk};x_{nk})}{\Pr(e_{nk};x_{nk})} \times h\left(e_{n;k};x_{n;k}\right) + h\left(e_{n;k}^u;x_{n;k}\right) \right) f_{n-1}(q_n - x_{nk}) \right\} \end{aligned} \tag{6.43}$$

Equation (6.43) represents the embedded recursive equation for n subscenarios of the main scenario s. Figure 6.12 depicts the embedding of Bayes' theorem in Bellman's principle of optimality.

As we have discussed, similar to maximizing the confidence in the occurrence of scenario s_1, it is also important to gain evidence that maximizes the probability of nonoccurrence of the scenario (i.e. using intelligence to support the nonoccurrence hypothesis in order to reduce the likelihood of false positives). We may seek to invest in gathering intelligence in order to achieve a better balanced analysis that would minimize both the false negative and false positive predictions. Both approaches follow the same formulation; thus, while this formulation pertains only to maximizing the confidence in the occurrence of the scenario, it can be generalized to deal with both cases.

6.4.6 Efficacy Versus Cost of New Intelligence

It is always helpful and desirable to quantify, to the extent possible, all relevant costs, benefits, and risks associated with resource allocation, especially for public projects and activities. One of the challenges facing the deployment of the methodology presented in this chapter is in relating the efficacy of new intelligence evidence, e, to its associated cost of collection and analysis, $x(e)$. Clearly, such an endeavor is case specific, depending on the type of intelligence being collected, under what conditions, whether it is human or technology based, and on the reliability and credibility of the gathering and analysis processes. Certain aspects of intelligence gathering were scheduled in the past well in advance, for example, satellite coverage and control dictates this to be true. To avoid pitfalls, yet to demonstrate the applicability of the proposed methodology, we draw from the US experience during the Cold War and the nuclear threat that clouded relations between the United States and the Soviet Union. The central importance of this analogy is that in that period, the decision-making process for resource allocation was *prescribed*. Namely, an a priori decision is made as to what percentage of the budget should be allocated for intelligence gathering and analysis for a given (main) scenario. Many factors lead to a slow response time for change from prescribed resource allocation. These include (i) the cost of change (be it new technology or analysts with new specialization areas); (ii) the sufficient agreement on current indicators that an expensive change is needed (e.g. the Iraq situation with WMD); and (iii) the time to complete the implementation of changes can be very long. These indicate that gaining an early agreement on the need for changes can be very important and a systemic method for determining new needs would be helpful.

An important starting point is identifying the main scenario or scenarios at which to direct the intelligence gathering activity; this is followed by the decision to allocate the appropriate percentage of the budget for that purpose. Having identified scenarios, there are three areas that when taken together constitute the intelligence collection effort: (i) intelligence related to the potential targets of attack, (ii) intelligence about people that might organize and carry out the attack, and (iii) intelligence about the potential timing and methods for conducting attacks. Each of these areas (target, people, and method) has its own costs and techniques for collecting information that must be assessed as part of determining the resources that could be applied. The next challenge is to associate a quantitative value to the efficacy of the evidence gained from the collected intelligence. Two major factors may be used for this purpose: *target* and *agent*. For the *target* factor, the consequences and irreversibility associated with the scenario may be weighed highly: the greater and more irreversible the consequences, the larger the percentage of the intelligence budget allocated for the given scenario. As an aid for evaluating this factor, analysts and decision makers may compare the relative importance of consequence and irreversibility among various scenarios. For the *agent* factor, analysts build on the multiple dimensions or categories of quality identified by Garvin (1988): *performance*, *features*, *reliability*, *conformance*, *durability*, *serviceability*, *aesthetics*, *and perceived quality*. For the intelligence collected by agents, the following dimensions or categories of quality may be considered: relevance, reliability, credibility, cost, and timeliness.

6.4.7 Bayesian Analysis of Intelligence Sensitivity and Evidence Ratio

Daily, security agencies receive a plethora of data, information, and other intelligence reports on threats to any homeland. This calls for a search for connectedness, motives, patterns, hidden terrorist plans, and ultimately for a roadmap of the terrorist networks. There is a crucial need for quantitative and systemic intelligence analyses. In this section, we use the abridged notation $\Pr(s_{jk}|e_{jk}; x_{jk}) = \Pr(s|e)$ for simplicity.

Assume that a set of evidence, e, is collected about a specific terrorist attack scenario, s. In general, we expect that the likelihood of s will be quite small. Assume, for example, that an attack scenario s has a probability of 0.0001 of actually occurring within a given time frame of concern. From Bayes' theorem we see that the probability $\Pr(s|e)$ of the scenario s occurring given the evidence, e, is

$$\Pr(s|e) = \frac{\Pr(e|s)\Pr(s)}{\Pr(e|s)\Pr(s) + \Pr(e|s')\Pr(s')} \tag{6.44}$$

where $\Pr(s')$ is the likelihood of no terrorist attack occurring. Note that $\Pr(s')$ is equal to 0.9999 for this situation.

If we divide the numerator and denominator in Eq. (6.44) by the numerator, we can transform the equation for $\Pr(s|e)$ into the following form:

$$\Pr(s|e) = \frac{1}{1 + \{\Pr(e|s')/\Pr(e|s)\}\{\Pr(s')/\Pr(s)\}} \tag{6.45}$$

For the example numerical values of $\Pr(s) = 0.0001$ and $\Pr(s') = 0.9999$, this equation becomes

$$\Pr(s|e) = \frac{1}{1 + \{\Pr(e|s')/\Pr(e|s)\}9999} \tag{6.46}$$

Note that unless $\{\Pr(e|s')/\Pr(e|s)\}$, which we will call the *evidence ratio*, is a small enough number to significantly offset the ratio of the initial value of $\{\Pr(s')/\Pr(s)\}$, the value of $\Pr(s|e)$ will remain small. That is, the likelihood of the attack will remain small unless there is some other mitigating evidence. Sample values are illustrated in Table 6.9.

This example emphasizes the point that the evidence ratio must be very large in order to offset an initial estimate that a specific terrorist attack is unlikely. This could require substantial evidence-collection

TABLE 6.9 Evidence Ratio and Likelihood of Attack

Evidence ratio	Pr(s| e)
1.0	0.00 010
10.0	0.000 999
100.0	0.0099
1000.0	0.0909
10 000.0	0.50

TABLE 6.10 Subscenarios for the Food Poisoning Example

Subscenarios		Description
s_1	Formalized intent	Earliest stage, where the terrorist group agrees upon a specific type of food poisoning attack *focused on meat*. Details of how and where the attack will take place are not yet set
s_2	Identified target	The terror group has chosen a specific set of *slaughterhouse* target(s). At this stage, they will most likely have an idea of which slaughterhouses have the most impact, least chance of being detected, least security, etc. They might also already have a rough idea of how to infiltrate the facilities, and the types of poisons that can be utilized
s_3	Developed plan	At this point, they have developed the specifics on how they will carry out the attack, including the terror cell(s) involved, alternative options, funding sources, poison to be used, etc.
s_4	Executed plan	The terror cell(s) involved have started executing the plan. They may have started to infiltrate the slaughterhouse facilities by applying for employment. They will most likely wait until they gain the trust and confidence of their employers before commencing any suspicious activities. Also, they might be gradually manufacturing poisons, using commercial materials. It could take a significant amount of time before they accomplish their main objective

efforts related to each of the three focus areas identified above (target-related intelligence, people-related intelligence, and method of attack-related intelligence). As a result, scenarios with higher prior likelihoods provide sharp reductions in the results required from new intelligence. Of course, the cost for increased evidence ratios are not necessarily uniform across collection techniques, and the combined Bayesian analysis and dynamic programming approach provides the optimal distribution accounting for both the prior and intelligence collection factors.

6.4.8 General Formulation of Food Poisoning Scenario

The following specific example illustrates how the methodology can be used for scenario-based tracking for terrorism. This involves a main scenario of food poisoning initiated in a meat slaughterhouse by a terrorist organization. Four subscenarios, s_j, $j = 1, 2, 3, 4$ (see Tables 6.9 and 6.10), are considered. These subscenarios can be subdivided into more detailed subscenarios as needed – e.g. a subscenario under s_4 can be the employment of a person associated with a known terrorist organization. However, for the purposes of this chapter, we are concerned only with these four subscenarios and one main scenario, s, the threat of mass food poisoning.

Assume that the government needs to gather new information and additional evidence to update the probability of occurrence of a mass food poisoning, S. This is accomplished by pursuing four subscenarios. The total resources available to invest in intelligence gathering is Q = \$10 million for all subscenarios, s_j, j = 1, 2, 3, 4. For practical management purposes the \$10 million is divided into 10 sublevels of investments.

Define the following variables:

x_{jk} = level of investment k at stage j (\$ millions)
q_j = dummy state variable designating level of investment allowed at stage j
Q = total level of investment allowed for all stages (\$ millions)
e_{jk} = new evidence we aim to collect by investing x_{jk}

6.4.9 Example of Scenarios A and B

The embedded Bayesian and dynamic programming model was run for two scenarios, A and B. We assume A and B are two unrelated scenarios and their subscenarios are also unrelated from one other. For the

TABLE 6.11 Parameters of Scenario A

	Full description of return given allocation decisions (scenario A)											
	Stage 1 ($j = 1$)			Stage 2 ($j = 2$)			Stage 3 ($j = 3$)			Stage 4 ($j = 4$)		
x	$\Pr(e_1; x_1)$	$\Pr(e_1 \mid s_1, x_1)$	$h(e_1; x_1)$	$\Pr(e_2; x_2)$	$\Pr(e_2 \mid s_2, x_2)$	$h(e_2; x_2)$	$\Pr(e_3; x_3)$	$\Pr(e_3 \mid s_3, x_3)$	$h(e_3; x_3)$	$\Pr(e_4; x_4)$	$\Pr(e_4 \mid s_4, x_4)$	$h(e_4; x_4)$
0	1.00E-04	1.00E-04	0.00E+00	1.00E-04	1.00E-04	0.00E+00	1.00E-04	1.00E-04	0.00E+00	1.00E-04	1.00E-04	0.00E+00
1	5.00E-04	2.00E-03	1.00E-02	1.50E-04	6.80E-04	2.00E-03	2.50E-04	1.00E-03	3.00E-02	3.00E-04	1.20E-03	1.00E-01
2	1.00E-03	4.80E-03	8.00E-02	2.00E-04	1.00E-03	1.00E-01	3.00E-04	1.60E-03	1.00E-01	3.50E-04	1.44E-03	5.00E-02
3	1.50E-03	7.20E-03	5.00E-03	2.50E-04	1.00E-03	1.40E-01	3.50E-04	1.80E-03	1.60E-01	4.00E-04	1.80E-03	1.00E-01
4	2.00E-03	8.40E-03	1.30E-01	3.00E-04	1.28E-03	1.00E-03	4.00E-04	2.00E-03	1.80E-01	4.50E-04	2.08E-03	1.00E-01
5	2.50E-03	1.12E-02	5.00E-02	3.50E-04	1.56E-03	3.00E-02	4.50E-04	2.20E-03	5.00E-04	5.00E-04	2.20E-03	1.00E-01
6	3.00E-03	1.40E-02	2.00E-01	4.00E-04	1.72E-03	1.00E-03	5.00E-04	2.24E-03	1.00E-01	5.50E-04	2.40E-03	7.00E-02
7	3.50E-03	1.60E-02	1.00E-01	4.50E-04	2.00E-03	1.00E-03	5.50E-04	2.36E-03	5.00E-02	6.00E-04	2.60E-03	1.00E-01
8	4.00E-03	1.76E-02	1.00E-01	5.00E-04	2.40E-03	2.00E-01	6.00E-04	2.64E-03	1.00E-01	6.50E-04	2.64E-03	5.00E-04
9	4.50E-03	2.00E-02	7.00E-02	5.50E-04	2.52E-03	1.00E-02	6.50E-04	2.80E-03	5.00E-04	7.00E-04	2.76E-03	1.00E-01
10	5.00E-03	2.40E-02	1.00E-03	6.00E-04	2.56E-03	2.00E-02	7.00E-04	4.00E-03	1.00E-02	7.50E-04	3.60E-03	3.00E-03

example demonstration, the parameters are assumed in (Table 6.10). Each run used the following assumptions: there are four subscenarios (stages) for the main scenario. We assume that there are 11 levels of funding for each subscenario. (Zero is also a possible funding level. With zero investment, the probability of obtaining the evidence is zero.)

Table 6.11 can be abridged by showing instead the calculated g values by Eq. (6.32).

Abridged parameters are provided for scenarios A and B (we did not show the full parameters table of scenario B) in Tables 6.12 and 6.13.

Solving the dynamic programming yields the following results for optimal allocation of resources for scenarios A and B (Tables 6.14 and 6.15).

Table 6.16 shows by example that the sequencing of stages (or subscenarios) will result in the same optimal allocation of resources for scenario A, in comparison with the optimal values found in Table 6.14. The optimal expected posterior probability during the last stage is calculated to be 4.29×10^{-4}, which is independent of how the order of the stages is defined. It is easy to understand if we look back at the objective function in Eq. (6.21). Based on the Bayes' theorem, the overall return function we formulated is the products of return functions in each stage. Since multiplication is a symmetric operator, the sequencing of stages has no influence on the final results.

TABLE 6.12 Abridged Description of Return (Scenario A)

x	$g_1(x_1)$	$g_2(x_2)$	$g_3(x_3)$	$g_4(x_4)$
0	1.000 00	1.000 00	1.000 00	1.000 00
1	1.030 00	1.007 07	1.090 00	1.300 00
2	1.304 00	1.400 00	1.433 33	1.155 71
3	1.019 00	1.420 00	1.662 86	1.350 00
4	1.416 00	1.003 27	1.720 00	1.362 22
5	1.174 00	1.103 71	1.001 94	1.340 00
6	1.733 33	1.003 30	1.348 00	1.235 45
7	1.357 14	1.003 44	1.164 55	1.333 33
8	1.340 00	1.760 00	1.340 00	1.001 53
9	1.241 11	1.035 82	1.001 65	1.294 29
10	1.003 80	1.065 33	1.047 14	1.011 40

TABLE 6.13 Abridged Description of Return (Scenario B)

x	$g_1(x_1)$	$g_2(x_2)$	$g_3(x_3)$	$g_4(x_4)$
0	1.00	1.00	1.00	1.00
1	1.06	1.04	1.03	1.08
2	1.10	1.11	1.06	1.01
3	1.22	1.13	1.12	1.09
4	1.36	1.18	1.20	1.13
5	1.40	1.26	1.21	1.21
6	1.30	1.32	1.28	1.25
7	1.42	1.51	1.32	1.36
8	1.56	1.40	1.49	1.54
9	1.67	1.50	1.54	1.30
10	1.72	1.65	1.78	1.70

TABLE 6.14 Optimal Allocation Results for Scenario A

Scenario A results, with optimal allocation highlighted (f_1 values assume prior 0.0001)

q	x_1	$f_1(x_1)$	x_2	$f_2(f_1, x_2)$	x_3	$f_3(f_2, x_3)$	x_4	$f_4(f_3, x_4)$
0	0	1.00E-04	0	1.00E-04	0	1.00E-04	0	1.00E-04
1	1	1.03E-04	0	1.03E-04	1	1.09E-04	1	1.30E-04
2	2	1.30E-04	2	1.40E-04	2	1.43E-04	0	1.43E-04
3	2	1.30E-04	2	1.44E-04	3	1.66E-04	1	1.86E-04
4	4	1.42E-04	2	1.83E-04	2	2.01E-04	1	2.16E-04
5	4	1.42E-04	3	1.85E-04	3	2.33E-04	1	2.61E-04
6	6	1.73E-04	2	1.98E-04	2	2.62E-04	1	3.03E-04
7	6	1.73E-04	3	2.01E-04	3	3.04E-04	1	3.40E-04
8	6	1.73E-04	2	2.43E-04	4	3.14E-04	1	3.95E-04
9	6	1.73E-04	3	2.46E-04	3	3.30E-04	1	4.08E-04
10	6	1.73E-04	3	2.46E-04	2	3.48E-04	1	4.29E-04

TABLE 6.15 Optimal Allocation Results for Scenario B

Scenario B results, with optimal allocation highlighted (f_1 values assume prior 0.0001)

q	x_1	$f_1(x_1)$	x_2	$f_2(f_1, x_2)$	x_3	$f_3(f_2, x_3)$	x_4	$f_4(f_3, x_4)$
0	0	1.00E-04	0	1.00E-04	0	1.00E-04	0	1.00E-04
1	1	1.06E-04	0	1.06E-04	0	1.06E-04	1	1.08E-04
2	2	1.10E-04	2	1.11E-04	0	1.11E-04	1	1.14E-04
3	3	1.22E-04	0	1.22E-04	0	1.22E-04	0	1.22E-04
[4]	4	1.36E-04	0	1.36E-04	0	1.36E-04	0	1.36E-04
5	5	1.40E-04	1	1.41E-04	0	1.41E-04	1	1.47E-04
6	5	1.40E-04	2	1.51E-04	0	1.51E-04	1	1.53E-04
7	7	1.42E-04	2	1.55E-04	1	1.55E-04	1	1.63E-04
8	8	1.56E-04	4	1.60E-04	4	1.63E-04	1	1.68E-04
9	9	1.67E-04	[5]	1.71E-04	[0]	1.71E-04	[1]	[1.76E-04]
10	10	1.72E-04	7	1.84E-04	0	1.84E-04	1	1.85E-04

TABLE 6.16 Optimal Allocations for Scenario a Resulting from Reordering of Stages: $s_a = s_4$, $s_b = s_3$, $s_c = s_2$, and $s_d = s_1$

q	x_a	$f_a(x_a)$	x_b	$f_b(f_a, x_b)$	x_c	$f_c(f_b, x_c)$	x_d	$f_d(f_c, x_d)$
0	0	1.00E-04	0	1.00E-04	0	1.00E-04	0	1.00E-04
1	[1]	1.30E-04	0	1.30E-04	0	1.30E-04	0	1.30E-04
2	1	1.30E-04	2	1.43E-04	0	1.43E-04	0	1.43E-04
3	3	1.35E-04	2	1.86E-04	0	1.86E-04	0	1.86E-04
4	4	1.36E-04	[3]	2.16E-04	0	2.16E-04	0	2.16E-04
5	4	1.36E-04	4	2.24E-04	2	2.61E-04	0	2.61E-04
6	4	1.36E-04	3	2.24E-04	[2]	3.03E-04	0	3.03E-04
7	4	1.36E-04	4	2.32E-04	2	3.13E-04	2	3.40E-04
8	4	1.36E-04	4	2.34E-04	3	3.18E-04	2	3.95E-04
9	4	1.36E-04	4	2.34E-04	2	3.25E-04	2	4.08E-04
10	4	1.36E-04	4	2.34E-04	3	3.30E-04	[4]	[4.29E-04]

6.4.10 Combined Scenarios A and B in a Multiobjective Tradeoff Analysis

A multiobjective problem can be formulated if the two scenarios use the same resources. Note that Chapter 5 is devoted in its entirety to multiple objectives and tradeoff analyses. As we have assumed before, A and B are two independent scenarios, where the only relation between A and B is that they compete for the limited resources. In such combined scenarios, the sum of the resources allocated to each scenario must not exceed the total available resources, i.e. $Q_A + Q_B \le Q$. Tables 6.14 and 6.15 summarized the result for the optimal allocation of Q when we consider A and B separately, where $Q_A = Q = 10$ and $Q_B = Q = 10$, respectively. Thus $Q_A + Q_B = 20$. Now we assume that the total resources for the two scenarios are 10 (if 20, then we can use the results from Tables 6.14 and 6.15 directly). Table 6.17 displays the optimal posterior probabilities for scenarios A and B with the allocation of resources varying from 0 to 10. For each investment level, we ran the dynamic program twice, once for scenario A and once for scenario B. For example, for investment level (9, 1), we set $Q_A = 9$ and $Q_B = 1$ for the dynamic program. Note that the sum of $Q_A + Q_B = 10$ for all entries in Table 6.17. This integration of scenarios A and B yields Pareto-optimal solutions of all the 11 combinations as depicted in Figure 6.13. For example, the combined allocation of three units to scenario A and seven units to scenario B $(Q_A, Q_B) = (3, 7)$ is a Pareto-optimal solution, in which the expected posterior probability of scenario A is 1.86×10^{-4} and the expected posterior probability of scenario B is 1.63×10^{-4}.

TABLE 6.17 Pareto-Optimal Frontier for Integrated Scenarios A and B

Scenario A (Q_A)	Scenario B (Q_B)	Posterior A $f_4(Q_A)$	Posterior B $f_4(Q_B)$
10	0	4.29E-04[a]	1.00E-04[a]
9	1	4.08E-04[a]	1.08E-04[a]
8	2	3.95E-04[a]	1.14E-04[a]
7	3	3.40E-04[a]	1.22E-04[a]
6	4	3.03E-04[a]	1.36E-04[a]
5	5	2.61E-04[a]	1.47E-04[a]
4	6	2.16E-04[a]	1.53E-04[a]
3	7	1.86E-04[a]	1.63E-04[a]
2	8	1.43E-04[a]	1.68E-04[a]
1	9	1.30E-04[a]	1.76E-04[a]
0	10	1.00E-04[a]	1.85E-04[a]

[a] Pareto-optimal solution.

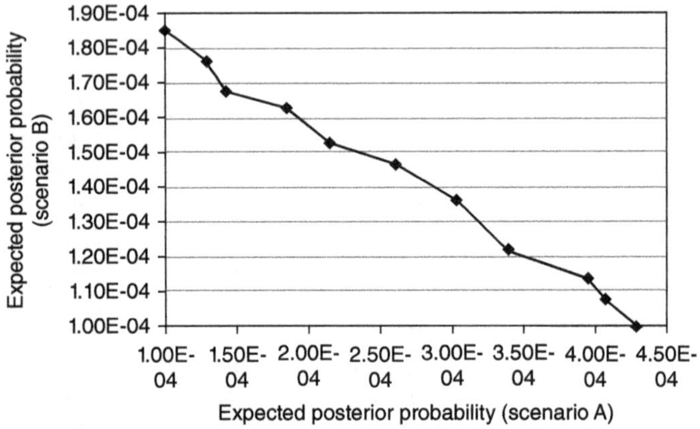

Figure 6.13 Pareto-optimal frontier in objective functional space for integrated scenarios A and B.

Invariably, a large number of scenarios may be candidates for further intelligence gathering and analysis. The ability to optimally distribute limited intelligence resources to those scenarios that yield the most valuable information can be determined within the context of Pareto-optimality, where the ultimate tradeoffs are made by the decision makers in charge.

6.4.11 Value of Discounting Evidence

The objective function in scenarios A and B was to maximize the expected confidence in their occurrence. An important characteristic in the maximization process of these scenarios is manifested by a ratio of $g = \Pr(e|s)/\Pr(e) \cdot h(e) + h(e^u)$ (where $h(e) + h(e^u) = 1$) that is greater than or equal to one. In other words, by virtue of seeking evidence in support of the hypothesis, the model (i.e. the recursive equation) maximizes the expected posterior probabilities. From this, one can conclude that all the collected evidence in which $g > 1$ is confirming the hypothesis, i.e. it increases expected confidence in the occurrence of the main scenarios. It is plausible, however, that some evidence may be available, which in fact discounts the likelihood of the main scenario. For such evidence, $\Pr(e|s)/\Pr(e)$ is less than 1, (i.e. $g = \Pr(e|s)/\Pr(e) \cdot h(e) + h(e^u) < 1$). Under the objective function that maximizes the expected confidence of occurrence of the scenario, this kind of evidence will be excluded from the optimal policy automatically by the dynamic programming algorithm. However, they can be employed to maximize the expected confidence of the nonoccurrence of the scenarios, that is, to minimize the false positive. As we have discussed before, similar scenario and formulation can be developed to illustrate the behavior of the model when such evidence is available.

6.4.12 Summary

Embedding Bayes' theorem with Bellman's principle of optimality in dynamic programming for the purpose of resource allocation for intelligence gathering in countering terrorism has been demonstrated methodologically and computationally. Two overlapping and complementary formulations are presented. One maximizes the expected confidence in the hypothesis that a terrorist scenario *is underway*. The other minimizes the expected confidence in the hypothesis, namely, a terrorist scenario is *not* underway. In this chapter, we mainly studied the first objective function, which is maximizing the expected confidence in the hypothesis and solved it with dynamic programming. Since the two objectives are dual and both approaches follow the same formulation, it can be easily generalized to deal with both cases. We also introduced a two-track intelligence collection strategy: One team aims to maximize the posterior probabilities of occurrence and one team aims to minimize the posterior probabilities of occurrence. The two resulted posterior probabilities from the two teams will serve as the estimated upper and lower bound likelihoods of the risk scenario.

REFERENCES

Asbeck, E.L. and Haimes, Y.Y. (1984). The partitioned multiobjective risk method. *Large Scale Systems* **6**(1): 13–38.

Bellman, R.E. and Dreyfus, S.E. (1962). *Applied Dynamic Programming*. Princeton, NJ: Princeton University Press.

Bender, J.F. (1962). Partitioning procedures for solving mixed-variables programming problems. *Numerische Mathematick* **4**: 238–252.

Bernardo, J.M. (1994). *Bayesian Theory*. New York: Wiley.

Dantzig, G.B. and Wolfe, P. (1960). Decomposition principle for linear programs. *Operations Research* **8**(1): 101–111.

Garvin, D.A. (1988). *Managing Quality: The Strategic and Competitive Edge*. New York: The Free Press.

Gelman, A., Carlin, J., Stern, H. et al. (2013). *Bayesian Data Analysis*. Boca Raton, FL: CRC Press.

Haimes, Y.Y. (1977). *Hierarchical Analyses of Water Resources Systems: Modeling and Optimization of Large-Scale Systems*. New York: McGraw-Hill.

Haimes, Y.Y. (1981). Hierarchical holographic modeling. *IEEE Transactions on Systems, Man, and Cybernetics* **SMC-11**(9): 606–617.

Haimes, Y.Y. (1991). Total risk management. *Risk Analysis* **11**(2): 169–171.

Haimes, Y.Y. (2016). *Risk Modeling, Assessment, and Management*, 4e. Hoboken, NJ: Wiley.

Haimes, Y.Y. and Macko, D. (1978). Hierarchical structures in water resources systems management. *IEEE Transactions on Systems, Man, and Cybernetics* **SMC-3** (4): 396–402.

Haimes, Y.Y., Tarvainen, K., Shima, T., and Thadathil, J. (1990). *Hierarchical Multiobjective Analysis of Large-Scale Systems*. New York: Hemisphere.

Lasdon, L.S. (1970). *Optimization Theory for Large-Scale Systems*. New York: The Macmillan Company.

Lasdon, L.S. and Schoeffler, J.D. (1966). Decentralized plant control. *ISA Transactions* **5**(2): 175–183.

Paté-Cornell, E. (2002). Fusion of intelligence information: a Bayesian approach. *Risk Analysis* **22**(3): 445–454. doi: 10.1111/0272-4332.00056.

Yan, Z. and Haimes, Y.Y. (2010). Cross-classified hierarchical Bayesian models for risk-based analysis of complex systems under sparse data. *Reliability Engineering and System Safety* **95**(7): 764–776.

Yan, Z. and Haimes, Y.Y. (2011). Risk-based multiobjective resource allocation in hierarchical systems with multiple decision makers, Part I: Theory and methodology. *Systems Engineering* **14**(1): 1–16. http://www3.interscience.wiley.com/cgi-bin/fulltext/123270626/PDFSTART

7

Hierarchical Multiobjective Modeling and Decision Making for Complex Systems of Systems

PART I: EXPLORING SYSTEMIC RISK TO SYSTEMS OF SYSTEMS WITH MULTIPLE OBJECTIVES

7.I.1 INTRODUCTION

Physical infrastructures, serving as the foundations of society's well-being, encompass the entire private and public sectors and are the driving force of today's social and economic development. By virtue of their multifarious reach into our lives, they have also become the focal interest of diverse stakeholders that span local, state, and federal governments. The need to better understand and improve the current status of physical infrastructure systems is evidenced in reports spanning the last two and a half decades. For example, as early as three decades ago in 1988, a US national commission issued a report titled *Fragile Foundations,* citing: "The National Council on Public Works Improvement ... has found convincing evidence that the quality of America's infrastructure is barely adequate to fulfill current requirements, and insufficient to meet the demands of future economic growth and development" (National Council on Public Works Improvement, 1988). The practice of persistent infrastructure underinvestment, coupled with a significant growth in the demand for commercial and noncommercial transportation, has left the transportation infrastructure of many countries ill equipped for the needs of a churning global economy. Meanwhile, there is a likelihood of potential adverse consequences to infrastructure caused by *emergent forced changes (EFCs)*, which affect their performance. EFCs connote external or internal positive or negative sources of risk that may adversely affect one system or the entire complex systems of systems (SoS) (Complex SoS). Examples of these EFCs, which originate from both within and outside of the system, include global climate change, global and national economic crises, terrorist activities, cyber war and crimes, increasing user demands, physical deterioration, and the lack of available maintenance resources. The need to better maintain US infrastructure systems is real and urgent. However, the complexity of these systems and the lack of theories and methodological approaches to understand and analyze the risks associated with their deterioration significantly complicate this effort.

Modeling and Managing Interdependent Complex Systems of Systems, First Edition. Yacov Y. Haimes.

Many of the world's physical infrastructure systems fall within the category of Complex SoS. As discussed in earlier chapters of this book, these are commonly composed of *intra- and interdependent and interconnected* subsystems, which in their essence constitute Complex SoS with multiple functions, operations, and stakeholders. These emergent, large-scale engineering systems, such as ground transportation, aviation, supply chains, power grid, and cyber-infrastructure systems, pose great challenges in their risk modeling and management. The complexity of physical infrastructure Complex SoS is characterized by highly interdependent and interconnected physical, economic, and social components, which contribute a major source of EFCs to the system. To meet the increasing needs for reliable services provided by these infrastructure systems, their owners and decision makers need tools to foresee potential EFCs from within or outside the system, identify interdependencies among its different system components, and understand the impacts to the systems so that efficient risk management strategies, including preparedness and response plans, can be developed. Risk assessment, management, and communication are indispensable tools with which to evaluate the states of the system, reduce its vulnerability, and increase its resilience to any EFCs. This chapter presents the development and application of key risk analysis theories and methodologies for managing this physical infrastructure Complex SoS effectively and efficiently. In this chapter the entity Bridge Complex SoS will refer to the organizational and physical bridge infrastructure Complex SoS (Guo, 2015; Guo and Haimes, 2016, 2017).

The ideas developed in this chapter and throughout the entire book are driven by the need to understand, model, and thus manage their interdependencies and interconnectedness (I-I) and also to perform risk analysis for their effective performance. More specifically, the I-I are manifested by systems and subsystems that Complex SoS share – common states and other essential entities. The term "essential entities" connotes shared/common decisions, decision makers, stakeholders, resources, organizational behavior and norms, policies and procedures, management, culture, and others.

We first explore some systemic risks to Complex SoS by analyzing a unique failure mechanism caused by interdependencies among its subsystems through shared state variables from a control perspective in Part I. We then discuss a general approach with which to identify potential precursors to system failure through systemically exploring a metamodel of system failure mechanisms in Part II. The aim of Part II is to develop a systemic precursor analysis framework for Complex SoS that (i) builds on a proactive, dynamic anticipatory analysis methodology; (ii) identifies, prioritizes, detects, and evaluates different sources of EFCs; and (iii) evaluates the likelihood of different failure modes of the system based on information from identified precursors so that decision makers and other stakeholders can make more timely and informed decisions in response to emergent risks to prevent severe future consequences. In Part III of this chapter, we present two levels of metamodels used to model various aspects of Bridge Complex SoS. At a lower level, metamodels are used to extract the functional components of subsystems from their constituent physical components, such that the behavior of a subsystem can be modeled as the result of I-I *functions instead of physical components*. This metamodeling approach enables a modeler to analyze the functions of a "complicated subsystem" without understanding how these functions are realized by various physical components. This abstraction of functions is useful for a model to capture different failure modes of the system and their conditions. At a higher level, metamodels are commonly used to model the behaviors of subsystems based on different decompositions. Subsystems having shared states can be converted via pseudo-variables into subsystems with independent states and inputs from other related subsystems (see Chapters 2 and 4). This process enables modelers at the SoS level to coordinate multiple subsystems without knowing the internal states, structures, and working mechanism of each subsystem. This level of metamodeling has important realistic implications because of the operational and managerial independence property of the organizations as Complex SoS.

This methodology is demonstrated with a case study on highway bridge infrastructure Complex SoS, which constitute an important part of transportation infrastructure and are the lifelines of commerce in every country. The dismal state of many of the world's bridges and other physical infrastructure has been well documented in the literature. For example, highway bridges in the United States are continuously deteriorating. This is due to the lack of appropriate maintenance, with a significant percentage of bridges structurally deficient or functionally obsolete. Many bridges receive insufficient inspection and maintenance due to limited funding, equipment, manpower, and available technology. A bridge infrastructure system is indeed a Complex SoS with many interdependent functional subsystems, which are also managed by different decision-making entities and organizations. The developed methodologies in this chapter and throughout this book demonstrate the effectiveness of modeling and managing Bridge Complex SoS. More importantly, they are designed to help government agencies and bridge owners efficiently prioritize and plan for inspection, maintenance, and rehabilitation activities based on precursors and to reduce the risk of bridge failure.

7.I.1.1 Precursor Analysis in the Risk Assessment of Physical Infrastructure Complex Systems of Systems

Government agencies and infrastructure system owners seek to understand the trends of risks associated with EFCs that affect the states of their systems and Complex SoS to prevent, mitigate, or prepare for future undesired occurrences. Earlier, we defined EFCs as emergent sources of risk (i) to a system, or (ii) to SoS that can originate inside or outside a system within the SoS, or (iii) from the entire Complex SoS with the potential to cause system failure. Unanticipated, undetected, misunderstood, or ignored EFCs, whether they originate from within or from outside a system, are likely to affect a multitude of states of that system with potentially adverse consequences. Therefore, it is imperative to be able to envision, discover, and track precursors, collect data, and measure EFCs to Complex SoS. This can be accomplished through scenario structuring, modeling, and risk analysis discussed in more detail in Chapter 9.

Probabilistic risk assessment (PRA) (Haimes (2016)) is a systematic methodology to evaluate risks associated with a complicated engineered technological system. It has been successfully performed at all phases of a project life cycle from concept definition and predesign through safe removal from operation. As discussed in Chapter 9, Kaplan and Garrick (1981) introduced the theory of scenario structuring (TSS) and within it the triplet questions in the risk assessment process: *(i) What can go wrong? (ii) What is the likelihood that it would go wrong? (iii) What are the consequences?* Kaplan et al. (2001) subsequently modified the TSS by stating that the set of all scenarios cannot be a complete set. Risk assessment methodologies based on the concept of PRA systematically develop risk scenarios of the initiating events and attempt to quantify the probability and consequences for each scenario. This approach is used in event trees and event sequence diagrams. Principal examples of inductive approaches include failure mode and effects analysis (FMEA), failure mode, effects, and criticality analysis (FMECA), and preliminary hazard analysis (PHA). A decade later, Haimes (1991) added the following question for the risk assessment process: *(iv) What is the time frame?* And then he added the following three questions for risk management: *(i) What can be done and what options are available? (ii) What are the tradeoffs in terms of all costs, benefits, and risks? (iii) What are the impacts of current decisions on future options?*

Applying the above inductive risk assessment approach to Complex SoS faces new challenges. The term "emergent" denotes that the forced changes to SoS are usually dynamic, evolving, and possibly unexpected. Investigations of several accidents to complex systems show that the causes of their failures usually include multiple component failure and their unexpected interactions. The Three Mile Island accident is one example of this. For instance, multiple initiating events posing no risk to each individual subsystem may have possibly unknown complex interdependencies and causal relationships, subsequently causing significant adverse consequences at Complex SoS level. In these cases, it is practically impossible to enumerate an all-inclusive set of potential risk scenarios. This is due to the large number of scenarios resulting from the combination of multiple initiating events as well as the analyst's limited knowledge about the subsystem interdependency. Also, *an EFC might not be directly observable.* Furthermore, there might not be any available historical nor expert evidence data to estimate the probability of risk scenarios for emergent Complex SoS, nor to support efficient allocation of limited risk management resources. In addition to the above issues, short of actually observing triggering changes in the state space of the system by collecting and analyzing information and other evidence, the static and passive approach lacks the capability to (i) track and monitor different risk scenarios over time and (ii) forecast indicators and warnings on evolving forced changes and EFCs. Risk assessment that depends solely on inductive and static methods might fail to detect emergent risks to Complex SoS, resulting in inefficient allocation of risk management resources.

For example, a comprehensive risk assessment of Bridge Complex SoS should employ both inductive and deductive approaches to augment each other. The basic difference between deductive and inductive methods is the direction of the analysis: (i) *Inductive method* is the appropriate analysis to carry out if a given set of initiating causes is identified and the goal is to determine the resulting consequences, and (ii) *deductive method*, on the other hand, is the appropriate analysis to carry out if a given undesired event is defined and the goal is to determine its basic causes. The inductive approach is useful in assuring that the analysis is broad enough to encompass all possible scenarios, while the deductive approach has the benefit of focusing the analysis on the undesired event. Principal examples of inductive approaches include fault-tree analysis deployed throughout this book (also see Haimes (2016)).

Deductive risk assessment of infrastructure Complex SoS requires an understanding of *why* and *how* risk propagates through the system and results in significant adverse consequences. Such adverse consequences usually manifest themselves in the form of system failures or accidents. Thus, the questions of why Complex SoS fail and how to detect and evaluate signs prior to such failures, i.e. via precursors, become the focus of a deductive risk assessment. Identifying specific failure modes and their failure mechanisms is a key step toward extending current risk assessment methods to analyze Complex SoS.

Indeed, effective risk assessment and management of Complex SoS depend on understanding their failure mechanisms and designing mitigation strategies to manage these risks. Failure analysis tools based on reliability theory, such as sequential (event-based) accident models and epidemiological models, work well for accidents caused by failures of physical components or human errors in relatively simple systems. However, they are limited in their capability to explain accident causation in the more complex systems that were developed in the last half of the twentieth century (Lundberg et al., 2009). Several theories have been proposed to understand failures and accidents in Complex SoS. Based on systems theory, systemic accident models endeavor to describe the characteristic performance at the level of the system as a whole, rather than at the level of specific cause–effect "mechanisms."

Risk analysis is about the future. Thus, a proactive risk analysis of EFCs calls for a continuous process of designing a data collection mechanism, developing metrics with which to measure changes in the system, assessing whether observed changes are sufficiently significant, and determining criteria for actions. All of these are requisites for effective risk modeling, assessment, management, and communication for future EFCs. With the advancements in sensing, communication, and information processing technology, automated detecting and monitoring devices generate a high volume of information flow regarding past and current states of the system. This trend in technology provides a foundation for a proactive risk assessment that utilizes information to detect signs prior to system failure, through (i) observing them and (ii) noting the changes in the relationship among different states of the system. Precursors are important signs prior to system failures and thus deserve our further investigation.

Precursor analysis identifies factors that increase likelihood and detects indicators and warnings of possible future system failures (Guo, 2015; Guo and Haimes, 2016, 2017). Precursors can be used to determine whether adversarial events are either occurring or expected and provide opportunities for decision makers to take preparedness and response actions to avoid projected adverse consequences. Analyzing precursors to system failures based on available information and expert evidence is a deductive and proactive risk assessment approach. Thus, this is an integral part in risk assessment of cyber–physical infrastructure Complex SoS.

Precursor analysis, which has been widely adopted in nuclear power, aviation, and health-care industries, has been proven an effective approach to identify potential risk factors leading to accidents and other adverse consequences. Although literature across different disciplines has discussed the potential use of precursors to anticipate system failures and accidents, they lack a formal process of precursor analysis based on systems theory and a rigid, systemic, and justifiable framework. The literature commonly identifies precursors on the basis of accident, or near-miss incident reports and analysis of available data and of expert evidence and knowledge. These methods for precursor analysis neither account for the internal causal relationships nor capture the essence of the system dynamics. Without a systems model capable of describing the mechanism of system failure, it is very difficult to identify and evaluate precursors and provide useful insights for risk management.

Another major issue in precursor analysis is the frequent discrepancy between pre- and post-accident risk assessment using precursors, where in the former case the effectiveness of using precursors is quite limited. This perception may be the result of *hindsight bias*; namely, after an accident, individuals often believe that the event should have been considered highly likely, if not inevitable, by those who observed the precursors prior to the event. Consider the following three major facts that contribute to the existence of hindsight bias:

1. Precursors are usually weak signals to forecast system failure, and in many cases the likelihood of system failure before and after the observation of a specific precursor is not objectively quantified. Also, response actions are not justified based on such vague indicators.
2. There are one-to-many relationships between a precursor and various failure modes of the system. Observing a specific precursor may increase the likelihood of multiple failure modes simultaneously, without necessarily increasing the situational awareness of decision makers.
3. Uncertainties in precursor detection and prediction are not well accounted for in a nonmodel-based and informal analysis (Guo and Haimes, 2016, 2017).

Addressing these problems in precursor analysis for Complex SoS would require (i) a systemic and quantitative approach to identify essential system building components, (ii) understanding complex system behaviors and failure mechanisms, (iii) quantifying uncertainties in detection and prediction, and (iv) integrating information from multiple precursors.

A basic adage posits that models should be as simple as possible but as complex as required. The challenges facing systems modelers of Complex SoS are nontrivial. The fact that Complex SoS usually consist of multiple subsystems, sub-subsystems, and numerous components with multiple time frames, decision makers, and stakeholders further complicates the modeling process. In Chapters 2 and 4 of this book we introduced the concepts of decomposition and higher-level hierarchical coordination – a common modeling approach applied to Complex SoS. Furthermore, the finer the granularity of the model, (i) the more complex the modeling process, (ii) the more the increase in the number of variables and equations, (iii) the more time and cost are required to develop the model, and (iv) the greater the difficulty in finding the solutions. The level of complexity of the model needs to be determined properly for a model to answer specific questions. For example, a model of bridge infrastructure systems, built from detailed physical models for each bridge element, may not be very suitable for the purpose of focusing on strategic policy formulation. However, without understanding the behaviors of each component of the system, a higher-level model cannot be built from a bottom-up approach.

Metamodeling, introduced in Chapter 4, connotes a framework that (i) builds on systems-based theory and methodology; (ii) enables modelers to relate, coordinate, and integrate submodels of multiple models of SoS subsystems; and (iii) provides a better understanding and modeling of the SoS as a whole. Metamodels are usually constructed using a data-driven, bottom-up approach. The exact inner working of the system is not assumed to be known or even understood, solely the input–output behaviors that are important to the upper level.

7.I.2 MODELING PHYSICAL INFRASTRUCTURE COMPLEX SYSTEMS OF SYSTEMS

Physical infrastructures in their essence constitute Complex SoS with multiple functions, operations, and stakeholders. Their complexity is characterized by highly interconnected, intra- and interdependent physical, cyber, organizational, and economic subsystems through shared resources, decisions, and states. Modeling an entire Complex SoS, including all its components, relationships, functions and behaviors, and data collection among other tasks, is by any measure a daunting job, involving a continuous iterative process of learning, discovery, modification, and validation. The complexity of the multiple interdependent and interconnected subsystems constitutes a major challenge in modeling and managing Complex SoS. These challenges emerge from systems modeling as well as risk assessment, management, and communication. From the modeling perspective, no single model is able to represent the multiple perspectives of Complex SoS. One key issue in modeling Complex SoS is how to identify and quantify the causal intra- and interconnected and interdependent relationships within and among the subsystems. Modeling that fails to consider the I-I among the subsystems would be unable to capture the impacts of decisions made on one subsystem on other related subsystems. Thus it would also fail to uncover intricate complex interactions and causal relationships that constitute Complex SoS.

Risk analysts of Complex SoS must (i) analyze and quantify the interdependencies among the subsystems, (ii) foresee the EFCs from within or from outside the system, and (iii) evaluate their impacts to the system. It requires understanding the paths through which sources of risk propagate through the system and result in adverse consequences. The latter usually manifest themselves in the form of system failures, such as power outages, cyber incidents, and bridge collapses, among others. These failures must be understood in advance, monitored in real time, and, if possible, predicted based on available information so that resulting adverse consequences can be effectively discovered and managed. Identifying specific failure modes and their failure mechanisms are key steps in extending current risk assessment methodologies to analyze Complex SoS.

This chapter draws on Chapter 5 on multiple objectives in Complex SoS. Here, we aim to identify internal systemic risks specific to infrastructure Complex SoS through (i) formulating infrastructure maintenance as a multiobjective decision problem under a control structure, (ii) identifying characteristics of Complex SoS and exploring unique system failure modes resulting from these characteristics, and (iii) developing methods to mitigate and manage these risks. An analysis of a multiobjective sequential decision process of a nonlinear SoS reveals some inherent systemic risks among its subsystems.

7.I.2.1 Decomposition of Subsystems with Shared States

No single model is capable of representing the multiple perspectives of a Complex SoS; thus a realistic model of Complex SoS necessarily consists of multiple subsystems models and their relationships. Decomposition is a common approach to modeling large-scale systems. When Complex SoS are decomposed into their subsystems and sub-subsystems, the I-I among must be preserved. The intrinsic modeling approach, discussed in Chapters 2 and 4, is to model subsystem interdependency. This assumes that the output from one subsystem is the input to another subsystem. This input–output modeling approach (Leontief, 1951a, b; Haimes and Jiang, 2001; Lian and Haimes, 2006; Haimes, 2016) the subject of Chapter 8, has been widely used to model systems where the exchange of I-I among physical commodities constitutes a major source of complexity. Examples include production processes, supply chains, and the overall economy. However, the characteristics of Complex SoS and subsystems suggest that their interdependencies rely more on sharing common states, resources, and information as well as decisions and constraints termed earlier in this book as *shared states and other essential entities* (defined earlier). These interdependencies cannot be directly represented as input–output relationships. Information about these interdependent and interconnected sectors of the economy requires a different approach that builds on state-space theory for modeling Complex SoS. The phantom systems model (PSM) introduced in Chapter 4 is an intrinsic modeling approach. It posits that some specific commonalities, interdependencies, interconnectedness, or other relationships must exist between any two subsystems within any Complex SoS. The essence of each subsystem can be represented by a finite number of essential state variables. The term "essential state variables" connotes the minimum number of state variables required to represent the essence of the system, or of Complex SoS. Shared state variables are defined as state variables common to two or more subsystems models; therefore they play a central role in modeling systems and their interdependencies. Decomposing Complex SoS, and a decentralized approach to subsystem coordination and control, are more common in reality when subsystems share state variables due to their operational and managerial independence. As discussed in Chapter 2, the decentralized approach treats each subsystem independently at a lower level of the hierarchy and coordinates them at a higher level, where only limited information is exchanged between subsystems and the higher-level coordinator. We build on a decomposition method that transforms subsystems that have shared state variables in Complex SoS into subsystems that are connected through inputs and outputs among them.

Building on Chapter 2 of this book, we introduce here a method to decompose subsystems with shared state variables. We then coordinate individual subsystems to achieve the overall objective of the Complex SoS through a two-level hierarchical control structure. The decomposition method plays a fundamental role in understanding systemic risks caused by subsystem interdependencies.

A simple example of the linear state-space system, shown in Figure 7.I.1, demonstrates the decomposition and coordination of two separate subsystems sharing one state variable (Guo and Haimes, 2016, 2017).

Consider a linear system with three state variables that can be described by Eq. (7.1).

$$s(k+1)=\begin{bmatrix} s_1(k+1) \\ s_2(k+1) \\ s_3(k+1) \end{bmatrix} = \mathrm{As}(k)+\mathrm{Bu}(k)=\begin{bmatrix} a_{11} & 0 & a_{13} \\ 0 & a_{22} & a_{23} \\ a_{31} & a_{32} & a_{33} \end{bmatrix} \begin{bmatrix} s_1(k) \\ s_2(k) \\ s_3(k) \end{bmatrix} + \begin{bmatrix} b_{11} & 0 \\ 0 & b_{22} \\ b_{31} & b_{32} \end{bmatrix} \begin{bmatrix} u_1(k) \\ u_2(k) \end{bmatrix}$$

$$y(k+1)=\begin{bmatrix} y_1(k) \\ y_2(k) \end{bmatrix} = \mathrm{Cs}(k)=\begin{bmatrix} c_{11} & 0 & c_{13} \\ 0 & c_{22} & c_{23} \end{bmatrix} \begin{bmatrix} s_1(k) \\ s_2(k) \\ s_3(k) \end{bmatrix} \tag{7.1}$$

According to the structure of matrix $\mathrm{A}=\begin{bmatrix} a_{11} & 0 & a_{13} \\ 0 & a_{22} & a_{23} \\ a_{31} & a_{32} & a_{33} \end{bmatrix}$, it is feasible to decompose it into two separate subsystems where state variables s_1 and s_2 each belong to one of the subsystems and the state variable s_3 is shared between the two subsystems.

The formulation of the model guarantees that the interdependencies between the two subsystems are manifested solely on the shared state variable s_3. Based on Eq. (7.1), a system diagram is constructed and illustrated in Figure 7.I.1. During the decomposition, the shared state variable belongs to, and thus remains in, both subsystems. By analyzing the shared state variable s_3, it can be shown that

$$s_3(k+1)=a_{31}s_1(k)+a_{32}s_2(k)+a_{33}s_3(k)+b_{31}u_1(k) + b_{32}u_2(k)=a_{33}s_3(k)+z_1(k)+z_2(k) \tag{7.2}$$

where $z_1(k)=a_{31}s_1(k)+b_{31}u_1(k)$ contains all necessary information from subsystem 1 at $t=k$ and $z_2(k)=a_{32}s_2(k)+b_{32}u_2(k)$ contains all necessary information from subsystem 2 at $t=k$.

A decomposition scheme based on this property is shown in Figure 7.I.2.

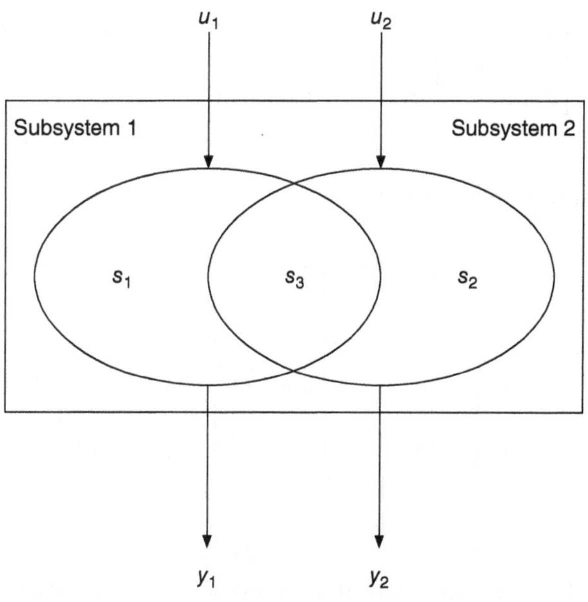

Figure 7.I.1 *Two subsystems sharing one state variable.*

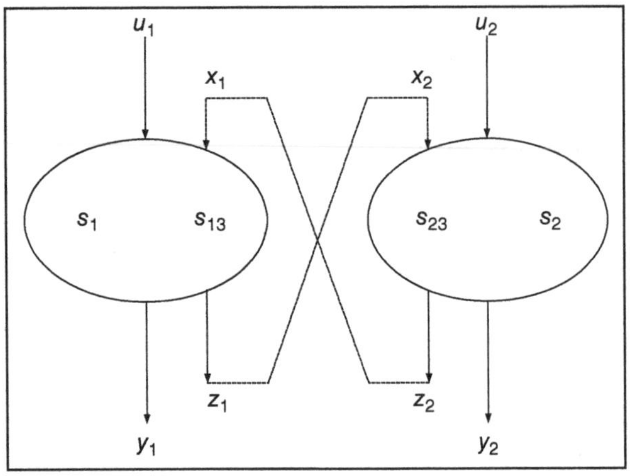

Figure 7.I.2 *Decomposition of systems sharing one state variable.*

The shared state variable s_3 remains in both subsystems as s_{13} for subsystem 1 and s_{23} for subsystem 2.

For subsystem 1, $s_{13}(k+1) = a_{33}s_{13}(k) + a_{31}s_1(k) + b_{31}u_1(k) + z_2(k)$, and for subsystem 2, $s_{23}(k+1) = a_{33}s_{23}(k) + a_{32}s_2(k) + b_{32}u_2(k) + z_1(k)$.

One extra input x and one extra output z are added to each subsystem to incorporate the information needed to update the value of the shared state variable from the other subsystem. The systems models for the decomposed subsystems are derived as follows:

For subsystem 1:

$$\begin{bmatrix} s_1(k+1) \\ s_{13}(k+1) \end{bmatrix} = \begin{bmatrix} a_{11} & a_{13} \\ a_{31} & a_{33} \end{bmatrix} \begin{bmatrix} s_1(k) \\ s_{13}(k) \end{bmatrix} + \begin{bmatrix} b_{11} & 0 \\ b_{31} & 1 \end{bmatrix} \begin{bmatrix} u_1(k) \\ x_1(k) \end{bmatrix}$$

$$\begin{bmatrix} y_1(k) \\ z_1(k) \end{bmatrix} = \begin{bmatrix} c_{11} & c_{13} \\ a_{31} & 0 \end{bmatrix} \begin{bmatrix} s_1(k) \\ s_{13}(k) \end{bmatrix} + \begin{bmatrix} 0 & 0 \\ b_{31} & 0 \end{bmatrix} \begin{bmatrix} u_1(k) \\ x_1(k) \end{bmatrix} \quad (7.3)$$

For subsystem 2:

$$\begin{bmatrix} s_2(k+1) \\ s_{23}(k+1) \end{bmatrix} = \begin{bmatrix} a_{22} & a_{23} \\ a_{32} & a_{33} \end{bmatrix} \begin{bmatrix} s_2(k) \\ s_{23}(k) \end{bmatrix} + \begin{bmatrix} b_{22} & 0 \\ b_{32} & 1 \end{bmatrix} \begin{bmatrix} u_2(k) \\ x_2(k) \end{bmatrix}$$

$$\begin{bmatrix} y_2(k) \\ z_2(k) \end{bmatrix} = \begin{bmatrix} c_{22} & c_{23} \\ a_{32} & 0 \end{bmatrix} \begin{bmatrix} s_2(k) \\ s_{23}(k) \end{bmatrix} + \begin{bmatrix} 0 & 0 \\ b_{32} & 0 \end{bmatrix} \begin{bmatrix} u_2(k) \\ x_2(k) \end{bmatrix} \quad (7.4)$$

subject to $x_1(k) = z_2(k)$ and $x_2(k) = z_1(k)$.

This decomposition method can be extended to nonlinear systems as long as the transition equation of the shared state variable has the form $s_3(k+1) = g_3(s_3(k)) + g_1(s_1(k), u_1(k)) + g_2(s_2(k), u_2(k))$, where $g_1(\cdot)$, $g_2(\cdot)$, and $g_3(\cdot)$ are arbitrary real functions.

7.I.2.2 Coordinating Subsystems with Shared States

The interdependencies caused by shared states introduce significant challenges for the decision makers to understand the impact of their decisions as well as the decisions of others on all of the connected subsystems. Consider the following questions: How do decision makers of a subsystem achieve their goals when some of their subsystem's state variables are shared with other subsystems and are partially uncontrollable? What would be the strategy of the decision makers to optimize subsystem performance?

Fundamental to this discussion is the decentralization principle: A set of subsystems is optimally controllable in a decentralized manner if and only if there are no variables common to two or more subsystems. The presence of common or shared variables is equivalent to the existence of constraints between the optimization problems, requiring an integrated or simultaneous solution of two or more subproblems. The decentralized control problem with input–output structure has been formulated and discussed in Chapter 2. A two-level hierarchical control structure is used to derive the solution to the problem. At the first level, each subsystem aims to achieve its own objectives by varying its input and decision variables; at the second level, a controller is assigned the task of coordinating the first-level subsystems so that the overall system goal is achieved. The decomposition method developed in Section 7.I.2.1 transforms the interdependency caused by shared states into input–output relations and provides a foundation for coordinating the subsystems with shared states.

We demonstrate through an example problem that any single-objective optimization problem of a subsystem with shared states is equivalent to an optimization problem of an independent subsystem with two competing objectives, with an extra objective of gaining maximum controllability of the shared states. In our example, it is the second-level controller's role to set an appropriate tradeoff level on the Pareto-optimal frontier for the subsystem decision maker in order to optimize the overall performance of the Complex SoS.

Consider here the optimization of a single-objective, one-stage state-space system using a multilevel nonfeasible method introduced in Chapter 2. As an example, we use the linear state-space model described in Eq. (7.1) and Figure 7.I.1. We assume that the performance of the overall system is a function of the performances of individual subsystems, and we choose an additive performance function of the form $F = y_1(k+1) + y_2(k+1)$ to be minimized for time period $k+1$ by the selection of manipulated/control variables $u_i(k)$.

We formulate the problem as

$$\min_{u_i(k)} F = \sum_{i=1}^{2} y_i(k+1) \tag{7.5}$$

$$s.t. \quad z_i(k) = x_j(k), \quad \text{for } i,j = 1,2,\ i \neq j$$

To incorporate the constraints, a Lagrangian equation is formed as

$$L = \sum_{i=1}^{2} y_i(k+1) + \sum_{\substack{i,j=1,2 \\ i \neq j}} \lambda_i \left[x_j(k) - z_i(k) \right] \tag{7.6}$$

where λ_i are Lagrange multipliers. Note that the Lagrangian function is discussed in the Appendix of this book. By manipulating and reorganizing the Lagrangian function, we decompose it into 2 subobjectives. Define $\widetilde{y}_i$ as the adjusted objective function of subsystem $L = \sum_{i=1}^{2} \widetilde{y}_i(k+1)$ where $\widetilde{y}_i(k+1) = y_i(k+1) - \lambda_i z_i(k) + \lambda_i x_i$.

During the course of computation, the constraint that the "cut" variables must be equal is not satisfied in general, and we define the errors as $E_i(k) = z_i(k) - x_j(k)$. A two-level nonfeasible approach can be used to solve the integrated problem in a completely decentralized manner, where at the lower level the ith subproblem is to determine $u_i(k)$ and $x_i(k)$ such that $\widetilde{y}_i(k+1)$ is minimized for fixed λ_i and λ_j. And, at the higher level, the goal of the coordination process is to find λ_i^* and λ_j^* such that $E_i(k) = 0$ for all i. The optimal solution must be such that the constraints are met.

Notice that each $\widetilde{y}_i(k+1)$ is an unconstrained function only of the independent variables associated with subsystem i for fixed multipliers. For each subsystem, given certain values of the Lagrangian multipliers, the decision maker tries to minimize a new unconstrained objective function $\widetilde{y}_i(k+1)$, which is the sum of the original objective function $y_i(k+1)$ and an adjusted factor: $-\lambda_i z_i(k) + \lambda_j x_i(k)$, where $x_i(k)$ defines the inputs from connected subsystems and $z_i(k)$ the outputs to connected subsystems. Minimizing this adjusted factor means that from an individual subsystem's perspective, the decision maker needs to minimize the input from connected subsystems and to maximize the output to connected subsystems. This can be interpreted as maximizing the subsystem's controllability on the shared state variable. In such case, gaining controllability of the shared state variables is another important objective to be considered in the decision-making strategy. The subsystem decision maker's original single-objective decision problem now becomes a decision problem with two competing objectives, and the Lagrangian multipliers serve as weighting factors between the subsystem's original objective and the controllability of the shared state variable.

In a real system, the physical form of the Lagrange multipliers includes incentive, subsidy, policy, and regulation, all of which have an influence on the decision maker's preference between the two objectives. The coordinator's role is to find the optimal values of the multipliers such that the optimal decision of subsystem 1 will also help to optimize subsystem 2 through shared states, and vice versa. In the next section, we apply the above decomposition and coordination methods in order to understand the risks of a subsystem in the Complex SoS under external perturbations caused by shared state variables.

7.I.3 SYSTEMIC RISKS IN COMPLEX SYSTEMS OF SYSTEMS

This section builds on the theories of modeling, decomposing, and coordinating SoS discussed in Section 7.I.2. Here we develop a framework to explore systemic risks inherent in Complex SoS in order to understand their unique sources of failure and to enable decision makers to mitigate and manage these risks. The term "systemic risk" refers to risks resulting from the structure, topology, and characteristics of the SoS rather than of an individual subsystem. We demonstrate that sources of risk may come from (i) the decision maker's preference among multiple competing objectives and (ii) unexpected perturbations due to potential couplings (interdependencies) among interconnected subsystems. Specifically, it demonstrates that a Complex SoS can fail even if all its components are functioning as they are designed; an optimal decision strategy does not necessarily guarantee system safety.

In subsequent discussion we introduce some special characteristics of Complex SoS as potential sources of systemic risks, explore a unique failure model in a stand-alone SoS subsystem, and illustrate the theory with a numerical example. We then discuss the implications of our findings and discuss both subsystem risks caused by interdependencies through shared states and a method to estimate system safety margins. Finally, we discuss how subsystem coordination helps to mitigate these risks.

7.I.3.1 Characteristics of Physical Infrastructure Represented via Bridge Complex Systems of Systems

Each physical infrastructure SoS has a unique feature that distinguishes it from conventional systems. However, the following set of characteristics is common to many infrastructure Complex SoS: (i) multiple stakeholders and decision makers, (ii) multiple goals and objectives, (iii) multiple interdependent and interconnected subsystems through shared states and decisions, (iv) unknown interactions among subsystems, (v) nonlinear system dynamics, and (vi) adaptive system behaviors. Knowing these characteristics is useful in identifying some internal risks to Complex SoS. Specifically, we focus on how the interactions among the system's functional components become sources of risk to system failure, rather than focusing on the physical reliability of each system component.

Consider a highway bridge as an example. Bridges are traditionally considered merely as physical systems from an engineering perspective. However, if we view bridges as part of an overall transportation infrastructure, we must address the various functionalities, objectives, and stakeholders, as well as the decision-making process of the encompassing system. This chapter considers a bridge as an infrastructure Complex SoS with broad social and economic impact due to the possibility of bridge failure. Bridges possess many of the characteristics of a Complex SoS. For example, a physical bridge includes all of its physical subsystems, such as deck, beam, bearing, abutment, and pier. A bridge system has multiple stakeholders, including its owner such as the Department of Transportation (DoT), the bridge users, and inspection and maintenance teams, among others. Stakeholders make decisions according to their goals and objectives, which are usually noncommensurable and competing with each other. Decisions made regarding one subsystem are likely to have direct or indirect impact on other subsystems due to their interdependencies. For example, deicing bridges during the winter season improves driving safety on the deck subsystem. However, deicing chemicals will also accelerate the deterioration process of the rebar in the reinforced concrete beam of the superstructure subsystem, thus reducing the load capacity of a bridge over time. Due to the complexity and uncertainty of the system dynamics, decision makers would need to adjust and adapt their decisions based on current states of the bridge, in an effort to keep it at or near the desired condition. The decision process resembles a closed-loop feedback control system, and a sequential set of decisions is made every 2 years through the life span of the bridge. Finally, the deterioration process is nonlinear in nature, as is the material behavior beyond the elastic limit. Thus, bridge infrastructure must be viewed and modeled from a Complex SoS perspective.

Regardless of the type, structure, or material from which a bridge is built, a bridge must meet two basic requirements – (i) reliability and (ii) effectiveness – for the traffic crossing it. We consider two basic subsystems of a Bridge Complex SoS: the maintenance subsystem and the traffic engineering subsystem. A subsystem here consists of both the physical elements of the bridge and all of the organizations with their functions, decisions, and control processes.

The decision maker of the bridge subsystem is usually the maintenance division in the state DoT, with the goal of maintaining the reliability and structural integrity of both superstructure and deck at an acceptable level. The decisions they make include planning the maintenance budget and the types of repairs to be made. The decision maker of the traffic engineering subsystem is usually the traffic engineering division in the state DoT, with the goal of ensuring the efficiency of traffic across the bridge. The decisions made by this division include deicing, lane allocation, work zone design, signal and illumination placement, and speed and tonnage posting. Figure 7.I.2 illustrates the functional components and control structure of the two interdependent subsystems of a Bridge Complex SoS. To model the interdependency between these two subsystems, three state variables are chosen to represent the essential states of the two subsystems. The condition rating of the superstructure s_s is a state variable of the maintenance subsystem, the average daily traffic of the bridge s_t is a state variable of the traffic engineering subsystem, and the condition rating of the deck s_d is a shared state variable of both subsystems.

In reality Complex SoS are usually formed through a process of adaption and evolution rather than a process of purposeful design and implementation. As a result, an individual subsystem's decision maker may not be aware of the coupling introduced by the shared state variable and does not necessarily manage to control it. We develop a descriptive model to mimic the optimal decision process that a rational decision maker would take without knowing the coupling to other subsystems and identify the systemic risks under this situation.

7.I.3.2 Risks in a Nonlinear Dynamic Multiobjective Sequential Decision-Making Process

The decomposition method developed in Chapter 2 and Section 7.I.2 in this chapter provide an opportunity to focus on each individual subsystem's behaviors and dynamics. To understand the systemic risks of a Bridge Complex SoS, we first developed a simple mathematical model for an individual subsystem within the SoS. This illustrative model has two state variables, two objective functions, and one decision maker. We assume one of the subsystem's objectives is related to the performance of the subsystem and the other objective is related to its safety. These two objectives are usually noncommensurable and competing with each other such that tradeoffs between objectives must be considered. The decision maker makes a set of sequential decisions to optimize both objectives based on the observed states of the subsystem over a time horizon. A set of constraints determine the operational safety boundary of the subsystem, and system failure occurs when the safety constraints are no longer satisfied. This type of system structure is a common representation of many real-world systems. We hypothesize that the decision maker's preference to increase the system performance objective against the safety objective constitutes a driving force that propels the system to migrate toward the boundary of unsafe operation. We test our hypothesis using the following subsystems model.

Recall the notation used in Section 7.I.2, and let $s_1(k)$ and $s_3(k)$ be two state variables of one subsystem at time k, where $s_3(k)$ is a shared state variable with other subsystems. Let $u(k)$ be the decision made by the decision maker at time k. At this stage, we temporarily ignore the coupling between this subsystem and the connected subsystem. The effect of the coupling will be discussed in Section 7.I.3.5.

A second-order state transition function is used to represent the nonlinear dynamics of this subsystem, and a mathematical model is shown in Eq. (7.7):

$$\begin{bmatrix} s_1(k+1) \\ s_3(k+1) \end{bmatrix} = \begin{bmatrix} a_{11} & a_{12} & a_{13} & a_{14} \\ a_{21} & a_{22} & a_{23} & a_{24} \end{bmatrix} \begin{bmatrix} s_1^2(k) \\ s_3^2(k) \\ s_1(k) \\ s_3(k) \end{bmatrix} + \begin{bmatrix} b_1 \\ b_2 \end{bmatrix} u(k)$$

$$k = 0, \ldots, T-1 \tag{7.7}$$

To simplify the discussion without a loss of generality, the second-order interaction terms (e.g. $s_1(k)$ $s_3(k)$) are omitted in the model. Parameters a_{ij} and b_i are system parameters for the state variables and control, respectively. Let y_{11} and y_{12} be two objective functions of the subsystem representing system safety and performance, respectively. In many cases, such as quality control, the decision maker aims to maintain the outputs of the system within certain boundaries and to minimize deviation from a nominal value. The safety objective can also take this form, where either a higher or lower shift in the level of the output would cause system failure. For simplicity, the two objectives assume the form $\begin{bmatrix} y_{11}(k) \\ y_{12}(k) \end{bmatrix} = \begin{bmatrix} (s_1(k)-c_1)^2 \\ (s_3(k)-c_3)^2 \end{bmatrix}$ for $k = 0, \ldots, T-1$ where c_1 and c_3

are control objectives of the state variables and represent the target level of safety and performance, respectively. Due to sources of uncertainties from within or outside of the subsystem, the decision maker decides at each time step to compensate for possible deviations and adjust the system's states. The decision problem is formulated as

$$\min_{u(k)} \{y_{11}(k+1),\ y_{12}(k+1)\} \quad \text{for each} \quad k \in \{0,\ldots,T-1\}.$$

Among various approaches to multiobjective decision-making problems, the weighting method is the simplest one, as long as the objective functions are convex. Let θ be the weight for these two objectives, $\theta \in (0,1)$, and the above problem becomes a single-objective decision-making problem with θ as a system parameter:

$$\min_{u(k)} \{\theta y_{11}(k+1) + (1-\theta) y_{12}(k+1)\} \quad \text{for each} \quad k \in \{0,\ldots,T-1\}. \tag{7.8}$$

To find the optimal value of $u(k)$, we set the derivative of y with respect to u to zero, and get

$$u^*(k) = \frac{\theta b_1 \left[c_1 - a_{11} s_1^2(k) - a_{12} s_3^2(k) - a_{13} s_1(k) - a_{14} s_3(k)\right]}{\theta b_1^2 + (1-\theta) b_2^2} + \frac{(1-\theta) b_2 \left[c_3 - a_{21} s_1^2(k) - a_{22} s_3^2(k) - a_{23} s_1(k) - a_{24} s_3(k)\right]}{\theta b_1^2 + (1-\theta) b_2^2} \tag{7.9}$$

Substitute $u^*(k)$ into the state transition matrix in Eq. (7.7), and $s_1(k+1)$ and $s_3(k+1)$ can be solved accordingly.

To analyze the dynamics of the state variable $s = [s_1; s_3]$ in the system state space, we are interested in the function

$$f(\theta, \boldsymbol{s}) = \boldsymbol{s}(k+1) - \boldsymbol{s}(k) = \begin{bmatrix} s_1(k+1) - s_1(k) \\ s_3(k+1) - s_3(k) \end{bmatrix} = \beta_2 \begin{bmatrix} s_1^2(k) \\ s_3^2(k) \end{bmatrix} + \beta_1 \begin{bmatrix} s_1(k) \\ s_3(k) \end{bmatrix} + \beta_0 \tag{7.10}$$

where parameter β_2, β_1, β_0 can be expressed as functions of a, b, c, and θ.

When $f(\theta, \boldsymbol{s}) = 0$, the system state $\boldsymbol{s}$ is in a steady state. Otherwise, it is in a transient state and may or may not reach the steady state, depending on its initial states:

$$f(\theta, \boldsymbol{s}) = \beta_2 \begin{bmatrix} s_1^2(k) \\ s_3^2(k) \end{bmatrix} + \beta_1 \begin{bmatrix} s_1(k) \\ s_3(k) \end{bmatrix} + \beta_0 = 0 \tag{7.11}$$

The solution to Eq. (7.11) for different values of θ provides information about the characteristics of stationary points and stable and unstable regions of the state space. The existence of solutions depends on the values of a_{ij} and b_i, as well as on the value of θ. When no solution exists, the system has no steady state and will very likely drift away from the control objectives and cause system failure. Thus, the question we want to answer here is: Given coefficients a_{ij} and b_i, what is the boundary of θ such that Eq. (7.11) has at least one solution?

7.1.3.3 A Numerical Example and Analysis of Results

Consider a numerical example; let $A = \begin{bmatrix} -0.06 & 0 & 0.93 & -0.01 \\ 0 & -0.06 & 0.01 & 0.91 \end{bmatrix}$, $B = \begin{bmatrix} 1 \\ 5 \end{bmatrix}$.

These values are derived from a bridge maintenance subsystems model with two state variables representing the condition rating of the superstructure and the condition rating of the deck, c_1 and c_3 representing the control target values of the condition rating of superstructure (safety) and deck (performance), and $u(k)$ being the decision (total maintenance spending in million dollars) made for the maintenance subsystem.

An analysis of this specific example reveals that Eq. (7.11) represents two hyperbolae in the two-dimensional state space. The problem is then equivalent to determining whether these two hyperbolae intersect or not. A solution to this problem has been proposed by Wang et al. (2001) and shows that when $\theta < 0.103$, Eq. (7.11) has two different solutions, thus the stationary points of the system. Among the two solutions, one is a stable stationary point, and the other is an unstable stationary point. When $\theta = 0.103$, Eq. (7.11) has only one solution, and when θ is decreased further, there is no solution to

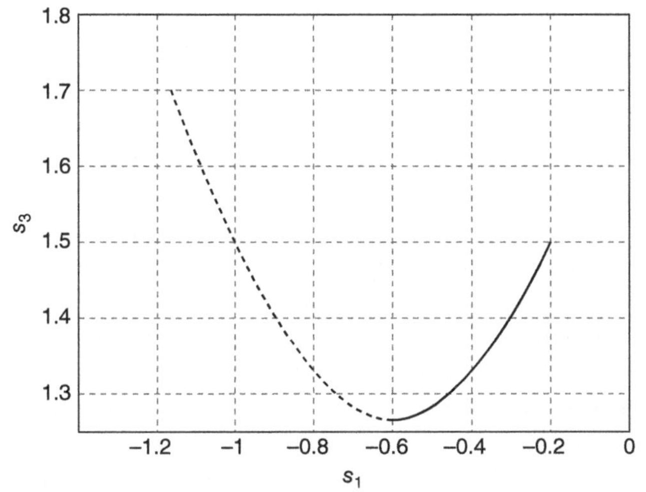

Figure 7.I.3 *Theoretical trajectory of two stationary points as a function of decreasing θ.*

Eq. (7.11). This means that when the value of θ is less than a threshold determined by system parameters, there would be no steady state in the state space and the values of the state variables may drift (or oscillate) beyond the safe operation boundaries and cause system failures. We must emphasize that even in this case, the system is controlled under an "optimal" control strategy $u^*(k)$, where the decision maker aims to optimize two objective functions given a certain value of θ. As we can see, this type of optimal control strategy does not necessarily ensure the safe operation of the system.

Figure 7.I.3 shows the trajectory of both stationary points in the state space as the value of θ decreases from 0.5 to 0.103. The solid line is the trajectory of the stable stationary point, and the dashed line is the trajectory of the unstable stationary point. As θ decreases, they eventually converge to one point, and beyond a threshold of θ, there is no stationary point in the state space.

The threshold of θ is a crossover point of the system between its stable and unstable regions of operation. If the decision maker's preference for the system performance objective over the safety objective becomes so high that θ moves across this crossover point, the system will not be able to return to a stable region even with a feedback mechanism. The crossover only occurs due to the decision maker's preferences and not to any external forces; thus this is a systemic risk to the system.

7.I.3.4 Implications

Results from the above analysis deserve further discussion. In a multiobjective decision problem, all solutions on the Pareto-optimal frontier are considered both feasible and optimal. Decision makers can choose any solution on the Pareto-optimal frontier based on their preferences. (Note that Chapter 5 is devoted in its entirety to multiobjectives, Pareto-optimality, and the associated tradeoffs.) Our analysis reveals that changing the preference does impact system stability and safety. When the decision maker has a very high preference for one objective over another and thus chooses the solutions on the extreme sides of the Pareto-optimal frontier, a nonlinear system may eventually enter an unstable region. The decision maker's change in preference, either intentionally or spontaneously, may be the result of pursuing better system performance or reacting to aggressive competitive market conditions. In the case of the bridge maintenance subsystem, the decision maker's decreasing preference for the superstructure over the deck may be the result of growing traffic demand, higher user cost for delays due to poor deck condition, or higher maintenance costs for the superstructure.

The system's behavior is also similar to a bifurcation structure as they are both dynamic and nonlinear with slowly changing parameters. The threshold of θ is a boundary between a stable and unstable region of operation and is analogous to a bifurcation point or critical point in a bifurcation system, as shown in Figure 7.I.3. Different types of bifurcation structures in complex systems have been explained through the principle of universality. In this chapter, we posit that the decision maker's preferred structure in a multiobjective decision process might offer another mechanism for the system to have such bifurcation structures.

This analysis also provides insights into identifying precursors of failure and designing warning systems for physical infrastructure Complex SoS. Traditional control paradigms monitor trends and patterns in controlled processes and use hypothesis testing to identify anomalies in the processes. This approach is reasonable for linear systems as the changes in the state space of linear systems are usually predictable through its trends. However, in a nonlinear system, state changes are usually abrupt, and trend-based technologies are limited in predicting these changes. This analysis calls for other metrics to be used as signals to predict abrupt state changes in nonlinear systems. Our further experiments with various values of system parameters indicate that a highly nonlinear system exhibits behaviors similar to critical slowing down (CSD) before abrupt state changes. Thus, some statistical measures may be practically useful for predicting this type of system failure.

The parameter θ, which is the Lagrangian multiplier, or literally the weight a decision maker places on objectives, is a measure of the decision maker's subjective preference between objectives. It is a latent variable that is difficult to measure or quantify even by the decision maker. However, as we can see from the example, when the value of θ is very low, the system becomes unstable and introduces significant risk of system failure. If a preference elicitation process to track and monitor a decision maker's preference along the sequential decision-making process can be designed, it may also provide useful information to predict potential system failure.

7.I.3.5 Risks Caused by Interdependencies Through Shared States

The previous section discusses systemic risks within a single subsystem. Recall that Complex SoS are usually large scale and consist of many interdependent and interconnected subsystems, which usually take the form of shared states. When a state variable is common to two or more subsystems, it necessarily introduces couplings between these subsystems. Therefore, any decision in one subsystem may propagate through the coupling and cause intended or unintended consequences to other interdependent subsystems. The system's stability under external perturbations induced by shared states is an important issue in understanding systemic risks to Complex SoS.

As shown in Section 7.I.3.2, the subsystem operates in a stable region as long as the value of θ does not exceed a certain threshold. In practice, the interdependencies among subsystems are often difficult to identify and quantify when the subsystems are not coordinated or they do not share any information. When one of the state variables is shared with other subsystems, the impact from other subsystems on the shared state variables usually can be treated as perturbations that force system states to deviate from a stable region. Depending on the magnitude of that perturbation and the margin around the stable region in the state space, the system may or may not return to the steady state. If the perturbations are strong enough, the system will probably become unstable and experience sudden state changes. In other words, due to the perturbations introduced into the subsystem, interdependencies among subsystems further reduce the boundary of θ within which the subsystem can operate safely. The ability of a subsystem to withstand this perturbation depends on the size of the stable region around the system operation point. When subsystems are not coordinated or they do not share any information, a decision maker of a subsystem has no information on the shared state variable, and Eq. (7.7) becomes Eq. (7.12):

$$\begin{bmatrix} s_1(k+1) \\ s_3(k+1) \end{bmatrix} = \begin{bmatrix} a_{11} & a_{12} & a_{13} & a_{14} \\ a_{21} & a_{22} & a_{23} & a_{24} \end{bmatrix} \begin{bmatrix} s_1^2(k) \\ s_3^2(k) \\ s_1(k) \\ s_3(k) \end{bmatrix} + \begin{bmatrix} b_1 \\ b_2 \end{bmatrix} u(k) + \begin{bmatrix} 0 \\ \Delta(k) \end{bmatrix} \tag{7.12}$$

where $\Delta(k)$ represents the perturbation caused by the interdependent subsystems with unpredictable magnitude. The state of the subsystem will remain at the stable stationary point for a given level of θ until the perturbation forces the state away from the stationary point by a distance of $\Delta(k)$. A fourth-order Runge–Kutta method can be used to determine whether the new state is in the unstable region of the state space. See, for example, Press et al. (1992).

The distance between a stable stationary point and the boundary of a stable region is called the *safety margin* of the subsystem. If the perturbation is strong enough to push the states beyond the safety margin, the system will not return to the stable stationary point, eventually causing system failure. Using the Runge–Kutta method, we can calculate the trajectory of system states under different levels of perturbation Δ. Figure 7.I.4 shows 10 different trajectories with Δ ranging from 0.05 to 0.5 with increment steps of 0.05, with $\theta = 0.2$.

The solid trajectories are those states that have returned to the stable stationary point, and the dashed

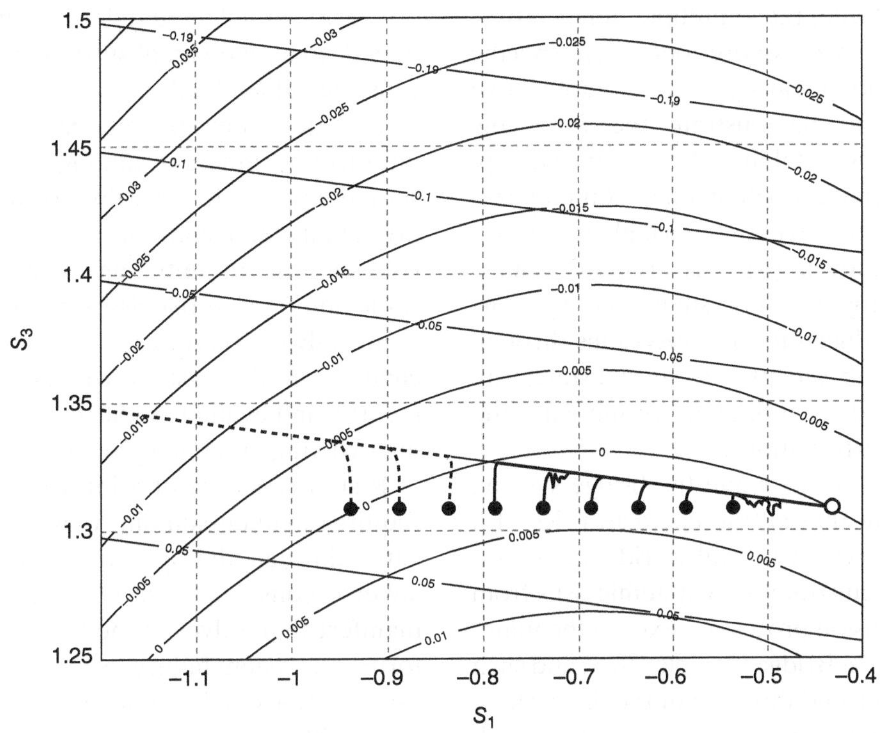

Figure 7.I.4 The theoretical trajectories of system states under different levels of perturbation, with black dots indicating system states immediately after different levels' perturbation. Solid lines show system trajectories returning to stationary point (indicated by the hollow circle at the right), and dashed lines show system trajectory leaving stationary point.

trajectories are those states that have not returned to the stable stationary point. In this case, when $\Delta > 0.35$, the system will leave the stable region, indicating that the safety margin is around 0.35.

Figure 7.I.3 shows that the safety margin is also a function of θ. When θ decreases, the safety margin of the system also decreases, which reduces the system's ability to withstand unexpected impact from other interdependent subsystems. For the subsystem to withstand the perturbation, the safety margin of the system must be greater than the potential maximum perturbations caused by shared states.

The risks caused by subsystem interdependencies can be mitigated in two ways: (i) The decision maker may operate the system around a region in the state space where the safety margin is sufficiently large to withstand most unexpected perturbations. However, in most cases this solution is at the expense of degrading system performance. (ii) An alternative approach is through system coordination, where impact from other interdependent subsystems can be anticipated and analyzed. If the decision maker of the subsystem is able to foresee that the level of the perturbation resulting from an optimal control perspective will exceed the safety margin, necessary communication can be made to the decision maker of the connected subsystem or to the second-level coordinator to review and modify the decisions.

The coordination method introduced in Section 7.I.2.2 ensures that the optimal decision for one subsystem generates exactly the required output z to optimize the other subsystem, and vice versa. In other words, the two decision makers are coordinating with each other to optimize their individual objectives (within a multiobjective tradeoff analysis), and the "perturbations" through shared states now become part of the anticipated inputs necessary to optimize the system. Knowing the impact on shared states from other connected subsystems reduces uncertainties in the shared states and provides the subsystem decision maker an opportunity to safeguard its operation.

Adapting risk analysis theories and methodologies to Complex SoS, such as large-scale infrastructure systems, is a necessary requirement for their owners, decision makers, and users. This adaptation requires a systemic and holistic approach that is responsive to shared states and other interdependencies among the subsystems and integrates multiple perspectives, models, and tools. We explore one such specific systemic risk of a nonlinear dynamic multiobjective decision process and demonstrate the following: (i) The system may lack stability and become susceptible to small perturbations when a decision maker is faced with multiple objectives or has a high preference for one of the objectives. (ii) It is prudent to have additional safety margins in subsystems where perturbations can be introduced to the subsystems through shared states. (iii) Subsystems with shared states can be decomposed and coordinated to mitigate the impact from such perturbations.

In this chapter, as we examine the bridge system, we also demonstrate that all the characteristics of a Complex SoS are present in real-world bridge systems. This example illustrates that systemic risks from a multiobjective decision process do exist in the maintenance subsystem of a Bridge Complex SoS, and such risks should be monitored throughout the life cycle of the system. Other applications of the model developed in this chapter may include high frequency trading systems, power grids, and some of the supervisory control and data acquisition (SCADA) systems.

Finally, the emerging behavior resulting from the adaptive and nonlinear interactions among individual subsystems may enable the performance of the SoS to exceed the sum of its parts. Although this chapter uses an additive form of performance function for simplicity, the overall Complex SoS performance function may be further generalized with coefficients or nonlinear functions that could manifest as either performance hindrances or boosters. The formulation of this problem and the related control and optimization methods deserve further study.

PART II: RISK MODELING OF CYBER–PHYSICAL INFRASTRUCTURE WITH PRECURSOR ANALYSIS

7.II.1 PRECURSOR ANALYSIS FOR PHYSICAL INFRASTRUCTURE COMPLEX SYSTEMS OF SYSTEMS

The ambiguity and uncertainty in the highly interdependent and interconnected subsystems constitute additional sources of risks to Infrastructure Complex SoS. For example, multiple initiating events that pose no risk to an individual subsystem might interact and cause significant adverse consequences at the SoS level. Due to multiple feedback loops and nonlinear interactions inherent among the subsystems, it is sometimes not feasible in practice to estimate the probability and consequences of each hypothetical risk scenario or initiating event. Furthermore, any static and passive risk assessment approach that fails to actually collect information and monitor triggering changes in the system will lack the capability to identify and evaluate indicators and warnings regarding evolving and emergent risks to the system.

Risk assessment of a infrastructure Complex SoS requires an understanding of why and how risk propagates through the system and results in significant adverse consequences. These consequences usually manifest themselves in the form of degradation, performance system failure, or accidents. Precursor analysis identifies and evaluates indicators and warnings of possible system failure and provides opportunities for proactive risk mitigation.

Accident models show the relationships between causes and effects and explain why accidents occur. Sequential accident models describe an accident as a chain of discrete events that occur in a particular temporal order. Epidemiological models regard events leading to accidents as the outcome of a combination of factors, some evident and some latent, that happen to exist together in space and time.

The use of control theory to analyze precursors to infrastructure failure is a natural result from a systems view of the organizational and structural preconditions to technological accidents and disasters in Infrastructure Complex SoS. There are several reasons that a *control-based* accident model may enhance the risk analysis of complex systems in addition to the reliability-based model. First, the system is usually modeled at a desired level of functional components instead of physical components. This abstraction makes it easier to identify potential failure modes caused by functional interdependencies rather than component failures. Modeling the functional components at a higher level also makes the systems modeling and risk filtering processes manageable for Infrastructure Complex SoS without acquiring detailed information on lower-level physical components. Unlike the reliability model that has only two states (failure/not failure) for each system component, the control model provides opportunities for capturing and monitoring essential system state variables in the functional space and thus increase situational awareness of system failure. This aspect is also essential for defining and analyzing a system's functional failures such as bridge structural deficiencies and obsolete functions that do not involve any failure of physical components. Finally, formulating system safety as a control problem enables the model to capture the behaviors of Complex SoS by including a dynamic and adaptive human decision-making process in the model and discovering deep and latent causal factors at organizational levels.

Common use of precursor identification and monitoring methods falls into three general categories: (i) trend prediction, (ii) statistical correlation, and (iii) causal relationship. In a *trend-based approach*, some safety-related states of the system are controlled and continuously measured or estimated, and their trends are used to predict if and when the safety threshold will be reached. While easy to implement, the trend-based prediction is not very reliable, and the selection of thresholds has to balance both false negative and false positive errors. In addition, this approach only identifies the "sharp-end" factors (Reason, 1990) leading to system failure, resulting in situations with limited response times. Approaches based on *statistical correlation* identify statistical measures that are significant prior to system failure. This approach is useful when the causal factors of system failure are difficult to understand. However, a training data set including the data captured prior to actual system failure is needed to enable model learning, and, as a result, signals related to unknown failure modes might not be identified. Finally, approaches based on *causal relationships* systematically explore the failure mechanism of a system and are useful in identifying precursors that have a long-term effect on the safety of a system. However, understanding the causal relationships in a complex system is often daunting; thus, a high-level functional model is usually used.

In the following sections we develop a systemic precursor analysis framework for a Bridge Complex SoS as a representative of infrastructures. Here, a proactive, dynamic, and anticipatory analysis tool is designed to identify, prioritize, and evaluate various sources of emergent risks that have the potential to cause system failure. We present three indispensable phases in a formal precursor analysis process: (i) precursor identification, (ii) prioritization, and (iii) evaluation. The *identification* phase utilizes a generic systems model based on control theory to model the failure mechanism of an Infrastructure Complex SoS and identify potential factors that cause system failure. To design an efficient monitoring system, the *prioritization* phase selects the most important causal factors as precursors based on (i) their likelihood to cause system failure and (ii) the expected remaining service life of the system, given that a particular causal factor occurs. The *evaluation* phase incorporates the uncertainties in precursor detection and provides improved situational awareness through a holistic system-level understanding of the risks of system failure.

This methodology is explained by building on the Bridge Complex SoS introduced previously. The conditions of highway bridges constitute an important part of the transportation infrastructure and are the lifeline of commerce. However, they are continuously deteriorating due to a lack of appropriate maintenance. For example, 25% of US bridges are structurally deficient or functionally obsolete. The following methodology provides insight into efficiently prioritizing and planning for inspection, maintenance, and rehabilitation activities based on precursors that could reduce the risk of bridge failure (Guo and Haimes, 2016, 2017).

7.II.2 PRECURSOR ANALYSIS FRAMEWORK USING AN EXAMPLE OF A BRIDGE COMPLEX SYSTEMS OF SYSTEMS

In this section, we introduce a systemic precursor analysis framework for a Bridge Complex SoS, with the goal to (i) understand its failure mechanisms, (ii) develop an effective and efficient monitoring process, and (iii) support informed risk management decision making while reducing hindsight bias in using precursors. To achieve the above goals, we (i) develop a functional model of a Bridge Complex SoS based on control theory to identify causal factors of system failure, (ii) prioritize the identified causal factors based on their impacts on system failure, and (iii) improve situation awareness for the risk analyst. We do this by incorporating multiple detected precursors and evaluating the likelihood of all identified failure modes of the system, with the uncertainties in the precursor detection process accounted for.

First, we introduce a set of terms that will be used in the following discussion:

1. *System failure* is the state or condition of not meeting a desirable or intended system objective.
2. A *failure mode* is the way in which a system fails functionally. Often a system has multiple failure modes.
3. A *causal factor* to system failure is any combination of events, conditions, malfunctions, anomalies, or deviations from normal operation internal to the system that may increase the probability of a system failure.

An infrastructure, together with a particular set of worst-case environmental conditions has the chance to lead to a system failure. A *precursor* is also a causal factor: Given a precursor event, the conditional probability of system failure is significantly higher than when the precursor has not occurred.

Causal factors are different from external initiating events, such as scour, earthquake, and ship collision with a bridge, among others. In our modeling approach, these initiating events are considered external perturbations to the system with potential adverse consequences. One important aspect of the modeling approach is to model and understand why the internal control structure, both physical and organizational, are unable to maintain the system's resilience to withstand various external perturbations. Although these initiating events are not included in the model, their impact on the bridge can still be observed and assessed during bridge inspection and be used as inputs to the evaluation process. Also, no risk analysis methodology can guarantee capturing all sources of risk of system failure. Risk analysis is inherently a continuous learning process that explores the vulnerability of the system to specific internal or external perturbations over time. This precursor analysis framework is an iterative process to accommodate the dynamic and adaptive characteristics of the Bridge Complex SoS.

7.II.3 SUBSYSTEMS OF A HIGHWAY BRIDGE COMPLEX SYSTEMS OF SYSTEMS AND ITS FAILURE MODES

To understand the causes of large-scale infrastructure systems failure, we need to take an SoS perspective of the technical, organizational, and structural preconditions to failures. This precursor identification process is based on exploring the causal relationships among functional system components that can lead to system failure. To demonstrate the applicability of this approach in a physical infrastructure Complex SoS, we continue to use a highway bridge as an example.

A typical highway bridge in its physical form is usually composed of deck, superstructure, substructure, and other auxiliary systems. Bridges are traditionally considered merely as physical systems, and bridge maintenance has been a research topic mainly within the civil engineering domain. However, bridge failure is usually the result of a complex process involving not only the physicochemical deterioration process of its physical elements but also the effectiveness of inspection and repair activities as well as the policy, standard, and operation decision-making processes. Multiple factors – technical, managerial, economic, political, and institutional – have brought about a situation in which highway bridges do not provide adequate service life. There is a need for a paradigm shift in modeling, understanding, and managing the risk and the life cycle of the Infrastructure Complex SoS. This SoS view of an Infrastructure can be understood by identifying various functional and organizational decision-making entities and processes that interact with each other and determine the bridge's reliability and service life span as illustrated in Figure 7.II.1 for a bridge.

As the elements of a bridge deteriorate under its operating environment, the owner of the bridge, usually the DoT, schedules maintenance and repair activities based on inspection results. The state DoT also manages the traffic pattern to control load impact on the bridge. When bridge load capacity is below a certain limit, the bridge will be considered structurally deficient, and bridge collapse may be triggered by some uncontrollable external events. In the United States, if an accident occurs, the National Transportation Safety Board (NTSB) will investigate the root causes. It will then provide

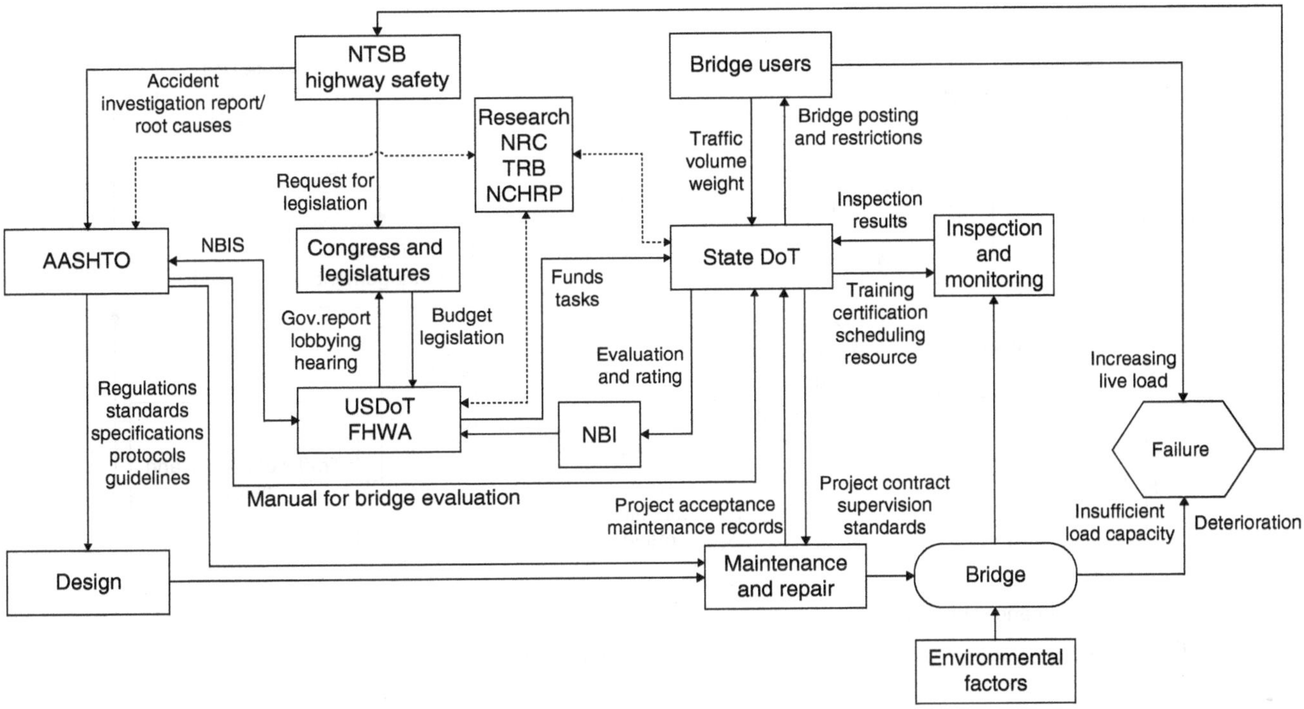

Figure 7.II.1 *An overview of a bridge SoS.*

information to both the Congress for safety-related legislation and the American Association of State Highway and Transportation Officials (AASHTO) to set new technical standards for bridge construction and maintenance. The Federal Highway Administration (FHWA) requests budgets and provides stewardship and oversight of highway programs and funding at both the federal and state level. Research organizations and programs such as the US National Research Council (NRC), the Transportation Research Board (TRB), and the National Cooperative Highway Research Program (NCHRP) provide theoretical and technical support for the above activities. From this perspective, a bridge possesses many of the characteristics of a Infrastructure Complex SoS.

Our focus in this analysis is on bridge maintenance at an operational level, which covers a subset of the encompassing Complex SoS shown in Figure 7.II.2 in terms of Hierarchical Holographic Modeling (HHM), to which Chapter 3 was devoted in its entirety. It includes the state DoT, bridge users, and inspection and maintenance teams. Bridge inspection and monitoring functions provide information on the current state of the bridge to the state DoT, and the DoT uses this information as a basis for its decision making. Bridge maintenance functions implement maintenance and repair jobs on the bridge. Bridge management functions include the goal of bridge owners to maintain the reliability and functionality of the bridge. Bridge users include private and commercial vehicles commuting across the bridge.

There are generally two types of system failure – physical failure and functional failure – in a Bridge Complex SoS. Physical failures include damage to major structure elements or even the collapse of the bridge, especially during natural disasters. The focus here is to understand the mechanism of bridge functional failures, which is a more common and prevailing issue. The definition of functional failure usually depends on the context of the problem. One of the major functional failures is bridge deficiency, which does not usually cause bridge collapse. However, it reduces the safety margin of the bridge and accelerates the deterioration process.

Identifying different failure modes is a first step to a basic understanding of how systems fail. This process is usually based on knowledge, experience, and expert judgment. There are three common failure modes for

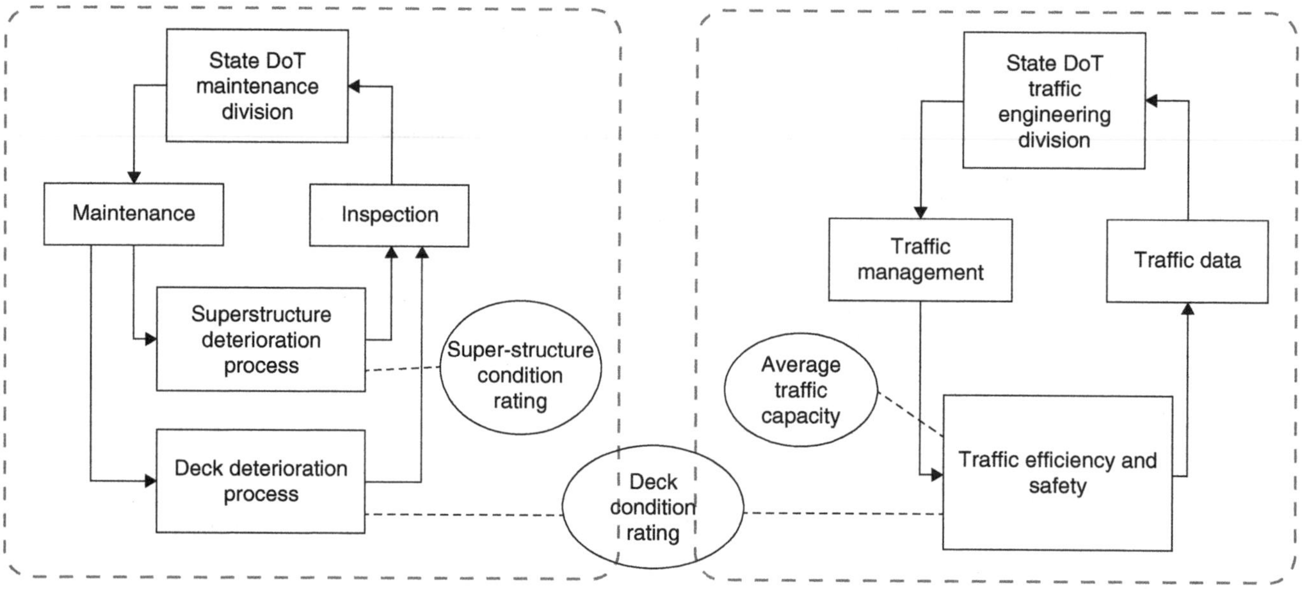

Figure 7.II.2 *A bridge SoS with interdependent maintenance and traffic engineering subsystems.*

bridge deficiency according to the *FHWA Bridge Preservation Guide* (2011):

1. Structurally Deficient (SD) – Significant load-carrying elements are found to be in poor condition due to deterioration or damage.
2. Functionally Obsolete (FO) – The deck geometry, load carrying capacity, clearance, or approach road alignment no longer meets the usual criteria for the system of which it is an integral part.
3. Overload (OL) – Live load and dead load exceed the load capacity of the bridge.

The next step in the analysis is to identify system constraints for each failure mode. This is achieved by expressing a failure mode in terms of the relationships among system state variables using a set of equality or inequality equations. If a failure mode needs to be expressed by more than one constraint, Boolean logic can be used to describe the relationships among these constraints.

7.II.4 MODELING AND CONTROL INFRASTRUCTURE SYSTEMS OF SYSTEMS

The system failure conditions identified in previous sections determine the safe operation boundaries of the system, and well-designed engineering systems, regardless of their structures and purposes, should contain a control mechanism to enforce system safety.

There are five major functional components in a typical feedback control system: the controlled process, process model, actuator, sensor, and controller. Each functional component normally consists of hardware, software, human operator or some other organizational entity, and control processes. Identifying these functional components enables us to understand how the safety constraints are enforced in the system and to develop precursors for each failure mode. In a Bridge Complex SoS, the deterioration process of bridge elements and traffic flow are the controlled processes. The state DoT usually plays the role of a controller through making bridge maintenance and other related decisions. The bridge maintenance team is the functional unit that performs maintenance on the bridge to maintain its rating condition; thus it plays the role of actuator. The bridge inspection team provides the DoT with actual bridge condition information and functions as a sensor.

Consider two basic subsystems of a bridge SoS: (i) the maintenance subsystem and (ii) the traffic engineering subsystem. The decision makers in the maintenance subsystem are usually the maintenance division in the state DoT, with the goal to maintain the reliability and structural integrity of both superstructure and deck at an acceptable level. The decisions they make must consider the amount of maintenance budget and the type of repairs to be made. The decision maker in the traffic engineering subsystem is usually the traffic engineering division in the state DoT, with the goal of ensuring the safety and efficiency of traffic across the bridge. The decisions this division makes include deicing, lane allocation, work zone design, signal and illumination placement, and speed and tonnage posting. Each subsystem has its own functional components and control structure. Figure 7.II.2 illustrates the control structure of these two subsystems.

However, interactions between these two subsystems through the condition of the deck further complicate the system behaviors and introduce emerging risks to the Bridge Complex SoS. For example, a heavy traffic load accelerates the deterioration process of the deck, while maintenance activities on the deck create work zones and reduce the traffic flow of the bridge. The consequences resulting from decisions in the traffic engineering subsystem will propagate through the change in the deck condition and impact the superstructure, and vice versa. To capture these interdependencies, essential state variables of both subsystems need to be identified, with special attention paid to state variables that are common to both subsystems. Three conceptual state variables are chosen here to represent the essential states of the two subsystems. The condition rating of the superstructure s_s is a state variable of the maintenance subsystem, the average daily traffic of the bridge s_t is a state variable of the traffic engineering subsystem, and the condition rating of the deck s_d is a shared state variable of both subsystems.

The identified control structure in Figure 7.II.2 is a qualitative systems model that will be used to identify causal factors of system failure. However, being able to project the likelihood of system failure into the future given a specific causal factor requires the use of a quantitative bridge model so that analytical analysis or numerical simulation can be performed. Various advanced bridge modeling tools and commercial software are available for condition prediction and maintenance scheduling. It is not the purpose here to develop a better or more realistic bridge model. *Instead, a simplified representative mathematical model is used to illustrate how this precursor methodology can be applied to a bridge system so that any practitioner can apply the methodology with more advanced models to a specific bridge of interest.* Due to the lack of sufficient data from a specific bridge spanning its whole service cycle, the national-level bridge data in the National Bridge Inventory has been used to estimate the major parameters of the bridge model. We assume that the dynamics of a bridge are relatively smooth and the nonlinearity of the deterioration process can be adequately captured by a second-order polynomial over the relevant range of bridge conditional rating. To illustrate the process, a quadratic state-space model is used to represent the identified structures. The resulting model provides an acceptable approximation of the deterioration characteristics of a hypothetical "national average bridge." The purpose of this model is not to predict the probability of future system failure of a specific bridge. Instead, it creates a base from which to compare the impact of different precursor scenarios and to demonstrate a precursor filtering and prioritization process that will be discussed in Section 7.II.5.

Let $s_s(k)$ and $s_d(k)$ represent the logit transformation of the condition rating of the superstructure and deck, respectively, at time stage k and $s_t(k)$ represent the average traffic at time stage k. Let $u_m(k)$ represent the decision (total maintenance spending in million dollars) made for the maintenance subsystem and $u_t(k)$ represent the decision (total traffic engineering spending in million dollars) made for the traffic engineering subsystem. Equation (7.13) describes the deterioration process of the superstructure and deck and the traffic flow. Some of these parameters, such as deterioration coefficients, can be estimated from bridge inspection data, but parameters such as effectiveness of maintenance activities vary case by case, and a nominal value is used based on expert estimations:

$$\begin{bmatrix} s_s(k+1) \\ s_d(k+1) \\ s_t(k+1) \end{bmatrix} = \begin{bmatrix} a_{11} & a_{12} & 0 \\ a_{21} & a_{22} & 0 \\ 0 & 0 & 0 \end{bmatrix} \begin{bmatrix} s_s^2(k) \\ s_d^2(k) \\ s_t^2(k) \end{bmatrix} + \begin{bmatrix} a_{14} & a_{15} & 0 \\ a_{24} & a_{25} & a_{26} \\ 0 & a_{35} & a_{36} \end{bmatrix} \begin{bmatrix} s_s(k) \\ s_d(k) \\ s_t(k) \end{bmatrix} + \begin{bmatrix} b_{11} & 0 \\ b_{21} & b_{22} \\ 0 & b_{32} \end{bmatrix} \begin{bmatrix} u_m(k) \\ u_t(k) \end{bmatrix}$$

$$\text{s.t.} \quad u_1(k) \geq 0, \quad u_2(k) \geq 0, \quad \text{for } k = 0, \ldots, T-1 \tag{7.13}$$

Let c_s, c_d, and c_t be the control targets of each state variable. The decision maker of the maintenance subsystem aims to maintain appropriate superstructure and deck condition ratings at the levels of c_s and c_d by controlling the maintenance spending. The decision maker of the traffic engineering subsystem aims to maintain the traffic level of the bridge at c_t. Let the functions $f_s(k)$ and $f_d(k)$ be the output (objective) functions of the maintenance subsystem and $f_t(k)$ be the output (objective) function of the traffic engineering subsystem:

$$\begin{bmatrix} f_s(k) \\ f_d(k) \\ f_t(k) \end{bmatrix} = \begin{bmatrix} (s_s(k)-c_s)^2 \\ (s_d(k)-c_d)^2 \\ (s_t(k)-c_t)^2 \end{bmatrix} \tag{7.14}$$

For the decision makers of the maintenance subsystem, the objective is to

$$\min_{u_m(k)} \theta f_s(k) + (1-\theta) f_d(k), \quad \text{for } k = 0, \ldots, T-1 \tag{7.15}$$

where θ is the preference or tradeoff between the two competing objectives. For the decision makers of the traffic engineering subsystem, the objective is to

$$\min_{u_t(k)} f_t(k), \quad \text{for } k = 0, \ldots, T-1 \tag{7.16}$$

where k represents the time interval between each decision period. As new information on the condition of the bridge is obtained every 2 years through bridge inspection, k is selected for a 2-year period in this case.

7.II.5 IDENTIFYING CAUSAL FACTORS OF SYSTEM FAILURE

Each functional component in a control system may malfunction in certain ways, which may lead to the failure of the entire system. Thus, a possible defect in a functional component or deviation from a controlled process can be seen as a causal factor of system failure. Five common types of defects for each functional component in a bridge system are summarized below:

1. Controller (bridge maintenance decision maker):
   - Missing or wrong bridge inspection results.
   - Inadequate or flawed planning method.
   - Conflicting or competing decision objectives.
2. Process model (bridge deterioration model used by the decision maker):
   - Deterioration model is unable to describe and predict actual deterioration process.
3. Actuator (maintenance team):
   - Inappropriate, ineffective, or missing maintenance activities.
   - Delayed or inadequate maintenance.
4. Sensor (inspection team):
   - Inspection errors.
   - Inadequate, delayed, or missing information reported to DoT.
   - Inadequate or incorrect inspection method is used.
5. Controlled process (bridge deterioration):
   - Failure of bridge elements.
   - Deterioration process changes over time.
   - Extreme external conditions.
   - Multiple processes with unknown interdependencies. For example, materials used for maintenance increase bridge dead-load overtime.

These common issues can be used as a checklist for each functional component identified in the system control structure in Figure 7.II.2. If evidence shows that certain defects or potential issues may exist in a functional component, this can be considered a causal factor to system failure. As we proceed with the checklist for all the functional components in the identified system control model, a comprehensive set of causal factors leading to system failure can be identified.

Investigations of industrial accidents reveal that it is rare that a complex engineering system fails due to a single cause. *In most cases, it is a combination of multiple causes with their unexpected interactions that lead to the failure of the system.* In the identification process, we should not only be concerned with the relationship of a single causal factor to a single failure mode but should also consider (i) the combination of multiple causal factors leading to a common failure mode and (ii) common causal factors leading to multiple failure modes. This process requires both expertise in the specific field and creative thinking. In order to facilitate this discovery process, the Hierarchical Holographic Modeling (HHM) introduced in Chapter 3 is constructed and used to organize causal

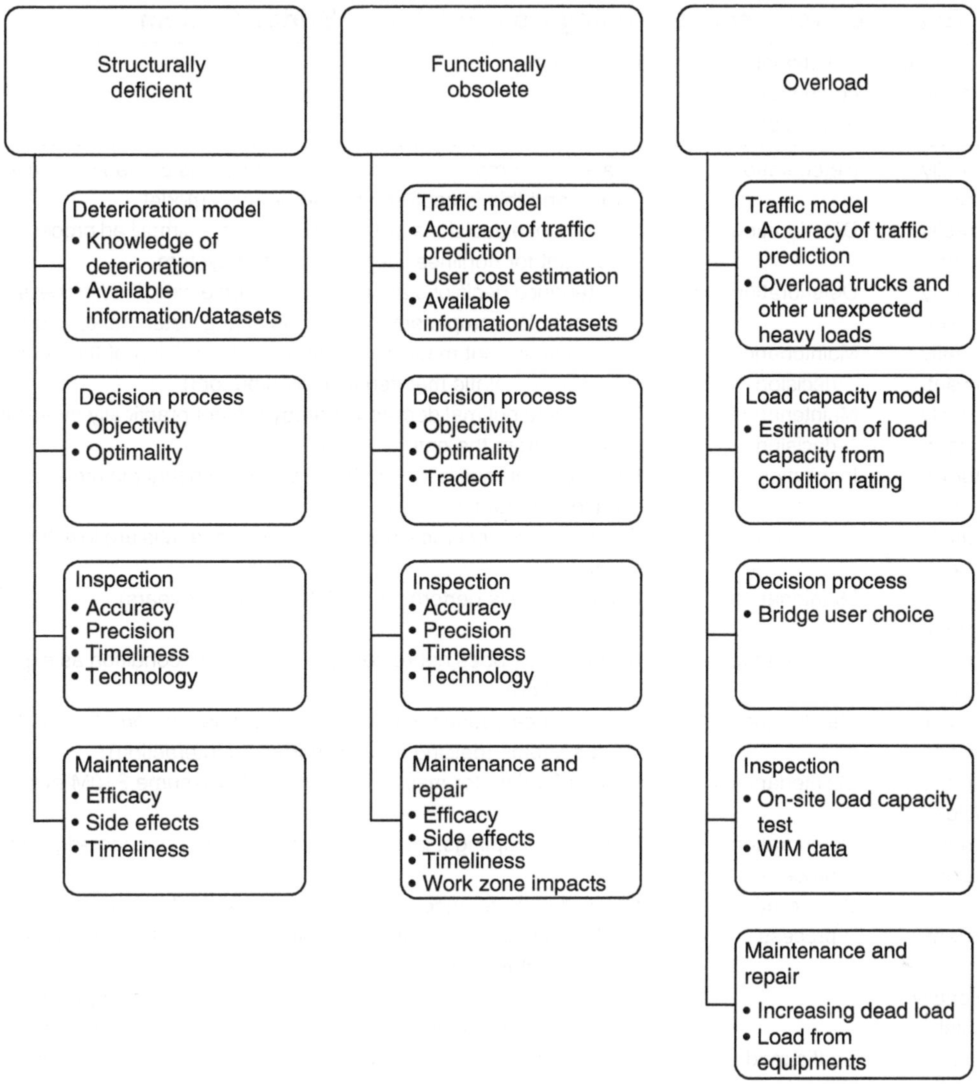

Figure 7.II.3 *Example HHM for the bridge SoS.*

factors under different failure modes and functional components, as shown in Figure 7.II.3.

In this HHM, each failure mode is listed as a head topic, and functional components of the control structure constitute the subtopics under each head topic. Then, the different causal factors can be organized under each functional component. Exploring the HHM along the horizontal axis provides an intuitive way to identify common causal factors of the system's multiple failure modes. For example, potential defects in all aspects of the inspection function may become causal factors for failure modes of structurally deficient and functionally obsolete components. Exploring the HHM along the vertical axis reveals all causal factors and their possible combinations for a single failure mode of the system. An example subset of the identified causal precursor factors is summarized in Table 7.II.1.

7.II.6 PRECURSOR PRIORITIZATION

The identification process of precursors to the failure of Complex SoS may generate a large number of factors that have the potential of causing system failure. Due to limited resources, it is impractical for a monitoring system, either automated or human operated, to monitor and track all these causal factors. However, not all the identified factors have the same likelihood of causing system failure. The purpose of the prioritization process is to systemically select a manageable subset of the most important causal factors

TABLE 7.II.1 Example Precursors Resulting from HHM for a Bridge System

Precursor code	Failure mode (head topic)	Functional component (subtopic)	Description
1.1.1	Structurally deficient	Deterioration model	The actual component deterioration process is 2% faster than the national average rate, and this is captured by the process model
1.1.2	Structurally deficient	Deterioration model	Maintenance decisions are made based on a simplified process model that does not account for nonlinear behaviors of the system
1.1.3	Structurally deficient	Deterioration model	Maintenance decisions are made based on a simplified process model that does not account for component interactions during deterioration process
1.2.1	Structurally deficient	Maintenance decision	Decisions are not made based on actual condition of the component (e.g. using a predefined static maintenance plan/budget)
1.2.2	Structurally deficient	Maintenance decision	Solutions to optimal decision strategy are not practically available. Actual decisions deviate from the optimal ones
1.3.1	Structurally deficient	Inspection	Inspection interval is longer than 2 years such that maintenance decisions are often based on outdated data
1.3.2	Structurally deficient	Inspection	Precision and accuracy in visual inspection results are low. Inspection error $\sim N(0.5, 0.8^2)$
1.4.1	Structurally deficient	Maintenance	Maintenance is performed too late (lagging 2 years)
1.4.2	Structurally deficient	Maintenance	Maintenance does not restore the component condition as expected (assume 75% restoration)
1.4.3	Structurally deficient	Maintenance	Maintenance causes unexpected consequences on other bridge components or subsystems (deicing causes increased interaction)
1.4.4	Structurally deficient	Maintenance	Limited funding for maintenance projects (assume $10M every 2 years)
1.5.1	Structurally deficient	Deterioration process	Unknown couplings, interdependencies, and unexpected perturbations exist
1.5.2	Structurally deficient	Deterioration process	Process change under different conditions and not captured by the process model (chloride concentration level at the interface of rebar exceeds the corrosion reaction threshold)
	Functionally obsolete		All precursor scenarios for structurally deficient are applicable to functionally obsolete. Only unique precursors to functionally obsolete are listed below
2.1.1	Functionally obsolete	Traffic model	The model does not capture the increasing trend in traffic volume over the bridge (assume an additional 50% increase)
2.1.2	Functionally obsolete	Traffic model	Maintenance work zones on deck have a higher than expected impact on traffic and user cost
2.2.1	Functionally obsolete	Maintenance decision	In a multiobjective decision process, the priority (preference) on some objectives is too low

as precursors for the monitoring system. The risk filtering and ranking method (RFRM) (Haimes, 2016) is used to prioritize the identified causal factors. The output of the RFRM is usually a manageable set of precursors to be monitored (Haimes, 2016).

A quantitative bi-criteria filtering and prioritization process is used with the two criteria being the (i) likelihood of failure and (ii) expected remaining service life of a system. *Given the existence of a causal factor*, (a) the first criterion is a measure of the maximum conditional probability of system failure within a specific time domain, and (b) the expected remaining service life is a measure of the expected time to reach a specific failure probability. Each causal factor is evaluated by these two criteria, respectively, and the causal factors causing higher likelihood and/or less expected remaining service life will be identified as precursors to system failure and receive higher priority for further monitoring. This prioritization process is not limited to the above two criteria. Additional criteria that matter to a risk analyst should be added when necessary.

The system control model presented in Eqs. (7.13)–(7.16) describes the system's behavior under normal operations. The key step in evaluating each causal factor is to incorporate the deviation from normal operation into a modified systems model, such that variation between the normal operation (baseline) scenario and the causal factor scenario can be compared. Below is an example to demonstrate this process of modifying the model according to a causal factor scenario. Furthermore, the system control model in Eqs. (7.13)–(7.16) is the ideal model without considering the practical issues of each functional component. For example, it is assumed that the bridge inspection team that functions as a sensor is able to measure the true deterioration state of the bridge elements and provide accurate and precise measurements of condition ratings to the decision maker in the state DoT. In this ideal scenario, any rational decision maker of the maintenance subsystem will choose an optimal maintenance spending (assuming no interaction from the traffic subsystem):

$$u_{\mathrm{m}}^{*}(k)=\rho\theta b_{11}\left[c_{\mathrm{s}}-a_{11}s_{\mathrm{s}}^{2}(k)-a_{12}s_{\mathrm{d}}^{2}(k)-a_{14}s_{\mathrm{s}}(k)-a_{15}s_{\mathrm{d}}(k)\right]$$
$$+\rho(1-\theta)b_{21}\left[c_{\mathrm{d}}-a_{21}s_{\mathrm{s}}^{2}(k)-a_{22}s_{\mathrm{d}}^{2}(k)-a_{24}s_{\mathrm{s}}(k)\right.$$
$$\left.-a_{25}s_{\mathrm{d}}(k)\right]\ \text{for}\ k=1,\ldots,T-1,\ \text{where}$$
$$\rho=\frac{1}{\theta\, b_{11}^{2}+(1-\theta)\, b_{21}^{2}} \tag{7.17}$$

However, the true state of the bridge element is usually not directly observable, and current practice relies heavily on visual inspections due to limited resources for other types of inspection methods. This practice increases the bias and uncertainty in the results of observed bridge states and may affect the reliability of the bridge in the long run. To evaluate the impact of this causal factor on the likelihood and timing of bridge failure, we modified Eq. (7.17) to incorporate errors in bridge inspection by adding a random variable $\omega(k)$, $k=1, \ldots, T$. The mean of $\omega(k)$ represents the accuracy of the inspection results, and the variance of $\omega(k)$ represents the precision of the inspection results. Based on an empirical study (Phares et al., 2004), the errors in the superstructure visual inspection results follow approximately a normal distribution of $N(0.5, 0.8^2)$. The new solution incorporating inspection uncertainty becomes

$$u_{\mathrm{m}}'(k)=\rho\theta b_{11}$$
$$\left[c_{\mathrm{s}}-a_{11}(s_{\mathrm{s}}(k)+\omega(k))^{2}-a_{12}s_{\mathrm{d}}^{2}(k)-a_{14}(s_{\mathrm{s}}(k)+\omega(k))-a_{15}s_{\mathrm{d}}(k)\right]$$
$$+\rho(1-\theta)b_{21}\left[c_{\mathrm{d}}-a_{21}(s_{\mathrm{s}}(k)+\omega(k))^{2}-a_{22}s_{\mathrm{d}}^{2}(k)-a_{24}(s_{\mathrm{s}}(k)\right.$$
$$\left.+\omega(k))-a_{25}s_{\mathrm{d}}(k)\right] \tag{7.18}$$

Equations (7.17) and (7.18) enable us to compare the inspection error causal factor scenario (in Table 7.II.1) with the baseline scenario by evaluating the likelihood of failure and the expected remaining service life of a system, either by analytical methods or by numerical simulation.

To account for practical constraints and our limited knowledge of the bridge deterioration process, certain levels of rule-based decision processes and uncertainties in model parameters have been incorporated into the model for both scenarios. Figure 7.II.4 shows a typical simulation result of the baseline scenario and the inspection error causal factor scenario. The maximum probability of system failure within 50 years and

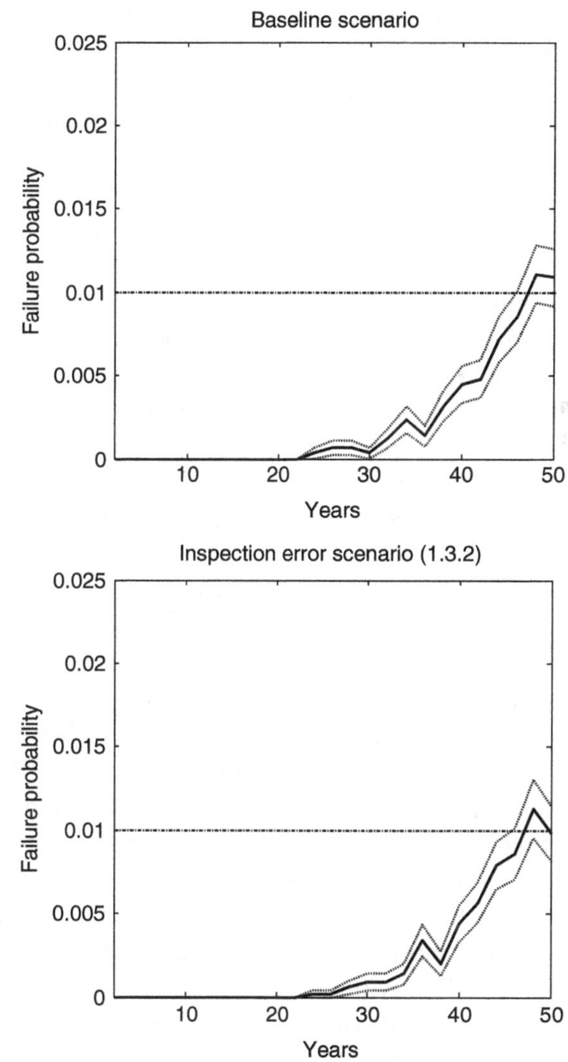

Figure 7.II.4 *Comparison of simulation results of the estimated system failure probability with 90% confidence interval between the baseline and precursor inspection error scenarios.*

the time to reach a failure probability of 0.01 are used to compare these two scenarios.

As the figure shows, the errors in bridge inspection have almost no impact on either criterion. This result shows that inspection errors at this level do not have a significant impact on system failure and this causal factor is not justified as a precursor and may not deserve further monitoring.

7.II.7 PRECURSOR EVALUATION: DETECTING UNCERTAINTY

The capability to detect precursors is a critical task in failure detection. If the identified precursors cannot be detected in a timely manner with certain confidence, their contribution to risk management will be significantly limited. However, in practice the existence of a precursor may not be directly observable, and indirect or imperfect information is needed to detect a specific precursor. We use the term *evidence* to refer to the information needed to detect the existence of a precursor. This evidence may include but not be limited to physical sensor measurements, inspections and tests, expert judgment, other forms of intelligence, and estimation or prediction from certain data analysis algorithms.

The detection of precursors depends on the definition and characteristics of each precursor, and there is no standard procedure for that task. Some precursors are well defined such that detection is straightforward, but the detection of many others is more involved. In a real working environment, noises and errors are inevitably embedded in the detection results. Therefore, it is important to account for the uncertainties in the detection process.

For example, using a physical sensor with on/off measurements, the level of evidence can be quantified as type I and type II errors. (A type I error connotes detecting an effect that is not present; a type II error is failing to detect an effect that is present.) In precursor analysis, we are interested in knowing the probability of system failure given an observed piece of evidence: Pr(system failure within a specific time domain | evidence), which can be calculated using precursor detection probability Pr(precursor | evidence) *and* the theorem of total probability as

$$\begin{aligned}
&\Pr(\text{system failure within a specific time domain}\\
&|\text{evidence}) =\\
&\Pr(\text{system failure within a specific time domain}\\
&|\text{existence of precursor})\cdot\\
&\Pr(\text{existence of precursor}|\text{evidence}) +\\
&\Pr(\text{system failure within a specific time domain}\\
&|\text{nonexistence of precursor})\cdot\\
&\Pr(\text{nonexistence of precursor}|\text{evidence}) \qquad (7.19)
\end{aligned}$$

When multiple precursors have been detected along the life cycle of the Complex SoS, it is important to understand their impact on current states of the system and improve the risk analyst's contextual understanding and situational awareness in order to achieve a more informed decision-making process for risk management. The precursor evaluation process integrates information from detected precursors at different time stages in the past to update and track the current likelihood of multiple failure modes of the system. This is a real-time, online process, which is different from the projection of future system failure probability in the filtering process where the existence of a hypothetical precursor is assumed.

The system control model developed in Eqs. (7.13)–(7.16) can be readily used to incorporate multiple precursor scenarios whenever a precursor is detected along the timeline. Instead of calculating the failure probability of the whole system, the model can calculate the failure probability of each failure mode (Structurally Deficient (SD), Functionally Obsolete (FO), and Overload (OL) in this example). When the likelihood of each failure mode and the likelihood of no failure are quantified along the timeline, the risk analysts should have a better understanding of the current risk situation. Risk management plans targeted on a specific failure mode can be developed and executed. An illustrative example is used to demonstrate this precursor evaluation process.

In this example, the probabilities of the three failure modes – SD, FO, and OL – as well as the probability of no failure are calculated and monitored for the past 20 years. The initial superstructure and deck condition ratings are assumed to be 6. Due to the fact that failure modes are not necessarily mutually exclusive, the probability of no failure is calculated as:

$$\begin{aligned}\Pr(\text{no failure}) = 1 - &\Pr(\text{SD}) - \Pr(\text{FO}) - \Pr(\text{OL}) \\ &+ \Pr(\text{SD} \cap \text{FO}) + \Pr(\text{FO} \cap \text{OL}) \\ &+ \Pr(\text{OL} \cap \text{SD}) - \Pr(\text{SD} \cap \text{FO} \cap \text{OL})\end{aligned} \tag{7.20}$$

We assume four precursors are detected along the 20-year time span:

1. Faster deterioration (2% increase in the deterioration rate) started at year 4, with Pr(existence of precursor | evidence) = 0.5.
2. Rehabilitation project was not performed at year 16. (This is a known fact with no detection uncertainty.)
3. No inspection was performed at year 18. (This is a known fact with no detection uncertainty.)
4. Increasing dead load due to maintenance throughout the 20-year period, with Pr(existence of precursor | evidence) = 0.8.

The state-space systems model in Eqs. (7.5)–(7.8) is modified to incorporate these precursor scenarios from the time they are detected, and the failure probabilities of each failure mode for all time stages in the past are simulated and plotted in Figure 7.II.5. For example, from year 4, the parameters determining the deterioration rate are increased; at year 16, $u(16) = 0$; and at year 18, a maintenance decision is based on $s(16)$ instead of $s(18)$; the dead load is assumed to have a linear increase throughout the 20-year period. Does this make sense if there is a linear increase *throughout* the 20-year period?

Figure 7.II.5 shows that the first precursor at year 4 does not significantly increase the failure probability of any failure mode. The probabilities of all three failure modes remain relatively low in the first 14 years. Then the failure probabilities of SD and OL gain a sharp increase and surpass FO after the detection of the second precursor at year 16. The detection of the third precursor raises the failure probability of OL well above that of the SD. At the end of the 20-year time period, the failure probability of OL is highest among the three failure modes. The results also show that none of the precursors have a significant impact on the failure probability of FO. In this case, the failure due to overload should be considered most likely at the current time stage, and risk management actions aimed to reduce overload should be planned and executed with higher priority. Figure 7.II.5 plots and directly compares the probability of all failure modes before and after the detection of each precursor. It provides an intuitive yet objective way to communicate the emergent forced system changes to the risk analysts to supplement their decision-making process.

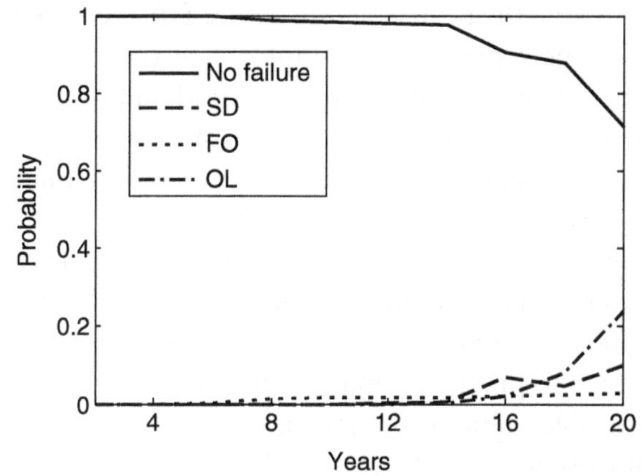

Figure 7.II.5 *Evaluation of failure probabilities of each failure mode with multiple detected precursors.*

This part of Section 7.II.7 demonstrates that through a systems engineering approach to understanding system failure mechanisms and designing efficient monitoring systems, this precursor analysis framework is able to provide a sound theoretical base for risk management of a physical infrastructure Complex SoS.

The unique value of this precursor analysis framework is that it recognizes the Bridge Complex SoS where the physical elements, human decision processes, control and detection actions, and the interactions among these components all contribute to the dynamics of the overall SoS. We demonstrate that all the characteristics of a Bridge Complex SoS exist in real-world bridge systems. Risk management methods focusing only on *part* of the encompassing SoS, such as the physical bridge elements only, will eventually miss critical factors leading to system failure.

Although the bridge infrastructure SoS is complex, it is not at the same level of complexity as other complex systems, such as aviation or power grid, in terms of the number and sophistication of stakeholders and decision makers, the frequency of decision making, the complexity in system structure, the way information is processed and exchanged, the level of system nonlinearity, and the speed at which the system responds and adapts. With an emphasis on holism, the approach to precursor analysis presented here can add measurable value to the risk assessment and management of other infrastructure SoS.

PART III: METAMODELING OF INTERDEPENDENT SYSTEMS: AN APPLICATION TO BRIDGE INFRASTRUCTURE MANAGEMENT

7.III.1 INTRODUCTION

As we have demonstrated in the previous sections, a critical infrastructure in the Transportation Complex SoS is the bridge system, which in its essence is itself Complex SoS. In this part, we explore the many manifestations of the Bridge Complex SoS and its unique I-I with many other Complex SoS. By its connotation, the bridge infrastructure plays a major role in connecting myriad assembly of population, cities, and physical and cyber–physical infrastructures, among others.

The longtime life span of bridges and the associated cost of their construction render their operational life span to commonly exceed the original planned design and construction reality. This is probably one of the major factors contributing to the current critical situation in many countries in the world, where most Bridge Complex SoS have not been designed and constructed to provide the current level of service required. It is reasonable to posit that approaches to the planning, design, construction, maintenance and overall management varied among countries, traffic loads, weather, and myriad other distinguishing factors, including condition and performance measures that are very subjective and heavily weighted toward ordinal assessments that are based upon nonquantitative criteria and methods. Past and current practice is to design and construct bridges with minimal regard for adequate inspection, maintenance, preservation, and renewal. As an example, the most common type of highway bridge being placed into service today is a prestressed concrete multigirder bridge with a reinforced concrete deck. However, the condition and integrity of the prestressing strands that provide the strength for these bridges cannot be determined once the bridge is placed in service. Additional examples of common bridge elements that cannot be inspected include grouted post-tensioning tendons, reinforcing steel, and driven piles. It is also the case that these critical elements cannot be maintained, replaced, or renewed. The current common approach to assessing the condition and integrity of such bridge elements is to rely upon visible indications of deterioration. This approach, by design, is reactive and limited to deterioration that has advanced to the point where the element is damaged. Deterioration could, and probably does, exist that has not progressed to the point where visible indications are present. There have been many sudden failures that have been traced back to hidden or latent defects, many of which were directly related to bridges that were not constructed as detailed and with materials that did not meet specifications. In addition, there could be deterioration processes occurring that are not yet understood or recognized. These challenges must be understood, modeled, and managed by considering the bridge infrastructure as Bridge Complex SoS and by applying appropriate risk management principles to identify those factors critical to the goal (Haimes, 2016).

Bridges epitomize the criticality of physical infrastructures: They are visible, and we all utilize them, as we cross many bridges while traveling from one location to another. It would be virtually impossible to effectively address the intricate life cycle of Bridge Complex SoS (or other physical infrastructures) without models that build on the indispensable domain knowledge of the complexities that stem from the multiscale composition of bridges, spanning the physical molecular and macro dimensions, as well as the myriad parties, stakeholders, organizations, the public, and political forces. All of the above determine the fate of bridges worldwide. In this sense, not only bridges but also other physical infrastructures constitute Complex SoS (Andrijcic, 2016; Andrijcic and Haimes, 2017).

7.III.2 MODELING INTERNAL AND EXTERNAL INTERDEPENDENCIES AND INTERCONNECTEDNESS CHARACTERIZING BRIDGE COMPLEX SYSTEMS OF SYSTEMS

The characterization of the Bridge Complex SoS is unique, because bridges representing Complex SoS, are continuously bombarded by internal forces (stress, fatigue, rust, and aging, among others) and external forces (weather, excessive loads, lack of appropriate maintenance, and timely replacement of deteriorated parts, among others).

What follows are major contributing sources to the deterioration of the Bridge Complex SoS due to the I-I within and outside the bridge infrastructure: (i) lack of discovery of the primary factor(s) limiting the service life of highway bridges; (ii) failure to deliver bridges that were constructed as planned, designed, and properly financed; (iii) lack of knowledge about deterioration processes; (iv) failure to detect damage before failure occurs; (v) failure to adequately maintain bridges; (vi) failure of current technical, management, and institutional systems to prevent these failures; and (vii) failure to apply a holistic approach to the modeling, construction, operation, and maintenance of Bridge Complex SoS infrastructure.

The latter includes the lack of a holistic approach that studies and models the encompassing Bridge Complex SoS, including its planning, design, construction, maintenance, management, and replacement.

As presented in earlier sections, the above multiple sources of risk to the viability of the bridge enterprise can be best modeled, assessed, managed, and communicated by studying and analyzing the bridge enterprise as Bridge Complex SoS. Thus, multiple models are required to represent the essence of the multiple perspectives characterizing the subsystems (though their shared and other essential entities) and can only be harmonized and coordinated via a metamodeling schema (Andrijcic, 2016).

To achieve an effective understanding, modeling, and management of the internal and external structural and operational forces placed on Bridge Complex SoS, a paradigm shift is needed from the commonly present lack of vision by not addressing them holistically. To achieve the above, there is a need for identifying, documenting, and quantifying the technical and institutional factors that limit the service life of existing Bridge Complex SoS. This can be at least partially achieved by studying a representative subset of regional bridge infrastructures, thus adding to the body of knowledge and understanding of the factors that limit the useful service lives of Bridge Complex SoS. The collection of appropriate database on current functional bridges, and the reliance on past data obtained by studying decommissioned bridges, are essential. Such a systemic process must identify, characterize, quantify, and model internal and external forces affecting the integrity of the bridge infrastructure. These initiatives would enable a more encompassing, systemic, rational, and scientifically based approach to the design, fabrication, construction, operation, maintenance, and preservation of Bridge Complex SoS. Furthermore, such knowledge can be used to develop maintenance and preservation strategies that would enable a bridge infrastructure to be kept in service and in a state of good repair for as long as it is needed. This knowledge can also be used to influence engineering curricula at major universities to help bring about a change of focus from newly designed and constructed bridges to the challenges of managing and maintaining the existing Bridge Complex SoS.

Consider the following factors that have caused the dismal state of countless number of bridges around the world and the need for the adoption of a paradigm shift grounded on systems-based risk management guiding principles to inhibit their development (see Chapter 9): (i) Most civil infrastructures have not been designed and constructed to provide the level of service required. (ii) Bridge infrastructure has been designed without regard for adequate maintenance and renewal for the required service levels. (iii) Current approach to bridge management is founded on data obtained by subjective visual condition and nonquantitative performance measures. (iv) Current practice is to design and construct bridges with limited regard for adequate inspection, maintenance, and preservation. There have been many sudden failures that have been traced back to hidden or latent defects, many of which were directly related to bridges that were not constructed as detailed and with materials that did not meet specifications. Data must continuously be collected on the deterioration processes and mechanisms affecting Bridge Complex SoS, including (i) corrosion, (ii) fatigue, (iii) extreme loads, (iv) impact, (v) concrete degradation, and (vi) coating degradation, among others. Quantitative data collected on the actual factors that have limited the service lives of specific decommissioned highway bridges may include (i) the actual concrete cover on *in situ* bridges, (ii) *in situ* condition and performance of bridge bearings, (iii) *in situ* condition and performance of bridge joints, (iv) quantitative measurements of diffusion and corrosion rates and metal loss, and (v) quantitative measurements of residual stress. Such database can support objective identification, characterization, and definition of these factors and allow the identification of the systemic failures at the nexus of Bridge Complex SoS.

7.III.3 THE ROLE OF STATE SPACE IN THE MODELING OF BRIDGE COMPLEX SYSTEMS OF SYSTEMS

Recall that the complexity of SoS stems primarily from the I-I within and among the systems and subsystems of SoS. And Chen's (2012) conceptual definition of state variables posits that the state $x(t_o)$ of a system at time t_o is the information at time t_o that, together with the input $u(t)$, for $t \geq t_o$, determines uniquely the output $y(t)$ for all $t \geq t_o$.

Consider the internal and external I-I within and among the numerous systems and subsystems that constitute Bridge Complex SoS. A major initiating event, e.g. an earthquake, that may cause the failure of the one system, e.g. structural integrity, would necessarily propagate and adversely affect other functions of the bridge because of I-I characterizing the Bridge Complex SoS. *Clearly, there is a need to understand and thus model the pathway through which the failure of one system propagates to other interconnected systems.* An effective risk analysis of Bridge Complex SoS (as a representative of complex infrastructure) requires a clear understanding of the respective configurations to enable the identification of specific critical failure modes. In previous chapters, also see Chapter 10, we built on fault-tree analysis to identify and quantify the I-I characterizing Complex SoS. This enabled the development and deployment of effective risk mitigation and other management strategies. Current risk models of systems, which do not consider the inherent interdependencies among other systems, are likely to be inferior to those models that do.

Models can help answer limited questions about the behavior of these systems under both steady-state conditions and dynamic forced changes. Because each Bridge Complex SoS is composed of multiple subsystems, each of which, small or large, is characterized by multiple perspectives, it is necessary to develop a set of models to represent this multiplicity. Such set of models must reflect the essence of the multiplicity of perspectives, attributes, functions, and dimensions of the Bridge Complex SoS. Physical, chemical, biological, and other natural laws serve as the first principles and the foundations of such models. Although they are most necessary, these natural laws are not sufficient for model construction. This is because the abovementioned multiple societal and political perspectives constitute critical driving forces for which all decision-making models must account. Furthermore, the above multiple perspectives characterizing Bridge Complex SoS cannot be adequately modeled in a single model, a fact that presents a challenge to modelers. Thus what is needed is a mechanism, a systemic framework, with which to augment the natural and physical laws with imagination, inventions, innovation, entrepreneurship, out-of-the-box thinking, and boundless experimentation. The PSM and the PSM laboratory (PSML), introduced in Chapter 4, will be briefly reviewed here (Haimes, 2007, 2012a).

The essence of metamodel coordination and integration is to build on all relevant direct and indirect sources of information to gain insight into the I-I among the submodels and on the basis of which to develop representative models of the Complex SoS. The coordination and integration of the results of the multiple models are achieved at the metamodeling phase within the PSM, thereby yielding a better understanding of the system as a whole. More specifically, modeling the I-I within and among the subsystems of Complex SoS requires an understanding of the intricate relationships that characterize the dynamics within and among the states of the subsystems. This very important task can be achieved at the metamodeling level of the PSM by observing, estimating, and assessing the outputs for given inputs and by building on the intrinsic common states within and among the subsystems. Note that although the intrinsic common states constitute a key element of the PSM, the extrinsic (input–output) relationships are also very important and support the intrinsic one. Indeed, the selection of the trial inputs to the model and the inquisitive process of making sense of the corresponding outputs are at the heart of system identification and parameter estimation. This is not a one-shot process; rather, it can be best characterized by tireless experimentation, trial and error, and parameter estimation and adjustments, as well as by questioning whether the assumed model's topology is representative of the system being modeled.

The PSM-based intrinsic metamodeling of Bridge Complex SoS stems from the basic assumption that some specific commonalities, I-I, or other relationships must exist between and among any two systems within any Complex SoS. More specifically, a metamodel represents the overall coordinated and integrated submodels of the Bridge Complex SoS. We define a metamodel as a family of submodels, each representing specific aspects of the subsystem for the purpose of gaining knowledge and understanding of the multiple I-I among the submodels and thus to comprehend the Complex SoS as a whole. For example, the physical structure of a Bridge Complex SoS is composed of multiple subsystems, e.g. a deck and wearing surface, which provides the riding surface and distributes load to the superstructure, the main structural support of the bridge, and transfers load to the substructure and distributes the load to the foundation soils. Each of these subsystems in turn contains and relies upon multiple subsystems such as bearings and joints and various materials, eg. concrete, structural steel, reinforcing steel, prestressing steel, and coatings – thus the hierarchy of systems and subsystems. Recall that the essence of each subsystem can be represented by a finite number of state variables. (The term essence of a system connotes the quintessence of the system, the heart of the system – everything critical about the system.) Given that a system may have a large number of state variables, the term essential states of a system connotes the minimal number of state variables in a model with which to represent the system in a manner that permits the questions at hand to be effectively answered. Thus, these state variables become fundamental for an acceptable model representation.

Consider the following sample of the intricate modeling of the multifarious characteristics of infrastructures that characterize Bridge Complex SoS: (i) sociotechnical SoS composed of physical and engineered components, as well as commercial and noncommercial users and decision and policy makers; (ii) presently managed solely as physical systems; (iii) maintained with decisions that are usually reactive and with fixes that are often short-term solutions; (iv) little to no consideration devoted to the impacts of current decisions on future maintenance options; (v) lack of essential proactive inspection and maintenance; (vi) prevailing political and societal environment of the community within which the bridges are located; (vii) lack of public accountability; and (viii) psychological perception of denial of risk to Bridge Complex SoS.

7.III.4 METAMODELING BRIDGE COMPLEX SYSTEMS OF SYSTEMS

Bridges are primarily modeled and managed for maintenance purposes solely as physical infrastructure systems. And bridge maintenance fixes are often short term and exclude a serious consideration of the impact of current decisions on future maintenance options and on other stakeholders (e.g. commercial and noncommercial infrastructure users). We present a case study to illustrate the use of a metamodeling framework that recognizes that the bridge infrastructure constitutes a Complex SoS, which includes sociotechnical systems composed of more than their various physical subsystems. In other words, the Bridge Complex SoS are characterized by a diverse set of managerially and operationally independent stakeholders whose changing needs and constraints evolve over different time frames. Furthermore, the distributed and decentralized decision-making environment often results in bridge infrastructure management strategies and outcomes that are "optimal" from the perspective of a single subsystem of the Bridge Complex SoS but might be harmful or suboptimal from a global holistic SoS perspective. The key to modeling Bridge Complex SoS lies in understanding the genesis of characterizing I-I of systems manifested through shared states and other *essential entities* within and among the systems that constitute Bridge Complex SoS. Recall that the term "essential entities" includes shared/common decisions, decision makers, stakeholders, resources, organizational behavior and norms, policies and procedures, management, culture, and others. This undertaking can be accomplished by building on state-space theory that is fundamental to systems engineering and process control. This modeling of Bridge Complex SoS should enable different bridge infrastructure stakeholders to identify, quantify, and explicitly model some common and conflicting interests and needs, with the ultimate goal of creating more sustainable, responsible, and satisficing bridge management policies that are cognizant of and acceptable to a diverse array of involved stakeholders.

Similar to most, if not all, Complex SoS, the development of representative models and submodels of the engineering, social, and economic submodels and the identification of the intricate I-I within the Bridge Complex SoS require multiple tasks, including (i) better accounting for the impact of travel delays, travel cost reactions, progressive increases in vehicle weights and sizes, and substitutions of work and shopping with telecommuting and online commerce; (ii) augmenting analytical submodels with, for example, agent-based simulations; and (iii) developing more streamlined and detailed engineering, economic, and social submodels that will be more scalable and applicable to a larger number of bridges.

Consider the following guiding premises: (i) The Bridge Complex SoS cannot be modeled through a single model. (ii) Multidisciplinary submodels must be developed to represent different stakeholder-level perspectives. (iii) The I-I among the bridge subsystems (submodels) are modeled primarily through shared state and other essential entities (also including physical condition of structural components, unrestricted load capacity, maintenance budget, traffic demand). (iv) A metamodel must be developed to coordinate and integrate the various interdependent bridge submodels. This metamodel, given its integrative capacity, should provide a deeper insight into the I-I among the Bridge Complex SoS.

In addition, the metamodel must take into account the wide range of factors that may affect the bridge subsystems and their interdependencies. These may include (i) load, capacity, and foundations; (ii) the nature of bridge design itself, in particular the fact that bridges are built from bottom up, whereas the load comes from top down; (iii) heavy trucks, corrosion, combined forces, and changes in technology; and (iv) load distribution, change in capacity, extreme events, risk, economics, and stakeholder expectations based on today's standards.

Traditional bridge management models, such as load rating using girder line, as well as more complex methods such as finite element analysis and statistical deterioration, are well known to and applied by the bridge engineering community. However, these models provide a very limited picture of the bridge maintenance and management problem, and they often ignore other factors (e.g. socioeconomic factors, regional development factors, historical importance) that could influence the choice of an optimal bridge maintenance plan. A more comprehensive maintenance approach would be more holistic and would consider the objectives and needs of all stakeholders, ranging from system owners to system users.

The Bridge Complex SoS encompasses not only physical subsystems but also human, social, and economic systems; their I-I is of critical importance, given the broad impacts of bridge failures. Concurrent use of multidisciplinary models is essential for the exploration of the I-I among the various systems and subsystems, with the goal of discovering how current bridge maintenance practices affect, and are affected by, social and economic systems. The essential development and deployment of multidisciplinary models allow us to examine the operation of the Bridge Complex SoS from its various perspectives.

7.III.5 TEN SYSTEMS-BASED GUIDING PRINCIPLES FOR THE NEXT GENERATION OF THE DESIGN, CONSTRUCTION, OPERATION, MAINTENANCE AND REPAIR OF BRIDGE COMPLEX SYSTEMS OF SYSTEMS

In this section we apply the guiding principles to the Bridge Complex SoS.[1] By its nature, risk analysis is an intricate, dynamic process – an amalgamation of the arts and sciences – tailored to each specific set

[1]Chapter 9 is devoted in its entirety to risk analysis of complex SoS, including the 10 systems-based guiding principles.

of sources of the risks to specific systems. It follows, then, that for any subsystem of the Bridge Complex SoS, the balance between quantitative and qualitative risk analysis will be problem and domain specific. The 10 principles introduced in Chapter 9 are intended as an efficacious, valuable risk analysis instrument for guiding the quantitative and qualitative decisions made by bridge engineers and other professionals (Haimes, 2012b, 2016). Indeed, violations of these basic principles have led in the past to disastrous results in Bridge Complex SoS worldwide. The idea behind these guiding principles is to provide engineers and other professionals with a checklist that will help them to avoid the most common errors while still providing them with the flexibility to make appropriate tradeoffs in constructing, deploying, and executing various functions of the Bridge Complex SoS. Further general discussion of the effectiveness of this approach is the subject of *The Checklist Manifesto* by Atul Gawande (2011).

A bridge engineer or a manager might ask: "Why do we need a set of principles to perform risk analysis?" Although no single answer can do justice to this broad question, one may posit that any decision-making process that addresses important probable, including dire, consequences resulting from changes emanating from within or without the Bridge Complex SoS ought not to be performed on an ad hoc basis. Thus, the life cycle of the bridge, to be studied and modeled as Complex SoS, must adhere to fundamental guiding principles that are cognizant of and responsive to the inevitable EFCs. Namely, trends in external or internal sources of risk to the Bridge Complex SoS may adversely affect its specific states. Examples of EFCs may include elements of the *Evolving Base* (introduced in Chapter 9), which is characterized by dynamic shifting rules and realities, including (i) goals and objectives; (ii) stakeholders, decision makers, and interest groups; (iii) organizational, political, and budgetary baselines; (iv) reorganization and reallocation of key personnel; (v) emergent new technology; (vi) requirements, specifications, delivery, and users; and (vii) natural hazards.

First Principle: Holism is the common denominator that encompasses systems engineering, risk analysis, and effective project management. The Bridge Complex SoS includes the physical bridge system; the multiple users – residential and commercial – that add to the load on the bridge system; the constant variability of weather that affects the structural integrity of the bridge system; and the political-budgetary constraints that affect the effective maintenance, operation, and rehabilitation of the physical bridge system. Analyzing the bridge as a Bridge Complex SoS cannot be a selective process, driven by limited perspectives; rather, a holistic systems-based approach must be embraced to account for EFCs; multiple databases and numerous technological and organizational subsystems; multiple objectives, agencies, stakeholders, and decision makers; and multiple time horizons associated with each phase in the system life cycle. Modeling and managing the risk to, and the life cycle of, bridge infrastructure require understanding the individual behaviors of these components as well as their expected and unexpected interactions.

Second Principle: The process of risk modeling, assessment, management, and communication must be methodical, disciplined, systemic, integrated, and commensurate in its comprehensiveness with the criticality of the Bridge Complex SoS and of its associated risks. There are numerous sources of risk emanating from the above subsystems. Furthermore, this process must consider the centrality of the time frame in risk analysis and, more specifically, the impacts of the *Evolving Base* on the ultimate integrity and functionality of the physical bridge system. As an example, consider five categories of EFCs as risk factors associated with the physical Bridge Complex SoS, which constitutes one subsystem of the bridge infrastructure SoS: (i) human factors (personnel, culture, and collaboration), (ii) technological factors (materials, manufacturing, inspection methods), (iii) infrastructure factors (traffic network and corridors, control facilities, and cyber infrastructure), (iv) operational factors (maintenance, traffic management, planning, and scheduling), and (v) policy factors (regulations, budget, and liability). Because safety is the level of acceptable risk, and no risk to a threatened bridge subsystem can be completely eliminated, tradeoffs are made between the cost of reducing one or more components of the risk vector and that of the remaining residual risk. Furthermore, risk analysts must be capable of communicating the above reality to the respective stakeholders.

Third Principle: Models and state variables are central to quantitative risk analysis. The integrity and functionality of the physical Bridge Complex SoS are directly related to its states – the levels of corrosion, physical and other structural conditions, and available funds to ensure proper operation, maintenance, and rehabilitation. State variables are essential building blocks of mathematical models, and all decisions aim to change the outputs of the system by changing (or maintaining the current) states of the system. In addition, vulnerability and resilience (Haimes, 2016) of the system are also the manifestation of the states of the system. The designed functional objectives and life cycle management of a bridge can be addressed and ultimately realized if and only if the multiple states of each subsystem of the Bridge Complex SoS can be identified, modeled, evaluated, and ultimately controlled to meet these objectives. These states include the quality of trained professionals, the quality and reliability of the bridge elements, and the effectiveness of the states of coordination, to cite a few.

Fourth Principle: Multiple models are required to represent the essence of the multiple perspectives of Bridge Complex SoS. Clearly, each of the multiple subsystems that constitute the Bridge Complex SoS must each be modeled with a specific model to represent the essence of the associated subsystems. These models include but are not limited to physical models, statistical models, and simulation models. Consider a sample of the dozens of subsystems that constitute the Bridge Complex SoS: the physical bridge, designers, manufacturers, inspectors, and maintenance teams. Each requires one or more specific models to represent the essence of each subsystem. Then at a higher level, a holistic metamodeling approach is used to coordinate different subsystems to achieve effective and efficient management of Bridge Complex SoS.

Fifth Principle: Metamodeling and subsystems integration must derive from the intrinsic states of the Bridge Complex SoS. The extrinsic modeling approach assumes that the output of a subsystem becomes the input of another subsystem, while the intrinsic modeling approach assumes that subsystem interdependencies originate from shared states and decisions. Identifying states that are common to multiple subsystems is an important way to understand inter-subsystem interdependencies and interactions. For example, the traffic capacity is a state variable of both engineering subsystem and socioeconomic subsystem. The metamodeling effort takes advantage of the subsystems models by relating the shared states and decisions among the subsystems and subsequently coordinating and integrating the information generated from the submodels into a cohesive representation of the entire Bridge Complex SoS. To maintain a reliable and sustainable bridge infrastructure SoS, all its subsystems must function as harmoniously as possible. This challenge can be achieved if and only if there is a clear understanding of the I-I among the intrinsic states of the entire Complex SoS.

Sixth Principle: Multiple conflicting and competing objectives are inherent in Bridge Complex SoS. Systems involving stakeholders and decision makers are characterized and driven by multiple conflicting, competing, and noncommensurate objectives; the Bridge Complex SoS is no exception. For example, (i) minimize all costs associated with the bridge life cycle; but (ii) maximize quality, reliability, and availability of the bridge system by controlling the traffic load over the bridge and allocating all required resources for its maintenance and rehabilitation. Decision makers must quantify and understand the tradeoffs among different competing objectives and identify nondominated options. Furthermore, diverse state, federal, and private sector stakeholders have an interest in, and an impact on, these objectives.

Seventh Principle: Life cycle management of Bridge Complex SoS must account for epistemic and aleatory uncertainties. Uncertainty characterizes most if not all subsystems of the Bridge Complex SoS. Both epistemic (knowledge) and aleatory (variability) uncertainties must be addressed in modeling, assessing, managing, and communicating the risks associated with the Bridge Complex SoS. These uncertainties come from both within and outside the system. Examples of epistemic uncertainties include the unknown deterioration process, bridge design error and unknown structural weakness, and prediction for future traffic load. Examples of aleatory uncertainties include the variabilities in (i) environmental factors, (ii) vision inspection results, and (iii) daily traffic.

Eighth Principle: Risk analysis must account for risks of low probability with extreme consequences.

Safety-critical systems, especially those that serve the public at large, must account for all sources of risk, in particular risks associated with a low probability of initiating events with dire and catastrophic consequences. The Bridge Complex SoS epitomize safety-critical systems. Bridge failure, such as the collapse of the bridge, is a relatively rare event; thus the expected value metric of risk cannot be applied to Bridge Complex SoS. The failure of bridge not only causes direct property damages and loss of human lives but also has broader socioeconomic consequences due to the interdependencies across multiple industry sectors. The cost of these cascading ripple effects must be estimated and taken into account so that sustainable bridge maintenance activities can be planned under a cost–benefit framework.

Ninth Principle: The time frame is central to life cycle management of SoS. The deterioration of subsystems of Bridge Complex SoS, affecting the bridge load capacity, the traffic load, and other EFCs, is a function of time. Understanding the impact of current decisions on future options is an important part of risk analysis. The dynamic nature of all the events – natural and human made – that affect and impact the integrity and functionality of the Bridge Complex SoS renders the time frame pivotal to risk analysis. It is also critical to consider the variability of the time frames associated with different subsystems. Multi-year bridge life cycle planning that spans several decades is fraught with unexpected surprises associated with every subsystem. More specifically, the bridge life cycle planning process is partitioned in sequences of different stages: design, construction, operation, and so on. Clearly, decisions made during design period would constrain and affect subsequent future decisions during construction and operation. In particular, "wrong" or inflexible decisions at any stage could result in programmatic risk (project's cost overrun and time delay in its completion) and technical risk (not meeting performance criteria).

Tenth Principle: Risk analysis of Bridge Complex SoS must be holistic, adaptive, incremental, and sustainable, and it must be supported with appropriate data collection, metrics with which to measure efficacious progress, and criteria on the basis of which to act. To ensure an effective risk assessment, management, and communication process, a continuous adaptive feedback mechanism must be put in place. Prudent risk management calls for a continuous process of designing a data collection mechanism, measuring the performance of a system, developing metrics with which to measure performance, assessing whether observed changes are sufficiently significant, and determining criteria for actions – all are requisites for effective risk modeling, assessment, management, and communication. Understanding bridge system behaviors must based upon scientific evidence and credible database. A holistic risk assessment, management, and communication process also guide the need for future data collection.

7.III.6 A CASE STUDY ON MAINTENANCE OF BRIDGE COMPLEX SYSTEMS OF SYSTEMS

7.III.6.1 Introduction to the Case Study

There are many reasons that have led to the consistent underspending for transportation infrastructure maintenance in many other parts of the world.[2] Some are discussed in Andrijcic et al. (2013) including the short electoral cycle for elected officials in the United States, which encourages investments in projects that have tangible short-term outcomes and results, and are thus easier to "campaign around for their funding." Furthermore, the benefits associated with regular long-term investments in infrastructure maintenance are largely "unseen" by the public, making it harder to justify those expenditures politically. Commonly, politicians have an immediate gain from building new infrastructures and working on "banner" projects that receive a lot of public attention. The lack of political will to allocate necessary funds for transportation infrastructure improvement stems, in part, from the disharmonious goals and objectives among the various stakeholders and decision makers; as well as from the lack of understanding and appreciation for the critical linkages and I-I among the myriad subsystems of the Transportation Complex SoS.

[2]This case study draws from research performed by Y.Y. Haimes, Z. Guo, and E. Andrijcic. See works cited in Part II and III introductory material for citations.

In this case study we propose a forward-looking and collaborative modeling approach that recognizes that Transportation Complex SoS are more than just built constructs; they must account for the diversity of the stakeholders and their vision, needs and objectives, and financial and other limitations and constraints. We illustrate the potential usefulness of the approach by modeling the bridge infrastructure as Bridge Complex SoS, composed not only of physical subsystems (i.e. the engineered components) but also of subsystems that reflect the socioeconomic and political environment within which the infrastructure operates. This representation is appropriate, given that these stakeholders are held accountable to the viability and functionality of the bridge infrastructure support and management.

The aim of this case study is neither to provide concrete maintenance solutions for Bridge Complex SoS nor to develop detailed mathematical models representing different subsystems of the infrastructure within the Bridge Complex SoS. Rather, our purpose is to illustrate the extent to which the metamodeling of Bridge Complex SoS can contribute to (i) making decisions on bridge infrastructure management in a decentralized environment characterized by multiple and conflicting stakeholders and objectives; (ii) enabling decision makers from representative DoTs and metropolitan planning organizations (MPOs) to assess the broader socioeconomic impacts of their maintenance investment alternatives; (iii) allowing other stakeholders of the Bridge Complex SoS, including commercial and noncommercial users of the infrastructure, to evaluate the impacts that infrastructure maintenance decisions have on them, given their changing needs, demands, and constraints; and (iv) leading to bridge maintenance policies, procedures, and maintenance schedules and solutions that are socially satisficing for the involved stakeholders. This collaborative modeling approach can lead to a more transparent understanding of the I-I characterizing Bridge Complex SoS and might in turn increase the level of accountability that the general population (i.e. infrastructure users) places on infrastructure management, thus transferring risk ownership to all involved stakeholders.

7.III.6.2 Metamodeling of the Bridge Infrastructure via Shared States and Other Essential Entities

Recall that Complex SoS constitute a collection of interdependent and interconnected subsystems, each of which is connected to at least one other subsystem through one or more shared states and other essential entities. Furthermore, shared states can be, over some period of time, directly impacted by decisions made by decentralized stakeholders representing different subsystems. Also, shared states do not have to represent shared *physical* aspects of the subsystems; they can also represent shared virtual, social, economic, or other properties that explicitly or implicitly connect two or more modeling perspectives. When considered from a single modeling perspective, shared states and other essential entities might not be of prime interest; however, when considered from the Complex SoS perspective, these entities are critical for the explicit modeling of the subsystem I-I and for our better understanding of how decisions made in a single subsystem might impact the shared states of interdependent subsystems.

In this case study we consider the decision makers of the Bridge Complex SoS infrastructure to constitute the DoTs and MPOs. They are primarily concerned with changes to the bridge condition and load ratings, as well as the cost of maintenance to address deterioration issues. Their mission is (i) maximizing bridge condition and load ratings while (ii) minimizing the net present value (NPV) of bridge maintenance investments and (iii) maximizing the return on the same investments. Noncommercial bridge users are generally concerned about maximizing their ease of access to places of interest and minimizing their cost of travel. Similarly, commercial users rely on the bridge infrastructure for efficient and affordable supply chain management; hence, they are concerned with maximizing their access to places of business, minimizing their cost of travel, and maximizing the loads they are allowed to carry.

For maintenance purposes, bridges are primarily modeled and managed solely as physical systems. We posit that existing bridge management practices, focused primarily on a consideration of engineering factors, often inadvertently identify policies that are "optimal" from the engineering perspective, but are not cognizant of, or responsive to, the changes stemming from the environment within which the Bridge Complex SoS infrastructures operate. While many models and approaches have been developed to assess some of the socioeconomic impacts of transportation infrastructure investments, many of the existing approaches are complex and require extensive data, preventing them from being commonly put into practice. We illustrate the potential benefits of studying the bridge maintenance problem through the use of a metamodel in which the different modeling perspectives are connected via a (set of) shared state variable(s). The specific details of the metamodeling process are described in Andrijcic and Haimes (2017).

In short, the metamodeling process enables a set of interdependent and interconnected subsystems to be modeled from multiple structural, organizational, or stakeholder perspectives. Each model addresses the problem from a limited perspective, and the combination of limited perspectives provides a better overall understanding and appreciation of the Bridge Complex SoS. Recall that all models are connected via a single or multiple shared states and other essential entities and the modeling process often requires a metamodeling coordination (see Chapters 2 and 4). Our metamodeling coordination is an iterative process of manipulating adjustable decisions and constraints associated with interdependent and interconnected subsystems in response to changes in shared state and other essential entities. Model coordination is accomplished by comparing the ranges of values for shared states and other essential entities. Similar to our model decomposition and higher-level coordination introduced in Chapters 2–4, the higher level coordinates the subsystems' specific preferred decisions and whether the resulting ranges are acceptable to all relevant stakeholders over the entire planning horizon. A level of shared states rendered to be unacceptable to subsets of relevant stakeholders necessitates the identification of those shared states and other essential entities that might be negotiable at the level of an individual subsystem or through a collaborative effort of several subsystems. The purpose of such a negotiation process is to achieve a range of values of shared states and other essential entities that would be more acceptable to relevant stakeholders over the entire planning horizon. This iterative process continues until solutions/policies are achieved that satisfy local subsystem objectives while also identifying levels of shared states and other essential entities that deem to be acceptable to all stakeholders over a longer planning horizon. A specific example of the modeling process is provided in the following section.

In general, a failure of subsystems to agree on globally acceptable ranges of values of shared state variables could manifest in local subsystem difficulties that could, over time, propagate and become detrimental to the entire system. Therefore, satisficing, that is, accepting suboptimal but adequate solutions (Simon, 1956), is critical from a broader public policy standpoint because "going after critical [global] requirements is usually a better bet for 'survival' than going after what seems optimal" from an individual subsystem perspective (term in brackets added by authors) (Ben-Haim, 2012, p. 1328).

7.III.6.3 The Case Study

We focus our case study on the Hampton Roads Bridge–Tunnel (HRBT) Complex SoS in the state of Virginia, USA, due to its geographic and strategic importance. The Hampton Roads Bridge–Tunnel (HRBT) is a 3.5-mile crossing for US Interstate 64. It is a four-lane facility including bridges, man-made islands, and tunnels under the main shipping channels for Hampton Roads harbor. It is the primary connection between a highly populated region in south Hampton Roads with the inland areas of the Commonwealth of Virginia. The HRBT Complex SoS has two 12-ft (3.7 m)-wide lanes each way, on separately built bridge–tunnel structures. Close to over 75% of the Hampton Roads bridges are "load posted," suggesting that as their condition deteriorates further, it is likely that additional weight restrictions will continue to limit the number of trucks that are allowed to cross the bridge and speed limit restrictions will increase general congestion levels. These two actions can create adverse consequences for the regional economy and residents. Therefore, to be effective, bridge management in the Hampton Roads area is considering changing demands placed on the infrastructures of the HRBT Complex SoS and the evolving needs and constraints of the infrastructure users.

We assume that major maintenance decisions for the HRBT Complex SoS are made every two years, following a bridge inspection, and we are interested in which of the available long-term maintenance policies are most acceptable to all stakeholders over a 22-year horizon (starting in 2011, ending in 2033). The stakeholders under consideration include decision makers from the Virginia DoT (VDoT) and Hampton Roads Metropolitan Planning Organization (HRMPO), as well as noncommercial and commercial bridge users. For illustrative purposes, we have created five alternative maintenance policies shown in Table 7.III.1. These policies are not based on actual maintenance plans for the HRBT, but are illustrative examples that range from an end-loaded policy in which maintenance is performed only at the end of the planning horizon when the condition of the bridge has deteriorated significantly. Furthermore, preventive measurements are taken immediately, and no follow-up maintenance is performed to several types of distributed policies, where investments are made at different points in time during the planning horizon.

7.III.6.3.1 Three Modeling Perspectives

For this case study we are concerned with the long-term macro-level view that considers how technical maintenance decisions made today affect the bridge structure, as well as the users of the infrastructure, in the future. We address this problem from three modeling perspectives, namely, engineering, social, and economic, which are coordinated at the metalevel through a single shared state variable of the bridge traffic capacity (as will be discussed in detail in subsequent paragraphs). We acknowledge that there are other modeling perspectives that could be of interest to the stakeholders (e.g. modeling perspectives reflecting different physical or functional subsystems of the bridge infrastructure or the natural environment), but our choice is sufficient to demonstrate the value of the modeling approach. Each of our three modeling perspectives represents a collection of supporting submodels that are mathematical representations of the subsystems under study.

7.III.6.3.2 The Engineering Perspective and Submodel

From a traditional engineering perspective, decision makers from DoTs and the HRMPO would like to spend minimum funds to ensure that a bridge is structurally sound. The determination of structural health is not an easy one, as there are numerous physical components to a bridge that are affected by different deterioration processes. Numerous civil engineering models have been developed to study these processes in detail, and the goal of this case study is not to try to replicate those. Instead, we focus on the process of deterioration as it ultimately relates to changes in bridge load rating, as this will play a key role in determining traffic flow over the bridge.

We identify several major factors (variables) that would enable DoT and MPO decision makers to track the changes in the condition of the bridge over some planning horizon of interest. These include bridge age, frequency of salting, climate/environmental factors, maintenance investment amount, deterioration rate (i.e. change in bridge condition rating), and bridge load capacity (i.e. load rating). These are shown in Figure 7.III.1.

Condition ratings indicate the level of physical deterioration (e.g. cracking, scour, section loss, etc.) caused by environmental effects and traffic wear on the superstructure, substructure, and deck of a bridge (Dunker and Rabbat, 1995). Load rating indicates the live load capacity that a bridge can safely carry. As bridge condition and load rating describe the state of deterioration and state of capacity of a bridge at any point in time, given the other factors, they represent the essential state variables of the engineering submodel. In this section we do not yet introduce the shared states and other essential entities that connect the three modeling perspectives; however, we simply build the engineering submodel from existing engineering models to represent the perspective of the DoT and MPO decision makers.

Our intent is to assess how different long-term maintenance policies selected in period i affect the superstructure condition and load ratings of the bridge at period $i+k$ ($k=0, 1, \ldots, T$) and how they frame the extent of future maintenance options. Additionally, we aim to understand the impact of maintenance decisions on commercial and noncommercial users of the

TABLE 7.III.1 Maintenance Decisions are Made Every 2 Years, Following a Bridge Inspection

Policy	Maintenance investment (in $ million)											
	$u(1)$	$u(2)$	$u(3)$	$u(4)$	$u(5)$	$u(6)$	$u(7)$	$u(8)$	$u(9)$	$u(10)$	$u(11)$	$u(12)$
1. End loaded	0.00	0.00	0.00	0.00	0.00	0.00	0.00	0.00	0.00	0.00	0.00	10.00
2. Equal annual distribution	0.83	0.83	0.83	0.83	0.83	0.83	0.83	0.83	0.83	0.83	0.83	0.83
3. Front loaded	10.00	0.00	0.00	0.00	0.00	0.00	0.00	0.00	0.00	0.00	0.00	0.00
4. Distributed front loaded	5.00	0.00	5.00	0.00	0.00	0.00	0.00	0.00	0.00	0.00	0.00	0.00
5. Distributed	2.50	0.00	0.00	2.50	0.00	0.00	2.50	0.00	0.00	2.50	0.00	0.00

This table indicates investment amounts (in $ million) at each decision point, $u(1) = 2011$, $u(12) = 2033$. The total amount of investment over the planning horizon is the same for all alternatives, $10 million.

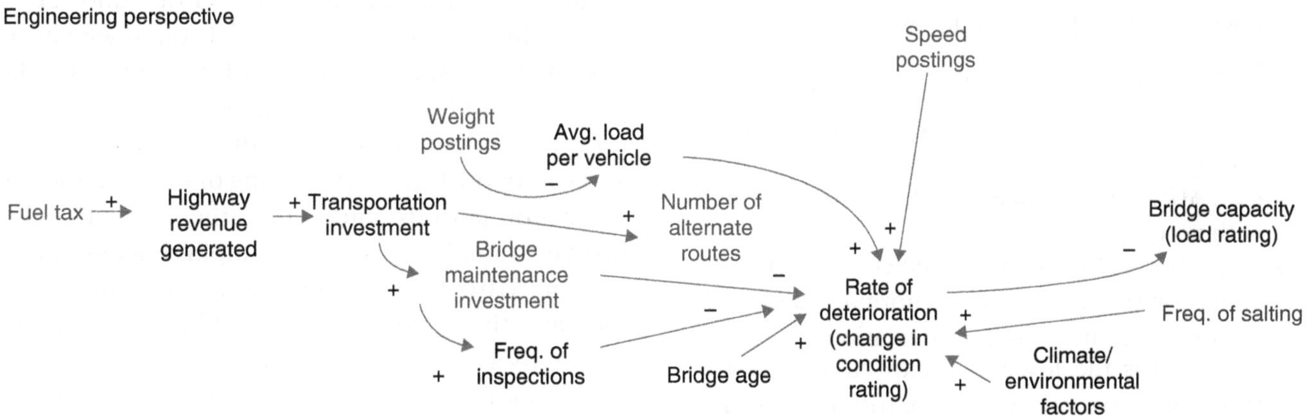

Figure 7.III.1 System dynamics (SD) diagram for the engineering perspective enables us to envision the dynamic relationships among the different factors under consideration.

bridge. We adopt a linear time invariant (LTI) system formulation for our engineering submodel. While there are certain drawbacks to using LTI formulations, there are also certain benefits. In particular, LTI formulation guarantees the existence of closed-form solutions, it implies constant parameters, and it reflects stakeholders' concerns with the performance domain, which is critically important in our problem.

For the purpose of this case study, we present only high-level information on how the factors identified in Figure 7.III.1 were included in our engineering submodel shown in Eq. (7.21). To construct an LTI representation of the engineering submodel, we utilize two existing engineering models by Chase et al. (1999) and Chase and Gáspár (2000). These models enable us to assess the changes in the bridge condition and load ratings conditional on bridge age, changes in climate, and maintenance actions. We assume that these two models provide a good starting point to evaluate the macro-level changes in the superstructure condition and load ratings over our planning horizon. We develop the following modeling construct:

$$\begin{aligned} s_{\text{eng},1}(k+1) &= a_{\text{eng}} \cdot s_{\text{eng},1}(k) + b_{\text{eng}} \cdot u_{\text{eng},1}(k) + c_{\text{eng}} \cdot u_{\text{eng},2}(k) \\ &\quad + d_{\text{eng}} \cdot u_{\text{eng},3}(k) + \omega_{\text{eng}}(k) \\ s_{\text{eng},2}(k+1) &= e_{\text{eng}} \cdot s_{\text{eng},1}(k+1) \\ y_{\text{eng},1}(k) &= E\left[s_{\text{eng},1}(k)\right] \\ y_{\text{eng},2}(k) &= E\left[s_{\text{eng},1}(k)/\beta \le \mu - \sigma\right] \end{aligned} \tag{7.21}$$

where

- k is the bridge inspection period (inspections occurring every 2 years: hence $\Delta k = 2$ years)
- $s_{\text{eng},\ 1}(k)$ is the superstructure condition rating (this variable ranges from 0 to 9, with 9 representing a new bridge and 0 representing a nonoperational bridge), and $s_{\text{eng},\ 1}(0)$ is estimated for

HRBT from the Chase et al. (1999) superstructure generalized linear regression model that estimates changes to bridge condition rating based on local climatic conditions, bridge age, and average daily traffic

- $s_{\text{eng},\,2}(k)$ is the bridge load rating (in thousands of pounds), and $s_{\text{eng},\,2}(0)$ is estimated for HRBT from the Chase and Gáspár (2000) model that relates the superstructure condition rating to the bridge load rating
- $u_{\text{eng},\,1}(k)$ is the bridge maintenance investment (in \$)
- $u_{\text{eng},\,2}(k)$ is the decision to implement a speed posting – a scalar variable indicating yes (1) or no (0)
- $u_{\text{eng},\,3}(k)$ is the decision to implement a weight posting – a scalar variable indicating yes (1) or no (0)
- $\omega_{\text{eng},\,3}(k)$ is the normally distributed white noise with zero mean
- $y_{\text{eng},\,1}(k)$ is the mean expected superstructure condition rating ($s_{\text{eng},\,1}(k)$ is normally distributed with mean μ and standard deviation σ)
- $y_{\text{eng},\,2}(k)$ is the low-range expected superstructure condition rating with partitioning β taken at one standard deviation below the mean

Coefficient a_{eng} represents the linearized yearly change in the superstructure condition rating based on bridge age, amount of precipitation, frequency of deicing, and number of freeze–thaw cycles. Its initial estimate ($a_{\text{eng}} = 0.989$) was obtained for the HRBT (for years 2011 to 2033) from the superstructure condition rating model presented by Chase et al. (1999). Under the assumptions of the above model, this initial estimate suggests that in each 2-year period the superstructure condition rating for the HRBT changes by approximately 1.1%. As this estimate of the coefficient does not account for possible extreme changes in the global and regional climate resulting in harsher winters and more precipitation, a sensitivity analysis was conducted using the following values for this coefficient – 0.989, 0.980, and 0.970 – representing a 1.1, 2, and 3% change in the superstructure condition rating over a 2-year period, respectively. To provide a worst-case scenario, the results presented in this article reflect a 3% change in the superstructure condition rating over a 2-year period.

Coefficient b_{eng} denotes the change in the bridge condition rating based on the maintenance investment made during the previous planning period of 2 years. It estimates the effectiveness of maintenance activities. This parameter would be identified by conducting a time-series analysis comparing the changes in superstructure condition rating for a particular bridge to actual maintenance expenditures for the same bridge. However, bridge maintenance investments for individual bridges are not recorded in a standard fashion, and many state DoTs do not track in house spending adequately. The Long-Term Bridge Performance Program (Friedland et al., 2007) will provide a framework through which such data will be collected, but as of now accurate data are unavailable. We acknowledge that this parameter is dependent on many factors, including the type and size of the bridge, the timing of maintenance and inspection periods, and many others. For this illustrative example, we make a simplifying assumption that a \$10 000 000 maintenance investment will increase the superstructure condition rating of a bridge by 1 point; hence $b_{\text{eng}} = 10^{-7}$. With the addition of more accurate data, this parameter could be reassessed by following existing approaches in the pavement maintenance literature (Fwa and Sinha, 1987; Mouaket and Sinha, 1990; Ramaswamy and Ben-Akiva, 1990; Madanat et al., 1995; Labi and Sinha, 2003).

Coefficient c_{eng} reflects the impact of a speed posting on the superstructure condition rating. If the DoT decision makers determine that a speed posting is not warranted, then $c_{\text{eng}} = 0$. If a speed posting is put in place, then $c_{\text{eng}} > 0$, and the actual value would depend on the actual speed limit. Coefficient d_{eng} reflects the impact of a weight posting on the superstructure condition rating. If the DoT decision makers determine that a weight posting is not warranted, then $d_{\text{eng}} = 0$. If a weight posting is put in place, then $d_{\text{eng}} > 0$, and the actual value would depend on the actual weight limit. The identification of c_{eng} and d_{eng} is not a trivial task as the effects of speed and weight postings are dependent on many factors. The identification of these parameters could be conducted by performing time-series analysis to compare superstructure condition ratings on bridges that are comparable in structure, size, and traffic loadings, some of which require speed and/or weight postings and some of which do not.

Coefficient $e_{eng} = 4.37$ relates the bridge condition rating to the bridge load capacity, and the value was adopted from Chase and Gáspár (2000). Additionally, for Eq. (7.22) we assume that $\omega(k)$ is normally distributed and purely random with mean 0 and variance 0.01. We also assume that the initial state $s_{eng,1}(0)$ for the HRBT is normally distributed and that $s_{eng,1}(0) = 5.86$ (obtained from the Chase et al. (1999) model for year 2011).

We assume that from a traditional engineering perspective, DoT and MPO decision makers have three major objectives, namely, maximizing the mean and low-range conditional expected bridge condition rating (or in this case superstructure condition rating) at the end of the planning horizon, which inevitably maximizes the mean and low-range conditional expected load rating at the end of the planning horizon as well, and minimizing the NPV of maintenance investments at a discount rate $r\%$, if total investment in period k is constrained by budget B_k (Eq. (7.22b)). We recognize that the NPV might not be the best metric for valuing long-term projects that might have considerable social and economic impact on the region in question and that certain metrics from the theory of welfare economics might be more suitable for assessing future worthiness of engineering projects (OECD, 2008). However, for this example we use the NPV as it is still widely used for transportation infrastructure projects. Furthermore, we recognize that in addition to these objectives, DoT and MPO decision makers have other less quantifiable and more politics-driven objectives, like satisfying local or state politicians.

$$\begin{aligned} \phi_{eng,1} &= \max_{u \in U_{eng}} \left(y_{eng,1}(k)\right) \quad \text{when} \quad k = T \\ \phi_{eng,2} &= \max_{u \in U_{eng}} \left(y_{eng,2}(k)\right) \quad \text{when} \quad k = T \\ \phi_{eng,3} &= \min\left(\sum_{k=0}^{T} \frac{u_{eng,1}(k)}{(1+r\%)^k}\right) \end{aligned} \qquad (7.22a)$$

$$\begin{aligned} & s.t. \\ & 0 \le s_{eng,1}(k) \le 9 \\ & u_{eng,1}(k) \le B_k \\ & u_{eng,2}(k) = 0 \ \text{or} \ u_{eng,2}(k) = 1 \\ & u_{eng,3}(k) = 0 \ \text{or} \ u_{eng,3}(k) = 1 \end{aligned} \qquad (7.22b)$$

7.III.6.3.3 Constructing the Social Submodel

From the perspective of noncommercial bridge users (i.e. the social modeling perspective), untimely and insufficient bridge maintenance is of concern only to the extent that it affects the ability of individuals to travel over the bridge of interest. The inability of noncommercial travelers to travel over the bridge can result in many ripple effects, including effects on workforce commute and earnings and changes in real estate values, among others. There are many factors responsible for determining travel patterns of noncommercial bridge users, among which are natural changes in the general population, impact of regional economy on work and nonwork car trips, and the impact of travel time and cost of travel (Figure 7.III.2).

Detailed quantitative and qualitative approaches have been developed to study the effects of some of these factors on travel patterns (see NCHRP (2001) for details). However, for the purpose of this case study, we focus on quantifying only the total number of noncommercial bridge users and the average travel time cost. The total number of noncommercial bridge users serves as a surrogate measure for social accessibility.

We again adopt an LTI representation that allows us, over time, to compare the anticipated changes in noncommercial traffic with anticipated changes in the structural condition of the bridge caused by selected maintenance decisions. A single essential state variable represents the state of the system at any point in time, namely, the average annual daily noncommercial traffic that can safely cross the bridge. This number depends on the size and growth patterns of the overall population, and it is impacted by decisions made in the social subsystem, which are implicitly accounted for in the social submodel (e.g. decision to travel for work or nonwork reasons, decision to utilize the HRBT) as well as the decisions made in the engineering subsystem (e.g. speed postings, weight postings). Our LTI representation of the social submodel is given in Eq. (7.23):

$$\begin{aligned} s_{soc,1}(k+1) &= a_{soc} \cdot s_{soc,1}(k) - b_{soc} \cdot u_{soc,1}(k) \cdot s_{soc,1}(k) + \omega_{soc}(k) \\ y_{soc,1}(k) &= E[s_{soc,1}(k)] \end{aligned} \qquad (7.23)$$

where

- k is the discrete assessment point ($\Delta k = 2$ years)
- $s_{soc,1}(k)$ is the normally distributed average annual daily noncommercial traffic over the

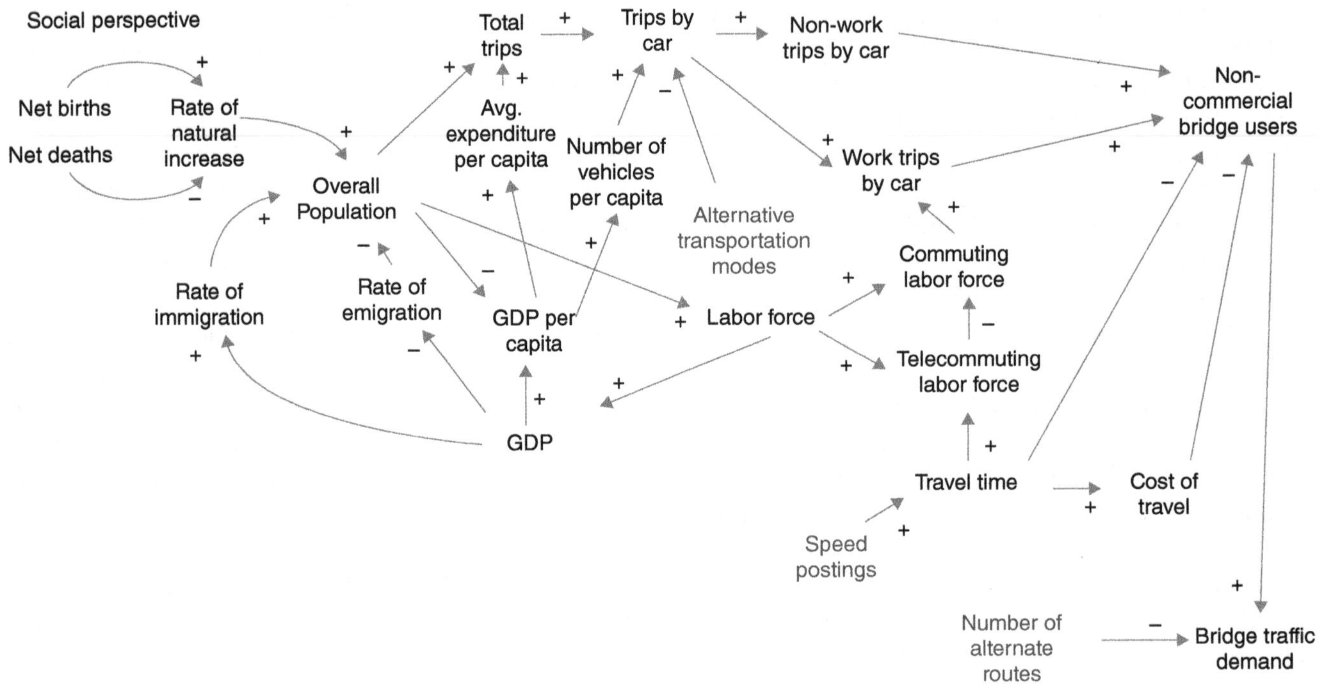

Figure 7.III.2 SD diagram for the social perspective.

bridge (we are making an assumption that each trip is conducted by a single individual; thus this state variable reflects the size of the population that is dependent on the bridge for work-commuting, social, cultural, and personal activities)

- $u_{soc,\ 1}(k)$ is the input into the model from the engineering perspective that indicates a speed posting flag (0 if no speed posting was introduced in period k, or 1 if a speed posting was put into place in period k)
- $\omega_{soc}(k)$ is the normally distributed white noise with zero mean
- $y_{soc,\ 1}(k)$ is the mean expected average annual daily noncommercial traffic over HRBT

The goal of the stakeholders representing the social subsystem is to perform their daily work and nonwork (discretionary) activities with as few disruptions as possible and at a lowest possible cost. Hence, from the social perspective, stakeholders wish to maximize the average annual daily noncommercial traffic over the bridge at any period of time (as this number serves as a surrogate measure of the ability of individuals to conduct their daily activities), as well as minimize the average total travel cost per noncommercial traveler (TC_{soc}). We assume that the decisions made by the noncommercial stakeholders are implicitly accounted for in the social submodel. These decisions refer to making a choice to conduct or not conduct the trip and making the choice to conduct or not conduct the trip over the HRBT versus an alternative. Hence the number of alternative route factor indicated in Figure 7.III.2 is not explicitly accounted for in this simplified model, but is assumed to be an implicit factor in the decision-making process. The objectives of the stakeholders from the social perspective are expressed in Eq. (7.24):

$$\phi_{soc,1} = \max_{u \in U_{soc}} (y_{soc,1}(k)) \quad \forall k = 0, \ldots, T$$

$$\phi_{soc,2} = \min_{u \in U_{soc}} \left(\sum_{k=0}^{T} TC_{soc}(k) \right) \tag{7.24a}$$

s.t.

$$TC_{soc} = TTC_{soc} + SIMC_{soc} \tag{7.24b}$$

$$TTC_{soc} = \frac{100}{AES} \cdot TTVAL_{soc}$$

We represent the average total cost per noncommercial traveler (in \$ per thousand vehicle miles). TTC_{soc} represents the average travel time cost (in \$ per thousand vehicle miles). It is a metric commonly used in the FHWA's Highway Economic Requirements System State Version (HERST-ST). Although the HERS-ST model has many component costs that are used to estimate TTC_{soc}, we use a simplified version of this model to obtain a general estimate of direct costs to the traveler. AES represents the average effective speed of a vehicle on the bridge. We assume that the introduction of a speed limit slightly reduces the average effective speed of the traffic on the bridge and thus increases the TTC_{soc}. For illustrative purpose in this example, we assume that the introduction of a speed limit reduces the overall speed flow by one mile per hour. $TTVAL_{soc}$ represents the average value of time (in \$) per occupant of a noncommercial vehicle.

7.III.6.3.4 Constructing the Economic Submodel

From the perspective of commercial bridge Complex SoS users (i.e. the economic modeling perspective), untimely and insufficient bridge maintenance is of concern only to the extent that it affects the ability of freight carriers to travel over the bridge of interest. The inability of commercial travelers to use the bridge can cause disruptions in manufacturing production schedules and sales patterns. There are many factors responsible for determining travel patterns of commercial bridge users, among which are changes in the size of the regional industry and legally allowable weight loads (Figure 7.III.3). For the purpose of this example, we focus on quantifying the total number of commercial travelers, the average annual daily commodity load transported, and the average travel time cost. The total number of commercial travelers serves as a surrogate measure for economic accessibility.

In our LTI representation, we identified two essential state variables, namely, the average annual daily commercial traffic that can safely use the bridge and the average annual daily commodity load transported over the bridge. The LTI representation of the economic subsystem is given in Eq. (7.25). Similar assumptions hold for this model as for the social model, and as in the social model, some of the factors identified in Figure 7.III.3, including the number of alternate routes, are implicitly modeled as a part of the decision-making process.

$$s_{\text{econ},1}(k+1) = a_{\text{econ}} \cdot s_{\text{econ},1}(k) + \omega_{\text{econ},1}(k)$$

$$s_{\text{econ},2}(k+1) = b_{\text{econ}} \cdot (1 - c_{\text{econ}} \cdot u_{\text{econ},1}(k)) \cdot s_{\text{econ},1}(k+1) + \omega_{\text{econ},2}(k)$$

$$y_{\text{econ},1}(k) = E[s_{\text{econ},1}(k)] \tag{7.25}$$

where

- K is the discrete assessment point ($\Delta k = 2$ years)
- $s_{\text{econ},\ 1}(k)$ is the normally distributed average annual daily commercial traffic over the bridge

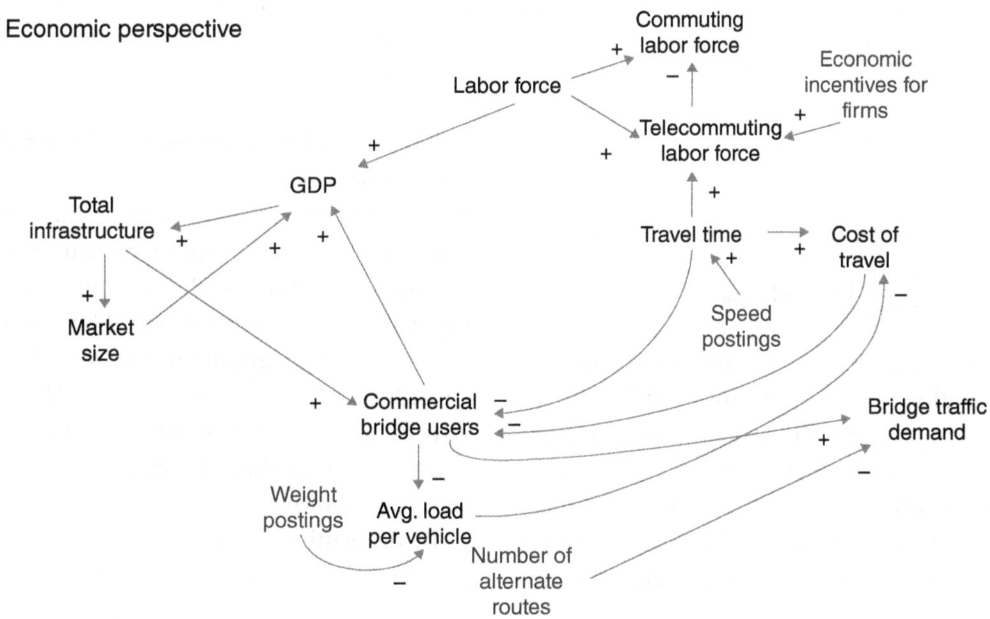

Figure 7.III.3 *SD diagram for the economic perspective.*

- $s_{\text{econ},\,2}(k)$ is the average annual daily commodity load transported over the bridge
- $u_{\text{econ},\,1}(k)$ is the input into the model that indicates a weight posting flag (0 if no weight posting was introduced in period k, or 1 if a weight posting was put into place in period k)
- $\omega_{\text{econ},\,i}(k)$ is the normally distributed white noise with zero mean
- $y_{\text{econ},\,1}(k)$ is the mean expected average annual daily commercial traffic over HRBT

The goal of the stakeholders representing the economic subsystem is to perform their scheduled commercial activities (i.e. delivery of raw and intermediate supplies and finished goods, etc.) with as few disruptions as possible and at the lowest possible cost. Hence, from the economic perspective, stakeholders wish to maximize the average annual daily commercial traffic over the bridge at any period of time (as this number serves as a surrogate measure of the ability of companies, industrial or service sectors, to conduct their business), as well as minimize the average total travel cost per commercial traveler (TC_{econ}). Similar assumptions hold for this model as for the social model. The objectives of the stakeholders from the economic perspective are expressed in Eq. (7.26):

$$\phi_{\text{econ},1} = \max_{u \in U_{\text{econ}}} \left(y_{\text{econ},1}(k)\right) \ \forall k = 0, \ldots, T$$
$$\phi_{\text{econ},2} = \min_{u \in U_{\text{econ}}} \left(\sum_{k=0}^{T} TC_{\text{econ}}(k)\right) \tag{7.26a}$$

$$s.t.$$
$$\text{TC}_{\text{econ}} = \text{TTC}_{\text{econ}} + \text{SIMC}_{\text{econ}}$$
$$\text{TTC}_{\text{econ}} = \frac{1000}{\text{AES}} \cdot \text{TTVAL}_{\text{econ}} \tag{7.26b}$$

In Eq. (7.26), TC_{econ} represents the total cost per commercial traveler (in \$ per thousand vehicle miles). TTC_{econ} represents the average travel time cost (in \$ per thousand vehicle miles). AES represents the average effective speed of a vehicle on the bridge. $\text{TTVAL}_{\text{econ}}$ represents the average value of time (in \$) per occupant of a commercial vehicle. The US DoT–HERS-ST (2005) provides values of 1 h of travel time for commercial and noncommercial users traveling in different types of vehicles (in 1995 \$). We assume that a five-axle Comb truck can serve as a representative commercial vehicle for the HRBT, and we use for the average value of 1 hour of travel time per vehicle $\text{TTVAL}_{\text{econ}}$ = \$37.95 (1995 values from US DoT–HERS-ST (2005) were adjusted for inflation).

In Eq. (7.26a), parameters a_{econ} and b_{econ} are estimated from a regional freight study conducted by the Hampton Roads Planning District Commission (2007). This report estimated the freight traffic growth at 4.88% per year, based on tonnage. We assume that the yearly growth rate of freight traffic over the bridge of interest is the same as the growth rate for all commercial traffic passing through the area, hence assuming no compounding effects; for a 2-year period this growth rate is 9.76%. Then, $a_{\text{econ}} = 1.0976$. Parameter $b_{\text{econ}} = 66\,055$ represents the average commodity load (in pounds) transported on a single commercial trip, and it is obtained from the same study.

We assume that a weight posting placed on the HRBT will decrease the average load that can be transported by a single truck. We introduce a reduction in the average commodity load transported on a single commercial trip at time $k+1$ if a weight posting is put in place at time k. For illustrative purposes we set $c_{\text{econ}} = 0.25\%$, indicating that when a weight posting is introduced, the average commodity load transported on a single commercial trip is reduced by 0.25%.

7.III.6.3.5 Harmonizing the Three Submodels into a Metamodel

The final model integrates the three modeling perspectives into a single metamodel that provides us with a more holistic view of the problem. Figure 7.III.4 shows an SD diagram that represents the integrated engineering, social, and economic considerations. Figures 7.III.1–7.III.3 are integrated into Figure 7.III.4 through the introduction of a state variable that is shared among all three perspectives, described below.

In determining what connects the three modeling perspectives, we note that the number of commercial

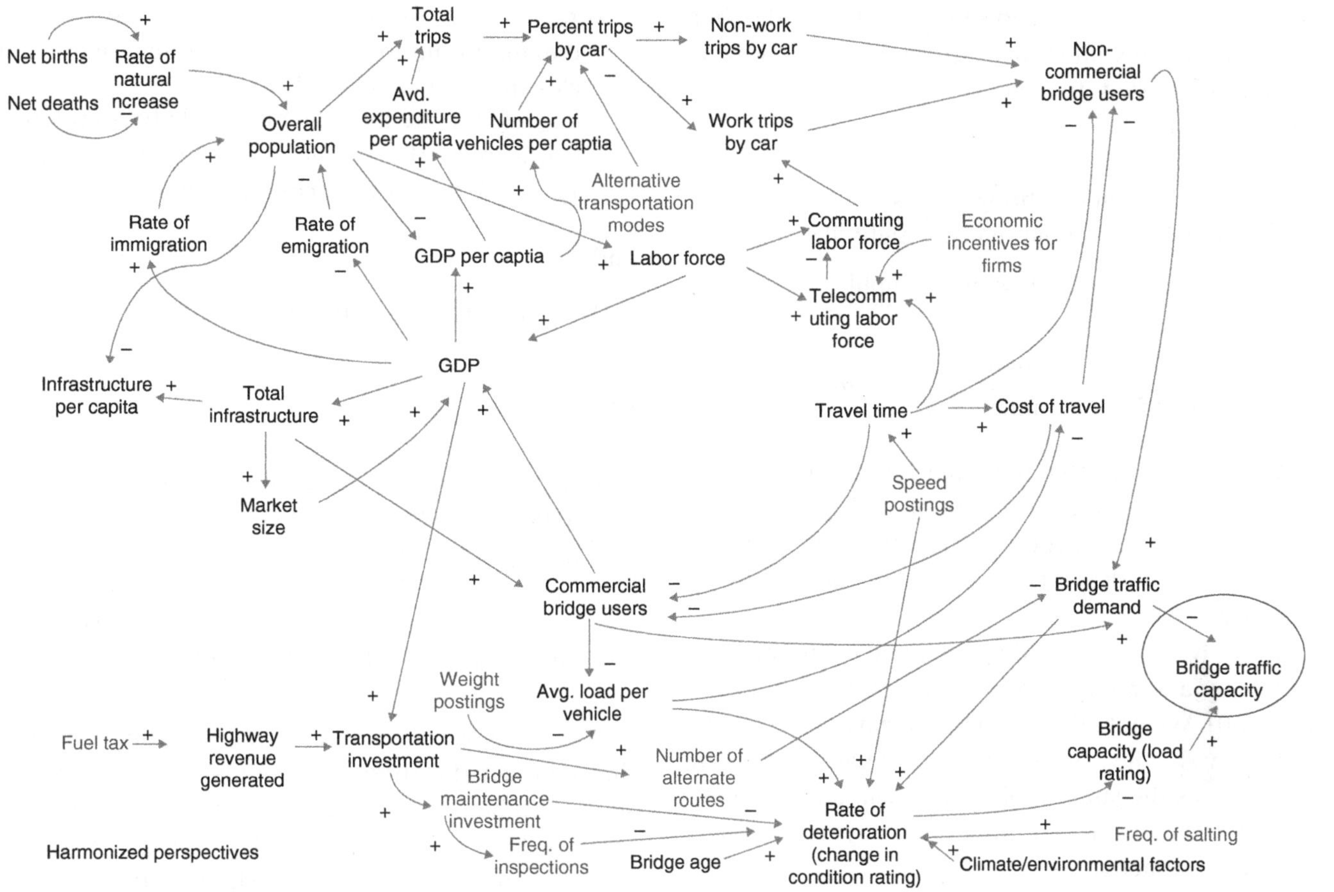

Figure 7.III.4 *SD diagram for integrated considerations. The shared state variable connecting the three modeling perspectives (bridge traffic capacity) is circled.*

and noncommercial users together represents the total traffic demand placed on the bridge. From the engineering perspective the bridge load capacity determines the maximum live load that a bridge can safely carry. Hence, we introduce a physical measure of the number of vehicles that can utilize the bridge during a specific period of time, which we term bridge traffic capacity. This variable describes the state of the overall system at some point in time, and it relates the number of commercial and noncommercial users of the bridge to the structural characteristics of the bridge and its load rating. This shared state variable essentially gives an indication of the additional traffic capacity that may be available on the bridge at a certain period of time, given current and projected commercial and noncommercial demand, and physical changes of the bridge structure. In order to maintain an adequate level of service for the HRBT over the entire planning horizon, the stakeholders must ensure that the additional traffic capacity is not negative. Thus, they must investigate this additional capacity for each of the five maintenance alternatives.

We harmonize the three submodels into a single model by introducing coupling functions for each subsystem. These coupling functions link bridge traffic capacity to a subset of the unshared state variables in each subsystem. For example, in the engineering submodel, we link bridge traffic capacity to bridge load rating (Eq. (7.27)):

$$x_{\text{eng},3}(k+1) = d_{\text{eng}} \cdot x_{\text{eng},2}(k+1) \tag{7.27}$$

In this coupling function, parameter $d_{\text{eng}} = 5.036$ converts bridge load carrying capacity ($x_{\text{eng},2}$) to the number of vehicles that can safely use the bridge in a given day. In this illustrative example, the following formula was used to calculate this parameter for HRBT (Eq. (7.28)):

$$d_{\text{eng}} = \frac{\text{number bridge spans}}{\text{weighted average vehicle weight [lbs]}} \cdot \frac{\text{average traffic flow speed [mph]}}{\text{length of bridge [miles]}} \cdot 24\text{ h} \quad (7.28)$$
$$= \frac{123}{5861.65\text{ lbs}} \cdot \frac{35\text{ mph}}{3.5\text{ miles}} \cdot 24\text{ h} = 5.036\text{ lbs}^{-1}$$

We acknowledge that this equation is simplistic and does not account for distribution of loads per axle of a moving vehicle; however, we argue that it is sufficient for illustrating the effectiveness of the modeling approach.

Similarly we augment the social and economic submodels by adding a coupling function to Eqs. (7.29) and (7.30). We note that from the social and economic perspectives, the bridge traffic capacity can be expressed as the sum of all commercial and noncommercial travelers, in addition to the yet unutilized additional capacity:

$$x_{\text{soc},2}(k+1) = x_{\text{soc},1}(k+1) + u_{\text{soc},2}(k+1) + \text{additional capacity} \quad (7.29)$$

where

- $x_{\text{soc},\,2}(k)$ is the bridge traffic capacity (i.e. the maximum number of average annual daily vehicles that can cross the bridge)
- $u_{\text{soc},\,2}(k)$ is the average annual daily commercial traffic over the bridge (this is a state variable $x_{\text{econ},\,1}(k+1)$ in the economic submodel)

and

$$x_{\text{econ},3}(k+1) = x_{\text{econ},1}(k+1) + u_{\text{econ},2}(k+1) + \text{additional capacity} \quad (7.30)$$

where

- $x_{\text{econ},\,3}(k)$ is the bridge traffic capacity (i.e. the maximum number of average annual daily vehicles that can cross the bridge)
- $u_{\text{econ},\,2}(k)$ is the average annual daily noncommercial traffic over the bridge (this is a state variable $x_{\text{soc},\,1}(k+1)$ in the social submodel)

Now that we have introduced coupling functions that connect the three submodels through a shared state variable, we can develop a single harmonizing metamodel (Eq. (7.31)), in which $\underline{x}_{\text{eng}} \cap \underline{x}_{\text{soc}} \cap \underline{x}_{\text{econ}} \neq \emptyset$ indicates the existence of a shared state variable (where $\underline{x}_i$ indicates a vector of all essential state variables of subsystem i):

$$\text{Metamodel} = \cup(\text{engineering, social, economic perspectives})$$

$$\text{Metamodel} = \cup \left(\begin{array}{l} \begin{cases} s_{\text{eng},1}(k+1) = 0.97 \cdot s_{\text{eng},1}(k) + 10^{-7} \cdot u_{\text{eng},1}(k) + \omega(k) \\ s_{\text{eng},2}(k+1) = 4.37 \cdot s_{\text{eng},1}(k+1) \\ s_{\text{eng},3}(k+1) = 5.037 \cdot s_{\text{eng},2}(k+1) \end{cases} \\ \begin{cases} s_{\text{soc},1}(k+1) = \left[1.0152 - 2.5 \cdot 10^{-5} \cdot u_{\text{soc},1}(k)\right] \cdot s_{\text{soc},1}(k) \\ s_{\text{soc},2}(k+1) = s_{\text{soc},1}(k+1) + u_{\text{soc},2}(k+1) + \text{additional capacity} \end{cases} \\ \begin{cases} s_{\text{econ},1}(k+1) = 1.0976 \cdot s_{\text{econ},1}(k) \\ s_{\text{econ},2}(k+1) = 66\,055 \cdot \left(1 - 2.5 \cdot 10^{-5}\right) \cdot u_{\text{econ},1}(k) \cdot s_{\text{econ},1}(k+1) \\ s_{\text{econ},3}(k+1) = s_{\text{econ},1}(k+1) + u_{\text{econ},2}(k+1) + \text{additional capacity} \end{cases} \end{array} \right) \quad (7.31)$$

such that $x_{eng,\ 3}(k) = x_{soc,\ 2}(k) = x_{econ,\ 3}(k) \quad \forall k = 0, \ldots, T-1$ represents the shared state variable.

7.III.6.3.6 Metamodel Coordination

The decentralized decision-making process that takes place within a bridge system, or to other infrastructure Complex SoS often leads to global solutions that are not acceptable to all stakeholders. Since decisions made in interdependent subsystems affect shared states, an uncoordinated decision-making process could result in levels of shared states that are not acceptable to all involved parties over a longer planning horizon.

We define coordination related to our modeling process as the iterative process of manipulating adjustable decisions and constraints to respond to changes in shared states, with the purpose of achieving more satisficing global objectives.

In the bridge maintenance decision making, decentralized stakeholders represent the three subsystems (engineering, social, economic) generally make decisions without considering how their decisions might impact other interdependent subsystems. As an example, we consider which of the maintenance policies is most acceptable to DoT and MPO decision makers, assuming that they have no knowledge of the subsystem interdependencies. In this case, policy 1 (end-loaded policy in which maintenance investment is not made until the end of the planning horizon) is most preferred, because it results in highest end-of-planning-period mean and conditional superstructure condition and load ratings, and the lowest NPV (Figure 7.III.5).

If we now consider that DoT and MPO decision makers have knowledge of the shared states (bridge traffic capacity) and thus have some insight into the subsystem interdependencies, then we can assume that in addition to the end-of-planning-horizon bridge condition and load ratings, they might also be interested in two other performance measures: (i) consistency of condition ratings over the entire planning horizon (consistency is defined as the ratio between the maximum and minimum superstructure condition rating achievable over the planning horizon under a specific policy (Figure 7.III.6)) and (ii) available additional traffic capacity resulting from a specific maintenance policy, aggregated over the entire planning horizon (Figure 7.III.7).

Assuming no speed or weight limitation postings have been issued for the HRBT, we assess the values of the shared states (bridge traffic capacity) resulting from the five maintenance policies at each period k, where $k = 0, 1, \ldots, T$. At each period k the bridge traffic capacity is compared to the traffic demand, and an additional traffic capacity value is recorded. A positive value indicates that the bridge is able to support additional traffic. A negative capacity indicates that the bridge is unable to support projected demand.

Recall that the objectives of the stakeholders from the social and economic perspectives are to maximize

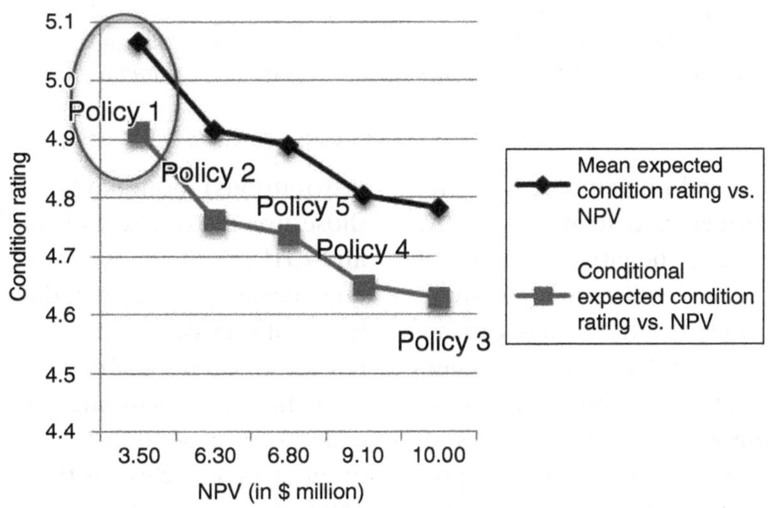

Figure 7.III.5 End-of-planning-horizon superstructure condition rating versus NPV for five maintenance policies (discounted at 10%). Conditional expected superstructure condition rating is calculated as mean +1.525σ (Haimes, 2016. Reproduced with the permission of John Wiley & Sons).

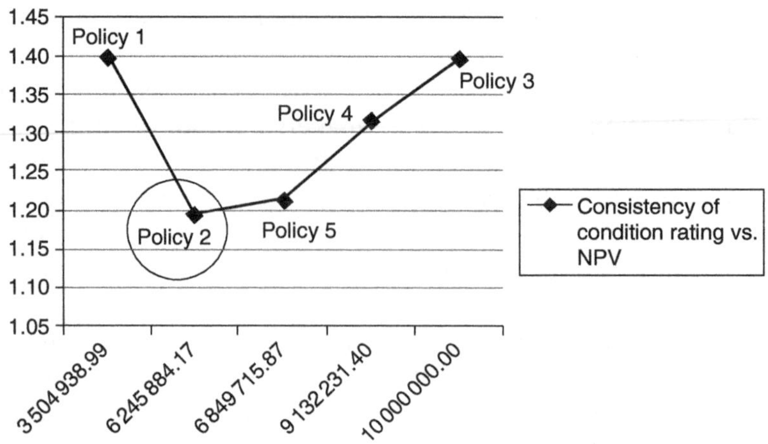

Figure 7.III.6 *Consistency of superstructure condition rating versus NPV for the five maintenance policies. Consistency is calculated as the ratio between the maximum and minimum superstructure condition rating achievable under a certain policy over the entire planning time frame. We prefer values closer to 1.*

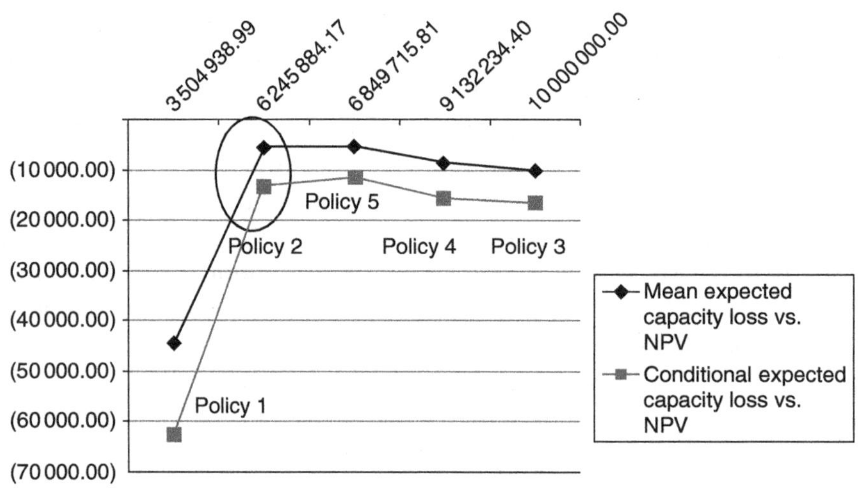

Figure 7.III.7 *Aggregate values of the loss of bridge traffic capacity at period* k = *T. Bridge traffic capacity is compared at each* k *to bridge demand, and additional capacity is recorded. This graph shows the sum of the additional bridge traffic capacity. We see that all five maintenance policies result in a lack of traffic capacity over the entire planning time frame.*

the number of commercial and noncommercial vehicles traveling over the bridge and to minimize the overall cost of travel. Since negative values for expected bridge traffic capacity will reduce the number of vehicles that can safely use the bridge causing delays and increasing TTC_{soc} and TTC_{econ}, commercial and noncommercial users would prefer a maintenance policy that minimizes the potential loss of capacity and that maximizes the consistency of provided service. In this example, policy 2 (equal annual distributions) dominates all others with regard to those two measures, as is indicated in Figures 7.III.6 and 7.III.7. Policy 1, while preferred from a traditional engineering perspective, does not result in consistent levels of service over the entire planning horizon and results in the largest expected loss of traffic capacity over the entire planning horizon. For this illustrative example, we assume that the average annual daily number of vehicles on the bridge at $k = 0$ is 90 000, of which $x_{soc,1}(0) = 87\,300$ (97%) are cars and personal vehicles (noncommercial) and $x_{\text{econ},1}(0) = 2700$ (3%) are commercial trucks. The regional freight study conducted by the Hampton Roads Planning District Commission (2007) suggests that these values are realistic in the urban area of interest.

When considering results from Figures 7.III.5, 7.III.6, and 7.III.7, we note that there is no maintenance policy that seems optimal from all three modeling perspectives. However, intrinsic metamodel can be used to guide the coordination of decentralized, yet interdependent decisions made by the different stakeholders. Because of the coupling functions linking the shared states (bridge traffic capacity) to other essential variables of each submodel, stakeholders from the three subsystems can more easily identify those decisions and constraints that might be adjusted either at the individual subsystem level or through a collaborative effort in order to reach a range of values of the bridge traffic capacity that would be more acceptable to all stakeholders over the entire planning horizon while still satisficing subsystem objectives.

DoT and MPO decision makers understand that their most preferred policy may not be acceptable from a social and economic perspective. They understand that a more proactive maintenance plan, as proposed in policy 2, is expected to produce more consistent superstructure condition and load ratings, thus ensuring a more consistent and adequate level of service over the entire planning horizon. However, when considered from a cost perspective, the NPV of policy 2 is nearly double that of policy 1; hence policy 2 will not be sustainable if additional innovative funding sources are not supplied.

To reduce the traffic demand on a deteriorated bridge and address the issue of insufficient capacity caused by their preferred policy 1, the DoT and MPO decision makers could put into place speed or weight postings. Speed or weight postings would increase the travel costs for commercial and noncommercial travelers, making policy 1 even less attractive to stakeholders from the social and economic perspectives. In order to make policy 1 more acceptable from the social and economic perspectives, representative stakeholders could consider options that would reduce the number of commercial and noncommercial bridge users, including alternative transportation modes (e.g. buses) and telecommuting work options.

To make policy 2 more acceptable from the engineering perspective, stakeholders must consider how to implement innovative funding mechanisms to raise the required additional investments. Both commercial and noncommercial users could consider participating in shared infrastructure maintenance costs (SIMC_{soc} and $\text{SIMC}_{\text{econ}}$) by paying increased fuel taxes, usage tolls, and/or fees for transporting excessive weights over the bridge. While the participation in the shared infrastructure maintenance costs would increase the total cost per traveler, stakeholders could assess tradeoffs between increased SIMC and decreased TTC. For example, while the current level of maintenance investments made in policy 2 are insufficient to maintain an excess bridge traffic capacity, an increase of \$367K at each period k ensures the existence of the excess traffic capacity at each period k. If we naively assume that the \$367K could be equally divided among all commercial and noncommercial bridge users at each stage k, then we see that a fairly small investment on behalf of infrastructure users (~7.4% of TTC_{soc} or 4.6% of TTC_{econ} every 2 years) completely solves the bridge traffic capacity problem while resulting in a sustainable and proactive bridge maintenance plan. On the other hand, assume that policy 1 is implemented, resulting in a speed and a weight posting. Based on the assumptions of our metamodel, such a policy could increase the TTC_{soc} and TTC_{econ} by approximately 3% for every 1000 miles traveled. Stakeholders representing the social and economic perspectives could then evaluate tradeoffs between these alternatives to determine which option results in more acceptable long-term system of systems solutions.

Every time a new effort is made to adjust a local decision/constraint, new Pareto-optimal curves must be computed for each subsystem representing the suggested changes, and based on those local optimization results, new projected ranges of shared states must be reevaluated. This process continues until a solution is found that satisfices local subsystem objectives while also resulting in levels of shared states that are acceptable to all stakeholders over a longer planning horizon. A failure of subsystem stakeholders to agree on a globally acceptable range of values of shared states could result in local subsystem difficulties that could, over time, propagate and become detrimental to the entire system.

7.III.7 CONCLUSION

The submodels and assumptions presented in this case study provide a first step in understanding the potential value of metamodeling through shared states. The implemented modeling process encourages the different stakeholders to identify, quantify, and explicitly model some common and conflicting interests and needs, and it provides guidance about the extent to which the knowledge of shared states/objectives/interests can help in making decisions in a decentralized environment characterized by multiple and conflicting stakeholders and objectives.

In particular, the metamodeling process enables DoT and MPO decision makers to get a macro-level estimate of the potential socioeconomic impacts of their maintenance alternatives, estimates that can then be assessed in more detail through more specific models. Hence, their ultimate maintenance choices are informed and guided by considerations of how the maintenance decisions made today impact the choice of future maintenance options and how they affect other system stakeholders. Based on this knowledge, other system stakeholders, including commercial and noncommercial users of the infrastructure, can evaluate the impacts that infrastructure maintenance decisions have on them, given their changing needs, demands, and constraints. Maintenance plans can be collaboratively reevaluated by considering, among other things, whether (i) other less damaging alternatives exist, (ii) innovative funding mechanisms (partially supported by infrastructure users) could be designed to support alternative projects, and (iii) consequences to infrastructure users could be temporarily assuaged (e.g. by utilizing more alternate routes or different public modes of transportation). This type of a collaborative, forward-looking management approach maintains and potentially increases, within a limited budget, the number of available maintenance options in the future, thus transforming an operational, short-term approach into a more strategic, long-term, and collaborative approach.

REFERENCES

Andrijcic, E. (2016). Intrinsic meta-modeling of systems of systems: a study in bridge infrastructure management. Ph.D. dissertation. University of Virginia.

Andrijcic, E. and Haimes, Y.Y. (2017). Metamodeling of interdependent systems: application to bridge infrastructure management. *Journal of Infrastructure Systems* **23**(2): 04016028. doi: 10.1061/(ASCE)IS.1943-555X.0000322.

Andrijcic, E., Haimes, Y.Y., and Beatley, T. (2013). Public policy implications of harmonizing engineering technology with socio-economic modeling: application to transportation infrastructure management. *Transportation Research Part A: Policy and Practice* **50**: 62–73.

Ben-Haim, Y. (2012). Doing our best: optimization and the management of risk. *Risk Analysis* **32**(8): 1326–1332.

Chase, S. and Gáspár, L. (2000). Modeling the reduction in load capacity of highway bridges with age. *Journal of Bridge Engineering* **5**(4): 331–336.

Chase, S., Small, E., and Nutakor, C. (1999). An in-depth analysis of the national bridge inventory database utilizing data mining, GIS and advanced statistical methods. *TRB Transportation Research Circular* **498**: 1–17.

Chen, C. (2012). *Linear System Theory and Design*, 4e. New York: Oxford University Press.

Dunker, F.K. and Rabbat, B.G. (1995). Assessing infrastructure deficiencies: the case of highway bridges. *ASCE Journal of Infrastructure Systems* **1**(2): 100–119.

FHWA (2011). *Bridge Preservation Guide: Maintaining a State of Good Repair Using Cost Effective Investment Strategies*. Washington, DC: FHWA Publication Number: FHWA-HIF-11042.

Friedland, I.M., Ghasemi, H., and Chase, S.B. (2007). *The FHWA Long-Term Bridge Performance Program*. McLean, VA: Federal Highway Administration, Turner-Fairbank Highway Research Center.

Fwa, T.F. and Sinha, K.C. (1987). An Aggregate Damage Approach for Highway Pavement Performance Analysis. *Tech. Rep*. FHWA/JHRP-87-15, Joint Highway Research Project, Purdue University, West Lafayette, IN.

Gawande, A. (2011). *The Checklist Manifesto: How to Get Things Right*. New York: Henry Holt and Company.

Guo, Z. (2015). A systematic approach to risk analysis of infrastructure systems of systems. Ph.D. dissertation. University of Virginia.

Guo, Z. and Haimes, Y.Y. (2016). Risk assessment of infrastructure system of systems with precursor analysis. *Risk Analysis* **36**(8): 1630–1643. doi: 10.1111/risa.12559.

Guo, Z. and Haimes, Y.Y. (2017). Exploring systemic risks in systems-of-systems within a multiobjective decision framework. *IEEE-SMC Systems* **47**(6): 906–915.

Haimes, Y.Y. (1981). Hierarchical holographic modeling. *IEEE Transactions on Systems, Man, and Cybernetics* **11**(9): 606–617.

Haimes, Y.Y. (1991). Total risk management. *Risk Analysis* **11**(2): 169–171.

Haimes, Y.Y. (2007). Phantom system models for emergent multiscale systems. *Journal of Infrastructure Systems* **13**(2): 81–87.

Haimes, Y.Y. (2012a). Modeling complex systems of systems with phantom system models. *Systems Engineering* **15**(3): 333–346.

Haimes, Y.Y. (2012b). Systems-based guiding principles for risk modeling, planning, assessment, management and communication. *Risk Analysis* **32**(9): 1451–1467.

Haimes, Y.Y. (2016). *Risk Modeling, Assessment, and Management*, 4e. Hoboken, NJ: Wiley.

Haimes, Y.Y. and Jiang, P. (2001). Leontief-based model of risk in complex interconnected infrastructures. *ACSE Journal of Infrastructure Systems* **7**(1): 1–12. 111–117.

Hampton Roads Planning District Commission (2007). Intermodal management system regional freight study. Hampton Roads, VA. http://www.hrtpo.org/uploads/docs/IMS2007%20Final%20Report_reduced.pdf Accessed 30 May 2016.

Kaplan, S. and Garrick, B.J. (1981). On the quantitative definition of risk. *Risk Analysis* **1**(1): 11–27.

Kaplan, S., Haimes, Y.Y., and Garrick, B. (2001). Fitting hierarchical holographic modeling into the theory of scenario structuring and a resulting refinement of the quantitative definition of risk. *Risk Analysis* **21**(5): 807–815.

Labi, S. and Sinha, K. (2003). The Effectiveness of Maintenance and Its Impact on Capital Expenditures. *Tech. Rep*. FHWA/IN/JTRP-2002/27. Joint Transportation Research Program, West Lafayette, IN.

Leontief, W.W. (1951a). Input/output economics. *Scientific American* **185**(4): 15–21.

Leontief, W.W. (1951b). *The Structure of the American Economy, 1919–1939*, 2e. New York: Oxford University Press.

Lian, C. and Haimes, Y.Y. (2006). Managing the risk of terrorism to interdependent infrastructure systems through the dynamic inoperability input-output model. *Systems Engineering* **9**: 241–258. doi: 10.1002/sys.20051.

Lundberg, J., Rollenhagen, C., and Hollnagel, E. (2009). What-you-look-for-is-what-you-find-the consequences of underlying accident models in eight accident investigation manuals. *Safety Science* **47**: 297–1311.

Madanat, S., Bulusu, S., and Mahmoud, A. (1995). Estimation of infrastructure distress initiation and progression models. *ASCE Journal of Infrastructure Systems* **1**(3): 146–150.

Mouaket, I.M. and Sinha, K.C. (1990). Cost Effectiveness of Rigid and Composite Highway Pavement Routine Maintenance. *Tech. Rep. FHWA/JHRP-90-15*, Joint Highway Research Project, Purdue University, West Lafayette, IN.

National Council on Public Works Improvement (1988). *Fragile Foundations: A Report on America's Public Works*. Washington, DC: Government Printing Office.

National Cooperative Highway Research Program (NCHRP) (2001). Guidebook for Assessing the Social and Economic Effects of Transportation Projects. *NCHRP Rep*. 456.

OECD (2008). International Transport Forum. Joint Transport Research Centre Discussion Paper No. 2008-6. Round Table, 25–26 October 2007, Boston. http://www.internationaltransportforum.org/jtrc/discussionpapers/DP200806.pdf.

Phares, B.M., Washer, G., Rolander, D. et al. (2004). Routine highway bridge inspection condition documentation accuracy and reliability. *Journal of Bridge Engineering* **9**(4): 403–413.

Press, W.H., Flannery, B.P., Teukolsky, S.A., and Vetterling, W.T. (1992). Runge-Kutta method. In: *Numerical Recipes. FORTRAN: The Art of Scientific Computing*, 704–716. Cambridge: Cambridge University Press.

Ramaswamy, R. and Ben-Akiva, M. (1990). *Estimation of Highway Pavement Deterioration from In-Service Pavements Data*, Transportation Research Record, vol. **1272**. Washington, DC: Transportation Research Board, National Research Council.

Reason, J. (1990). The contribution of latent human failures to the breakdown of complex systems. *Philosophical Transactions of the Royal Society of London Series B Biological Sciences* **327**(1241): 475–484.

Simon, H.A. (1956). Rational choice and the structure of the environment. *Psychological Review* **63**(2): 129–138.

U.S. Department of Transportation (US DoT) – Federal Highway Administration. (2005). Highway Economic Requirements.

Wang, W., Wang, J., and Kim, M. (2001). An algebraic condition for the separation of two ellipsoids. *Computer Aided Geometric Design* **18**(6): 531–539.

8

Modeling Economic Interdependencies among Complex Systems of Systems

OVERVIEW

Electricity, communications, and water are three safety-critical sectors of every country's economy and of its population's well-being. They literarily constitute the lifeline of every modern community that transcends cultural, societal, and political borders. Together, they comprise interdependent and interconnected emergent complex systems of systems (SoS) (Complex SoS). Furthermore, their interdependencies and interconnectedness (I-I) represent a significant universal source of risk. To model the impacts of their disruption to the regional and national economy, we select the supply chain, which constitutes a Complex SoS, and to which we have devoted Chapter 11 in its entirety to this critical global sector of the economy. Furthermore, we will introduce (via four case studies in Chapter 10) the I-I that exist among electricity, communications, and water Complex SoS. Note, however, that in Chapter 10, we quantified the reinvention of fault-tree analyses via their I-I by building on OR Gates (systems connected in series) and on AND Gates (systems connected in parallel). In this chapter we quantify the consequences resulting from the I-I of all sectors of a country's economy via the inoperability input–output model (IIM) (Haimes and Jiang, 2001; Haimes et al. 2005a, b; Haimes, 2016). The IIM builds on and extends the input–output model developed by the Nobel laureate Wassily Leontief (1951a, b). Applications of the IIM and its extensions with case studies are presented in Chapter 14.

8.1 INOPERABILITY INPUT–OUTPUT MODEL (IIM)

We start with electricity, as an electric power utility, which in its essence is a complex interdependent and interconnected indispensible sector of every economy of the world. This sector is a large-scale, hierarchical, and interconnected Complex SoS. At the US national level, it consists of three interconnected networks: the (i) Eastern Interconnected System, covering the eastern two-thirds of the United States; (ii) Western Interconnected System, covering the Southwest and areas west of the Rocky Mountains; and (iii) Texas Interconnected System, consisting mainly of Texas. Indeed, each constitutes Complex SoS, and, certainly at the network level,

Modeling and Managing Interdependent Complex Systems of Systems, First Edition. Yacov Y. Haimes.

each network, as its name implies, is an interconnected system in itself, comprising numerous generators, distribution and control centers, transmission lines, converters, and other elements. Proper functioning of these interacting interdependent and interconnected systems and Complex SoS is crucial to the continuous operation of the entire power system. In addition to its essential internal dependency, the US electric power Complex SoS is externally dependent upon other infrastructure systems, notably telecommunications, water (primarily for cooling towers) fuel supply, and transportation. For example, its operation is heavily dependent upon voice and data communications. Data communications provide real-time updates (i.e. every few seconds) of electrical system status to supervisory control and data acquisition (SCADA) systems in distribution and bulk electric control centers. Data communications are also used for the remote control of devices in the field, such as circuit breakers, switches, transformer taps, and capacitors. Moreover, data communications allow generating units to follow the real-time signals from the control center that are necessary to balance electricity generation with consumer demand instantaneously. Although the power industry commonly owns and operates the majority of its communications equipment, a substantial portion is dependent upon local telephone carriers, long-distance carriers, satellites, cellular systems, paging systems, networking service providers, Internet service providers, and others.

Historically, many critical infrastructures around the world were physically and logically separate systems with little interdependence. This situation has rapidly changed, and close relationships among infrastructures can now take many forms. For example, telecommunications, power, transportation, banking, and others are marked by immense complexity, characterized predominantly by strong I-I as well as hierarchies. This complexity takes many forms, including flows of information, shared security, and physical flows of commodities, among others. In this chapter we introduce a high-level, overarching modeling framework capable of describing the risks to a country's critical infrastructures and industry sectors – focusing on risks arising from the I-I characterizing the four safety-critical sectors of the economy, each of which is Complex SoS: electricity, communications, water, and supply chain.

In assessing the vulnerability of Complex SoS, it is important to analyze both the I-I of the subsystems that compose them and the I-I with other external Complex SoS. Addressing the importance of I-I can be achieved by modeling the way "inoperability" or "dysfunctionality" propagates throughout critical infrastructure systems or industry sectors of the economy. The inoperability caused by the lack of effective maintenance, willful attacks, accidental events, or natural causes can set off a complex chain of cascading impacts on other interconnected systems. For example, similar to other critical infrastructures, water resources systems – surface and groundwater sources, water transport, treatment, distribution, storage, and wastewater collection and treatment – heretofore have been designed, built, and operated without a threat to their integrity. Today, the I-I among infrastructures pose a "silent threat" to entire sectors of the economy.

8.2 THE ORIGINAL LEONTIEF I–O MODEL

Wassily Leontief received the 1973 Nobel Prize in Economics for developing what came to be known as the Leontief input–output (I–O) model of the economy (Leontief, 1951a, b, 1986). The economy (and thus the model) consists of a number of subsystems, or individual economic sectors or industries, and is a framework for studying the equilibrium behavior of an economy. The model enables understanding and evaluating the interconnectedness among the various sectors of an economy and forecasting the effect on one segment of a change in another. Leontief's I–O model describes the equilibrium behavior of both regional and national economies (Isard, 1960; Liew, 2000; Lahr and Stevens, 2002), and the I–O model is a useful tool in the economic decision-making processes used in many countries (Miller et al., 1989).

The Leontief model enables accounting for the I-I within each and among all critical systems that compose Complex SoS as well as the interconnectedness among them. Miller and Blair (1985) provide a comprehensive overview of I–O analysis with deep insights into the Leontief economic model and its applications. Other literature in the area of cascading failures through interconnected sectors can be found in US DOC (2003) and Embrechts et al. (1997). Many notable extensions were later created based on the original Leontief model, including the nonlinear Leontief model (Krause, 1992), energy I–O analysis (Griffin, 1976; Proops, 1984), and environmental I–O analysis (Converse, 1971; Lee, 1982). Haimes and Nainis (1974) and Haimes (1977, 2016) developed an I–O model of supply and demand in a regional water resources system. Olsen et al. (1997) developed an I–O model for risk analysis of distributed flood protection. Extensions of I–O analysis were described by Lahr and Dietzenbacher (2001).

The brief outline below is based on Intriligator (1971), Haimes (1977, 2016), and Haimes et al. (2005a, b). It provides a simplified version of Leontief's (1951a) I–O model to trace resources and products within an economy. The economy Complex SoS is assumed to consist of a group of n interacting sectors or industries, where each "industry" produces one product (commodity). A given industry requires labor, input from the outside, and also goods from interacting industries. Each industry must produce enough goods to meet both interacting demands (from other industries in the group) plus external demands (e.g. foreign trade and industries outside the group). A static (equilibrium-competitive) economy, with constant coefficients for a fixed unit of time (1 year), is assumed. Define the following notation:

- x_j is the output (for the total economy) of jth goods, $j = 1, 2, \ldots, n$.
- r_k is the input (for the total economy) of kth resource, $k = 1, 2, \ldots, m$.
- x_{ij} is the amount of the ith goods used in the production of the jth goods.
- r_{kj} is the amount of the kth resource input used in the production of the jth goods.

Leontief's model assumes that the inputs of both goods and resources required to produce any commodity are proportional to the output of that commodity:

$$x_{kj} = a_{kj}x_j, \quad j,k, = 1,2,\ldots,n \tag{8.1}$$

$$r_{ij} = b_{ij}x_j, \quad k = 1,2,\ldots,m, \quad j = 1,2,\ldots,n \tag{8.2}$$

Furthermore, the output of any commodity is used either as input for the production of other commodities or as final demands, c_k. The balance Eq. (8.1) is a key to the subsequent development of the Leontief-based Eq. (8.3):

$$x_k = \sum x_{kj} + c_k, \quad k = 1,2,\ldots,n \tag{8.3}$$

Combining Eqs. (8.1) and (8.3) yields the Leontief equation:

$$x_k = \sum_j a_{kj}x_j + c_k, \quad k = 1,2,\ldots,n \tag{8.4}$$

Similarly, the proportionality assumption applies to the resources:

$$r_{ij} = b_{ij}x_j \tag{8.5}$$

$$\sum_j r_{ij} = \sum_i b_{ij}x_j \tag{8.6}$$

Since the demand for the ith resource cannot exceed its supply, then

$$\sum_i b_{ij}x_j \le r_i, \quad r_i \ge 0, \quad i = 1,2,\ldots,m \tag{8.7}$$

The above basic model of the economy is written in a compact matrix notation in Eq. (8.8):

$$\mathbf{x} = \mathbf{A}\mathbf{x} + \mathbf{c} \Leftrightarrow \left\{ x_i = \sum_j a_{ij}x_j + c_i \right\} \forall i \tag{8.8}$$

8.3 INOPERABILITY INPUT–OUTPUT MODEL (IIM)

Grounded on Leontief's work, a first-generation inoperability I–O model (IIM) of interconnected Complex SoS was developed by Haimes and Jiang (2001). This *physical-based* model considers multiple intra- and interconnected systems. The primary purpose of the model is to improve understanding of the impact of complexity on the continued and sustained operability of systems under adverse conditions. Other related works on infrastructure interdependencies and risks of terrorism are presented in Haimes (2002, 2004); Haimes and Horowitz (2004); Santos and Haimes (2004); Crowther and Haimes (2005); Jiang and Haimes (2004); and Haimes (2016).

Note that the "supply" and "demand" concepts in the Leontief economy model assume a different interpretation and have been inverted to some extent in the IIM risk model. Although the mathematical construct of the two models is similar, the interpretation of the model parameters is fundamentally different. *Dollars* are the units used in the Leontief I–O model for the economy. Our infrastructure model uses units of *risk of inoperability* [0, 1], defined above as a measure of the probability (likelihood) and degree (percentage) of the inoperability (dysfunctionality) of a system, or of the entire Complex SoS. An inoperability of 1 would mean that an infrastructure is totally out of commission. As stated earlier, inoperability may take various forms according to the nature of the system. When the model is applied to study any infrastructure system, one of the very first tasks is to define the specific inoperability and the associated risks.

This model addresses the equilibrium state of economic Complex SoS in the event of an attack, provided that the interdependency matrix is known. The *input* to the system is an initial perturbation triggered by an attack of terrorism, an accidental event, or a natural disaster. The *outputs* of the system are the resulting risks of inoperability of different infrastructures due to their connections to one another. The output can be triggered by one or multiple failures due to their inherent complexity, or to external perturbations (e.g. natural hazards, accidents, or acts of terrorism).

In his basic I–O model, Leontief considered an economy that produces n goods as output and uses m primary resources as input. For the IIM we consider a system consisting of n critical complex intra- and interconnected infrastructures (Haimes and Jiang, 2001; Haimes, 2016). Although the equations are similar, there is a major difference in the interpretation of the variables. In other words, the basic Leontief Eqs. (8.1)–(8.8) are similar to the IIM Eqs. (8.9)–(8.12) that will be introduced subsequently; however, they connote different meanings.

In the IIM the output is the infrastructure's *risk of inoperability* or simply *inoperability* that can be triggered by one or multiple failures due to complexity, accidents, or acts of terror. *Inoperability* is defined as *the inability of the system or the entire Complex SoS to perform its intended natural or engineered functions*. In the model, the term *inoperability* can denote the level of the system's, or Complex SoS dysfunction, expressed as a percentage of the system's "as-planned" level of operation. Alternatively, inoperability can be interpreted as a degradation of a system's capacity to deliver its intended output (or supply). Although inoperability in its current scope applies to physical and economic losses, it can be extended to assess impacts due to information failure. In addition, other factors for assessing failures, such as loss of lives, environmental quality, and others, can supplement the economic factors used in the context of inoperability.

Inoperability is assumed to be a continuous variable evaluated between 0 and 1, with 0 corresponding to a flawlessly operable system/SoS state and 1 corresponding to the Complex SoS being completely inoperable. Inoperability may take different forms, depending upon the nature of the problem and the type of the Complex SoS. When the production level is of major concern, inoperability may well be defined as the unrealized production (i.e. the actual production level subtracted from the desired production level). For instance, if the Complex SoS under consideration is a power plant, then the inoperability may be defined as the ratio of the actual amount of power produced (in appropriate units) to the desired amount. Furthermore, the notion of inoperability also attempts to capture the quality of a system's function. Assuming that quality can be measured numerically, a defective Complex SoS whose performance is of degenerate functional quality is considered partially operable and thus has inoperability greater than zero. For instance, a television set that has a picture but no sound is only partially operable and thus has inoperability greater than zero. By the same token, a water supply Complex SoS producing slightly contaminated water is also considered partially operable and thus has inoperability greater than 0. Finally, inoperability of a system is not necessarily a continuous variable. Under certain circumstances, it may take discrete values such as binary values. Here, we focus our discussion on the continuous case.

The risk of inoperability can also be viewed as an extension of the concept of *unreliability*. Unreliability is the conditional probability that a system will fail during a specified period of time t, given that it operates perfectly at $t = 0$. In fact, the system may not fail completely during this time span; it may fail partially with certain probability. For instance, during this period of time, it may fail 100% with probability 0.1, it may lose 50% of its functionality with probability 0.4, or it may lose 10% of its functionality with probability 0.8 and so forth (provided that the functionality is quantifiable). Thus a natural extension of the notion of unreliability is to average out all these possibilities by considering both the failure level and the likelihood. In so doing, we end up with a quantity that represents *the expected value of the failure level during a certain period of time.* In other words, if the expected-value metric is adopted in the definition of risk, then the risk of inoperability can be viewed as the expected inoperability. A conditional expected-value metric, to supplement the expected-value metric (Haimes, 2016). Hence, for the sake of brevity, in the following discussion, we sometimes use "inoperability" in lieu of "risk of inoperability."

The inoperability of interdependent and interconnected infrastructure Complex SoS may be manifested in several dimensions, e.g. geographical, functional, temporal, or political. On the one hand, these and other perspectives markedly influence the values assigned to the probability (coefficient) of inoperability in the model. On the other hand, each may justify the construction of a different inoperability model addressing a specific dimension. An example would be inoperability that spans regional or statewide, short-term, or long-term, and one-function failure, or multiple failures of infrastructure Complex SoS. In such cases, each model will require specific and different probabilities of inoperability. In addition, one such inoperability model might evaluate, and measure in monetary terms, the risk of inoperability or damage to property, production, service, or injury under extreme natural and accidental conditions, or due to acts of terrorism.

In the following discussions, we assume that each of the four infrastructure Complex SoS performs a uniquely defined function, that is, no two systems perform the same function. In other words, in this preliminary model, we do not consider the issue of redundancy.

Let x_j, $j = 1, 2, \ldots, n$ be the overall risk of inoperability of the jth I-I infrastructure that can be triggered by one or multiple failures caused by accidents or acts of terrorism.

Let x_{kj} be the degree of inoperability triggered by one or multiple failures that the jth infrastructure can contribute to the kth infrastructure due to their complex I-I.

Let a_{kj} be the probability of inoperability that the jth infrastructure contributes to the kth infrastructure. In our model, a_{kj} describes the degree of dependence of the kth infrastructure on the jth infrastructure. For example, if $a_{kj} = 1$, then this means a complete failure of the jth infrastructure will lead to a complete failure of the kth infrastructure. A value of $a_{kj} = 0$, on the other hand, indicates that the failure of the jth infrastructure has no effect on kth infrastructure.

Let c_k be the natural or man-made perturbation into the kth critical infrastructure.

At this stage, the proportionality assumption that underpins Leontief's economy model is assumed to hold for the inoperability I–O risk model as well; then we have

$$x_{kj} = a_{kj} x_j \quad j,k = 1,2,\ldots,n \tag{8.9a}$$

The following balance equation is a key to the subsequent development of the linear model:

$$x_k = \sum_j x_{kj} + c_k, \quad j,k = 1,2,\ldots,n \tag{8.9b}$$

Combining the balance equation with the proportionality equation yields the inoperability equation for the infrastructure model:

$$x_k = \sum_j a_{kj} x_j + c_k, \quad j,k = 1,2,\ldots,n \tag{8.10}$$

The above equation can be written in matrix notation as follows:

$$\mathbf{x} = \mathbf{A}\mathbf{x} + \mathbf{c} \tag{8.11}$$

where $\mathbf{x} = [x_1, x_2, \ldots, x_n]^T$, $\mathbf{c} = [c_1, c_2, \ldots, c_n]^T$, $\mathbf{r} = [r_1, r_2, \ldots, r_m]^T$, $[.]^T$ = a column vector, and $\mathrm{A} = [a_{kj}]$ $n \times n$ matrix.

Defining $\mathrm{I} = n \times n$ identity matrix and assuming that $(\mathrm{I} - \mathrm{A})$ is nonsingular, the vector of inoperability $\mathbf{x}$ in Eq. (8.11) can be solved using the following matrix operation:

$$\mathbf{x} = (\mathrm{I} - \mathrm{A})^{-1}\mathbf{c} \tag{8.12}$$

Determining the values of the **A**-matrix during the modeling process is a very challenging undertaking, and extensive data collection and data mining may be required to complete this step. The following are general guiding principles for determining the **A**-matrix:

- Explore the potential use of publicly available I–O tables to enable understanding of the transactions among various sectors in the economy. Prior to conducting actual field surveys and interviews, these I–O tables can provide valuable insights into the interdependencies among various infrastructures.
- Define the level of resolution and the boundary conditions of each infrastructure, because a system may be analyzed at different levels of resolution. Note, however, that the level of resolution adopted in the analysis must be harmonious with the accuracy of the data and the analytical tractability (including the determination of the I–O relationships). The realism that exists at high granularity should not be sacrificed in the process of aggregation.
- Identify physical connections among the infrastructures. In general, if there are no physical connections between infrastructures i and j, then $a_{ij} = a_{ji} = 0$. Physical boundary conditions are very critical in identifying the physical connections among different infrastructures.
- If there are any deterministic correlations among any infrastructures, then these relationships should be singled out first. For instance, if the failure of infrastructure i will definitely lead to the failure of infrastructure j, then $a_{ji} = 1$. By the same token, if the failure of infrastructure i will definitely lead to the failure of one of the two subsystems of infrastructure j, which performs 50% of that infrastructure's functions, then $a_{ji} = 0.5$.
- If the correlation between two infrastructures (e.g. infrastructures i and j) is of a stochastic nature, then all conceivable scenarios must be analyzed, and a statistical average has to be taken to obtain a_{ij} and a_{ji}. For example, if the failure of infrastructure i leads, with probability 0.3, to complete failure of infrastructure j, and with probability 0.7, leads infrastructure j to be 50% inoperable, then $a_{ji} = (0.3)\ (1) + (0.7)\ (0.5) = 0.65$. If the real data are not sufficient, a simulation may be helpful in order to obtain data for the probability distributions.

8.4 REGIMES OF RECOVERY

Several time frames, or regimes, exhibit different features of interdependencies following an attack or other extreme events affecting infrastructure. We will use the term emergent forced changes (EFCs) for any or for all of the above disruptions. We define EFCs as internal or external forces that adversely affect the operational or functionality of a system, or of the entire Complex SoS, or other extreme event affecting infrastructure. The nature and extent of sector interactions will vary from one time frame to the next. Moreover, the metrics of outcomes will be allowed to vary from time frame to time frame. Within each time frame, the inoperability I–O risk model can describe a conceptual situation of equilibrium. Before equilibrium is reached, the system will have evolved to a distinct and new frame of interactions. A sample of several time frames that will be addressed by IIM is presented in Figure 8.1.

Further uses of the regimes include comparing the physical versus psychological effects of an EFC. While the physical-based inoperability I–O risk model analyzes the physical losses caused by either natural or human-caused disasters, it is important to consider psychological factors as well. Specific empirical studies, such as those by Susser et al. (2002) and Galea et al. (2002), show the significance of the "fear factor" induced by the 11 September 2001 terrorist attack in the United States. Fear can cause the public to reduce their demand for the goods and services produced by an attacked industry.

For example, public apprehension after the 9/11 attack in the United States about the safety of air transportation caused a drastic reduction in the operations of the airlines and other airline-dependent industries. These retrenchments and changes in demand can have large economic repercussions that compound the physical losses (e.g. degraded production capacity). Both physical and psychological considerations ought to be accounted for in

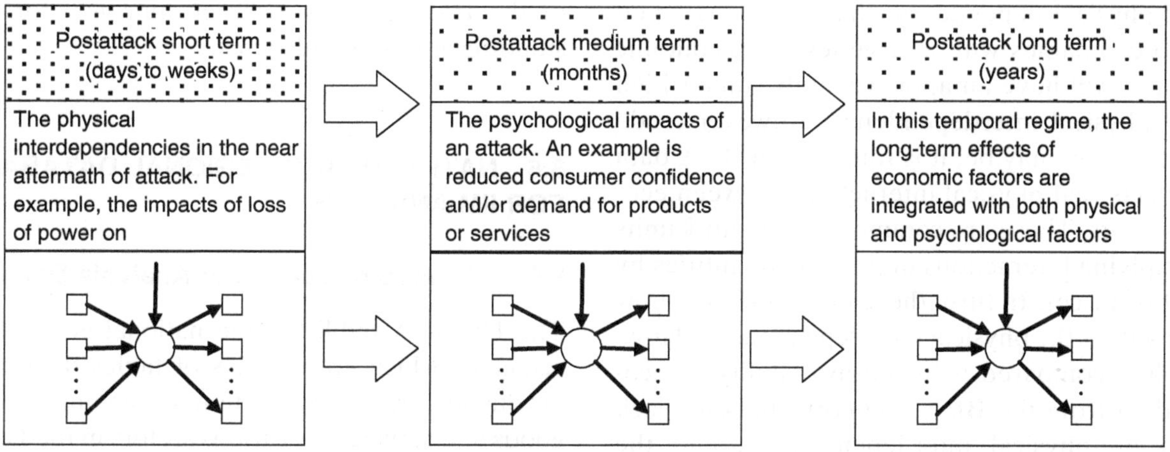

Figure 8.1 *Three temporal regimes of recovery that are considered in IIM analysis of impacts resulting from EFCs.*

analyzing the long-term adverse economic impacts on the "as-planned" operation levels of interconnected sectors.

8.5 SUPPORTING DATABASES FOR IIM ANALYSIS

An advantage of building on the Leontief I–O model is that it is supported by major ongoing data collection efforts. These available databases of interdependency statistics provide an essential foundation for applying the IIM to model the economic consequences of EFCs or a terrorist attack. In this section we review two main data resources available in the United States. The (i) Bureau of Economic Analysis (BEA) database of national I–O accounts and (ii) Regional Input–Output Multiplier System (RIMS II) accounts. The BEA database provides an overview of the national economic I–O accounts; this is a series of tables depicting the production and consumption of commodities (i.e. goods and services) by various sectors in the US economy. The BEA consumption and production tables are combined to calculate the Leontief technical coefficient matrix for over 500 industry sectors of the US economy and their corresponding interdependencies with the workforce sector. RIMS II is a set of regional data maintained by the BEA Regional Economic Analysis Division. Empirical tests suggest that regional multipliers can be used as surrogates for time-consuming and expensive surveys without compromising accuracy.

Utilizing the BEA database, the demand reduction IIM augments the physical-based model developed by (Haimes and Jiang, 2001; Haimes, 2016). While this model quantifies inoperability in terms of degraded capacity to deliver the intended outputs, the demand-based model addresses the demand reductions that can potentially stem from perturbations (Santos, 2003; Santos and Haimes, 2004). Logically, the demand reduction of a perturbed sector produces further adverse impacts on the operations of other dependent sectors. For example, the demand reduction of the airline industry – an industry primarily affected by the 9/11 terrorism – caused the demand for other dependent industries to decline as well (e.g. travel and hotel industries). Specifically, a 33.2% reduction in passenger enplanements (FAA, 2002) and a 19.2% reduction in hotel occupancy (Ernst and Young, 2002) were realized in the aftermath of 9/11, relative to 2000 figures. Integrating the concept of inoperability into Leontief's economic I–O model makes it possible to analyze how demand reduction inoperability affects other interdependent infrastructures.

Two motivations have driven the use of an economic model to study physical interactions. One deals with the general issue of translating physical to economic values, while the other accounts for EFCs (e.g. terrorist attacks) on power sources as well as on equipment operated by using sectors (e.g. computers, control systems). An assumption made when applying the IIM is that the level of economic dependency constitutes a surrogate measure of the level of physical dependency, that is, it is assumed that two companies with a large level of economic interaction will have an approximately similar high level of physical interdependency. However crude this assumption may be, it is founded on BEA data that reflects real physical interactions between economic sectors. These are commensurate into $ units by multiplying interactions of physical quantities by producers' prices. In turn, these prices indicate how a sector values the physical interdependencies. However, when compared with the availability of economic data from the BEA, the corresponding lack of data on physical interdependencies, and the extraordinary cost required to collect such information on the scale of the economic data collections, the degree of inaccuracy in IIM results becomes a question. Other studies determined the rank order of interdependent sectors, the loss in their production capacities, and the corresponding economic impact. This can be used to determine the size of the risk and where to invest to reduce it. One possible way to add confidence in the results is to carry out a study of the top sectors resulting from the IIM analysis to determine how close the physical ties are relative to economic ties. Such a study might be bounded enough to be carried out at an acceptable cost when compared with costs of poor risk management and could result in modifying prioritizations.

When applying the IIM to consequences from EFCs or a potential terrorist attack, the BEA's data can be used to determine the expenditures of all economic sectors on items that use electricity (i.e. how much a sector spends on computers and other electrical equipment). Using the percentage of each sector's total resources that are spent on electrical equipment to estimate the production-focused level of dependence on electric power, we can estimate the percentage loss in production level that each sector would suffer due to its own electrical devices failing. This permits us to create an input vector for inoperability that includes not only the unavailability of power sources but also the production losses of power-dependent sectors even with power restored (e.g. due to dysfunctional equipment). However limited it may be by substituting economic for physical data, this use of the IIM provides a direct approach for understanding interdependencies.

8.6 NATIONAL AND REGIONAL DATABASES FOR IIM ANALYSIS

8.6.1 Bureau of Economic Analysis Database

The US BEA publishes the national economic I–O accounts, which are a series of tables depicting the production and consumption of commodities (i.e. goods and services) of various sectors in the US economy. The detailed national tables are composed of hundreds of industries, organized according to the North American Industry Classification System (NAICS) codes.

In the original Leontief model formulation, each industry is assumed to produce a distinct commodity. The term *commodity* here refers to the output of an industry, which can be in the form of goods or services. Realistically, however, it is possible that a given industry can produce more than one commodity. On the other hand, a given commodity may not be a unique output of a given industry. The BEA recognizes that the one-to-one correspondence assumption between an industry and commodity is generally not true. The BEA makes a distinction between an industry and a commodity in its published I–O data via the *industry-by-commodity* and *commodity-by-industry* matrices. Figure 8.2, adapted from Miller and Blair (1985), shows a summary of the types of national I–O accounts maintained by the BEA.

The *make* matrix in Figure 8.2, denoted by **V**, would show the monetary values of the different column commodities *produced* by the different row industries. A sample of *make* matrix data is shown in Table 8.1. The *use* matrix on the other hand, denoted by **U**, would show the monetary values of the different row commodities *consumed* by the different column industries. A sample of *use* matrix data is shown in Table 8.2. Note that Figure 8.2 does not directly specify the I–O matrix representing the industry-by-industry transactions. This matrix, denoted by **A** in the Leontief formulation, is called the industry-by-industry technical coefficient matrix in Leontief parlance. It would show the input of industry i to

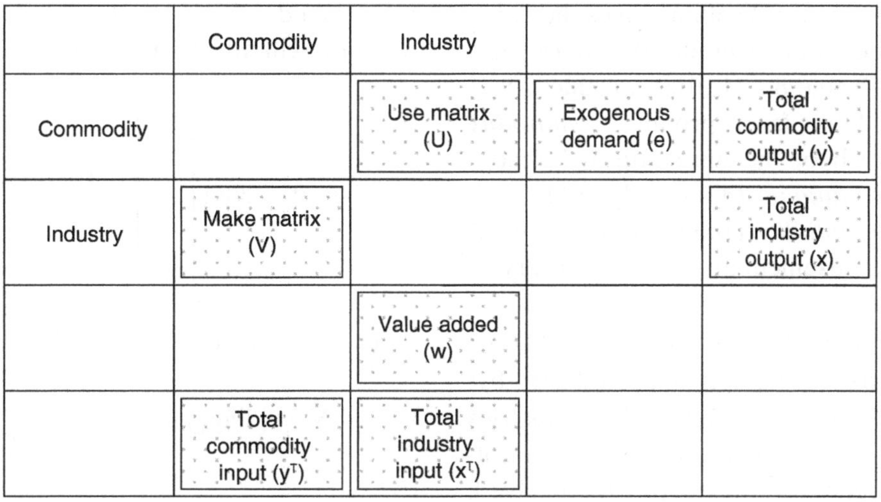

Figure 8.2 *Summary of economic input–output accounts.*

TABLE 8.1 Sample *Make* Matrix for 1992 US Economy

Industry (SIC code)	Commodity output (SIC code)	Value (in million $)
1.0100	....	20 285
	1.0100	19 646
	4.0001	86
	14.0600	365
	76.0206	188

Excerpt from US Department of Commerce, p. 47 (1998).

TABLE 8.2 Sample *Use* Matrix for 1992 US Economy

Commodity	Using industry	Value (in million $)
2.0502	....	2162
	2.0502	55
	14.1900	2099
	93.0000	4
	94.0000	4

Excerpt from US Department of Commerce, p. 83 (1998).

j, expressed as a proportion of the total production inputs to industry j. BEA does not publish the elements of the **A**-matrix because this task is left to the analyst. Typically, the **A**-matrix is established from the *make* and *use* matrices using various assumptions (e.g. commodity–technology assumption (CTA) and industry–technology assumption (ITA). One approach is carried out by first normalizing the values of the *make* and *use* matrices. The following sections discuss the operations for deriving the *normalized make* matrix ($\widehat{\mathbf{V}}$) from the *make* matrix (**V**) and the *normalized use* matrix ($\widehat{\mathbf{U}}$) from the *use* matrix (**U**).

8.6.2 Coefficients of Production in National and Regional Economies

The *make* matrix (**V**) in BEA I–O reports shows the itemized production of commodities by various industries. Each element of the *make* matrix (v_{ij}) shows industry i's production of commodity j (typically measured in millions of dollars). If there are m commodities and n industries, then the total industry output for the ith industry (x_i) must follow the balance equation below (see Figure 8.2):

$$x_i = v_{i1} + v_{i2} + \cdots + v_{im} = \sum_{j \leq m} v_{ij}; \quad \forall i = 1,2,\ldots,n \tag{8.13}$$

Denoting **x** as the vector of total industry outputs, **Σ** as a unity vector (i.e. a vector whose elements are all 1s, also known as a summation vector), and **V** as the make matrix, it can be shown that Eq. (8.14) can be written in the following matrix form:

$$\mathbf{x} = \mathbf{V\Sigma} \tag{8.14}$$

Due to the volume of data, the BEA does not present the *make* matrix in the format of v_{ij} (i.e. with the industries arranged along the rows and commodities along the columns). Rather, referring to Table 8.1, one industry is given at a time (see first column), the second column enumerates the commodities produced by that industry, and the third column gives the value of those commodities. For example, the dairy farm product *industry* in Table 8.1 (1.0100) produces \$20 M worth of the dairy farm product *commodity* (1.0100); \$86 M worth of the agricultural, forestry, and fishery service *commodity* (4.0001); \$365 M worth of the fluid milk *commodity* (14.0600); and \$188 M worth of the other amusement and recreation service *commodity* (76.0206).

Equation (8.19) shows the formulation for the *normalized make* matrix $\hat{\mathbf{V}} = [\hat{v}_{ij}]$. It is an industry-by-commodity matrix because it shows the industries along the rows and the commodities along the columns. To better understand how Eq. (8.19) is derived, we dissect the elements of the underlying *make* matrix ($\mathbf{V}$) and the total commodity output vector ($\mathbf{y}^T$) as follows (see Figure 8.2):

$$\mathbf{V} = \begin{bmatrix} v_{11} & \cdots & v_{1j} & \cdots & v_{1m} \\ \vdots & & \vdots & & \vdots \\ v_{i1} & \cdots & v_{ij} & \cdots & v_{im} \\ \vdots & & \vdots & & \vdots \\ v_{n1} & \cdots & v_{nj} & \cdots & v_{nm} \end{bmatrix} \tag{8.15}$$

$$\mathbf{y}^T = \left[y_1 = \sum_i v_{i1} \cdots y_j = \sum_i v_{ij} \cdots y_m = \sum_i v_{im} \right] \tag{8.16}$$

The *normalized make* matrix, whose elements are denoted by $\hat{v}_{ij}$, can be obtained by dividing each element of the make matrix (v_{ij}) by the respective column sum (y_j) as follows:

$$\hat{\mathbf{V}} = \begin{bmatrix} v_{11}/y_1 & \cdots & v_{1j}/y_j & \cdots & v_{1m}/y_m \\ \vdots & & \vdots & & \vdots \\ v_{i1}/y_1 & \cdots & v_{ij}/y_j & \cdots & v_{im}/y_m \\ \vdots & & \vdots & & \vdots \\ v_{n1}/y_1 & \cdots & v_{nj}/y_j & \cdots & v_{nm}/y_m \end{bmatrix} \tag{8.17}$$

As Eq. (8.18) shows, Eq. (8.17) can be written in a compact matrix notation by first denoting the operator $\text{diag}(\boldsymbol{\theta})$ as the resulting diagonal matrix constructed from a given vector $\boldsymbol{\theta}$. (Note that this notation will also be used later.)

$$\text{diag}(\boldsymbol{\theta}) = \text{diag}\begin{bmatrix} \theta_1 \\ \theta_2 \\ \vdots \\ \theta_m \end{bmatrix} = \begin{bmatrix} \theta_1 & 0 & \cdots & 0 \\ 0 & \theta_2 & \ddots & \vdots \\ \vdots & \ddots & \ddots & 0 \\ 0 & \cdots & 0 & \theta_m \end{bmatrix} \tag{8.18}$$

Thus, from Eqs. (8.17) and (8.18), yields:

$$\hat{\mathbf{V}} = \mathbf{v}[\text{diag}(\mathbf{y})]^{-1} \Leftrightarrow \left\{ \hat{v}_{ij} = \frac{v_{ij}}{y_j} \right\} \forall i, j \tag{8.19}$$

8.6.3 Coefficients of Consumption in National and Regional Economies

The *use* matrix ($\mathbf{U}$) in BEA I–O reports shows the itemized consumption of commodities by various industries. Each element of the *use* matrix (u_{ij}) shows industry j's consumption of the ith commodity (typically measured in millions of dollars). Suppose there are m commodities and n industries. Denoting e_i as exogenous consumptions for commodity i (or final commodity demands), the total commodity output for the ith commodity (y_i) must follow balance Eq. (8.20). (The notation $\mathbf{c}$ or c_i throughout this chapter refers to final *industry* demand. It should be distinguished from $\mathbf{e}$ or e_i, which refers to the exogenous or final *commodity* demand.) (See Figure 8.2.)

$$y_i = u_{i1} + u_{i2} + \cdots + u_{im} + e_i = \sum_{j \le n} u_{ij} + e_i; \quad \forall i = 1, 2, \ldots, m \tag{8.20}$$

Denoting the total commodity output vector by $\mathbf{y}$, the summation vector by $\boldsymbol{\Sigma}$, and the *use* matrix by $\mathbf{U}$, Eq. (8.20) can be written in the following matrix notation. (The notation $\mathbf{x}$ or x_i throughout this chapter refers to total *industry* output. It should be distinguished from $\mathbf{y}$ or y_i, which refers to the total *commodity* output.)

$$\mathbf{y} = \mathbf{U}\sum + \mathbf{e} \tag{8.21}$$

Sample data from the *use* matrix is depicted in Table 8.2. Due to the volume of data, BEA does not present the *use* matrix in the format of u_{ij} (i.e.

the commodities arranged along the rows and industries along the columns). Rather, one commodity is listed at a time (see the first column of Table 8.2), the second column enumerates the industries that use that commodity, and the third column gives the amount of that commodity used by the industries. For example, the usage of the sugar crops *commodity* in Table 8.2 (2.0502) is as follows: \$55 M by the sugar crops *industry* (2.0502); \$2099 M by the sugar *industry* (14.1900); \$4 M as change in business inventories (93.0000); and \$4 M as exports of goods and services (94.0000). Note that the last two codes, 93.0000 and 94.0000, are not industries per se. Rather, they are the final commodity consumptions (e_i) in the balance Eq. (8.20).

Equation (8.25) shows the formulation for the normalized *use* matrix $\hat{\mathbf{U}} = [\hat{u}_{ij}]$. This is a commodity-by-industry matrix because it shows the commodities along the rows and the industries along the columns. To better understand how Eq. (8.25) is derived, we dissect the elements of the underlying *use* matrix ($\mathbf{U}$) and the total industry output vector ($\mathbf{x}$) as follows (see Figure 8.2):

$$\mathbf{U} = \begin{bmatrix} u_{11} & \cdots & u_{1j} & \cdots & u_{1n} \\ \vdots & & \vdots & & \vdots \\ u_{i1} & \cdots & u_{ij} & \cdots & u_{in} \\ \vdots & & \vdots & & \vdots \\ u_{m1} & \cdots & u_{mj} & \cdots & u_{mn} \end{bmatrix} \quad (8.22)$$

$$\mathbf{x} = \begin{bmatrix} x_1 = \sum_i v_{1j} \\ \vdots \\ x_i = \sum_i v_{ij} \\ \vdots \\ x_n = \sum_i v_{nj} \end{bmatrix} \Leftrightarrow \mathbf{x}^T = [x_1 \ \cdots \ x_j \ \cdots \ x_n] \quad (8.23)$$

The *normalized use* matrix, whose elements are denoted by $\hat{u}_{ij}$, can be obtained by dividing each element of the *use* matrix (u_{ij}) by its respective column sum, which happens to be x_j (see Figure 8.2). The *normalized use* matrix is represented by the following matrix notations:

$$\hat{\mathbf{U}} = \begin{bmatrix} u_{11}/x_1 & \cdots & u_{1j}/x_j & \cdots & u_{1n}/x_n \\ \vdots & & \vdots & & \vdots \\ u_{i1}/x_1 & \cdots & u_{ij}/x_j & \cdots & u_{in}/x_n \\ \vdots & & \vdots & & \vdots \\ u_{m1}/x_1 & \cdots & u_{mj}/x_j & \cdots & u_{mn}/x_n \end{bmatrix} \quad (8.24)$$

Thus,

$$\hat{\mathbf{U}} = \mathbf{U}[\text{diag}(\mathbf{x})]^{-1} \Leftrightarrow \left\{ \hat{u}_{ij} = \frac{u_{ij}}{x_j} \right\} \forall i, j \quad (8.25)$$

8.6.4 Technical Coefficient Matrix

The *technical coefficient* matrix, denoted by $\mathbf{A}$, has industries along the rows as well as the columns. It can be shown that $\mathbf{A}$ is the product of the *normalized make* and the *normalized use* matrices:

$$\mathbf{A} = \hat{\mathbf{V}}\hat{\mathbf{U}} \Leftrightarrow \left\{ a_{ij} = \sum_k \hat{v}_{ik}\hat{u}_{kj} \right\} \forall i,j \quad (8.26)$$

On the other hand, the vector of industry final demands ($\mathbf{c}$) can be shown to be the product of the *normalized make* matrix and the *exogenous commodity demand* vector:

$$\mathbf{c} = \hat{\mathbf{V}}\mathbf{e} \Leftrightarrow \left\{ c_i = \sum_k \hat{v}_{ik} e_k \right\} \forall i \quad (8.27)$$

Deriving Eqs. (8.26) and (8.27) involves the following steps:

Substituting Eq. (8.19) for Eq. (8.14) we have

$$\mathbf{x} = \mathbf{V}\Sigma = \left(\hat{\mathbf{V}}\text{diag}(\mathbf{y})\right)\Sigma \quad (8.28)$$

Equation (8.28) can be simplified further by using the fact that diag($\mathbf{y}$)) $\sum = \mathbf{y}$

$$\mathbf{x} = \hat{\mathbf{V}}\mathbf{y} \quad (8.29)$$

Similarly, we substitute Eq. (8.26) for Eq. (8.21) to form the following equation:

$$\mathbf{y} = \left(\hat{\mathbf{U}}\text{diag}(\mathbf{x})\right)\Sigma + \mathbf{e} \quad (8.30)$$

Equation (8.30) can be simplified further by using the fact that $\text{diag}(\mathbf{x})\sum = \mathbf{x}$

$$\mathbf{y} = \widehat{\mathbf{U}}\mathbf{x} + \mathbf{e} \tag{8.31}$$

Premultiply $\hat{\text{v}}$ to Eq. (8.31):

$$\widehat{\mathbf{V}}\mathbf{y} = \widehat{\mathbf{V}}\widehat{\mathbf{U}}\mathbf{x} + \widehat{\mathbf{V}}\mathbf{e} \tag{8.32}$$

Substitute Eq. (8.29) for Eq. (8.32):

$$\mathbf{x} = \widehat{\mathbf{V}}\widehat{\mathbf{U}}\mathbf{x} + \widehat{\mathbf{V}}\mathbf{e} \tag{8.33}$$

For Eq. (8.33) to become equivalent to the usual Leontief balance Eq. (8.8), then Eqs. (8.26) and (8.27) must be true. Thus, we have shown that the Leontief industry-by-industry coefficient matrix (**A**) can be calculated on the bases of the *normalized make* and *normalized use* matrices as described in Eq. (8.26). In addition, the industry final demand can be constructed from the exogenous commodity demand by premultiplying it by the *normalized make* matrix, as described in Eq. (8.27).

8.6.5 Relevant Data for Workforce Sector Vulnerability Analysis

We have added a new *workforce* row and column to the original national technical coefficient matrix (**A**). By extracting the household portion of the exogenous demand (measured in terms of personal consumption expenditures) and the household portion of the value added (measured in terms of personnel compensations), we were able to generate an updated **A**-matrix. This integrates information on additional interdependency impacts contributed by the household sector. The extraction of household portions from exogenous demand and value-added vectors is described in Figure 8.3. The household sector, a standard BEA sector classification, is the source of labor inputs in various sectors of the economy. Thus, from here on, we refer to it as the "workforce sector."

8.7 REGIONAL INPUT–OUTPUT MULTIPLIER SYSTEM (RIMS II)

Regional decomposition enables a more focused and thus more accurate analysis of interdependencies for regions of interest in the United States. Miller et al. (1989) and Lahr and Dietzenbacher (2001) discuss the validity of "closing" the I–O analysis to a particular region (i.e. a single regional I–O framework as opposed to multiregional) since interregional feedbacks empirically are found to be "small."

Crowther (2007) developed multiregional IIM model (MRIIM) for strategic preparedness. Other contributors on regional IIM and its extensions include Haimes and Jiang, 2001; Jiang, 2003; Haimes et al., 2005a, b; Crowther 2007; Barker, 2008; and Barker and Haimes, 2008. The RIMS II division of the US Department of Commerce is responsible for releasing multipliers for various regions in the United States. Empirical tests suggest that regional multipliers can be used as surrogates for time-consuming

	Commodity	Sector			
Commodity		Use matrix (U)	Workforce (e_1)	Exogenous demand (e_2)	Total commodity output (y)
Sector	Make matrix (V)				Total sector output (x)
		Workforce (z_1^T)			
		Value Added (z_2^T)			
	Total commodity input (y^T)	Total sector input (x^T)			

Figure 8.3 *Economic input–output accounts reconfigured for workforce analysis.*

and expensive surveys without compromising accuracy. With the availability of national I–O tables and location quotients, analysts can convert and customize the national data according to the region of interest.

The RIMS II utilizes location quotients derived from "personal income data" and "wage and salary data" to regionalize the national Leontief technical coefficient matrix (i.e. the **A**-matrix). A location quotient indicates how well an industry's production capacity satisfies the regional local demand. In addition, as the value of an industry's location quotient tends to 1, its relative concentration in the regional approaches that of the national level:

$$l_i = \frac{\hat{x}_i^R / \hat{x}_s^R}{\hat{x}_i / \hat{x}_s} \quad (8.34)$$

where

$\hat{x}_i^R$ is the regional output for the ith industry
$\hat{x}_s^R$ is the total regional output for all regional-level industries
$\hat{x}_i$ is the national output for the ith industry
$\hat{x}_s$ is the total national output for all national-level industries

The regional industry-by-industry technical coefficient matrix $\mathbf{A}^R$, whose elements are denoted by a_{ij}^R, is then established as follows:

$$a_{ij}^R = \begin{cases} (a_{ij}) & l_i < 1 \\ a_{ij} & l_i \geq 1 \end{cases} \quad (8.35)$$

When **l** is used to denote a vector of location quotients and Σ a unity vector, Eq. (8.35) can be written in the following matrix notation:

$$\mathbf{A}^R = \text{diag}[\text{Min}(\mathbf{l}, \Sigma)]\mathbf{A} \Leftrightarrow \left\{a_{ij}^R = \text{Min}(l_i, 1)a_{ij}\right\} \quad \forall i, j \quad (8.36)$$

RIMS II issues a series of multipliers for various sectors of a specified region, generated via the region's location quotients (see Eq. (8.34)). Some examples are as follows:

- *Output multiplier* – gives the change in the *production output* of a sector resulting from a \$1 change in the demand for another sector's output.
- *Earning multiplier* – gives the change in the *workforce earnings* of a sector resulting from a \$1 change in the demand for another sector's output.
- *Employment multiplier* – gives the change in the *number of workers* of a sector resulting from a \$1 M change in the demand for another sector's output.

RIMS II multipliers are presented in the form of 38 × 490 matrices. The columns in Figure 8.4 represent detailed sectors (e.g. column 420 (C420), electric services/utilities). On the other hand, the rows in the matrix of RIMS II multipliers represent an aggregation of several column sectors (e.g. R26 (electric, gas, and sanitary services)). Thus, this specific row corresponds to the aggregated version of C420–C424, which includes C421, natural gas transportation; C422, natural gas distribution; and so on.

An extreme event such as a terrorist attack degrades the capability of a sector to supply its "as-planned" level of output. A sector's supply reduction necessarily leads to demand reduction (e.g. consumption adjusts when available supply is below the "as-planned" demand level). The RIMS II multipliers can be utilized for predicting the impact of reduced demand or supply on various interconnected sectors of a region due to extreme events.

8.8 DEVELOPMENT OF IIM AND ITS EXTENSIONS

8.8.1 Physical-Based IIM

A first-generation physical-based inoperability I–O model (or physical IIM, for simplicity) describes how the impact of willful attacks can cascade through a system of interconnected infrastructures. Inoperability connotes degradation in the system's functionality (expressed as a percentage relative to the intended state of the system). The formulation of the physical-based model is as follows:

$$x_i^P = \sum_j a_{ij}^P x_j^P + c_i^P \Leftrightarrow \mathbf{x}^P = \mathbf{A}^P\mathbf{x}^P + \mathbf{c}^P \quad (8.37)$$

Here we add the superscript P in Eq. (8.37) to the original formulation to distinguish it from Leontief's model. Although the mathematical construct of the

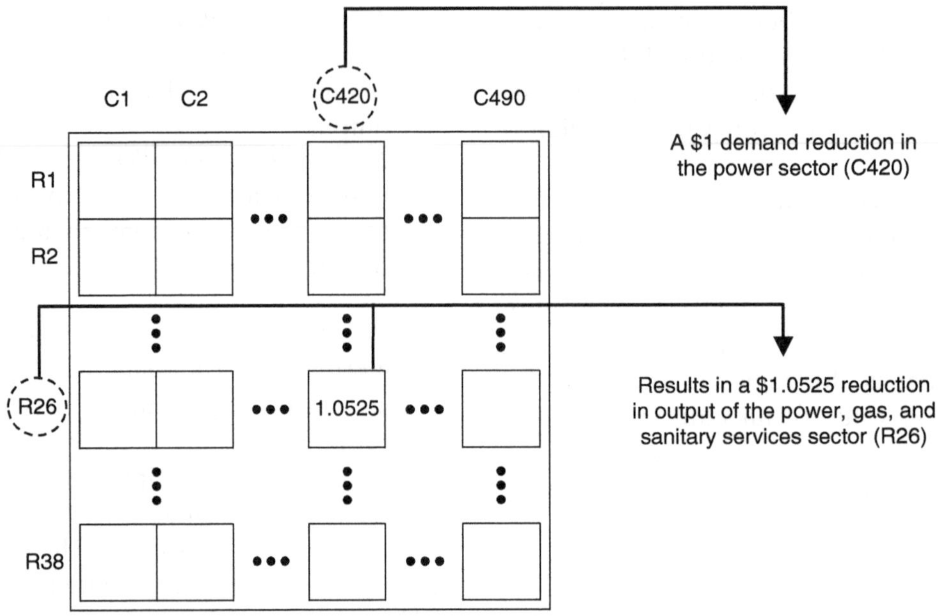

Figure 8.4 *Sample interpretation of RIMS II multipliers.*

two models is similar, the interpretation of the model parameters is fundamentally different. The "supply" and "demand" concepts in the Leontief economy model now assume different interpretations and have been inverted to some extent in the physical-based IIM. In Leontief's model, **c** and **x** represent commodities typically measured in production or monetary units. In the physical-based model, the vector $\mathbf{c}^P$ represents the input to the interconnected infrastructures – perturbations in the form of natural events, accidents, or willful attacks. The output is defined as the resulting vector of inoperability of the different infrastructures, denoted by $\mathbf{x}^P$, due to their connections to the perturbed infrastructure and to one another. The long-run inoperabilities of the interconnected infrastructures following an attack can be calculated using Eq. (8.37).

The inoperability vector $\mathbf{x}^P$ describes the degree of functionality of I-I economic Complex SoS. Thus, it takes on values between 0 and 1, where flawless operation corresponds to $\mathbf{x}^P = 0$ or $x_1^P = x_2^P = \cdots = x_n^P = 0$ for n interconnected infrastructures. When this condition is in effect, the infrastructures are said to be at their "as-planned" or *ground state.* A perturbation input $\mathbf{c}^P$ will cause a departure from this "as-planned" state. In addition, a perturbation can intuitively set off a chain of effects, leading to higher-order inoperabilities. For example, a power infrastructure (the kth infrastructure) would initially lose 10% of its functionality due to an attack that delivers a perturbation of $c_k^P = 0$. This means that the perturbation can be interpreted as the resulting inoperability of the power infrastructure *right after* an attack. In addition, the inoperability propagated by the power infrastructure to other power-dependent infrastructures will in turn cause more inoperabilities and ultimately may cause additional inoperability in the power infrastructure itself. In general, we expect the long-run inoperability of an attacked infrastructure to increase from its postattack value (i.e. the perturbation).

8.8.2 Demand Reduction IIM

The demand reduction IIM is derived by combining the insight and intuition gained from the physical I–O model with the rigor and proven BEA databases that accompany the original Leontief model. The BEA data is a record of the physical exchange of commodities between various interconnected industry sectors of the economy that have been scaled by producers' prices into one common unit of dollars. Therefore, this will be the foundation for our measure of interdependency.

Using the definition of normalized production loss, we derive the demand-based model on the basis of the Leontief model. We first define an "as-planned" production scenario based on the Leontief balance:

$$\hat{\mathbf{x}} = \mathbf{A}\hat{\mathbf{x}} + \hat{\mathbf{c}} \tag{8.38}$$

The variables in Eq. (8.38) are defined as follows:

$\hat{\mathbf{x}}$: "As-planned" total production vector.
$\mathbf{A}$: Leontief coefficient matrix.
$\hat{\mathbf{c}}$: "As-planned" final demand vector.

We also define a degraded production scenario based on the Leontief balance equation:

$$\tilde{\mathbf{x}} = \mathbf{A}\tilde{\mathbf{x}} + \tilde{\mathbf{c}} \tag{8.39}$$

The variables in Eq. (8.39) are defined as follows:

$\tilde{\mathbf{x}}$: Degraded total production vector
$\mathbf{A}$: Leontief coefficient matrix
$\tilde{\mathbf{c}}$: Degraded final demand vector

A reduction in the final demand (denoted by $\delta\mathbf{c}$ in Eq. (8.41)) is defined to be the difference between the "as-planned" and degraded final demands. This reduction in final demand consequently triggers a reduction in production (denoted by $\delta\mathbf{x}$ in Eq. (8.40)), which is defined to be the difference between the "as-planned" and degraded productions:

$$\delta\mathbf{x} = \hat{\mathbf{x}} - \tilde{\mathbf{x}} \tag{8.40}$$

$$\delta\mathbf{c} = \hat{\mathbf{c}} - \tilde{\mathbf{c}} \tag{8.41}$$

Subtracting Eq. (8.39) from Eq. (8.38) will result in the following relationship between $\delta\mathbf{x}$ and $\delta\mathbf{c}$:

$$(\hat{\mathbf{x}} - \tilde{\mathbf{x}}) = \mathbf{A}(\hat{\mathbf{x}} - \tilde{\mathbf{x}}) + (\hat{\mathbf{c}} - \tilde{\mathbf{c}}) \Leftrightarrow \delta\mathbf{x} = \mathbf{A}\delta\mathbf{x} + \delta\mathbf{c} \tag{8.42}$$

The transformations in Eqs. (8.43), (8.44), and (8.45) are needed to derive the demand-based model in a form analogous to the balance equation of the Leontief model:

$$\mathbf{c}^* = \left[\left(\operatorname{diag}(\mathbf{x})^{-1}\right)\delta\mathbf{c}\right] \tag{8.43}$$

$$\mathbf{A}^* = \left[(\operatorname{diag}(\hat{\mathbf{x}}))^{-1}\mathbf{A}(\operatorname{diag}(\hat{\mathbf{x}}))\right] \tag{8.44}$$

$$\mathbf{q} = \left[(\operatorname{diag}(\hat{\mathbf{x}}))^{-1}\delta\mathbf{x}\right] \tag{8.45}$$

Define the transformation matrix:

$$\mathbf{P} = \left[\operatorname{diag}(\hat{\mathbf{x}})^{-1}\right] \tag{8.46}$$

Using the transformation matrix in Eq. (8.46), Eq. (8.42) becomes Eq. (8.48) by the transformation defined in Eq. (8.47):

$$[\mathbf{P}\delta\mathbf{x}] = \left[\mathbf{PAP}^{-1}\right][\mathbf{P}\delta\mathbf{x}] + [\mathbf{P}\delta\mathbf{c}] \tag{8.47}$$

$$\mathbf{q} = \mathbf{A}^*\mathbf{q} + \mathbf{c}^* \tag{8.48}$$

Assuming that the demand-based interdependency matrix $\mathbf{A}^*$ is nonsingular and stable, the demand-based inoperability $\mathbf{q}$ can be calculated as follows:

$$\mathbf{q} = [\mathbf{I} - \mathbf{A}^*]^{-1}\mathbf{c}^* \tag{8.49}$$

8.8.3 Regional IIM

At the national level, the derived form of the demand reduction IIM is $\mathbf{q} = \mathbf{A}^*\mathbf{q} + \mathbf{c}^*$. The regional model takes a similar form:

$$\mathbf{q}^R = \mathbf{A}^{*R}\mathbf{q} + {}^R\mathbf{c}^{*R} \tag{8.50}$$

The system of equations corresponding to Eq. (8.50) is as follows:

$$\begin{aligned} q_1^R &= a_{11}^{*R}q_1^R + a_{12}^{*R}q_2^R + \cdots + a_{1n}^{*R}q_n^R + c_1^{*R} \\ q_2^R &= a_{21}^{*R}q_2^R + a_{22}^{*R}q_2^R + \cdots + a_{2n}^{*R}q_n^R + c_2^{*R} \\ &\vdots \\ q_n^R &= a_{n1}^{*R}q_1^R + a_{n2}^{*R}q_2^R + \cdots + a_{nn}^{*R}q_n^R + c_n^{*R} \end{aligned} \tag{8.51}$$

The term a_{ij}^{*R} in Eq. (8.51) can be expressed in terms of the regional technical a_{ij}^R coefficient using the identity shown in Eq. (8.52). This identity is analogous to the corresponding national-level formula $\mathbf{A}^* = \left[(\operatorname{diag}(\hat{\mathbf{x}}))^{-1}\mathbf{A}(\operatorname{diag}(\hat{\mathbf{x}}))\right]$:

$$\begin{aligned} &\mathbf{A}^{*R} = \left[\operatorname{diag}(\hat{\mathbf{x}}^R)\right]^{-1}\mathbf{A}^R\left[\operatorname{diag}(\hat{\mathbf{x}}^R)\right] \\ &\Leftrightarrow \left\{a_{ij}^{*R} = a_{ij}^R\left(\frac{\hat{x}_j^R}{\hat{x}_i^R}\right)\right\}\forall i,j \end{aligned} \tag{8.52}$$

Now, we express the regional industry-by-industry technical coefficient matrix ($\mathbf{A}^R$) in terms of the counterpart national matrix ($\mathbf{A}$). The resulting $\mathbf{A}^R$ matrix in Eq. (8.53) is obtained by substituting Eq. (8.36) for Eq. (8.52). Thus, the regional interdependency matrix $\mathbf{A}^{*R}$ can be established on the bases of the location quotients, the national industry-by-industry technical coefficients, and the "as-planned" production outputs of the regional industries:

$$\mathbf{A}^{*R} = \left[\text{diag}(\hat{\mathbf{x}}^R)\right]^{-1}[\text{diag}[\text{Min}(\mathbf{l})]\mathbf{A}]\left[\text{diag}(\hat{\mathbf{x}}^R)\right]$$
$$\Leftrightarrow \left\{ a_{ij}^{*R} = \min(l_i,1)a_{ij}\left(\frac{\hat{x}_j^R}{\hat{x}_i^R}\right)\right\}\forall i,j \tag{8.53}$$

8.8.4 Multiregional IIM

Regional IIMs can be interconnected to develop a multiregional version that improves spatial explicitness, model flexibility, and analysis coverage. The construction of the MRIIM builds on the regionalized IIM from the previous section by accounting for cross-regional flows of goods and services that interconnect regions (Crowther, 2007; Haimes, 2016). Accounting for cross-regional flows enables calculating multiregional coefficients, which in turn adjust the intraregional interdependency matrices $\mathbf{A}^*$. A spatially explicit interdependency matrix can be formed as a block diagonal matrix in Eq. (8.54), where $\mathbf{A}^s$ is a matrix containing all intraregional technical coefficients for region s calculated above:

$$\mathbf{A} = \begin{bmatrix} \mathbf{A}^1 & & & \\ & \mathbf{A}^2 & & \\ & & \ddots & \\ & & & \mathbf{A}^p \end{bmatrix} \tag{8.54}$$

Multiregional coefficients are calculated using commodity and service flow data. These coefficients describe the way that multiple regions are interconnected as larger regional systems due to their economic transactions of goods and services across geographical areas. To decision makers in "large regional" systems, these coefficients provide a measure of economic I-I across smaller (sub)regions that can result in either cascades of impacts or sources of resilience following a disaster scenario. To the decision makers in smaller (sub)regions, they systemically provide (i) a demand "footprint" describing other regions from which they purchase goods and services and (ii) a supply "footprint" describing other regions to which they deliver goods and services. Such decision makers can adapt strategic preparedness to mitigate risks against disaster scenarios that produce (i) supply perturbations in their demand footprint and (ii) demand perturbations in their supply footprint. We will henceforth refer to commodities and services as commodities. Let z_i^{rs} be the value of commodity i produced in region r and consumed in region s. For each commodity, we form an origin–destination matrix similar to the matrix in Table 8.3.

TABLE 8.3 Multiregional *Origin–Destination* Table for Commodity *i*

		Region of destination			
		1	2	...	p
Region of origin	1	z_i^{11}	z_i^{12}	...	z_i^{1p}
	2	z_i^{21}	z_i^{22}		
	⋮	⋮		⋱	
	p	z_i^{p1}			z_i^{pp}
Column sums:		s_i^1	s_i^2	...	s_i^p

The ratio of commodity flow z_i^{rs} to the total consumed commodities at the final destination s_i^s represents the portion of commodities consumed in region s that arrived from region r.

Equation (8.55) estimates the interregional technical coefficient, given the demand pooling assumption:

$$a_{ij}^{rs} = \left(\frac{z_i^{rs}}{s_i^s}\right) z_{ij}^{\dot{s}}/x_i^s = t_i^{rs} z_{ij}^{\dot{s}}/x_j^s = t_i^{rs} a_{ij}^{\dot{s}} \tag{8.55}$$

where $t_i^{rs} = z_i^{rs}/s_i^s$ is the proportion of commodity i consumed by region s that originated in region r.

Equation (8.56) defines the spatially explicit interregional flow matrix $\mathbf{T}$, and Eqs. (8.57) and (8.58) define $\mathbf{x}$ and $\mathbf{f}$, respectively, for a p-region economy. Note that each block matrix $\mathbf{T}^{rs}$ in $\mathbf{T}$ is a diagonal matrix by construction:

$$\mathbf{T} = \begin{bmatrix} \mathbf{T}^{11} & \cdots & \mathbf{T}^{1p} \\ \vdots & \ddots & \vdots \\ \mathbf{T}^{p1} & \cdots & \mathbf{T}^{pp} \end{bmatrix} \tag{8.56}$$

$$\mathbf{x} = \begin{bmatrix} \mathbf{x}^1 \\ \vdots \\ \mathbf{x}^p \end{bmatrix} \tag{8.57}$$

$$\mathbf{f} = \begin{bmatrix} \mathbf{f}^1 \\ \vdots \\ \mathbf{f}^p \end{bmatrix} \tag{8.58}$$

This composition of the various components results in a multiregional Leontief-based model for p regions with n sectors per region. Equation (8.59) shows the multiregional Leontief-based model used to construct the MRIIM. Constructions similar to Eq. (8.59) can be found in Miller and Blair (1985) and Isard et al. (1998).

$$\mathbf{x} = \mathbf{TAx} + \mathbf{Tf} \Leftrightarrow \left\{ x_i^r = \sum_{js} t_i^{rs} a_{ij}^{s} x_j^s + \sum_{s} t_i^{rs} f_j^s \right\}, \forall i, r \tag{8.59}$$

Each component of the multiregional Leontief-based model can be transformed according to the equations above. Following the same derivation, $\mathbf{T}$ is transformed similarly as shown in Eq. (8.60):

$$\mathbf{T}^* = \left[\hat{\mathbf{x}}^{-1}\mathbf{T}\hat{\mathbf{x}}\right] = \begin{bmatrix} \left[\hat{\mathbf{x}}^1\right]^{-1}\mathbf{T}^{11}\hat{\mathbf{x}}^1 & \cdots & \left[\hat{\mathbf{x}}^1\right]^{-1}\mathbf{T}^{1p}\hat{\mathbf{x}}^p \\ \vdots & \ddots & \vdots \\ [\hat{\mathbf{x}}^p]^{-1}\mathbf{T}^{p1}\hat{\mathbf{x}}^1 & \cdots & [\hat{\mathbf{x}}^p]^{-1}\mathbf{T}^{pp}\hat{\mathbf{x}}^p \end{bmatrix} \tag{8.60}$$

8.9 DYNAMIC IIM

To address more effectively the temporal dynamic behavior of industry recoveries in the static IIM, a dynamic IIM (DIIM) is proposed and formulated. In this section, the concept of an *industry resilience coefficient* is introduced as a key element in the dynamic extension that supplements and complements the static IIM. Fundamentals on how to define a resilience coefficient and its connection to parameters of recovery are also discussed. A comparison of dynamic and static models at the end of this section shows the consistency of the two models.

8.9.1 Introduction to the Dynamic IIM (DIIM)

In I–O literature, the classic Leontief DIIM takes the following form (see Miller and Blair (1985)):

$$\mathbf{x}(t) = \mathbf{Ax}(t) + \mathbf{c}(t) + \mathbf{B}\dot{\mathbf{x}}(t) \tag{8.61}$$

Matrix $\mathbf{B}$ in Eq. (8.61) is a square matrix of capital coefficients. It represents the willingness of the economy to invest in capital resources. Blanc and Ramos (2002) argue that the elements of $\mathbf{B}$ must either be zero or negative for an economic system to be stable. Such a condition will produce an economic behavior consistent with the static model, independent of initial conditions and final demand. Therefore, the capital coefficient matrix $\mathbf{B}$ can be interpreted as an expression of short-term countercyclical policy instead of long-term growth. For intuition about $\mathbf{B}$, consider the case investigated by Blanc and Ramos (2002) where $\mathbf{B} = -\mathbf{I}$, which represents an economy that quickly adjusts its production levels following information about mismatches in supply and demand:

$$\dot{\mathbf{x}}(t) = \mathbf{Ax}(t) + \mathbf{c}(t) - \mathbf{x}(t) \tag{8.62}$$

Using the classic Leontief I–O model and the results above, we can extend IIM to model the industry sectors' dynamic recovery behaviors and dynamic interactions caused by demand reduction or terrorist attacks on industry sectors. Consider a diagonal matrix form of the capital coefficient matrix $\mathbf{B}$:

$$\mathbf{B} = \text{diag}(b_i) \quad \forall i = 1, 2, \ldots, n \tag{8.63}$$

Furthermore, we define a $\mathbf{K}$ matrix as follows:

$$\mathbf{K} = \text{diag}(k_i) \quad \forall i = 1, 2, \ldots, n \tag{8.64a}$$

We relate Eqs. (8.64a) and ((8.64b)) as follows:

$$\mathbf{K} = -\mathbf{B}^{-1} \Leftrightarrow k_i = \frac{1}{b_i} \quad \forall i = 1, 2, \ldots, n \tag{8.64b}$$

Substituting Eq. (8.64b) for Eq. (8.61) and rearranging the terms will yield the following equation:

$$\dot{\mathbf{x}}(t) = \mathbf{K}[\mathbf{Ax}(t) + \mathbf{c}(t) - \mathbf{x}(t)] \tag{8.65}$$

Or in discrete form:

$$\mathbf{x}(k+1) - \mathbf{x}(k) = \mathbf{K}[\mathbf{Ax}(k) + \mathbf{c}(k) - \mathbf{x}(k)] \tag{8.66}$$

Transforming Eqs. (8.65) and (8.66) into normalized inoperability form will yield the following equations:

$$\dot{\mathbf{q}}(t) = \mathbf{K}[\mathbf{A}^*\mathbf{q}(t) + \mathbf{c}^*(t) - \mathbf{q}(t)] \tag{8.67}$$

$$\mathbf{q}(k+1) - \mathbf{q}(k) = \mathbf{K}[\mathbf{A}^*\mathbf{q}(k) + \mathbf{c}^*(k) - \mathbf{q}(k)] \tag{8.68}$$

In Eqs. (8.65) and (8.66), matrix $\mathbf{A}$ is the Leontief technical coefficient matrix; vector $\mathbf{c}(t)$ is the final demand vector at time t; and vector $\mathbf{x}(t)$ represents the total output of sectors at time t. In Eqs. (8.67) and (8.68), matrix $\mathbf{A}^*$ is the normalized interdependency matrix; vector $\mathbf{c}^*(t)$ is the normalized final demand vector at time t; and $\mathbf{q}(t)$ is the inoperability vector at time t. Collectively, Eqs. (8.65)–(8.68) give the formulation for the DIIM.

Matrix $\mathbf{K}$ will be referred to as the *industry resilience coefficient matrix*; each element k_i in the matrix measures the resilience of sector i, given an imbalance between supply and demand. In the case of a terrorist attack or other catastrophic event, it measures the recovery rate of the industry sectors. In the case of demand reduction, k_i measures the production adjustment rate of the sector.

The resilience coefficient k_i can be controlled and managed. Each resilience coefficient k_i in the matrix $\mathbf{K}$ is determined by the nature of the individual sector itself as well as by the controls on it via risk management policies. Hardening and other risk mitigation efforts in the industry sectors increase k_i during the recovery. Consequently, economic losses and other adverse impacts are minimized with shorter recovery times. This would enable policymakers to assess the return on investments associated with candidate risk management actions for expediting recovery. A general solution to Eq. (8.61) is

$$\mathbf{q}(t) = e^{-\mathbf{K}(\mathbf{I}-\mathbf{A}^*)t}\mathbf{q}(0) + \int_0^t \mathbf{K}e^{-\mathbf{K}(\mathbf{I}-\mathbf{A}^*)(t-z)}\mathbf{c}^*(z)dz \tag{8.69}$$

If the final demand $\mathbf{c}^*(t)$ is stationary, Eq. (8.69) can be further simplified to

$$\mathbf{q}(t) = (\mathbf{I}-\mathbf{A}^*)^{-1}\mathbf{c}^* + e^{-\mathbf{K}(\mathbf{I}-\mathbf{A}^*)t}\left[\mathbf{q}(0) - (\mathbf{I}-\mathbf{A}^*)^{-1}\mathbf{c}^*\right] \tag{8.70}$$

or

$$\mathbf{q}(t) = \mathbf{q}_\infty + e^{-\mathbf{K}(\mathbf{I}-\mathbf{A}^*)t}[\mathbf{q}(0) - \mathbf{q}_\infty] \tag{8.71}$$

In the equation above, $\mathbf{q}_\infty$ stands for the equilibrium inoperability, determined by the final demand vector. The exponential term $e^{-\mathbf{K}(\mathbf{I}-\mathbf{A}^*)t}\,[\mathbf{q}(0)-\mathbf{q}_\infty]$ is the temporal term that is decaying with time. When Eq. (8.70) reaches its equilibrium, it becomes

$$\mathbf{q}(t) = (\mathbf{I}-\mathbf{A}^*)^{-1}\mathbf{c}^* \tag{8.72}$$

In the equilibrium state, the DIIM reduces to the form of the static IIM. It can be viewed as a more general extension of the static IIM, and/or the static model can be viewed as a description of the dynamic model at its equilibrium condition.

8.9.2 Assessing the Industry Resilience Coefficient

As discussed in the previous section, the industry resilience coefficient is the key to modeling the DIIM. The resilience coefficient reflects the output response of each individual industry sector to an imbalance of supply and demand. For a detailed assessment of the industry resilience coefficients, consider an economy consisting of n sectors. It is assumed that initially sector i is attacked by terrorists. Based on the postattack economic response, two sets of sectors should be analyzed.

The first is sector i. After an attack, sector i will start the recovery process (e.g. rebuild the factories, machines, and so on) with a recovery rate k_i, $0 \le k_i < 1$. Depending on the risk mitigation efforts and the damage, the faster sector i recovers, the larger the value of k_i will be. The second set of sectors encompasses all the others in the economy that are affected by the attack due to their dependence on sector i. To be able to respond efficiently to the attack scenario, the production outputs of these sectors must be immediately adjusted relative to the new level of demand. Such immediate adjustments to mismatches in supply and demand correspond to the maximum recovery rates $k_i = 1, j \neq 1$.

For a special case where $k_i = 0$, and momentarily neglecting the dependence of i on j, $a_{ij}^* = 0\ \forall j \neq i$, and if final demand stays constant, the ith row in Eq. (8.68) will read as follows:

$$q_i(k+1) - q_i(k) = 0 \tag{8.73}$$

from which follows:

$$q_i(k+1) = q_i(0)\ \ \forall k = 0,1,2,\ldots,T \tag{8.74}$$

In other words, during the period of time under consideration, sector i has a constant inoperability equal to the initial perturbation.

In the following discussion, the assessment of the recovery rate of the attacked sector, corresponding to k_i, $0 < k_i < 1$, is addressed and formulated in greater detail.

In Eq. (8.69), if $k_i > 0$, $a_{ij}^* = 0\ \forall j \neq i$, and final demand stays constant, then the inoperability equation for sector i becomes

$$1-q_i(t)=1-e^{-k_i\left(1-a_{ii}^*\right)t}q_i(0) \quad (8.75)$$

Equation (8.75) is called an *individual sector recovery trajectory*. Similar to the concept of inoperability, $q_i(t)$, the term $1-q_i(t)$ is defined as the operability of sector i at time t. From this we conclude that a recovery trajectory that follows an exponential curve in temporal space will have a recovery parameter $k_i\left(1-a_{ii}^*\right)$.

The recovery trajectory of a sector can also be written in the following form typically found in reliability literature. (Note: The ratio λ/τ will be clarified in the forthcoming example.)

$$1-q_i(t)=1-e^{-(\lambda/\tau)t}q_i(0) \quad (8.76)$$

Comparing Eqs. (8.75) and (8.76) generates the following formula that can be used to estimate the resilience coefficient of sector i:

$$k_i=\frac{\lambda}{\tau\left(1-a_{ii}^*\right)} \quad (8.77)$$

when $a_{ii}^* \ll 1$, Eq. (8.77) can be approximated further, as follows:

$$k_i \approx \frac{\lambda}{\tau} \quad (8.78)$$

This equation provides the connection between the resilience coefficient (recovery rate) and recovery parameter. It justifies the definition of k_i in the DIIM as a *sector resilience coefficient* or *recovery rate* (λ/τ). As an example, the derivation of the recovery rate for the electric power generation and supply sector is shown to illustrate the process. Consider a power blackout scenario that follows an exponential recovery such that 99% recovery is achieved in 60 days. The resilience coefficient of the power sector (denoted by the subscript p) can be derived as shown below. According to Eq. (8.76) the recovery parameter can be calculated as follows:

$$\frac{\lambda}{\tau}=\frac{-\ln\left[\frac{q_p(60)}{q_p(0)}\right]}{60}=0.0768/\text{day}$$

Through the BEA data, we determined for the power sector that $a_{pp}^*=1.217\times10^{-4}$. From Eq. (8.77) we can calculate the recovery rate to be $k_p \approx 0.0768$/day. Therefore, the individual recovery trajectory for the power sector has the following function, which is depicted in Figure 8.5:

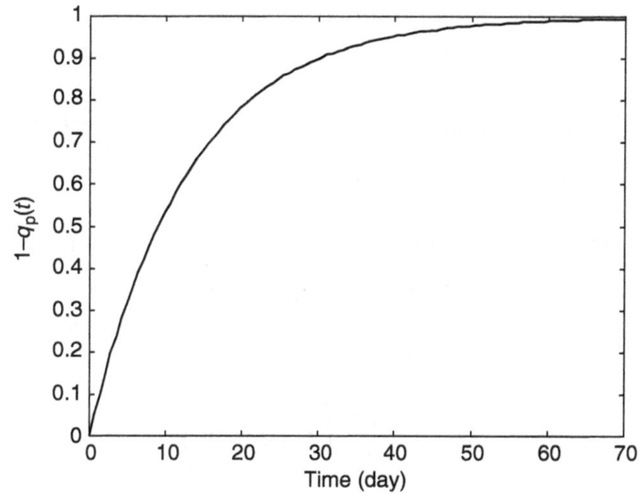

Figure 8.5 *Individual recovery trajectory of power sector.*

$$1-q_p(t)=1-e^{-k_p\left(1-a_{pp}^*\right)t}q_p(0)=1-e^{-0.0768t}q_p(0)$$

8.9.3 Assessing Economic Loss During the Recovery Through the Dynamic IIM

To understand better what the impacts of the attack will be and facilitate the tradeoff analysis in the risk management decision making, it is imperative that the economic loss during the recovery from each individual industry sector be estimated in quantitative dollar amounts for all kinds of possible scenarios. During the recovery process, it is important to know not only a sector's own loss compared with the "as-planned" level; all the indirect losses from its interdependent industry sectors should be quantified and taken into account as well. The national and regional case studies both consider the two measures in dollar amounts: (i) economic losses of the attacked sector and (ii) economic losses (direct and indirect) from all sectors. According to the dynamic model, in the continuous form the cumulative economic loss for each individual industry i is given by

$$Q_i(t)=\hat{x}_i\int_{t=0}^{T}q_i(t)\mathrm{d}t \quad (8.79)$$

$\hat{x}_i$: The "as-planned" output rate of industry i (\$/time unit).

$q_i(t)$: The inoperability of industry i at time t.
$Q_i(t)$: The cumulative economic loss of industry i by time t.
$q_i(t)$ is subject to $\mathbf{q}(t) = (\mathbf{I}-\mathbf{A}^*)^{-1}\mathbf{c}^* + e^{-\mathbf{k}(\mathbf{I}-\mathbf{A}^*)t}\left[\mathbf{q}(0) - (\mathbf{I}-\mathbf{A}^*)^{-1}\mathbf{c}^*\right]$.

Therefore $Q_i(t)$ will also be exponential due to the exponential recovery trajectory of sector i.

Similarly, the total economic loss from all n sectors by time t (denoted by $Q(t)$) is assessed as

$$Q(T) = \sum_{i=1}^{n} \left(\hat{x}_i \int_0^t q_i(t)\mathrm{d}t \right) \tag{8.80}$$

8.9.4 Comparing the Static IIM and Dynamic IIM

The static and dynamic models are consistent under equilibrium conditions, and the dynamic model can be transformed into a static model through the concept of *equivalent static inoperability*.

As noted in the previous section, the DIIM takes the form of $\dot{\mathbf{q}}(t) = \mathbf{K}[\mathbf{A}^*\mathbf{q}(t) + \mathbf{c}^*(t) - \mathbf{q}(t)$. When equilibrium is reached, $\dot{\mathbf{q}}(t) = 0$. It follows that $\mathbf{A}^*\mathbf{q}(t) + \mathbf{c}^*(t) - \mathbf{q}(t) = 0$ or $\mathbf{q}(t) = [\mathbf{I} - \mathbf{A}^*\mathbf{q}(t)]\mathbf{c}^*(t)$. Therefore, under equilibrium conditions, the dynamic model becomes the static model.

8.9.5 Specializing the Static IIM to the Dynamic IIM

In the dynamic model, suppose that sector i follows a dynamic inoperability function $q_i(t)$ from time $t = 0$ to T. A static inoperability $\bar{q}_i$ exists, defined as follows:

$$\bar{q}_i = \frac{1}{T} \int_{t=0}^{T} q_i(t)\mathrm{d}t \tag{8.81}$$

$\bar{q}_i$ is called equivalent static inoperability during $[0, T]$. Through the equivalent static inoperability, the economic loss accumulated during the dynamic recovery can be estimated statically using the following equation derived from Eq. (8.79):

$$Q_i(t) = \hat{x}_i \int_{t=0}^{T} q_i(t)\mathrm{d}t = (\hat{x}_i)(\bar{q}_i T) \tag{8.82}$$

As depicted in Figure 8.6, the scenario where the power sector recovers from 100% inoperability to 1% in 60 days has an equivalent constant static inoperability of 22% for the 60-day period.

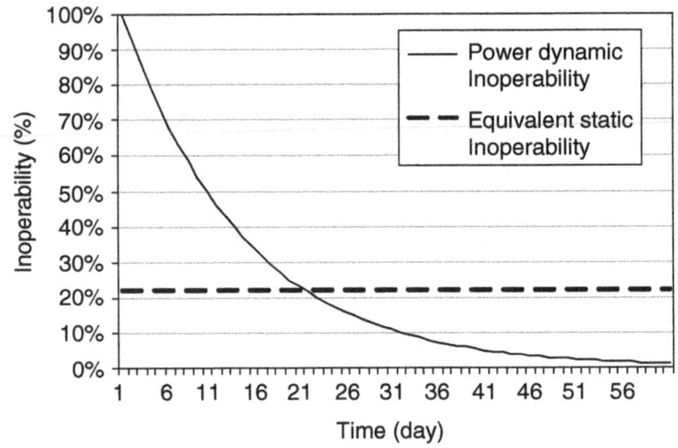

Figure 8.6 *Dynamic inoperability and equivalent static inoperability.*

8.10 PRACTICAL USES OF THE IIM

The IIM provides a computation base for risk–impact analysis that, as noted earlier, utilizes I–O data from the BEA – the agency responsible for documenting the transactions of approximately 500 producing and consuming sectors within the US economy. Through our direct use of the detailed national I–O tables published by BEA, we benefit from their intensive data collection efforts and resource base. In addition, we utilize the data available through the RIMS II for conducting regional-level analysis. This provides a solid foundation for any analysis, especially one as sensitive to the unknown as the analysis of a terrorist attack.

Given that BEA data provides each producing sector's requirements or support from other sectors (i.e. production inputs such as products and services), IIM is capable of:

- Computing the propagating impacts of diverse perturbation scenarios for various regions.
- Computing the impact of varying recovery rates for interdependent sectors.
- Computing various perspectives of impact, including *inoperability* and *economic loss*. This yields insight into societal consequences and provides a quantitative method for resource allocation.

As part of using economic-based data for analyzing a terrorist attack situation, the IIM application is based upon the assumption that the level of economic interdependencies between sectors is also representative of physical interconnectedness (i.e. in general, two sectors that have a large number of economic transactions similarly have a large degree of physical linkage). Therefore, utilizing economic interdependencies made accessible to us through BEA and RIMS II is an efficient and cost-effective alternative for comprehensively accounting for physical linkages between national sectors. (Otherwise, a similar or even greater special data collection effort would be required.) By allowing a holistic integration of sectors, IIM provides analysts with a tool for systemically prioritizing sectors deemed to be economically and physically critical, in addition to identifying those sectors whose products are critical during recovery operations. The IIM's prioritization capability also serves to avoid erroneous assumptions that might otherwise occur in preselecting "most vulnerable" sectors or commodities.

Specifically for a power sector analysis, for example, the IIM could provide the following information essential for assessing and managing the propagating impacts of a terrorist attack:

- Direct economic and power production impacts of a terrorist attack on the power generation and power supply sectors.
- Economic and production capacity impacts to electrical power users (manufacturing, commerce, household, and others) due to terrorist destruction of vulnerable electronic equipment.
- Tradeoffs between the possible reductions in economic losses and the corresponding cost of investment required for carrying out various equipment recovery/resource allocation options.
- Labor requirements to support production, delivery, and use of "as-planned" power outputs.
- Economic and production impacts due to the possible psychological effects of a terrorist attack.

8.10.1 Assumptions and Limitations of the IIM

Several assumptions from the original Leontief economic structure are retained in the IIM formulation. Many of these remain unchanged because of the need to capitalize on the vast BEA databases, which were designed specifically for Leontief's linear, deterministic, equilibrium model. It is important to address the underlying model assumptions for optimal understanding and interpretation of IIM analysis results.

8.10.2 Equilibrium Modeling of the Static IIM

The equilibrium assumption of the IIM is perhaps the hardest to manage in situations where it is highly possible to experience nonequilibrium conditions. Equilibrium implies that industry inputs and outputs will find balance with the final consumption of the sectors' outputs. In the long run such a condition is evidently true. Moreover, during a recovery process equilibrium conditions will also dominate, as industries are constantly improving their states in an interdependent fashion, as illustrated by the DIIM. However, in the short time immediately following scenarios that impose large, widespread perturbations, nonequilibrium conditions could dominate, and the IIM results would not exactly reflect real recovery production rates or economic losses.

A terrorist attack would most likely impact only a defined region of the country while fortunately leaving surrounding regions intact. The specific attributes of the attack scenario determine the size and location of the impacted region and which regional economies are categorized by equilibrium economic data. Where the impacted region is relatively small, the consequences, while large within that region, can potentially be dealt with either by importing resources from the rest of the country or exporting resources or problems out (e.g. hospital patients or unusable inventory to support increased production in other regions). These transfers would complement other activities to restore normal operations. When applying the IIM, we anticipate that the national impact on the economy and production capacity due to a terrorist attack is important, but not approaching anything like 100%, even during the time period immediately following an attack. The smaller the fraction of inoperability (e.g. <10%), the more applicable are the results obtained from IIM. This is because the overall capacity of the country to produce goods plays a significant role. It does this through redistributions that could be feasible without drastic economic adjustments that would go beyond the IIM model's assumptions of constant technology and overall economic structure. At the regional level, the results from using IIM would be more suspect, since the region under attack would likely have very large disruptions. However, the redistribution of national resources to the regional and the recovery process could occur quickly, bringing the region into a condition that is within the IIM boundaries. The initial period of redeployment and recovery would not be suitable for IIM analysis of inoperability or economic loss, but once within a close fraction of normal (e.g. 10%), the model could provide results for the remaining periods, leading up to full recovery.

For cases encompassing a relatively small region, a brief time period (days to weeks), or slight inoperability (<10%), the bulk of economic losses accumulate in the long period of final recovery, where companies operate normally, quarter by quarter, with constantly improving states dominated by equilibrium conditions. Since the bulk of economic losses are accumulated later, the costs in data and time required for building highly accurate transient models are prohibitively large. Therefore, although the initial period of redeployment and recovery is not suitable for IIM analysis of inoperability or economic loss, once within a close fraction of normal (e.g. 10%), the model could provide results for the remaining periods, leading up to full recovery. Additionally, it can provide an optimal prioritization strategy for recovery during the uncertain transient periods to minimize the overall losses when the equilibrium condition is reached.

8.10.3 Stability of the Technical Coefficient Matrix

At the core of Leontief's model and the IIM is the technical coefficient matrix A, derived from BEA's databases. These equilibrium data define deterministic guidelines for sector interactions based on the assumptions of constant technology and economic structure. The properties of these data provide the remaining assumptions that lead to model limitations.

Again, the BEA decomposes the US economy into about 500 sectors whose outputs contribute to the input of other sectors. Each element in this matrix gives a constant, deterministic value that represents the contributions to one sector, say j, from any other sector, say i, which is proportional to the output of infrastructure j. There are obvious examples where this assumption is exact and valid. Moreover, as the number of industry subdivisions increases, a first-order approximation becomes increasingly accurate. (500 sectors create a matrix defining 250 000 interdependencies.) For example, if infrastructure i is tire production and infrastructure j is automobile production, then the value of tires used by infrastructure j increases linearly with the value of automobiles produced. On the other hand, there are also cases where the linearity assumption may not be valid. For example, consider any process where human innovation is involved. As production increases proportionally, producers seek to increase resource sharing, among other measures, to consume fewer manufacturing materials. For example, an increase in the production of cars may not necessarily lead to a proportional increase in the requirement for alloy wheels (i.e. basic car models often use simple lug bolt rims with hubcaps). Nonlinearity in the use of input requirements may also be observed in the case of shared resources (e.g. multiple computers sharing external drives or peripheral devices). Thus, in addition to the extreme cost of such added modeling, the database would become overwhelming and probably less meaningful to the analyst.

It is also important to note that the constants that define the relationships between industries are time invariant and deterministic. This stems from the fact that the BEA generates data every 5 years directly from US Census data, because of economic momentum that causes values to change very little from year to year. Therefore the variance of commodity flows changes very little from year to year and in most industry sectors. (This is intuitive because production procedures change very little over short periods of time.) Returning to the car example, four tires for every one car will certainly vary negligibly over time. However, this obviously will not always be the case. In order to protect against major flaws, studies have compared various years of economic data. Analysis results have shown only negligible changes between two recent 5-year periods.

8.11 EXAMPLE PROBLEMS

8.11.1 Example Problem 1

The following example demonstrates the workings of the inoperability I–O risk model. Twelve infrastructure sectors have been selected (Santos, 2003): (i) coal, (ii) petroleum refining, (iii) railroads and related services, (iv) trucking and courier, (v) water transportation, (vi) air transportation, (vii) telephone and telegraph, communication services, (viii) electric services, (ix) water supply and sewerage systems, (x) banking, (xi) eating and drinking places, and (xii) hospitals. For illustration purposes, the interdependency matrix (***A***) for these sectors shown in Table 8.4 has been generated through a transformation of the I–O matrices published by the BEA.

Assume that the electric service infrastructure ($k = 8$) is attacked, causing its operability to be reduced by 20% (i.e. $c_8 = 0.20$). Also assume that the rest of the elements of the **c** vector are zeroes (i.e. electric service is the only infrastructure that was attacked). Applying Eq. (8.11), $\mathbf{x} = \mathbf{Ax} + \mathbf{c}$, Table 8.5 presents the inoperability vector (**x**) after the attack, arising from infrastructure interconnectedness.

The framework imbedded in our infrastructure inoperability I–O risk model is represented by selected infrastructures as depicted in Figure 8.7, where $\{x_{kj}\}$ and c_k are the inputs to infrastructure k, and $\{x_{jk}\}$ is the output of infrastructure k. The matrix $\boldsymbol{A} = \{a_{kj}\}$ plays a central role in the problem definition. When the system is in a perfect condition – a condition where all components are operating flawlessly (i.e. $x_k = 0$ for all k and $\sum_k x_k = 0$) – then it is said to be in *ground state*.

In risk assessment, we ask the triplet questions posed in Chapter 9: (i) What can go wrong? (ii) What is the likelihood that it would go wrong? (iii) What are the consequences? (Kaplan and Garrick, 1981). Applying these questions to the analysis of risks inherent in a system of interconnected infrastructures requires in-depth definition and understanding of S_0 – the system's *as-planned scenario*. Any deviations from S_0 are *risk scenarios* (S_i's), which can have adverse impacts on the functionality of a system. Hierarchical Holographic Modeling (HHM) (Haimes, 1981, 2016) (see Chapter 3) and the theory of scenario structuring (TSS) (Kaplan, 1996) (see Chapter 9) are risk assessment methods appropriate for defining the S_0, and the corresponding S_is, of a particular system. Furthermore, the fusion of HHM and TSS, as described by Kaplan et al. (2001), offers a risk assessment framework for systematically identifying the myriad S_is that can perturb a system's S_0.

TABLE 8.4 Interdependency Matrix (*A*)

	$j=1$	$j=2$	$j=3$	$j=4$	$j=5$	$j=6$	$j=7$	$j=8$	$j=9$	$j=10$	$j=11$	$j=12$
$i=1$	0.1130	0.0826	0.2236	0.0667	0.0060	0.0118	0.0090	0.1204	0.0000	0.0644	0.0370	0.0000
$i=2$	0.0000	0.0618	0.0026	0.0050	0.0010	0.0003	0.0015	0.0120	0.0000	0.0134	0.0036	0.0000
$i=3$	0.0000	0.0308	0.0617	0.0020	0.0002	0.0019	0.0010	0.0013	0.0000	0.0252	0.0067	0.0000
$i=4$	0.0000	0.0385	0.0025	0.1569	0.0003	0.0021	0.0148	0.0050	0.0000	0.0112	0.0080	0.0000
$i=5$	0.0002	0.0848	0.0020	0.0169	0.1247	0.0066	0.0046	0.0170	0.0000	0.1203	0.0151	0.0020
$i=6$	0.0000	0.1307	0.0014	0.0047	0.0006	0.0614	0.0245	0.0050	0.0000	0.0185	0.0576	0.0000
$i=7$	0.0000	0.0006	0.0001	0.0012	0.0000	0.0016	0.1236	0.0030	0.0000	0.0137	0.0055	0.0000
$i=8$	0.0144	0.0093	0.0338	0.0035	0.0008	0.0012	0.0018	0.0001	0.0000	0.0194	0.0046	0.0000
$i=9$	0.0000	0.1635	0.0313	0.4487	0.0000	0.0455	1.0060	0.7701	0.0000	0.8386	0.2021	0.0006
$i=10$	0.0000	0.0005	0.0001	0.0064	0.0000	0.0013	0.0062	0.0033	0.0000	0.0474	0.0038	0.0000
$i=11$	0.0000	0.0011	0.0008	0.0054	0.0000	0.0010	0.0023	0.0118	0.0000	0.0074	0.0150	0.0000
$i=12$	0.0000	0.0015	0.0009	0.0040	0.0000	0.0014	0.0047	0.0058	0.0000	0.0040	0.0168	0.0000

TABLE 8.5 Inoperabilities Resulting from 20% Degraded Functionality of the Electric Infrastructure

Infrastructure (i)	$i=1$	$i=2$	$i=3$	$i=4$	$i=5$	$i=6$	$i=7$	$i=8$	$i=9$	$i=10$	$i=11$	$i=12$
Inoperability (x_i)	0.0279	0.0026	0.0004	0.0014	0.0044	0.0016	0.0007	0.2005	0.1574	0.0007	0.0024	0.0012

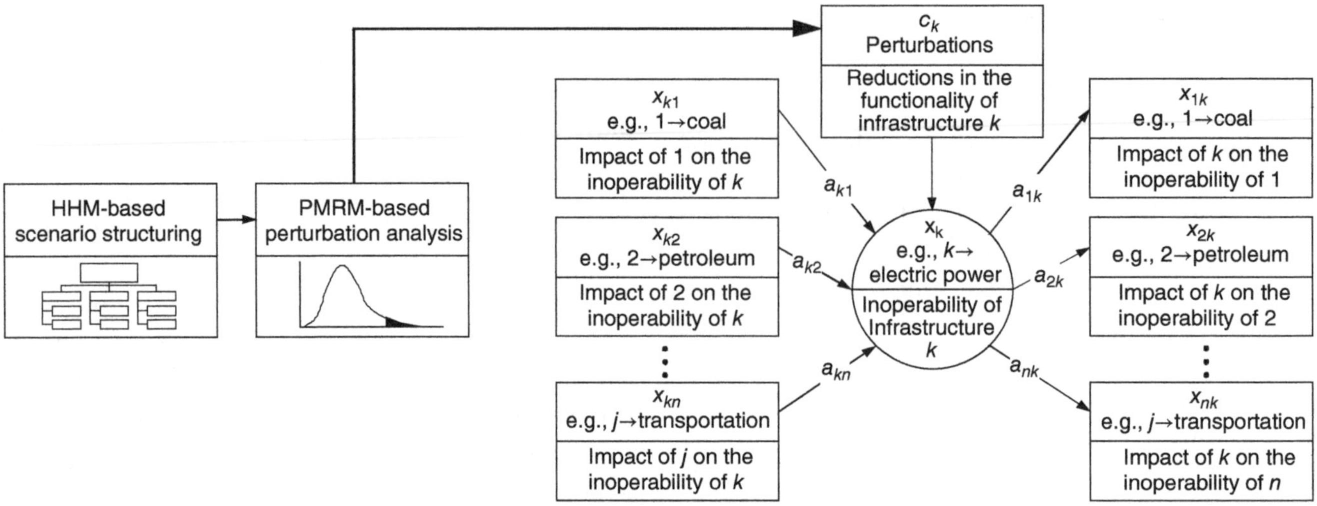

Figure 8.7 *Framework for the infrastructure inoperability I–O model.*

So far, the treatment of the term *perturbation* (i.e. events that trigger a reduction in the functionality of a system) is compatible with the above risk assessment framework. In the 12-infrastructure example, however, we specifically analyze a scenario corresponding to a reduction of 20% in the functionality of the electric power infrastructure (i.e. $c_k = 0.2$). In doing so, we skip the important process underlying the quantification of this perturbation (c_k). To adhere to the risk assessment framework in the infrastructure inoperability I–O risk model, identifying the system's risk scenarios should be a prerequisite to generating specific c_ks. For instance, a bombing incident could manifest as a 0.2 (or 20%) reduction in the operability of the kth infrastructure. Through the use of an elaborate HHM-based scenario structuring process, we posit c_k to be a preliminary *output* (or *throughput*), which is rooted in a risk scenario. The perturbation c_k resulting from a specific risk scenario then becomes the *input* to our IIM model. Additionally, the Partitioned Multiobjective Risk Method (PMRM) (see Appendix) (Asbeck and Haimes, 1984; Haimes, 2016) will be utilized to generate the magnitude of the perturbation. Since perturbations are generally nondeterministic, a distribution may be more appropriate to use than point estimates. For example, a specific risk scenario may degrade at least 1% of a system's functionality, or worse, 50%. The PMRM enables analyzing various "what ifs" based on a risk scenario's perceived distribution. The conditional expectations used in the PMRM can effectively distinguish low-consequence/high-probability events from high-consequence/low-probability events (i.e. extreme events).

8.11.2 Example Problem 2

To show how to apply the Leontief equation, we solve the following example. Suppose we have a system with two subsystems (Jiang, 2003). The inoperability of these two subsystems is represented as x_1 and x_2, respectively. Now suppose a failure at subsystem 2 will lead subsystem 1 to be 80% inoperable, and a failure at subsystem 1 will lead subsystem 2 to be 20% inoperable.

Thus, the A-matrix reads

$$A = \begin{pmatrix} 0 & 0.8 \\ 0.2 & 0 \end{pmatrix} \tag{8.83}$$

Now suppose that subsystem 2 loses 60% of its functionality due to an external perturbation (such as an attack by terrorists). We want to know the inoperability of the two subsystems. Plugging the A-matrix into the Leontief equation, we have

$$\begin{pmatrix} x_1 \\ x_2 \end{pmatrix} = \begin{pmatrix} 0 & 0.8 \\ 0.2 & 0 \end{pmatrix} \begin{pmatrix} x_1 \\ x_2 \end{pmatrix} + \begin{pmatrix} 0 \\ 0.6 \end{pmatrix} = \begin{pmatrix} 0.8x_2 \\ 0.2x_1 + 0.6 \end{pmatrix} \tag{8.84}$$

Solving the equation, we get $x_1 = 0.571$ and $x_2 = 0.714$. This means the inoperability of subsystem

1 is 0.571, even though it was *not* attacked directly by terrorists. This effect is due purely to the interconnectedness between the two subsystems. The inoperability of subsystem 2 also increases by 0.114 due to its connection to subsystem 1.

We can also evaluate the impact of an attack of varying intensity on the operability of the system. Suppose the intensity of the attack is $h \times 100\%$, meaning that $h \times 100\%$ of the operability of subsystem 2 is lost due to the attack alone. The Leontief equation reads

$$\begin{pmatrix} x_1 \\ x_2 \end{pmatrix} = \begin{pmatrix} 0 & 0.8 \\ 0.2 & 0 \end{pmatrix} \begin{pmatrix} x_1 \\ x_2 \end{pmatrix} + \begin{pmatrix} 0 \\ h \end{pmatrix} = \begin{pmatrix} 0.8x_2 \\ 0.2x_1 + h \end{pmatrix} \tag{8.85}$$

The solution is $x_1 = 0.952h$, $x_2 = 1.190h$, for $0 \leq h \leq 0.84$, and $x_1 = 0.8$, $x_2 = 1.0$, for $0.84 < h \leq 1$. Note that due to the constraints $0 \leq x_1, x_2 \leq 1$, subsystem 2 fails completely when the external attack brings down 84% of its operability. The remaining 16% is taken away by its dependency on subsystem 1.

8.11.3 Example Problem 3

Consider a system consisting of four subsystems:

Subsystem 1: A power plant.
Subsystem 2: A transportation system (roads, signs, signaling facilities, etc.).
Subsystem 3: A hospital.
Subsystem 4: A grocery store.

Suppose the A-matrix for this system is as simple as

$$A = \begin{pmatrix} 0 & 0.9 & 0 & 0 \\ 0.4 & 0 & 0 & 0 \\ 1 & 0.8 & 0 & 0 \\ 1 & 0.9 & 0 & 0 \end{pmatrix} \tag{8.86}$$

In this example, the A-matrix can be interpreted in the following manner: If the power plant fails completely, then the transportation system can perform only 60% of its functionality, whereas both the hospital and the grocery store cannot operate at all. If the transportation system completely fails so that the workers cannot get to work and trucking delivery becomes impossible, then the power plant and the grocery store can perform only 10% of their full functionality, and the hospital can perform only 20% of its functionality. On the other hand, the inoperability of the hospital or the grocery store has no impact on the operation of the power plant or the transportation system, nor do they have any appreciable impact on each other.

Now suppose a major hurricane hits the area and destroys 50% of the functionality of the transportation system. Due to this disaster, many workers are not able to get to work, and the delivery trucks are not able to arrive as scheduled. Using the given A-matrix, we have the following Leontief equation:

$$\begin{pmatrix} x_1 \\ x_2 \\ x_3 \\ x_4 \end{pmatrix} = \begin{pmatrix} 0 & 0.9 & 0 & 0 \\ 0.4 & 0 & 0 & 0 \\ 1 & 0.8 & 0 & 0 \\ 1 & 0.9 & 0 & 0 \end{pmatrix} \begin{pmatrix} x_1 \\ x_2 \\ x_3 \\ x_4 \end{pmatrix} + \begin{pmatrix} 0 \\ 0.5 \\ 0 \\ 0 \end{pmatrix} \tag{8.87}$$

The solution to this equation is $(x_1, x_2, x_3, x_4) = (0.78, 0.70, 1, 1)$. Thus, due to the ravages of the hurricane, the power plant loses 78% of its functionality, the transportation system loses 70% of its operability, and the hospital and the grocery store are not able to operate at all.

As with the previous example, we can also evaluate the impact of a hurricane attack of varying intensity on the operability of the system. Suppose the intensity of the attack is $h \times 100\%$ with respect to the transportation system, meaning that $h \times 100\%$ of the operability of the transportation system is lost due to the hurricane alone. Solving the Leontief equation yields

$$(x_1, x_2, x_3, x_4) = (1.41h, 1.56h, 2.66h, 2.80h), \quad \text{for } 0 \leq h \leq 0.357 \tag{8.88}$$

When $h = 0.357$, the value of x_4 reaches 1. The value of x_3 reaches 1 when h increases to 0.376. This means that when the magnitude of the attack on the transportation system is 0.376, the actual inoperability of the power plant is 0.53, and the actual inoperability of the transportation system is 0.59 due to the compound effect of the direct attack and the inoperability of the power plant. Both the grocery store and the hospital are out of commission at this point due to the inoperability of both the transportation system and the power plant. When h reaches the level of 0.641, x_2 becomes 1, which means the transportation system completely fails.

8.12 SUMMARY OF THE INOPERABILITY INPUT–OUTPUT MODEL (IIM)

The IIM is based on Leontief's I–O model to study the effects that intra- and interconnectedness can have on the inoperability of a system composed of a large number of components (subsystems). We proposed viewing the generic Leontief-based model as a first-order approximation of a general model, and we discussed the approximation of linearization under various circumstances.

The underlying economic data utilized in the IIM provide each sector's requirements of support from other sectors (i.e. production inputs such as products and services). The IIM is capable of calculating the propagating impacts of diverse perturbation scenarios for various regions. In using economic-based data for analyzing a terrorist attack, the application of the IIM is based upon the observation that the level of economic interdependency between sectors is often representative of physical interconnectedness (i.e. in general, two sectors that have a large volume of economic transactions have a similarly large degree of physical linkages). Utilizing interdependencies based on economic data gives IIM the capability to comprehensively assess the vulnerability of approximately 500 nationwide sectors given an attack to one or multiple sectors. By allowing the holistic integration of sectors, the IIM (i) provides analysts with a tool for systemically prioritizing sectors deemed to be economically critical and (ii) identifies those sectors whose continued operability is critical during recovery operations.

The following features and capabilities of the IIM make it particularly useful for conducting an impact analysis of a terrorist attack:

- The IIM considers numerous sectors of the economy. This is to avoid misleading results that can stem from studying only those sectors that are selected because they are related to direct vulnerabilities without also accounting for indirect ripple effects.
- The IIM provides a comprehensive ranking of approximately 500 BEA sectors according to inoperability and economic loss impact metrics. It does this in a graphic format (e.g. histograms) that is relevant to risk management of terrorist attacks.
- The IIM is capable of modeling workforce recovery. This enables the modeling of critical and electronically vulnerable sectors such as the power and health service sectors and identifying the most essential personnel for response to a terrorist attack.
- The IIM provides various geographic resolutions (e.g. counties, states, and economic regions) that can be customized via RIMS II data to closely approximate different attack strengths and intensities.
- The DIIM is capable of modeling different temporal frames of recovery (e.g. recovery rates for different sectors) and establishing various interdependent adjustments to levels of equilibrium that may occur after a terrorist attack.

Developing a meaningful model capable of capturing the complex essence of the intra- and interconnectedness of our critical infrastructures is by any account a daunting task that requires the contributions of many individuals from several disciplines. The IIM contributes to the gigantic effort needed to better our understanding of these dependencies and subsequently to manage more cost effectively the threats and risks that critical infrastructures encounter today. Thus, this model is another important tool in a comprehensive risk assessment and management framework for ensuring the integrity and continued operability of our complex critical infrastructures.

However, before the model is brought to bear on any complex problems, there are three issues meriting special attention. First and foremost, we must define inoperability for each of the subsystems. This is important to capture the essence of the problem and to be sure that the characteristics of all subsystems pertinent to the objectives of the problem are appropriately and effectively represented. Second, we must make sure that the assumption of linearization is justified. If the circumstance suggests that nonlinearity reigns, then starting from Eq. (8.9a), we must unearth the underlying relationships (f-functions) by looking into the detailed structure and coupling of the system. Last but not least, the matrices A, B, and C play overarching roles in both formulating and solving the problem. To determine the elements of any of these matrices could become an overwhelming task and may entail very extensive data collecting and data mining. The extent to which we can project the principles of the Leontief I–O model into analyzing the I-I among critical infrastructures constitutes an opportunity to those in charge of ensuring their continued operation, as well as a challenge to the research community. In this sense, the Leontief-based I–O infrastructure model is intended to be used for at least two purposes:

- The primary and dominant purpose is to improve our understanding of the effects of impacts on the continued and sustained operability of our critical infrastructures under all plausible conditions.
- The secondary purpose is to serve as a tool to allocate resources for an effective process of risk assessment and risk management. In particular, the added security, survivability, assurance, and integrity of our critical infrastructures can be best accomplished through a multiobjective-based risk management framework, where all important costs, benefits, and risks are traded off in a systemic way.

The challenge in realizing both purposes undoubtedly lies in further developing the needed theoretical foundations, methodological instruments, and essential vast appropriate database. Countries worldwide, including the United States, have successfully developed and deployed Leontief I–O models for their economies. Their accomplishments in meeting similar data collection challenges (for a large number of coefficients in the Leontief model) should be a source of encouragement and optimism in our quest to ensure the operability of our complex critical infrastructures.

8.13 COMMUNICATIONS, ELECTRICITY, WATER, AND SUPPLY CHAIN AS INTERDEPENDENT AND INTERCONNECTED COMPLEX SYSTEMS OF SYSTEMS

We present in Chapter 10 four studies, which addressed, among others, the following three infrastructures: (i) communication, (ii) electricity, and (iii) water. In this chapter, we add a fourth critical sector of the economy, (iv) supply chain, to which Chapter 11 is devoted. To augment the contributions of the IIM in quantifying the impacts of the disruption of one sector of the economy on all other interdependent and interconnected sectors, we model in this section the I-I that exist among all four safety-critical systems by (i) identifying shared states and other essential entities among all four Complex SoS; (ii) developing appropriate fault trees, which translate the above into systems connected in series (OR Gates) and in parallel (AND Gates); and (iii) interpreting the results from (i) and (ii), as well as drawing conclusions on the ensuing risks from the event of disabling any one of the four Complex SoS (communication, electricity, water, and supply chain) to the other three. In many ways, deploying the IIM introduced in this chapter on the same four Complex SoS would also generate pedagogically valuable results, albeit in different forms and quantification.

Fault-tree analysis would crystalize the I-I among the four systems in terms of subsystems/systems connected in series and in parallel, thus highlighting the criticality of systems connected in series (via OR Gates). In Chapter 10 we provided a detailed introduction to fault trees, minimal cut sets, and the criticality of the latter to the failure of the entire Complex SoS (with the knowledge if any member of that the minimal cut-set connection fails, then the entire SoS would also fail). The uninitiated reader with fault trees is encouraged to consult the Appendix of this book and the case studies in Chapter 10.

There is a significant difference between the level of granularity of the consequences resulting from an initiating event provided by the IIM versus the shared states and other essential entities.

The fault tree depicted in Figure 8.8 highlights the I-I in terms of OR Gates. In other words, if any one of the four (water, electricity, communications, and supply chain) fails, then the entire communication, electricity, water, and supply chain Complex SoS fails.

The above fundamental differences between results provided by the (i) IIM and (ii) shared states and other essential entities, supported with fault-tree analysis, provide stakeholders and modelers two entire different avenues with which to model and assess the propagation of adverse consequences to sectors of the economy.

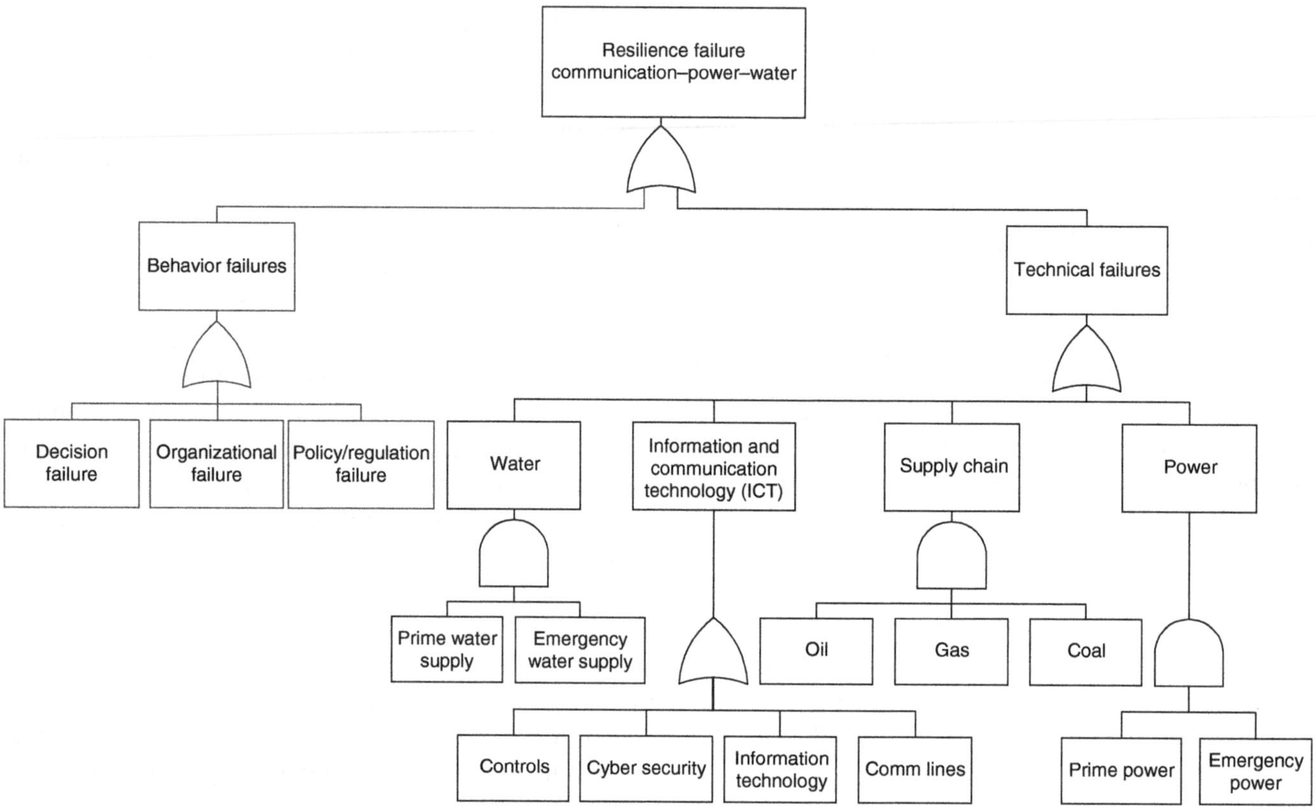

Figure 8.8 *Communication, electricity, water, and supply chain fault tree.*

REFERENCES

Asbeck, E.L. and Haimes, Y.Y. (1984). The partitioned multiobjective risk method (PMRM). *Large Scale Systems* **6**(1): 13–38.

Barker, K. (2008). Extensions of inoperability input-output modeling for preparedness decisionmaking: uncertainty and inventory. Ph.D. dissertation. University of Virginia.

Barker, K. and Haimes, Y.Y. (2008). Uncertainty Analysis of Interdependencies in Dynamic Infrastructure Recovery: Applications in Risk-Based Decisionmaking. *Tech. Report-08-2008*. Center for Risk Management of Engineering Systems. University of Virginia, Charlottesville, VA.

Blanc, M. and Ramos, C. (2002). *The Foundations of Dynamic Input-Output Revisited: Does Dynamic Input-Output Belong to Growth Theory?* Oviedo: University of Oviedo. Department of Applied Economics. Faculty of Economic and Business Sciences https://www.researchgate.net/profile/Carmen_Ramos7/publication/28092604_The_foundations_of_dynamic_input-output_revisited_Does_dynamic_input-output_belong_to_growth_theory/links/5465da1a0cf2052b509fb3c3.pdf (accessed 6 February 2018.

Converse, A.O. (1971). On the extension of input–output analysis to account for environmental externalities. *American Economic Review* **61**: 197–198.

Crowther, K.G. (2007). Development and deployment of the multiregional inoperability input-output model (MRIIM) for strategic preparedness of interdependent regions. Ph.D. dissertation. University of Virginia.

Crowther, K.G. and Haimes, Y.Y. (2005). Application of the inoperability input–output model (IIM) for systemic risk assessment and management of interdependent infrastructures. *Systems Engineering* **8**(4): 323–341.

Embrechts, P., Kluppelberg, C., and Mikosch, T. (1997). *Modeling External Events for Insurance and Finance*. New York: Springer.

Ernst & Young (2002). *Manhattan Lodging Forecast*. New York: Ernst & Young Real Estate Advisory Group.

Galea, S., Ahern, J., Resnick, H. et al. (2002). Psychological sequelae of the September 11 terrorist attacks in New York City. *The New England Journal of Medicine* **346**(13): 982–987.

Griffin, J. (1976). *Energy Input–Output Modeling*. Palo Alto, CA: Electric Power Research Institute.

Haimes, Y.Y. (1977). *Hierarchical Analyses of Water Resources Systems: Modeling and Optimization of Large-Scale Systems*. New York: McGraw-Hill.

Haimes, Y.Y. (1981). Hierarchical holographic modeling. *IEEE Transactions on Systems, Man and Cybernetics* **11**(9): 606–617.

Haimes, Y.Y. (2002). Roadmap for modeling risks of terrorism to the homeland. *ASCE Journal of Infrastructure Systems* **8**(2): 35–41.

Haimes, Y.Y. (2016). *Risk Modeling, Assessment, and Management*, 4e. Hoboken, NJ: Wiley.

Haimes, Y.Y. and Horowitz, B.M. (2004). Adaptive two-player hierarchical holographic modeling game for counterterrorism intelligence analysis. *Journal of Homeland Security and Emergency Management* **1**(3): 1038.

Haimes, Y.Y. and Jiang, P. (2001). Leontief-based model of risk in complex interconnected infrastructures. *ASCE Journal of Infrastructure Systems* **7**(1): 1–12.

Haimes, Y.Y. and Nainis, W.S. (1974). Coordination of regional water resource supply and demand planning models. *Water Resources Research* **10**(6): 1051–1059.

Haimes, Y.Y., Horowitz, B.M., Lambert, J.H. et al. (2005a). Inoperability input-output model (IIM) for interdependent infrastructure sectors. I: Theory and methodology. *ASCE Journal of Infrastructure Systems* **11**: 67–79.

Haimes, Y.Y., Horowitz, B.M., Lambert, J.H. et al. (2005b). Inoperability input–output model (IIM) for interdependent infrastructure sectors, II: Case study. *ASCE Journal of Infrastructure Systems* **11**: 80–92.

Intriligator, M.D. (1971). *Mathematical Optimization and Economic Theory*. Englewood Cliffs, NJ: Prentice-Hall.

Isard, W. (1960). *Methods of Regional Analysis: An Introduction to Regional Science*. Cambridge, MA: MIT Press.

Isard, W., Azis, I.J., Drennan, M.P. et al. (1998). *Methods of Interregional and Regional Analysis*. Brookfield, VT: Ashgate.

Jiang, P. (2003). Input–output inoperability risk model and beyond: a holistic approach. Ph.D. dissertation. University of Virginia.

Jiang, P. and Haimes, Y.Y. (2004). Risk management for Leontief-based interdependent systems. *Risk Analysis* **24**(5): 1215–1229.

Kaplan, S. (1996). *An Introduction to TRIZ, the Russian Theory of Inventive Problem Solving*. Southfield, MI: Ideation International, Inc.

Kaplan, S. and Garrick, B.J. (1981). On the quantitative definition of risk. *Risk Analysis* **1**(1): 11–27.

Kaplan, S., Haimes, Y.Y., and Garrick, B.J. (2001). Fitting hierarchical holographic modeling (HHM) into the theory of scenario structuring and a refinement to the quantitative definition of risk. *Risk Analysis* **21**(5): 807–819.

Krause, U. (1992). Path stability of prices in a nonlinear Leontief model. *Annals of Operations Research* **37**: 141–148.

Lahr, M.L. and Dietzenbacher, E. ed. (2001). *Input–Output Analysis: Frontiers and Extensions*. New York: Palgrave.

Lahr, M.L. and Stevens, B.H. (2002). A study of regionalization in the generation of aggregation error in regional input–output models. *Journal of Regional Science* **42**(3): 477–507.

Lee, K.S. (1982). A generalized input–output model of an economy with environmental protection. *Review of Economics and Statistics* **64**: 466–473.

Leontief, W.W. (1951a). Input/output economics. *Scientific American* **185**(4): 15–21.

Leontief, W.W. (1951b). *The Structure of the American Economy, 1919–1939*, 2e. New York: Oxford University Press.

Leontief, W.W. (1986). *Input–Output Economics*, 2e. New York: Oxford University Press.

Liew, C.J. (2000). The dynamic variable input-output model: an advancement from the Leontief dynamic input–output. *The Annals of Regional Science* **34**: 591–614.

Miller, R.E. and Blair, P.D. (1985). *Input-Output Analysis: Foundations and Extensions*. Englewood Cliffs, NJ: Prentice-Hall.

Miller, R.E., Polenske, K.R., and Rose, A.Z. (1989). *Frontiers of Input–Output Analysis*. New York: Oxford University Press.

Olsen, J.R., Beling, P.A., Lambert, J.H., and Haimes, Y.Y. (1997). Leontief input-output model applied to optimal deployment of flood protection. *Journal of Water Resources Planning and Management* **124**(5): 237–245.

Proops, J.L. (1984). Modeling the energy-output ratio. *Energy Economics* **6**: 47–51.

Santos, J.R. (2003). Interdependency analysis: extension to demand reduction input-output inoperability modeling and portfolio selection. Ph.D. dissertation. University of Virginia, Charlottesville, VA.

Santos, J.R. and Haimes, Y.Y. (2004). Modeling the demand reduction input–output (I–O) inoperability due to terrorism of interconnected infrastructures. *Risk Analysis* **24**(6): 1437–1451.

Susser, E.S., Herman, D.B., and Aaron, B. (2002). Combating the terror of terrorism. *Scientific American* **287**(2): 70–77.

U.S. Department of Commerce (2003). *Digital Economy*. Washington, DC: US Government Printing Office http://www.esa.doc.gov/DigitalEconomy2003.cfm (accessed 1 November 2017.

U.S. Department of Commerce, Bureau of Economic Analysis (1998). *Bureau Benchmark Input–Output Accounts of the United States, 1992*. Washington, DC: US Government Printing Office.

U.S. Federal Aviation Administration (2002). *Aviation Industry Overview Fiscal Year 2001*. Washington, DC: Federal Aviation Administration, Office of Aviation Policy and Plans.

9

Guiding Principles for Modeling and Managing Complex Systems of Systems

PART I: RISK MODELING OF COMPLEX SYSTEMS OF SYSTEMS: AN OVERVIEW

9.I.1 INTRODUCTION

The terms complexity and complex systems have evolved and assumed different connotations during the last 50 years and thus the way to model and manage complexity. For example, decision makers and stakeholders by *necessity aspire to a reliable and trustworthy risk analysis, not only for mundane single systems but also especially for interdependent and interconnected complex systems of systems (SoS) (Complex SoS)*. Indeed, in their essence, most cyber–physical, organizational, and government enterprises, now and in the future, belong to this category. New theory and methodology in risk analysis must respond to and accommodate the above legitimate aspiration. In this chapter, we communicate with systems analysts and prospective decision makers and stakeholders by translating the essence of complexity through a case study.

Two major fundamentals guide our reevaluation of current practices of risk analysis when addressing the risk to interdependent and interconnected Complex SoS. These are grounded on (i) understanding and evaluating the interdependencies and interconnectedness (I-I) within and among the Complex SoS manifested through shared/common states and other *essential entities* within and among the systems and subsystems that make up Complex SoS. The term "essential entities" connotes shared/common decisions, decision makers, stakeholders, resources, organizational behavior and norms, policies and procedures, management, culture, and others; and (ii) being cognizant of and responsive to the emergent nature of Complex SoS, given the *Evolving Base*.

The *Evolving Base* is represented by the following dynamic shifting rules and realities for each subsystem and for the entire Complex SoS: (i) goals and objectives; (ii) stakeholders, decision makers, and interest groups; (iii) organizational, political, and budgetary baselines; (iv) reorganization and reallocation of key personnel; (v) emergent technology

Modeling and Managing Interdependent Complex Systems of Systems, First Edition. Yacov Y. Haimes.

and its deployment; and (vi) requirements, specifications, delivery, users, and clients (Haimes, 2016). Whether large or small, natural or in a constructed environment, most systems are subject to myriad sources of risk, originating from within or outside the systems themselves. These risks are even more prevalent in interdependent and interconnected Complex SoS.

In the past decades, risk analysis has become a dominant and requisite area of expertise in almost every discipline as well as in government, public, and private organizations. By its nature, risk analysis is an intricate, dynamic process – a marriage of the arts and sciences – tailored to address specific sources of risk to Complex SoS. It follows, then, that the balance between quantitative and qualitative risk analysis will depend on the problem at hand and the type of the system. This is especially true for Complex SoS, which are unique in their complexity. The ten principles set forth in this chapter are thus intended to guide both quantitative- and qualitative-centered risk analyses. Furthermore, the approach to best define, model, and quantify the multidimensional risk function of a single system and Complex SoS will depend on the discipline and the expertise of its risk modelers and decision makers. For example, while the disciplines of systems engineering and risk analysis share multiple precepts in philosophy, theory, methodology, and practice, each discipline has evolved through separate paths and thus has garnered distinctive followers. Although the principles we set forth in this chapter may be applied to a range of disciplines, to retain focus we draw from, and are guided by, the theory, methodology, and practice of both risk analysis and systems engineering.

9.I.2 FUNDAMENTAL DIFFERENCES BETWEEN RISK ANALYSES OF SINGLE SYSTEMS AND COMPLEX SYSTEMS OF SYSTEMS

Complex SoS are commonly composed of myriad interdependent and interconnected subsystems; each of which can be characterized by a hierarchy of shared or interacting components and each with multiple functions, operations, databases, costs, and stakeholders. All systems, natural and the constructed environment, are commonly subjected to, and thus affected by, multiple adverse initiating events that could originate from outside or from within one or more subsystems that constitute Complex SoS.

The fundamental differences between emergent forced changes (EFCs) (that connote internal or external positive or adverse events affecting some systems or the entire Complex SoS positively or negatively) on a single system, as opposed to the impacts on Complex SoS, stem from the following intrinsic characteristics of Complex SoS: (i) They are commonly composed of innumerable subsystems, which in their essence constitute *systems of systems*; (ii) each subsystem can be characterized by a hierarchy of shared or interacting components; and (iii) each with multiple functions, operations, databases, costs, and stakeholders. Clearly, no single model can ever successfully capture the essence of such systems – their multiple dimensions and perspectives. Furthermore, almost every living entity, all infrastructures, and both the natural and constructed environment are in their essence Complex SoS, because they are composed of multiple interdependent and interconnected systems that *share* (*have in common*) most, if not all, of the following building blocks (of models and reality): (i) states that constitute the fundamental entities of all systems (more will be written subsequently on state-space theory); (ii) decision makers and stakeholders; (iii) resources; (iv) organizational setups; (v) emergent behavior and evolution; (vi) often competing and conflicting goals and objectives; (vii) historical and cultural identity; and (viii) time frame, among others. These inherent natures and characteristics of the systems and subsystems that make up Complex SoS, which differentiate them from single systems, are commonly subjected to fundamentally different sources of risk to each system/subsystem and thus to the entire Complex SoS as a whole. The need to review and recalibrate the intricate process of risk modeling, assessment, management and communication for Complex SoS is imperative.

The evolution of the field of risk analysis, since the founding of the Society for Risk Analysis (SRA) and its journal *Risk Analysis*, requires continuous reassessment of its prevailing theory, methodology, and practice. Thus, current risk analysis applied to a single system must be fundamentally extended when applied to Complex SoS, which constitute multiple systems with emergent behaviors. The complexity of these systems results primarily from the I-I within and among the systems and subsystems of Complex SoS. Consider, for example, the interdependencies and interconnections among three common cyber–physical infrastructures: electricity, communications, and water (and the communities they serve) presented in the case studied in Chapter 10. A major initiating event that may cause the failure of the electrical system would result in adverse consequences to the other two cyber–physical infrastructures, because of their dependency on electricity; and neither one can operate without electricity. *Clearly, there is a need to understand and thus model the pathway through which the failure of one system propagates to other interconnected systems.* An effective risk analysis of Complex SoS requires a clear understanding of the SoS configurations to enable the identification of specific critical failure modes. This enables the development and deployment of effective risk mitigation and other management strategies. Current risk models of systems, which do not consider the inherent interdependencies among other systems, are likely to be inferior to those models that do. The premise of this chapter and book is that the interdependencies and interconnections are best understood and thus modeled through the *shared states* within the SoS (and other *shared essential entities*) and of the SoS as a whole. (The term "essential entities," which was defined earlier, connotes shared/common decisions, decision makers, stakeholders, resources, organizational behavior and norms, policies and procedures, management, culture, and others.) Scientific literature has many definitions of state variables in modeling, because they play a significant role in estimating the future behavior of the systems for any given inputs. Recall Chen's (2012) conceptual definition of state variables posits that the state $x(t_o)$ of a system at time t_o is the information at time t_o that, together with the input $u(t)$, for $t \geq t_o$ determines uniquely the output $y(t)$ for all $t \geq t_o$. Nise also emphasizes that state variables represent the smallest set of linearly independent system variables such that the values of the members of the set at time t_o along with known forcing functions completely determine the value of all system variables for all $t \geq t_o$ (2015). Thus, the behavior of the states of the system, as a function of time and other inputs, enables modelers and risk analysts to determine, under certain conditions, its future behavior for any given *inputs* or *initiating events*. Indeed, in industrial production and management, the feedback control process (as an integral part of risk management and quality control) is predicated on the fact that the smooth operations of the multiple interconnected subsystems are built on the knowledge of the operational states of each subsystem at any instant. As a trivial example, to determine the reliability and functionality of a car, one must know the *states* of the fuel, oil, tire pressure, and the states of other mechanical and electrical components (due to the dynamic nature of all states).

Systems modelers face nontrivial challenges when selecting the minimum number of states to adequately and effectively represent (model) the subsystems under consideration and, thus, the risk thereto. Consider, for example, the challenges in selecting the states in the risk modeling process for sustained years of drought, including (i) availability and quantity (states) of ground and surface water; (ii) quality (states) of ground and surface water; and (iii) human and livestock population and industrial needs for water. The above states of water quality and quantity have direct impacts on (i) rural or urban populations and livestock; (ii) agriculture and industry; and (iii) future recovery of the states of water quality and quantity for the entire River Basin.

From the perspective of the *reliability of Complex SoS*, shared states and other essential entities within and among the subsystems represent the essence of the I-I that characterize Complex SoS: *Initiating events affecting one or more of the interdependent and interconnected subsystems would necessarily increase the probability of failure of the entire Complex SoS*. This fundamental fact could be viewed in the parlance of fault-tree analysis (NRC, 1981) as converting subsystems from being "marginally connected," or in "parallel," to becoming "directly connected" or in "series." Nuclear reactors, which epitomize Complex SoS, have always relied on fault-tree analysis to ensure their safety. Thus, from that perspective, subsystems that share states and other essential entities *are most likely to* be affected by adverse initiating events. For example, this phenomenon is most evident in the emergent extensive use of cloud computing technology, where numerous hardware/software subsystems are shared among multiple users (Haimes et al., 2015). An initiating event may not affect shared states of different subsystems in the same way. This fact necessarily implies the following scenarios when analyzing risk to Complex SoS: When (i) each subsystem has different decision makers, then decisions made to control subsystem A may affect positively or negatively subsystem B that shares subsystem A's states and other essential entities; (ii) decision makers collaborate among themselves and coordinate their decisions and this can have a positive effect on Complex SoS and improve their overall effectiveness and management; and (iii) different initiating events affect one or more subsystems and the level of shared states and other essential entities among them can positively or negatively affect the risk management process of the entire Complex SoS. Therefore, modelers and risk analysts should recognize and exploit these attributes of shared states and other essential entities among subsystems. As a natural example of interdependent subsystems, the states of the earth and its moon relative to each other and to the sun are known at any instant due to shared gravitational forces. In their case, without any initiating events affecting any of them, the states of their celestial coordinates in space and time can be determined.

Uncertainty analysis becomes even more imperative in risk analysis of Complex SoS. Two major sources of uncertainty in modeling affect risk analysis (Paté-Cornell, 1990, 1996; Apostolakis, 1999). *Knowledge (epistemic) uncertainty* may manifest itself in faulty selection of appropriate model topology (structure) and model parameters due to ignorance of the system or SoS under consideration. *Variability (aleatory) uncertainty* applies to all relevant and important events and must not be overlooked. Sources of uncertainty dominate most decision-making processes, especially Complex SoS, and are the Achilles heel for all deterministic and probabilistic models. Sources of uncertainty and lack of understanding of the complexity associated with one subsystem of SoS would likely result in (i) adherence to unrealistic assumptions, (ii) lack of awareness of and accountability to critical interdependencies and interconnections of the SoS under consideration, (iii) poorly selected representative model topology and comprehensiveness, (iv) a dated or insufficient database to populate and calibrate the multiple subsystems models, and (v) essential risk scenarios being poorly represented or structured for all interdependent and interconnected systems. In addition, model uncertainties are often introduced through human errors of both commission and omission. In sum, uncertainty analysis associated with Complex SoS is probably one of the most difficult, albeit important, tasks in the broader risk analysis process.

9.I.3 UPDATING THE BASIC QUESTIONS LEADING TO RISK ASSESSMENT AND MANAGEMENT

In the inaugural issue of *Risk Analysis*, Kaplan and Garrick (1981) presented the following three questions in risk assessment: (i) What can go wrong? (ii) What is the likelihood? (iii) What are the consequences? Here we add: What is the time frame? (Haimes, 1991). It is constructive to paraphrase the above fundamental questions in the context of I-I Complex SoS. The first question becomes much more encompassing, and thus challenging, when applied to Complex SoS. Let us assume that we are integrating the three current major systems – communications, navigation, and surveillance (CNS) into a single Complex SoS. NextGen (FAA, 2014) is the term used by the FAA for "the next generation" of air traffic control and the reconfiguration of the US airspace. The answer to the question "What can go wrong in the high frequency signal integrity in the communications system?" is now much more encompassing, as shown in the sample of shared states and other essential entities among the three systems CNS, as one Complex SoS, depicted in Tables 9.I.1 and 9.I.2. Note that the state of the "signal effective range" is shared by all three systems. Thus, its failure would directly affect the CNS Complex SoS, which could be catastrophic. Clearly, a comprehensive answer would necessarily address the effect of the above initiating event not only on the communications system but also on the navigation system as well, because of their multiple shared states and other essential entities. Similarly, this analysis could be applied to the other shared states among the CNS as Complex SoS as depicted in Table 9.I.1.

TABLE 9.I.1 Shared States (System View)–Dynamic RNP Configuration

States	Communication	Navigation	Surveillance
HF signal integrity	✓	✓	
VHF signal integrity	✓	✓	
UHF signal integrity		✓	✓
SATCOM signal integrity	✓	✓	
GPS signal integrity		✓	✓
GPS signal availability		✓	✓
GPS signal accuracy		✓	✓
Radar signal integrity			✓
Comprehension	✓		
Plane authenticity/identification	✓		✓
Signal effective range	✓	✓	✓
Station authenticity/identification		✓	
Distance calculation accuracy		✓	
Position drift		✓	
Detection availability			✓
Compliance	✓	✓	✓

TABLE 9.I.2 Shared Decisions–Dynamic RNP Configuration–Approach

Decision	Communication	Navigation	Surveillance
Type of approach			
Instrument flight rules	✓	✓	✓
Visual flight rules	✓	✓	
Approach system			
ILS		✓	
Vectored approach	✓		✓
GPS (RNAV) approach		✓	
VOR approach		✓	
NDB approach		✓	
Runway clearance	✓		
Aircraft separation	✓		✓
Approach route	✓	✓	
Holding pattern	✓	✓	

The second and third questions posed by Kaplan and Garrick, "What is the likelihood?" and "What are the consequences?" *must now also be addressed for each of the three systems separately, as well as in terms of their shared states and other essential entities.* For example, due to the unique hardware–software configurations and the organizational setup, culture, and structure of each of the three CNS systems described in this example, it is reasonable to assume that the (i) probabilities of adverse initiating events affecting each system of the CNS, (ii) subsequent probabilities of the consequences associated with each system (when affected by either a single or multiple initiating events), and (iii) propagating consequences from one system to another would all likely be different. *Essentially, the deployment of the three basic risk assessment questions to any Complex SoS requires a fundamental adjustment from the prevailing risk analysis process commonly applied to single systems when deployed to Complex SoS.*

The question "What are the consequences?" must also be extended to all affected systems of the Complex SoS that share states and other essential entities. *Although the time frame is at best assumed implicit in the above three questions when addressing a single system, it must be made explicit in the context of Complex SoS.* The reason stems from the different characteristics and missions of each system within the Complex

SoS, thus, of different expected responses in time and space of each system within the Complex SoS to the same initiating event. It is easy to project this assertion to the different temporal behavior of each of the three safety-critical cyber–physical Complex SoS in the four case studies discussed in Chapter 10. The extension of the above three questions, along with the time frame, in the risk assessment process from one system to Complex SoS would require risk analysts to learn more about and address the I-I within and among the systems that make up Complex SoS. Finally, we ask: What is the time frame?

Similar adjustments must be made when answering the following three risk management questions (Haimes, 1991, 2016, 2018) for Complex SoS: (i) What can be done and what options are available? (ii) What are the tradeoffs in terms of all associated costs, benefits, and risks? and (iii) What are the impacts of current decisions on future options? In addressing the first question, *What can be done and what options are available?* risk analysts, decision makers, and stakeholders must realize that any risk management action tailored specifically to any affected system would likely impact directly or indirectly other interdependent and interconnected systems within the Complex SoS. Thus, such risk management can be successful and effective only to the extent that the broader consequences resulting from acting on one system can be well understood and projected on all other interdependent and interconnected systems within the Complex SoS.

Addressing the second question in risk management: *What are the tradeoffs in terms of all associated costs, benefits, and risks?* is even more encompassing, because the tradeoffs in terms of all associated costs, benefits, and risks would necessarily be different for each system of the Complex SoS. *Indeed, assessing such intricate tradeoffs associated with the affected systems, and projecting them over time to all systems of the Complex SoS, constitutes a measurably more intricate and demanding task than dealing with a single system.* The above discussion of the first two questions in risk management was presented in relation to shared states among the systems that constitute the Complex SoS. A similar modified risk management process must also be applied to *shared essential entities*, e.g. *shared decisions*, as depicted in Table 9.I.2. Note, for example, that the communications and navigation systems share four decisions and that the communications and surveillance systems share three decisions within and among the three CNS systems. One important shared decision, which we lay travelers can appreciate from Table 9.I.2, is that all three CNS systems share the decisions on instrument flight rules.

Addressing the third question in risk management – "What are the impacts of current decisions on future options?" – requires explicit dynamic projections of the commonly multiple objectives that characterize a single system and Complex SoS. Pareto-optimal risk management policies made at time (t) would certainly impact future options during future periods ($t + \Delta$) (introduced in Part II of this chapter). A decision is defined as Pareto-optimal, efficient, noninferior, or nondominated if improving one objective function necessarily degrades another. (Recall that Chapter 5 is devoted in its entirety to multiple objective decision making.) Thus, systems that are interdependent and interconnected necessarily imply the propagation of dynamic tradeoffs among the multiple objectives that characterize most systems, especially Complex SoS.

Thus, the essential requirement of updating and appropriately modifying currently applied risk modeling, assessment, management, and communication to a single system when applied to Complex SoS.

9.I.4 RISK OF LOW PROBABILITY AND EXTREME CONSEQUENCES TO COMPLEX SYSTEMS OF SYSTEMS

Risk of low probability but with extreme and dire consequences affecting one subsystem may result in catastrophic consequences to Complex SoS that share critical states and other *essential entities* with an affected subsystem. The catastrophic collapse of the Fukushima Daiichi nuclear plant in Japan on 11 March 2011 is a classic example. The use, and often misuse, of the expected-value metric of risk remains one of the most unfortunate practices of risk analysis, especially when it is used as the sole representation and quantification of risk to a system. The concerns of the public and most decision makers focus on events with dire consequences, even with low probabilities. Thus, the importance of studying "extreme" events must be acknowledged, especially for Complex SoS.

Moreover, the expected value of risk *masks* the criticality of catastrophic events by intrinsically equating a low probability of high-consequence events with a high probability of low-consequence events. Indeed, relying on the commonly used expected-value metric, when it is used as the *sole measure of risk*, is likely to lead to very bad choices (Asbeck and Haimes, 1984, Haimes, 2016).

The scope and extent of catastrophic risk to Complex SoS are magnified due to the nature of their interdependencies and interconnections. Planners and stakeholders expect neither miracles nor precise hindsight from risk modelers and analysts, nor are the latter capable of delivering such knowledge or information. Yet modeling and simulation of Complex SoS enable the exploration of what might evolve in the future, given the prevalence of adverse initiating events, and help avoid likely negative or disastrous impacts of current policies on future options. Initiating events, whether internal or external to Complex SoS, may be manifested by the dynamically shifting rules and realities highlighted in the following dynamically *Evolving Base* (Haimes, 2012a, 2016): (i) goals and objectives; (ii) stakeholders, decision makers, and interest groups; (iii) organizational, political, and budgetary considerations; (iv) reorganization and reallocation of key personnel; and (v) requirements, specifications, and delivery dates.

PART II: SYSTEMS-BASED GUIDING PRINCIPLES FOR RISK TO COMPLEX SYSTEMS OF SYSTEMS

9.II.1 THE JOURNEY

Part II of this chapter is grounded on the premise that the intricate process of risk assessment, management, and communication, when applied to Complex SoS, ought to be guided by universal systems-based principles. It is written from the perspective of systems engineering with the hope and expectation that the principles introduced here will be supplemented and complemented by principles from the perspectives of other disciplines. Indeed, there is no claim that the following ten guiding principles constitute a complete set; rather, the intent is to initiate a discussion on this important subject that will incrementally lead us to a more complete set of guiding principles. The 10 principles are as follows:

1. *First Principle*: Holism is the common denominator that bridges risk analysis and systems engineering.
2. *Second Principle*: The process of risk modeling, assessment, management, and communication must be systemic and integrated.
3. *Third Principle*: Models and state variables are central to quantitative risk analysis.
4. *Fourth Principle*: Multiple models are required to represent the essence of the multiple perspectives of complex systems of systems.
5. *Fifth Principle*: Metamodeling and subsystems integration must be derived from the intrinsic states of the complex system of systems.
6. *Sixth Principle*: Multiple conflicting and competing objectives are inherent in risk management.
7. *Seventh Principle*: Risk analysis must account for epistemic and aleatory uncertainties.
8. *Eighth Principle*: Risk analysis must account for risks of low probability with extreme consequences.
9. *Ninth Principle*: The time frame is central to quantitative risk analysis.
10. *Tenth Principle*: Risk analysis must be holistic, adaptive, incremental, and sustainable, and it must be supported with appropriate data collection, metrics with which to measure efficacious progress, and criteria on the basis of which to act.

The relevance and efficacy of each guiding principle is demonstrated by applying it to the FAA Next Generation (NextGen) Complex SoS.

Risk analysis has become a dominant interest and a requisite area of expertise in almost every discipline as well as in government, public, and private organizations. By its nature, risk analysis is an intricate, dynamic process – an amalgamation of the arts and sciences – tailored to each specific set of sources of the risks to specific systems. It follows, then, that for any system, or Complex SoS, the balance between quantitative and qualitative risk analysis will be problem and domain specific. The ten principles set forth in this part are thus intended to guide both quantitative- and qualitative-centered risk analyses. Furthermore, meeting the challenges associated with defining, modeling, and quantifying the multidimensional risk function of a system and of Complex SoS will likely be guided by the specific discipline performing such tasks and influenced by the specific experiences and expertise of its risk modelers and decision makers. For example, although the disciplines of systems engineering and risk analysis share a wide common denominator of philosophy, theory, methodology, and practice, each discipline has historically evolved through separate paths and thus has distinctive followers. Furthermore, the principles we set forth may in fact be applied to a range of disciplines, to retain focus this part draws from and is guided by both risk analysis and systems engineering theory, methodology, and practice.

It is constructive first to define the relationship between systems engineering and risk analysis. As a starting point, one may ask: Are risk analysis and systems engineering/analysis grounded on similar principles? Do they represent two distinct fields or disciplines? Or, in reality, do they reinforce and add synergy to each other and thus constitute a unified approach to problem solving and an effective decision-making process? On the one hand, the two fields have a common philosophical approach to decision making; on the other hand, they differ in their historical evolution and philosophical and technical maturity. Yet both groups aspire to a gestalt–holistic philosophy in their problem-solving approaches. In particular, systems modeling, which builds on a plethora of theory, methods, tools, and techniques, offers the instruments with which problems are studied, assessed, understood, managed, and solved.

Systems engineering is distinguished by a practical philosophy that advocates holism in cognition and in decision making. This philosophy is grounded on the arts, natural and behavioral sciences, and engineering, and it is supported by a complement of modeling methodologies, optimization and simulation techniques, data management procedures, and decision-making approaches. The ultimate purpose of systems engineering is to (i) build an understanding of a given system's nature, functional behavior, and interaction with its environment; (ii) ensure a holistic multiperspective representation and systems integration; (iii) improve the decision-making process (e.g. in planning, design, development, operation, and management); and (iv) identify, quantify, and evaluate risks, uncertainties, and variability within the decision-making process.

The entire process of risk assessment, management, and communication is essentially a synthesis and amalgamation of the empirical and the normative, the quantitative, and the qualitative and of objective and subjective evidence. It is a process, in fact, that has been built over the years on the contributions of individuals from diverse disciplines. Many of the theories, quantitative and qualitative tools, and methods employed by risk analysts today have been developed by mathematicians, statisticians, biostatisticians, social and behavioral scientists, health scientists, economists, and engineers, among others. In particular, social, behavioral, and organizational scientists have markedly contributed to our understanding and appreciation of the human and social dimensions of risk analysis, e.g. human perception, organizational and institutional barriers, communication, trust, and conflict resolution. In 1980, in recognition of the evolving field of risk analysis and its growing constituencies, the SRA was founded. According to its charter, the SRA "brings together individuals from diverse disciplines and from different countries and provides them opportunities to exchange information, ideas, and methodologies for risk analysis and risk problem-solving."

A reader might ask: "Why do we need a set of principles to perform risk analysis?" Although no answer can do justice to this encompassing question, one may posit that any decision-making process addressing important probable, including dire, consequences resulting from EFCs ought not to be performed on an ad hoc basis. Beyond this, however, because risk analysis has grown to embrace a wide range of specialists from myriad fields, a governing set of principles will provide a conceptual and theoretical framework – a roadmap – that will enable these individuals to work in concert on modeling and managing risks to Complex SoS.

9.II.2 THE JOURNEY: GUIDING PRINCIPLES IN THE BROADER CONTEXT OF THE FAA'S NEXTGEN

To demonstrate the efficacy of the 10 principles in the risk modeling, planning, assessment, management, and communication process for Complex SoS, this section offers a conceptual journey, outlining each principle, and then applying these principles to a new initiative of the US National Airspace System (NAS). At present, the FAA is planning the most significant overhaul of the NAS since its establishment in 1982. This is an ambitious, multiyear, and multibillion dollar enterprise, termed by the FAA as the Next Generation or the FAA NextGen (FAA, 2014). The successful integration of new technologies into the existing operational structure is meant to contribute to aviation safety. Indeed, safety, aircraft-centric operations, and aircraft equipage are keys to NextGen's successes. (Aircraft centric operations revolve around the idea of relying on autonomous aircraft movements, interconnected with other aircraft, with less reliance on external control.) In other words, the emergent NextGen Complex SoS is a collection of programs and initiatives meant to improve the capacity, efficiency, safety, security, and positive environmental impact of aviation in the United States. Indeed, the NextGen epitomizes the I-I of Complex SoS, where risk modeling, assessment, management, and communication are central to the successful planning, development, and deployment of solutions. The planning of a large and complex project usually faces various sources of risk. In this book, these sources of risk are termed EFCs, connote *trends in sources of risk, whether external or internal, that may adversely affect specific states of the subsystem or of the Complex SoS.*

The conceptual journey (see Figure 9.II.1) starts at Part I, the "Evolving Base." There follows Part II, "The Journey." Part III, "The Compass," outlines the modeling process that is carried out under the guidance of the 10 Principles. This journey continues with Part IV, "Interim Destination" – which revisits the process of risk assessment, management, and communication. Then we return to Part I, the "Evolving Base." This cyclical process is at the heart of the intricate and complex risk analysis process that interdependent and interconnected Complex SoS require (Haimes, 2012a).

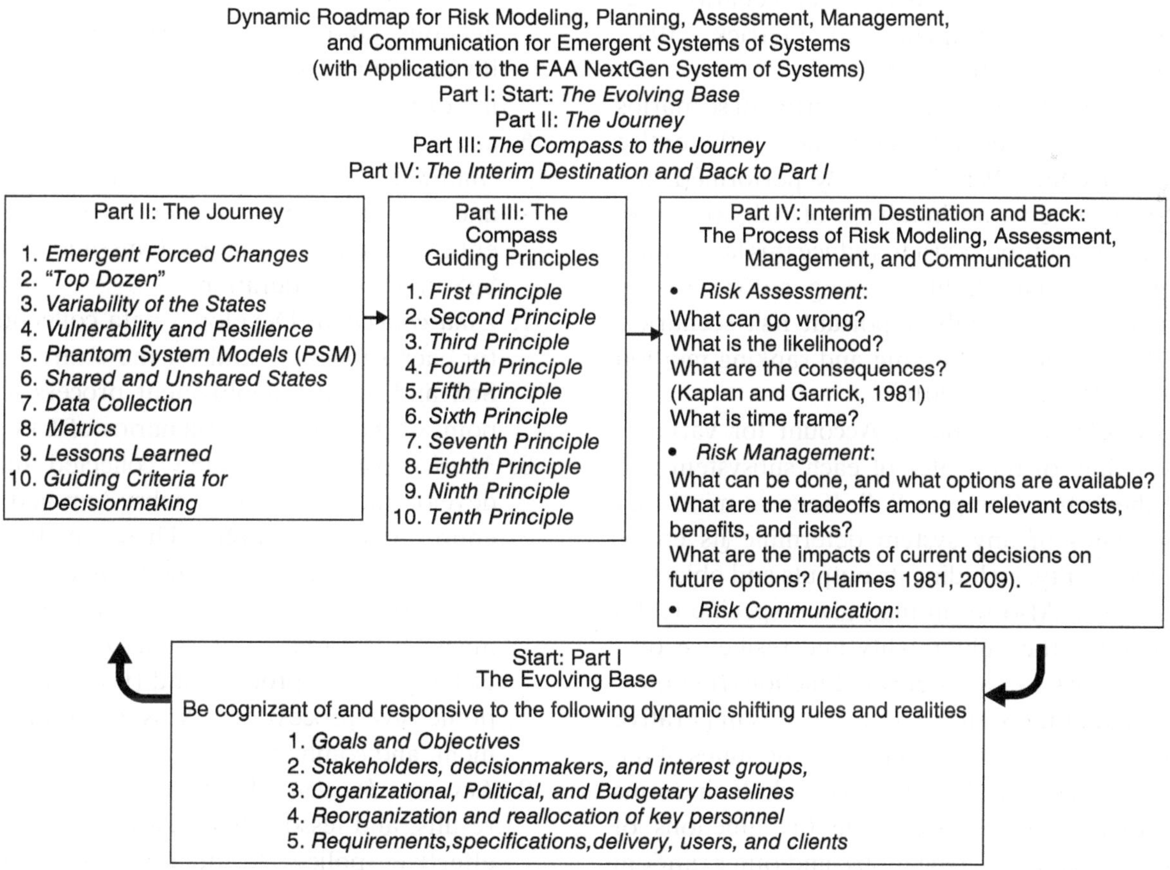

Figure 9.II.1 *Dynamic Roadmap for risk modeling, assessment, management, and communication.*

9.II.2.1 The *Evolving Base*

Risk analysts and decision makers, especially those associated with Complex SoS, must be cognizant of and responsive to the following sample of dynamic, shifting considerations for each subsystem and for the entire Complex SoS: (i) goals and objectives; (ii) stakeholders, decision makers, and interest groups; (iii) organizational, political, and budgetary baselines; (iv) reorganization and reallocation of key personnel; and (v) requirements, specifications, delivery, users, and clients.

9.II.2.2 Part II: The Journey

The *Journey* provided by the roadmap of the 10 principles implies that successful risk analysis of all subsystems – and of the Complex SoS as whole – adheres to the following guidelines:

1. *Emergent Forced Changes (EFCs)*: Develop a comprehensive and complete list of EFCs for each subsystem and for the entire SoS, by embracing the modified theory of scenario structuring (TSS) (Kaplan and Garrick, 1981; Haimes, 1981, 2011, 2016).
2. *Top dozen*: Identify the most pertinent scenarios of EFCs for each subsystem and for the entire Complex SoS. When properly performed, the TSS may generate hundreds of risk scenarios. Systemic methods exist to reduce this large number into a manageable one (e.g. a dozen) without missing critically important risk scenarios, including the risk filtering and ranking method (RFRM) (Haimes, 2016).
3. *Variability of the states*: Account for variance over time of the states of each subsystem and of the entire Complex SoS. Keep in mind that the states of any system determine its entire functionality, including its outputs and objective functions. Also recall that, given specific EFCs and time, the vulnerability and resilience of – and the risk to – a system are functions (manifestations) of the states. Therefore, it is imperative that we focus on the behavior of the states of the subsystems and of the SoS as a whole, in terms of their variance over time (as functions of inputs, decisions, exogenous, and other random variables).
4. *Vulnerability and resilience*: Assess the vulnerability and resilience of each subsystem and the SoS as a whole to each member of the top dozen scenarios of EFCs, keeping in mind that both the vulnerability and resilience of each subsystem are manifestations of their states and can be completely defined only as functions of specific EFCs.
5. *Phantom System Models (PSM)*: Explore the PSM-based intrinsic metamodeling of SoS introduced in Chapter 4. This modeling rests on the basic assumption that some specific commonalities, interdependencies, or other relationships must exist between and among any two systems within any Complex SoS, through shared and other essential entities (also see the *Fifth Principle*). Recall that the term "essential entities" includes *shared/common decisions*, *decision makers*, *stakeholders*, *resources*, *organizational behavior and norms*, *policies and procedures*, *management*, *culture*, among others. Note that shared states and other essential entities, which represent the interdependencies and interconnections among the subsystems, provide an invaluable mechanism for the enhanced modeling of SoS and of their ultimate management.
6. *Shared and unshared states and other essential entities*: Develop a comprehensive list of shared and unshared states, as well as other essential entities, among the subsystems and the entire SoS under consideration.
7. *Data collection*: Develop a data collection plan for each subsystem and for the entire Complex SoS and for the deployment process for each point of the top dozen scenarios of EFCs.
8. *Metrics*: Building on the collected database, develop metrics for each subsystem and for the entire Complex SoS. These metrics help (i) evaluate the progress being made; (ii) track precursors to emerging risks; (iii) assess the impacts of current progress on succeeding phases of the planning process; and (iv) initiate a comprehensive process of risk assessment, management, and communication (see Section 9.II.3).
9. *Lessons learned*: Evaluate the experiences, results, and lessons learned as a prelude to an effective policy formulation and decision-making process.

10. *Guiding criteria for decision making*: Develop guiding criteria for required actions. Keep in mind Lowrance (1976) epistemological question: "Who should decide on the acceptability of what risk, for whom, in what terms, and why?"

9.II.3 THE COMPASS – FUNDAMENTAL GUIDING PRINCIPLES FOR ANALYSIS OF COMPLEX SYSTEMS OF SYSTEMS

The following set of guiding principles constitutes a theoretical and conceptual roadmap for an effective process of risk modeling, assessment, management, and communication for Complex SoS. These principles are not offered as a complete or definitive set; rather, they are intended as the basis for a holistic, adaptive, incremental, and sustainable risk-based decision-making process.

9.II.3.1 First Principle: Holism Is the Common Denominator That Bridges Risk Analysis and Systems Engineering

The first principle posits that holism is the hallmark of an effective systems modeling, systems engineering, and risk analysis for single systems and Complex SoS. Risk by definition, a measure of the probability and severity of adverse effects (Lowrance, 1976), encompasses future probable adverse events and probable adverse consequences resulting from EFCs. This commonly adopted definition has some ambiguity to it. The term "probability and severity of adverse effects" can be interpreted in multiple ways: (i) in terms of the *probability of the occurrence of the adverse* event to select subsystems and/or to the entire Complex SoS and (ii) in terms of the *probability of the severity* of the adverse event, given each of the above occurrences. The latter may be further interpreted in terms of a conditional probability distribution function representing the severity of the adverse effects. All interpretations are valid; however, each carries different conceptual and theoretical connotations, interpretations, and challenges.

Since risk is the likelihood of future adverse consequences, the analyst needs an encompassing holistic approach to identify all perceived and imaginable adverse consequences, as advocated by TSS. Moreover, the holistic character of the risk analysis process – the modeling, assessment, management, and communication of risk – does not stop at the TSS level, because each of the elements of risk analysis, alone and collectively, calls for a systemic, comprehensive, and encompassing process. For example, quantifying the risk function requires determining the causal relationship between the timing and severity of the initiating event and the probability of each of its consequences (as an element of the multidimensional consequences) and thus the resulting risk function – for each subsystem and for the entire Complex SoS. This conceptually, philosophically, and analytically intensive risk analysis task can be effectively achieved through further reliance on and support of the practical philosophy of systems engineering on its supporting tools and methodologies and on the guiding principles introduced in this chapter.

In many ways, the gestalt–holistic philosophy can be viewed as the common denominator that bridges risk analysis with systems engineering. For example, the quest for, and indeed requisite to, identifying all conceivable risk scenarios depends on a holistic approach; otherwise, the risk assessment process, and its subsequent management and communication, would be partial, biased, and ineffective at best. To ensure such a comprehensive identification of all conceivable risk scenarios, a systems-based methodology is required.

In 1981, both the TSS and Hierarchical Holographic Modeling (HHM), introduced in Chapter 3, were independently published to address the multiple perspectives of a system and to promote the consideration of every conceivable EFC. Subsequently, a joint paper streamlined the TSS and the HHM, and the revised definition of risk made explicit what was only implicit before (Kaplan et al., 2001). The TSS and its extension constitute a cornerstone of risk analysis, which requires the generation of all conceivable risk scenarios for each subsystem and for the entire SoS. This comprehensive process necessarily requires subsequent analysis, filtering, and prioritization of risk scenarios. Other methods that serve the purpose of analyzing risk scenarios include failure modes and effects analysis (FMEA), event trees, and fault trees, among others (Haimes, 2016).

***Relevance to the NextGen SoS**: Modeling, analyzing, and managing systems of systems cannot be a selective process, driven by limited perspectives; rather, a holistic systems-based approach must be embraced to account for EFCs, including multiple shared states and other essential entities, including multiple databases and numerous technological and organizational subsystems; multiple objectives, agencies, stakeholders and decision makers; multiple time horizons associated with each subsystem and the entire Complex SoS; and the phased planning and development process time. These considerations, among others, characterize the challenges in modeling Complex SoS, such as the NextGen enterprise.*

9.II.3.2 Second Principle: The Process of Risk Modeling, Assessment, Management, and Communication Must Be Methodical, Disciplined, Systemic, Integrated, and Equally Critical of Each Subsystem and the Entire Systems of Systems

Given the importance of the *Second Principle*, it is useful to revisit some key definitions: *Risk management* is commonly distinguished from *risk assessment*, even though some may use the term *risk management* to connote the entire process of risk assessment, management, and communication.

Risk Assessment

In risk assessment, the analyst often attempts to answer the following four questions (Kaplan and Garrick, 1981): What can go wrong? What is the likelihood that it would go wrong? What are the consequences? And, the fourth question: "What is the time frame?" (Haimes, 1991). However, when addressing Complex SoS, we must modify and paraphrase the above questions as follows. For risk assessment we ask: (i) What can go wrong for each subsystem and for the entire SoS and when? (ii) What is the likelihood that it would go wrong for each subsystem and for the entire Complex SoS? (iii) What are the consequences for each subsystem and for the entire Complex SoS? (iv) What is the duration of each consequence for each subsystem and for the entire Complex SoS? Answers to these questions help risk analysts to identify, quantify, and evaluate myriad categories of risks and their consequences.

To prevent, mitigate, and prepare for undesirable future occurrences, all government agencies as well as the public and private sectors seek to understand the trends of risks associated with EFCs. If trends/changes are unanticipated, undetected, misunderstood, or ignored, they are likely to hurt a multitude of states of that subsystem and/or the entire Complex SoS. Thus, it is imperative to be able – *through scenario structuring, modeling, and risk analysis* – to anticipate, discover, and track EFCs and their crossovers. Recall that *crossovers connote situations where a projected EFC, e.g. an impending failure or a terrorist threat, can cross from a state of uncertainty to a forecasted state of high likelihood of occurrence.* EFCs may be caused by natural phenomena, malevolent human actions, or accidents. Identifying such *precursors (based on tangible or intangible information) is most effective when the information is used in conjunction with other sources of knowledge and when they are followed with vigilance.* Often, a crossover is ignored during the tracking of EFCs and is recognized too late, after it becomes a disaster. *In the context of modeling, crossovers may also connote situations where, during the simulated scenarios of a dynamic system, the projected EFC crosses the state of knowledge from uncertainty to a more likely eventuality of its occurrence.* Duderstadt et al. (2005) emphasize that "Change is coming, and the biggest mistake could be underestimating how extensive it will be." Of course, there are also positive EFCs, which may bring with them opportunities in terms of new technological innovations and discoveries.

Bayesian analysis is a crucial element in the process of modeling, estimating, and tracking over time precursors to the emergence of forced changes affecting each subsystem and the entire SoS. Indeed, Bayes' theorem, introduced in Chapter 6, incorporates new, probabilistically acquired knowledge, whether gained by direct observations or through expert evidence obtained through prior probabilistic work. Scenario development constitutes a basis upon which to track and estimate precursors. In scenario tracking, both the probabilities and consequences are considered when evaluating the importance of any specific scenario. The uncertain nature of EFCs requires that both aleatory and epistemic sources of uncertainty be considered in the risk analysis process (see *Seventh Principle*). Without such comprehensive considerations, the risk-based decision-making process would be less representative, effective, and credible.

One important demarcation line that characterizes sound risk-based decision making is the extent to which it evaluates the known and known–unknown future developments and EFCs affecting subsystems and/or the entire Complex SoS. Anticipating and planning for EFCs, through preparedness and other risk management actions, will improve the resilience of the system. Inherently, challenging tradeoffs must be made among (i) certain (nonprobabilistic) investments in preparedness in the quest to address probable EFCs and (ii) added resilience to the system in anticipation of the uncertain nature of the realization of the EFCs.

The often injudicious allocation of precious limited resources (today) in the quest to protect against (future) probabilistic adverse consequences has commonly marred the deployment of risk management options. Lack of investment in cyber security is one example. Indeed, responding to each and every credible scenario of critical EFCs is unrealistic and could lead to bankruptcy. A holistic, adaptive, incremental, and sustainable risk management process, introduced in the *Tenth Principle*, alleviates the need for such massive and unsustainable investments in risk management.

Risk Management

Similarly, risk management builds on the risk assessment process by seeking answers to a second set of three questions (Haimes, 1991, 2016); thus, we ought also to modify the three questions in risk management, when addressing Complex SoS: (i) What can be done and what options are available for each subsystem and for the entire Complex SoS? (ii) What are the tradeoffs among all relevant costs, benefits, and risks (both mitigated and residual) associated with each subsystem and with the entire Complex SoS? (iii) What are the impacts of current decisions on future options on each subsystem and on the entire Complex SoS? Note that the last question is the most critical one for any sound and effective managerial decision making. Unless the future impact of current decisions are assessed and evaluated to the greatest extent possible; these policy decisions cannot be said to be effective or, indeed, desirable.

Risk management is effective only when the three questions are addressed for each individual system and for the entire Complex SoS in the broader context of management, where all options and their associated tradeoffs are considered within the common hierarchical organizational structures of most, if not all, Complex SoS. Of course, this task cannot be done seriously and meaningfully without modeling the SoS and allocating appropriate resources. In other words, risk management must be an integral part of the overall organization's policies and procedures. Furthermore, an effective and sustainable risk management approach must address the four major sources of failure: (Haimes, 1991, 2016) *hardware, software, human, and organizational.*

The four sources of failure of Complex SoS are not necessarily independent of each other. For example, the distinction between software and hardware is not always straightforward, given that cyber–physical Complex SoS dominate our world today. Furthermore, human and organizational failures are often inseparable. Nevertheless, these four categories provide a meaningful foundation upon which to build a comprehensive risk management framework.

Organizational errors are often at the root of failures of critical engineering systems and SoS. Yet when searching for risk management strategies, engineers often tend to focus on technical solutions, in part because of the way risks and failures have been analyzed in the past. For example, in her study of offshore drilling rigs, Paté-Cornell (1990) found that over 90% of documented failures were caused by organizational errors. To better account for such oversights, the risk assessment and management process must be comprehensive. It must involve all relevant life cycle aspects of each subsystem and the entire SoS, such as including the requirements, specifications, planning, design, integration, construction, deployment, operation, and management of the Complex SoS. This requires the total involvement of all the principals. Blue- and white-collar workers and managers at all levels of the organizational hierarchy must engage with the process.

Risk Communication

The risk assessment and management process aim to answer the above specific questions in order to make better decisions under uncertain conditions. Similar to systems modeling, where a model must be as simple as possible and as complex as required, *the risk assessment and management process must follow this same basic maxim*. These seemingly conflicting attributes – simplicity and complexity – can be harmonized through an effective *risk communication process*.

Invariably, questions posed to risk analysts during the risk assessment and management process originate from decision makers and other stakeholders of varying levels of responsibility, including managers, designers, journalists and other media professionals, politicians, proprietors, and government, or other officials. These decision makers must be able to communicate the multidimensional perspectives of the risks and the associated challenges facing them – risks and challenges that they wish to understand better and for which they seek answers. In turn, risk analysts must be able to translate complex technical analysis into answers that decision makers and other stakeholders can understand and incorporate into actions. The questions raised are likely to be intricate and may also require the translation of similarly complex sets of answers in a language that can be understood by intelligent laypersons.

Relevance to the NextGen: *A systemic approach to risk modeling, planning, assessment, management, and communication is an absolute necessity for the credibility and effectiveness of the NextGen multifarious Complex SoS, affecting their planning and development process, as well as the quantification of the complex multidimensional risk functions associated with its multiple subsystems (Haimes, 2016). As an example, consider five categories of EFCs as risk factors associated with the unmanned aircraft system (UAS), which constitutes only one subsystem of the NextGen: (i) human factors (personnel, culture, and collaboration), (ii) technological factors (aircraft type, manufacturer, communication), (iii) infrastructure factors (launch and recovery site, control facilities, and cyber-infrastructure), (iv) operational factors (aircraft operations, organizational operations, support and schedule), and (v) policy factors (regulations, security, and liability). Since safety is the level of acceptable risk (Lowrance, 1976) and no risk to a threatened NextGen subsystem can be completely eliminated, tradeoffs must be made for each subsystem and the entire Complex SoS. Furthermore, risk analysts must be capable of communicating the above reality and must have the tools and training to address the questions raised in the intricate, yet essential, risk assessment, management, and communication process.*

9.II.3.3 Third Principle: Models and State Variables Are Central to Quantitative Risk Analysis

Consider the following simple problem: an engineer is asked to determine the safety of a municipality's drinking water. Can the task be performed without determining the *state* of acidity of the water, and its turbidity, dissolved oxygen level, and amount of bacteria and other pathogens? This example shares the following systems-based fact with almost all other challenges facing risk analysts: *all systems are characterized at any moment, and at all times, by their respective states (termed state variables in modeling)*. These states are affected and shaped by the other building blocks of mathematical models (e.g. decision, random, and exogenous variables, and inputs and outputs). In reality, all states are under continuous natural, desired, or EFCs (positive and negative). This does not mean that all models must be built with time-dependent state variables, despite the fact that in reality the states of all subsystems and the entire Complex SoS are usually dynamic and that all risk-based problems are time dependent. Recall that models ought to be as simple as possible and as complex as required.

To further highlight the centrality of the states of the system in modeling and in risk analysis, we pose the following question: *Why do farmers irrigate their crops in dry seasons?* Indeed, the answer is fundamental to understanding both the role of state variables in modeling and the definitions of vulnerability, resilience, and therefore risk. The farmer irrigates the crops to maintain an appropriate state of soil moisture during the specific growth period. Furthermore, the farmer adds fertilizer to maintain an appropriate state of nutrients. In other words, the farmer and all decision makers endeavor to control or maintain the states of the system under their management at acceptable levels by deploying appropriate risk management policy options in the case of risk-based decision making.

Recall Chen's (2012) following succinct definition of *state variable*: "The state $x(t_o)$ of a system at time t_o is the information at time t_o that, together with the input $u(t)$, for $t \geq t_o$, determines uniquely the output $y(t)$ for all $t \geq t_o$." The states of a system constitute the main building blocks of both analytical and simulation models. Of course, by their nature and composition, Complex SoS are characterized by a hierarchy of multiple states and substates and by the interdependencies and interconnections among the subsystems that comprise it. Recall that the essence of these interdependencies and interconnections is grounded on and characterized by shared states and other essential entities. Also, models are built to answer specific questions and must be made as simple as possible and as complex as required. Without models, risk analysts would be traveling blindly in unknown and often perilous terrains with potentially dire consequences to those whom they serve. The centrality of state variables in decision making, and particularly in risk analysis, has not been sufficiently emphasized. Indeed, a representative model can effectively provide answers only by first identifying and incorporating all relevant and critical states of the system. Statistical models, which may not suggest a direct cause–effect relationship among the states of the system and the other building blocks of the mathematical model(s), can also play a valuable role in quantitative risk analysis. For example, insurance policies, based on statistical accounting of accidents, distribute high-cost events across the population through which they occur, thereby spreading the cost through insurance premiums.

As noted in Chapter 1 and other earlier chapters, vulnerability and resilience are key concepts in risk analysis. Their systems-based definitions improve our understanding of risk and help make them operational for modeling. *Vulnerability is the manifestation of the inherent states of the system (e.g. physical, technical, organizational, and cultural) that can be subjected to EFCs – e.g. natural hazard – or be exploited to adversely affect (cause harm or damage to) that system* (Haimes, 1981, 2016). Moreover, the vulnerability of a subsystem or of Complex SoS is multidimensional, a *vector* in mathematical terms. Consider the risk of a major hurricane to a community. A sample of the states of the community, as Complex SoS, that represents its vulnerabilities to natural hazards includes the functionality/availability of telecommunications, electric power, water supply, schools and other public facilities, and transportation and other essential services, all of which are critical to the overall functioning of the community and to public safety. Furthermore, note that these states are not static; they are dynamic, changing, and evolving continuously.

Similar to vulnerability, the resilience of Complex SoS or any of their subsystems is also *a manifestation of the states of the system, and it is a vector that is time and threat (initiating event) dependent* (Haimes, 2011, 2012b, 2016). More specifically, *resilience* represents *the ability of the system to withstand a major disruption within acceptable degradation parameters and to recover within an acceptable frame of cost and time.* In other words, resilience is a vector state of each subsystem and of the entire SoS, which is neither abstract, static, nor deterministic. Resilience is similar to vulnerability in that it cannot simply be measured in a single unit metric. Any specific EFCs affecting a subsystem, or the entire Complex SoS, would ultimately cause multidimensional outputs to each subsystem and to the entire SoS, based on the corresponding states at the time of the event.

Relevance to the NextGen: *Consider the four major initiatives and objectives of the NextGen SoS enterprise: Improving the capacity, safety, security, and environmental impacts of aviation in the United States. Each one of these objectives can be addressed and ultimately realized if and only if the multiple states of each subsystem of the NextGen Complex SoS can be identified, modeled, evaluated, and ultimately controlled to meet the above objectives. A sample of these states includes the quality of an appropriate number of trained professionals, the quality and reliability of the supporting physical, cyber, and cyber–physical infrastructures, and the effectiveness of the states of communication, to cite a few. For example, consider the safety and security objective. Since the vulnerability and resilience of each subsystem are manifestations of its states at the time of a specific initiating event (e.g. an EFC), then the complex multidimensional risk function can be assessed only when sufficient knowledge is available of the states of each subsystem at the time of the event.*

9.II.3.4 Fourth Principle: Multiple Models Are Required to Represent the Multiple Perspectives of Complex SoS

When modeling cyber–physical infrastructure Complex SoS including their operation and management, such as dams and levees or the FAA's NextGen project, multiple mathematical, conceptual, simulation, and other types of models are likely to emerge, each of which may focus on a specific aspect of a subsystem or on the entire Complex SoS. And only when these multiples models are continuously reevaluated with the support and integration of effective data would they constitute an effective representation of each subsystem and of the entire Complex SoS (Haimes, 1981, 2016). This phenomenon is particularly common in organizational and cyber–physical infrastructure SoS. For example, a River Basin may be modeled from at least four different perspectives: (i) hydrology (e.g. controlling floods and soil erosion for different watersheds), (ii) geographical–political boundaries (e.g. state or county boundaries), (iii) temporal decomposition (e.g. different planning horizons), or (iv) functional decomposition (e.g. flood control and hydropower generation). Hence, it is impossible to represent within a single model all perspectives and functionalities and configurations of Complex SoS (Haimes, 1981, 2016).

Indeed, most cyber–physical infrastructures and most government, private, and public organizations are Complex SoS composed of multiple subsystems, which in their essence constitute SoS. Each subsystem, characterized by its own states (and thus vulnerability and resilience), functionality, and organizational structure, will necessarily respond differently to different EFCs. Hence, when faced with an adverse event, Complex SoS are likely to experience multidimensional consequences, which commensurately require multiple models to represent the effects within each subsystem.

Despite the comprehensive and representative characterization of Complex SoS offered by Maier (1998), the precise definition of Complex SoS remains elusive. In response to the question "What is a system of systems?" Sage and Biemer (2007) provide the following answer: "No universally accepted definition of a system of systems is available at this time." However, systems analysts must quantify causal intra- and interconnected and interdependent relationships when modeling cyber–physical infrastructures. On the other hand, as we noted in Chapter 4, identifying shared states and other essential entities among interdependent and interconnected subsystems helps analysts to model and understand Complex SoS as a whole. The growing recognition that cyber–physical infrastructures, health care, the military, the environment, and many other entities are Complex SoS has inspired a commensurate awareness of the need for their modeling – which in turn will foster their better understanding and management.

Relevance to the NextGen: *Consider a sample of the dozens of subsystems that constitute the NextGen: The national air space UAS; airports and their subsystems, e.g. runways, airliners, air traffic controllers; and communications. Each requires one or more specific models to represent the essence of the subsystems. The next principle addresses the metamodeling of Complex systems of systems.*

9.II.3.5 Fifth Principle: Metamodeling and the Integration of Subsystems Must Derive from the Intrinsic States of the Systems of Systems

Modeling a simple system, or Complex SoS, requires determining its properties, constructing the relationships among its inputs and outputs through its state variables and other variables and parameters (e.g. random inputs, decision, and exogenous variables), quantifying intra- and interdependencies within and among its various components and subsystems, and determining the model topology (structure) and parameters that best represent its essence and functionality. Models enable us to experiment and test hypotheses, to explore different design options, or to generate responses to or impacts of varied scenarios and policy options. Inversely, by their nature, Complex SoS constitute black holes to modelers by forcing us to acknowledge our inability to directly uncover, understand, or predict behavior of the Complex SoS under different scenarios of EFCs. We commonly lack sufficient knowledge to assess the causal relationships among the subsystems, and to compensate for this shortfall, we revert to multiperspective experimentation aided by the ingenuity, creativity, and the domain knowledge of experts, supported by available databases. There is no assurance that modelers will be able to explain the reasons behind every variability among submodels; nevertheless, the very intricacy of modeling such as variability may highlight limited databases, inconsistent assumptions, unrecognized epistemic and aleatory uncertainties, and a host of other technical or perceptual shortcomings that ought not to be dismissed.

The PSM presented in Chapter 4 builds on the centrality of the states of the system in modeling and in risk analysis, it is designed to model inter- and intradependencies between and among the subsystems of Complex SoS by exploiting vital knowledge and information embedded in the shared states and other essential entities among the subsystems.

Relevance to the NextGen: *To plan, develop, and ultimately deploy the NextGen system of systems, all its subsystems must function as harmoniously as possible. This can be achieved if and only if there is a clear understanding of the interdependencies and interconnections among the intrinsic states of the entire Complex SoS. Thus, the metamodeling and integration of subsystems cited in the Fifth Principle must be performed with diligence and patience through a process of trial and error.*

9.II.3.6 Sixth Principle: Multiple Conflicting and Competing Objectives Are Inherent in Risk Management

All risk-based decision-making problems, especially for Complex SoS, are characterized by multiple, noncommensurate and often conflicting objectives (Keeney and Raiffa, 1976; Chankong and Haimes, 1983, 2008; Haimes, 2016). Recall that Chapter 5 is devoted to multiple objectives. For example, risk management actions – such as increasing preparedness, reducing vulnerability, or enhancing the resilience of a system, or of the entire Complex SoS – require the expenditure of monetary and human resources in exchange for protecting lives, property, and the well-being of the public. Furthermore, for most risk-based decision making, there exists a hierarchy of objectives, subobjectives, sub-subobjectives, and so on. In risk management, and in decision making in general, it is important to identify this hierarchy of objectives and to be cognizant of the tradeoffs among the objectives within the same hierarchical level of importance, and to avoid trading off objectives that belong to different hierarchical levels. Consider, for example, an electric power Complex SoS that includes power generation turbines, transformers, supervisory control and data acquisition (SCADA), and transmission and electric power distribution systems. Multiple objectives of various levels of importance are associated with each of the above subsystems – a fact that would make meaningless the tradeoffs among the above needs and objectives. Ensuring the continuation of government operations following an attack is much more important than repairing a minor infrastructure. This priority-driven principle is significant because risk to a system is commonly multifaceted and time dependent, represented in noncommensurate units.

Pareto-Optimality and Tradeoffs

The concept of Pareto-optimality or identifying a noninferior solution to a multiobjective decision-making problem (see Chapter 5) was introduced in 1896 by Vilfredo Pareto, a prominent economist. But it has only been since 1950, when Kuhn and Tucker (1951) published their seminal work on the necessary and sufficient conditions for (proper) noninferiority, that considerable effort has been devoted to developing procedures for generating noninferior solutions to multiobjective problems. The Kuhn–Tucker conditions are discussed in the Appendix. As noted in Chapter 5, a solution to a two-objective function is termed Pareto-optimal if improving one objective function can be achieved only by degrading another. Indeed, Pareto-optimality (also termed noninferior solution, nondominated solution, and efficient solution) is central to determining the additional expenditure required for a further risk reduction. To paraphrase Lowrance (1976) "safety is the level of acceptable risk." Thus, an acceptable level of safety is a Pareto-optimal level of the associated cost and residual risk.

Building on the Kuhn–Tucker necessary and sufficiency conditions for Pareto-optimality, the Epsilon-constraint method (Haimes, 2016) and the surrogate worth tradeoff (SWT) method (Haimes and Hall, 1974) can generate all required Pareto-optimal policies and their associated tradeoffs for any multiobjective optimization problem, without the need for any commensuration among the objective functions; also see Appendix. The Epsilon-constraint method (Haimes et al., 1971; Haimes, 1973, 2016), introduced in Chapter 5, maintains one objective function and converts the remaining objectives into constraints bounded by variable epsilons. Then, the new constrained optimization problem, for each subsystem and for the entire Complex SoS, is further converted into an unconstrained problem through the Lagrangian function. This enables risk analysts and other modelers to generate risk management options and better understand the associated tradeoffs for two or more objectives functions, e.g. in terms of monetary cost, human lives, and continuity of operation following a failure (time in hours, days, etc.). Furthermore, the Lagrange multiplier(s) λ_{ij} as functions of ε_j – namely, $\lambda_{ij}(\varepsilon_j)$ – represent the tradeoffs among the ith and jth objective functions and can be generated between any two objective functions f_i and f_j at the Pareto-optimal frontier, where $\lambda_{1j}(\varepsilon_j) = -\frac{\partial f_1(\cdot)}{\partial \varepsilon_j}$, $j = 2, 3, \ldots, n$ for $\lambda_{1j}(\varepsilon_j) > 0$. Tradeoff analyses are central to all decision making processes, because ultimate decisions among multiple noncommensurate objectives are rarely made solely on the absolute values of each objective (whether cost, risk, or benefit); rather, they are also determined on the basis of the marginal increase or decrease among the corresponding objectives. This fact is particularly salient when two Pareto-optimal policies are closely related but have significantly different associated tradeoffs.

In sum, the way to maximize the efficiency of risk analysis is to develop sound risk management policy options with acceptable tradeoffs among cost, residual risk, and benefit. This important task can be best accomplished, when possible and practical, through quantitative tradeoffs among all relevant competing and commonly noncommensurate objectives among the subsystems and within the entire Complex SoS. Note, however, that the number of selected objectives in risk management introduces an implicit bias in the ultimate subsequent Pareto-optimal policies generated. For example, answers to the second question in risk management "What are the tradeoffs among all relevant costs, benefits, and risks?" (see the *Second Principle*) would squarely depend on the preselection of the specific objective functions. Furthermore, adding or deleting one objective function from consideration in the risk management process would influence the decision-making process, because it would also a priori modify the set of Pareto-optimal frontiers. To address this reality, concerted effort must be devoted to assessing the sensitivity of the resulting policy options to the addition or deletion of one or more objective functions associated with risk management (whether from specific subsystems or from the Complex SoS). Note that removing one of the above three objective functions would eliminate a significant number of Pareto-optimal solutions and vice versa. (For a more comprehensive discussion on Pareto-optimality consult Chapter 5, and Chankong and Haimes, 1983, 2008.)

Relevance to the NextGen: *The NextGen enterprise epitomizes the reliance on multiple, noncommensurate, and competing objectives associated with most, if not all, real systems. The multiple objectives cited earlier to which the NextGen aspires – improving the capacity, safety, security, and environmental impacts of aviation in the United States – are a case in point. Furthermore, diverse state, federal, and international organizations as well as the private sector have an interest in, and an impact on, these objectives. Thus, understanding and responding to (i) the dynamics and interplay among the subsystem's objectives; (ii) the objectives of the entire Complex SoS; (iii) the objectives of individual stakeholders who are advancing them; and (iv) ultimately achieving acceptable tradeoffs among them through effective multiobjective risk modeling, assessment, management, and communication constitute an imperative, though involved, task.*

9.II.3.7 Seventh Principle: Risk Analysis Must Account for Epistemic and Aleatory Uncertainties

Uncertainty, which is commonly defined as *the inability to determine the true state of a system*, depends on factors that affect the modeling of systems and thus risk analysis (Paté-Cornell, 1996; Apostolakis, 1999; Haimes, 2016): (i) incomplete *knowledge (epistemic) uncertainty* that manifests itself in the selection of model topology (structure) and model parameters, among other sources of ignorance (e.g. lack of knowledge of important interdependencies within the system and among other systems) and (ii) *stochastic variability (aleatory) uncertainty* that includes all relevant and important random processes and events, as well as EFCs. Uncertainty dominates most decision-making processes and is the Achilles heel for all deterministic – and most probabilistic – models. Clearly, both categories of uncertainty markedly affect the quality and effectiveness of the risk analysis efforts, and ultimately, the decision-making process.

All too often ignored in the quest for quantitative predictive models, a significant source of uncertainty is the limited ability to develop models that would represent the future behavior of a system. In addition, human errors of both commission and omission introduce more model uncertainties. A sample of sources of uncertainties and errors in modeling includes the following six major categories: *model topology, model parameters, model scope or focus, data collection, decision-making methods*, and *human subjectivity*.

Modelers' subjectivity strongly influences how subsystems are modeled as well as the overall model of the entire Complex SoS and thus the associated risk assessment, management, and communication process. "Subjectivity" includes the background, training, competence, personal preference, self-interest, proficiency, and experience of the analyst. Human subjectivity also plays a very important role in selecting (by omission or commission) the specific building blocks of the model or models. Uncertainty, stemming from stochastic variability, is associated with the inherent fluctuations, or differences in the numerous parameters that populate all models. In other words, variability occurs when the quantity of concern is not of a specific value, but rather a population of values. Also, the construction, calibration, identification, testing, "validation," and, ultimately, implementation – which could reduce epistemic uncertainty – of a model depend in large part on the ability of the analyst to access sufficiently representative data. Finally, an inaccurate, or insufficient database (due to deficiencies in collection, acquisition, processing, and analysis) may cause substantial errors in modeling and thus in the risk analysis process.

Relevance to the NextGen: *Epistemic and aleatory uncertainties influence and impact all other principles. This is particularly the case with the NextGen due to the numerous sources of uncertainties associated with planning for and deploying new technology: (i) radical utilization of the restructured multilayer air space for different flights and aircrafts; (ii) heavier reliance on GPS and diminished reliance on radar to navigate airplanes through narrower airspace corridors; training of professionals at almost all levels, as well as their adjustment to new responsibilities and functionality; (iii) and new rules and procedures that must be adapted internationally for all carriers that fly to the United States. These myriad sources of aleatory and epistemic uncertainty are only a scant sample among a host of others, new organizational and budgetary environments to which multiple players must confront a long host of unknowns.*

9.II.3.8 Eighth Principle: Risk Analysis Must Account for Risks with Low Probability but Extreme Consequences

The translation of the definition of risk into quantitative terms is extremely difficult and has been a major source of misuse by many disciplines, often leading to erroneous conclusions (Haimes, 2016). The most common quantification of risk is through the mathematical construct known as the *expected-value-of-risk* metric, which holds equal low-probability, high-consequence events commensurate with high-probability, low-consequence events. This practice has played a decisive role in *dangerously masking* the criticality of extreme and catastrophic events. It is important to recognize the misuse of the expected value – the averaging of risk – when *it is used as the sole criterion for risk in decision making*. The Partitioned Multiobjective Risk Method (PMRM) (Asbeck and Haimes, 1984; Haimes, 2016) supplements the expected value measure of risk with a conditional expected value of risk of extreme events. *A conditional expectation* is defined as *the expected value of a random variable, given that this value lies within some prespecified probability or consequences range*. During the last four decades, the interest in risk of extreme events has been the subject of many researchers and practitioners (Haimes, 2016; Paté-Cornell, 1990; Perrow, 1999; Taleb, 2007). The perception of risk plays an important role in the extent to which individuals, the public, and policymakers are willing to invest resources in preparedness, response, and recovery, or in the mitigation of low-probability risks of extreme events that exist far in the future (Slovic, 2000, 2002).

Statistics of Extremes: The theory of the statistics of extremes is concerned with studying the largest (e.g. a flood) or smallest (e.g. a drought) value realized from a sample of *n* independent, identically distributed random variables (Ang and Tang, 1984). The asymptotic theory of statistical extremes was developed early in the twentieth century by a number of statisticians, including Fisher and Tippett (1927) and Gnedenko (1963). This is an analytic theory concerned with the forms of probability functions with n observations, and as $n \to \infty$, such probabilities may converge to one of three particular forms, which Gumbel (1954) termed as Types I, II, and III. It can be shown that the behavior of the tail of the initial probability distribution (variant) determines to which type the extreme distribution corresponds. If the tail of the initial distribution decays exponentially, then the largest value is of Type I; if the tail decays in a polynomial way, then the largest value is of Type II. The third form, Type III, is only reserved for initial variants that have a finite upper bound. The importance of extreme and catastrophic events in risk analysis makes the theory of statistics of extremes an invaluable tool. In particular, the incorporation during the last three decades of the PMRM and the statistics of extremes in risk analysis has proven to be effective and informative (Haimes, 2016).

Relevance to the NextGen: *Civilian air travel is probably the most scrutinized and protected mode of transportation around the world, and historically it has had an impressive record of safety. Thus, the expected-value metric of risk cannot be applied to air transportation. Consider that there are over 35 000 annual fatalities from car accidents in the United States. This staggering number of deaths has become the "expected value" in surface transportation; rarely does the media report on such calamities on a regular basis. However, a collision of two airliners would be considered catastrophic by the media and the public, which gives a false sense of the danger of air travel over surface travel. Thus, the risk of low probability and extreme consequences for both single systems and for Complex SoS must be modeled and quantified using proper mathematics and statistics as discussed in the Eighth Principle.*

9.II.3.9 Ninth Principle: The Time Frame Is Central to Quantitative Risk Analysis

Can risk analysis be performed effectively without considering future EFCs (affecting every subsystem and the entire Complex SoS) within the TSS and within an appropriate time frame? The fact that the risk function is composed of the likelihood of adverse initiating events and of their resulting consequences, both of which are time dependent, speaks to the centrality of the time frame in risk analysis. Thus, a systemic approach to risk analysis should (i) embrace the continuous and timely assessment and evaluation of precursors to potential forced changes, (ii) balance present resource allocation for preparedness (to reduce the states of vulnerability and enhance the states of resilience of the system to specific adverse initiating events or threats), (iii) update the likelihood of the tracked initiating events (on each subsystem and on the entire Complex SoS) and of the resulting consequences (given that investments in preparedness affect the states of the system, as may be represented by changes in its vulnerability and resilience), and (iv) update, based on the above, the entire process of risk assessment, management, and communication associated with the Complex SoS in question. In its essence, the Bayesian theory and analysis are predicated on the premise that new information can update and enhance prior information. In particular, flexibility in the formulation of risk management policy, with a focus on its future impacts, would make the subsystems and the entire Complex SoS more agile and provide a buffer against possible catastrophic or irreversible consequences. Furthermore, a systemic risk analysis should stress the importance of impact analysis due to the dynamics that affect all systems – societal, organizational, and technological. Examples include demographic changes, advancements in technology, and natural deterioration over time of all physical infrastructures. Furthermore, as discussed in the *Sixth Principle*, each stage in this evolving dynamic world is characterized by multiple objectives, and the tradeoffs among the attainment of present objectives and future flexibility ought to be incorporated within the overall risk-informed decision-making process.

Dynamic Complex SoS, where time is of paramount importance, introduce another challenge to the multiple objectives presented in the *Sixth Principle*. For example, to ignore the changes in the states of a dynamic safety-critical system as time progresses would be to render static-based models and decisions to be misleading, if not disastrous. Recognizing this fact, event trees, decision trees, and process control, among other methods, account for the impacts of current decisions on future options. Figure 9.II.2 presents

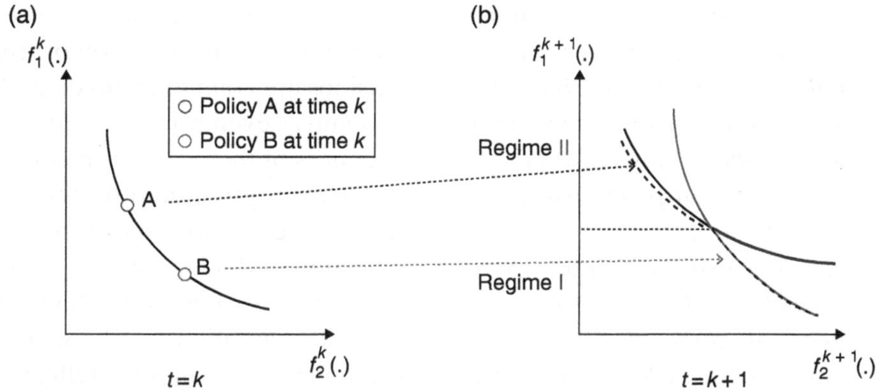

Figure 9.II.2 *Impact of policies at time* t = k *on future options at time* k + 1.

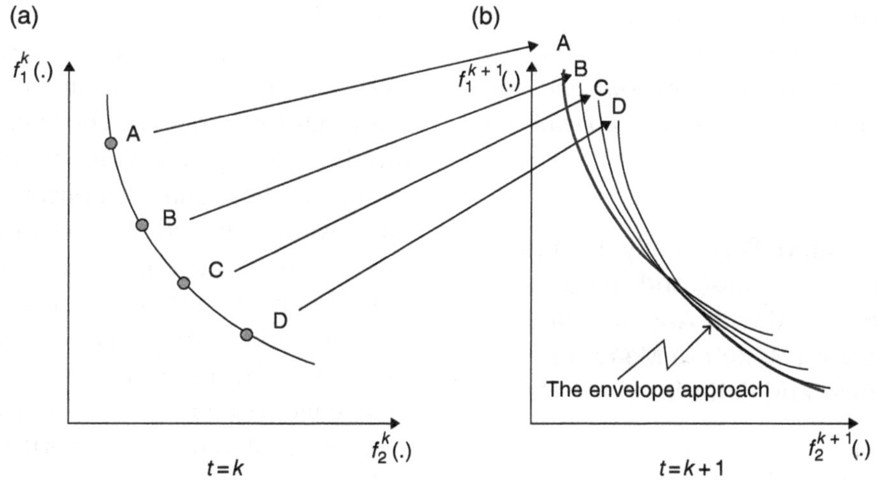

Figure 9.II.3 *(a) and (b) The envelope of the combined Pareto-optimal frontier of policies A, B, C, and D for* k + *1st period.*

the impact of policies at time $t = k$ on future options at time $k + 1$, where the dotted line represents a new Pareto-optimal frontier generated at time $k + 1$.

Note that if the decision maker(s) were to prefer to operate during period $k + 1$ in regime I, then policy B would be inferior to A. Conversely, if the decision maker(s) were to prefer to operate during period $k + 1$ in regime II, then policy A would be inferior to B. The envelope approach was developed to account for the propagation of Pareto-optimal solutions from one period to another (Li and Haimes, 1987). Figure 9.II.3a extends this concept through the envelope approach, where the Pareto-optimal frontier can be generated for any n periods. Note that the solid-line outer curve represents the envelope of the combined Pareto-optimal frontier of policies A, B, C, and D for the $k + 1$st period. Note that analytically, the Envelope Approach is applicable to multiple objectives; however, graphical displays are more restricted.

Finally, the dynamic tradeoffs among multiple objectives must augment the determination of Pareto-optimality in risk management, given the dynamic nature of evolving events. Similarly, the process of eliciting the decision makers' preferences must be systemic and harmonious with the realization that conditions, and thus policies, are dynamic and subject to update and modification.

Relevance to the NextGen: *The time frame has been commonly implicit in risk analysis. Yet its centrality in risk and decision making is so dominant that it must be addressed explicitly by decision makers, modelers, and other practitioners. This notion is imperative for Complex SoS. The two sets of questions in risk assessment and risk management introduced in the Second*

Principle would lose their impetus without a model that adhered to the time frame. Reconsider, in the context of the NextGen, the third question in risk management: What are the impacts of current decisions on future options? Planning that spans more than a decade is fraught with unexpected surprises associated with every subsystem within the over two-dozen major subsystems that make up the NextGen Complex SoS. More specifically, the NextGen planning process is partitioned in sequences of three five-year time periods: Alpha, Bravo, and Charlie. Clearly, decisions made during period Alpha would constrain and affect subsequent future decisions during Bravo and Charlie. In particular, "wrong" or inflexible decisions at any planning period (e.g. period Alpha) could result, during periods Bravo or Charlie, in both programmatic risk (project's cost overrun and completion delayed) and technical risk (not meeting performance criteria).

9.II.3.10 Tenth Principle: Risk Analysis Must Be Holistic, Adaptive, Incremental, and Sustainable, and It Must Be Supported with Appropriate Data Collection, Metrics with Which to Measure Efficacious Progress, and Criteria on the Basis of Which to Act

Addressing the question "What are the impacts of current decision on future options" requires adhering to the first nine principles, which together constitute *a requisite to a holistic, adaptive, incremental, and sustainable risk analysis process*. A holistic risk assessment, management, and communication process of a system of systems must identify all conceivable sources of risk by adhering to the TSS, focusing on and tracking the most critical EFCs. At a minimum, this process must recognize that a Complex SoS encompasses many of the following attributes: it is dynamic; it is fraught with myriad sources of risk and uncertainty, driven by EFCs; it involves a hierarchy of multiple decision makers; it often must address multiple conflicting, competing, and noncommensurate objectives and aspirations; it is driven by ever-evolving requirements, demands, and targets; and, not least of all, it protects, adheres to, and promotes organizational core values.

Fundamental to the risk analysis of single systems and interdependent and interconnected Complex SoS is the consideration of present and future risks associated with any decision or tradeoff. Since critical decisions are commonly made under uncertain conditions and as events are tracked over time and the consequences of the decisions made (positive or negative on each subsystem and on the entire Complex SoS) and assessed, some appropriate *incremental* course correction and adaptation is needed. Thus, the term *incremental* is to be understood in the context of the dynamic adaptation to EFCs. As more knowledge is accumulated and more intelligence and information become available, making incremental investments in risk management becomes an effective policy with which to minimize potential shortfalls with budgeted resources.

A sustainable risk-based decision-making process must share many of the traits of a holistic process, in that it must be strategic, encompassing, and well balanced, involving and communicating with all concerned parties and responsive to the uncertain world within which the results of the decision will be realized. In sum, a holistic, adaptive, incremental, and sustainable risk assessment, management, and communication process has the following attributes: (i) is driven by and responsive to EFCs; (ii) rests on a systemic, comprehensive, and encompassing doctrine; (iii) continuously updates the states of the system and the corresponding decisions; (iv) incrementally updates all appropriate policy decisions in a step-by-step manner that builds on new knowledge, ensuring a viable and repeatable process; and (v) is sustainable – that is, "maintained at length without interruption, weakening, or losing power or quality" (*Webster*, 1976).

Finally, risk analysis associated with a single system or with Complex SoS, must be supported by:

Data collection: Develop a data collection plan and deployment process for each component of the tracked top EFCs and precursors thereto.

Metrics: Develop metrics with which to (i) measure and evaluate, on the basis of the collected database, the progress being made in the risk analysis process; (ii) track and evaluate precursors to emerging risks; (iii) assess the impacts of current decisions and progress on future options and on succeeding phases in the risk analysis process; and (iv) evaluate the experiences, results, and lessons learned as a prelude to an effective decision-making process.

Other guiding criteria for decision making: Develop guiding criteria for required actions when a tracked risk scenario, or other risks, emerge as imminent and thus must be confronted and acted upon.

Relevance to the NextGen: *The necessity for adherence to the Tenth Principle is grounded on the intricacy of all Complex SoS (including the NextGen) and on all nine principles and other premises discussed in this chapter. More specifically, (i) the long-term planning horizon and complexity of most such systems of systems require a holistic comprehensive approach; (ii) the myriad EFCs and the dynamic changes within and outside the subsystems and entire Complex SoS call for an adaptive process; (iii) the subsequent responses to these internal and external forces must be incremental (unless a catastrophic failure is discovered); (iv) all decisions must be sustainable; and (v) a continuous adaptive feedback mechanism must be put in place to ensure an effective risk assessment, management, and communication process. This mechanism must be supported with appropriate data collection, metrics with which to measure progress, and criteria on the basis of which to act, when actionable decisions must be made.*

9.II.4 THE CONTINUOUS PROCESS OF RISK ASSESSMENT, MANAGEMENT, AND COMMUNICATION

Risk assessment, management, and communication is a continuous, never-ending process. It constitutes the ultimate incorporation and deployment of the ten guiding principles within the complex process of risk assessment and management (answering the two sets of questions, respectively) as well as risk communication.

9.II.5 EPILOGUE

The multidisciplinary nature of the field of risk analysis necessarily implies that each discipline has over time developed, adopted, and adapted appropriate theories, methodologies, and practices from other disciplines. This cross-fertilization has been the hallmark of this field as is manifested through the diverse membership of the SRA and by its flagship journal *Risk Analysis: An International Journal*. In this context, this chapter is written from the perspectives of systems engineering with the hope and expectation that the principles it introduce will be augmented by others.

REFERENCES

Ang, A. and Tang, W. (1984). *Probability Concepts in Engineering Planning and Design*, vol. **2**. New York: Wiley.

Apostolakis, G. (1999). The distinction between aleatory and epistemic uncertainties is important: an example from the inclusion of aging effects into probabilistic safety assessment. *Proceedings of PSA'99* (22–25 August 1999). Washington, DC: American Nuclear Society.

Asbeck, E. and Haimes, Y.Y. (1984). The partitioned multiobjective risk method (PMRM). *Large Scale Systems* **6**(1): 13–38.

Chankong, V. and Haimes, Y.Y. (1983). *Multiobjective Decision Making: Theory and Methodology*. New York: North Holland.

Chankong, V. and Haimes, Y. (2008). *Multiobjective Decision Making: Theory and Methodology*. New York: Dover.

Chen, C. (2012). *Linear System Theory and Design*, 3e. New York: Oxford University Press.

Duderstadt, J., Wulf, W., and Zemsky, R. (2005). Envisioning a transformed university. *Issues in Science and Technology* **22**(1): 35–41.

Fisher, R.A. and Tippett, L.H.C. (1927). Limiting forms of the frequency distribution of the largest of smallest numbers of a sample. *Proceedings of the Cambridge Philosophical Society* **24**: 180–190.

Gnedenko, B. (1963). *The Theory of Probability* (Trans. B. Seckler). New York: Chelsea Publishing Company.

Gumbel, E. (1954). *Statistical Theory of Extreme Values and Some Practical Applications*, Applied Mathematical Series, vol. **3**. Washington, DC: National Bureau of Standards.

Haimes, Y.Y. (1973). Integrated system identification and optimization. In: *Advances in Control System Theory and Applications*, vol. **10** (ed. C.T. Leondes), 435–518. New York: Academic Press.

Haimes, Y.Y. (1981). Hierarchical holographic modeling. *IEEE Transactions on Systems, Man, and Cybernetics Part A: Systems and Humans* **11**(9): 606–617.

Haimes, Y.Y. (1991). Total risk management. *Risk Analysis* **11**(2): 169–171.

Haimes, Y.Y. (2011). On the complex quantification of risk: systems-based perspective on terrorism. *Risk Analysis* **31**(8): 1175–1186.

Haimes, Y.Y. (2012a). Systems-based guiding principles for risk modeling, planning, assessment, management and communication. *Risk Analysis* **32**(9): 1451–1467.

Haimes, Y.Y. (2012b). Modeling complex systems of systems with phantom system models. *Systems Engineering* **15**(3): 333–346.

Haimes, Y.Y. (2016). *Risk Modeling, Assessment, and Management*, 4e. Hoboken, NJ: Wiley.

Haimes, Y.Y. (2018). Risk modeling of interdependent complex systems of systems: Theory and practice. *Risk Analysis*, **38**(1): 84–98. doi: 10.1111/risa.12804.

Haimes, Y.Y. and Anderegg, A. (2015). Sequential Pareto-optimal decisions made during emergent complex systems of systems: An application to the FAA NextGen. **18**(1): 28–44.

Haimes, Y.Y. and Hall, W.A. (1974). Multiobjectives in water resources systems analysis: the surrogate worth trade-off method. *Water Resources Research* **10**(4): 615–624.

Haimes, Y.Y., Lasdon, L., and Wismer, D. (1971). On the bicriterion formulation of the integrated system identification and systems optimization. *IEEE Transactions on Systems, Man, and Cybernetics SMC* **1**: 296–297.

Haimes, Y.Y., Horowitz, B.M., Guo, Z. et al. (2015). Assessing systemic risk to cloud-computing technology as complex interconnected systems of systems. *Systems Engineering* **18**(3): 284–299.

Kaplan, S., Haimes, Y.Y., and Garrick, B. (2001). Fitting hierarchical holographic modeling into the theory of scenario structuring and a resulting refinement of the quantitative definition of risk. *Risk Analysis* **21**(5): 807–815.

Keeney, R. and Raiffa, H. (1976). *Decisions with Multiple Objectives*. New York: Wiley.

Kuhn, H.W. and Tucker, A.W. (1951). Nonlinear programming. In: *Proceedings, 2nd Berkeley Symposium on Mathematical Statistics and Probability* (ed. J. Neyman), 481–492. Berkeley, CA: University of California Press.

Li, D. and Haimes, Y.Y. (1987). The envelope approach for multiobjective optimization problems. *IEEE Transactions: Systems, Man and Cybernetics* **17**(6): 1026–1038.

Lowrance, W. (1976). *Of Acceptable Risk*. Los Altos, CA: William Kaufmann.

Maier, M. (1998). Architecting principle for systems-of-systems. *Systems Engineering* **1**(4): 267–284.

Nise, N. (2015). *Control Systems Engineering*, 7e. Pomona, CA: California State Polytechnic University.

Pareto, V. (1896). *Cours d'économie Politique professé a l'université de Lausanne*, vol. **I**, in-8 de v-430 pages. Rouge: editeur a Lausanne Pichon, libraire a Paris.

Paté-Cornell, E. (1990). Organizational aspects of engineering system safety: the case of offshore platforms. *Science* **250**: 1210–1217.

Paté-Cornell, E. (1996). Uncertainties in risk analysis: six levels of treatment. *Reliab Eng Syst Safe* **54**: 95–111.

Perrow, C. (1999). *Normal Accidents: Living with High-Risk Technologies*. Princeton, NJ: Princeton University Press.

Sage, A.P. and Biemer, S.M. (2007). Processes for system family architecting, design and integration. *IEEE Systems Journal* **1**(1): 5–16.

Slovic, P. (2000). *The Perception of Risk*. Sterling, VA: Earthscan Publications Ltd.

Slovic, P. (2002). Terrorism as hazard: a new species of trouble. *Risk Analysis* **22**(3): 425–426.

Taleb, N. (2007). *The Black Swan: The Impact of the Highly Improbable*. New York: Random House.

U.S. Federal Aviation Administration (2014). *FAA's NextGen: Implementation Plan*. NextGen Integration and Implementation Office: Washington, DC.

U.S. Nuclear Regulatory Commission (1981). *Fault Tree Handbook*, NUREG-81/0492. Washington, DC: United States Government Printing Office.

Webster (1976). *Webster's Third International Unabridged Dictionary*. Springfield, MA: Merriam-Webster.

10

Modeling Cyber–Physical Complex Systems of Systems: Four Case Studies

OVERVIEW

This chapter develops models for risk to interdependent cyber–physical complex systems of systems (SoS) (Complex SoS) that builds on the shared/common states and other essential entities among the systems and subsystems that constitute Complex SoS (the term "essential entities" include *shared/common decisions*, *decision makers*, *stakeholders*, *resources*, *organizational behavior and norms*, *policies and procedures*, *management*, *culture*, and others.)

Four case studies focus on safety-critical cyber–physical systems and infrastructures are presented; and they utilize state-space theory and methodology of modeling and managing Complex SoS. The interdependencies and interconnectedness (I-I) that characterize Complex SoS, which are modeled throughout this book via shared states and other essential entities, constitute the common denominator of all four case studies. They represent an operational sample of modeling, theory, methodology, and practice espoused in this book. Building on fault-tree analysis enables us to accomplish the identification and quantification of critical shared states and other essential entities within the Complex SoS and their role as critical sources of risk thereto.

Modeling cyber–physical infrastructures, or other Complex SoS, necessarily implies determining their properties, constructing the relationships among the inputs and outputs through their state variables and other variables and parameters (e.g. random, decision, and exogenous variables), quantifying intra- and interdependencies within and among their various components and subsystems, and determining the appropriate model topology (structure) and parameters that best represent their essence and functionality. To achieve this, we commonly rely on data collection through testing, observation, experimentation, and measurement and through a tedious learning process about the Complex SoS, including the use of simulation. However, studying, modeling, and managing the myriad sources of risk to Complex SoS require an augmented schema that addresses the essence of what constitutes this complexity. In our case studies we addressed this undertaking by building on state-space theory, which has enabled us to understand and model the necessary defining characteristic of complexity, manifested through the I-I within and among the systems that make up Complex SoS.

Modeling and Managing Interdependent Complex Systems of Systems, First Edition. Yacov Y. Haimes.

In sum, the four cyber–physical case studies share the following common denominator, namely, building on and exploiting (i) state-space theory and methodology, (ii) modeling I-I via shared states and other essential entities (shared/common decisions, decision makers, stakeholders, resources, organizational behavior and norms, policies and procedures, among others), (iii) analytically quantifying via our reinvention of fault-tree analysis (without reliance on any reliability information) the I-I within and among the subsystems that compose Complex SoS, and (iv) sharing with the reader the detailed methodological process and lessons learned.

PART I: MODELING OF INTERDEPENDENT AND INTERCONNECTED AVIATION COMPLEX SYSTEMS OF SYSTEMS

10.I.1 INTRODUCTION AND OVERVIEW

The Federal Aviation Administration (FAA) is planning the integration of three major systems – communications (C), navigation (N), and surveillance (S) (which today operate semi-independently with coordination) – into one complex systems of systems (CNS Complex SoS). The genesis of the complexity stems from the newly created I-I among the integrated CNS Complex SoS. We trace the I-I among the integrated CNS in terms of the newly shared/common state (variables) and other "essential entities" within and among the three C, N, and S systems. We subsequently develop a sample of comprehensive tables identifying shared/common states and other essential entities, (decisions, decision makers, and resources) among all three members of the CNS Complex SoS. We accomplish this critical/technical task with the help and contributions of FAA and MITRE Corporation domain experts in the operations and interface among members of the CNS Complex SoS. The tables that summarize the I-I among the (planned) integration of the C, N, and S systems provide the essential step in modeling and quantifying the I-I resulting from the (planned) systems integration. Subsequent to tabulating the shared states (Table 10.I.1), shared decisions (Table 10.I.2), shared decision makers (Table 10.I.3), and shared resources (Table 10.I.4) within and among the three CNS Complex SoS, we translated the tables into the reinvented fault trees, without the need for, nor reliance on, the reliability of the C, N, and S systems. Our tables enabled us to determine (with the advice and expertise of FAA operators and technical experts) whether the shared systems and/or their subsystems are connected in (i) series (via OR Gates), or connected, in (ii) parallel (via AND Gates). These essential designations of "AND" or "OR" Gates in the construction of fault trees, which were guided by the expertise and experience of FAA engineers, technicians, and managers, were reconfirmed at a second round. Note that

TABLE 10.I.1 Shared States (System View) – Required Navigation Performance (RNP) Configuration

States	Communication	Navigation	Surveillance
HF signal integrity	✓	✓	
VHF signal integrity	✓	✓	
UHF signal integrity		✓	✓
SATCOM signal integrity	✓	✓	
GPS signal integrity		✓	✓
GPS signal availability		✓	✓
GPS signal accuracy		✓	✓
Radar signal integrity			✓
Comprehension	✓		
Plane authenticity/identification	✓		✓
Signal effective range	✓	✓	✓
Station authenticity/identification		✓	
Distance calculation accuracy		✓	
Position drift		✓	
Detection availability			✓
Compliance	✓	✓	✓

TABLE 10.I.2 Shared Decision–Dynamic RNP Configuration–Approach

Decision	Communication	Navigation	Surveillance
Type of approach			
Instrument flight rules	✓	✓	✓
Visual flight rules	✓	✓	
Approach system			
ILS		✓	
Vectored approach	✓		✓
GPS (RNAV) approach		✓	
VOR approach		✓	
NDB approach		✓	
Runway clearance	✓		
Aircraft separation	✓		✓
Approach route	✓	✓	
Holding pattern	✓	✓	

TABLE 10.I.3 Shared Decision Makers–Dynamic RNP Configuration–Approach Phase

Decision\decision maker	Pilot	ATC	Airport
Type of approach			
Instrument flight rules	✓	✓	
Visual flight rules	✓	✓	
Approach system			
ILS	✓	✓	✓
Vectored approach	✓	✓	✓
GPS (RNAV) approach	✓	✓	✓
VOR approach	✓	✓	✓
NDB approach	✓	✓	✓
Runway clearance		✓	✓
Aircraft separation		✓	
Approach route	✓	✓	
Holding pattern		✓	

TABLE 10.I.4 Shared Resources–Dynamic RNP Configuration–Approach Phase

Resource	Communication	Navigation	Surveillance
Radar sites			✓
DME substations		✓	
Satellites	✓	✓	
Air traffic controllers	✓		✓
Pilots	✓	✓	
Airplane	✓	✓	✓
NACO charts	✓	✓	✓
Flight management computer	✓	✓	
ATN–B2 receiver/transceiver	✓		
NAVAID receiver		✓	
Weather stations	✓	✓	✓

if all systems (subsystems) in the minimal cut set fail, then the entire CNS Complex SoS would fail. Stated differently, the failure of any subsystem connected in series (as a member of the minimal cut set) would cause the failure of all other connected systems. For more detailed discussion on fault trees, the reader may consult the Appendix of this book.

We introduce a theoretically-based methodology to model and prioritize safety-critical interdependent and interconnected Complex SoS. Limiting our modeling efforts to only those prospective faults/risks initiated at one subsystem that can propagate to other subsystems of the Complex SoS (due to shared states and other essential entities), captures the most significant risks to the integrated CNS Complex SoS (Lewis and Haimes, 2018).

With future technologies and a fully integrated CNS Complex SoS, the current level of assumed safety would significantly change. Clearly, such an unprecedented systems integration of three safety-critical systems into a single CNS Complex SoS must undergo extensive risk modeling, assessment, management, and communication review prior to any partial or full deployment of the integrated system. The following is an overview of the theoretical and methodological processes for modeling CNS Complex SoS that were used by the MITRE/University of Virginia team on this FAA case study.

10.I.2 THEORETICAL AND METHODOLOGICAL APPROACH

This part develops a quantitative modeling methodology for interdependent and interconnected cyber–physical Complex SoS. We build on two major tenets of systems modeling and risk analysis, namely, (i) the central role of state variables in modeling and (ii) reinventing the traditional use of fault-tree analysis, without the reliance on reliability. We achieve this task, first by identifying the I-I that characterize the interface between and among the planned integrated C, N, and S systems and their associated subsystems. We exploit the fact that all outputs from a system are functions of the states of that system and thus also of the decisions and exogenous variables and all other inputs to the system. This fact is of particular significance to modeling Complex SoS. Chen (2012) offers the following succinct definition of *state variable*: "The state $x(t_o)$ of a system at time t_o is the information at t_o that together with the input $u(t)$, for $t \geq t_o$, determines uniquely the output $y(t)$ for all $t \geq t_o$." The states of a system, commonly a multidimensional vector that characterizes the system as a whole, play a major role in estimating its future behavior for any given input. Thus, the behavior of the states of the system as a function of time enables modelers to determine, under certain conditions, the system's future behavior for any given *input* or *initiating event*. (For example, to determine the reliability and functionality of a car, one must know the *state* of the fuel, oil, tire pressure, and other mechanical and electrical components.) All systems are characterized at any moment by their respective state variables and the conditions thereof, and these conditions are subject to continuous variation and fluctuation. We embrace the fact that complex systems cannot, by their essence and definition, be understood or modeled using one "straight-jacket" modeling schema. Most importantly, by understanding what makes I-I so central to the complexity of SoS and challenging to both understanding and modeling them, we will ultimately improve future management of Complex SoS.

We discover and model the essence of their complexities by tracing their I-I via shared/common states and other essential entities among the individual systems and their subsystems. The time frame of a system is a major driver of what we term complexity, because all systems continue to evolve, emerge, and change, while our ability to keep pace with these changes, via modeling, continues to lag. Although static analytical analyses assume a steady-state model, our domain-expert FAA/MITRE teams did project likelihood of future evolving I-I within and among the subsystems of the planned integrated CNS Complex SoS.

A key step in the modeling process was the identification (with several aviation domain experts) of critical shared (i) states, (ii) decisions, (iii) decision makers, and (iv) resources among the three systems and their subsystems (see Tables 10.I.1–10.I.4, respectively). Clearly, without projecting, understanding, and thus tabulating the consequences resulting from the ensuing shared states and other essential entities, the planned integration of the three semiautonomous systems could result in disastrous consequences. To focus on the most consequential shared states and other essential entities, we deployed risk filtering, ranking, and management (RFRM) (Haimes et al., 2002; Haimes, 2016a), the inclusion of which is beyond the scope of this part. Specific details on fault trees and their essential contributions to the theory and methodology for quantifying the I-I among the three CNS systems are introduced in Section 10.I.4. We achieve the quantification of the I-I among the three CNS systems by reinventing the use fault trees without the need of knowledge of the reliability of the systems under consideration. Thus, we were able to (i) reinvent the use of fault trees (without a required knowledge of the reliability of each system or subsystem) through modeling using AND Gates, and OR Gates; discover consequential interdependencies and interconnections that would emerge from the planned integration of the three major safety-critical systems into a CNS Complex SoS by developing Tables 10.I.1–10.I.4 of shared/common states, decisions, decision makers, and resources within and among the three CNS systems and their subsystems; (ii) quantify the risk of dire consequences resulting from the I-I of shared states and other essential entities by modeling shared states and other essential entities via AND Gates (systems connected in parallel) and OR Gates (systems connected in series); and (iii) discover consequential interdependencies and interconnections that would emerge from the planned integration of the three major safety-critical systems into a CNS Complex SoS by developing Tables 10.I.1–10.I.4 of shared/common states, decisions, decision makers, and resources within and among the three CNS systems and their subsystems.

An effective modeling of any system, particularly Complex SoS, requires a clear understanding of their configuration. This methodological approach enabled us to identify and track the essence of the I-I characterizing the system's complexity via the common/shared states and other essential entities among the systems that compose the CNS Complex SoS.

10.I.3 MODELING SHARED STATES AND ESSENTIAL ENTITIES IN COMPLEX SYSTEMS OF SYSTEMS: SPECIFIC LESSONS LEARNED

Modeling the I-I among the three systems is centered on identifying shared states and other essential entities between at least two of the three C, N, and S systems. Tables (10.I.1–10.I.4) provide safety-critical knowledge about the added sources of risk due to the I-I among the components of the CNS Complex SoS. According to the following tables, (i) five states are shared between communications and navigation, (ii) three states are shared between communications and surveillance, (iii) six states are shared between navigation and surveillance, and (iv) one state (compliance) is shared with all three systems. Using Tables 10.I.1–10.I.4, we are able to determine the most critical shared states and other essential entities. Together, they further our understanding of each of the C, N, and S systems, as well as their interdependencies.

For example, the increased utilization of global positioning system (GPS) technology has enabled the performance of advanced activities, such as dynamic required navigation performance (dynamic RNP), resulting in a significant gain in efficiency and savings. At the same time, there is an added source of risk associated with using a dynamic RNP configuration when relying on GPS technology. This is because GPS and dynamic RNP efficiencies significantly increase the interdependencies within and among the CNS Complex SoS components (FAA, 2014). More specifically, Table 10.I.1 represents shared safety-critical states and their place within the integrated CNS Complex SoS. *In a subsequent discussion we will refer to Tables 10.I.1–10.I.4 as critical precursors to catastrophic failures due to the (i) shared states (Table 10.I.1), (ii) shared decisions (Table 10.I.2), (iii) shared decision makers (Table 10.I.3), and (iv) shared resources (Table 10.I.4).*

The use of GPS in both the navigation and surveillance systems makes GPS a vital resource, if not an essential entity. Indeed, the integrity, availability, and accuracy of the GPS become vital to the integrated CNS Complex SoS. Furthermore, via the shared states and other essential entities, we are able to discover the criticality and integrity of the very high frequency (VHF) signal, which connects the communications and navigation systems through the signals used in both voice and data communication and navigation. The effective signal range is vital to all three systems and indicates the importance of complete coverage and an understanding of when and where to switch from one system to another. Furthermore, navigation is largely considered the pilot's responsibility and is aided by the flight management system (FMS). Surveillance, considered the traffic controller's (ATC) responsibility, is the ability to determine the aircraft's position within the airspace and maintain correct separation among all aircraft. The communications system represents the connection between the navigation and surveillance systems and can be separated into two categories – voice communication and data communication – and modeled accordingly. The current FAA system relies heavily on voice communication between the pilot and the ATC. While this link will never fully disappear with the CNS integration, the new system will rely much more heavily on data communication. Voice communication is used to vector aircraft, avoid weather, maintain specific separation, ask for and receive altitude corrections, and other pilot–controller communication. In the current data communications system, the aircraft communications addressing and reporting system is used to push small text data to and from the aircraft. The communication is limited to initial flight plan loading, text messages, weather updates, and terminal conditions.

Furthermore, the integration of three safety-critical systems into one CNS Complex SoS also require systematic, comprehensive answers to the following basic questions in risk analysis *for each of the three C, N, and S systems and for the entire integrated CNS Complex SoS*: (i) What can go wrong? (ii) What is the likelihood? (iii) What are the consequences? and (iv) What is the time frame? Although our research project did not address the above basic explicit questions in risk analysis (they were delegated to other teams), our results identified critically important sources of risk (via the fault-tree analysis) resulting from the planned integration of the CNS systems into a single Complex SoS.

Assessing and managing the impact of current decisions on future options are necessary in risk modeling and analysis in general and in particular for safety-critical systems. Recall that each of the three C, N, and S systems currently operates semiautonomously, with integration planned to exceed two decades. Thus, the risk-based ramifications of this integration being carried out by a federal agency that has the mandate to *operate and regulate* the entire aviation Complex SoS cannot be underestimated. A sample of one of the most consequential findings resulting from our modeling the I-I resulting from the systems integration is the need to systematically examine the shared states and other essential entities with an eye toward "what not to do." *That is, the consequences resulting from the shared states and other essential entities of the C, N, and S systems must be carefully studied and their integration evaluated to avoid "one-point failure"* of the aviation Complex SoS. Our study identified specific aspects of the communications, navigation, and surveillance systems and their subsystems that should never be integrated because their merging could lead to disastrous consequences (Tables 10.I.1–10.I.4 represent a sample of our findings). Our theoretical and methodological premises regarding modeling Complex SoS published earlier (Haimes, 2008, 2012a) have been measurably clarified, solidified, and substantiated by this and other case studies. Indeed, this modeling process provided an opportunity for the MITRE/FAA/University of Virginia team to better understand the intricacies resulting from the integration of the C, N, and S systems. This investigative and learn-as-you-go process has lead to (i) developing a better understanding of the consequences resulting from the planned systems integration, (ii) identifying specific safety-critical failure modes associated with the planned integration, and (iii) alerting the decision makers and stakeholders regarding "what not to do," probably one of the most important and immediate benefits of the FAA case study.

10.I.4 REINVENTING THE USE OF FAULT TREES FOR MODELING COMPLEXITY VIA SHARED STATES AND OTHER ESSENTIAL ENTITIES

The extensive reliance on fault trees and their reinvention call for the following succinct description of the fault-tree model (U.S. Nuclear Regulatory Commission, 1981), which is most relevant to modeling Complex SoS:

> A fault tree analysis can be simply described as an analytical technique, whereby an undesired state of the system is specified (usually a state that is critical from a safety standpoint), and the system is then analyzed in the context of its environment and operation

to find all credible ways in which the undesired event can occur (in the case of aviation could mean disastrous consequences). The fault tree itself is a graphic model of the various parallel and sequential combinations of faults that result in the occurrence of the predefined undesired event. The faults can be events that are associated with component hardware failures, human errors, or any other pertinent events, which can lead to the undesired event. A fault tree thus depicts the logical interrelationships of basic events that lead to the undesired event – which is the top event of the fault tree. It is important to understand that a fault tree is not a model of all possible system failures or all possible causes for system failure. A fault tree is tailored to its top event, which corresponds to some particular system failure mode, and the fault tree thus includes only those faults that contribute to this top event. Moreover, these faults are not exhaustive – they commonly cover only the most credible faults as assessed by the analyst. (pp. IV–1)

Failures in interdependent and interconnected engineered systems are primarily triggered by the failure of one or more components (subsystems). Three of the most safety-critical systems of the National Airspace System (NAS) are addressed in this case study – communications, navigation, and surveillance. The formulation of policies and procedures (driven by and represented in the "common/shared essential entities") guide the NAS, the ATCs, and the pilots, among myriad other "systems/parties." Therefore, complete compliance with them by all parties is essential. Indeed, ignoring the projected consequences resulting from the planned integration of C, N, and S systems can trigger failures that could lead to unacceptable and catastrophic consequences. It is imperative that knowledgeable engineers, operators, and other experts in the NAS identify all conceivable new sources of risk (due to the planned integration) that could cause the failure of the safety-critical NAS system. In this case study we address only a sample of shared/common states and other essential entities, each of which could trigger a catastrophic failure. The shared states and other essential entities provide an answer only to the first question in risk analysis, "what can go wrong," within any of the three CNS systems and with the entire CNS Complex SoS. We introduce two fault trees, which relate the I-I between communications and navigation systems. Figure 10.I.1 is an expanded fault tree with failure modes caused by data link failure in the data link fault tree. Figure 10.I.2, an expanded fault tree with failure modes caused by data link failure in the positioning system fault tree, also highlights the interdependencies between communications and navigation systems, albeit with a focus on the positioning systems. All fault trees represent major sources of risks resulting from the I-I among the three C, N, and S systems via the minimum cut sets. For a detailed discussion on fault trees with a large number of example problems, the reader is referred to the Fourth edition of the book *Risk Modeling, Management and Assessment* (Haimes, 2016a).

For example, in Figure 10.I.1 "on ground failure," one minimum cut set includes the following three systems: (i) radio receiver network failure, (ii) central computer failure, and (iii) data link service provider failure. Similarly, in Figure 10.I.2, the minimum cut set "accuracy degradation" includes the following five systems: (i) intentional signal blocking, (ii) receiver issues, (iii) lack of continuity, (iv) satellite issues, and (v) signal interference.

The four remaining fault trees (presented in Figures 10.I.3–10.I.6), built on the four tables, respectively, illustrate specific relationships of commonalities and I-I among the subsystems of the CNS Complex SoS. The two more comprehensive fault trees (Figures 10.I.1 and 10.I.2) will provide a more complete view of the encompassing complexity associated with the planned integration of the C, N, and S systems.

Exploiting our knowledge of the shared states and other essential entities among the C, N, and S systems is manifested in the development of the four tables and the construction of the above figures of fault trees. Recall that we structured fault trees to determine and track failure paths associated with shared states, decisions, decision makers, and resources among the integrated CNS Complex SoS. Indeed, fault-tree analysis enabled us to focus on the extreme consequences associated with the planned integration and on the emergence of new sources of risk of catastrophic failure. We began with the large number of subsystems within the C, N, and S systems, each with its own unique functionality and performance, the reliability of which was neither publically available nor required for this study. Then, we built on the most valuable attribute of fault-tree analysis, namely, identifying and focusing on the newly formed I-I among the three integrated CNS systems. We identify the

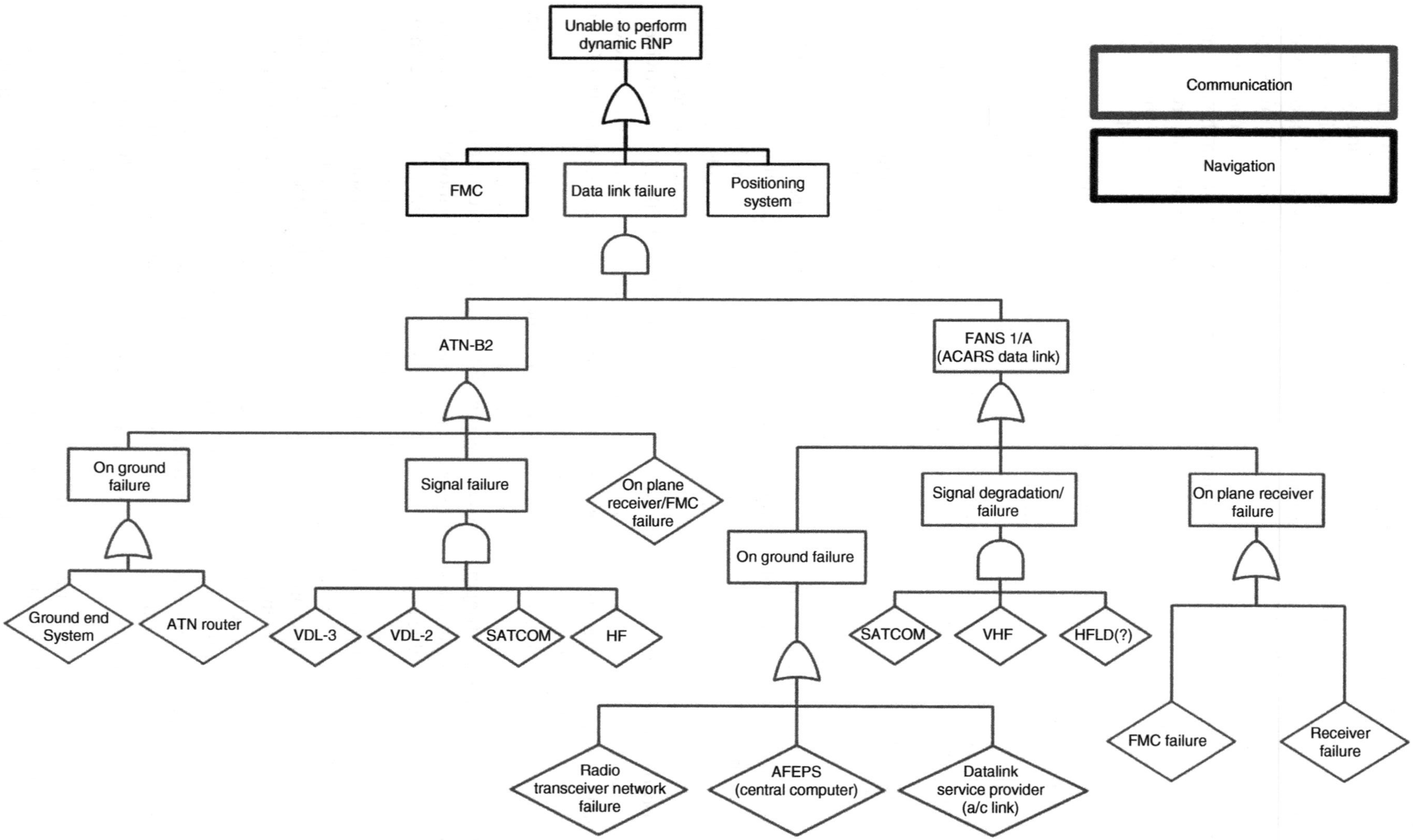

Figure 10.I.1 *Data link fault tree: expanded fault tree with failure modes caused by data link failure.*

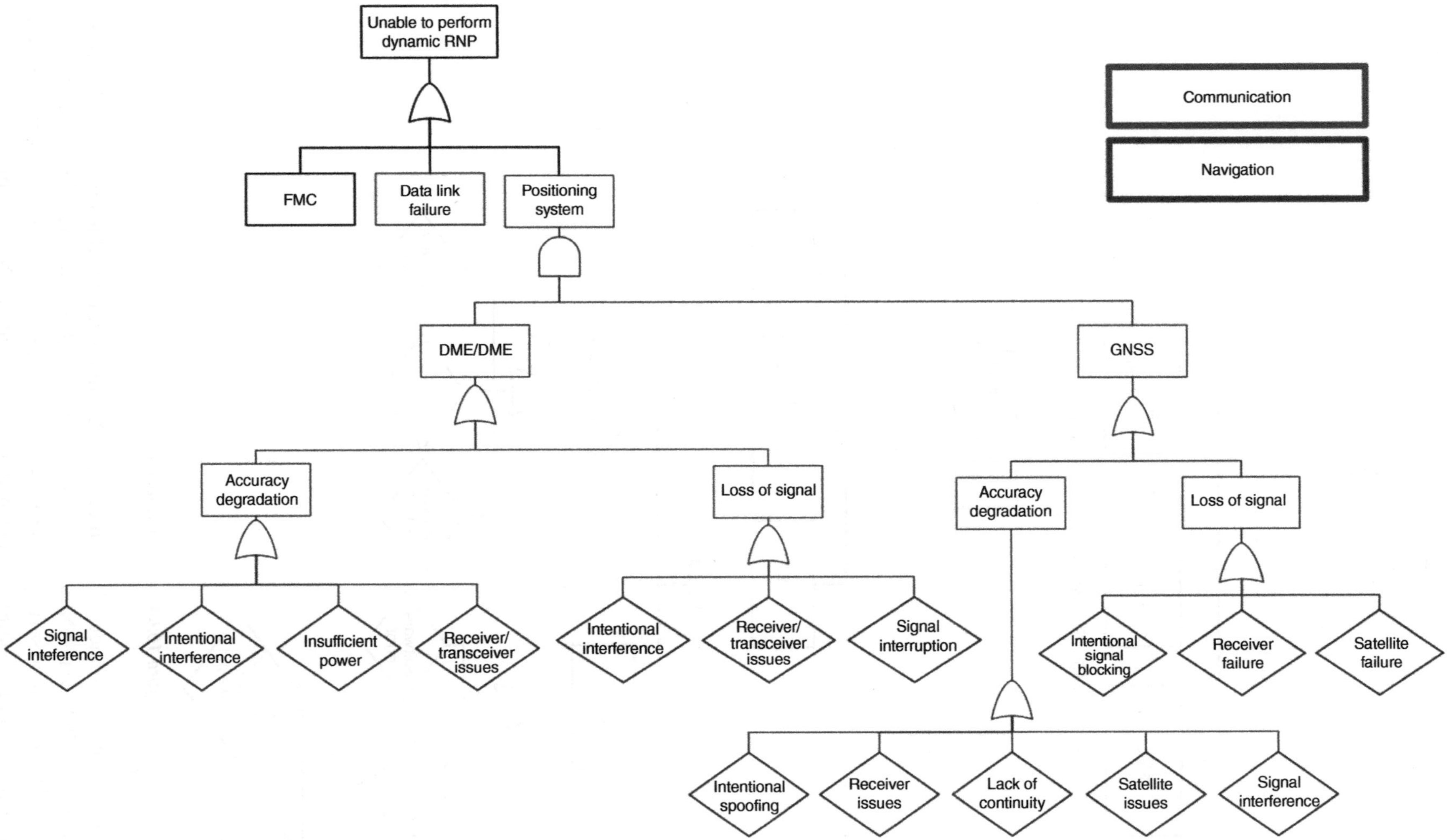

Figure 10.I.2 *Positioning system fault tree: expanded fault tree with failure modes caused by data link failure.*

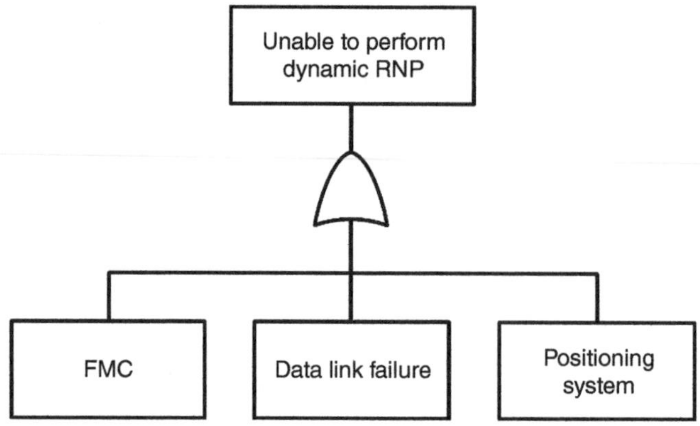

Figure 10.I.3 *Simple dynamic RNP fault tree.*

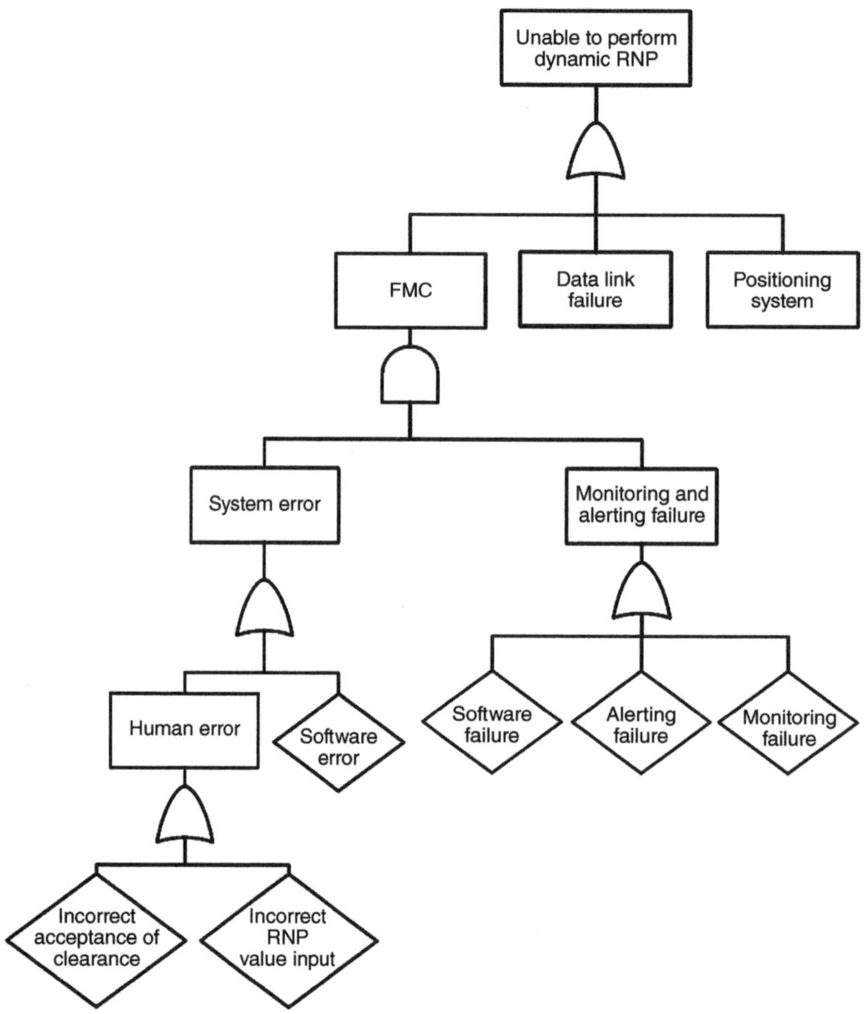

Figure 10.I.4 *The FMC fault tree.*

interdependencies and interconnections as connected in series (via OR Gates), or in parallel (via AND Gates). Clearly, newly connected CNS systems in series would introduce significant risks to the entire aviation enterprise. As noted earlier, this important discovery was enabled by the FAA domain experts' identification of safety-critical shared states and other essential entities within the integrated CNS system.

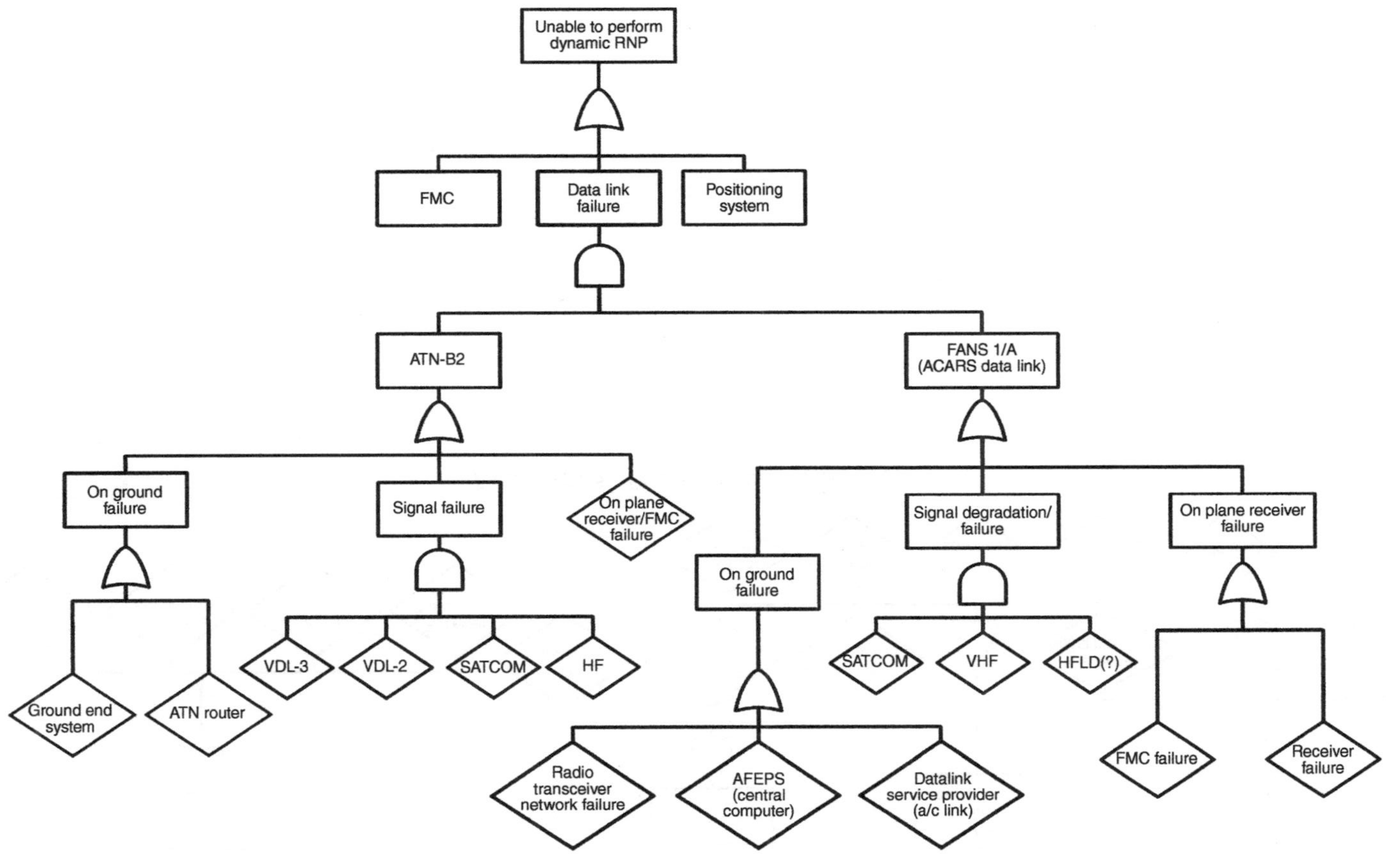

Figure 10.I.5 *Data link failure fault tree.*

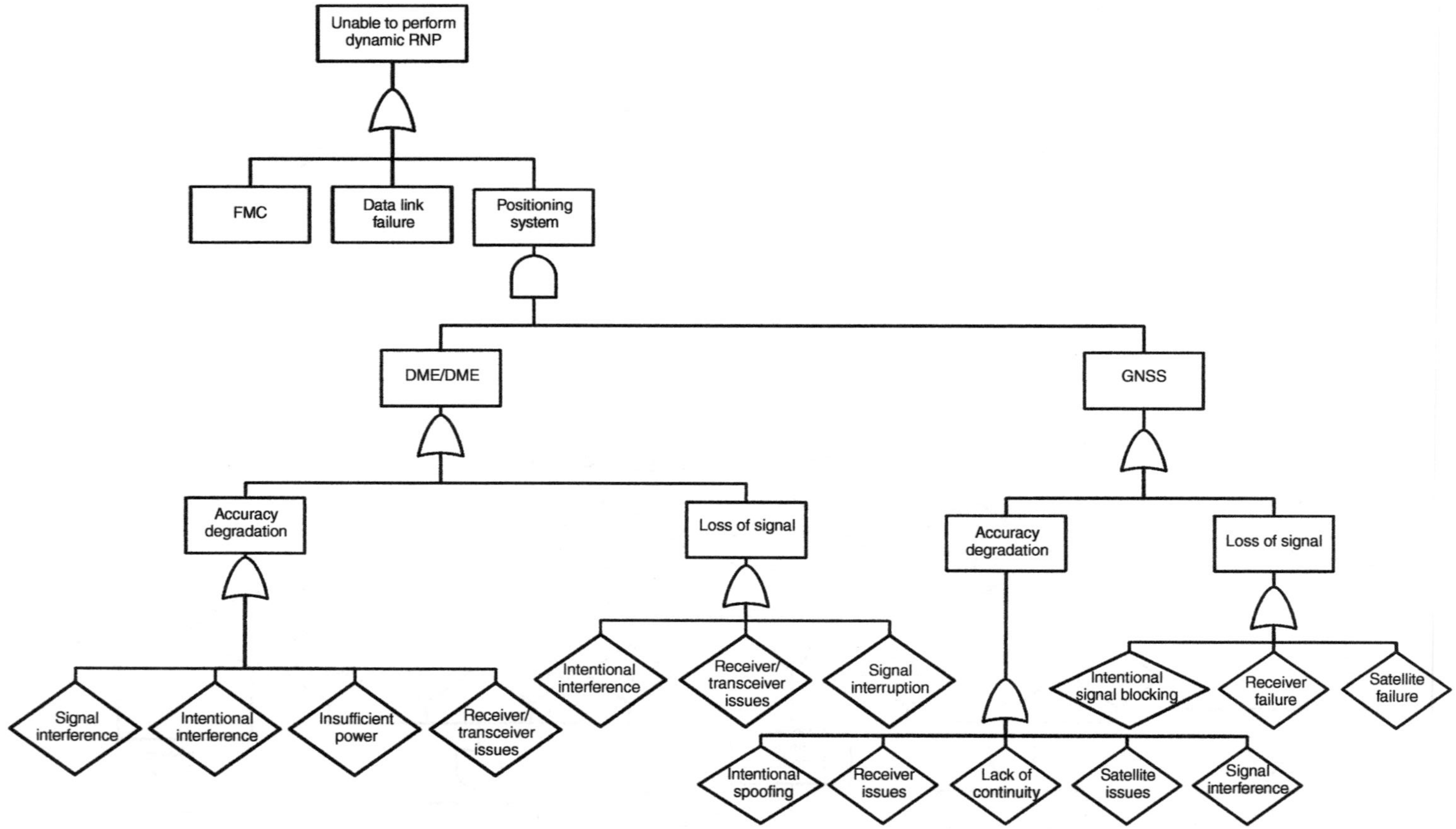

Figure 10.I.6 *Positioning system fault tree.*

Another important advantage resulting from translating shared states and other essential entities into fault trees is the identification of potential single-point failure caused by the newly formed I-I resulting from the integration of C, N, and S as a Complex SoS. Critical paths to failure were discovered using *minimal cut sets*. By definition, a minimal cut set is a combination of intersections of primary events in parallel sufficient for the top event to occur (if all parallel components fail). This combination is the "smallest" combination in that all the failures in the minimal cut set need to occur for the top event (system failure) to occur. If any one component in the parallel combination does not occur, then the top event will not occur (by this combination).

Each fault tree provides valuable insight into the interdependencies and interconnections that could lead to system failure. For example, the fault tree in Figure 10.I.1 is fully expanded in Figure 10.I.2; it shows that the three main systems – the flight management computer (FMC), the data link, and the positioning system – are connected in series, which is necessary for the operation of dynamic RNP, the FMC, data link, and the positioning system. The dynamic RNP capability enhances present and future trajectory-based operations by expanding RNP operations to live uplink of RNP values on a leg-by-leg basis, radius-to-fix legs, and fixed radius transitions. The key is the ability to adjust the route or RNP value in real time versus a fixed predetermined value and clearance. Each of the three systems – the FMC, the data link system, and the positioning system – when integrated into the CNS Complex SoS, contributes to the minimal cut set of the simple fault tree presented in Figure 10.I.3. To determine a minimal cut set for each system, we constructed individual fault trees: the FMC (Figure 10.I.4), data link (Figure 10.I.5), and positioning (Figure 10.I.6). This analysis provides a more comprehensive understanding of the structural perspective of the dynamic RNP system and, therefore, of the CNS Complex SoS. Furthermore, the fault tree for the FMC (Figure 10.I.4) reveals that two subsystems ("system error and monitoring" and "alerting failure"), which are vital to the health of the FMC (also known as the flight management system), are connected in parallel via AND Gates.

One of the main benefits of dynamic RNP is its critical function as a monitoring system to alert and inform the pilot of problems within the system and with the possible inability to maintain the required RNP value for a specific airspace. The multiple OR Gates represented in Figure 10.I.4 imply that the minimal cut sets are composed of many combinations of subevents. For example, one minimal cut set is "incorrect acceptance of clearance and software failure," and another would be "incorrect RNP value input and an alerting failure." The fault tree for the data link system (Figure 10.I.5) is similar to that of the FMC with two main data links available to perform dynamic RNP. The positioning system fault tree (Figure 10.I.6) is similar to the other two main subsystems for the use of the dynamic RNP capability, where the GPS global navigation satellite system is considered the main positioning system for dynamic RNP. The fault tree for the FMC (Figure 10.I.4) shows two important subsystems vital to the health of the FMC. As these events, "system error and monitoring" and "alerting failure," are connected with an AND Gate, it will take failures of both events to cause the top event to occur. The OR Gates represent both the "system error and monitoring" and "alerting failure" events and mean that the minimal cut sets are made up of many combinations of subevents underneath "system error and monitoring" and "alerting failure" events. Thus, the significance of the minimal cut set and the synergy between this structural perspective and the operational perspective is evident.

In sum, using fault-tree analysis to examine shared states and other essential entities enabled us to determine minimal cut sets, and highlight critical I-I within the CNS Complex SoS that would play a key role in the failure of safety-critical systems. Finally, fault-tree analysis and minimal cut sets advance our understanding of, and add insight into, the myriad sources of risk associated with interdependent and interconnected Complex SoS.

10.I.5 RISK MODELING OF THE CYBER–PHYSICAL CNS COMPLEX SYSTEM OF SYSTEMS

If the adage "To manage risk, one must measure it" constitutes the compass for risk assessment and management, then modeling constitutes the roadmap that guides the analyst throughout the entire journey of risk analysis. The process of risk assessment, management, and communication may be viewed through many lenses depending on one's perspectives, vision, and circumstances. Kaplan and Garrick (1981) argued that for each risk scenario in risk assessment, we must ask: "What can go wrong? What is the likelihood? What are the consequences?" In the risk management process, we ask: "What can be done and what options are available? What are the tradeoffs in terms of all relevant costs, benefits, and risks?" "What are the impacts of current decisions on future options?" "What is the time frame?" (Haimes, 1991). In this case study, not all questions in the risk assessment and management were relevant to the FAA. This position is reasonable because the focus was only on the first question, namely, "What can go wrong as a result from merging the three separate systems into a single CNS Complex SoS?" Thus, although all other questions are important, they were too early to be addressed, and these questions were premature and not an integral part of the study. However, if the FAA were interested in pursuing the risk modeling, assessment, management, and communications, then we should have modified the above set of questions from addressing a single system to interdependent and interconnected Complex SoS: In risk assessment, "What can go wrong for each and every C, N, and S system," and "What can go wrong for the entire CNS Complex SoS? The same adjustment should be applied to all other questions. For example, answers to the last question of risk management – "What are the tradeoffs in terms of all costs, and benefits?" – would require immense effort, given the multiple competing and conflicting objectives associated with the FAA's Next Generation, of which the integration of CNS is only one part.

10.I.6 SUMMARY AND CONCLUSIONS

We identified two innovations in this case study. First we modeled the I-I characterizing Complex SoS by building on shared/common states, decisions, decision makers, and resources, among other entities within and among the communications, navigation, and surveillance systems. Due to the resulting I-I within the planned integration of the three systems, the likelihood of catastrophic failures must be addressed and cannot be ignored. We subsequently identified and developed comprehensive tables of shared states and other essential entities within any two of, and among all three, members of the CNS Complex SoS. In this part we focused on (i) shared states (Table 10.I.1), (ii) shared decisions (Table 10.I.2), (iii) shared decision makers (Table 10.I.3), and (iv) shared resources (Table 10.I.4). We accomplished this critical/technical task with the support and contributions of FAA domain experts in the operations and interface among members of the CNS Complex SoS. The tables that summarize the I-I among the three (planned) integration of the C, N, and S systems provide the essential step in modeling and quantifying the I-I resulting from the (planned) integration of the three systems. The second innovation in the quantification of the I-I within and among the three C, N, and S systems, as an integrated CNS Complex SoS, is the reinvention of the use of fault trees, without the need for, nor reliance on, the reliability on the C, N, and S systems. The required reliability information for each subsystem in the fault tree is replaced with the shared states and other essential entities, representing the I-I that characterizes the CNS Complex SoS. More specifically, we populated the fault trees from the tables generated by FAA domain-expert engineers. We quantified the I-I by determining whether the systems are connected in (i) series (via OR Gates), or connected, or in (ii) parallel (via AND Gates). This essential designation of "AND" or "OR" Gates in the construction of fault trees is also guided by the expertise and experience of FAA engineers, technicians, and managers (as well as by MITRE's team). Recall that the failure of any subsystem connected in series would cause the failure of all other connected systems, whereas in systems connected in parallel, all subsystems must fail for the entire system to fail.

This fundamental quantitative modeling of the interdependencies and interconnections by the reinvention of fault trees enabled the translation of safety-critical I-I into AND Gates and OR Gates and prevent avoidable disastrous consequences that would result from risky integration of subsystems of the CNS systems. Recall that the genesis of shared states and other essential entities occurs through the conversion of the three semiautonomous systems into a single CNS Complex SoS. In the above analysis, we also discovered a synergy between the operational and structural perspectives. Clearly these two perspectives, while different, are closely linked as demonstrated through the fault-tree analysis.

On the one hand, shared states and other essential entities represent the risks through functional/operational and time frame considerations that result from decisions made within the CNS Complex SoS. On the other hand, fault-tree analysis represents the structural/cyber–physical dimensions of the I-I within and among the CNS Complex SoS components. Furthermore, we also explored the link between shared states and other essential entities and fault-tree analysis that were highlighted and analytically quantified the I-I within the CNS Complex SoS. This knowledge can be used to (i) identify, prevent if possible, and prepare for potentially adverse initiating events and their potential consequences and (ii) contribute to key risk assessment and management decisions in future operations of the C,N, and S systems as an integrated Complex SoS.

In sum, *knowing what not to do via modeling Complex SoS, especially for safety-critical Complex SoS, constitutes an invaluable contribution of systems engineering to public safety.*

PART II: QUANTITATIVE MODELING OF INTERDEPENDENT CYBER–PHYSICAL COMPLEX SYSTEMS OF SYSTEMS

OVERVIEW

This case study similarly presents a quantitative method for risk modeling of Global Positioning System (GPS) Complex SoS, guided by the work performed for the US government by the MITRE Corporation (Bogdanor, 2014; Bogdanor and Haimes, 2018). The modeling methodology is also grounded on state-space theory, risk analysis, and analytical systems modeling. We document adverse consequences resulting from spoofing the GPS and consequential risks to the national electricity and communications systems. This case study reinforces the methodological contributions made to the FAA project. One is the innovative modeling of the three cyber–physical systems accomplished by building on state-space theory to model and quantify the I-I manifested by shared/common states and other essential entities within and among the interdependent and interconnected GPS, Electricity, and Communications Complex SoS. A sample of essential entities includes shared/common decisions, decision makers, stakeholders, resources, organizational behavior and norms, policies and procedures, management, and culture, among others. The second innovation is reinventing the use of fault-tree analysis with neither the need for nor a reliance on the reliability data for each subsystem of the fault tree. And, similar to the FAA case study, the normally required reliability data is replaced by the shared states and other essential entities characterizing the Complex SOS. We use expert evidence from MITRE engineers and technicians to populate multiple scenarios with low, medium, and high probabilities for the event-tree analyses and develop a simulation model with event trees to gain more insight and to further improve our understanding of the interdependencies among GPS, the electrical, and the communications sectors. Similar to the FAA case study, the main contributions of this case study are (i) building on and extending the modeling of interdependent and interconnected Complex SoS by identifying shared/common states and other essential entities (ii) reinventing the use of fault trees without the need for the reliability data of any one of the three cyber–physical systems; and (iii) quantifying the I-I within and among the GPS, Electricity, and Communications Complex SoS.

10.II.1 INTRODUCTION

The US Department of Homeland Security (DHS) commissioned a research project with the MITRE Corporation, in collaboration with the University of Virginia, to study projected dire impacts on the electrical grid and on the communications sector should the GPS be spoofed, adversely affected, or impaired. The adage in risk modeling, assessment, management, and communication is to build appropriate models capable of helping risk analysts and stakeholders answer the following fundamental questions: (i) What can go wrong? (ii) What is the likelihood? (iii) What are the consequences? (iv) What is the time frame? (Kaplan and Garrick, 1981; Haimes 1991, 2016a; Kaplan et al. 2001). Risk management was premature; thus, it was beyond the scope of this research. The scope of the study was further narrowed by addressing only two of the above risk-based questions, namely, (i) What can go wrong? and (ii) What are the consequences? This decision was based on the fact that there was limited technical information publically available on these three safety-critical interdependent and interconnected cyber–physical infrastructure systems.

Electricity is generated and can be transferred in two forms: (i) direct current and (ii) alternating current. The interdependence and interconnections between the electricity and communication sectors are due to the fact that to operate and transfer information, the communications sector requires energy from the electrical sector. Similarly, the electrical sector requires the availability of communications media to operate and to effectively and efficiently transfer electricity from point A to point B using timely information and knowledge-based decisions. The transfer of electricity via alternating current is safer and more cost effective than with direct current. Furthermore, the dependency of the efficient and effective generation of electrical power and its transfer to long distances requires inputs from the GPS.

To minimize energy losses, electric power generated in local power generation stations is transferred to long distances through the electrical grid with *very high voltage*. Three-phase electric power is a common method of generating and transmitting alternating current electric power. One major role of the GPS system is to provide, via pulse measurement units, accurate information on the *timing and phase of the alternating current*. The critical dependency of the electricity sector on accurate pulse measurement units, which are based on GPS timing and location, constitutes the essence of the interdependencies between electricity and GPS – thus the interdependencies and interconnections among the three cyber–physical systems. For completeness, we introduce another enabling system – supervisory control and data acquisition (SCADA) system – that enables the joint operations of the three electricity, communications, and GPS systems. Note, for example, that GPS signal availability affects the synchrophasor of the electricity system and the voltage phasor of the SCADA system and so forth.

The above technical information provides the basis for the interdependencies and interconnections that characterize the electricity, communications, and GPS systems as Complex SoS. For clarity and to avoid overcomplexity in this part, we consider the SCADA system as a supporting system.

10.II.2 THEORETICAL AND METHODOLOGICAL CONTRIBUTIONS

The introduction presented the physical and technical relationships among the three cyber–physical systems. Indeed, to be able to model any system, we must understand it from as many perspectives and viewpoints as possible. To *model* the existing relationships within and among the GPS, Electricity, and Communications systems, we rely on state-space theory and methodology, which provides the foundation of systems and control engineering, among other fields. As noted earlier, analytical modeling of simple systems, or of Complex SoS, must build on state-space theory and methodology that guided this research project. Recall that all outputs from a system are functions of the states of that system and the latter are functions of all decisions and are exogenous to all other inputs to the system. This notion is also of particular significance and central to modeling Complex SoS. Chen (2012) offers the following succinct definition of *state variable*: "The state $x(t_o)$ of a system at time t_o is the information at time t_o that, together with the input $u(t)$, for $t \geq t_o$, determines uniquely the output $y(t)$ for all $t \geq t_o$." The states of a system, commonly a multidimensional vector, characterize the system as a whole and play a major role in estimating its future behavior for any given input. Thus, the behavior of the states of the system, as a function of time, enables modelers to determine, under certain conditions, its future behavior for any given *input* or *initiating event*. All systems are characterized at any moment by their respective state variables and the conditions thereof, and these conditions are subject to continuous variations and fluctuations. Similarly, the state of health of an individual is multifaceted and includes blood composition, blood pressure, and pulse rate, among myriad others, and the relationships among the states of all human biological systems must be considered. It is important to learn from the evolution of current thinking about interdependent and interconnected Complex SoS and seek new approaches for their modeling and management. We embrace the fact that complexities cannot, by their essence and definition, be compounded, packaged, understood, or modeled using a single modeling schema. Most importantly, by seeking to understand the relationships that are central to the complexity of SoS, which challenge their modeling and their understanding, we will ultimately manage them in better ways than we have done in the past.

As noted earlier, complexity, by definition is multifaceted, cannot be understood from any single perspective, nor be represented within any single model. Similar to the previous FAA case study, our approach, described in this case study, to modeling Complex SoS, builds on *shared/common states and other essential entities* within and among the three systems. (Recall that the term shared "essential entities" connotes shared/common decisions, decision makers, stakeholders, resources, organizational behavior and norms, policies and procedures, management, and culture, among others.) The time frame has always been recognized as a major driver of what we term complexity. This is due to the fact that all systems continue to evolve, emerge, and thus change, while our ability to keep pace with these changes via our modeling tools continues to lag behind, and our inability to model the dynamic changes that characterize Complex SoS remains the Achilles' heel that characterizes and drives our understanding and our modeling and managing interdependent and interconnected Complex SoS. We must acknowledge the importance of the time frame in modeling; however, the added complexity of the evolving states of the three systems was beyond the scope of this case study. This modeling process, which builds on shared/common states and other essential entities within and among the three systems, enables us to quantify the relationships and to uncover the resulting sources of risk to the Complex SoS. Table 10.II.1 and Figures 10.II.1–10.II.5, presented in subsequent sections, provide a sample of shared states and other essential entities that characterize the interdependencies and interconnections among the GPS, Electrical, and Communications Complex SoS and highlight the multiple paths to failure that result from the relationships that characterize the systems' complexity.

In sum, this introduction explores and recaps the shared states and other essential entities among the three systems, which enabled us to quantify the sources of risk resulting from the interdependencies

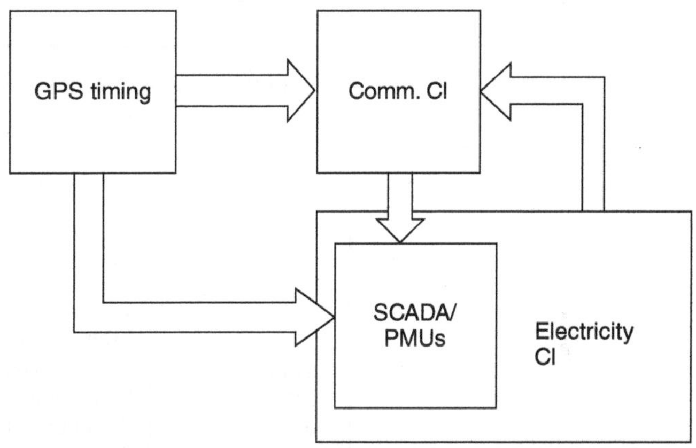

Figure 10.II.1 *Interdependencies and interconnectedness among GPS timing, electricity, communications, and SCADA/PMUs.*

TABLE 10.II.1 Compilation of Shared States

GPS timing	Communications	Electricity	SCADA/PMU
1. Signal accuracy	1. *Data transmission capacity*	1. **Current synchrophasor**	1. **Current phasor**
2. Signal availability	2. *Comm. availability*	2. **Voltage synchrophasor**	2. **Voltage phasor**
3. Signal integrity	3. *Comm. integrity*	3. Load on transmission lines	3. *State of comm.*
	4. GPS timing		PMU
	5. STATE OF POWER SOURCE		1. **Current synchrophasor**
			2. **Voltage synchrophasor**
			3. *State of comm.*
			4. GPS timing

Four letter fonts – (i) regular, (ii) bold faced, (iii) italics, and (iv) upper case – are used to signify shared/common states among the four systems.

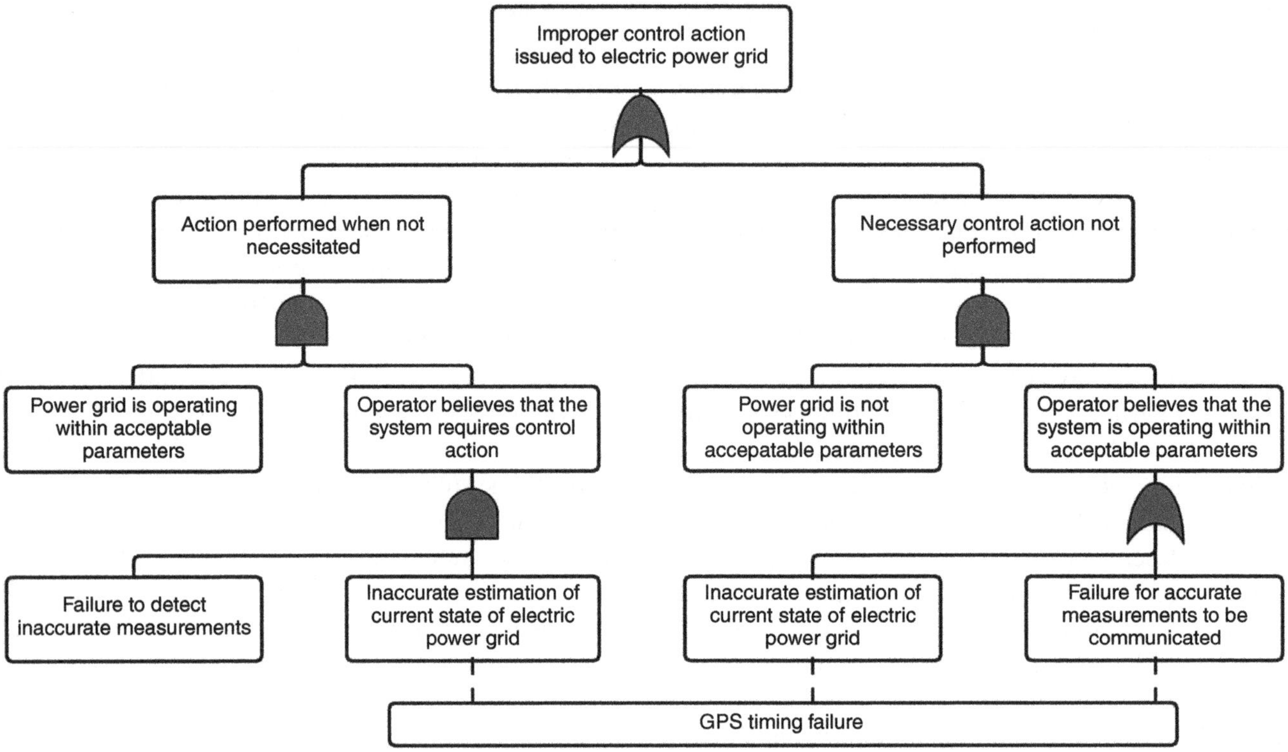

Figure 10.II.2 *Fault tree of smart electrical power grid.*

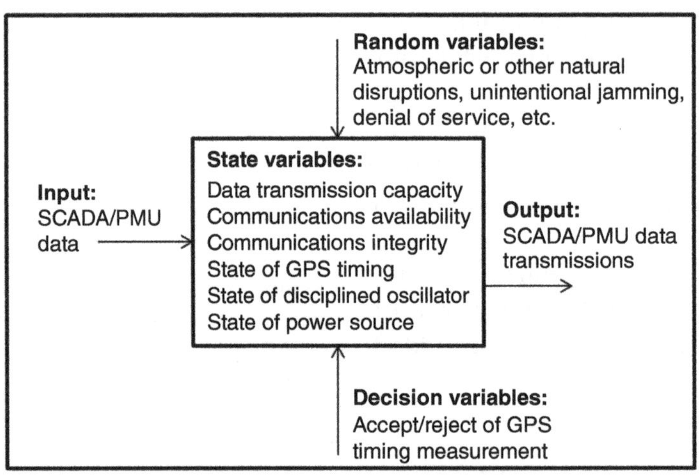

Figure 10.II.3 *State variable model for GPS timing-dependent communications CI used by the electrical sector.*

and interconnections among the GPS, Electrical, and Communications Complex SoS. Thus, once we have identified a representative number of shared states and other essential entities within and among the three cyber–physical systems, we proceed to the next step. In this second step we quantify the interdependencies and interconnections within and among the Complex SoS, and by reinventing the selective use of fault trees, we are able to identify and characterize the relationships among the three systems in terms of "OR" Gates (systems connected in series) and "AND" Gates (systems connected in parallel). Modeling the I-I of the GPS, Electrical, and Communications Complex SoS via shared states and other essential entities and exploiting and translating this knowledge into fault trees allow us to

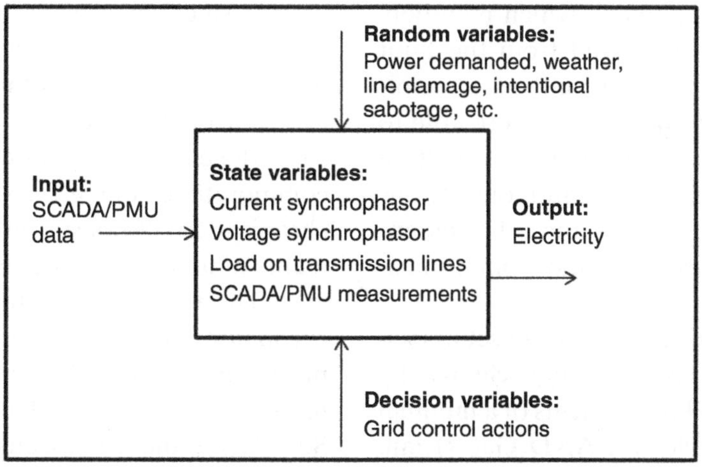

Figure 10.II.4 *State variable model for the electrical sector.*

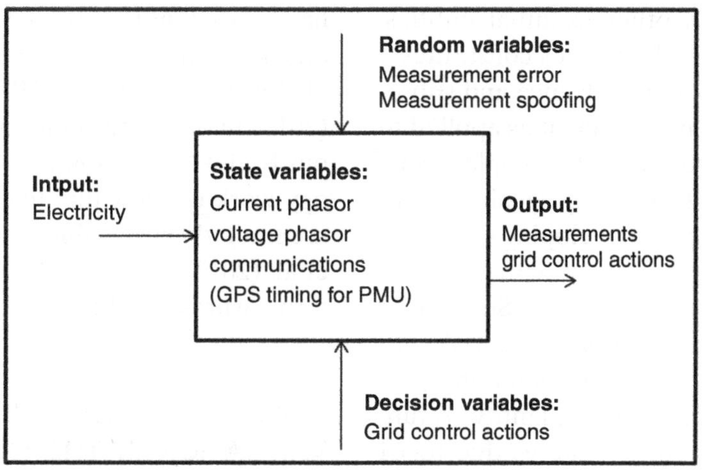

Figure 10.II.5 *State variable model for SCADA/PMU systems.*

(i) translate and model the I-I among the three systems in terms of the shared states and other essential entities; (ii) quantify the risk of failure of the GPS, Electrical, and Communications Complex SoS in terms of "OR" Gates and "AND" Gates, and ultimately the generation of "minimal cut sets"; and (iii) discover and thus recommend actions that should not be taken, given the new knowledge gained via the modeling process.

10.II.3 REINVENTING THE USE OF FAULT-TREE ANALYSIS

This case study, similar to the previous one, reinvents the traditional use of fault trees, one that neither needs nor relies on the reliabilities of any of the three GPS, Electrical, and Communications Complex SoS. We consider fault trees as a "technology" with various useable parts appropriate for different purposes. The reliability of each system and subsystem of the fault tree, which are the main drivers of the fault-tree analysis, is neither of concern to our modeling efforts nor of interest in this DHS-commissioned project. Here we (i) explore the specific connections among all parts of the GPS, Electrical, and Communications Complex SoS and (ii) use fault trees to transfer/translate the common/shared states and other essential entities into a depiction of systems being connected in series (OR Gates) and in parallel (AND Gates). Subsequently, we discover which of these connections are on the "critical path," namely, which members of the minimal cut sets are derived from the fault tree. We made use of the "minimal cut set technology" in an innovative way. Namely, we were able to quantify prospective dire consequences resulting from the interdependencies and interconnections that exist among the shared states and other essential entities of the GPS, Electrical, and Communications Complex SoS.

The process of calculating a minimal cut set associated with the three systems is commonly achieved by using Boolean algebra. If all components of a minimal cut set (connected in parallel via AND Gates) fail, then there would be a complete failure of the entire GPS, Electrical, and Communications Complex SoS modeled by that fault tree. In this sense, when modeled via shared states and other essential entities, the reinvention of the use of fault trees constitutes a significant quantitative modeling advance and subsequent improvement in risk management as applied to interdependent and interconnected Complex SoS. In this part we are not avoiding dealing with "complexity," but rather we are developing a theoretically-based and tested methodology with which to model and quantify the risks to Complex SoS due to their I-I using shared states and other essential entities. Furthermore, for "planning-for-operation" purposes, the ability to model and quantify future projected, unintended, disastrous failures due to I-I constitutes another valuable contribution to the quantitative modeling of Complex SoS.

10.II.4 WHAT HAVE WE LEARNED FROM THE LITERATURE ON GPS

Several modeling philosophies and methods have been developed over the last five decades to address the modeling of Complex SoS and to offer various modeling schema. With respect to modeling a cyber–physical Complex SoS such as GPS, Electrical, and Communications Complex SoS, the subject of our case study, Ross et al. (2014) address the interdependencies between and the integration of social and physical perspectives and decision makers and the resulting positive synergistic effects. Smart grid initiatives from the Department of Energy (2013) call for continuous real-time monitoring of the electric power grid. The North American SynchroPhasor Initiative (NASPI) is proposing to use a nationwide system of phasor measurement units to attain wide-area measurement and control. A phasor measurement unit (PMU) measures the electrical waves on an electrical grid using a common time source for synchronization. Time synchronization allows synchronized real-time measurements of multiple remote measurement points on the grid. The resulting measurement is known as a synchrophasor. It is important to note that appropriate measurement of PMUs provided by GPS is essential for proper operation of the national electrical grid SoS. While there is extensive literature praising the benefits of implementing PMUs in the electric power grid (Zhong et al., 2005; Carta et al., 2009; Das et al., 2012), more recent work from Shepard et al. (2012) has shown that PMUs are vulnerable to GPS spoofing attacks. Humphreys et al. (2009) addressed the simplicity of executing a GPS spoofing attack. Wesson (2013) has applied similar methodologies from previous Radio Navigation Laboratory of United Technology work to show that telecommunications networks are similarly vulnerable to GPS spoofing attacks. Jiang et al. (2013) demonstrated the need to develop countermeasures to GPS spoofing threats.

10.II.5 COMPUTATIONAL RESULTS: VALIDATION OF THE THEORETICAL AND METHODOLOGICAL CONTRIBUTIONS

In this case study and similar to the previous one, we model the I-I that exist among the three cyber–physical infrastructures – GPS, Electrical, and Communications – as Complex SoS through their shared/common states and other *essential entities* within and among their subsystems. The GPS system shares states, e.g. PMUs, with electricity, which implies that *a lack of accuracy in the PMU (provided by GPS) would jeopardize the viability of the electrical grid.* This relatively recent dependency between GPS and electricity emerged as a cost reduction effort in electricity transmission, among other factors. *It is with this in mind that the interdependencies among the GPS and PMU timing, the electricity sector, the communications sector, and the SCADA system as SoS are subsequently explored and modeled via multiple fault trees.*

In our modeling efforts we have been able to *quantify the sources of risk to the electricity sector (and subsequently to the communication sector), if the GPS system were spoofed or otherwise disabled, thereby affecting GPS timing and the electrical grid. Furthermore, because the* communications sector cannot operate without electricity, *a compromised GPS would also impact the communications sector.* Table 10.II.1 presents a sample of the interdependencies among the communications and electricity systems and SCADA based on GPS timing. Note, for example, that GPS signal availability affects the synchrophasor of the electricity system and the voltage phasor of the SCADA system. Four different fonts in Table 10.II.1 present a sample of shared states and other essential entities among the GPS timing, communications, electricity, and SCADA PMU systems: (i) regular, (ii) **bold faced**, (iii) *italics*, and (iv) UPPER CASE. (For simplicity, we included the PMU with the SCADA system.)

Building on both the major shared states and other essential entities that have been identified, and on the interrelationships among subsystems of the GPS, Electricity, and Communications Complex SoS, enables us to construct, in subsequent sections, fault trees for analyzing the dependency of the electricity sector on GPS timing accuracy. The fault-tree analysis provides us with a quantitative evaluation of the specific configurations among the subsystems and the consequences resulting from shared states and other essential entities among the three systems (excluding SCADA). *Most importantly, recall that our reinvention of fault-tree analysis performed here is used not to develop the specific interdependencies and interconnections among the GPS timing, electricity, communications, and SCADA/PMU. Rather, the analysis is performed solely to identify and model those specific I-I relationships in terms of the AND Gates and OR Gates characterizing the GPS, Electricity, and Communications Complex SoS.*

It is through these interdependencies that threats impacting the GPS timing system (e.g. via spoofing satellites) can propagate into the communications sector. Similarly, consider the interdependencies between the communications and electricity sectors, a sample of which is presented in Figure 10.II.1. As the extended smart grid implementation expands, it is expected that interdependencies between the communications and electricity sectors will increase, with electric utilities relying less upon private communications infrastructures, and more on communications carriers to provide their data networks.

Presently, PMUs are used for monitoring the state of the electrical grid. To highlight the risk of failure within the electricity sector following a failure with GPS timing, Figure 10.II.1 depicts the interdependencies and interconnections among GPS timing, the electrical sector, the communications sector, and the SCADA/PMU systems – all of which constitute an emergent Complex SoS. In its essence, Figure 10.II.1 epitomizes the complexities that characterize the GPS, Electrical, and Communications Complex SoS manifested through their interdependencies and interconnections. Also note that in Figure 10.II.1, GPS timing has a one-way relationship with communications as well as with PMUs. Within SCADA/PMUs, only PMUs are directly dependent on GPS timing for synchronization. Because this case study deals only with those networks that provide support to SCADA/PMU systems, the communications sector is shown connected to the electricity sector as an input to SCADA/PMUs, which are modeled as a subsystem of the electricity sector. Clearly, additional interdependencies within the GPS, Electrical, and Communications Complex SoS under consideration exist, but they are beyond the scope of this case study.

SCADA devices measure and control the state of the electric power grid and serve as a subsystem of the electricity sector. Many types of communications support data transmission for SCADA systems, such as privately owned microwave transmission networks and fiber-optic networks. Those networks that are owned and operated by communications carriers also rely upon GPS timing for synchronization. Because of their shared states and other essential entities with communications carriers, these networks carry risks of power loss and blackouts affecting the electricity sector. In particular, PMUs share similar relationships with both the electricity and communications sectors. Presently, PMUs are used for monitoring the state of the electric grid. In the future, smart grid initiatives plan to utilize PMUs to perform real-time autonomous/semiautomated control.

Recall that fault-tree analysis is used to demonstrate its efficacious contributions to the risk modeling and analysis of the *GPS, Electrical, and*

Communications Complex SoS through graphical and analytical representation. Figure 10.II.2 depicts the fault tree of a smart electric power grid and represents a future risk scenario of smart grid PMU autonomous control. In this fault tree, the structure of the cyber–physical Complex SoS is comprised of GPS timing, the electricity sector, the communications sector, and the SCADA/PMU systems. Total failure is precipitated by some improper control action being issued to the electric power grid. This can occur either when an action is performed that was not necessary or when a necessary control action was not performed. When an unnecessary control action is performed, it occurs because the grid is operating within acceptable parameters, yet the controlling system believes the grid requires a control to be implemented. This occurs when there is an inaccurate assessment of the current state of the electric power grid and there is a simultaneous failure to detect the inaccuracy. When considering the risk that GPS timing attacks pose to the electrical sector, it is important to realize that this inaccurate assessment can stem from GPS timing failures that introduce error to measurements taken from PMUs.

Consider the following realistic scenario: When the power grid is operating in a range that requires a control to be implemented, but the control system believes that the grid is operating in an acceptable range, the necessary control action is not performed. This occurs either when there is an error in the measurements, which could be the result of GPS timing failure, or the state of the grid is properly measured yet the communication of this information fails to be received by the control system. Once again, a GPS timing failure may be to blame. By tracing up the tree in Figure 10.II.2, we note a clear path to electricity sector failure initiated by failure in GPS timing systems. Signal availability impacts the GPS disciplined oscillator (GPSDO) system; when GPS timing is not available, the oscillator enters holdover mode.

The states depicted in Figure 10.II.3 are impacted by random occurrences both natural and intentional. Atmospheric or other natural disruptions related to weather can impact the success of SCADA/PMU data transmissions, as can unintentional jamming that disables GPS updates within the GPSDO subsystems. Figure 10.II.4 presents the state variables for the electricity sector, and Figure 10.II.5 presents the state variables for the SCADA/PMU systems. Figure 10.II.5 presents the final state variable model of SCADA/PMU systems.

The states of signal accuracy, signal availability, and signal integrity within GPS timing are shared with the communications sector as well as with SCADA/PMU systems (see Table 10.II.1). Data transmission capacity, communications availability, and communications integrity represent shared state variables between the communications sector and SCADA/PMU systems. Current and voltage phasors are shared between the electrical sector and SCADA systems, while their synchrophasors are shared states between the electrical sector and PMUs. Lastly, load on power lines is a shared state variable between the electrical sector and communications sector.

Major natural and malevolent sources of risk to the SoS under consideration impact signal accuracy, availability, and integrity. The risks include a signal that differs from the GPS signal sent, spoofing attacks, solar or other space or weather events, or atmospheric anomalies. Note that the distinction between accuracy and integrity, the lack of signal accuracy, and a lack of signal integrity are not necessarily correlated. Brief random events may impact genuine signals originating from GPS satellites. While the signal received may not be accurate, it is still a genuine signal and can be trusted, and most likely will be correct. Additionally, an accurate signal does not necessarily equate to a signal with high integrity. This distinction is important because sophisticated spoofing attacks may convince receivers to bypass checks for integrity when the accuracy of the incoming signal appears to be high. Bypassing a check for integrity can be a major source of risk to the SoS. Furthermore, these attacks gradually decrease the accuracy of the signal over time with increased system tolerance for the error until a critical point is reached.

Data transmission capacity is a major concern moving forward with smart grid initiatives. It is essential for communications networks to be able to keep up with the increasing amount of data transmitted by PMUs as they significantly exceed that of current SCADA systems. If portions of a network become unavailable due to timing discrepancies initiated by a GPS timing attack, this change in the state of data transmission capacity could negatively impact the data output from these communication networks. If measurement data is delayed arriving at control centers, improper control actions might be implemented within the grid. If time-sensitive control data is delayed in arriving at the proper systems, impacts such as blackouts might occur more rapidly. *The state of communications availability has two distinct states similar to the availability of GPS signals: available or not available.* The availability state of communications is the likelihood that, at any given time, the communications network is available. This is also the most common case with GPS signals. Communications integrity is a measure of the trustworthiness of communications at a given point in time.

Building on the above, we represent the I-I within and among the multiple systems of the cyber–physical Complex SoS through fault-tree analysis. As noted in the previous case study, fault trees are an essential process by which to represent the inherent complexity of Complex SoS through their interconnections (i) in parallel through AND Gates, (ii) in series through OR Gates, or (iii) by a combination thereof.

10.II.6 CONCLUSIONS

We modeled the complex interdependencies and interconnections of critical infrastructure sectors of the economy as Complex SoS. This challenge has been on the national agendas of many countries, including the United States. In this case study, we quantified the dependence of the communications and electrical sectors on GPS timing and the potential impact of the US smart grid initiative. We also represented the inherent I-I among GPS, the electrical, and communications sectors (coordinated through SCADA systems) through their shared states and other essential entities. We have demonstrated our risk modeling and analysis by (i) building on shared states among the subsystems, (ii) using fault trees coupled with event-tree analysis, and (iii) guided by systems-based risk analysis, thus highlighting the impact of the dependencies within the cyber–physical Complex SoS. The fault-tree analysis enabled us to demonstrate the I-I among the subsystems through AND Gates (subsystems connected in parallel) and OR Gates (subsystems connected in series). We demonstrated that the electricity subsector becomes dependent upon GPS timing through the proposed use of PMUs for determining control actions on the grid. We also noted the existing dependence on GPS timing through the communications infrastructure, which in turn is used by the electrical subsector for the transmission of PMU and SCADA data.

In this part, we have (i) demonstrated the nature of the I-I among the GPS, Electrical, and Communications systems through our modeling and analyses that builds on shared states and other essential entities among the subsystems and (ii) reinvented the use of fault trees without the need for, nor a reliance on, the reliabilities of any of the three cyber–physical systems. Furthermore, this case study was guided by systems-based risk analysis, which highlights the consequential impacts of the I-I within the three cyber–physical Complex SoS. Fault-tree analysis enabled us to demonstrate the interdependencies and interconnections among the subsystems through *AND* Gates and *OR* Gates. We demonstrated that the electricity subsector becomes dependent on GPS timing through the proposed use of PMUs for determining control actions on the electrical grid and noted the already existing dependence of electricity on GPS timing through the communications infrastructure, which in turn is used by the electricity subsector for the transmission of PMU and SCADA data.

To study and model the risk of cyber attack to the proper functionality of GPS, on which both the electrical and communications sectors are dependent, we built on shared states and other essential entities among these three interdependent SoS. The time frame has always been recognized as a major driver of complexity. Given that spoofed cyber attacks on the GPS system *is not unlikely*, these shared states necessarily increase the risk to the electrical and communications systems should the GPS system get attacked. We translate the dependencies of shared states and other essential entities among the three systems by reinventing the use of fault trees to model these interdependencies as OR Gates (connected in series; namely, if one member of the OR Gate fails, then all other members of that Gate will also fail), or as AND Gates (connected in parallel; namely, where all members of the AND Gate must fail for that Gate to fail). To circumvent the fact that *no database* about the vulnerabilities of the three cyber–physical systems to any specific threat is publically available and yet be able to gain more insight into the risks to these critical SoS, we experimented with event trees. Here we used expert evidence from individuals at

MITRE to populate multiple scenarios with low, medium, and high probabilities for the event-tree analyses. We developed a simulation model with event trees to gain more insight and to further improve our understanding of the interdependencies among GPS and the electrical and communications sectors.

PART III: REGIONAL INFRASTRUCTURES AS COMPLEX SYSTEMS OF SYSTEMS: A SHARED STATE MODEL FOR REGIONAL RESILIENCE

OVERVIEW

This part of Chapter 10 introduces an alternative methodology to model, manage, and prioritize safety-critical infrastructure systems as Complex SoS for resilience planning and operation (Haberlin and Haimes, 2018). The genesis of the complexity addressed in this case study – similar to the two previous case studies, stems from the I-I among the subsystems and systems. Our modeling methodology traces all critical I-I in terms of shared/common decisions, decision makers, stakeholders, resources, organizational behavior and norms, policies and procedures, management, and culture, among others. This elaborate and tedious process provides modelers, planners, and managers with an invaluable mechanism for enhanced modeling and ultimate management of Complex SoS. Moreover, present modeling and risk management for resilience of interdependent and interconnected physical and cyber–physical infrastructure SoS often rely on the same risk functions, models, and algorithms used for single systems. Furthermore, to understand, model, and manage Complex SoS, it is fundamental to study, focus on, and account for the mutual influence resulting from two or more subsystems that share states and other essential entities. Recall that the term shared "essential entities" connotes shared/common decisions, decision makers, stakeholders, resources, organizational behavior and norms, policies and procedures, management, and culture, among others. Limiting our modeling efforts to only those prospective faults/risks initiated at one subsystem that can propagate to other subsystems of the Complex SoS (due to shared states and other essential entities) captures the most significant risks to the entire Complex SoS. To further prioritize the above sources of risk as in the two previous case studies, we have again reinvented the use of fault trees, without the common need for or knowledge of the associated reliabilities. We "quantify" the I-I among the subsystems, which constitute a significant source of risk, by modeling them as connected in series (via OR Gates), or connected, or in parallel (via AND Gates). Recall that the failure of a subsystem connected in series would cause the failure of all other connected systems, whereas in systems connected in parallel, all subsystems must fail for the entire system to fail.

10.III.1 INTRODUCTION: REGIONAL AND COMMUNITY RESILIENCE

Consider the centrality of an effective and uninterrupted water supply to a regional community, which is dependent on electricity and communications, the subject of this part. Now consider the following daily critical needs of water resources by a community, where most, if not all, of which are provided delivered with SCADA Complex SoS: (i) drinking water; (ii) water for the generation of electricity required for cooling towers; (iii) water for firefighting; (iv) pumping water from groundwater, especially from deep aquifers; (v) industrial water use; and (vi) agricultural water use, among many others. None of the above sample of safety-critical needs and functions can be performed timely and effectively without electricity, nor without communications. The reason is simple: (i) Today's communications and electricity sectors cannot perform their intended functions without each other, (ii) SCADA systems cannot operate without both electricity and communications, and (iii) water cannot be delivered without electricity, communications, and SCADA. In sum, there can be no effective and safe resilience to the three interdependent and interconnected electricity, communications, and water Complex SoS without identifying and managing the inherent I-I among the three above systems.

The methodological framework introduced in this part and deployed in the case study highlights the inescapable interdependent and interconnected reality within and among the three – water, electricity, and communications (along with SCADA) – which constitute Complex SoS. Thus, by methodically and technically tracing the above critical interdependencies and modeling them via our reinvented fault trees, we succeed in discovering those I-I among them that are on the critical path to failure.

10.III.2 INNOVATIVE USE OF FAULT-TREE ANALYSIS

The extensive reliance on fault trees and their reinvention in this chapter call for the following succinct description of the fault-tree model (U.S. Nuclear Regulatory Commission, 1981)[1]:

> A fault tree analysis can be simply described as an analytical technique, whereby an undesired state of the system is specified (usually a state that is critical from a safety standpoint), and the system is then analyzed in the context of its environment and operation to find all credible ways in which the undesired event can occur. The fault tree itself is a graphic model of the various parallel and sequential combinations of faults that will result in the occurrence of the predefined undesired event. The faults can be events that are associated with component hardware failures, human errors, or any other pertinent events, which can lead to the undesired event. A fault tree thus depicts the logical interrelationships of basic events that lead to the undesired event – which is the top event of the fault tree. It is important to understand that a fault tree is not a model of all possible system failures or all possible causes for system failure. A fault tree is tailored to its top event, which corresponds to some particular system failure mode, and the fault tree thus includes only those faults that contribute to this top event. Moreover, these faults are not exhaustive – they cover only the most credible faults as assessed by the analyst. (pp. IV–1)

Fault trees have been deployed in almost all safety-critical systems from nuclear reactors to space shuttles and to numerous other Complex SoS. In all of these applications, the reliability of each system and the probability of its failure within the Complex SoS were central to the contribution of fault-tree analyses. The novelty of relying on fault trees in our study stems from the ability to translate the interdependencies and interconnections inherent in the shared states and other essential entities (that characterize Complex SoS) into AND Gates (systems connected in parallel) and into OR Gates (systems connected in series). This is done without the need of knowing the reliability of each system or of the electricity, communications, and water Complex SoS. More specifically, we were able to build on the experience of domain-knowledge experts and use other relevant technical information on the I-I among the three systems. We integrated the above into quantitative metrics via AND Gates and OR Gates, repeating the process for multiple permutations of shared states and other essential entities among the integrated electricity/communications/water Complex SoS.

A community or a region is characterized as interdependent and interconnected Complex SoS. Consider the following definition of a region: *A region is a geographic area with social and economic ties, and interdependent infrastructure, that require coordination to build quality of life. As such, a region can be a town, county, city, metropolitan statistical area (MSA), planning district, state, or even a multistate area.* For this part, the above definition is adopted for a region and characterizes it as interdependent and interconnected cyber–physical Complex SoS. We interpret and define I-I among the systems that constitute Complex SoS as shared states and other essential entities among them. Recall that the term shared "essential entities" connotes shared/common decisions, decision makers, stakeholders, resources, organizational behavior and norms, policies and procedures, management, and culture, among others. A sample of subsystems that characterize a regional community includes multiple infrastructures, decision makers and stakeholders, and multiple organizational entities that span local, state, and federal agencies, along with private, non-for-profit public organizations, and so on.

In risk assessment for single systems we ask: (i) What can go wrong? (ii) What is the likelihood?

[1] For each case study in this chapter to be self-sufficient, we opted to keep the introduction to fault trees at the expense of repetition.

(iii) What are the consequences? (Kaplan and Garrick, 1981); here we add, (iv) What is the time frame? Answering each question is much more intricate; they must be addressed for each system of the Complex SoS in terms of their shared states and other essential entities. In risk management for single systems, we ask the following questions – (i) What can be done and what options are available? (ii) What are the tradeoffs in terms of associated costs, benefits, and risks? (iii) What are the impacts of current decisions on future options? (Haimes, 1991, 2016a, 2018). Similarly, each of the above three questions in risk management must also be addressed more comprehensively: Risk analysts, decision makers, and stakeholders must realize that any risk management action tailored specifically to any affected system would likely impact directly or indirectly other interdependent and interconnected systems within the Complex SoS. Thus, due to the I-I within the Complex SoS, such risk management can be successful and effective only to the extent that the broader consequences resulting from acting on one system can be well understood and projected on all other interdependent and interconnected systems within the Complex SoS.

To understand and then model Complex SoS, it is fundamental to focus on and account for the dynamic changes resulting from two or more subsystems that share states and other essential entities, keeping in mind that all states are functions of time and all inputs and decisions. In addressing the risk to and the resilience of infrastructures, modelers and decision makers should be cognizant of and responsive to the *Evolving Base* that affect most if not all Complex SoS (introduced in Chapter 9). All systems, including and especially Complex SoS, are continuously affected by shifting rules and realities highlighted in the following dynamically *Evolving Base* of moving (i) goals and objectives; (ii) stakeholders, decision makers, and interest groups; (iii) organizational, political, and budgetary considerations; (iv) reorganization and reallocation of key personnel; and (v) requirements, specifications, and delivery dates (Haimes, 2012b, 2016a, 2018). Although all infrastructure subsystems are likely to be affected by the *Evolving Base*, each one may respond differently and consequently the Complex SoS as a whole. The *Evolving Base* recognizes a shift away from an environment in which "event-based models developed to explain physical phenomena (which they do well) are inadequate to explain accidents involving organizational and social factors and human decisions and software design errors in highly adaptive, tightly-coupled, interactively complex socio-technical systems" (Leveson, 2012).

This part introduces a systemic methodology to prioritize critical infrastructure components for resilience planning. Specifically, enhancing the resilience of critical infrastructures that support regional communities in the face of major natural or malevolent disturbances necessarily implies accounting for and understanding the I-I within and among three critical infrastructures that provide life support to the community – water, electricity, and communication – supported by SCADA systems and identifying critical shared states and other essential entities among the three cyber infrastructures (see a sample in Figures 10.III.1–10.III.3). As a control system architecture that uses computers and networked data communications for supervisory management of infrastructure control systems, SCADA is logically a source of interconnectivity between regional infrastructure systems and must be considered in any analysis of shared states.

10.III.3 SHARED STATES AND ESSENTIAL ENTITIES

One novelty in modeling the complexity of SoS in this part stems from deciphering the genesis of the I-I characterizing them. The fundamental fact is that if the interdependencies and interconnections between systems A and B are connected in series and systems A is subjected to failures due to an initiating event, then system B will also be affected by the same initiating event and be subject to failure as well. Thus to model this I-I characterizing Complex SoS, we (i) explore the I-I among the systems of Complex SoS via the help and advice of technical experts, who possess technical, managerial, and other expertise in the systems that constitute the Complex SoS under consideration. (ii) We identify shared/common decisions, decision makers, stakeholders, resources, organizational behavior and norms, policies and procedures, management, and culture between and among the systems that make up Complex SoS. For brevity, the term "shared states and other

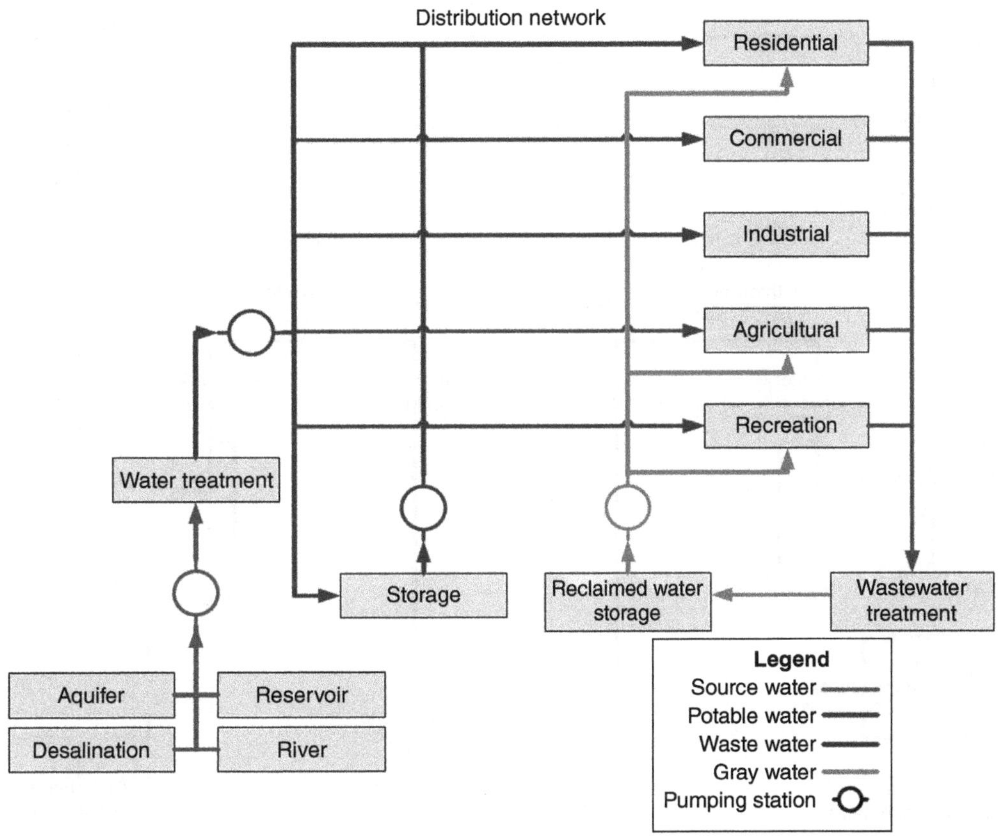

Figure 10.III.1 *Representative regional water distribution.*

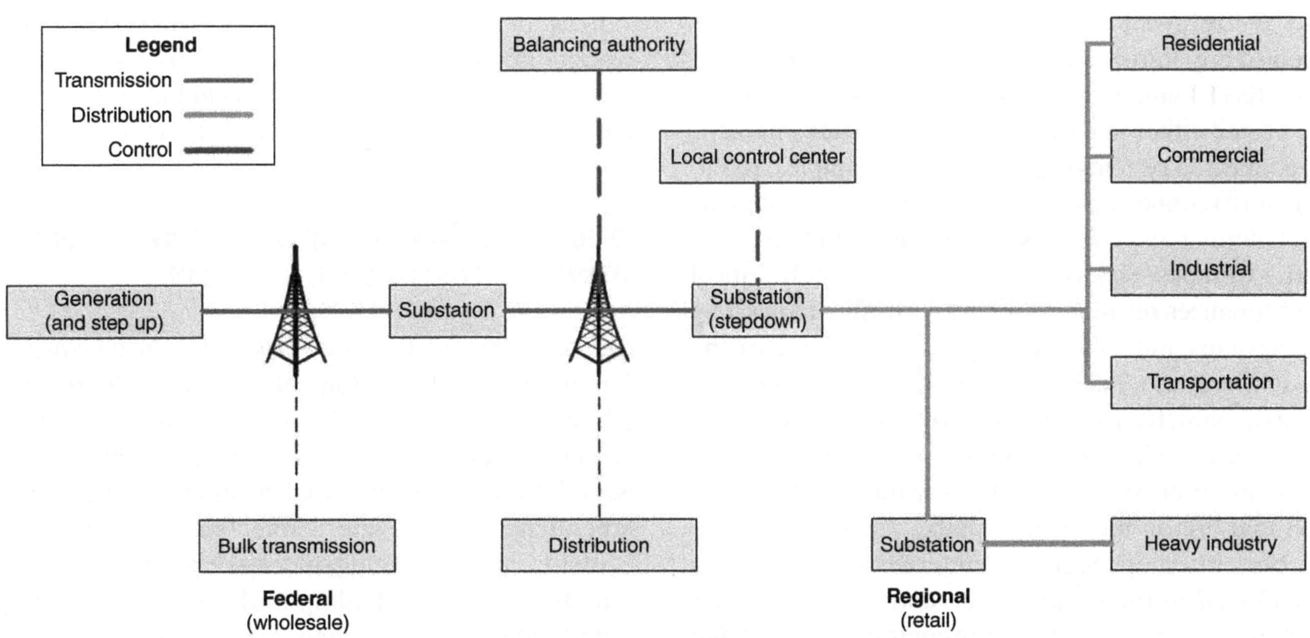

Figure 10.III.2 *Representative regional electric power distribution.*

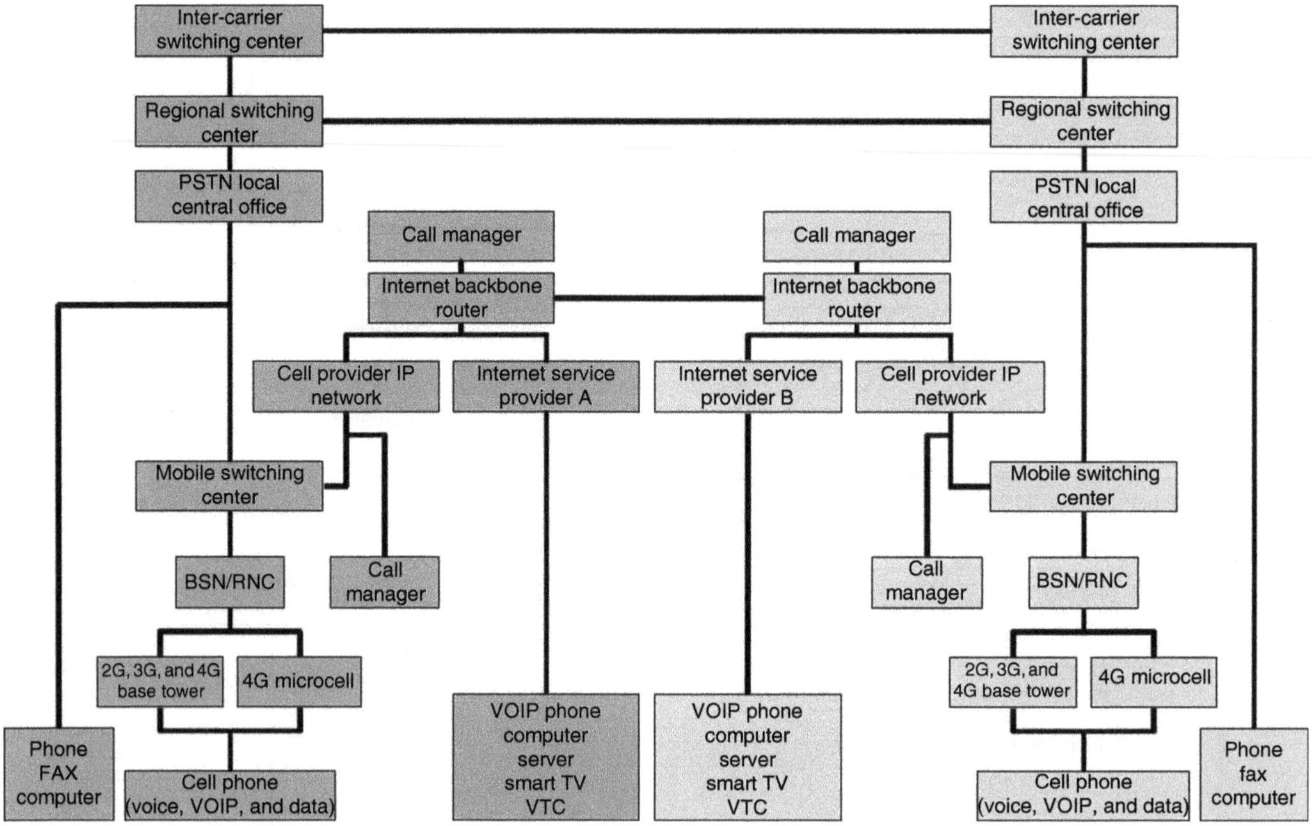

Figure 10.III.3 *Representative regional communications (Adapted from MITRE (2015)).*

essential entities" will be used. We further arrange the shared states and other essential entities (within two or more systems) into tables, leading to quantifying the I-I characterizing the Complex SoS. (iii) We reinvent the use of fault trees (without the need of any information on their reliability) by classifying the I-I among the systems and subsystems as (a) connected either in parallel (via AND Gates), namely, all connected systems must fail for the Complex SoS to fail, or (b) connected in Series (via OR Gates), namely, the failure of any system would cause the failure of the entire Complex SoS. (iv) We analyze the likely critical consequences resulting from the I-I within and among the systems and subsystems and discover one of the most important benefits from the above modeling efforts, namely, *what not to do*. More specifically, we discover via our systemic and methodological modeling process select systems within the Complex SoS that are on the "critical path," where the failure of one interdependent or interconnected system would lead to the failure of the entire Complex SoS and thus to catastrophic consequences. Quantifying via fault trees the critical shared states and other essential entities, whether the critical interdependent or interconnected systems and subsystems are connected in parallel or in series, will be introduced in subsequent sections. Future work will incorporate sources of behavioral risk to the subsystems and expand upon the initial model to integrate faults shared with other subsystems, including transportation, shelter, food, sewage, and health care.

10.III.4 CENTRALITY OF STATES IN MODELING COMPLEX SYSTEMS OF SYSTEMS

The centrality of the states of a system constitutes the fundamental building blocks of modeling and risk analysis in general and of Complex SoS in particular. Focusing on shared states and other essential entities is essential to modeling the interdependencies and interconnections and thus to assessing and managing the associated risks stemming from their I-I. Failing to do so would severely curtail the risk modeling and analysis process. Recall Chen's definition of the states of the systems

(Chen, 2012): "The state $x(t_o)$ of a system at time t_o is the information at time t_o that, together with the input $u(t)$, for $t \geq t_o$, determines uniquely the output $y(t)$ for all $t \geq t_o$." Thus, in the context of the above definition, the states of a system play an important, significant, and critical role in systems modeling and especially in modeling and understanding and managing Complex SoS. More specifically, no model can ever predict the future behavior of a dynamic system without building on the states of that system.

From the perspective of state-space control theory, the states of a system constitute the most important building blocks of mathematical models. This is because the states of a system are functions of all decisions, inputs, random variables, exogenous variables, and time. Invoking Chen in the same context, the vulnerability and resilience of a single system are both manifestations of the states of the system and may be represented by a state transition function that captures their physical, technical, organizational, and cultural attributes – all as functions of the states of the system, the initiating events, and the time frame. Note that the vulnerability of any subsystem of Complex SoS to a specific initiating event is likely to yield a multidimensional vector of consequences, each with its specific associated probability of occurrence and consequences. Resilience also represents the ability of a system to withstand a major disruption within acceptable degradation parameters and to recover within an acceptable composite cost and time (Haimes, 2016a). This necessitates a shift in the state of the system from one of productive operation to intolerable degradation caused by a precipitating event. Shared states of systems are points of commonality in these building blocks between two or more systems. The shared state model of critical infrastructure subsystems promotes understanding of this Complex SoS by focusing attention on the most critical sources of technical risk without the distraction of the myriad potential faults that affect only a single subsystem. Faults of a single system are prioritized lower than those affecting multiple systems when it comes to applying limited resources to resilience planning. Risk and vulnerability are assessed against a specific threat and its timing to the SoS or subsystems. The "behavior" of systems can be understood and modeled via the states of the system as functions of the input/action/forces placed on them, namely, (i) all inputs to the system, e.g. decisions and inputs from other systems or from outside the system, and (ii) all random and other exogenous variables. This is due to the fact that the states of the system are functions of above (i) and (ii). Furthermore, all outputs and objectives of all models are functions of all states (thus, we have function of functions). In other words, for Complex SoS, all outputs (including objective functions) are functions of all states and all other essential entities.

The reliability of Complex SoS performing functions beyond those of its individual component functions is strongly dependent on the specific configuration of the complete network control structure. Both immediate and latent degradations to functionality are exacerbated through the I-I of its subsystems. Thus as noted earlier, this fundamental fact could be viewed in the parlance of fault-tree analysis (NRC, 1981) as converting subsystems from being "marginally unconnected," or in "parallel," to becoming "heavily connected," namely, in "series." This implies that an initiating event that affects one or more states and other essential entities of the electricity sector would necessarily affect other interdependent and interconnected infrastructure subsystems, e.g. water and communications. Thus, from that perspective, subsystems that share states and other essential entities may experience cascading effects across multiple systems when exposed to adverse initiating events, which is a central characteristic to Complex SoS. However, different initiating events may not affect shared states and other essential entities of different subsystems in the same way. For example, when each subsystem has different decision makers, then decisions made to control subsystem A may affect positively or negatively subsystem B that shares A's states and other essential entities (e.g. the decision to incorporate electrical power redundancy positively affects the water and communications subsystems). On the other hand, coordinated decisions and collaboration among these decision makers could positively affect the Complex SoS and improve its overall resilience and management (e.g. memoranda of understanding for mutual disaster support). Positive impacts can be attributed to collaborative management by the principal stakeholders of the infrastructure subsystems that make up the Complex SoS. Thus, risk analysts should recognize and exploit these attributes of shared states and other essential entities among subsystems. This process should ultimately account for most of the resulting critical risks to each subsystem and to the Complex SoS as a whole.

10.III.5 REGIONAL INFRASTRUCTURE SUBSYSTEMS

Three critical infrastructure subsystems are assessed in this part: water, power, and communications. Focusing on these systems enables the important process of identifying the complexity of shared states and other essential entities and the prioritization of mitigation policies and risk management measures.

An introductory description of subsystem components is provided to enable the understanding of the critical features common to (shared by) a region's subsystems.

Water

- The mission of the water subsystem is to provide safe, potable, and industrial water to public and commercial users. In fact, this can be summarized in the Water Sector Specific Plan (EPA, 2010) as "A secure and resilient drinking water and wastewater infrastructure that provides clean and safe water [....] as an integral part of daily life, ensuring the economic vitality of and public confidence in the Nation's drinking water and wastewater services through a layered defense of effective preparedness and security practices in the sector." Implied tasks for the water subsystem include: (i) accessing water resources, (ii) performing water treatment, (iii) performing desalination, (iv) providing drinking water distribution, (v) providing industrial water distribution, and (vi) performing water quality monitoring/measuring and performing water rationing.

Figure 10.III.1 provides a conceptual representation of a regional water resources, distribution, and treatment system. Water is drawn from a variety of resources, including underground aquifers, rivers, surface reservoirs, and oceans through desalination. The water is treated to produce potable water before storage or distribution to customers, such as residential, commercial, industrial, agricultural, and recreational entities. Waste from these customers is collected and treated before being stored and distributed for reuse.

Throughout the distribution and collection network, electrically driven pumps transfer water among components at the required level, coordinated by SCADA systems.

Power

- The mission of the electric power subsystem is to provide clean, stable power to the public, business, and government entities. Implied tasks include (i) providing power generation, (ii) providing power transmission and distribution, (iii) keeping power safe and reliable, and (iv) maintaining competitive prices.

Figure 10.III.2 is a general representation of the major components of a regional electric power cyber-infrastructure subsystem. Electric generation is provided via many possible sources including nuclear, hydroelectric, solar, coal, gas, and wind technologies. This electric power is stepped up to allow bulk transmission across long distances. A balancing authority coordinates the wholesale production and transmission of electric power across multiple regions. Within a region, power is stepped down for retail delivery to many different types of customers, including residential, commercial, industrial, and transportation. Coordination is performed at the local control center of the regional cyber-infrastructure Complex SoS.

Generation and bulk transmission of wholesale electric power is coordinated at the federal level through the balancing authorities. Distribution of retail electric power is coordinated at the regional level by the local control center of a provider organization.

Communications

- The mission of the communications system is to enable information sharing among federal, state, regional, and local public and private entities. Implied tasks (i) providing communications via voice and data, (ii) providing data exchange, (iii) ensuring necessary bandwidth to support users, (iv) supporting emergency services,

(v) providing emergency notification to the public, (vi) supporting intraregion commerce, (vii) supporting interregion commerce, and (viii) supporting communications needs of public.

Figure 10.III.3 provides an illustration of representative communications systems for ground-based communications among multiple public switched telephone networks (PSTN) and Internet service providers. Landline routing of communications occurs among carriers at the intercarrier switching, regional switching, and Internet backbone router level. Mobile switching centers provide commonality between traditional PSTN and mobile networks. Traditional communications media (phone, fax, and computer) are routed through the PSTN local central office. Mobile communications are routed through base towers and microcells to a mobile switching center, which routes the signal through traditional PSTN or through an Internet backbone router. Similarly, communications such as VoIP phone, computers, servers, smart television, and video teleconferencing (VTC) are routed onto the Internet backbone though the Internet service provider.

Communications networks are a complex mix of switches, traditional copper cable, Ethernet copper cable, and fiber-optic cable. Providers are also working to improve performance by shortening the distance of copper runs and using fiber cable at the local level.

10.III.6 METHODOLOGICAL APPROACH

The current regional infrastructure Complex SoS model incorporates three subsystems –electric power, water, and communications (including PSTN) – and SCADA system; the latter commonly provides the medium through which all subsystems communicate. These subsystems are subsequently studied and modeled by building on their shared states and other essential entities. In the future, this modeling process will be further extended and applied to all other subsystems (e.g. transportation, food, health care, etc.).

A major subsystem is selected for modeling and evaluation, followed by others. Electric power is selected because it is commonly assumed to be critical to many other subsystems. We identify states and other essential entities that represent the interdependencies and interconnectedness among the subsystems that constitute the regional cyber–physical Complex SoS. For planning-for-operation purposes, this task identifies failures that would cause the power subsystem to not perform its intended functions of providing electricity to other subsystems, namely, failure of primary power, failure of secondary power, or failure of cyber security.

The interdependent and interconnected subsystems are then subjected to fault-tree analysis to identify critical subsystems connected in series, where the failures of a subset therefrom would cause a degradation of performance for the entire Complex SoS. We populate the fault trees with the help of domain experts: First, all major shared states and other essential entities are tabulated, and subsequently are transferred into the fault trees by translating, as appropriate, connections that are made in series (into OR Gates) and connections that are made in parallel (into AND Gates). A completed fault tree (see Figure 10.III.4) illustrates the critical path to failure for the regional cyber–physical Complex SoS. The fault-tree analysis reaches its conclusion with the calculation of its minimal cut set. A minimal cut set is defined as the smallest combination of component failures, which, if they all occur, will cause the top event to occur (NRC, 1981; Haimes, 2016b). The process of calculating a minimal cut set associated with the Complex SoS is commonly achieved by using Boolean algebra. In the context of this case study, a minimal cut set is calculated from the fault tree to identify the smallest combination of component failures of the three infrastructures – water, electric power, and communication – which, if they occur, will result in a degradation of SoS performance to meet its regional objectives. Note that the shared states and other essential entities among the three subsystems (and the SCADA system introduced above) constitute the building blocks of the generated fault tree illustrated in Figure 10.III.4.

10.III.7 SHARED STATE MODEL

Modeling the I-I characterizing Complex SoS is captured in a fault tree (Figure 10.III.4) which highlights

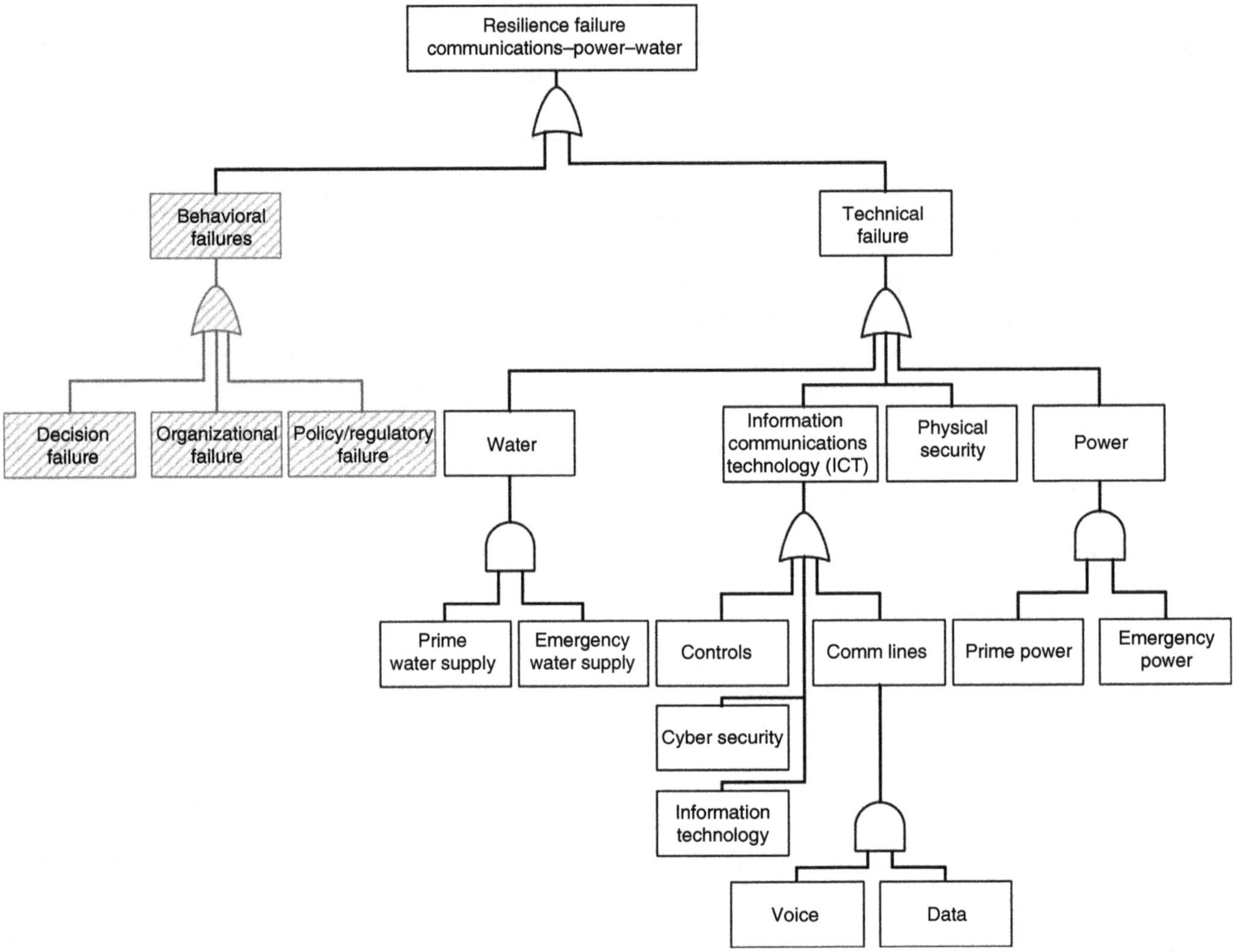

Figure 10.III.4 *Power–water–communications fault tree.*

the shared states and other essential entities among the subsystems– a fact that can cause cascading failures and degrade the entire Complex SoS. The fault tree is interpreted as a failure of a "child node" (or combinations of child nodes) that results in a failure of its "parent." For example, failure of both prime power and emergency power result in an overall failure of power to the Complex SoS.

As previously introduced and expounded upon below, the failures were categorized into those resulting either from behavioral issues, or from technical assets. The components of the minimal cut set should be prioritized for the purpose of improving the resilience of the Complex SoS for specific critical initiating events developed through the scenario structuring. The large number of shared states and other essential entities generated through the risk scenario structuring process enabled by the HHM is further subjected to the Risk Filtering, Ranking and Management (RFRM) process (Haimes, 2016a). In this way, the fault tree is subjected only to a set of initiating events, whose occurrence could ultimately lead to a degradation in performance of the Complex SoS.

10.III.8 A SAMPLE OF BEHAVIORAL FAILURES

A brief introduction of each failure category is provided followed by further detail on technical failures.

Behavioral failures (shaded region of Figure 10.III.4) are those caused by the actions of individuals (or groups of individuals) due to their choices of actions, or failure to act, in a way that would minimize the effect of a hazard to the subsystem. The behavioral failures were loosely organized into three broad categories introduced below; detail of their contents and resulting fault trees are left for future work. Note that the criticality of the time frame associated with actions or inaction is often as critical as the decisions themselves.

Decision Failures. Decision failures define conscious choices made by decision makers, or decision-making bodies, that negatively affect the subsystem. These may include decisions regarding operations, coordination, planning, resourcing, and maintenance.

Organizational Failures. Organizational failures are caused by dysfunction within and among subsystems' leadership and staff that lead to inter-subsystem failures. These may result from such attributes as command climate, tribal culture, command structure, and poor leadership.

Policy/Regulatory Failures. Policy and regulatory failures are the result of misinformed or missing policy that results in the failure of a subsystem. Examples of policy/regulatory failures include rationing priorities, quality monitoring, reserve margin, and peak demand calculation.

Understanding the effect of behavioral failures on subsystems is an area of research that crosses multiple disciplines and includes many levels of organizational stakeholders. Future work will further decompose these failures to better understand their interrelationships and mitigation strategies.

Technical Failures

Technical failures (nonshaded region of Figure 10.III.4) in the fault tree describe resources and hardware components that cause multiple subsystem failures and ultimately a failure of the system of systems.

Water Failure. A failure of the water subsystem limits access to treated water and elimination of wastewater by consumers with varying requirements on quality and quantity. Regions also have internal water stockpiles that will absorb supply limitations for a short period of time.

Figure 10.III.5 illustrates a further decomposition of the prime water supply into failure of its major components based on the conceptual representation in Figure 10.III.1.

Figure 10.III.5 illustrates the three primary components of the prime water supply: source, treatment, and the pumping/distribution network. A failure of any one of those components results in the Complex SoS shifting to a backup source of water as shown in Figure 10.III.4. Further, for the source component of the prime water supply system to fail, all possible water sources must fail, as indicated by the presence of the *AND* Gate. A fully realized, region-specific prime

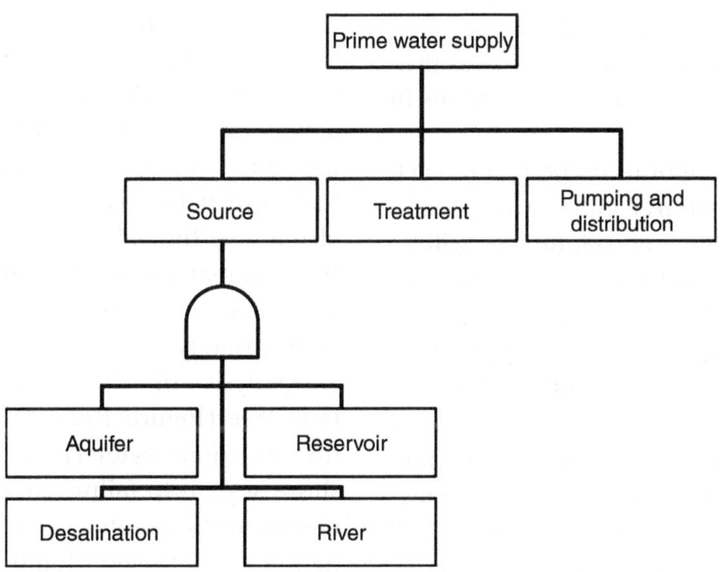

Figure 10.III.5 *Prime water supply fault tree.*

water supply model would continue decomposition of each of these nodes to a level of granularity suitable for assessment of the entire system.

Information and Communications Technology (ICT) failure. A failure in information and communications technology (ICT) prevents the flow of communications controls and supporting data throughout the Complex SoS, thus preventing efficient coordination and flow of resources. There are four major components within the ICT: controls, cyber security, information technology, and communications lines. A loss of any one of these components fails the system and degrades the Complex SoS.

Physical Security Failure. Physical security protects people, data, equipment, systems, and facilities. It includes such resources as site design and layout, environmental components, emergency response, training, access control, intrusion detection, power, and fire protection (NRC, 1975, 1981). A failure of physical security exposes subsystem components to environmental and human threats that could result in failure of the subsystem. For example, failure of security at a water treatment plant could result in contamination of potable water resources within a local area.

Power Failure. Failure of the power subsystems affects multiple interconnected subsystems and would have severe impact on the Complex SoS. For this reason, subsystems maintain their own backup power resources to continue electrical service for a limited duration until prime power is restored. Memoranda of agreement may also exist that allow electrical providers to utilize power generated by organizations in adjacent regions.

There is necessarily a temporal component to each of these failures, and the fault tree must be assessed in the context of time. Therefore, each fault tree will be relevant to a specific point in time. That is, what are the shared states and other essential entities that introduce risk to the Complex SoS at the 12 h, 24 h, 72 h, etc. point? The fault tree may be different as time passes. For example, if a hazard results in a 12 h loss in potable water production, are there sufficient resources in the emergency water supply to meet demand? Will the same answer be true after 72 h? Similarly, from a planning context, the question is: How many hours of demand for potable water can be met by the existing emergency water supply? Each of these scenarios must be assessed when planning resilience measures for critical components.

Shared States Identification

Identification of subsystem shared states and other essential entities is commonly a collaborative effort among domain experts in which participants explore potential failures within each of the categories that could lead to an individual subsystem failure, categorized as behavioral or technical as introduced above. These are then compared across subsystems to determine common failures that would lead to a failure of multiple subsystems and eventually degradation to the entire Complex SoS. To maintain focus on the highest priority of resources affecting multiple subsystems, those failures that are not common across all three subsystems of interest are pruned from the fault tree. Thus, the shared states and other essential entities leading to the fault tree capture only those failures that affect all three subsystems: water, power, and communications.

Critical Component Identification

The fault-tree model adapted and deployed here constitutes an essential and logical representation of the failure criteria of the three I-I subsystems that constitute the Complex SoS. Technical and organizational domain knowledge of each subsystem and of the SoS as a whole is essential. It is important to note that because the failure of the Complex SoS is decomposed into contributing causes connected by logical gates, thus, the fault tree is limited to those constituent causal failures that are shared across three subsystems. Furthermore, a minimal cut set of the tree contains the failures across only the multiple subsystems that were included in the fault tree and can cause overall failure. Indeed, component failures identified in the minimal cut set are the most critical resources for resilience protection and should be resourced appropriately.

Analysis of the power–water–communications fault tree (Figure 10.III.4) including the decomposition of prime water (Figure 10.III.5) to identify the shortest path to failure identifies the minimal cut set of shared state fault tree, which is the smallest combination of component faults, which, if they occur, will result in a failure of the Complex SoS. Using Boolean

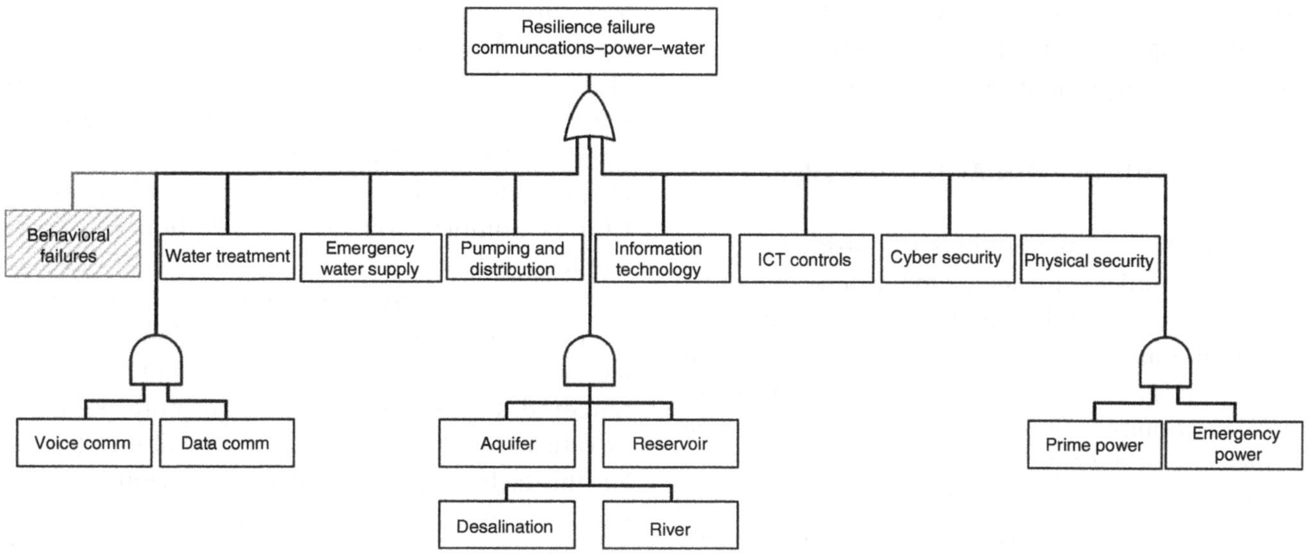

Figure 10.III.6 *Minimal cut set of power–water–communications fault tree.*

algebra to determine a minimal cut set of the technical failures resulted in the following resources being identified as critical (Figures 10.III.1 and 10.III.6).

This minimal cut set in the fault tree identifies the 15 components critical to operations of the power–water–communications regional infrastructure Complex SoS. Recall that the *AND* Gate indicates that all components must fail for an overall failure to occur. Therefore, a failure of any one of the following will result in a failure of the Complex SoS: (i) voice and data communications; (ii) water treatment; (iii) emergency water supply; (iv) water pumping and distribution; (v) source water from aquifer, reservoir, river, and desalination; (vi) information technology; (vii) ICT controls; (viii) cyber security; (ix) physical security; and (x) prime and emergency power (Haberlin and Haimes, 2018).

These components are shared across the three subsystems and are critical to the overall function of the regional infrastructure Complex SoS. Incorporation of additional systems into the analysis may further reduce the minimal cut set; it will not increase the set.

10.III.9 SUMMARY AND CONCLUSIONS

Part III provided another case study for modeling regional infrastructures as Complex SoS in which shared states and other essential entities are used to identify the highest priority for avoiding a disastrous failure, thus allocating appropriate resources for the resiliency planning associated with specific adverse initiating events. Three subsystems – water, electricity, and communication – are studied, and a fault tree representing the technical shared states and other essential entities is developed. An extended decomposition of the prime water supply node was provided as an example. A fully realized model would complete this decomposition for each node in the fault tree to a level of detail sufficient to assess the risk and resilience of each subsystem prior to calculation of the minimal cut set. For this case study, analysis of this fault tree resulted in a minimal cut set of 15 entities to be prioritized for resilience resourcing. The methodology employed can be extended to incorporate additional subsystems to establish technical components most critical to the entire Complex SoS.

A further pursuit of this case study includes at least two separate tasks: (i) expansion of the subsystems fault-tree model to incorporate additional subsystems and (ii) decomposition of the behavioral side of the fault tree. The first is relatively straightforward and may be accomplished using the methodology described above for the existing model. The latter is a new area of focus and will necessarily delve into social science topics including organizational behavior, policy development, and decision making–the theme of Chapter 12.

PART IV: ASSESSING SYSTEMIC RISK TO CLOUD-COMPUTING TECHNOLOGY AS COMPLEX SYSTEMS OF SYSTEMS

10.IV.1 INTRODUCTION AND OVERVIEW

Part IV of this chapter presents a fourth case study and a methodology for modeling cloud-computing technology (CCT) and its users as complex interdependent and interconnected Complex SoS. We posit and demonstrate that due to the inherent structure of CCT as Complex SoS, users of public *CCT are at a higher risk of unlawful intrusion than users of non-CCT systems*. At the same time, the flexibility and the agility of CCT Complex SoS provide advantages relative to most non-CCT systems. The need is for the CCT community to employ these advantages as an integral part of its cyber security strategy. Our analysis builds on the following theory and methodology: We consider CCT and its users as complex interconnected hardware and software subsystems that interact as Complex SoS through shared states and other essential entities. We defined earlier in this chapter the term "other essential entities. (The term "essential entities" includes *shared/common decisions*, *decision makers*, *stakeholders*, *resources*, *organizational behavior and norms*, *policies and procedures*, *management*, *culture*, and others.) We will further explore systems and subsystems that are connected in series (rather than in parallel, as is the case for most non-CCT systems). Exploiting the serial connectivity of CCT Complex SoS, we reinvent the use of fault-tree analysis to demonstrate their resulting unreliability. Furthermore, building on published literature, we incorporate Pareto-optimal frontiers among the multiple objectives associated with the users of CCT to compare the risks faced by security-conscious users of CCT (e.g. large corporations) versus those faced by cost-conscious users (e.g. small or start-up companies). We further show that *users of CCT as Complex SoS are at a higher risk than users of non-CCT systems for certain assumptions and types of cyber attacks*. Part IV of this chapter is structured as follows: Part I outlines the characteristics of CCT Complex SoS, addresses the inherent vulnerability of CCT Complex SoS to cyber intrusion and attack, and provides a brief literature review; Part II offers a rationale for modeling CCT as interdependent and interconnected Complex SoS; Part III demonstrates, through the reinvented use of fault-tree analysis, the higher risk faced by CCT and its users as Complex SoS; Part IV presents an economic analysis of the security of CCT as Complex SoS; and Part V offers conclusions and lessons learned.

Due to the considerable economic advantages that CCT Complex SoS offer, it is gaining a rapidly increasing user base. Its flexibility and agility have rendered it almost indispensible to both public and private enterprise. But this very flexibility has also left this technology vulnerable to malicious intrusion. Cyber-infrastructure systems (CIS) – public and private – are subjected to continuous attacks worldwide by adversaries and malicious intruders' intent on exploiting critical databases and the theft of intellectual property and other resources.

10.IV.2 CLOUD-COMPUTING TECHNOLOGY AS COMPLEX SYSTEMS OF SYSTEMS

One of the most salient – and for our purposes, one of the most critical – aspects of CCT is its continuous availability to a wide range of users. The National Institute of Standards and Technology (NIST) considers CCT as a model for enabling convenient, on-demand network access to a shared pool of configurable computing resources (e.g. networks, servers, storage, applications, and services) that can be rapidly provisioned and released with minimal management effort or service provider interaction. According to the NIST, CCT is composed of five *essential characteristics* ("on-demand self-service; broad network access; resource pooling; rapid elasticity or expansion; and measured service."), three *service models* ("software, platform and infrastructure") and four *deployment models* ("private, community, public and hybrid"). For the purposes of this chapter, we assume that these features – taken together with the decentralized power and authority of CCT – describe and constitute Complex SoS.

Drawing on this preliminary definition of CCT, in this chapter we address and provide answers to the following set of basic questions: (i) *Why* is studying and modeling CIS served by CCT (CIS-CCT) as Complex SoS critical to answering risk and security questions? (ii) *Why* is understanding the central role of the states of the CIS-CCT model critical to answering risk and security questions? (iii) *Why* is proper and effective hardware and software systems integration critical to answering risk and security questions? (iv) *Why*, and to what extent, are users of CIS-CCT at higher risk than those who do not use CCT? (v) *What* are the associated economic and security tradeoffs between using CCT and non-CCT technology?

While we highlight the added risks associated with CCT, we do not explicitly analyze the advantages that CCT systems have over non-CCT Complex SoS in terms of providing important, economically satisfying risk management opportunities. However, it is worth outlining some of the most important of these advantages: the (i) opportunity to employ the agility of CCT Complex SoS for moving target security solutions, (ii) opportunity to use the diverse geographical locations of CCT Complex SoS as part of moving target security solutions, and (iii) opportunity to employ different CCT Complex SoS to provide both diverse redundancy and moving target security solutions. In addition, the business mandate for CCT Complex SoS to serve a diverse customer base and rapidly respond to new needs necessarily results in agile system structures and the employment of configuration control processes and tools that are optimized for managing change – and certainly more so than for single users for whom the system might ordinarily be designed.

These new system structures and control processes make it possible to respond to new cyber threats more rapidly and with more assurance. Note that under certain assumptions, CCT might not be worse than traditional non-CCT computing environments for *all* types of cyber attacks. For example, recent cyber attacks have demonstrated that cloud computers can handle distributed denial-of-service attacks better than traditional servers can, precisely because they enable a quick instantiation of additional resources that can more efficiently handle increases in requests. Furthermore, non-cloud-computing environments (e.g. traditional enterprise systems) may be more vulnerable to cyber attacks that target specific IP data, since such attacks in the cloud would be ineffective unless the exact location of the information on the server was known.

Easy access to the cloud by anyone who has a valid email and credit card suggests that the cloud may be easier to infiltrate than a non-cloud-computing environment, in the sense that the initial access to the cloud requires a low level of technical ability and a small investment. Thus, the number of potential threats is increased. Furthermore, the structure of the cloud and its functionalities facilitate an environment in which it is easy for somebody, who is already in the cloud and has malicious intentions, to create and propagate a cyber attack with a relatively low budget. In some cases, it may be difficult for the cloud customer (in its role as data controller) to effectively check the data handling practices of the cloud provider and thus to be sure that the data is handled in a lawful way. Thus, the cloud is more vulnerable to ubiquitous, nonspecific cyber attacks in which the attackers have no desire to access specific information, e.g. intellectual property of a specific company, but are more interested in accessing any confidential or sensitive information they can find.

10.IV.3 CLOUD-COMPUTING TECHNOLOGY AS AN INTERDEPENDENT AND INTERCONNECTED COMPLEX SYSTEMS OF SYSTEMS

This section introduces and defines what constitutes Complex SoS and to demonstrate that CIS-CCT fits within this representation.

10.IV.3.1 Modeling Cloud-Computing Technology as an Interdependent and Interconnected Complex Systems of Systems

Clearly, no single model can ever attempt to capture the essence of Complex SoS – their multiple dimensions and perspectives. Instead, a variety of models must be built to represent the diverse Complex SoS perspectives. These submodels that constitute CCT Complex SoS are integrated (coordinated) into a metamodel by a process of intrinsic metamodel coordination through shared states and other essential entities. (The term "essential entities" includes *shared/common decisions, decision makers, stakeholders, resources, organizational behavior and norms, policies and procedures, management, culture*, and others.) The objective of metamodel coordination and integration is to build on all relevant direct and indirect sources of information to gain insight into the Intra- and interdependencies and interconnectedness characterizing the submodels and, on the basis of this insight, to develop representative models of the Complex SoS under consideration. The coordination and integration of the results of the multiple models are achieved in the metamodeling phase within the phantom systems model (PSM) methodology (introduced in Chapter 4), thereby yielding a better understanding of the system as a whole.

Recall the following systems-based building blocks of our CCT Complex SoS model:

1. Given a system's model, the states of the system are the smallest set of independent system variables whereby the values of the members of the set at time t_0, along with known inputs, decisions, and random and exogenous variables, completely determine the value of all system variables for all $t > t_0$.
2. Vulnerability is the manifestation of the inherent multidimensional states of the system (e.g. physical, technical, organizational, and cultural) that can be subjected to a natural hazard or be exploited to adversely affect (cause harm or damage to) that system, and it is a function of the specific threat to the system and of the time frame.
3. Resilience is also a manifestation of the inherent states of the system; it is a multidimensional vector that is time and threat dependent. More specifically, resilience represents the ability of the system to withstand a disruption within acceptable degradation parameters and to recover within acceptable loss and time parameters.
4. PSM are used for intrinsic metamodeling of Complex SoS using the basic assumption that some specific commonalities, interdependencies, interconnectedness, or other relationships exist through shared and unshared states, decisions, and inputs between any two systems within any system of systems. *The intrinsic shared states provide a powerful mechanism for understanding and exploiting the strong interdependencies among the subsystems of* Complex SoS.

The efficacy of shared states among subsystems may be manifested through (i) sharing databases, (ii) realizing that decisions made by the stakeholders of subsystem A can have a direct impact on subsystem B, and (iii) encouraging and enticing stakeholders of different subsystems to collaborate on inputs, constraints, and decisions that affect the shared states for win–win outcomes (Haimes 2012a, 2016a).

In Haimes (2012b) and earlier chapters we posited that *some specific commonalities, interdependencies, interconnectedness, or other relationships must exist between any two systems within any Complex SoS. The essence of each subsystem can be represented by a finite number of essential state variables.* Given that a system may have a large number of state variables, we define the term *essential states of a system* to connote the minimal number of state variables in a model with which to represent the system in a manner that permits the questions at hand to be effectively answered. Thus, these essential state variables become fundamental for an acceptable model representation.

For a properly defined Complex SoS, any interconnected subsystem will have essential state variable(s), common/shared essential entities (recall that the term was defined earlier), and objective(s) (shared with at least one other subsystem). This requirement constitutes a necessary and sufficient condition for modeling interdependent and interconnected subsystems as Complex SoS. This ensures an overlapping of states and other essential entities within the subsystems. Of course, the more we can identify and model shared (overlapping) state variables among the subsystems, the greater is the representativeness of the submodels and the metamodel coordination among Complex SoS.

Thus, by virtue of their structural design and organization, CIS-CCT are complex interdependent and interconnected Complex SoS. The above interdependencies and interconnectedness (I-I) within and among the subsystems and systems that compose CCT Complex SoS – as well as sharing users and databases – render their subsystems more vulnerable to the risk of malicious intrusion than those of non-CIS-CCT users, as discussed in the next section.

10.IV.3.2 Systemic Sources of Risk That Can Affect the Subsystems of the CIS-CCT Complex SoS

The following list represents a sample of systemic sources of risk that can affect the subsystems of the CIS-CCT Complex SoS:

1. Shared states, due to shared pools of configurable computing resources, e.g. networks, servers, storage, applications, client infrastructure, and server infrastructure.
2. Complexity of hardware and software systems integration resulting from the large scale of the systems with millions of lines of code.
3. Complexity related to the buffer (virtual machines) between the memory (hypervisor) and the operating system (OS). A hypervisor is one of many hardware virtualization techniques that allow multiple OS to run concurrently on a host computer. The hypervisor presents to the guest OS a virtual operating platform and manages the execution of the guest OS. Multiple instances of a variety of OS may share the virtualized hardware resources. The buffer can be closed by an adversary and affect the OS. At the core, there is a central (global) entity that has complete control, and when penetrated, problems arise and pose risk to the system.
4. Excess uniformity (through standards) in protocols. For example, all operating codes in Windows™ are the same; thus, if an intruder breaks one protocol, the rest would follow.
5. Tradeoffs between openness (exposure to attacks) and flexibility in CIS-CCT Complex SoS, which invariably have been driven by cost-cutting measures, resulting in increased exposure.
6. Digital fingerprints – messages and other data – left on electronic devices. All electronic devices have digital fingerprints. Although we cannot see the messages and data, we can follow their fingerprints, and so can would-be intruders.
7. Lack of system operations metrics. No known metrics are available for users to determine whether a CCT Complex SoS is working appropriately.
8. Readily available metrics on users and system states. Metrics available to users may allow access to the following vulnerable features of CCT Complex SoS: authentication, authorization, control over employees, state of patching; quality of failure (i.e. how resilient is the system, or how quickly can it recover from delay?), outcomes from audits, uses of audits, quality of audits and their frequency, quality of trust models (i.e. what trust models are used, and how can the owners' and users' trust models be compared?).
9. Hybrid CCTs. The interface between private and public CCTs – creating what is effectively a hybrid CCT – can provide opportunities for intrusion. A private cloud can be exploited by an adversary, who can pass it on to a public CCT.
10. Multiple layers of interacting parts designed for modularity and components acquired separately.
11. Hardware vulnerability. Although many systems operators are aware of the need to control access to data and software, hardware is also vulnerable. An intruder can modify hardware when obtaining access to the software; once the hardware is modified, both software and data can later be maliciously modified.
12. Uninsulated data. Data is commonly not insulated; thus, when it is in memory, a few algorithms can operate on other encrypted data and potentially compromise it.
13. Lack of knowledge about the origin of sources of data.
14. The ability to modify a chip.
15. Careless logging in, transferring files, etc. by users.
16. Difficulty in customizing the trust model in hybrid CCT systems, where the software for each customer is separate.
17. Unsafe methods for isolating concurrent programs.
18. Buffer overflow; a fixed size allocated for handling data may cause the buffer to fail, and an intruder can take control of the system by injecting a new code. This is a well-known problem, and there is no silver bullet against injecting malware or viruses into the system.
19. Lack of understanding of how many security layers are needed to bring risk down to an acceptable level.
20. Inability to safely update system configurations.

10.IV.4 HIGHER RISK TO CLOUD-COMPUTING TECHNOLOGY AND TO ITS USERS AS COMPLEX SYSTEMS OF SYSTEMS

10.IV.4.1 *Premise*: Users of Cloud-Computing Technology (as Complex SoS) Are at Higher Risk Than Users of Non-CCT Systems for Certain Assumptions and Types of Cyber Attacks

To validate the above premise, we compare cloud and non-cloud systems under similar conditions. We (i) explore public CCT as Complex SoS for a certain set of principles and assumptions detailed below; (ii) build on our earlier premises that the vulnerability and resilience of a system are manifestations of the states of that system and that they are functions of the specific intrusion/threat; (iii) recognize that the risk to a system is a function of all of the above; (iv) assert that the multidimensional probabilistic consequences resulting from a malevolent intrusion into a CCT Complex SoS necessarily yield a multidimensional risk function; (v) assert that the same type of intrusion/threat is applied to both systems; (vi) assert that both CCT Complex SoS and non-CCT systems operate under the same security standard; (vii) assert that the interconnected subsystems, which constitute CCT Complex SoS, share functional components and states and that the multiple users are mostly unknown to one another (e.g. some of the users could be insiders or intruders); and (viii) use the reliability of a system as a surrogate metric for the risk to CCT Complex SoS due to a malevolent intrusion.

The failure probabilities of subsystems of the CCT Complex SoS are dependent on the failure of other interconnected subsystems with shared states. The failure probabilities can be evaluated using conditional failure probabilities, because intrusion into any subsystem with shared states implies the ability to penetrate other subsystems with those shared states. *Thus, in comparison with non-CCT systems, the shared states are synonymous with replacing the probabilities of independent subsystem failure with conditional subsystem failure probabilities (in the parlance of fault-tree analysis, the subsystems connected in parallel become connected in series) and thus add vulnerability to the CCT Complex SoS. In sum, the reliability of interconnected subsystems with shared states is always lower than the reliability of subsystems that do not share states (i.e. modeled by independent failure probability).*

Let R_i be the independent reliability of subsystem i; $R_i|_j$ be the conditional reliability of subsystem i, given the failure of subsystem j; $R_{\text{s-CCT}}$ be the reliability of the entire CCT system; and $R_{\text{s-non-CCT}}$ be the reliability of the entire non-CCT system. The following inequality holds for subsystems connected in any form:

$$R_i|_j \leq R_i \text{ and } R_{\text{s}-\text{CCT}} \leq R_{\text{s}-\text{non}-\text{CCT}}.$$

This is because for a system whose subsystems share states with other subsystems, the failure of any one subsystem would cause the failure of other interconnected subsystems. Thus, from the perspective of reliability theory and fault-tree analysis, the CCT Complex SoS will have a lower reliability, i.e. a higher risk than non-CCT systems. In sum, the probability of adverse consequences to the interconnected CCT as Complex SoS, with the corresponding tangible and intangible risks, is higher than for the non-CCT systems.

10.IV.4.2 Demonstration of the Premise with a Mini-CCT Complex SoS

This section provides a demonstration of the premise through a fault-tree analysis on a virtual mini-CCT Complex SoS. The constructed virtual mini-CCT is harmonious with the literature on the architectures of CCT systems. To focus on the proof of concept, only essential components and functionalities are included in the mini-CCT. A fault-tree model is developed based on identified subsystems and their potential failure modes. The reliability of the CCT Complex SoS and the non-CCT system can be calculated and compared assuming conditional failure probability and independent failure probability of each subsystem.

The National Institute of Standards and Technology (NIST) classifies Complex SoS into different service models (e.g., Software as a Service [SaaS], Platform as a Service [PaaS], and Infrastructure as a Service [IaaS]) and deployment models (e.g., private, community, public, and hybrid) [NIST Special Publication 800-145]. Current CCT Complex SoS providers in the marketplace employ specific technologies and cloud architectures to meet the different needs of a broad range of cloud users. The literature

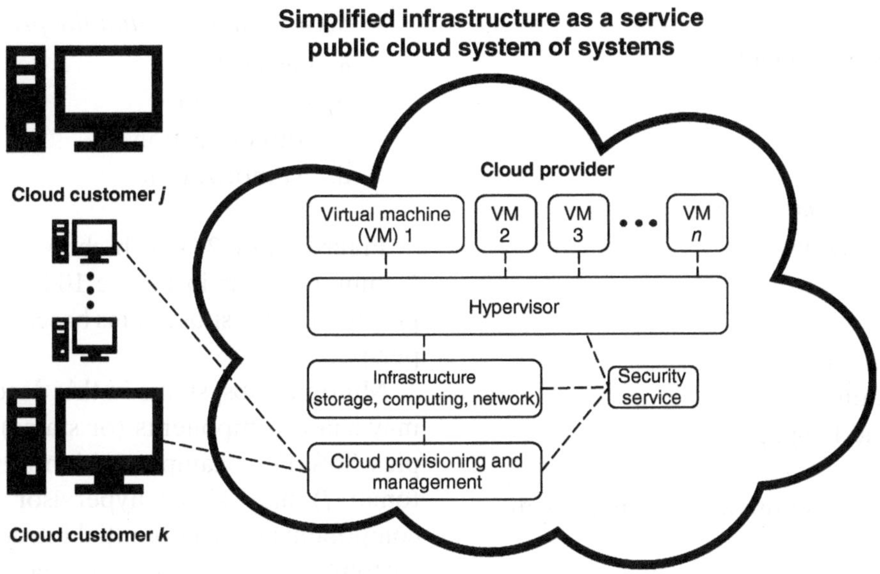

Figure 10.IV.1 *Essential components and functionalities of a simplified IaaS public cloud Complex SoS.*

identifies no standard technology or system architecture for CCT Complex SoS. For our purpose, a conceptual mini-IaaS public-CCT system is constructed in Figure 10.IV.1.

Each box in the figure represents an essential component or functionality of the CCT Complex SoS.

A specific CCT Complex SoS can be decomposed into multiple perspectives, including, for our purposes, physical, functional, and stakeholder perspectives. From a physical perspective, an IaaS public-CCT Complex SoS can be decomposed into the following subsystems and components:

1. *Hardware*
   a. Hard drives and memory
   b. CPU
   c. Switches and cables
2. *Software*
   a. System software (hypervisor, OS, middleware)
   b. Application software (applications and APIs)
   c. Communications software
   d. Security software
3. *Physical infrastructure*
   a. Data center
   b. Physical security system
   c. Power supply
4. *Human*
   a. Programmer/developers
   b. Administrator
   c. Operations and maintenance staff

This decomposition perspective provides limited insights into the security aspects of the CCT Complex SoS, because the overall security functionality cannot be deduced from the functionality of the individual components. In most cases, hardware, software, and human perspectives need to be integrated and work together to provide meaningful functionality to the CCT Complex SoS.

From a functional perspective, an IaaS public-CCT Complex SoS can be decomposed into the following subsystems and components:

1. *Infrastructure*
   a. Storage resources
   b. Computational resources
   c. Network resources
2. *Platform*
   a. Provisioning tool
   b. Monitoring and metering tool
   c. Systems management (performance, capacity, availability)
   d. Service catalog

e. Database
f. Run-time environment

3. *Application*
   a. User interface
   b. Machine interface
   c. Service management

4. *Security*
   a. Authentication
   b. Authorization
   c. Auditing and logging
   d. Firewall
   e. Encryption and entitlement management

The relationships among these functionalities constitute the architecture of the CCT Complex SoS. A combination of these functions is used to achieve the required characteristics of a CCT Complex SoS as defined by the NIST (on-demand self-service, broad network access, resource pooling, rapid elasticity, and measured service).

In the above two decomposition schemes, each subsystem has a specific function, and the failure of one subsystem will lead to the failure of the entire CCT Complex SoS. However, in this study we are not concerned with system failure, but in the security failure of a CCT Complex SoS due to internal or external emergent risks. We thus propose a decomposition scheme based on a stakeholder's perspective. From this perspective, an IaaS public-CCT Complex SoS can be decomposed into the following subsystems and components:

1. *Normal user*
   a. Virtual machine
   b. Physical machine
   c. Computing tasks
   d. Data

2. *Attacker*
   a. Virtual machine
   b. Physical machine
   c. Malware

3. *Operation agent within the provider*
   a. Physical machine
   b. Hypervisor
   c. Provisioning and system management

4. *Security agent within the provider*
   a. Firewall
   b. Encryption
   c. Authentication
   d. Authorization

Figure 10.IV.2, which shows the same mini-CCT Complex SoS as in Figure 10.IV.1, has been decomposed into four subsystems based on the stakeholder's perspective.

Multiple subsystems in this decomposition scheme may share components (or states) of the CCT Complex SoS. For example, an innocent user (cloud customer *j*) may share hypervisor and infrastructure components with a potential intruder (cloud phantom customer *k*); customer *k* may share the virtual machine, hypervisor, and infrastructure components with the cloud provider; and the cloud provider may share the hypervisor and infrastructure components with the security service. These shared components constitute the interdependency and interconnectedness among these subsystems and a major source of threat to the CCT Complex SoS.

10.IV.4.3 Building a Fault Tree for the Mini-CCT Complex SoS

In this section we demonstrate, through fault-tree analysis, that the I-I among the subsystems that characterize CCT Complex SoS increase its vulnerability to unlawful intrusion and thus the risk of financial and intellectual property losses. And, more specifically, we validate the premise that due to the inherent structure of CCT as I-I Complex SoS:

> Users of CCT are at a higher risk of unlawful intrusion than users of non-CCT systems.

In fault-tree analysis, when subsystems are connected through an OR Gate (in series), the system fails when at least one of its components fails. The reliability of the system is represented as $R_s(t) = \prod_{i=1}^{n} R_i(t)$, and $R_s(t) \leq \min_i\{R_i(t)\}$. On the other hand, when subsystems are connected through an AND Gate (in parallel), the system fails only when all of its components fail. The reliability of the system is represented as $R_s(t) = 1 - \prod_{i=1}^{n}[1 - R_i(t)]$, and $R_s(t) \geq \max_i\{R_i(t)\}$.

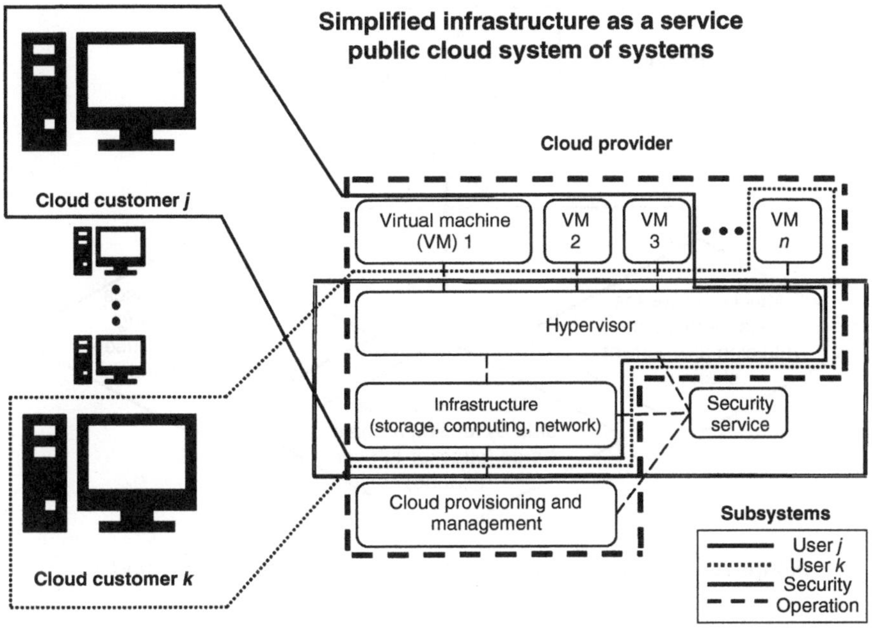

Figure 10.IV.2 *Identified subsystems and components of a simplified IaaS public cloud SoS.*

A minimal cut set is defined as the smallest combination of component failures, which, if they all occur, will cause the top event (i.e. failure) to occur (U.S. Nuclear Regulatory Commission, 1981). By definition, a minimal cut set is a combination of intersections of primary events in parallel sufficient for the top event to occur (if all parallel components fail). A fault tree will consist of a finite number of minimal cut sets, all of which are in series, which are unique for the top event to occur.

The literature suggests that a cloud intruder constitutes a major source of risk for cloud users; therefore, in this study the top event of the fault tree for the mini-CCT Complex SoS is defined as "attacker gains access to normal user's confidential data." A prime example of such an attack involved a researcher, Thomas H. Roth, using Amazon's cloud-computing service Elastic Compute Cloud (EC2) to fire 400 000 passwords a second at a secured Wi-Fi network. It took Roth 20 min to hack into the system (Homeland Security News Wire, 2011; The H Security, 2011). Epsilon, a cloud-based marketing service, experienced an actual cyber attack in which data (e.g. email and bank account details) of its customers, including JPMorgan Chase, Citibank, Barclays Bank, Marriott, and Hilton, was exposed to the hackers (Bhadauria and Sanyal, 2012).

Three potential events can individually cause the top event:

1. Event E1: Attacker accesses data by real-time intrusion (concurrent sharing).
2. Event E2: Attacker accesses data by leaving a Trojan horse (attacker-to-user sharing).
3. Event E3: Attacker accesses data by reading residual traces (user-to-attacker sharing).

Each of these events alone will cause the top event to occur (i.e. a complete systems failure) because all subsystems/components are connected through an OR Gate. The basic fault tree is shown in Figure 10.IV.3.

The necessary conditions for each of the three basic failure events are developed in detail in Figures 10.IV.4, 10.IV.5, and 10.IV.6: Figure 10.IV.4 presents an expanded fault tree with failure modes caused by real-time intrusion, Figure 10.IV.5 presents an expanded fault tree with failure modes caused by a Trojan horse, and Figure 10.IV.6 presents an expanded fault tree with failure modes caused by residual traces.

Each basic failure event in a circle represents a specific failure mode of a subsystem; all failures must occur to cause the top event (failure) to occur, because

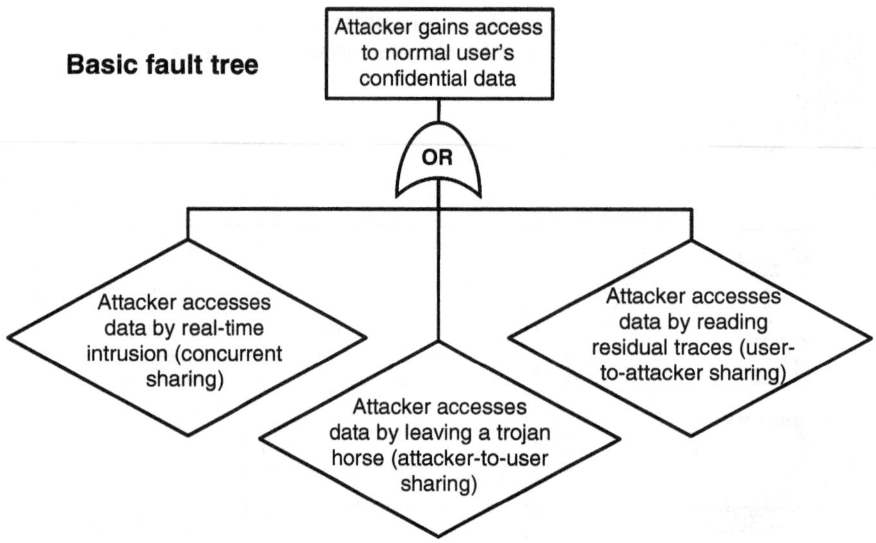

Figure 10.IV.3 *Fault tree top events and three potential failure modes.*

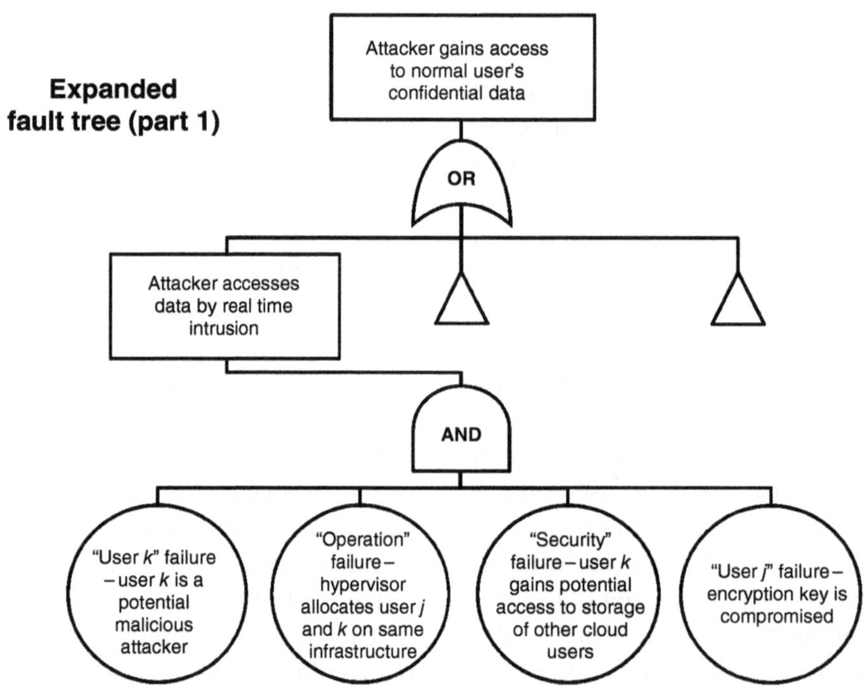

Figure 10.IV.4 *Expanded fault tree with failure modes caused by real-time intrusion.*

they are connected in parallel through an AND Gate. As demonstrated earlier, the failure probability of basic events (subsystems) in non-CCT systems are independent of each other, while the failure probability of basic events (subsystems) in a CCT Complex SoS are dependent on other interconnected subsystems through shared states. *In sum, the results demonstrate that the failure probabilities of the top event in a CCT Complex SoS are higher than those of non-CCT Complex SoS.*

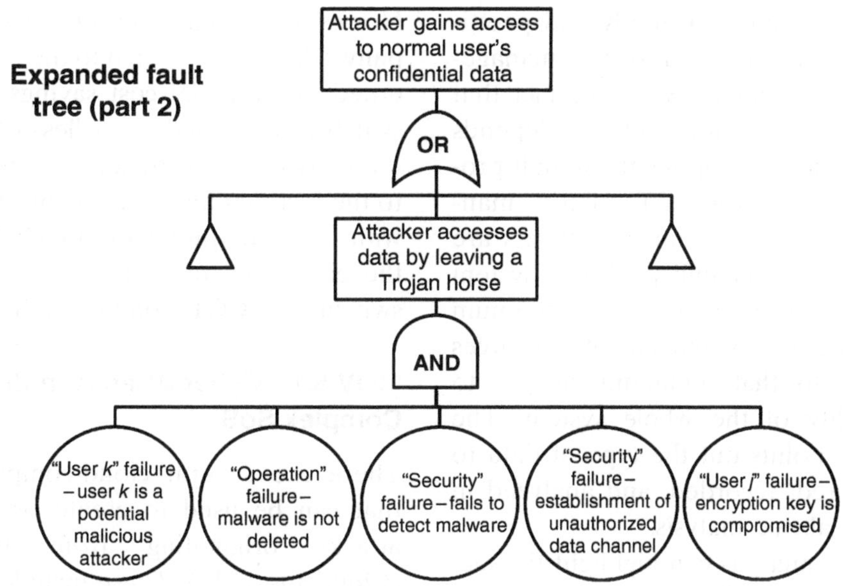

Figure 10.IV.5 Expanded fault tree with failure modes caused by Trojan Horse.

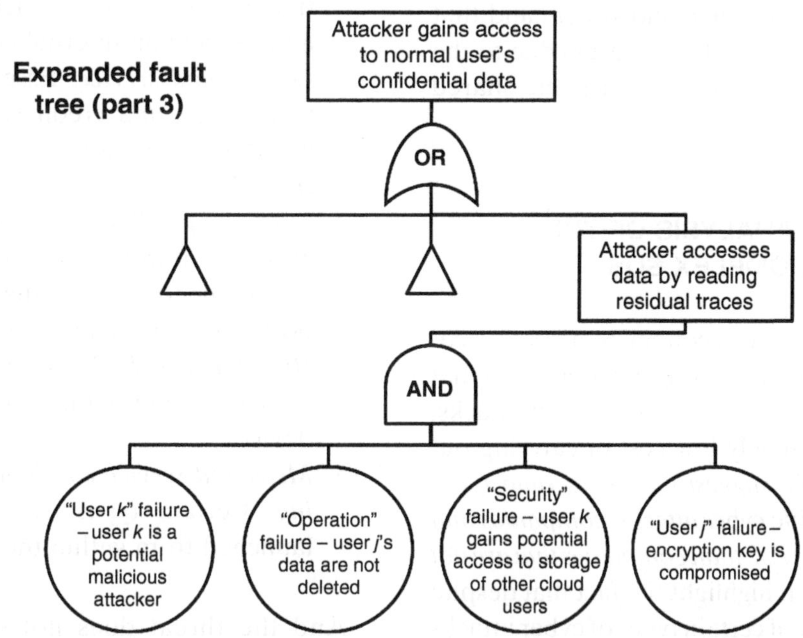

Figure 10.IV.6 Expanded fault tree with failure modes caused by residual traces.

10.IV.4.4 Summary of Fault-Tree Analyses

The above results demonstrate that the reliability of CCT Complex SoS is lower than non-CCT Complex SoS due to subsystem interdependency. This conclusion remains valid as long as the conditional failure probability is greater than the independent failure probability. Real CCT Complex SoS are of course much more complex than the virtual mini-CCT Complex SoS we have just examined; however, the general concept we have established regarding interdependent subsystems with shared states and the resulting interdependent failure probabilities can be used to analyze the reliability of more complex systems.

The fault-tree model can measurably and practically contribute to the development of risk management options for CCT Complex SoS. The fact that the reliability of each minimum cut set depends mainly on the most reliable component within it provides invaluable guidelines for CCT security managers. For example, since all minimum cut sets are connected in series, the reliability of the system depends primarily on the most unreliable minimum cut set. Thus, limited risk management resources should be allocated to that minimum cut set to improve the reliability of the whole system. The fault-tree model also points out the type of data to be observed, monitored, recorded, and analyzed in order to perform the above analysis.

In sum, a fault-tree analysis can highlight the fact that the cloud, being Complex SoS with multiple interdependent and interconnected subsystems and resources, is vulnerable to even one cloud user/intruder who has access to the cloud system and that user can also access the files of other users due to the shared states and other essential entities and shared cloud subsystems.

10.IV.5 ECONOMIC ANALYSIS OF THE SECURITY OF CCT COMPLEX SOS

In this section we address the economics of cloud security. We introduce cloud-computing functionalities that can be used to create, or to enable, cloud cyber attacks, and we discuss the relatively low cost of carrying out cloud cyber attacks. *We suggest that for certain types of ubiquitous, nonspecific cyber attacks, companies that use the cloud are more at risk than companies that are not using the cloud.* We then highlight the fact that despite the likely increased risk of certain types of cyber attacks targeting confidential and sensitive data stored in the cloud and despite the recognized fact that security may not be the highest priority for cloud providers, companies will continue to switch to the cloud to gain measurable economic benefits.

The prior sections of Part IV of this chapter indicate that CCT is more prone to certain types of cyber attacks, which suggests that companies that are considering switching to the cloud ought to be cognizant of the increased risk of losing confidential and sensitive information when using cloud-based applications. However, we suggest in this section that when a company's decision to switch to the cloud is primarily governed by potential cost savings, that company may switch to the cloud regardless of the increased risks if the savings are sufficiently high. Our analysis points to the fact that any assessment of the risks associated with cloud computing must include a consideration of the economic tradeoffs between switching and not switching to CCT Complex SoS.

10.IV.5.1 Cyber Attacks in the CCT Complex SoS

There are several cloud-computing functionalities that can be used to create or enable cloud cyber attacks originating from within the cloud (CloudBus, 2011). These include the following:

1. *Rapid elasticity*: A large amount of computational power can be quickly utilized to initiate and execute an internal cloud cyber attack.
2. *On-demand service*: Essentially anyone with a valid email and credit card can sign up for a cloud service and potentially launch attacks from the cloud.
3. *Resource pooling*: Malicious users launching an attack from within can utilize as many resources as they need because the service provider supports everyone's applications.
4. *Broad network access*: Cyber attacks can be launched from within a cloud from any digital device.
5. *Measured service combined with resource pooling*: Certain types of cyber attacks can be launched from within the cloud very cheaply.

And the threat does not stop there: Once the potential attacker is in the cloud, he or she has potential access to many companies/users that store their information on cloud servers. Thus, if an attacker is interested in ubiquitous nonspecific attacks, i.e. if he or she is not looking for specific IP data belonging to a specific company, but is more interested in obtaining any type of information, then, theoretically, with a single attack he or she can gain control of large amounts of confidential information. A theoretical example of such an attack is a hyperjacking attack in which a malicious user installs a rogue hypervisor that takes complete control of a company's servers. In such an attack, regular security measures are ineffective because the OS will not indicate that the machine has been compromised (McKay, 2011; Sarno and Rodriguez, 2011).

10.IV.5.2 The Cost of Cyber Attacks Internal to the Cloud

The Stratsec Winter School (Hayati, 2012) performed research to explore whether the cloud-computing environment provides certain benefits for cyber attackers and whether the cloud platform can be utilized to launch cyber attacks. Furthermore, they explored the ability of cloud providers to detect and report cyber security breaches in the cloud. They set up a botCloud (a group of cloud instances that are commanded and controlled by a malicious attacker), with which they attacked the victim hosts, which were also set up virtually in a controlled network environment. After performing four different experiments with different cases of malicious traffic and intrusion detection systems, the researchers concluded that, in general, cloud providers did not notify cloud users of the malicious activities occurring in the cloud and did not respond to the attacks by resetting or terminating connections. Given the cloud providers' inadequate response, malicious users could easily set up and exploit these attacks, since doing so required minimal knowledge of the internal architecture and system administration. Furthermore, the cloud environment made it effortless for intruders to create uncountable cloned opportunities from which to launch attacks; the cloud environment also meant that attackers needed no significant physical infrastructure of their own. Finally, the availability and elasticity of resources made such attacks very cheap. This example shows that the cloud-computing environment has opened doors to cyber attacks that can be significant in scope while requiring little physical infrastructure and costing little to design and execute.

10.IV.5.3 Security Implications for Cloud Users

Several questions deserve clarification: (i) What are cloud customers doing to protect themselves against such attacks? (ii) Are cloud customers aware that they might be exposed to a greater risk in the cloud? (iii) Are they requiring a higher level of security? and (iv) How much are they willing to pay for security in the cloud? Surprisingly, studies show that cloud customers are not well acquainted with what security they are paying for or what that security protects against. Furthermore, many admit to having lowered their security requirements after switching to the cloud.

A Ponemon Institute's study of cloud-computing service providers suggests that of the US providers surveyed, 73% believe that their cloud services do not secure and protect confidential information (Ponemon Institute, 2011). The report suggests that most cloud providers do not consider security as their priority and believe that protecting the customer's data is not their responsibility. The report observes: "[N]either the company that provides the service nor the company that uses cloud computing seem willing to assume responsibility for security in the cloud. In addition, cloud-computing users admit they are not vigilant in conducting audits or assessments of cloud-computing providers before deployment" (Ponemon Institute, 2010). The literature is sparse regarding data on the actual dollar value of security expenditures in the cloud, and most estimates are gathered at a higher level by considering total IT budgets of companies.

The question then becomes: How can we utilize this data, along with some general economic performance data for relevant economic sectors, to determine the way companies are implicitly trading increased profits for potentially decreased IT security? A majority of companies switch to the cloud for performance reasons, e.g. savings in capital and operating costs that can be directed toward increasing net profit margins, potentially enabling the cloud-using companies to remain economically competitive or increase their market share. Many of the companies that switch to the cloud believe that in the process they need to accept an overall lower level of security, yet many of them continue to place sensitive and confidential information in the cloud. *Even more importantly, many of the companies do not know what level of security service the cloud provides and are doubtful of the cloud providers' ability to protect their data.* Even some cloud providers are doubtful of their ability to protect the data that is stored on their servers. This suggests that some companies value the additional profit they might receive by switching to the cloud more than they value data security and potential loss of reputation and business.

In this analysis we examine what level of savings would justify companies of different sizes and different histories and from different sectors accepting a potentially lower or ambiguous level of security. Since we do not have data that would enable us to compare the levels of security of a specific company with respect to a specific threat to both cloud and non-cloud environments, *we will instead compare the increase in the net profit margin that companies might be able to achieve by reducing their IT expenditures by moving to the cloud environment.* In essence, through this analysis we are indirectly assessing how much security companies might be willing to give up in order to improve their net profit margin, i.e. to remain competitive. We assume the following:

1. Companies transfer sensitive and confidential information to the cloud. (According to a Ponemon Institute survey on encryption in the cloud (2012), 50% of US companies that participated in the survey transferred confidential and sensitive information to the cloud.)
2. Companies understand that their data might be more at risk in a cloud-computing environment. (According to a Ponemon Institute survey on encryption in the cloud (2012), 41% of US companies that participated in the survey believe that their security posture was decreased by moving to the cloud.)
3. Companies in the cloud are not fully aware of what the cloud provider is doing to protect their data. (Ponemon Institute's survey on encryption in the cloud (2012) suggests that 54% or US companies that responded to the survey were not aware of security features in the cloud.)
4. Companies in the cloud, as well as cloud providers themselves, are not confident about the security features provided by the cloud provider. (Ponemon's survey (2010) suggests that over 40% of cloud providers are not confident that they are meeting the customers' security requirements.)
5. Companies transfer IT operations to the cloud primarily to save money, and security concerns are not of primary interest to most companies. (According to a Ponemon Institute survey (2010), 73% of companies migrated to the cloud in order to reduce costs, and only 14% migrated to the cloud to improve security.)
6. On average, companies spend approximately 10% of their total IT budgets on IT security (Schwartz, 2011).
7. On average, companies (not including governmental agencies) save 20–30% in costs by switching to the cloud (Leung, 2010; Wright, 2012). We assume these savings come from reducing IT costs, and we assume scenarios in which the IT costs are reduced by 0–30% in increments of 5%.
8. We assume that companies are in business to make as much money as they can and that any reduction in costs will be redirected to increasing net profits. Thus, we assume that the IT costs that are saved by switching to the cloud are directly transferred into net profit. That is, for a company whose IT budget is 10% of revenues and whose net profit margin is 2%, net profit margin will be increased by 2% if IT costs are reduced by 20%.
9. Companies of different sizes and of longer or shorter histories will have varied concerns over the loss of reputation that might ensue from a cloud cyber breach. For example, bigger, more established companies will, in general, care more about their reputations than smaller, start-up companies and will thus require a larger net profit increase from an implied reduction in security. Thus, net profit increase vs. reduction in security cost curves will be different for companies in different sectors and of different sizes and histories.
10. The quantitative risk function of companies that are switching to the cloud is a multidimensional function composed of the following:
    a. Risk of cyber incident (loss of reputation, clients, and money). This may be of more concern for larger and more established companies that have created a high level of trust with their large client base than with smaller, newer companies that are more driven to increase their profits and that do not have an established client base.
    b. Risk of giving up existing IT staff, that is, not being able to return to the old configuration if the cloud environment proves inefficient.

This may be of more concern to larger and more established companies that have established legacy systems, and potentially large IT departments, than to smaller, newer companies that might not have extensive IT departments of their own.

c. Risk of not being able to switch from one cloud provider to another. Contract breakage fees may be high, and if a company does attempt to move to another provider, some personal or confidential information might remain with the original cloud provider.
d. Risk of no cost guarantee. Cloud costs are not guaranteed forever.

Based on estimates of net profit margins of different economic sectors (Butler Consultants; Yahoo Finance, 2012) and estimates of IT expenses as a percentage of revenues for different economic sectors (PricewaterhouseCoopers, 2008), we consider the following five economic sectors – banking and financial services, professional services, construction and engineering, manufacturing, and health care – shown in Table 10.IV.1.

For these five sectors, we consider the following six scenarios:

1. Migration into cloud will reduce IT expenses by 5%.
2. Migration into cloud will reduce IT expenses by 10%.
3. Migration into cloud will reduce IT expenses by 15%.
4. Migration into cloud will reduce IT expenses by 20%.
5. Migration into cloud will reduce IT expenses by 25%.
6. Migration into cloud will reduce IT expenses by 30%.

For each of these sectors, we explore the increase in the net profit margin that results from the six levels of IT cost reduction. We assume that all IT cost savings are directly redistributed to net profit and hence that net profit increases with any reduction of IT costs. Results for the five economic sectors are shown in Figure 10.IV.7, which indicates that under the aforementioned assumptions, a 30% reduction in IT costs for a company in the banking and finance sector could translate into a 1.6% increase in net profit. The same level of IT cost reduction could translate into a 0.5% increase in net profit for the construction and engineering sectors.

We can also consider how the potential increase in net profit translates into value for company shareholders. We do this by considering two financial metrics, namely, the *earnings per share* (EPS) and *return on equity* (ROE). The EPS metric indicates the amount of earnings for each outstanding share of a company's stock. *This metric is the single most important determinant of a company's stock price and is computed as a ratio of net income to average common shares.* Any increase in net income, while keeping average common shares constant, results in an equal increase in EPS. An increase in EPS then translates into higher stock prices, which are desirable to a company's shareholders and are thus desirable to the company itself. In other words, companies do what they can to improve their EPS and stock prices, and this can be done efficiently by increasing their net incomes. The second metric, ROE, indicates the rate

TABLE 10.IV.1 Net Profit Margins, IT Expenditure as a Percent of Revenues, and Security Expenditures as a Percent of IT for Five Economic Sectors in the United States

Industry	Net profit margin (%)	IT as % of revenues	Security as % of IT
Banking and financial services	10.08	5.40	10
Professional services	5.89	4.50	10
Construction and engineering	8.17	1.70	10
Manufacturing	8.08	3.60	10
Health care	15.13	2.90	10

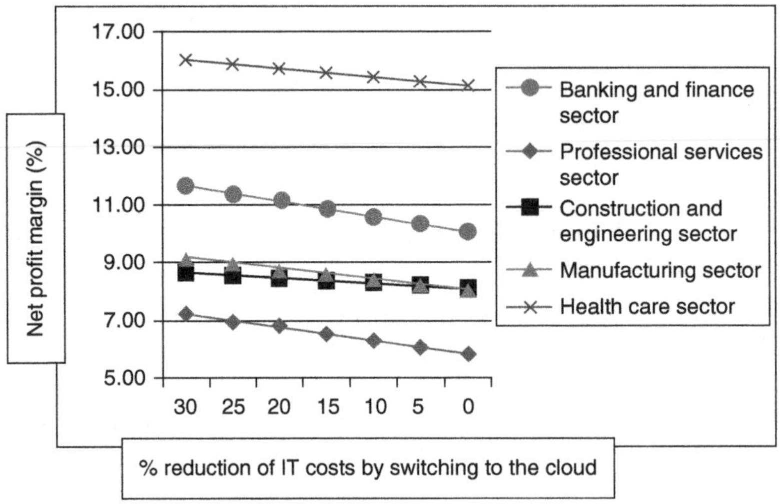

Figure 10.IV.7 *IT expenditures as % of revenues for US industries.*

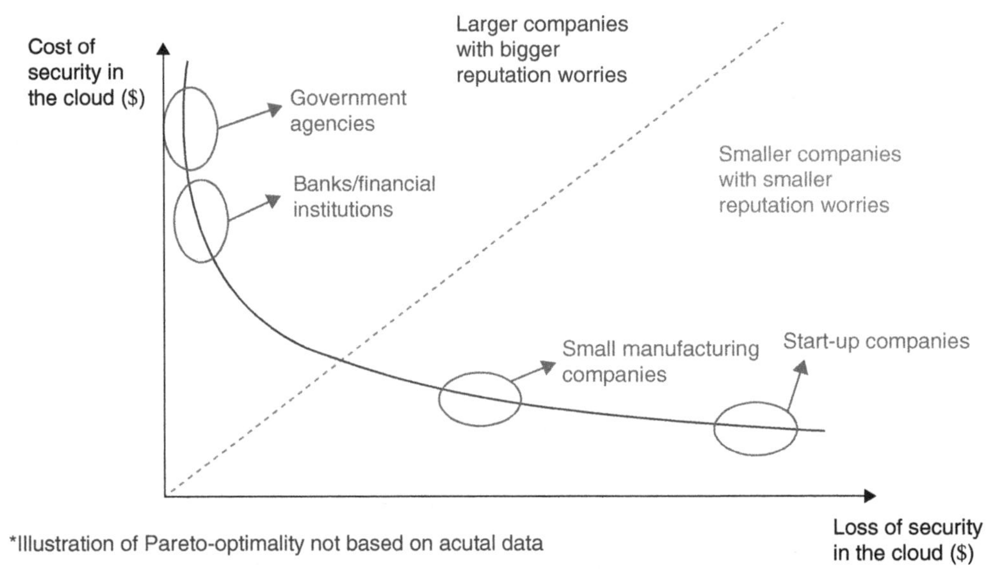

Figure 10.IV.8 *Illustration of a Pareto-optimal curve for different companies in the cloud.*

of return of ownership interest; as with EPS, a higher value is preferred. It is computed as the ratio of net income and average shareholder equity per period; thus any increase in net income, while keeping the average shareholder equity per period constant, results in an increase in the ROE. Figure 10.IV.8 could be modified to show a change in EPS or ROE instead of net profit. By relating the IT savings generated from switching to the cloud to improved EPS or ROE, companies may be persuaded to switch to the cloud in order to remain competitive. This will, of course, depend on the company's existing net profit, size, and concerns about reputation.

The generic curves in Figure 10.IV.8 do not account for the size of a company, its economic sector, or its relative concern with the four risk functions listed in Assumption 10. Recall that multiple objectives and Pareto-optimality are the subject of Chapter 5.

However, if we were to express these four risk factors in terms of multipliers, then we could apply those multipliers to these generic curves and get a better estimate of the actual relationship between net profit and reduction in IT costs for companies of different sizes and with different histories. For example, we could develop a reputation concern multiplier, which, as was already discussed, would generally be larger for larger companies having established histories and a large number of customers, as opposed to small start-ups that have not yet built credibility with their customers. Thus, a company more concerned with its reputation would require a larger increase in net profit gained by switching to the cloud than a company that is less concerned with reputation. Similarly, we could develop a loss of in-house IT staff risk multiplier, which again would generally be larger for larger companies. These multipliers could be broken down by industry type, company size, and history. They could then be applied to the generic curves shown in Figure 10.IV.8 to show the relative value of increases in net profit to different companies in different sectors and how these potential increases might convince a company to switch to the cloud – and therefore to accept a potentially lower level of security. While we currently do not have data to construct these multipliers, theoretically speaking such data could be obtained from case studies and surveys. With additional data about the cost of security provided in the cloud and more accurate data about financial losses from breached confidential and sensitive data in the cloud, we could start to create Pareto-optimal curves like the one in Figure 10.IV.8. This would enable us to say more accurately how much security various companies are willing to give up to reduce their costs.

When considering Figure 10.IV.8, note that individual companies face a variety of risks that are not related to cyber security; these risks in turn influence the level of investment the companies direct toward addressing cyber security risks. Figure 10.IV.8 takes into account the fact that the major challenge facing a start-up company is its ability to generate customers and revenue. In fact, using data from more than 2000 companies that received venture funding (generally at least $1 million) between 2004 and 2010, a Harvard researcher determined that three out of four start-ups fail. As a result, one can expect that most of the risks faced by start-up companies are associated with customer and revenue generation and that thus most of their investments will be directed toward those areas. *On the other hand, one can expect more significant investments in cyber security from larger companies that already have large profit margins and that are differentiated by their customers based on reputation.*

Our analysis demonstrates that the risk to a company of not switching to the cloud is not trivial – it could mean that a company could lose its competitive advantage as a result of an inability to reduce expenditures, to increase net profit, and to increase the overall value of the company. We argue that many companies opt to switch to the cloud based on this short-term view of the high economic risk of a failure to do so, without sufficiently considering, at the same time, the risk of the potential loss of confidential and sensitive information.

10.IV.5.4 Summary: Economic Perspectives of CCT Users

When assessing the risks of cloud computing, companies also must consider a myriad of other economic risks, including the consequences of not switching to the cloud and the impact that this would have on costs, price, market share, and profit. In reality, companies that are deciding whether to switch to the cloud must compare the risk of not switching to the risk of switching. We have provided a representative analysis to show that, in some cases, the potential economic losses of not switching to the cloud might be substantial. Our economic analysis is grounded on the basic business principle that companies are in the business of making money for shareholders. They do whatever they can to provide a higher return on investment for their shareholders, even if it means accepting a cloud-computing environment in which the level of security is not always clear to the company or to the provider. We do not have the data to fully determine the risk of switching to the cloud, but by conducting this type of analysis, we can simulate the type of thinking that occurs in companies when they are facing this decision. This enables us to assess how much security companies might be willing to give up in order to improve their net profit margin and remain competitive. Hence, this type of analysis can provide insight into what companies implicitly believe are the risks of cloud computing.

10.IV.6 CONCLUSIONS AND LESSONS LEARNED

Modeling CCT, its users, and would-be malicious intruders as interdependent and interconnected Complex SoS enabled us to apply theoretical, analytical, and methodological foundations from three fields: (i) advances in systems engineering in modeling Complex SoS by building on shared states and other essential entities among subsystems and use of the PSM; (ii) principles and guidelines in risk assessment, management, and communication; (iii) fault-tree analysis; and (iv) cyber security and CCT. Indeed, builders, users, and would-be malicious intruders of CCT Complex SoS are learning as they go through the process of discovering the emergent benefits as well as the potentially undesirable risks associated with the significant tangible and intangible losses and other consequences of switching to the cloud. This is true for government, public, and private organizations.

We posited the following premise: *Users of CCT (Complex SoS) are at higher risk than users of non-CCT systems for certain assumptions and types of cyber attacks*. The validity of this premise was demonstrated under similar non-CCT conditions through a quantitative fault-tree analysis on a virtual mini-CCT system. We further developed a fault tree of a virtual mini-CCT Complex SoS with the following components: (i) An intruder constitutes a major risk to cloud users. (ii) The top event of the fault tree for the mini-CCT system is defined as "attacker gains access to normal user's confidential data." (iii) Three potential events were considered to cause the top event, namely, a successful malicious intrusion:

1. Event E1: Attacker accesses data by real-time intrusion (concurrent sharing).
2. Event E2: Attacker accesses data by leaving malware (attacker-to-user sharing).
3. Event E3: Attacker accesses data by reading residual traces (user-to-attacker sharing).

The fault-tree model provided a demonstration of the premise that users of CCT Complex SoS are more at risk than users of non-CCT systems. The important finding from this demonstration Event C3 (establishment of unauthorized data channel) and D1 (encryption key is compromised) in minimum cut set 2 means that cut set 2 should have top priority for improvement. The concept of interdependent subsystems with shared states and the resulting interdependent failure probabilities can be used to analyze the reliability of other Complex SoS. Our results address what type of information is needed for future research and thus should what data be collected and analyzed.

In examining the economics of CCT Complex SoS, we found that detailed information is available regarding spending on security by cloud providers or regarding what cloud users are willing to pay for a certain level of security in the cloud. Importantly, companies make the decision to switch to the cloud by evaluating the potential profit increases that could result from reducing operating and capital costs, namely, in IT.

Finally, although we do not present a risk management analysis, we have highlighted the concept that CCT system designs are well aligned with providing risk management opportunities that non-CCT systems cannot provide within the same economic profile or with the same speed to implementation. It then becomes incumbent for those CCT providers who wish to maintain security at a level that can compete with that of non-CCT systems to implement solutions that exploit the inherent advantages of CCT system designs.

REFERENCES

Bhadauria, R. and Sanyal, S. (2012). Survey on security issues in cloud computing and associated mitigation techniques. *International Journal of Computer Applications* **47**(18): 47–66.

Bogdanor, J. (2014). Risk analysis of GPS-dependent communications critical infrastructure utilized by the U.S. Electric Power Grid. M.S. thesis. University of Virginia.

Bogdanor, J. and Haimes, Y.Y. (2018). Quantitative modeling of interdependent cyber-physical complex systems of systems. Submitted to *Reliability Engineering and System Safety*.

Butler Consultants. Free industry statistics: sorted by highest gross margin. http://research.financial-projections.com/IndustryStats-GrossMargin (accessed 26 January 2015).

Carta, A., Tocci, N., and Muscas, C. (2009). A PMU for the measurement of synchronized harmonic phasors in three-phase distribution networks. *IEEE Transactions on Instrumentation and Measurement* **58**(10): 3723–3730.

Chen, C. (2012). *Linear System Theory and Design*, 4e. New York: Oxford University Press.

CloudBus (2011). Cloud computing: a powerful tool for cyberattack!. http://cloudbus.blogspot.com/2011/05/cloud-computing-powerful-tool-for.html (accessed 26 January 2015).

Das, K., Hazra, J., Seetharam, D. et al. (2012). Real-time hybrid state estimation incorporating SCADA and PMU measurements. *2012 3rd IEEE PES International Conference and Exhibition on Innovative Smart Grid Technologies (ISGT Europe)* **1**(8): 14–17.

Haberlin, R. and Haimes, Y.Y. (2018). Regional infrastructures as complex systems of systems: a shared state model for regional resilience. *ASCE Journal of Infrastructure Systems* 24 (3) https://doi.org/10.1061/(ASCE)IS.1943-555X.0000421.

Haimes, Y.Y. (1991). Total risk management. *Risk Analysis* **11**(2): 169–171.

Haimes, Y.Y. (2008). Models for risk management of systems of systems. *International Journal of Systems of Systems Engineering* **1**(1/2): 222–236.

Haimes, Y.Y. (2012a). Modeling complex systems of systems with phantom system models. *Systems Engineering* **15**(3): 333–346.

Haimes, Y.Y. (2012b). Systems-based guiding principles for risk modeling, planning, assessment, management, and communication. *Risk Analysis* **32**(9): 1451–1467.

Haimes, Y.Y. (2016a). *Risk Modeling, Assessment, and Management*, 4e. Hoboken, NJ: Wiley.

Haimes, Y.Y. ed. (2016b). Fault trees. In: *Risk Modeling, Assessment, and Management*, 4e, 389–422. Hoboken, NJ: Wiley.

Haimes, Y.Y. (2018). Risk modeling of interdependent complex systems of systems: Theory and practice. *Risk Analysis* **38**(1): 84–98.

Haimes, Y.Y., Kaplan, S., and Lambert, J.H. (2002). Risk filtering, ranking, and management framework using hierarchical holographic modeling. *Risk Analysis* **22**: 383–397.

Hayati, P. (2012). BotCloud: an emerging platform for cyber-attacks. Cyber Warfare Intelligence (2 November 2012). http://0xicf.wordpress.com/2012/11/02/botcloud-an-emerging-platform-for-cyber-attacks/ (accessed 26 January 2015).

Homeland Security News Wire (2011). Hackers using cloud networks to launch powerful attacks. http://www.homelandsecuritynewswire.com/hackers-using-cloud-networks-launch-powerful-attacks (accessed 26 January 2015).

Humphreys, T.E., Ledvina, B.M., Psiaki, M.L. et al. (2009). Assessing the spoofing threat. *GPS World* **20**(1): 28–38.

Jiang, X., Zhang, J., Harding, B. et al. (2013). Spoofing GPS receiver clock offset of phasor measurement units. *IEEE Transactions on Power Systems* **28**(3): 3253–3262.

Kaplan, S. and Garrick, B.J. (1981). On the quantitative definition of risk. *Risk Analysis* **1**: 11–27.

Kaplan, S., Haimes, Y.Y., and Garrick, B. (2001). Fitting hierarchical holographic modeling into the theory of scenario structuring and a resulting refinement to the quantitative definition of risk. *Risk Analysis* **21**(5): 807–819.

Leung, L. (2010. Cloud customers report capital cost savings. Data Center Knowledge. http://www.datacenterknowledge.com/archives/2010/01/26/cloud-customers-report-capital-cost-savings/ (accessed 26 January 2010).

Leveson, N.G. (2012). *Engineering a Safer World: Systems Thinking Applied to Safety*. Cambridge MA: MIT Press.

Lewis, B. and Haimes Y.Y. (2018). Risk modeling of communications, navigation, and surveillance complex systems of systems for future aviation. Systems Engineering 2018: 1–10. doi: 10.1002/sys.21423.

McKay, D. (2011). A deep dive into hyperjacking. *Security Week*. http://www.securityweek.com/deep-dive-hyperjacking (accessed 3 February 2011).

MITRE (2015). Regional Resilience Continuum: Concepts, Constraints, and Proposed Approach, Version 1.0.

National Institute of Standards and Technologies (NIST) (2011). *The NIST Definition of Cloud Computing*, Special Publication, 800-145. Gaithersburg, MD: P. Mell and T. Grance. http://csrc.nist.gov/publications/-nistpubs/800-145/SP800-145.pdf (accessed 26 October 2011). http://www.nist.gov/itl/csd/cloud-102511.cfm.

Ponemon Institute (2010). Security of cloud computing users. http://www.ca.com/~/media/Files/IndustryResearch/security-cloud-computing-users_235659.pdf.

Ponemon Institute (2011). Security of cloud computing providers study. http://www.ca.com/~/media/Files/IndustryResearch/security-of-cloud-computing-providers-final-april-2011.pdf.

Ponemon Institute (2012). Encryption in the cloud. http://www.thales-esecurity.com/knowledge-base/gated-content? (accessed 26 January 2015).

PriceWaterhouseCoopers (2008). Why isn't IT spending creating more value? How to start a new cycle of value creation.

Ross, W., Gorod, A., and Ulieru, M. (2014). A socio-physical approach to systemic risk reduction in emergency response and preparedness. *IEEE Systems, Man and Cybernetics, Part A* **99**: 1–11.

Sarno, D. and Rodriguez, S. (2011). Hacker attacks show vulnerability of cloud computing. *Los Angeles Times* (17 June). http://articles.latimes.com/2011/jun/17/business/la-fi-cloud-security-20110617.

Schwartz, M. (2011). Security spending grabs greater share of IT budgets. *InformationWeek Security* (15 February).

Shepard, D., Humphreys, T., and Fansler, A. (2012). Evaluation of the vulnerability of phasor measurement units to GPS spoofing attacks. *Proceedings of the International Conference Critical Infrastructure Protection*, Washington, DC (19–21 March 2012).

The H Security (2011). Cracking WPA keys in the cloud. http://www.h-online.com/security/news/item/Cracking-WPA-keys-in-the-cloud-1168636.html (accessed 12 January 2011).

The SmartGrid (2013). An Introduction. www.energy.gov.

U.S. Federal Aviation Administration (2014). *Dynamic Required Navigational Performance Preliminary Concept of Operations*. Washington, DC: United States Government Printing Office.

U.S. Nuclear Regulatory Commission (1975). *Reactor Safety Study: An Assessment of Accident Risks in U.S. Commercial Nuclear Power Plants: Appendix I, WASH-1400 (NUREG 75/014)*. Washington, DC: U.S. Nuclear Regulatory Commission.

U.S. Nuclear Regulatory Commission (1981). *Fault Tree Handbook*, NUREG-81/0492. Washington, DC: U.S. Nuclear Regulatory Commission.

U.S. Water Sector Coordinating Council (2010). *Water Sector-Specific Plan: An Annex to the National Infrastructure Protection Plan*. U.S. Department of Homeland Security and U.S. Environmental Protection Agency.

Wesson, K. (2013). Spoofing and implications for telecom. http://rntfnd.org/wp-content/uploads/GPS-Spoofing-for-Telecom_UTex.pdf (accessed 18 September 2013).

Wright, T. (2012). How much can I save with cloud computing? Hazaa. http://hazaa.com.au/sharepoint-consultant/blog/how-much-can-i-save-with-cloud-computing/ (accessed 27 January 2012).

Yahoo Finance (2012). Industry index. http://biz.yahoo.com/p/sum_peed.html.

Zhong, Z., Chunchun, X., Billian, B. et al. (2005). Power system frequency monitoring network (FNET) implementation. *Power Systems, IEEE Transactions on* **20**(4): 1914–1921.

11

Global Supply Chain as Complex Systems of Systems

11.1 INTRODUCTION

The global supply chain complex systems of systems (Complex SoS) is the backbone of the U.S. and the economy of every country, and its success is paramount to the success of individual economies as well as of individual businesses with which we interact on a daily basis. Developing a strong understanding, modeling, and managing of the ever-evolving and growing global supply chain sector are paramount to keeping the functioning and safety of national as well as the global economies. A sample of the multiple facets of the supply chain includes economic, cultural, temporal, organizational, logistical, and the viability of a country's economic security. Furthermore, there are multiple organizations/entities that play a crucial role in the success of the supply chain; however, not all parties and players are committed to safeguarding its safety and integrity. In this chapter we will also address the adversarial entities of society that espouse to the disruption of the supply chain and the associated risk analysis to be considered to ensure its viability. The multifaceted nature of the supply chain sector requires its extensive modeling, and understanding of this safety-critical sector of every country's economy as interdependent and interconnected Complex SoS.

Denoting the "supply chain" as safety-critical Complex SoS must be understood from its myriad basic intertwined systems and subsystems throughout the U.S. and every country's economy, emanating from raw and scarce materials and heavy metals to abundant processed and manufactured commodities. The importance of the myriad resources and activities that contribute to the supply chain has been well recognized by the practitioners and the leadership of this important sector of the economy. For example, cyber security, which is a subset of the communication sector, is of paramount importance to the success of the supply chain sector, which is vulnerable to malevolent intrusions, leading to financial and other material losses, often compromising the entire country's and the global economy. Furthermore, the diverse consequences resulting from a cyber attack on the supply chain may adversely affect inventory, product quality, time of delivery, and multiple sequential chain effects of every country's economy.

The supply chain along with the Supervisory Control and Data Acquisition (SCADA) together constitute two entities essential to the integrity, sustainability,

Modeling and Managing Interdependent Complex Systems of Systems, First Edition. Yacov Y. Haimes.

and viability of the economy of every country in the world. Furthermore, SCADA comprises a critical part of a larger groupings of critical systems known as Industrial Control Systems (ICS), which include process control systems, or real-time systems. These technologies are used for controlling and monitoring the logistical processes, including physical operations on which the delivery of critical commodities is dependent. Failure of, or security breaches in these systems, could result in wide-reaching adverse impacts, not only to the community they serve, but also for the economy at large that depends on the quality, reliability, and timely delivery of goods and services. Several other organizations that depend on and control critical infrastructures include manufacturing, logistics and transportation, energy and utilities, telecommunication services, and agriculture and food distribution.

Of the many emergent technologies in support of the supply chain sector of the economy, consider the emergence of three-dimensional printer (3-D printer) and its on-site and timely capabilities which is becoming an important player. The usefulness of the 3-D printer has been extremely effective in the supply chain and global economy, especially with the ever-evolving technology and increasing consumer's on-time demand of delivery. However, this new technology, as with most innovations, can be used to serve the public needs, but also as an instrument of disruption and destruction via malevolent acts.

11.2 IMPORTANCE OF LEONTIEF INPUT–OUTPUT MODEL TO THE SUPPLY CHAIN

(Note: Chapter 8 is devoted in its entirety to Leontief input–output model and its extensions, the Inoperability Input–Output (I–O) Model (IIM).)

The advancement in information technology has markedly influenced and shaped the Supply Chain Complex SoS. It has also increased the interdependencies and interconnections of our critical infrastructures. Examples include the following critical sectors of the supply chain: manufacturing; agricultural products; produce, packaging, and storage facilities; multiple transportation modes; telecommunications and electrical power systems; gas and oil storage; banking and finance; groundwater and surface water supply systems; and emergency services and continuity of government. These supporting entities, which constitute vital infrastructure Complex SoS, are always subject to emergent forced changes (EFCs). The term EFCs connotes internal or external changes from within or outside the systems, or the entire Complex SoS, that may have correspondingly positive or negative effects. There is an emerging need to better understand and advance the art and science of modeling their complexity – the subject and theme of this entire book.

To illustrate this complexity, let us consider the electric power utility, which is commonly large-scale, hierarchical, and interconnected Complex SoS. This safety-critical utility is also vital to the effective operation of the Supply Chain Complex SoS. For example, at the national level in the United States, the electric grid consists of three interconnected networks: (i) the Eastern Interconnected System, covering the eastern two-thirds of the United States; (ii) Western Interconnected System, covering the southwest and areas west of the Rocky Mountains; and (iii) Texas Interconnected System, consisting mainly of Texas. At the network level, each network, as its name implies, is an interconnected system in itself, comprising numerous generators, distribution and control centers, transmission lines, converters, and other elements. Proper functioning of these interacting components is crucial to the continuous operations of the entire Electrical Power Complex SoS. In addition to its essential internal dependency, the distribution power infrastructure is externally dependent upon other infrastructure systems, notably fuel supply, telecommunications, water, and transportation. For example, its operation is heavily dependent upon voice and data communications. Data communications provide real-time updates (i.e. every few seconds) of electrical system status to SCADA systems in distribution and bulk electric control centers. Note that water is essential for cooling towers for the generation of electricity. Data communications, essential to effective operations of the supply chain, are also used for the remote control of devices in the field, such as circuit breakers, switches, transformer taps, and capacitors. Moreover, data communications allow generating units to follow the real-time signals from the control center that are necessary to balance electricity generation with consumer demand instantaneously. Although the power industry commonly owns and operates the majority of its communications equipment, a substantial portion is dependent upon local telephone carriers, long-distance carriers, satellites, cellular systems, paging systems, networking service providers, Internet service providers, and others.

Historically, many critical infrastructures around the world, whose support to the Supply Chain Complex SoS is vital, were physically and logically separate systems with little interdependence. This situation has been rapidly changing and close relationships among the supply chain and other sectors of the economy and infrastructures are now taking many forms. For example, telecommunications, power, transportation, and banking, which are critical drivers and facilitators of the effective functioning of the supply chain and the economy, are marked by immense complexity, characterized predominantly by strong intra- and inter-dependencies as well as multiple hierarchies. These interconnections take many forms, including flow of information, shared security, involvement of myriad levels of expertise, and functionality of labor of every discipline, skill and capability, and physical flows of commodities, among others. There is a need for a high-level, overarching modeling framework capable of describing the risks to critical infrastructures and industry sectors – focusing on risks arising from the interdependencies and interconnectedness (I-I) characterizing the Supply Chain Complex SoS, and on the vulnerability that exists within and among the subsystems of the supply chain. In assessing the vulnerability of the supply chain, it is important to analyze the interdependencies and interconnections that exist within and among the subsystems of the supply chain. Mitigating adverse consequences resulting from the I-I that characterize the supply chain can be achieved by modeling the way dysfunctionality and "inoperability" propagate throughout the supply chain by focusing on critical sectors of the economy. The inoperability whether caused by the lack of proper maintenance and effective management, or willful attacks, accidental events, or natural causes, can set off an intricate chain of cascading impacts on other Complex SoS. For example, similar to other critical infrastructures, e.g. water resources systems – surface and groundwater sources, water transport, treatment, distribution, storage, and wastewater collection and treatment – all have been heretofore designed, built, and operated without concerns to their vulnerability to EFCs and thus to their physical and operational integrity.

11.3 MODELING SUPPLY CHAIN INTERDEPENDENCIES VIA LEONTIEF INPUT–OUTPUT MODEL

The intricate I-I that characterize the Supply Chain Complex SoS can be best understood, and thus modeled, via the classic and most deployed economic models in the world – the Leontief input–output model. Following are two major objectives in incorporating the Leontief model in this chapter with the ability of the Leontief Inoperability I-O Model (IIM) (Haimes and Jiang, 2001; Haimes, 2016), (also consult Chapter 8) to (i) quantify the impacts of a disastrous failure of one major supply chain sector on a significant number of sectors of the Supply Chain Complex SoS; (ii) quantitatively evaluate such major impacts; and (iii) assess the efficacious contributions of remedial actions on the restoration of the affected supply chain sectors.

Wassily Leontief received the 1973 Nobel Prize in Economics for developing what came to be known as the Leontief input–output (I-O) model of the economy (Leontief, 1951a, b, 1986). The economy (and thus the model) consists of a number of subsystems, or individual economic sectors, or industries and is a framework for studying the equilibrium behavior of an economy. The model enables understanding and evaluating the interconnectedness among the various sectors of an economy and forecasting the effect on one segment of change on another. Leontief's I–O model describes the equilibrium behavior of both regional and national economies.

The brief outline of the Leontief I–O model below is based on Intriligator (1971) and Haimes (1977, 2005a, b, 2016). It provides a simplified version of Leontief's (1951a) I–O model to trace the economic impacts of resources and products within an economy. The economy (system) is assumed to consist of a group of n interacting sectors or industries, where each "industry" produces one product (commodity). A given industry requires labor, input from the outside, and also goods from interacting industries. Each industry must produce enough goods to meet both interacting demands (from other industries in the group) and external demands (e.g. foreign trade and industries outside the group). A static (equilibrium-competitive) economy, with constant coefficients for a fixed unit of time (one year), is assumed. Define the following notation:

x_j is the output (for the total economy) of jth goods, $j = 1, 2, \ldots, n$

r_k is the input (for the total economy) of kth resource, $k = 1, 2, \ldots, m$

x_{ij} is the amount of the ith goods used in the production of the jth goods

r_{kj} is the amount of the kth resource input used in the production of the jth goods

Leontief's model assumes that the inputs of both goods and resources required to produce any commodity are proportional to the output of that commodity:

$$x_{kj} = a_{kj} x_j, \quad j,k, = 1,2,\ldots,n \tag{11.1}$$

$$r_{ij} = b_{ij} x_j, \quad k = 1,2,\ldots,m, \quad j = 1,2,\ldots,n \tag{11.2}$$

Furthermore, the output of any commodity is used either as input for the production of other commodities or as final demands, c_k. The balance Eq. (11.1) is a key to the subsequent development of the Leontief-based Eq. (11.3):

$$x_k = \sum x_{kj} + c_k, \quad k = 1,2,\ldots,n \tag{11.3}$$

Combining Eqs. (11.1) and (11.3) yields the Leontief equation:

$$x_k = \sum_j a_{kj} x_j + c_k, \quad k = 1,2,\ldots,n \tag{11.4}$$

Similarly, the proportionality assumption applies to the resources:

$$r_{ij} = b_{ij} x_j \tag{11.5}$$

$$\sum_i r_{ij} = \sum_i b_{ij} x_j \tag{11.6}$$

Since the demand for the ith resource cannot exceed its supply, then

$$\sum_j b_{ij} x_j \le r_i, \quad r_i \ge 0, \quad i = 1,2,\ldots,m \tag{11.7}$$

The above basic model of the economy is written in a compact matrix notation in Eq. (11.8):

$$\mathbf{x} = \mathbf{A}\mathbf{x} + \mathbf{c} \Leftrightarrow \left\{ x_i = \sum_j a_{ij} x_j + c_i \right\} \forall i \tag{11.8}$$

11.4 INOPERABILITY INPUT–OUTPUT MODEL (IIM)

Grounded on Leontief's model introduced in Chapter 8, and in the previous section, a first-generation Inoperability I–O Model (IIM) of interconnected systems was developed by Haimes and Jiang (2001). Subsequently multiple Ph.D. dissertations and numerous applications extended and deployed the IIM (Haimes, 2016). The IIM enables the modeling of the interdependent and interconnected sectors of the economy (e.g. supply chain sectors) and quantifies the adverse consequences on all affected supply chain sectors. One primary purpose of the IIM is to improve understanding of the impact of complexity on the continued and sustained operability of the Supply Chain Complex SoS under adverse conditions. Other related works on infrastructure interdependencies and risks of EFCs are presented in Haimes (2002, 2016), Haimes and Horowitz (2004), Santos and Haimes (2004), Crowther and Haimes (2005), and Jiang and Haimes (2004).

Note that the "supply" and "demand" concepts in the Leontief economy model assume a different interpretation and have been inverted to some extent in the IIM. Although the mathematical construct of the two models is similar, the interpretation of the model parameters is fundamentally different. *Dollars* are the units used in the Leontief I–O model for the economy. Furthermore, the inoperability model uses units of *risk of inoperability* (Haimes et al., 2005b) defined above as a measure of the probability (likelihood) and degree (percentage) of the inoperability (dysfunctionality) of one supply chain sector, or of the entire Supply Chain Complex SoS (depending on the context of the sentence). An inoperability of 1 would mean that a supply chain sector or the entire Supply Chain Complex SoS (depending on the context) is totally out of commission. As stated earlier, inoperability may take various forms according to the nature of the system. When the model is applied to study any infrastructure system, one of the very first tasks is to define the specific inoperability and the associated risks.

The inoperability model addresses the equilibrium state of the system (i.e. a single supply chain sector, or the entire Supply Chain Complex SoS) in the event of an attack, or EFCs, provided that the I–O interdependency matrix is known. The *input* to the system is an initial perturbation triggered by EFCs, e.g. an attack of terrorism, an accidental event, or a natural disaster. The *outputs* of the system are the resulting risks of inoperability of different infrastructures due to their connections to one another. The output can be triggered by one or multiple failures due to their inherent complexity or to external perturbations (e.g. natural hazards, accidents, or acts of terrorism).

In the IIM the output is the infrastructure's *risk of inoperability*, or simply *inoperability* that can be triggered by one or multiple failures due to complexity, accidents, or acts of terror. *Inoperability* is defined as *the inability of the system to perform its intended natural or engineered functions*. In the model, the term *inoperability* can denote the level of the system's dysfunction, expressed as a percentage of the system's "as-planned" level of operation. Alternatively, inoperability can be interpreted as a degradation of a system's capacity to deliver its intended output (or supply). Although inoperability in its current scope applies to physical and economic losses, it can be extended to assess impacts due to information failure. In addition, other factors for assessing failures, such as loss of lives, environmental quality, and others, can supplement the economic factors used in the context of inoperability.

Inoperability is assumed to be a continuous variable evaluated between 0 and 1, with 0 corresponding to a flawlessly operable system's state and 1 corresponding to the system being completely inoperable. Inoperability may take different forms, depending upon the nature of the problem and the type of the system. When the production level is of major concern, inoperability may well be defined as the unrealized production (i.e. the actual production level subtracted from the desired production level). For instance, if the system under consideration is a power plant, then the inoperability may be defined as the ratio of the actual amount of power produced (in appropriate units) to the desired amount. Furthermore, the notion of inoperability also attempts to capture the quality of a system's function. Assuming that quality can be measured numerically, a defective system whose performance is of degenerate quality is considered partially operable and thus has inoperability greater than zero. For instance, a television set that has a picture but no sound is only partially operable and thus has inoperability greater than zero. By the same token, a water supply system producing slightly contaminated water is also considered partially operable and thus has inoperability greater than zero. Finally, inoperability of a system is not necessarily a continuous variable. Under certain circumstances, it may take discrete values such as binary values. Here, we focus our discussion on the continuous case.

Risk of inoperability can also be viewed as an extension of the concept of *unreliability*. Unreliability is the conditional probability that a system will fail during a specified period of time t, given that it operated perfectly at $t = 0$. In fact, the system may not fail completely during this time span; it may fail partially with certain probability. For instance, during this period of time, it may fail 100% with probability 0.1, it may lose 50% of its functionality with probability 0.4, or it may lose 10% of its functionality with probability 0.8, and so forth (provided that the functionality is quantifiable). Thus a natural extension of the notion of unreliability is to average out all these possibilities by considering both the failure level and the likelihood. In so doing, we end up with a quantity that represents the expected value of the failure level during a certain period of time. In other words, if the expected value metric is adopted in the definition of risk, then the risk of inoperability can be viewed as the expected inoperability. A conditional expected value metric, to supplement the expected value metric, is introduced in Haimes (2016). Hence, for the sake of brevity, in the following discussion, we sometimes use "inoperability" in lieu of "risk of inoperability."

The inoperability of an infrastructure may be manifested in several dimensions, e.g. geographical, functional, temporal, or political. On the one hand, these and other perspectives markedly influence the values assigned to the probability (coefficient) of inoperability in the model. Furthermore, each may justify the construction of a different inoperability model addressing a specific dimension. An example would be inoperability that spans regional or statewide, short-term or long-term, or one function failure or multiple failures of an infrastructure. In such cases, each model will require specific and different probabilities of inoperability. In addition, one such inoperability model might evaluate, and measure "dysfunctionality" in monetary terms, the risk of inoperability or damage to property, production, service, or injury under extreme natural and accidental conditions, or due to acts of terrorism.

In the following discussion, we assume that each infrastructure system performs a uniquely defined function, that is, no two systems perform the same function. In other words, in this preliminary model we do not consider the issue of redundancy. The systems that we consider here fall into the category of "unparallel" systems.

Let x_j, $j = 1, 2, \ldots, n$ be the overall risk of inoperability of the jth intra- and interconnected infrastructure that can be triggered by one or multiple failures caused by accidents or acts of terrorism.

Let x_{kj} be the degree of inoperability triggered by one or multiple failures that the jth infrastructure can contribute to the kth infrastructure due to their complex intra- and interconnectedness.

Let a_{kj} be the probability of inoperability that the jth infrastructure contributes to the kth infrastructure. In our model, a_{kj} describes the degree of dependence of the kth infrastructure on the jth infrastructure. For example, if $a_{kj} = 1$, then this means a complete failure of the jth infrastructure will lead to a complete failure of the kth infrastructure. A value of $a_{kj} = 0$, on the other hand, indicates that the failure of the jth infrastructure has no effect on kth infrastructure.

Let c_k be the natural or man-made perturbation into the kth critical infrastructure.

At this stage, the proportionality assumption that underpins Leontief's economy model is assumed to hold for the inoperability I–O risk model as well; then we have:

$$x_{kj} = a_{kj} x_j \quad j,k, = 1,2,\ldots,n \tag{11.9a}$$

The following balance equation is a key to the subsequent development of the linear model:

$$x_k = \sum_j x_{kj} + c_k, \quad j,k = 1,2,\ldots,n \tag{11.9b}$$

Combining the balance equation with the proportionality equation yields the inoperability equation for the infrastructure model:

$$x_k = \sum_j a_{kj} x_j + c_k, \quad j,k = 1,2,\ldots,n \tag{11.10}$$

The above equation can be written in matrix notation as follows:

$$\mathbf{x} = \mathbf{A}\mathbf{x} + \mathbf{c} \tag{11.11}$$

where $\mathbf{x} = [\,x_1, x_2, \ldots, x_n\,]^T$, $\mathbf{c} = [\,c_1, c_2, \ldots, c_n]^T$, $\mathbf{r} = [r_1, r_2, \ldots, r_m]^T$, $[.]^T$ = a column vector, and $A = [a_{kj}]$ $n \times n$ matrix.

Defining $I = n \times n$ identity matrix and assuming that (I–A) is nonsingular, the vector of inoperability $\mathbf{x}$ in Eq. (11.11) can be solved using the following matrix operation:

$$\mathbf{x} = (\mathrm{I} - \mathrm{A})^{-1}\mathbf{c} \tag{11.12}$$

Determining the values of the **A**-matrix during the modeling process is a very challenging undertaking, *and* extensive data collection; and data mining may be required to complete this step. The following are general guiding principles for determining the **A**-matrix:

- Explore the potential use of publicly available I–O Tables to enable understanding of the transactions among various sectors in the economy. Prior to conducting actual field surveys and interviews, these I–O Tables can provide valuable insights into the interdependencies among various infrastructures.
- Define the level of resolution and the boundary conditions of each infrastructure, because a system may be analyzed at different levels of resolution. Note, however, that the level of resolution adopted in the analysis must be harmonious with the accuracy of the data and the analytical tractability (including the determination of the I–O relationships). The realism that exists at high granularity should not be sacrificed in the process of aggregation.
- Identify physical connections among the infrastructures. In general, if there are no physical connections between infrastructures i and j, then $a_{ij} = a_{ji} = 0$. Physical boundary conditions are very critical in identifying the physical connections among different infrastructures.
- If there are any deterministic correlations among any infrastructures, then these relationships should be singled out first. For instance, if the failure of infrastructure i will definitely lead to failure of infrastructure j, then $a_{ji} = 1$. By the same token, if the failure of infrastructure i will definitely lead to failure of one of the two subsystems

of infrastructure j, which performs 50% of that infrastructure's functions, then $a_{ji} = 0.5$.

- If the correlation between two infrastructures (e.g. infrastructures i and j) is of a stochastic nature, then all conceivable scenarios must be analyzed and a statistical average has to be taken to obtain a_{ij} and a_{ji}. For example, if the failure of infrastructure i leads with probability 0.3 to complete failure of infrastructure j and with probability 0.7 leads infrastructure j to be 50% inoperable, then $a_{ji} = (0.3)(1) + (0.7)(0.5) = 0.65$. If the real data are not sufficient, a simulation may be helpful in order to obtain data for the probability distributions.

In his basic I-O model, Leontief considered an economy that produces n goods as output and uses m primary resources as input. For the IIM we consider a system consisting of n critical complex intra- and interconnected infrastructures. Although the equations are similar, there is a major difference in the interpretation of the variables. In other words, the basic Leontief Eqs. (11.1)–(11.8) are similar to the IIM Eqs. (11.9)–(11.12) however, they connote different meanings (also see Chapter 8).

11.5 THE CENTRALITY OF THE SUPPLY CHAIN IN THE GLOBAL ECONOMY

The interdependencies and interconnections within and among the myriad commodities that make up the supply chain are one of the most critical features that characterize the Supply Chain Complex SoS and the numerous sources of risk thereto. Moreover, almost all commodities that constitute the supply chain follow several alterations and transformations to meet specific requirements for different final demands. This endless process of product makeover is also fraught with numerous sources of risk and uncertainty affecting the ultimate quality of the final product in the supply chain. In his classic book *Managing Quality*, Professor David Garvin (1988) offers the following eight dimensions to quality; here we apply them to the supply chain: (i) performance, (ii) features, (iii) reliability, (iv) conformance, (v) durability, (vi) serviceability, (vii) aesthetics, and (viii) perceived quality. The time frame that is central to the performance of every Complex SoS plays a critical role and affects all of the above eight dimensions of quality. Innumerable as well as immeasurable sources of risk affect countless *time-sensitive* supply chain commodities, ranging from critical health and safety to industrial production and invariably commodities to each and every sector of the economy. The I-I within and among the Supply Chain Complex SoS constitute an Achilles' heel for many commodities, on which numerous other commodities depend and on a strict time frame. Indeed, these inherent attributes of the Supply Chain Complex SoS make it one of the most safety-critical sectors of the economy for every country in the world. Furthermore, the movement and management of supply chain commodities over vast distances, under different climate and weather conditions, at different time frames, all can have measurable effects on most of the eight dimensions of quality highlighted above by David Garvin. The almost immeasurable consequences on the fate and quality of the supply chain resulting from the above adverse effects can be disastrous. Here again we revert to Chapter 5, which addresses the multiple goals and objectives associated with each system that constitute interdependent and interconnected Complex SoS. Indeed, the supply chain epitomizes the inherent tradeoffs associated with almost every step in the life cycle of every commodity within it and every attribute of Garvin's eight dimensions of quality. Clearly these tradeoffs in terms of time of delivery, cost of the selected mode of transportation, and the quality assurance associated with the above, among other attributes, all must be considered in the ultimate multiobjective decision-making process and in terms of tradeoff analysis among cost, benefits, quality, and timely delivery, among many other attributes and considerations (Haimes et al., 1990).

An important player in quality management of the supply chain is the Council of Supply Chain Management Professionals (CSCMP). The following is its mission statement:

> To lead the supply chain profession by connecting, developing, and educating the world's logistics and supply chain management professionals throughout their careers.

The goal of CSCMP is to (i) provide opportunities for supply chain professionals to communicate in order to develop and improve their supply chain management skills, (ii) identify and conduct research that adds to the knowledge-base of supply chain theory and practice, and (iii) create awareness of the significance of supply chain to business and to the economy.

Chapter 12 is devoted in its entirety to understanding and managing the organizational dimension of Complex SoS. Indeed, there exist multiple organizational entities and associated hierarchies that ultimately manage, influence and impact, and determine the *essence* and fundamental nature of the supply chain. In many ways, such global management of the supply chain is uncountable, and its unique features are also its strengths as well as its weaknesses. The reason stems from the fact that core drivers that propel the supply chain are numerous, widely distributed geographically, supplied by uncounted number of commodities and resources, and span multiple time horizons, among many other attributes.

In Chapter 3, which is devoted to Hierarchical Holographic Modeling (HHM) (Haimes, 1981), we maintained that systems – small and large – must and can be modeled and understood from their multifarious functionality, organizational setups, and their web of I-I with other systems. In this chapter we encapsulate the multifarious intricacy of the supply chain as Complex SoS within the HHM philosophy and methodology. Consider the following sample of subsystems and perspectives of the supply chain, where all of which are interdependent and interconnected in numerous ways: (i) raw materials, (ii) finished products, (iii) time frame and just-in-time, (iv) quality, (v) cost, (vi) information technology, (vii) business relationships and partnerships, (viii) services, (ix) middleman enterprises, (x) myriad groups of customers, (xi) local and international rules and regulations, (xii) malevolent acts originating from inside or outside the supply chain, (xiii) management at all levels, and (xiv) multiple modes of transportation, among others.

11.6 CENTRALITY OF ORGANIZATIONAL INFRASTRUCTURE TO EFFECTIVE PERFORMANCE OF THE SUPPLY CHAIN COMPLEX SOS

In this section we address the role of organizational infrastructure to effective operations and delivery of the Supply Chain Complex SoS. Although the numerous organizational entities that serve billions of individuals globally and myriad institutions and other clienteles cannot be boxed into one unified structure or operational process, common fundamentals can be identified. *Note that every commodity in the Supply Chain Complex SoS commonly has an origin, whether raw materials or products from another supply chain. Thus the supply chain as Complex SoS is essentially composed of a hierarchy of myriad supply chain of classes of commodities, each with its customized organizational entity. And all together in this hierarchy of "subsupply chains" constitute one supply chain for every specific category of commodities.* This hierarchical organizational structure serves and enables the effective operations and delivery of the supply chain as Complex SoS. Note that as the organizational infrastructure expands and becomes more effective, so would the sources of risk to failures increase, unless the latter are identified, assessed, managed, and communicated explicitly and methodically. Indeed, there exist multiple hierarchies in the global Supply Chain Complex SoS. Chapters 2 and 4 of this book address decomposition and hierarchical-multilevel coordination of Complex SoS. An important source of risk to the global supply chain stems from damaged, defected, soiled, wrong, or maliciously tampered commodities on which other supply chain commodities depend.

Consider the following multiple and diverse, albeit augmented perspectives on organizational behavior that is perfectly applicable and appropriate for studying the complexity of managing the multinational supply chain. We are including a sample of perspectives from leading scholars and other experts in organizational behavior. For example, Drucker (2001) (cited in a different context Chapter 12) argues that each social sector institution exists to make difference in the lives of individuals and in society. Indeed, the supply chain affects individuals, families, communities, and society at all levels. Collins (2011) maintains that an organizational mission goes to the heart of the fundamental tension in any great institution due to the dynamic interplay between continuity and change. Indeed, it is almost impossible to identify another sector of the economy within the Supply Chain Complex SoS that is not continuously undergoing changes and is inescapably changing by such drivers as technology, competitiveness, consumers' demands and preferences, and the ever-evolving cultural and societal demands and preferences. Indeed, the mission of the organizations and the leadership involved in the Supply Chain Complex SoS is multidimensional and multifarious, involving diverse industries, organizations, suppliers, customers, and economies, among others. The globalization of the supply chain has also evolved as one of the major sectors of the economy that transcends boundaries and socioeconomic levels. In many ways, the supply chain as Complex SoS, and its international reach to all diverse countries and societies, has brought even more connectedness and acute interdependencies within and among countries. At the same time, these interdependencies and interconnections have also made the world smaller and more integrated, crossing geographical and political boundaries, with all the sociopolitical ramifications therefrom. The ramifications of the emergent globalization caused by the supply chain have yet to be assessed holistically.

11.7 HIERARCHICAL-MULTILEVEL COORDINATION AMONG SUBSUPPLY CHAIN COMMODITIES

The Supply Chain Complex SoS can be effectively viewed and modeled in terms of hierarchical systems or structures, which in their essence are families of supply chain subsystems arranged in a hierarchical fashion. Each subsystem is commonly concerned with a specific class of commodities of the overall Supply Chain Complex SoS, and each occupies a specific level in the overall hierarchy. Successful operations and management of any class of commodities, at any given level of the hierarchy, is directly and explicitly influenced by other commodities from higher-level subsystems. Policies and decisions of higher-level subsystems are commonly binding on the lower-level subsystems and reflect a priority of importance in the actions and goals of the higher-level subsystems. Note that due to the I-I that characterize the Supply Chain Complex SoS, fulfillment of the higher-level needs and objectives depend on the actions and quality products of the lower-level subsystems. We focus here on three types of hierarchical-multilevel structures for dealing with Supply Chain Complex SoS.

Consider the critical dependency of the supply chain's effective performance on transportation and on reliable communications. They can be categorized from at least five levels: (i) strategic, (ii) operational, (iii) technical, (iv) temporal, and (v) quality. To each of the above can be also associated at least the following four levels of I-I: (i) distributions, (ii) retails, (iii) warehouses, and (iv) demand and supply. At the temporal level, which will be further discussed, we can identify the following time frames: (i) online delivery, (ii) forecasting, (iii) quality control (and damage avoidance) along the entire journey, (iv) effective coordination between demand and supply, cost-effective inventory control, and (v) reliable forecasting of the demand of each of the myriad categories of commodities and the associated accountable supply thereof. The effectiveness of the above sample of critical sectors of the economy depends on a reliable, uninterrupted performance and timely delivery of goods and services of the Supply Chain Complex SoS, including safe and timely international, national, and regional distribution of innumerable categories of commodities.

In Section 11.11 we will present an analytical method for an optimal management of the supply–demand inventory control using dynamic programming.

11.7.1 Multistrata Hierarchy

The multistrata hierarchy is highly effective for modeling the Supply Chain Complex SoS. (More on multistrata hierarchy is discussed in Chapter 2.) The tradeoffs are between (i) simplicity, a requisite for understanding the I-I that characterize the intricate need of streamlining the organizational and logistical coordination of the numerous groups and classes of the Supply Chain Complex SoS, and consequently in the analysis, decision making, and management strategy applied, and (ii) multifarious modes of transportation associated with the scheduling and delivery of critical commodities, while accounting for the unpredictable, intricate behavior of the Supply Chain Complex SoS. These descriptive levels of a hierarchy are commonly referred to as strata. (A more detailed discussion on hierarchical modeling is presented in Chapters 2 and 4.) Lower strata involve more detailed and specialized descriptions of the system than the higher strata. Each stratum has its own concepts and principles and may deal with different aspects of Supply Chain Complex SoS.

11.7.2 Multilayer Decision-Making Hierarchy

A multilayer organizational hierarchy is one of the most critical challenges used in streamlining complex decision-making policies and protocols associated with the Supply Chain Complex SoS. Two trivial but important conditions of almost all real-life decision situations are to act without delay and therefore avoid decisions by default, and to take time for reflection and planning to better understand the situation. Layers of decision making are essentially levels of management tradeoffs. Parameters in lower-layer problems are fixed by "policies" of higher-layer systems, and a "solution" or understanding of the original problem is obtained when all subproblems and conflicts in the sequence are understood or "resolved." A complex decision-making hierarchy naturally emerges in reference to several essential phases of the decision-making process, e.g. gathering information, reducing uncertainties, and selecting a preferred course of action.

11.7.3 Multiechelon Hierarchy

The third kind of hierarchy, a multiechelon hierarchy, arises when Supply Chain Complex SoS are explicitly viewed as consisting of many interacting subsystems. Such hierarchies must deal with the mutual relationships among the various commodity subsystems comprising the Supply Chain Complex SoS. Each subsystem is considered as "goal seeking," that is, satisfying specific objectives and aspiration levels. In this multiechelon hierarchy, conflicts within subsystems on one echelon are resolved by higher echelon subsystems. Coordination and the resolution of conflicts are commonly accomplished by intervention, which can be achieved by including certain policy decisions in the subsystem's objectives to be considered and manipulated by higher echelon subsystems. For example, intervention may be of the following kinds: (i) goal intervention that affects goal-related factors, (ii) information gathering that affects outcome expectations, and (iii) constraint intervention that affects available alternative actions.

All three groups of hierarchies among others may emerge simultaneously in dealing with Supply Chain Complex SoS. Different hierarchies could serve different purposes and may be imbedded within one another: the multistrata hierarchies are for modeling purposes, the multilayer hierarchies are for resolving the complexities of the complex supply chain coordination and decision-making process, and the multiechelon hierarchies are for organizational purposes.

11.8 THE ROLE OF SHARED STATES IN RISK MODELING OF SUPPLY CHAIN COMPLEX SYSTEMS OF SYSTEMS

Risk analysis applied to a single system must be fundamentally extended when applied to the Supply Chain Complex SoS, which constitute multiple systems with emergent behaviors. As noted throughout this book, the complexity of these systems results primarily from the I-I within and among the systems and subsystems of SoS. An effective risk analysis of the Supply Chain Complex SoS requires a clear understanding of their configurations to enable the identification of the many critical failure modes, whether in the production, transportation, quality of the commodities, the time frame, and myriad other sources of risk. This would enable the development and deployment of effective risk mitigation and other management strategies. Risk models of systems, which do not consider the inherent interdependencies among other systems, are likely to be inferior to those models that do. Scientific literature has many definitions of state variables in modeling, because they play a significant role in estimating the future behavior of the systems for any given inputs. An extensive discussion on the centrality of state-space theory was introduced in Chapter 1 and beyond. Recall, Chen's (2012) conceptual definition of state variables posits that the state $x(t_o)$ of a system at time t_o is the information at time t_o that, together with the input $u(t)$, for $t \geq t_o$, determines uniquely the output $y(t)$ for all $t \geq t_o$. Similarly, Nise (2015) emphasizes that state variables represent the smallest set of linearly independent system variables such that the values of the members of the set at time t_o along with known forcing functions completely determine the value of all system variables for all $t \geq t_o$. Thus, the behavior of the states of the system, as a function of time and other inputs, enables modelers and risk analysts to determine, under certain conditions, its future behavior for any given *inputs* or *initiating events*.

From the perspective of the *reliability of the supply chain*, shared states and other essential entities within and among the subsystems represent the essence of the I-I that characterize SoS: *Initiating events affecting one or more of the interdependent and interconnected subsystems (commodities) would necessarily increase the probability of failure of the entire Supply Chain Complex SoS.* This fundamental fact could be viewed in the parlance of fault-tree analysis (NRC, 1981) as converting subsystems from being "marginally connected," or in "parallel," to becoming "directly connected," or in "series." Thus, from that perspective, most supply chain commodities (subsystems) that share states and other essential entities with other commodities *are most likely to* be affected by adverse initiating events. An initiating, e.g. a major snowstorm, may adversely affect only the local supply chain, but with no measurable effects on other regions.

Uncertainty analysis becomes even more imperative in risk analysis of the Supply Chain Complex SoS. Two major sources of uncertainty in modeling affect risk analysis (Paté-Cornell, 1990, 1996; Apostolakis, 1999): *Knowledge (epistemic) uncertainty* may manifest itself in faulty selection of appropriate model topology (structure) and model parameters, due to ignorance of the system or Complex SoS under consideration. *Variability (aleatory) uncertainty* applies to all relevant and important events and must not be overlooked. Sources of uncertainty dominate most decision-making processes, especially Supply Chain Complex SoS, and are the Achilles' heel for all deterministic and probabilistic models. Sources of uncertainty and lack of understanding of the complexity associated with one subsystem of the Supply Chain Complex SoS would likely adversely affect multiple regions and sectors of the economy.

The intricacy characterizing the Supply Chain Complex SoS requires a reevaluation of the way we model, assess, manage, communicate, and analyze the risk thereto. Several fundamentals must guide our reevaluation of current practices of risk analysis to the supply chain. We may start with our need of understanding and evaluating the I-I within and among the myriad systems and subsystems that constitute the supply chain. To do so we build on Chapter 3, which is devoted in its entirety to HHM. We maintained that all systems – small and large – must and can be modeled and understood from their multifarious functionality, organizational setups, and their web of I-I with other systems. In this chapter on the supply chain, we encapsulate the diversified intricacy of the supply chain as Complex SoS within the HHM philosophy and methodology. Consider the following sample of subsystems and perspectives of the supply chain, all of which are interdependent and interconnected in numerous ways: (i) raw materials, (ii) finished products, (iii) time frame and just-in-time, (iv) quality, (v) cost, (vi) information technology, (vii) business relationships and partnerships, (viii) services, (ix) middle-man enterprises, (x) groups of customers, (xi) local and international rules and regulations, (xii) malevolent acts originating from inside or outside the supply chain, (xiii) management at all levels, and (xiii) multiple modes of transportation, among others.

Recall that the I-I characterizing Complex SoS are manifested through shared/common states and other *essential entities* within and among the systems and subsystems that make up Complex SoS, and that the term *essential entities* connotes (i) decision makers and stakeholders, (ii) resources, (iii) organizational setups, (iv) emergent behavior, (v) cultural identity, (vi) time frame, and (vii) being cognizant of and responsive to the emergent nature of Complex SoS. To streamline our presentation, we reintroduce the *Evolving Base* from Section 9.II.2.1 which is represented by the following dynamic shifting rules and realities for each subsystem and for the entire Complex SoS: (i) goals and objectives; (ii) stakeholders, decision makers, and interest groups; (iii) organizational, political, and budgetary baselines; (iv) reorganization and reallocation of key personnel; (v) emergent technology and its deployment; and (vi) requirements, specifications, delivery, users, and clients.

Recall that Supply Chain Complex SoS are commonly composed of innumerable interdependent and interconnected subsystems, each of which can be characterized by a hierarchy of shared or interacting components and each with multiple functions, operations, databases, costs, and stakeholders. Indeed, all systems, natural and the constructed environment, are commonly subjected to, and thus affected by, multiple adverse initiating events that could be initiated from outside or from within one or more subsystems that constitute Complex SoS.

The fundamental differences between the impacts of adverse initiating events on a single system as opposed to the impacts on Complex SoS (e.g. the supply chain) stem from their intrinsic I-I characteristics. As noted in earlier chapters, Complex SoS are commonly composed of innumerable subsystems, which in their essence constitute *Complex SoS*. Furthermore, almost every living entity, all infrastructures, and both the natural and constructed environment are in their essence Complex SoS, because they are composed of multiple interdependent and interconnected systems that *share* (*have in common*) most, if not all, of the following building blocks (of models and reality): (i) states that constitute the fundamental entities of all systems (discussion on state-space theory can be found throughout the book), (ii) decision makers and stakeholders, (iii) resources, (iv) organizational setups, (v) emergent behavior and evolution, (vi) often competing and conflicting goals and objectives, and (vii) historical and cultural identity and time frame, among others.

11.9 THE ROLE OF ORGANIZATIONAL MANAGEMENT OF THE SUPPLY CHAIN COMPLEX SYSTEMS OF SYSTEMS

The production, supply, proper movement and allocation of efficient transportation modes, effective timely delivery of planned-for commodities – especially time-critical supplies – and an acceptable cost and quality, all are essential for effective Supply Chain Complex SoS. The above "chain" of critical tasks cannot necessarily be accomplished without a well-structured and well-operated organizational setup supported by effective cyber–physical infrastructures. Chapter 12 is devoted in its entirety to understanding and managing the organizational dimension of Complex SoS. The reader is strongly encouraged to consult that chapter for a more comprehensive appreciation of the pivotal role of management in the Supply Chain Complex SoS.

Today's cyber–physical systems (CPS) constitute a new generation of Complex SoS with integrated computational and physical capability that enables the most effective communications known to date. Cloud-computing technology (discussed in Chapter 10) has become a significant driver in the quest of the supply chain sector to becoming timely, efficient, and reliable. Chapter 13 is devoted in its entirety to cyber–physical Complex SoS. For example, Chan and Chieu (2010) have cogently outlined the present enthusiasm associated with the benefits (as well as the challenges) associated with cloud-computing technology:

> Cloud computing promises to provide high performance, flexible and yet low cost on-demand computing services with the benefits of speed, ease of deployment, scalability and service oriented architecture; it is extremely attractive; it offers a readily available and scalable computing environment without substantial capital investment and hardware administrative and maintaining cost. (p. 362)

The benefits gained from cloud-computing technology are augmented with new challenges, such as reliability, security, and accountability that must be properly addressed. Furthermore, while cloud computing is an enabling technology, it also dominates and essentially controls the parts of the communications sector that is vital and pivotal sector of the economy. Indeed, today's instant communications are provided in a measurable level by the cloud-computing technology within and among all parties involved in the life cycle of the Supply Chain Complex SoS: from raw materials to final production and from demand to instant supply.

Walmart: A Case Study

The following section is based on the website of Walmart Company – the largest Supply Chain Complex SoS supplier in the United States. The acquisition of commodities and their transportation, storage, processing, and their ultimate movement to the company's stores constitute one of the most elaborate and complex logistic enterprises. For example, cross-docking is a logistics practice that is the centerpiece of Walmart's strategy to replenish inventory efficiently. It means the direct transfer of products from inbound or outbound truck trailers without extra storage by unloading items from an incoming semitrailer truck or railroad car and loading these materials directly into outbound trucks, trailers, or rail cars (and vice versa), with no storage in between. Suppliers have been delivering products to Walmart's distribution centers where the product is cross-docked and then delivered to Walmart stores. Cross-docking keeps inventory and transportation costs down, reduces transportation time, and eliminates inefficiencies.

Walmart's fleet of truck drivers continuously deliver goods to distribution centers (located an average 130 miles from the store), where they are stored, repackaged, and distributed without sitting in inventory. Goods will cross from one loading dock to another, usually in 24 h or less, and company trucks that would otherwise return empty "backhaul" unsold merchandise. Using cross-docking, products are routed from suppliers to Walmart's warehouses, where they are then shipped to stores without sitting for long periods of time in inventory. This strategy reduced Walmart's costs significantly, and they passed those savings on to their customers with highly competitive pricing.

Walmart's supply chain management strategy has provided the company with several sustainable competitive advantages, including lower product costs, reduced inventory carrying costs, improved in-store variety and selection, and highly competitive pricing for the consumer. This strategy has helped Walmart become a dominant force in a competitive global market. As technology evolves, Walmart continues to focus on innovative processes and systems to improve its supply chain and achieve greater efficiency.

In his book *The World Is Flat*, Friedman (2005) presents the following succinct narrative on the "supply chain" enterprise epitomized by Walmart, the world's largest Supply Chain Complex SoS:

> I had never seen what a supply chain looked like in action until I visited Wal-Mart headquarters in Bentonville, Arkansas. My Wal-Mart hosts took me over to the 1.2-million-square-foot distribution center, where we climbed up to a viewing perch and watched the show. On one side of the building, scores of white Wal-Mart trailer trucks were dropping off boxes of merchandise from thousands of different suppliers. Boxes large and small were fed up a conveyor belt at each loading dock. These little conveyor belts fed into a bigger belt, like streams feeding into a powerful river. Twenty-four hours a day, seven days a week, the suppliers' trucks feed the twelve miles of conveyor streams, and the conveyor streams feed into a huge Wal-Mart river of boxed products. But that is just half the show. As the Wal-Mart river flows along, an electric eye reads the bar codes on each box on its way to the other side of the building. There, the river parts again into a hundred streams. Electric arms from each stream reach out and guide the boxes – ordered by particular Wal-Mart stores – off the main river and down its stream, where another conveyor belt sweeps them into a waiting Wal-Mart truck, which will rush these particular products onto the shelves of a particular Wal-Mart store somewhere in the country. There, a consumer will lift one of these products off the shelf, and the cashier will scan it in, and the moment that happens, a signal will be generated. That signal will go out across the Wal-Mart network to the supplier of that product – whether that supplier's factory is in coastal China or coastal Maine. That signal will pop up on the supplier's computer screen and prompt him to make another of that item and ship it via the Wal-Mart supply chain, and the whole cycle will start anew. So no sooner does your arm lift a product off the local Wal-Mart's shelf and onto the checkout counter than another mechanical arm starts making another one somewhere in the world. Call it "the Wal-Mart Symphony" in multiple movements – with no finale. It just plays over and over 24/7/365: delivery, sorting, packing, distribution, buying, manufacturing, reordering, delivery, sorting, packing .... (p.128)

11.10 RISK ANALYSIS OF THE SUPPLY CHAIN COMPLEX SYSTEMS OF SYSTEMS

11.10.1 Supply Chain, Risk Management, and the Future of Three-Dimensional Printing

The global Supply Chain Complex SoS, which is responsible for enabling and integrating all the supply and demand sectors of the worldwide economies, is continuously evolving and adopting new know-how, technologies, and capacity. As new innovations and technologies emerge and are adopted by the marketplace and sectors of the economy, other sectors must adapt to this unavoidable evolution, most notably manufacturing processes, transportation of commodities, and people's needs. Clearly, the infinite commodities of the supply chain necessarily generate more opportunities as well as more sources of risk to the supply chain itself, its users, and the public at large. Furthermore, any contamination or accident resulting from the introduction of untested new technology could have cascading effects on the entire supply chain and impact uncounted number of stakeholders. Figure 11.1 presents a sample of the

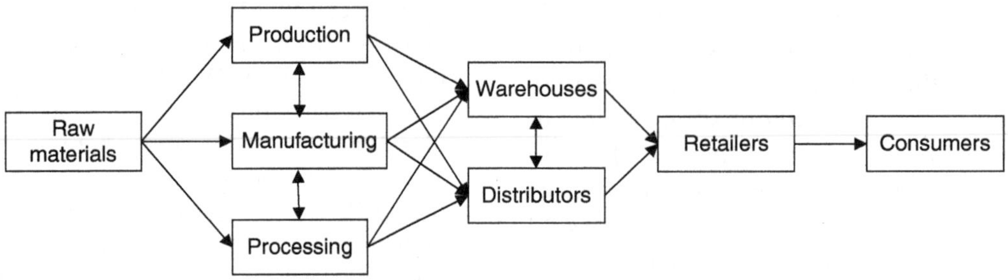

Figure 11.1 *The supply chain process.*

different phases that goods and services must pass through in the supply chain. While customers are generally not appreciative of every step of these processes in action, almost all products pass through the sample of steps depicted in Figure 11.1.

There is a direct correlation between the introduction of new technology, their adoption, and introductions to the Supply Chain Complex SoS, and to the marketplace. Often, most technological beneficial changes can be correlated with cost adjustments, given that the marketplace is constantly implementing new technologies to increase efficiency and decrease transportation time and costs. For example, the introduction of the 3-D printer to the interdependent and interconnected regional or the world economy has directly affected, in many ways, the supply chain sector of the economy. While 3-D printer is set to have a massive impact on the Supply Chain Complex SoS, it is too early to predict the ultimate impact of its malicious misuse. However, if creatively and positively incorporated into the supply chain, the 3-D printer will enhance the entire economy of a region, a country, or the world economy as Complex SoS.

All new technologies – and the 3-D printer is of no exception – introduce new sources of risk. The list below illustrates some of the important sources of risks of 3-D printers (O'Rourke, 2015):

Product risk. Since three-dimensional printing changes the traditional manufacturing model, industries will need to determine who owns a three-dimensional printed product and in the event of an accident how will liability be apportioned?
Operations risk. How will three-dimensional printing impact power supplies (the printers generate a lot of heat during operation), and how will the possible toxicity of ingredients and their by-products be addressed? In addition, what are the business interruption and transportation risks?
Cyber security risk. How would individuals and organizations protect their designs and formulas? How would they prevent counterfeiting?
Strategic risk. How to handle reputation and intellectual property issues? What happens to product development and its life cycle management?
Supply chain risk. Would the risk to the supply chain increase or decrease?
Market risk. What differentiates one product from another?

To model, assess, manage, and communicate the above sources of risk, we apply a systemic process (Haimes, 2016, 2018) by addressing the following set of questions.

In the risk assessment of Complex SoS, we ask the following questions: (i) What can go wrong for each subsystem of the supply chain and for the entire supply chain as Complex SoS? (ii) What is the likelihood associated with each subsystem of the supply chain and with the entire supply chain as Complex SoS? (iii) What are the consequences associated with each subsystem of the supply chain and with the entire supply chain as Complex SoS? (iv) What is the time frame associated with each subsystem of the supply chain and with the entire supply chain as Complex SoS?

Since the supply chain covers such a large number of processes and entities, there would be a very large range of responses to the above questions. Indeed, for each subsystem there can be hundreds of "things" that can go wrong, from small to large and from safety-critical to benign effects. Each source of risk would likely have different likelihoods, consequences, and time frames. However, we can address the above risk-based questions by building on HHM, introduced in Chapter 3. Below, we will briefly demonstrate the scope and effectiveness of the HHM methodology and philosophy by generating a sample list of threats to the supply chain across the different supply chain sectors. The next challenge is to examine ways and means with which to manage sources of risk (Haimes, 2016, 2018): (i) What can go wrong and what can be done with each subsystem of the supply chain and with the entire supply chain as Complex SoS? (ii) What are the tradeoffs in terms of all costs, benefits, and risks associated with each subsystem of the supply chain and with the entire supply chain as Complex SoS? (iii) What are the future impacts of management decisions associated with each subsystem of the supply chain and with the entire supply chain as Complex SoS?

The above risk assessment and management questions are discussed in detail and with examples in Chapter 9 of this book and in Haimes (2016).

11.11 ANALYTICAL METHOD FOR MODELING AND MANAGING THE SUPPLY CHAIN AS COMPLEX SYSTEMS OF SYSTEMS

11.11.1 Science/Engineering

Science and engineering can have significant impacts on the supply chain as Complex SoS. As time passes, science and engineering advance, creating new methods of production and consumption. Five key areas of science and engineering that call for more research are community, ecology, modern advancement, food, and 3-D printing. One effect that science and engineering can have on communities is that as technology develops, machines can do jobs more efficiently. This leads to less labor-intensive jobs and more technical jobs. One example of this is online streaming sites, e.g. Netflix, which have diminished the movie rental industry. Another effect on community is the advancement of transportation and shipping methods; communities no longer need to be self-sustaining. However, the reliance on imports makes the supply chain even more important, on which communities are more dependent.

Furthermore, we must be mindful of how our methods of production and consumption affect the environment and other organisms. For example, high levels of toxic chemicals have been found in the ocean as a result of hundreds of years of dumping waste. Traces of these toxins have been detected in fish and other oceanic organisms, and since animals are a significant food source, we should be careful when introducing contaminated animals and other byproducts to our supply chain. Figure 11.2 presents a

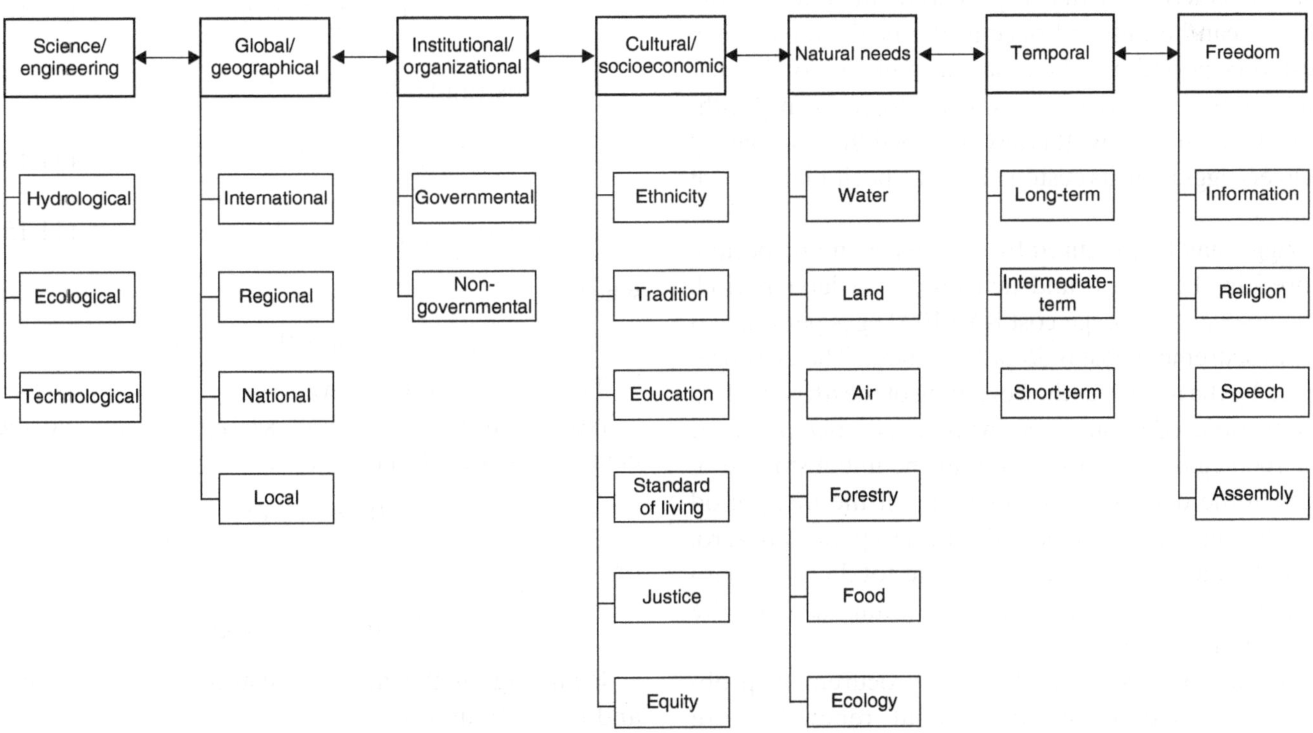

Figure 11.2 *A sample of sources of risk to the global supply chain complex systems of systems.*

sample of sources of risk to the global Supply Chain Complex SoS.

The advancement of technology can improve manufacturing, production, and transportation methods. Additionally, the advancement of the Internet and its use have led to substantial increased threat of cyber attack.

11.12 INVENTORY CONTROL OF SUPPLY CHAIN COMPLEX SYSTEMS OF SYSTEMS

A common problem in the supply chain industry is the optimal operation and management of storage facilities for commodities in transit, such as warehouses and depots. This section formulates and solves a prototype procurement problem via dynamic programming using a simplified poultry eggs inventory for illustration. (Note that Section A.1.3.1 of the Appendix addresses Bellman's principle of optimality.)

Assume that the agency is in charge of the poultry eggs supply to a region. *For simplicity, we will use the term "eggs" for "poultry eggs."* The agency must make this commodity available in sufficient quantities to meet all demands in N time periods. The agency procures eggs from various sources and stores them in its depot, which has a maximum capacity of Q 1 000 000 eggs. It is assumed that a procurement order by the agency can be initiated once at the beginning of each period (a period may be a day, a week, a month, etc.) and that the eggs are made available to the agency without a lead time delay. It is also assumed that the agency delivers eggs to all its customers at the beginning of each period.

Eggs may be procured by the agency in one period, stored in the depot, and delivered at a later period. The associated storage cost is \$$a$/1000 eggs/period, and the procurement cost is \$$b$/procurement. The eggs procured by the agency at the beginning of the ith period, x_i, can be ordered in quantities with integer increments, Δ (of 1000 eggs). It is assumed that the initial storage of eggs in the depots at the beginning of the first period and the final storage at the end of the last period are zero.

The objective is to minimize the total cost of supplying sufficient eggs to meet all demands over the entire planning horizon.

In the above statement of the procurement problem, simplified assumptions about renewals were made for pedagogical purposes. Note that these assumptions require that procurement lead time be zero, procurement orders be initiated only at the beginning of each period, inventory at the beginning or at the end of the last period be zero, demand occurs at the beginning of the period, and there be no shortages. All these constraints can be removed with proper modifications of the mathematical model developed here. The price of a unit of eggs (\$/1000 eggs) is not given (since it does not affect the optimization problem). Storage at the beginning and the end of the planning horizon is fixed.

11.12.1 Model Formulation

Let q_i be the stock level of eggs during period i, $i = 1, 2, \ldots, N$ (state variable), x_i be the quantity of eggs procured at the beginning of period i, $i = 1, 2, \ldots, N$ (decision variable), D_i be the demand of eggs at the beginning of period i, $i = 1, 2, \ldots, N$, and $g_i(x_i, q_i)$ be the procurement cost and holding cost of eggs for period i, $i = 1, 2, \ldots, N$.

Note that the number of stages in this dynamic programming formulation coincides with the number of periods, N, of the planning horizon.

The overall objective function is

$$\min_{x_i} \sum_{i=1}^{N} g_i(x_i, q_i) \tag{11.13}$$

The constraints are

$$x_i \geq 0, \quad i = 1, 2, \ldots, N \tag{11.14}$$

$$q_i = q_{i-1} + x_i - D_i, \quad i = 1, 2, \ldots, N \tag{11.15}$$

where

$$q_0 \equiv 0$$

Thus, $q_{i-1} = q_i + D_i - x_i$.

Note that the maximum storage at any period cannot exceed Q. Therefore,

$$0 \leq q_{i-1} \leq Q$$

or

$$0 \leq q_i + D_i - x_i \leq Q$$

Rearranging the above constraint yields a lower and upper bound on x_i:

$$q_i + D_i - Q \le x_i \le q_i + D_i, \quad i = 1,2,\dots,N$$

Define a new function $f_1(q_1)$ as follows:

$f_1(q_1)$ = the minimum cost of meeting water demand at the first period with a storage (stock) level of eggs at q_1.

Mathematically, the optimization problem for the first stage can be written as

$$f_1(q_1) = \min_{x_1} g_1(x_1, q_1) \tag{11.16}$$

$$q_1 + D_1 - Q \le x_1 \le q_1 + D_1 \tag{11.17}$$

Similarly, define the general function $f_n(q_n)$ to be

$f_n(q_n)$ = the minimum cost of meeting eggs demands all for all n previous periods, with eggs storage level q_1 during the nth period.

Mathematically, the general recursive equation for the dynamic programming formulation can be written as

$$f_n(q_n) = \min_{x_n} \{g_n(x_n, q_n) + f_{n-1}(q_{n-1})\} \tag{11.18}$$

$$q_n + D_n - Q \le x_n \le q_n + D_n, \quad n = 1,2,\dots,N \tag{11.19}$$

Substituting the value of q_{n-1} in $f_{n-1}(q_{n-1})$ yields

$$f_n(q_n) = \min_{x_n} \{g_n(x_n, q_n) + f_{n-1}(q_n + D_n - x_n)\} \tag{11.20}$$

The above recursive equation should be solved for all possible stock levels q_n for all planning periods, $n = 1, 2, \dots, N$. Then the optimal procurement policy, x_n^*, for all periods, $n = 1, 2, \dots, N$, can be determined using the state equation $q_{n-1} = q_n + D_n - x_n$, which relates the state variable at $(n-1)$st period to the state and decision variables at the nth period. A detailed discussion on the determination of the overall optimal procurement policy is given in the following numerical example.

11.12.2 An Example Problem

Given the following numerical values to the general procurement problem discussed above:

$N = 5$ (periods in the planning horizon)
$Q = 40$ (maximum storage capacity)
$\Delta = 10$ (integer units of procurement increments)
$a = \$0.10$ (holding cost per unit period based on the stock level at the end of the period)
$b = \$20$ (procurement cost per procurement)
$D_1 = 10$ (eggs demand at period 1)
$D_2 = 20$ (eggs demand at period 2)
$D_3 = 30$ (eggs demand at period 3)
$D_4 = 30$ (eggs demand at period 4)
$D_5 = 20$ (eggs demand at period 5)

Find the optimal procurement policy for all five periods at a minimum total cost.

Solution: First stage. The recursive equation for the first stage ($n = 1$) is

$$f_1(q_1) = \min_{x_1} \{g_1(x_1, q_1)\}$$
$$q_1 + 10 - 40 \le x_1 \le q_1 + 10$$

and

$$x_1 \ge 0$$

This recursive equation should be solved for all feasible incremental values of q_1 ($q_1 = 0$, 10, 20, 30, and 40):

$$f_1(0) = g_1(10,0) = 20 + 0 = 20$$
$$f_1(10) = g_1(20,10) = 20 + 1 = 21$$
$$f_1(20) = g_1(30,20) = 20 + 2 = 22$$
$$f_1(30) = g_1(40,30) = 20 + 3 = 23$$
$$f_1(40) = g_1(50,40) = 20 + 4 = 24$$

Note that the cost is composed of two parts: the fixed procurement cost of \$20 and the corresponding per-unit per-period holding cost.

Second stage. The recursive equation for the second stage ($n = 2$) is

$$f_2(q_2) = \min_{x_2} \{(g_2, q_2) + f_1(q_2 + 20 - x_2)\}$$
$$q_2 + 20 - 40 \le x_2 \le q_2 + 20$$
$$x_2 \ge 0$$

Again, the last recursive equation should be solved for $q_2 = 0$, 10, 20, 30, and 40.

1. For $q_2 = 0$:

$$f_2(0) = \min_{x_2} \{g_2(x_2,0) + f_1(0+20-x_2)\}$$

$$0 \le x_2 \le 20$$

$$f_2(0) = \min \begin{pmatrix} g_2(0,0) + f_1(0+20-0) = 0+0+22 = 22 \\ g_2(10,0) + f_1(0+20-10) = 20+0+21 = 41 \\ g_2(20,0) + f_1(0+20-20) = 20+0+20 = 40 \end{pmatrix}$$

$f_2(0) = 22.\ x_2^* = 0$

2. For $q_2 = 10$:

$$f_2(10) = \min_{x_2} \{g_2(x_2,10) + f_1(10+20-x_2)\}$$

$$0 \le x_2 \le 30$$

$$f_2(10) = \min \begin{pmatrix} g_2(0,10) + f_1(10+20-0) = 0+1+23 = 24 \\ g_2(10,10) + f_1(10+20-10) = 20+1+22 = 43 \\ g_2(20,10) + f_1(10+20-20) = 20+1+21 = 42 \\ g_2(30,10) + f_1(10+20-30) = 20+1+20 = 41 \end{pmatrix}$$

$f_2(10) = 24.\ x_2^* = 0$

3. For $q_2 = 20$:

$$f_2(20) = \min_{x_2} \{g_2(x_2,20) + f_1(20+20-x_2)\}$$

$$0 \le x_2 \le 40$$

$$f_2(20) = \min \begin{pmatrix} g_2(0,20) + f_1(20+20-0) = 0+2+24 = 26 \\ g_2(10,20) + f_1(20+20-10) = 20+2+23 = 45 \\ g_2(20,20) + f_1(20+20-20) = 20+2+22 = 44 \\ g_2(30,20) + f_1(20+20-30) = 20+2+21 = 43 \\ g_2(40,20) + f_1(20+20-40) = 20+2+20 = 42 \end{pmatrix}$$

$f_2(20) = 26.\ x_2^* = 0$

4. For $q_2 = 30$:

$$f_2(30) = \min_{x_2} \{g_2(x_2,30) + f_1(30+20-x_2)\}$$

$$10 \le x_2 \le 50$$

$$f_2(30) = \min \begin{pmatrix} g_2(10,30) + f_1(30+30-10) = 20+3+24 = 47 \\ g_2(20,30) + f_1(30+20-20) = 20+3+23 = 46 \\ g_2(30,30) + f_1(30+20-30) = 20+3+22 = 45 \\ g_2(40,30) + f_1(30+20-40) = 20+3+21 = 44 \\ g_2(50,30) + f_1(30+20-50) = 20+3+20 = 43 \end{pmatrix}$$

$f_2(30) = 43.\ x_2^* = 50$

5. For $q_2 = 40$:

$$f_2(40) = \min_{x_2} \{g_2(x_2,40) + f_1(40+20-x_2)\}$$

$$20 \le x_2 \le 60$$

$$f_2(40) = \min \begin{pmatrix} g_2(20,40) + f_1(40+30-20) = 20+4+24 = 48 \\ g_2(30,40) + f_1(40+20-30) = 20+4+23 = 47 \\ g_2(40,40) + f_1(40+20-40) = 20+4+22 = 46 \\ g_2(50,40) + f_1(40+20-50) = 20+4+21 = 45 \\ g_2(60,40) + f_1(40+20-60) = 20+4+20 = 44 \end{pmatrix}$$

$f_2(40) = 44.\ x_2^* = 60$

This concludes the calculations for the second stage.

Similar calculations have been made for the third, fourth, and fifth stages. The results of these calculations are summarized in Table 11.1. Note that at the fifth stage, there is no need to solve the recursive equation for all feasible values of $g_5(\cdot)$, since it was assumed that no inventory will be left at the end of

TABLE 11.1 Summary of Results for All Five Stages

	Column number									
q	1	2	3	4	5	6	7	8	9	10
	x_1	$f_1(q_1)$	x_2	$f_2(q_2)$	x_3	$f_3(q_3)$	x_4	$f_4(q_4)$	x_5	$f_5(q_5)$
0	10	20	0*	22	30*	42	0	45	0*	64
10	20	21	0	24	40	43	0	49	30	66
20	30*	22	0	26	50	44	50*	64	40	67
30	40	23	50	43	60	45	60	65	50	68
40	50	24	60	44	60	46	60	67	60	69

that period. Table 11.1, however, does give the corresponding values for g_5 at all feasible increments for pedagogical purposes.

The final step in solving the procurement problem is tracing the optimal procurement policies for all five stages or periods. This is done in a reverse order, starting with the fifth stage. All optimal values in Table 11.1 are identified by an asterisk.

The optimal procurement policy and the minimum cost for meeting water demand at all five stages are given in columns 9 and 10, respectively. These values are $x_5 = 0$, $f_5(0) = 64$. The corresponding optimal inventory level at $N = 5$ is $q_5 = 0$. It is now possible to find the optimal inventory level for stage 4:

$$q_4 = q_5 + D_5 - x_5$$

$$q_4 = 0 + 20 - 0 = 20$$

The optimal procurement policy at the fourth stage corresponds to $q_4 = 20$ and can be found in column 7 to be $x_4 = 50$.

The optimal inventory level for stage 3 is

$$q_3 = 20 + 30 - 50 = 0$$

The optimal procurement policy at the third stage corresponds to $q_3 = 0$, and it can be found in column 5 to be $x_3 = 30$.

The optimal inventory level for stage 2 is

$$q_2 = 0 + 30 - 30 = 0$$

The optimal procurement policy at the second stage corresponds to $q_2 = 0$, and it can be found in column 3 to be $x_2 = 0$.

Finally, the optimal inventory level for stage 1 is

$$q_1 = 0 + 20 - 0 = 20$$

The optimal procurement policy at the first stage corresponds to $q_1 = 20$, and it can be found in column 1 to be 30.

In summary, the optimal procurement policy vector, $\mathbf{x}*$, for all five periods is $\mathbf{x}* = (30, 0, 30, 50, 0)$, and the minimum cost is \$64.

11.13 SUMMARY

The supply chain epitomizes the complexity of interconnected and interdependent Complex SoS, with a sizable percentage of these systems being safety-critical. We identified the supply chain not only as Complex SoS but also the backbone of the global economy of every country. And its success is paramount to the success of individual economies as well as of individual businesses with which we interact on a daily basis. To achieve the above required harmonized operations and effective flow of commodities from point A to destination Z, organizational infrastructure becomes essential, while the commodities moved through many roads less traveled. Thus, a well-founded organizational infrastructure that supports and enables the supply chain to perform its intended functions effectively becomes an imperative. Chapter 12 addresses these and other fundamental questions on the role of organizations in effective management of Complex SoS.

REFERENCES

Apostolakis, G. (1999). The distinction between aleatory and epistemic uncertainties is important: an example from the inclusion of aging effects into probabilistic safety assessment. *Proceedings of PSA'99* (22–25 August 1999). Washington, DC: American Nuclear Society.

Chan, H. and Chieu, T. (2010). Ranking and mapping of applications to cloud computing services by SVD. *2010 IEEE/IFIP Network Operations and Management Symposium Workshops*, Osaka, Japan (19–20 April 2010), 362–369. doi: 10.1109/NOMSW.2010.5486550.

Chen, C. (2012). *Linear System Theory and Design*, 4e. New York: Oxford University Press.

Collins, J. (2011). *Good to Great: Why Some Companies Make the Leap and Others Don't.* New York: HarperBusiness.

Crowther, K.G. and Haimes, Y.Y. (2005). Application of the inoperability input–output model (IIM) for systemic risk assessment and management of interdependent infrastructures. *Systems Engineering* **8**(4): 323–341.

Drucker, P. (2001). *The Essential Drucker*. New York: HarperBusiness.

Friedman, T. (2005). *The World Is Flat*. New York: Farrar, Straus and Giroux.

Garvin, D. (1988). *Managing Quality*. New York: The Free Press.

Haimes, Y.Y. (1977). *Hierarchical Analyses of Water Resources Systems: Modeling and Optimization of Large-Scale Systems*. New York: McGraw-Hill.

Haimes, Y.Y. (1981). Hierarchical holographic modeling. *IEEE Transactions on Systems, Man and Cybernetics* **11**(9): 606–617.

Haimes, Y.Y. (2002). Roadmap for modeling risks of terrorism to the homeland. *Journal of Infrastructure Systems* **8**(2): 35–41.

Haimes, Y.Y. (2016). *Risk Modeling, Assessment, and Management*, 4e. Hoboken, NJ: Wiley.

Haimes, Y.Y. (2018). Risk modeling of interdependent complex systems of systems: theory and practice. *Risk Analysis* **38**: 84–98. doi: 10.1111/risa.12804.

Haimes, Y.Y. and Horowitz, B. (2004). Adaptive two-player hierarchical holographic modeling game for counterterrorism intelligence analysis. *Journal of Homeland Security and Emergency Management* **1**(3): Article 302.

Haimes, Y.Y., Li, D., and Tulsiani, V. (1990). Multiobjective decision-tree analysis. *Risk Analysis* **10**: 111–127.

Haimes, Y.Y., Horowitz, B., Lambert, J. et al. (2005a). Inoperability input-output model (IIM) for interdependent infrastructure sectors. I: Theory and methodology. *Journal of Infrastructure Systems* **11**(2): 67–79.

Haimes, Y.Y., Horowitz, B., Lambert, J. et al. (2005b). Inoperability input–output model (IIM) for interdependent infrastructure sectors. II: Case study. *Journal of Infrastructure Systems* **11**(2): 80–92.

Intriligator, M.D. (1971). *Mathematical Optimization and Economic Theory*. Englewood Cliffs, NJ: Prentice-Hall.

Jiang, P. and Haimes, Y.Y. (2004). Risk management for Leontief-based interdependent systems. *Risk Analysis* **24**(5): 1215–1229.

Leontief, W.W. (1951a). Input/output economics. *Scientific American* **185**(4): 15–21.

Leontief, W.W. (1951b). *The Structure of the American Economy, 1919–1939*, 2e. New York: Oxford University Press.

Leontief, W.W. (1986). *Input–Output Economics*, 2e. New York: Oxford University Press.

Nise, N. (2015). *Control Systems Engineering*, 7e. Pomona, CA: California State Polytechnic University.

O'Rourke, M. (2015). Risks and questions surround 3-D printing technology, Risk Management Monitor. http://www.riskmanagementmonitor.com/risks-and-questions-surround-3d-printing-technology (accessed 24 October 2017).

Paté-Cornell, E. (1990). Organizational aspects of engineering system safety: the case of offshore platforms. *Science* **250**: 1210–1217.

Paté-Cornell, E. (1996). Uncertainties in risk analysis: six levels of treatment. *Reliability Engineering and System Safety* **54**(2): 95–111.

Santos, J.R. and Haimes, Y.Y. (2004). Modeling the demand reduction input–output (I–O) inoperability due to terrorism of interconnected infrastructures. *Risk Analysis* **24**(6): 1437–1451.

U.S. Nuclear Regulatory Commission (1981). *Fault Tree Handbook*, NUREG-81/0492. Washington, DC: U.S. Nuclear Regulatory Commission.

12

Understanding and Managing the Organizational Dimension of Complex Systems of Systems

PART I: ORGANIZATIONAL CULTURE, VISION, AND QUALITY OF LEADERSHIP AS CRITICAL DRIVERS TO EFFECTIVE AND SUCCESSFUL PERFORMANCE OF COMPLEX SYSTEMS OF SYSTEMS

12.I.1 INTRODUCTION

Organizational structures of interdependent and interconnected complex systems of systems (Complex SoS) vary so widely, especially among the private and public sectors, that they may be branded from their modeling perspectives as an unbounded set. Furthermore, both types of structures are characterized by a hierarchy of multiple conflicting, noncommensurate, and competing objectives. Moreover, the hierarchy of objectives is associated with decision makers and stakeholders responsible for different levels of the organization's history, mandate, vision, structure, and operation. Fundamental differences can be associated with the nature and mission of each organization, based on several factors: (i) private versus public sector; (ii) local, regional, or global reach and functionality; (iii) mission orientation, e.g. secular, religious, or philanthropic, local, state, or federal government; (iv) commercial versus scientific and civilian versus military; and (v) profit versus nonprofit. Although each of the above organizational structures has very specific characteristics and would require customized modeling efforts, we will develop common denominators among them to enable modelers to exploit the essential attributes that most, if not all, organizational groups share. For example, all government organizations share very specific mandates, methods of operation, and missions. What makes understanding and modeling organizations as Complex SoS an especially challenging task is that the goals of the individuals who populate and run such organizations are so widely different, often resulting in internal organizational stress and conflicts.

Understanding, modeling, managing, and acting upon the organizational dimension of Complex SoS is a never-ending continuous enterprise. It is commonly characterized by a trial-and-error and learn-as-you-go process, best guided by the gestalt–holistic philosophy on which systems engineering is grounded. It is an intriguing process that when you think you have won, you have lost. In this chapter we relate

Modeling and Managing Interdependent Complex Systems of Systems, First Edition. Yacov Y. Haimes.

our experiences in the learn-as-you-go process of systems modeling, including data gathering and interpretation, and a continuous and endless process of modifying earlier assumptions, propositions, and perspectives. In many ways, modeling the intricacy of the ever-evolving organizational enterprises is an elaborate process. Such models are built to provide a mechanism with which to understand and answer specific questions, and where their effectiveness is highly dependent on well-thought and designed questions to which we seek answers, or at least gaining more clarification on what could be a "Black Box" target for intelligence gathering and interpretation to aid and contribute to the modeling process.

Although we build on previous chapters about the genesis of the interdependency and interconnectedness that characterize Complex SoS, it is constructive to recall that we can effectively trace these interdependencies through the shared/common *states* and to other shared *essential entities that characterize* Complex SoS. Note that the term "essential entities" connotes shared/common decisions, decision makers, stakeholders, resources, organizational behavior and norms, policies and procedures, management, culture, and others within and among the subsystems and systems that constitute Complex SoS.

The diverse drivers that propel organizational tradition and culture, the organization's evolving policies and procedures, its interpersonal relationships and organizational values, the inner forces that sustain the organization in good and difficult times, and their ability to sustain their leadership in their enterprise domain–all constitute an organizational litmus test for a sustained success or being marginalized.

Consider the following unfortunate common practices that engulf both the private and public sectors:

1. Embracing short-term band-aid policies and procedures versus adhering to long-term fundamentals and becoming agents of change.
2. Espousing to classic "decomposition" by conveniently addressing selective systems or subsystems of the overall enterprise versus adapting a more holistic approach that addresses the dynamic emergence of changes to the essentially interdependent and interconnected organization as Complex SoS.
3. Adapting a laissez-faire and business-as-usual attitude versus a proactive, futuristic, and systems-based planned actions approach.
4. Casually practicing risk analysis as an add-on task versus commitment of the leadership to proactively probing and searching for detrimental *precursors* that would likely negatively affect or alter the strategic plans and operations of the organization.

Effective organizations must apply the same diligence and systemic approaches to the discovery of *precursors* to major sources of risk. It is important to note the efficacy of developing scenarios for sources of risk and emergent forced changes (EFCs) that may originate from within or outside a subsystem or the entire Complex SoS. Organizations around the world, whether small, large, or Complex SoS, are structured in numerous ways, depending on the culture, tradition, and uniqueness of each country. In many ways the essence of each organizational structure is reflected by the history, tradition, culture, uniqueness, and the dominant sectors of the economy, among other distinctive factors in each country.

The structures and management of matrix organizations are commonly suitable for Complex SoS, because they promote integration and collaboration. Matrix structures are in particular common for engineering- and technology-driven organizations, because employees have dual reporting relationships, generally reporting to both a functional manager and a product manager. Of course, the choice of the organizational structures is not a random process; the latter evolve with the demands placed on them as they grow and develop new products, markets, or innovations. Nevertheless, the hierarchical or lateral interdependencies and interconnectedness (I-I) within and among the many units of organizational Complex SoS evolve and are restructured in response to, and through accommodation of, the strong shared states and other essential entities within and among the units/departments of the organizational Complex SoS. Improving interpersonal communication and collaboration are requisites for the stimulation and generation of new ideas, products, and markets to sustain organizational growth and prosperity. Although the myriad units, departments, and divisions that constitute organizations as Complex SoS are supposed to support and collaborate with other units, an effective synergistic collaboration is commonly predicated on conducive organizational culture and environments and, most importantly, on visionary leaders who possess a blend of humbleness and courage with determination.

To formally introduce the reader to the next section, we further augment the narratives presented by the following several pivotal books by creative authors.

12.I.2 PHILOSOPHICAL PERSPECTIVES ON ORGANIZATIONAL BEHAVIOR OF COMPLEX SYSTEMS OF SYSTEMS

A major tenet of the study of organizational behavior associated with Complex SoS is the imperative recognition that organizations are multifaceted in nature. They constitute (i) a hierarchy of multiple systems and subsystems (ii) involving multiple decision makers, stakeholders, and power brokers (iii) with multiple conflicting and competing objectives associated with different systems and subsystems of the organizational Complex SoS, and (iv) each diverse mission is commonly associated with different time frames. The following shared characteristics of Complex SoS, which are driven by human and organizational dynamics, are espoused by gifted authors of several classic books. We ought to keep in mind that although organizations have their own structures and dynamics, they are also commonly driven by their history and culture and ultimately by humans.

The *first thought-provoking seminal book*, titled *Thinking Fast and Slow* and written by the Nobel Laureate in Economic Sciences, Professor of Psychology Daniel Kahneman (2013), is a must read on human perception and thinking about probability and statistics. It highlights the way we respond to events, think, and act via two modes: (i) intuitively, spontaneously, and without deep thinking, and (ii) thoughtfully and methodically. We should appreciate that the principles and narratives of human decision making within an organization cannot be separated from the organizational culture and behavior. For decades the scholarly collaboration between Professor Daniel Kahneman and Professor Amos Tversky has mesmerized the professional community, transcending common divides between psychologists, economists, risk analysts, engineers, and other professionals and lay persons. Kahneman and Tversky's insightful and seminal publications continue to be taught and discussed in undergraduate and graduate courses, seminars, and professional conferences and workshops around the world.

It is worth noting in the context of organizational behavior the following perspectives that Kahneman provides on what he terms "the availability heuristic" (p. 130):

> The Availability Heuristic, like other heuristics of judgment, substitutes one question for another: you wish to estimate the size of a category or the frequency of an event, but you report an impression of the case with which instances come to mind. Substitution of questions inevitably produces systematic errors. You can discover how the heuristic leads to biases by following a simple procedure: List factors other than frequency that make it easy to come up with instances. Each factor in your list will be a potential source of bias.

Indeed, as we study organizational behavior as a prelude to also understanding Complex SoS, it is important to gain as much insight as possible on the multifarious structures, cultures, and successes of Complex SoS. The above discussion necessarily leads us to the gestalt–holistic philosophy, on which systems engineering is grounded, and further reflecting on the many ways with which one can decipher the complexity of organizational behavior, especially for Complex SoS. Recall that Chapter 3 is devoted to Hierarchical Holographic Modeling (HHM), which enables us to effectively study, model, and manage interdependent and interconnected Complex SoS from their multifarious perspectives.

In the *second book*, *Working Knowledge*, Davenport and Prusak (2000) suggest that knowledge moves through an organization via markets just as any other scarce resource does. There are buyers, sellers, and brokers of knowledge. Indeed, the multiple subsystems that make up Complex SoS provide ample opportunities for the exchange of knowledge and opportunities for higher achievements across the entire enterprise. Those who possess it will sell their knowledge if they are properly compensated with money, reciprocity, repute, bonuses, promotions, or other expected gain. Furthermore, if there is neither culture of belonging and of sharing knowledge and know-how nor sufficient compensation for those who share their knowledge, the organization will not achieve its excellence. This "market" for knowledge within a Complex SoS has some important implications for growth and high achievements.

Indeed, management and corporate culture are key influences and drivers that facilitate the stimulation and generation of knowledge and productivity. First and foremost, trust is required for the exchange of knowledge. Davenport and Prusak posit that knowledge "markets" are informal and lack the security of legal contracts and a system of courts with which to maintain the integrity of exchanges. *Therefore, trust is required so that sellers believe that they will receive appropriate compensation and buyers believe that the knowledge they receive is accurate.* The authors also maintain that management must create an environment that fosters trust. *When the factor of concern is risk, knowledge of failures and mistakes is usually the most useful knowledge of all.* Incidentally, knowledge of failures and mistakes is also the least likely to be divulged by an organization's members. Consequently, creating a culture of trust is imperative to obtaining the knowledge that is critical for an effective risk management. Punishing personnel for reporting mistakes and failures is certain to short-circuit the entire risk management process. Unfortunately, the large number of participants complicates building trust in a development or acquisition project. System knowledge must be obtained from all of the participating organizations. This means that trust must exist both within each organization and among organizations.

In the *third best-selling book*, *The Speed of Trust*, Stephen M.R. Covey (2008) quotes Jim Burke, Former Chairman and CEO of Johnson & Johnson, who echoes Davenport and Prusak on the importance of trust (p. 6):

> You can't have success without trust. The word trust embodies almost everything you can strive for that will help you to succeed. You tell me any human relationship that works without trust, whether it is a marriage or a friendship or a social interaction; in the long run, the same thing is true about business, especially business that deals with the public.

Underscoring the point, Covey also quotes the New York Times columnist Thomas Friedman (2005) from his book, *The World Is Flat* (p. 21):

> Without trust, there is no open society, because there are not enough police to patrol every opening in an open society. Without trust, there can be no flat world, because it is the trust that allows us to take down walls, remove barriers, and eliminate friction at borders. Trust is essential for a flat world…

Covey identifies four core values for credibility, which are foundational elements that make one trustworthy and believable: (i) *integrity*, which also connotes trust and honesty; (ii) *intent*, which has to do with motives, our agenda, and our resulting behavior; (iii) *capabilities*, which inspire confidence – our talents, attitudes, skills, knowledge, and style; and (iv) *results*, our track records, our performance, and our getting the right things done.

The above interpersonal relationships within and among the subsystems and systems that make up Complex SoS are indeed an imperative for effective and successful organizations, whether small systems or Complex SoS.

Covey also posits that establishing trust is not sufficient for an efficient knowledge market, however. Experts, who possess the knowledge, must feel that they are being compensated for the knowledge they are providing to the organization. When project managers provide open support for the exchange of knowledge and know-how, they enable an environment conducive for growth and high-quality products. This enabling will occur because the "knowers" are being compensated for the knowledge they bring to the table. The more knowledge is shared, the more likely it is that the overall organization and its management will succeed. When the link has been established that successful trust and the sharing of knowledge lead to a successful organization, then participants will also be compensated by their organization, with potential promotions, bonuses, pay increases, and other rewards.

In the *fourth groundbreaking book*, *Networks and Netwars*, Arquilla and Ronfeldt (2001) identify the following four major narratives that characterize organizations, social groups, tribes, and other closed communities. A sample of the multiple perspectives characterizing most Complex SoS is highlighted here (p. 324):

i. ***Narrative level***
Why have the members assumed a network form? Why do they remain in that form? Networks, like other forms of organization, are held together by the narratives, or stories, that people tell...These narratives provide a grounded expression of people's experiences, interests, and values.
ii. ***Doctrinal level***
This level of the analysis is very important for explaining what enables the members to operate strategically and tactically, without necessarily having to resort to a central command or leader.
iii. ***Technological infrastructure***
What is the pattern of and capacity for information and communications flow within an organizational network?
iv. ***Social underpinnings***
The full functioning of a network also depends on how well, and in what ways, the members are personally known and connected to each other.

Clearly, when we study organizational behavior and its direct, sustained, and consequential impacts on the successes and failures of Complex SoS, we discover that the characterization of organizations in terms of their *narrative level, doctrinal level, technological infrastructure, and social underpinnings shed a deeper light onto the life span of successful versus mediocre organizations.*

In the *fifth classic book*, the *Boundaryless Organization*, Ashkenas et al. (2002) state that organizations must confront their narratives and reshape the following four types of boundaries (p. 11):

i. ***Vertical***
Vertical boundaries represent layers within a company. They are the floors and ceilings that differentiate status, authority, and power. Traditional elements of vertical boundaries are spans of control, limits of authority, and other manifestations of hierarchy. In a hierarchical organization, roles are clearly defined, and more authority resides higher up in the organization than lower down.
ii. ***Horizontal***
Horizontal boundaries exist between organizational functions, product lines, or units. If vertical boundaries are floors and ceilings, horizontal boundaries are walls between rooms. Boundaries between functions exist in the traditional firms when each function has a singular agenda that may compete or conflict with other functional agenda.
iii. ***External***
External boundaries are barriers between firms and the outside world – principally suppliers and customers but also such entities as governmental agencies, special interest groups, and communities. In traditional organizations, clear differentiations exist between insiders and outsiders. Some of these differentiations are legal, but many are psychological, stemming from various senses of identity, strategic priorities, and cultures. These differences led most organizations to some form of we–they relationship with external constituents. Business is done through negotiation, haggling, pressure tactics, withholding of information, and the like. When there are multiple customers or suppliers, one may be played off against another.
iv. ***Geographic***
Geographic, or global, boundaries exist when complexly structured firms operate in different markets and countries. Often stemming from national pride, cultural differences, market peculiarity, or worldwide logistics, these boundaries may isolate innovative practices and good ideas within a single country, keeping the overall company from leveraging the learning from specific countries and markets to increase company success.

Indeed, these boundaries provide an excellent roadmap that helps decipher the many intriguing characteristics of organizational Complex SoS. The juxtaposition of, and the fundamental differences among the four boundaries reveal, if not expose, the inherent tension and often disharmony that characterize and dominate organizations as Complex SoS.

In the *sixth book*, *Fundamentals of Philosophy*, Spirkin (1985) reflects in the following quote on the I-I of what we term today as Complex SoS (p. 83):

> What is a connection? It is a dependence of one phenomenon on another in a certain relationship. The basic forms of connection may be classified as spatial, temporal, causal and consequential, necessary and accidental, law governed, immediate and mediate, internal and external, dynamic and static, direct and feedback, and so on.

Spirkin presents challenging dilemmas to the modelers of Complex SoS: How many of the above-cited perspectives ought to be modeled to answer any given set of questions, for which specific time frame, in what terms, and for what purpose? What causes the performance of an *emergent Complex SoS* to not meet the collective desired behavior? Is it due to unanticipated, undetected, misunderstood, or ignored EFCs from within or from outside a system? Recall that EFCs connote internal or external and positive or negative forces that may affect an organization positively or negatively. Or is it because changes in the states of Complex SoS may result in a negative emergent behavior? Would the premise that models ought to be as simple as possible and as complex as required apply equally to small systems as well as to Complex SoS? Answers to these and similar questions would provide helpful guidelines to modelers of Complex SoS.

Clearly, there can be no universal response to the challenging perspectives highlighted by Spirkin, given the experience and creative learn-as-you-go art- and science-based technical modeling process. It is worth noting that the states of a system are functions of time and of all decisions and inputs to it. Thus, the functionality of a structure as Complex SoS is also time dependent (due to the level of the states of its maintenance and operation) and, therefore, much more difficult to predict.

Accordingly, the path to studying and understanding the relationship between the structure of an organization and the functions that it performs necessarily depends on (i) its mission, culture, and history; (ii) leadership's vision and the trust that they cultivate and with which they imbue the organization; and (iii) dynamic interpersonal relationships and communications within and among the multiple layers of the hierarchical decision-making process embraced by the organization. Furthermore, the states of the organizational structure are often dynamically evolving as functions of all inputs and other EFCs. *In other words, the functionality of organizational Complex SoS is dynamically evolving over time.*

12.I.3 MULTIFARIOUS PERSPECTIVES ON ORGANIZATIONAL BEHAVIOR OF COMPLEX SYSTEMS OF SYSTEMS

In the *seventh book* edited by Robert I. Desourdis, Jr. (2012), we learn about the challenges, bordering upon utter failure, facing organizations that fail to share information on safety-critical Complex SoS. The author terms this “interoperability” and shares with the readers several cogent and informative definitions attributed to scholarly sources. The centrality of effective communications among multiple organizations – local, state, and federal government agencies, private and public sectors, and civil and other philanthropic organizations – has been well recognized by scholars and other experts in organizational behavior. The lack of information sharing and its devastating consequences have been demonstrated in numerous other scholarly publications. In particular, the lack of information sharing among organizations, which are the depository of safety-critical historical records, has been identified by several commissions investigating large and catastrophic failures of public safety.

Consider, for example, the author of the seminal IEEE paper (Desourdis, 2012, p. 221) on the 25 Pearl Harbor catastrophic deficiencies. Indeed, Desourdis Jr. posits that all of the 25 national disasters (see Table 12.I.1) that have struck the United States share characteristics of a lack of information sharing and ignorance of lessons learned from historical catastrophic events. These catastrophes include Pearl Harbor (Japan's attack on the US Navy), 9/11 attack on the United States and New York City, Hurricane Katrina, Virginia Tech shooting, where scores were killed, BP's disastrous Deepwater Horizon oil spill

TABLE 12.I.1 Pearl Harbor Planning Deficiencies.

No.	Deficiency of	Description
1	Organization	Multiple parallel organizations with ambiguous authority
2	Assumption	Action is taken upon belief, not verified facts
3	Omission	Information sharing distribution is incomplete, people and entities excluded
4	Verification	Commands/information sent, no follow-up to ensure understanding and action
5	Supervision	No supervision to verify understanding and predictable action or find capability gaps
6	Alertness	Heightened alert is undermined by repeated training and exercises
7	Complacency	Vigilance relaxes from the day-to-day lull of "business as usual"
8	Intelligence	Centralize intelligence services with tailored dissemination
9	Attitude	Superiors do not engage in open dialogue with peers and subordinates
10	Imagination	Worst-case scenarios not included in preparedness and response planning
11	Communications	Information exchanged is ambiguous, convoluted, or contradictory
12	Paraphrase	Messages altered according to assumption and no verification
13	Adaptability	Alert and response thresholds are not matched to the known threat environment
14	Disclosure	Intelligence so protected that it is inaccessible to those who urgently need it
15	Insight	Inadequate understanding of the threat make risks poorly estimated
16	Dissemination	Information is not provided to subordinates who need to know
17	Inspection	Leaders do not know or understand their personnel and critical systems
18	Preparedness	Prepare for consequences of what a threat might do, instead of what it can do
19	Consistency	Official direction is contradicted by unofficial speculation from authorities
20	Jealousy	One-upmanship for real or perceived self-benefit prevents collaboration
21	Relationship	Personal friendships define organizational coordination, but not institutionalized
22	Priority	Failure to prioritize critical needs over day-to-day activities and conveniences
23	Reporting	Subordinates fail to report information up or, if indicated, outside the command chain
24	Improvement	Failure to identify gaps, particularly in worst-case scenarios, and correct them
25	Delegation	Responsibility is delegated to individuals or groups without the authority to act

in the Gulf of Mexico, and similar initial disastrous response phases and the associated planning deficiencies.

In the *eighth seminal publication*, Crowther (2014) offers a list enumerating common failures of information sharing. He attributes these failings to the increasing complexity of government and defense systems that has resulted in a common set of information sharing failures that have caused and exacerbated the above disasters (Desourdis and Contestabile, 2011).

Crowther, who also cites Table 12.I.1, organizes the listed failures within four categories: (i) failure of partnership, (ii) failure of communication, (iii) failure of fusion/actions, and (iv) failure of feedback. He notes that "each failure is an example mechanism or observable shortcoming that is symptomatic of organizations or individuals not exchanging information in a form that leads to successful collaboration outcomes (i.e. shared understanding, consistent decision making, and coordinated activity)."

Figure 12.I.1 (after Crowther, 2014) generalizes and organizes the failures within four categories. For example, although *ambiguity of responsibility*, *unapproachable attitudes*, and *organizational rivalry* were separate recommendations in the *Pearl Harbor report* and separate failures documented in Prange et al. (1991) and Desourdis et al. (2009); they are all variations of a failure to build the partnerships necessary to move data and information in ways that could result in collaboration.

In the classic *ninth book*, *Pearl Harbor: The Verdict of History* (The Controversial Sequel to At Dawn We Slept), Gordon W. Prange, with Donald M. Goldstein and Katherine V. Dillon, present a comprehensive accounting of the tragedy that engulfed the entire population, including private, public, and government organizations, and agencies in the United States, and possibly around the world. In the introduction, the authors write (Pearl Harbor, p. xi):

> Over the years, many Americans have seemed less interested in asking themselves "How could Pearl Harbor have been prevented? What mistakes were made and how can we avoid them in the future?" than

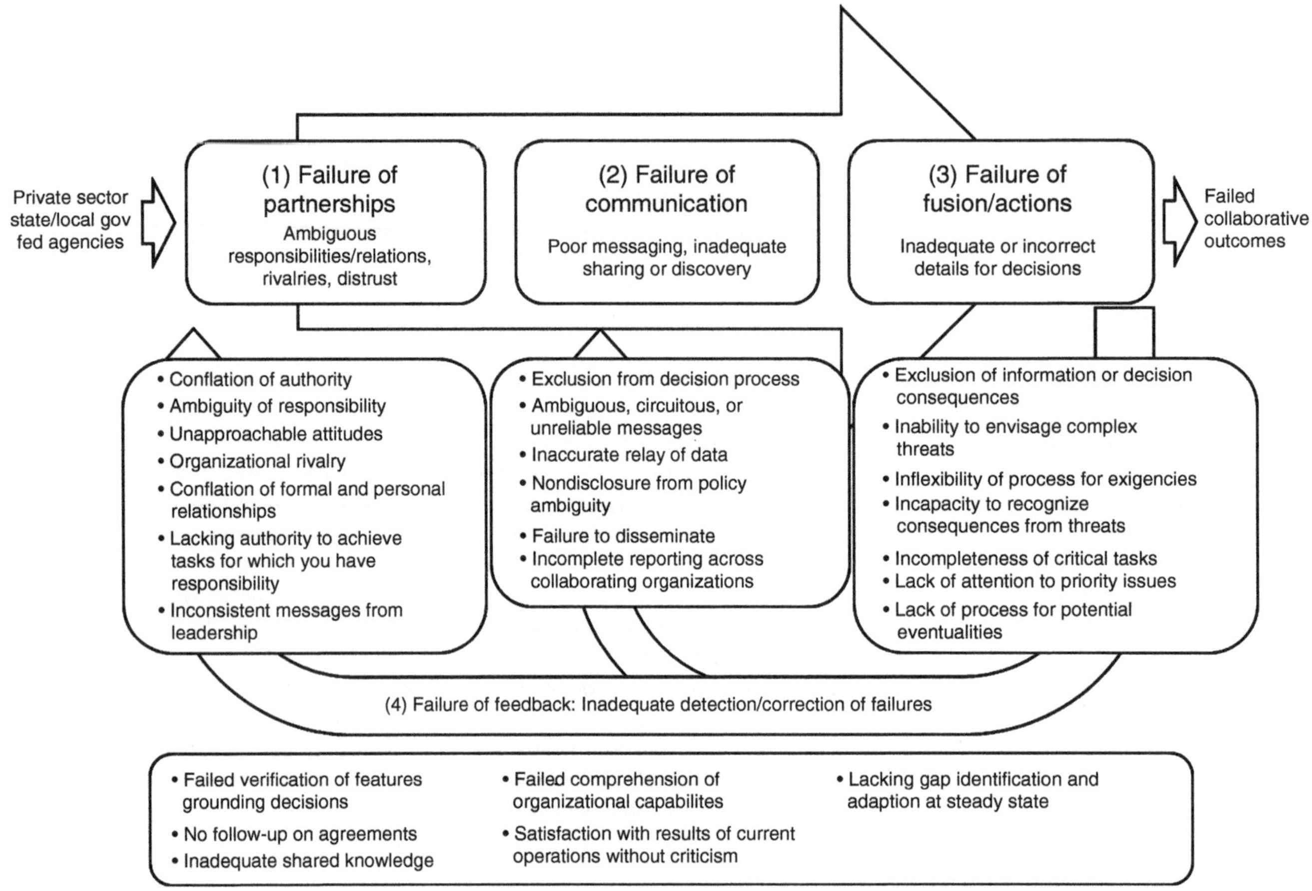

Figure 12.I.1 *Common information sharing failures, grouped in four categories. (Adapted from Crowther (2014). De Gruyter).*

in finding out "Whom can we hang?" What a relief it would be if one individual, or two, could take the rap so that everyone else could relax in the comfortable consciousness of "no fault"!

The Pearl Harbor national disaster can be attributed, among many other reasons highlighted in the most encompassing book, on organizational failures, leading among the culprits is the US Congress, which controlled the budget for national defense and security. The authors of the book *Pearl Harbor* share with us the following quote by Representative Jed Johnson of Oklahoma, who, following the attack, pointedly reminded his colleagues of the Congress's organizational failure (Pearl Harbor, p. 15):

> (t)hose Members of Congress who consistently and persistently opposed a two-ocean navy, also the Fortification of Hawaii, Guam, the Philippine Islands, and other American possessions, on the flimsy excuse that we could not afford to offend Japan, or upon the repeated assurance by them that there was not the remotest possibility of America ever being dragged into war on the Pacific, must also now realize that their advice and council [sic], if followed, would have meant total disaster, humiliation, and defeat for the United States.

Effective organizations mirror the personal quality and professional commitment to excellence of their leadership and reflect the organizational culture that has evolved over time. The following is a quotation from a Denver Post editorial (Pearl Harbor, p. 203):

> Why didn't Secretary of War Stimson and Secretary of the Navy Knox check up on Pearl Harbor defenses and find out whether their warnings had been heeded? Why didn't they ask Short and Kimmel to report on the precautionary steps taken to guard this outpost against attack? WHY DIDN'T THEY KNOW WHETHER THEY HAD ABLE AND EFFICIENT COMMANDERS IN CHARGE THERE? There must be some generals in the army

> and some admirals in the navy who haven't forgotten how to think and how to act.

The fight for organizational territorial turf, which is endemic for both government and private sector organizations, is one of the most debilitating of the factors that impede progress and efficiency. The following quote epitomizes this notion (Pearl Harbor, p. 370):

> One of the big arguments was: we wanted to take over the radar stations and get them set up and operating. The Signal Corps said no, that was their job; they wanted to set them up and get them operating and then turn them over to us for operational control. The Department headquarters decided in favor of the Signal Corps.

Flagrant expediency and disregard of established protocols, coupled with leadership's arrogance, can distract an organization from its mission, which is another common cause of organizational failure. Finally, we return to the ever-evolving conflict between the pride and self-interest of the leadership and the core organizational mission. Consider the following quote: (Pearl Harbor, p. 560).

> It is proper to suggest that, had both the commanding officers in Hawaii been less concerned between November 27 and December 7 about preserving their individual prerogatives with respect to Wake and Midway and more concerned about working together to defend the Hawaiian Coastal Frontier in the light of the warnings they had received, the defensive situation confronting the Japanese on the morning of December 7 might well have been entirely different.

In sum, the classic book *Pearl Harbor: The Verdict of History* presents a plethora of examples of organizational failures encompassing the endemic conflict between satiating the leadership's ego and sustaining the organization's mission, accomplishments, and reputation.

12.I.4 SUCCESSFUL HABITS OF VISIONARY ORGANIZATIONS

In the dust cover of the 10th *classic book, Built to Last* by James C. Collins and Jerry I. Porras (2004), the authors write:

> This is not a book about charismatic visionary leaders, it is not about visionary products or visionary market insights. Nor even is it about just having a corporate vision. This is a book about something far more important, enduring and more substantial. This is a book about visionary companies.

Indeed, in the parlance of this chapter, the above quote is about what we may learn about the drivers that ultimately engender organizational successes and failures. Collins and Porras present a comprehensive treatise on the essential attributes that characterize excellent organizations, focusing by and large on the attributes and indispensable characteristics of visionary and charismatic leaders, who enable sustained management of successful organizations that have been able to endure their dominance over decades.

The central tenet of Collins and Porras' book is that organizational success and failure is dominated by, and is a by-product of, the leadership's core values; their vision, quality, and commitment to excellence; their endurance and mentoring of homegrown management and the next generation of leadership; and their focus on "the end of the beginning." Indeed, the authors quote Winston S. Churchill: "This is not the end. It is not even the beginning of the end. But it is, perhaps, the end of the beginning." In the context of this chapter on organizational behavior associated with Complex SoS, it is constructive to benefit from the above authors' perspectives on the frequent practice of organizations that indulge in generating expansive plans without bringing them to fruition. Consider the following quote: *Just because a company has a 'vision statement' (or something like it) in no way guarantees that it will become a visionary company!* (Collins and Porras, 2004, p. 201).

The authors provide guidelines for CEOs, managers, and entrepreneurs. A key step in building a visionary company is to articulate a core ideology, equating the latter as the sum of core values plus purpose. According to Collins and Porras (p. 76): *In a visionary company, the core values need no rational or external justification. Nor do they sway with the trends and fads of the day. Nor even do they shift in response to changing market conditions.*

Clearly, the above core values and visionary approach that Collins and Porras articulate in leading and managing organizations are directly applicable to the core values and structure of organizations that in their essence constitute Complex SoS. They define core ideology as the sum of core values and purpose (Collins and Porras, p. 73):

> Core Values: The organization's essential and enduring tenets – a small set of general guiding principles; not to be confused with specific cultural or operating practices; not to be compromised for financial gain or short-term expediency.
> Purpose: The organization's fundamental reasons for existence beyond just making money – a perpetual guiding star on the horizon; not to be confused with specific goals or business strategies.

In sum, developing, adhering to, and enforcing the core values of organizations as Complex SoS are the *sine qua non* for their success and endurance.

The 11th *classic book* on organizational behavior, *Good to Great* (2011) by Jim Collins, who is also a coauthor of the book *Built to Last*, offers a panorama of requisites for successful organizations. Collins identifies what he terms *Level 5 leadership* and quotes Harry S. Truman: "You can accomplish anything in life, provided that you do not mind who gets the credit." To that end, he develops five levels of leadership (Collins, p. 20):

> Level 1: Highly Capable Individual: Makes productive contributions through talent, knowledge, skills, and good work habits.
> Level 2: Contributing Team Member: Contributes individual capabilities to the achievement of group objectives and works effectively with others in a group setting.
> Level 3: Competent Manager: Organizes people and resources toward the effective and efficient pursuit of predetermined objectives.
> Level 4: Effective Leader: Catalyzes commitment to, and vigorous of, pursuit of a clear and compelling vision, stimulating higher performance standards.
> Level 5: Executive: Builds enduring greatness through a paradoxical blend of personal humility and professional will.

On Level 5 Collins states on p. 21: Level 5 leaders channel their ego needs away from themselves and into the larger goal of building a great company. It's not that Level 5 leaders have no ego or self-interest. Indeed, they are incredibly ambitious – but their ambition is first and foremost for the institution, not themselves.

The 12th *book* on *Engineering Systems* (2016) is authored by three scholars: Olivier L. de Weck, Daniel Roos, and Christopher L. Magee. The genius of the three authors stems from their emergent vision about systems engineering versus engineering systems, encapsulated in the following brief statement in the preface:

> What we call Engineering Systems are the systems that fulfill important functions in society: transporting people and cargo, transmitting messages, providing energy, improving our health, and so forth. They are characterized by high levels of interwoven technological and social complexity.

On p. 31 the authors define engineering systems as: "A class of systems characterized by a high degree of technical complexity, social intricacy, and elaborate processes, aimed at fulfilling important functions in society."

The authors provide a succinct paragraph in the Appendix of the book on the "Complexity in Engineering Systems," which is the theme of this book. They write:

Arguably the key concept in engineering systems is complexity. Thus, we discuss it at greater length than the other terms that will follow. There are many definitions of complex systems, but we shall concentrate on just two.

> A system is behaviorally complex if its behavior is difficult to predict, analyze, describe or manage. In contrast, a system is structurally complex if the number of parts is large and the interconnections between its parts is intricate or hard to describe briefly. Systems that are structurally complex are usually behaviorally complex. Systems that have complex behavior need not have complex structure, since we know of relatively simple mechanical systems whose behavior is chaotic and hence complex. On the other hand, most behaviorally complex systems are structurally complex as well.
>
> Complex engineering systems are not simply technical in nature, but rely on people and their organizations for the design, manufacturing, and operation of the system, and are influenced by and influence the societal and physical context as well. (p. 185)

The above discussion necessarily leads us to the following premises that characterize organizational Complex SoS:

1. They have multiple systems and subsystems populated with similarly multiple hierarchies of personnel with competing but complementing goals and objectives.
2. Each level assumes distinct goals and objectives, which must be harmonious with those at the other levels of the organizational structure.
3. The I-I within and among the subsystems and the corresponding levels of the hierarchical structure necessitate harmonious collaboration and support within and among the organization.
4. Interpersonal communication and harmonious relationships (albeit, not always easy to achieve) within and among the players associated with subsystems and systems are imperative. These and other attributes are fundamental to successful organizations as Complex SoS.

12.I.5 HARMONY VERSUS DISHARMONY AMONG SUBSYSTEMS THAT CONSTITUTE COMPLEX SYSTEMS OF SYSTEMS

It is unrealistic to assume that a harmonious interplay always exists among the organizational subsystems that make up Complex SoS. For example, one might expect that the shared states, resources, decisions, decision makers, behavioral norms, and stakeholders would necessarily generate tensions and disagreements among the parties' stakeholders. Cognizant of this fact, the question that faces modelers of Complex SoS is the proper way to account for this reality when projecting the cause–effect relationship and outcomes associated with specific inputs and assumptions. For example, to what extent would harmonious as well as disharmonious relationships among the subsystems (what Arquilla and Ronfeldt (2001) term "tribes"), which often represent multiple factions among decision makers and stakeholders, affect the functionality of Complex SoS? And how should our modeling efforts reflect and account for this reality in order to achieve more realistic models? This observation necessarily requires that modelers, especially managers and other stakeholders, pay careful attention to the states, structures and functionality of the subsystems, and the relationships among the individual subsystems and the entire Organization Complex SoS. Furthermore, each subsystem is likely to represent diverse structural, organizational, functional, geographical, political, or cultural physiognomies. Thus, the intrinsic multifaceted characterizations of the diverse member subsystems that compose Complex SoS ought not to escape their would-be modelers and managers. Clearly, the above multiple organizational structures and the associated hierarchical levels of authority and decision making, render the role of organizations to be central in the overall understanding and modeling of Complex SoS and, ultimately, of their effective management.

Recall that Spirkin presents the following set of challenging dilemmas to the modelers of Complex SoS: (i) How many of the above-cited perspectives ought to be modeled to answer any given set of questions for a specific time frame? (ii) When the performance of the models of Complex SoS is disharmonious with the anticipated behavior, whether by unanticipated, undetected, misunderstood, or ignored EFCs from within or from outside a system, the state changes may result in a negative emergent behavior. Recall that EFCs connote *trends in external or internal sources of system risk that may adversely affect specific states of that system*. Would the premise that models ought to be as simple as possible and as complex as required to answer specific questions provide helpful guidelines to SoS modelers?

Clearly, there can be no universal response to the challenging perspectives highlighted by Spirkin, given the experience and creative learn-as-you-go art- and science-based technical modeling process. It is worth noting that the states of a system are functions of time and of all decisions and inputs to it. Thus, the functionality of a structure as Complex SoS is time dependent (due to the level of the states of its maintenance, operation, and management).

12.I.6 MULTIFARIOUS PERSPECTIVES ON ORGANIZATIONAL BEHAVIOR OF COMPLEX SYSTEMS OF SYSTEMS: REVISITED

In the 13th *seminal book* by the philosopher Jacob Bronowski (1978), *The Origin of Knowledge and Imagination*, we find some guiding principles for modeling and managing Complex SoS, building on the two groups of arts espoused by Bronowski that human beings practice by (i) arts, which are mediated by the sense of light, like sculpture and painting; and (ii) arts, which are mediated by speech and sound, e.g. the novel, drama, and music. Of the human senses, Bronowski argues that these two dominate our outlook! Although most of the time we use vision to give us information about the world and sound to give us information about other people in the world, we ought to embrace additional "mediations" for modeling Complex SoS. This leads us to the necessity of a multifarious characterization of Complex SoS. It is constructive to project the two media of communications (embraced by Bronowski) to individuals who populate and essentially manage the organizational Complex SoS environment. *The metaphysical message of Bronowski has deeper and broader implications than merely to the arts, because the interpersonal relationships and organizational behavior that are by and large manifested via communications among the very large number of workers, and which constitute the heart and soul of an organizational Complex SoS, dominate the culture and ambiance that prevail day in and day out.* The above implications have even more subtle impacts on the productivity, accomplishments, and successes of the organization.

In terms of modeling organizations as Complex SoS, each of the 13 perspectives enunciated by the authors of the 13 seminal publications necessarily could lead to a different modeling schema of organizational Complex SoS. Each could also lead to a different decomposition of the organization at hand through which to gain some degree of understanding of the corresponding subsystems and the organizational Complex SoS as a whole. Note that each decomposition would lead to different shared states, resources, decisions, and other essential entities. Since modeling is aimed at answering specific questions and to enhance our understanding of Complex SoS, then there is a very close relationship between the (i) selected decomposition, (ii) specific aspects of the organizational Complex SoS that ought to be modeled, (iii) decisions that have to be made, and (iv) specific time frames.

Here we revisit the "flipping the HHM" modeling schema already documented in Chapter 3 on Hierarchical Holographic Modeling to enable us to gain a better understanding of the I-I among the systems and subsystems and address and incorporate important shared states, decisions, and resources among the subsystems.

Flipping the HHM enables us to discover *important, hidden, nonobvious, or unperceived perspectives* heretofore through visible representation of inter- and intradependencies. Given that the modeling of organizational Complex SoS is essentially an unending and iterative, learn-as-you-go, and trial-and-error process, then flipping the HHM provides another visual conceptual-imaginative and powerful process through which to benefit from our inner conscious and subconscious ability to build and construct seemingly unrelated pieces/components/subsystems with new meanings and perspectives. This process is not dissimilar from the journey of discovery that enables painters, musicians, and architects, not to mention children playing with "Legos," to create new worlds and new realities.

As we continue in our never-ending process of understanding and modeling organizational Complex SoS through the HHM discovery journey, we also continue to identify shared states, decisions, and other essential entities. This discovery process, via flipping the HHM, seems imperative in our quest to build more representative models by capturing as many previously veiled interdependencies and interconnections within and among the subsystems and systems of Complex SoS.

Clearly, the multifarious perspectives, dimensions, and complexities associated with the I-I characterizing organizational Complex SoS can neither be effectively understood nor modeled by a single modeler. To be effective, the dynamical modeling process must be performed within a cross-disciplinary team whose members are well versed with specific perspectives and subsystems and collectively cover the required domain knowledge and experiences required in the complex modeling process.

To build on this idea, we defer to the 14th *book*, *The Wisdom of Teams* by Jon R. Katzenbach and Douglas K. Smith (2015), in which the authors identify four attributes of an effective team (p. 45):

> (i) A small number of team members; (ii) Multiple members with different experiences and perspectives; (iii) All members share the same goals and objectives; and (iv) All members are mutually accountable.

Indeed, there is no specific norm guiding the composition of modeling teams either for single systems or for any given Complex SoS. Rather, the selection of specific expertise and the levels of the team members' contributions to the modeling process are commonly built with core resident modelers and augmented with other expertise as needs arise over time.

A representative and comprehensive database constitutes the essential element/foundation for an effective and representative modeling of all systems; Complex SoS are of no exception. Flipping the HHM, introduced in Chapter 3, and within it, the discovery of the multiple layers of perspectives opens new and heretofore unimagined interdependent and interconnected subsystems that make up Complex SoS. Associated with this process of discovery is the imperative requirement to answer the following basic questions: What are the extent and the characteristics of the database that we ought to assemble that would help us to (i) understand the I-I within and among the systems, (ii) verify and validate, to the extent possible, the soundness of our modeling assumptions and constructions, and (iii) account for the dynamic variability of the subsystems and thus the Complex SoS, given the criticality of the time frame, especially in modeling organizational behavior of Complex SoS. Note that each subsystem of Complex SoS is likely to operate on a different time frame, ranging from seconds or minutes to years or decades. This is particularly true for the natural and the constructed environment and for organizational Complex SoS.

It is likely that in this era of big data, data collection and analysis and their associated updating process would evolve into improved levels of modeling sophistication and ingenuity. *One fundamental challenge associated with data collection and analyses is relating the dynamic evolution of the subsystems that make up Complex SoS to the respective time frame during which a database was collected.* This imperative correlation is not an easy process and is fraught with fundamental challenges, both technical and organizational.

How many perspectives of a system must be considered by modelers to achieve a close-to-a-holistic model(s)? And are we able to conceive and discover all the essential attributes, characteristics, and perspectives of Complex SoS? Such open-ended questions reinforce the notion that the modeling process is a journey of discovery, imagination, and creativity, entailing that when we think we have succeeded, we are likely to be proven failed. This assertion ought to be interpreted constructively and philosophically, but never fatalistically. In other words, we must recognize and acknowledge that the modeling process is an open-ended continuous journey of learning and exploration that is characterized by "successes" and "failures" during which progress is made, and eventually, models are declared representative and valuable. Furthermore, the fact that all systems – natural or the constructed environment – are dynamically changing and evolving necessarily requires the imperative consideration of the time frame in modeling, whether implicitly or explicitly, when possible.

Modeling dynamic systems necessarily requires comparably dynamic models. Such a requirement adds enormous challenges to modelers, who often revert to a steady-state model with the understanding of the adage that "models must be as simple as possible, but as complex as required." In other words, since the essence of modeling is an amalgamation of art and science, imagination, judgment and experience, and so on, then "assumptions" become an essential instrument that modelers use to navigate between the grace of static simplicity and the harshness of dynamic complexity, with the ever required challenging balance between the two. Of course, the choice between static and dynamic models is only one of the challenges facing models of Complex SoS. Not all submodels of subsystems necessarily require the same inherent characterization, e.g. linear versus nonlinear, static versus dynamic, deterministic versus stochastic–probabilistic, lumped parameter versus distributed parameter, or discrete versus continuous. Here again, modelers of organizational Complex SoS necessarily resort to the essential guidance provided by the arts and sciences, namely, the creativity and imagination that constitute the foundation of the modeling process. This never-ending process of tradeoffs necessarily gets resolved with justified assumptions by the modelers and by their ultimate users.

Interpersonal and intraorganizational relationships are dynamic and multifarious, a fact that makes the modeling of organizational Complex SoS a challenging enterprise, and from many perspectives opaque. Our observing, reading documentation, conversing with insiders, and exploring and exploiting all sources of information relevant to the organizational Complex SoS are tasks that are important and valuable. Nevertheless, this tedious and essential process does not remove the inherent mystery characterizing Complex SoS. Similar to an X-ray image of a body, the above solicited and collected information ought not to lead us to the illusion that what we observe is the entire reality. To augment our observations and added information with our understanding and modeling of the essence of real organizational Complex SoS is essential to their modeling. Indeed, we ought to embrace the notion that the modeling of organizational Complex SoS is an intricate amalgamation of the arts and sciences and engineering, guided by the ingenuity of the systems modelers.

There is a strong correlation between the (i) inner hierarchical structure and approach to governance and management and (ii) architectural design of the buildings and structures that house organizational Complex SoS. The physical–structural dynamics and the visible and invisible relationships that characterize cultural and intraorganizational relationships together are a manifestation of an organization's culture and core values. In many ways, the roles of the architects, who design and oversee the construction of the structures that house the organization, and that of the systems modelers, share inherent commonality, albeit differing in nature: The architect aspires to translating and bringing to reality the image, culture, and essence of the organization via the yet unborn multistory high-rise building, while the modeler attempts to represent the operational and core values of the organization via its modeling as organizational Complex SoS. Clearly, each discipline uses different crafts in its work. Systems modelers use the building block of mathematical and simulation models in the trial-and-error and a never-ending modeling process, while architects embed in their craft the art and science of architecture by augmenting their talent and imagination with the art and science of structural engineering.

12.I.7 WHAT HAVE WE LEARNED FROM PHILOSOPHERS ABOUT SYSTEMS

We read again in Spirkin (1983):

> A system consists of something more than structure: It is a structure with certain properties. When a structure is understood from the standpoint of its properties, it is understood as a system...So, structure is actually the law or set of laws that determine the system's composition and functioning, its properties and stability...The life of a structure manifests itself in its function, they condition each other. The structures of the organs of the body, for instance, are connected with their functions. Any breakdown in structure, any deformation of an organ leads to a distortion of the function. (p. 98)

To emphasize the I-I between the structure and functionality of a system, Bronowski (1978) uses the organs of the body as an example to relate the structure of a system to its functionality – different yet not unrelated perspectives on I-I between two essential human structures, eyes and ears, and their functionality (Bronowski, 1978):

> If you think of it, there are essentially only two groups of arts that human beings practice. One group includes arts, which are mediated by the sense of light, like sculpture and painting; the other group includes arts, which are mediated by speech and sound, like the poem and the novel and the drama and music. Of the human senses these two dominate our outlook. (p.10)

Fundamentally, this construal of the modeling process represents a mental translation that implies subjective cognitive understanding of each of the above multiple perspectives of the system as such and their integration as Complex SoS. Conceivably two different modelers would interpret and perceive the components (subsystems) and the integrated system of systems differently, given the amalgamation of the arts and sciences on which the modeling process is built. Furthermore, given the large number of state variables required to represent the multiple subsystems and their multitude of perspectives and the necessity for brevity yet representativeness, modelers from different disciplines and thus perspectives would naturally tend to be influenced by their backgrounds. Furthermore, the large number of states and other essential entities generated through the inquisitive and learn-as-you-go modeling process often leads to the selection of the subset of *essential states*. These would be the minimum number of state variables with which each subsystem model, and the Complex SoS as a whole, can be reasonably represented at an acceptable level. Here again we confront the necessary tradeoffs that dominate all systems modeling – representativeness that balances complexity with reasonableness.

Working collaboratively, modelers can develop a better model that espouses to combine the team's ingenuity and interpretation of the subsystems and of the entire Complex SoS. Alternatively, it is possible to envision separate modeling efforts by, say, two modelers and subsequently attempt to integrate the two models to yield better and more representative attributes of the overall Complex SoS. We ought to not overlook the inherent fact of the innovativeness of modeling that is characterized by a trial-and-error and learn-as-you-go process. In other words, the multipath exploration process that characterizes the modeling effort necessarily implies capturing the multiple perspectives of the Complex SoS, thus yielding better and more representative models. The above discussion is harmonious with Bronowski's (1978, p. 96) seminal statement:

> The world is totally connected. Whatever explanation we invent at any moment is a partial connection, and its richness derives from the richness of such connections as we are able to make.

In the 15th *book*, *Ageless Body, Timeless Mind*, the physician and philosopher, Deepak Chopra (1994) suggests the following three "models" of humans: physiology, mental capacity, and spirituality. No one would negate the notion that the human body is interdependent and interconnected Complex SoS. Indeed each organ is by itself a Complex SoS composed of multiple subsystems. The basic question is: Can we model (represent a complete understanding of a person) by ignoring any one of their attributes identified by Chopra? The same principle of completeness/representativeness must apply to other Complex SoS.

In the 16th *book* by Peter F. Drucker, who is considered one of the most influential philosopher and organizational behavior scholars, we quote from his book *The Essential Drucker* on "What is management?" (2001, pp. 10–11):

> Management is about human beings. Its task is to make people capable of joint performance, to make their strengths effective and their weaknesses irrelevant....Every enterprise requires commitment to common goals and shared values. Without such commitment there is no enterprise; there is only a mob. The enterprise must have simple, clear, and unifying objectives. The mission of the organization has to be clear enough and big enough to provide common vision. The goals that embody it have to be clear, public, and constantly reaffirmed. Management's first job is to think through, set, and exemplify those objectives, values, and goals...Management must also enable the enterprise and each of its members to grow and develop as needs and opportunities change....It must be built on communication and on individual responsibility. All members need to think through what they aim to accomplish – and make sure that their associates know and understand that aim...Performance has to be built into the enterprise and its management; it has to be measured – or at least judged – and it has to be continually improved... The result of a business is a satisfied customer.

On p. 273 Drucker summarizes what he calls "The Dos":

> (i) Purposeful, systematic innovation begins with the analysis of the opportunities. It begins with thinking through what I have called the seven sources of innovative opportunity. In different areas, different sources will have different importance at different times: (i) The organization's own unexpected successes and unexpected failures, but also the unexpected successes and unexpected failures of the organization's competitors; (ii) Incongruities, especially incongruities in the process, whether of production or distribution, or incongruities in customer behavior; (iii) Process needs; (iv) Changes in industry and market structures; (v) Changes in demographics; (vi) Changes in meaning and perception; and (vii) New knowledge.

On p. 275 Drucker summarizes what he calls "The Don'ts":

> (i) The first is simply not to try to be clever. Innovations have to be handled by ordinary human beings, and if they are to attain any size and importance at all, by morons or near-morons. Incompetence, after all, is the only thing in abundant and never-failing supply. Anything too clever, whether in design or execution, is almost bound to fail; (ii) Don't diversify;

don't splinter; don't try to do too many things at once. This is, of course, the corollary to the "do": be focused! Innovations that stray from a core are likely to become diffused; (iii) Finally, don't try to innovate for the future. Innovate for the present! An innovation may have long-range impact; it may not reach its full maturity until twenty years later.

12.I.8 THE ROLE OF POLICY FORMULATION IN ORGANIZATIONAL COMPLEX SYSTEMS OF SYSTEMS

There is a strong interplay between effective policy formulation and decision making that applies to, and equally propels, the success of both the public's harmonious aspirations and the organizational mission and its success. The decision-making process, for both the public and the private sectors, when it is properly grounded on visionary and well-formulated policies would commonly lead to more successes and tangible accomplishments to the affected constituencies. In many ways, policy formulation is one of the most creative, encompassing, and consequential propelling activities affecting all government and other public and private organizations. Indeed deliberate and thoughtful policy formulation can be expected to serve as consequential instrument in charting the future destiny, successes or failures, of the organization.[1]

Thus, if policy formulation must be a precursor to decision making, especially for public and organizational Complex SoS, then why do so many organizations fail over time? Answers might be attributed to the diminishing role of visionary players in the leadership, or unanticipated EFCs from within or outside the organization.

Contrary to the common deliberate formulation of corporate/organizational policies, public policy formulation must often accommodate multiple constituencies with competing and conflicting goals and objectives and with different advocated time frames.

In sum, policy formulation must be built on effective and representative models of the organizational Complex SoS as prelude to decision making. Given that decisions change the states of the system at many levels of the organization, it is imperative that effective policy analyses be performed prior to decisions to assess their impacts on the essential states of the system and subsequent impact on major goals and objectives. *In many ways, policy formulation constitutes the foundation upon which effective, sustainable, and viable decisions can be ultimately made, and it can serve as a potent instrument in risk reduction and prevention.* Note that building on a systemic modeling process enables the projection of impacts of current decisions on future options. Indeed, often decisions made in haste fail to assess their future impacts and ramifications on other important policies or states of specific systems, or on the entire Complex SoS. Furthermore, if the ultimate efficacy of prudent decision making and risk analysis is to not only manage future risk scenarios but also avoid the creation of unintended sources of risk and other consequences, *then building on policy formulation prior to decision making constitutes prudent risk avoidance and a precursor to risk management.*

Individual systems and subsystems of organizational Complex SoS commonly experience uncounted adversarial emergent forced tracking, and tracing precursors to such adverse events constitutes a proactive risk analysis. Such prudence calls for a continuous process of designing a data-collection mechanism: (i) developing metrics with which to measure changes in the Complex SoS, (ii) assessing whether observed changes are sufficiently significant, and (iii) determining criteria for actions. The advancements in sensing, communication, social networks, and information processing technology provide a foundation for a proactive risk assessment tool by building on information to detect signs prior to a system's failure. This can be achieved through the observation and/or prediction of changes in the state space of the system. Precursor analysis can track different failure modes of a specific unit of the organization, especially for cyber–physical (CP) systems, and it can forecast indicators and warnings of possible

[1] A personal note
During his sabbatical year as AAAS/AGU Congressional Science Fellow, the author of this book served for nine months on the House of Representatives' Subcommittee on the Environment, Committee on Science and Technology; and for three months in the Office of Science and Technology Policy, Executive Office of the President. In 1980, he completed the Executive Program (TEP) of the Darden School of Business at the University of Virginia.

system failures. Precursors to organizational failures can also be detected; however, this requires prudence and vigilant commitments from managers at all levels of the organization. Moreover, precursors can be used to determine whether adversarial events are either occurring or expected and provide opportunities for managers and decision makers to make more timely and informed decisions to respond to adverse consequences and to prevent future severe consequences.

12.I.9 ORGANIZATIONAL ROLE IN PLANNING AND MANAGEMENT OF WATER RESOURCES COMPLEX SYSTEMS OF SYSTEMS

River Basins epitomize the I-I characterizing Complex SoS. They commonly transcend hydrological and political boundaries, multiple planning time horizons, multiple conflicting and competing goals and objectives, involving diverse communities, and multiple decision makers and stakeholders. River Basin Complex SoS encompass human, agriculture, and environmental protection from flooding; and assurance of quality of water supply to population commonly exceeding millions, residing in multiple states that may span thousands of square miles of hydrologic area as well as City, County, or State boundaries. The I-I characterizing the organizational complexity of River Basins stem from the fact that State and County political boundaries commonly transcend hydrological boundaries, as well as all other political and social divisions. Clearly, more than one analytical or conceptual model is likely to emerge; each of which may focus on a specific aspect of a region or of the River Basin Complex SoS as a whole, yet all may be regarded as acceptable representations of the River Basin. This phenomenon is particularly common in hierarchical modeling of Complex SoS, because they often present a dilemma over the choice of the perceived critical subsystems that ought to be incorporated in the modeling and planning process. For example, an economic system may be decomposed into geographic regions or activity sectors. An electric power management system may be decomposed according to the various functions of the system (e.g. power generation units, energy storage units, transmission line units) or along geographic or political boundaries. Another decomposition might be a time-frame decomposition into planning periods. For an effective modeling of Complex SoS, several aspects of the subsystems must be addressed, because each adds value to our understanding and improves our modeling of the systems that make up Complex SoS.

For example, the Maumee River Basin may be modeled from four overlapping perspectives – (i) hydrological, (ii) geographical, (iii) organizational, and (iv) political (see Figures 12.I.2 and 12.I.3) – as well as multiple objectives and several time planning horizons.

The Maumee River Basin as Complex SoS requires human, agriculture, and environmental protection from flooding and assurance of quality of water supply to population exceeding four million, residing in three states (Indiana, Michigan, and Ohio), which span close to 7000 mi^2 of hydrologic area and close to 9000 mi^2 of county boundaries. The sources of risk to the River Basin Complex SoS can originate from any of the multiple overlapping perspectives depicted in Figure 12.I.3, characterizing its systems and subsystems. Thus there are intriguing challenges that systems modelers face in deciphering the myriad sources of risk to Complex SoS. In particular, the many sources of risk and their distribution within the multiple towns, municipalities, and states, as subsystems, often play a dominant role in the allocation of shared resources. Tradeoffs in risk management must be made among the multiple objectives associated with the affected systems and subsystems of the Complex SoS. Other examples of the multiple sources of risk to the River Basin include risk of drought and floods affecting populations in urban and agricultural areas along with their associated health and community well-being. Also, the River Basin spans an area of approximately 6000 mi^2 over parts of the states of Ohio, Michigan, and Indiana and has been divided into five planning subareas (PSAs), each one consisting of several counties (political/geographic decomposition).

From risk analysis and flood management perspectives, the basin can also be divided into eight watersheds crossing state and county boundaries (hydrological decomposition), as shown in Figure 12.I.3. Consider also the following seven major risk management objectives from the functional decomposition perspective: (i) protect agricultural land, (ii) reduce erosion and

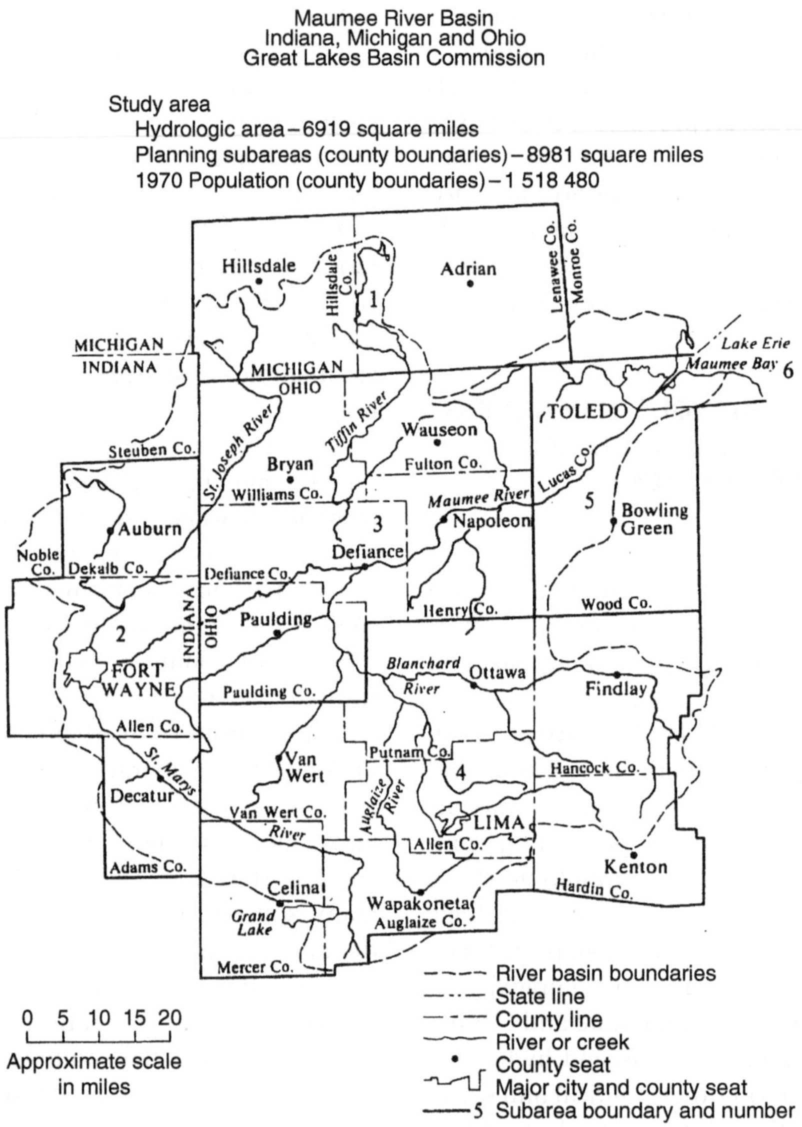

Figure 12.I.2 *Political–geographic decomposition of the Maumee River Basin.*

sedimentation, (iii) enhance water quality, (iv) protect fish and wildlife, (v) enhance outdoor recreational opportunities, (vi) reduce flood damage, and (vii) supply water needs.

Each modeling representation of a specific configuration of the subsystems of the River Basin Complex SoS represents and uncovers important sources of risk not evident using a single system model. In other words, Complex SoS, such as the Maumee River Basin, must be modeled and the associated sources of risk be assessed and studied from many perspectives and configurations. This intensive modeling process can be achieved and the risk be assessed from the following overlapping shared states and other essential entities: *(i) five PSAs each consisting of several counties (political/geographic risk modeling perspective), (ii) eight watersheds crossing state and county boundaries (hydrological risk modeling perspective), (iii) seven major objectives (functional risk modeling perspective), and (iv) three planning time horizons (temporal risk modeling perspective).*

In sum, we present the multidimensionality of the Maumee River Basin as a representative of organizational Complex SoS, along with the associated large number of decision makers and stakeholders, representing multiple constituencies, while their decisions affect cross-county and cross-state boundaries, laws, and regulations. *Clearly, the above intertwined natural*

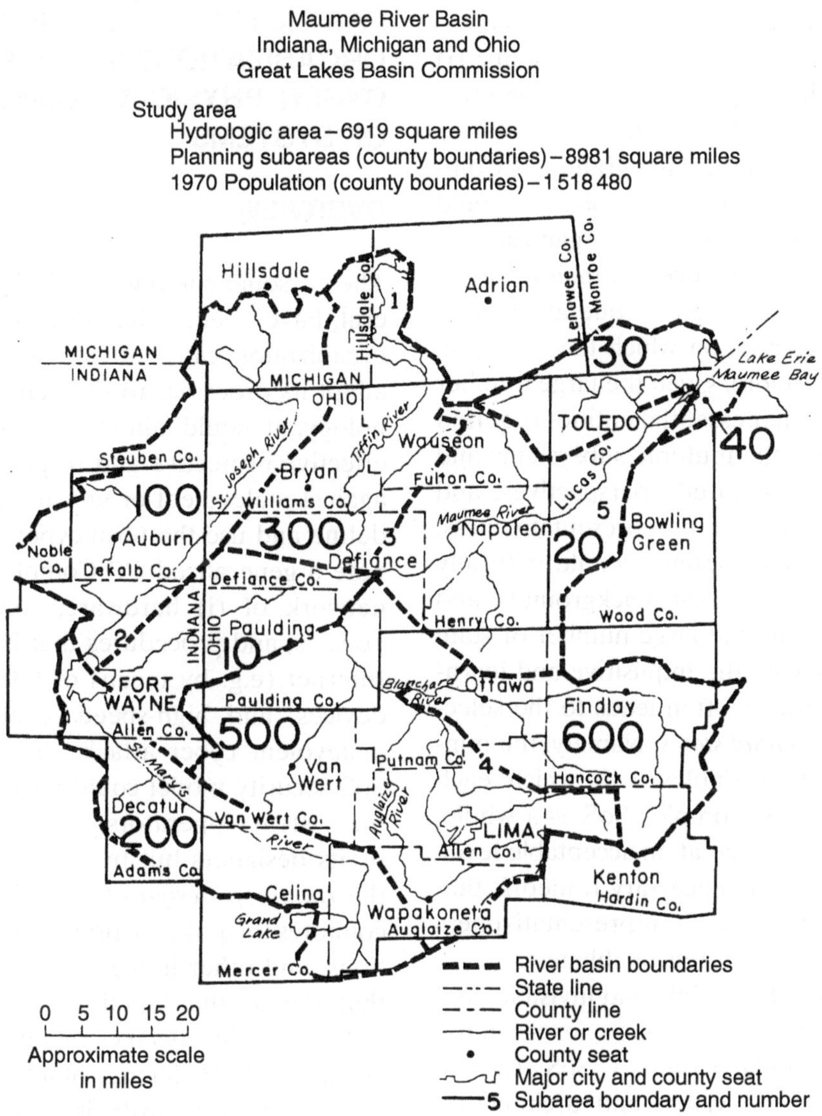

Figure 12.I.3 *Hydrological decomposition of the Maumee River Basin.*

environmental, legal, cultural, organizational, and political users and other entities defy and challenge the capability of any single model to represent the essence of the associated risk modeling and analysis process.

12.I.10 FINAL PHILOSOPHICAL REFLECTIONS ON PART I

In many ways, the two philosophers, Bronowski and Spirkin, provide us with at least two essential building blocks for modeling organizational Complex SoS – their *structure* and *functionality*. As we practice the art and science of systems modeling, we note that modelers commonly build on (i) domain knowledge of the subject, (ii) systems engineering theory and methodology, (iii) database, (iv) modeling experience, and (v) their imagination, ingenuity, and artistic creativity and philosophy, among others. One may view modeling as the perception and interpretation of multiple perspectives, including natural, physical, structural, organizational, or human behavior. An open-ended question remains unanswered: How much inference, perception, and ingenuity, among other capabilities, do effective modelers need to be able to translate the reality of organizational Complex SoS and the associated environment into an abstract vision that builds on state-space theory? Fundamentally, this construal modeling process represents: (i) a mental translation that implies subjective cognitive understanding of each of the multiple perspectives of the organization; and (ii) their integration as Complex SoS. Conceivably two different modelers would interpret, perceive, and model the organizational Complex SoS differently, influenced by their personal talent and experiences, and their unique amalgamation of the arts and sciences on which the modeling process is built. Furthermore, given the large number of state variables and other essential entities required to model and represent the multiple subsystems and of Complex SoS, their multitude perspectives, and the necessity for brevity yet representativeness, modelers from different disciplines, would naturally tend to be influenced by their backgrounds and experiences. Furthermore, the large number of state variables required through the inquisitive and learn-as-you-go modeling process often leads to the selection of the subset of *essential states*. Namely, the minimum number of state variables with which each subsystem model, and the Complex SoS as a whole, can be reasonably represented at an acceptable level. Here again we confront the necessary tradeoffs that dominate all systems modeling – representativeness that balances complexity with reasonableness.

The amalgamation of the visible and invisible, and the interplay between the arts and the sciences in the modeling of organizational Complex SoS, can be demonstrated through the vision and architecture involving the elaborate design of a 100-story high building. These images and the ultimate translation are envisioned by the architects in their drawings and following the tedious process through which the latter are materialized into the reality of a constructed building. In many ways, the role of the architects and that of the systems modelers share many similarities: Both translate and bring to reality the imaging of the "unborn yet" multistory high-rise building and models of Complex SoS. Clearly, each discipline uses different crafts in its work: Systems modelers use the building block of mathematical and simulation models, while architects use in their crafts in drawings, design, and other mechanisms to represent strength of materials, time frame, and the essential sequences through which they translate their models into the ultimate construction and utilization.

PART II: MODELING THE ROLE OF ORGANIZATIONS IN THE RESILIENCE OF CYBER–PHYSICAL COMPLEX SYSTEMS OF SYSTEMS

OVERVIEW

The personal computer and cyber technology in general, have opened the doors to the world. Unless an organization can cultivate the needed flexibility to adapt its processes to the demands of this new technology, it could diminish its capability to execute effective protective and responsive countermeasures against malevolent or unwanted cyber intrusion. We define and use the term cyber–physical (CP) to connote a generalized CP Complex SoS that includes a network of (i) hardware, software, organizational policies, and procedures that is (ii) connected to the Internet (e.g. by means of telephone lines, wireless devices, and high-speed technology). Following a malevolent cyber attack, the original CP Complex SoS security model could have been (unbeknown to the system's operators) perturbed and changed, not by its designers but by adversaries. In other words, the *roadmap design* of the CP Complex SoS, which connotes the policies, procedures, methods of connections, and other human and organizational activities that define the functionality of the CP Complex SoS, could have been changed following a malevolent cyber attack. Resilience of a CP Complex SoS, which is central to its security, is defined as the ability of the system to withstand a major disruption within acceptable degradation parameters and to recover within an acceptable composite cost and time. Recall from previous chapters that the resilience of a system is a manifestation of its states. The central role of systems integration in cyber security is highlighted because it not only connotes the integration of components but also involves an understanding of the functionality that emerges from that integration. Systems modeling is indispensable in developing awareness when a CP is intruded and is being violated. CP Complex SoS are essentially multidimensional and of multiscale compositions – from molecular to macro dimensions – and its wholeness is characterized by multiple and varied paths of interdependence. Thus, Part II of this chapter advocates a holistic approach as a requisite to understanding not only its myriad parts and functionality of CP Complex SoS, but also the forced changes imposed upon it both from within and without its domains. Furthermore, models can play a central role in intrusion detection, thus enabling information managers to develop protective and remedial courses of action. They can also be used to answer a wide range of questions, including how to deceive would-be intruders and lead them to believe they are inflicting damage on the real intended system. Through logically organized and systemically executed models, the Phantom System Models (PSM) laboratory, introduced in Chapter 4, is aimed at providing a reasoned experimental modeling framework with which to explore and thus understand the intricate relationships that characterize the nature of multiscale emergent systems.

12.II.1 INTRODUCTION

One of the strengths of any large and successful organization is that it has well-designed and -integrated policies, management, and culture with the ability to execute intricate procedures in a harmonious, fast, and repeatable manner. The foundation for this capability lies primarily in the vast array of organizational culture and collective doctrines, and procedures that the organization has cultivated and established over the years. These processes, moreover, are based on the education, training, and perhaps most importantly the cumulative experience of the organization's workforce. Namely, in any successful organization these processes, or sets of best practices, have developed as a direct result of its workforce learning what works and what does not. Lessons learned are invaluable; they cannot be duplicated without the organization going through similar experiences. However, once a successful corporation has arrived at the top of its game and has clearly emerged as the leader among its competitors, its management and workforce are often not willing to risk making any further significant process changes. This resistance to change has taken on more serious overtones over the last decade, with the advent of cyber technology. The emergence and adoption of any new technology has the potential to streamline an organization's processes, but it also has the potential to change these processes, and in the case of cyber technology, it also has the potential to radically disrupt them. Cyber technology in general opens the doors to the world, and unless an organization can cultivate the flexibility to adapt its processes to the demands of this new technology, it could diminish its capability to execute effective protective and responsive counter measures against malevolent or unwanted cyber intrusion.

Historically, the emergence of the personal computer industry offers an illustrative example. In the mid-1980s, there were several major corporations that led the minicomputer industry, but none of them was able to make the transition to the personal or desktop computer business. One premise was that these corporations were already successful and had all the right processes, methods, and tools to operate in that environment. However, when that environment changed, they were not flexible enough to change their processes in order to compete in the new marketplace. *We ask the following questions on CP Complex SoS in Part II of this chapter on organizational behavior: (i) Does Internet technology foster this type of revolutionary change? (ii) Do organizations need to examine their methods and processes in order to make effective use of this technology? (iii) Are they willing and/or able to do so?*

The Internet brings a wealth of tools and mechanisms that can provide an organization with almost unlimited access to their customers, suppliers, and the community at large. However, it also brings threats as well. The Internet brings with it innumerable explicit and implicit processes to which organizations must respond. Most organizations can and do react to changes in their businesses; however, the extent to which they are aware or want to believe in the critical security issues associated with the Internet is less clear. One significant problem in cyber security stems from the fact that the professional community, which eagerly adopted information technology and the Internet, has marginally failed to integrate the global computer network into its core operation, that is, the functionality of the new computer technology has allowed companies to replace many functions with speed and efficiency but without considering the security and safety of the new evolving Complex SoS. The result has been that companies around the world have realized tremendous capabilities but at the same time have had to contend with many serious problems related to cyber intrusion, ranging from loss of intellectual property to the denial of service and to corporate espionage (Chittister and Haimes, 2011; Haimes and Chittister, 2012).

The generic term "intrusion", as used in Part II of this chapter, connotes malevolence targeted against networked technology, including viruses, phishing, pharming, spam, and Trojans. One central reason for this state of affairs is what Clayton Christensen, in his book *The Innovator's Dilemma* (2003), has identified as the above "disruptive change" or "disruptive technology." Christensen's argument is that in some instances a piece of technology is introduced that is so new and revolutionary that it forcefully changes the very foundation of the business. In these situations, Christensen maintains, "the ultimate uses or applications for disruptive technologies are unknowable in advance" (p. 113). Christensen notes, in describing the innovator's dilemma that "the logical competent decisions of management that are critical to the success of their companies are also the reasons why they lose their positions and leadership" (p. xvi). Christensen's vision is especially pertinent when considering Internet technology: over the past decades and continuing to the present day, companies have consistently demonstrated that they do not understand the nature of the disruptive technology introduced by the Internet. These well-established companies and organizations possess the resources and energy to sustain their leadership in their field; however, they fail to make the necessary changes and to adapt to the new reality by integrating the new technology into their operation, and thus they fail to protect their data and other vital information (Christensen et al., 2006). For example, the best solution may be to isolate some information from the network to ensure its protection and adjust other processes to work around the limited access to specific data.

In sum, because well-established companies and organizations have been slow to adapt to new Internet technology (or have failed entirely to adapt to it) and because the disruptive innovators are not bound by the traditions and dynamics of these companies, cyber intruders have been able to penetrate the operations of these companies at a minimal cost to gain major benefits. At the same time, the inadequate responses of companies and organizations to the intruders suggest that in many ways they remain in denial of the realities described above. Indeed, no matter how well information is protected through firewalls and other patching measures, until companies adopt and embrace holistic policies and procedures on the use of the Internet as an integrative subsystem of the entire functionality of their overall operations, the risks associated with cyber intrusion will not only persist but also become more severe and more costly. The next section will address the complexities of the emerging realities resulting from the Internet disruptive technology syndrome. It should be noted that the reason this situation exists and is growing is because companies are getting tremendous benefit from the Internet. In essence they are accepting the risk of cyber intrusion.

12.II.2 THE COMPLEXITY OF THE SECURITY OF CYBER–PHYSICAL COMPLEX SYSTEMS OF SYSTEMS

Cyber security is a generic term that connotes the complex *state* of reliability and confidence in the sanctity of the information delivered by a network of computer Complex SoS. This complexity stems from the very fact that the network of any given computer system is itself comprised of integrated hardware–software organizational Complex SoS. Also, the focus on cyber security implies the existence of an element of insecurity that could be exploited by a variety of intruders, including state and state-sponsored hackers and spies. In Part II of this chapter we define and use the term *CP Complex SoS to connote a generalized complex, large-scale CP network that includes hardware, software, organizational policies, and procedures, and that is connected to the Internet* (e.g. by means of telephone lines, wireless devices, and high-speed technology). Note that each subset of this complex and multifaceted network is itself a composite of subsystems, developed and assembled together by diverse entities employing a variety of tools and technologies. Consider, for example, the software engineering component of the CP Complex SoS (Chittister and Haimes, 2004, 2011; Haimes and Chittister, 2012). The software is likely to include thousands (if not millions) of lines of code, a variety of subroutines and commercial off-the-shelf (COTS) technology, and to have been developed by teams that cross experiences, languages, continents, and areas of expertise. Yet this assembly and amalgamation of science, technology, and society is expected to deliver a product that provides *software assurance*; namely, a product with a performance that is expected to exceed "six-sigma quality", with a full proof protection against malevolent intrusion of viruses and related cyber threats. Indeed, it is by now axiomatic that the quality of a product is fundamentally dependent on, and a function of, the quality of its components. The assembly and amalgamation of these subsystems, known as systems integration, constitutes one of the most critical sources of risk to software assurance and ultimately to information assurance (Haimes and Chittister, 1996). *Information assurance* is the trust that information presented by the system is accurate and is properly represented; its measure of the level of acceptable risk depends on the critical nature of the system's mission. Furthermore, an improper systems integration of a number of intrusion-resistant software components is likely to introduce new vulnerability to malevolent intrusion, even if each component by itself was designed and developed to be intrusion resistant. (More will be discussed on quality control and the role of systems integration in systems' resilience in the subsequent section.)

Experience has shown that cyber intruders are developing their own technology and solutions with which to overcome and penetrate firewalls and inscription protection, among other defenses. In other words, not only have these intruders adversely changed the integrated design and deployment of CP Complex SoS, but they also have adversely altered the original roadmap designs of those systems (assuming they were actually well known to the systems integrators and properly modeled). Similarly, the original CP security *model* has in effect been unknowingly (to the system's operators) perturbed and changed, not by its designers, but by adversaries. The term *roadmap design of a CP Complex SoS connotes the policies, procedures, methods, connections, and other human and organizational activities that define the functionality of the CP Complex SoS*. The fact that attackers are continuously making their own software solutions and forcefully modifying the original roadmap designs of CP systems and subsystems that make up CP Complex SoS suggests that they have been able to introduce a chaotic and unstable reality. Clearly, the myriad-evolving sources of risk, which originate from this new, uncertain reality, require the development of adaptive and effective defenses.

This new reality for CP Complex SoS demands an innovative and fundamental remedy based upon an unconventional, out-of-the-box approach, not simply repeated patching. The new approach should have multiple dimensions and attributes, as will be discussed in subsequent sections. As a starting point, however, the model that constituted the original roadmap design of a CP Complex SoS that has been successfully breached must be reevaluated, and the forced changes introduced by the intruders must be discovered and studied prior to implementing any serious remedial action. Only then can proper, sustainable, and effective responses be developed and deployed.

12.II.3 ORGANIZATIONAL AND CYBER–PHYSICAL RESILIENCE

Resilience has been defined in a variety of ways for different systems (Haimes, 2009, 2016, 2018) and in Chapter 1 of this book. In this section, we define the resilience of a CP Complex SoS as the ability of the system to recover from a cyber intrusion and to assume close-to-normal operations within an acceptable time and at an acceptable total encompassing cost. This concept is equally applicable to the resilience of CP Complex SoS: established organizations, both public and private, which are not spending the required funds and other resources on cyber security; are ignoring the risks to their CP Complex SoS, as well as the imperative need for adding resilience to their systems, thereby exposing their enterprises to dire consequences. The fundamental lesson learned from the quest for the highest level of reliability possible for nuclear reactors was critical to avoiding one path to failure. Indeed, fault-tree analysis, which was perfected by the Nuclear Regulatory Commission (NRC, 1981), identifies all the paths connected in series (known as minimal cut sets) of which if any one path were to fail, the entire system would fail. Note that the fault trees are discussed in the Appendix of this book. Note that each such path may be composed of multiple components, all linked in parallel. Large organizations, however, in their efforts to lower costs, often fail to take into account all possible *minimal cut sets* (minimal cut sets are introduced in the Appendix in the section on fault trees) and as a result compromise risk management and reduce their systems' resilience to an unacceptable level. In other words, such organizations fail to fully address the four fundamental questions in risk management (Haimes, 1991, 2016): What can be done and what options are available? What are the tradeoffs in terms of all associated costs, benefits, and risks? What are the impacts of current decisions on future options? What is the time frame? Indeed, if *resilience* were an integral part of the organization's CP system's design and operations, where economic efficiency was appropriately incorporated and traded off with CP security, and all of the above four risk management questions were addressed, then the ability of the system to recover within an acceptable time and cost following a major malicious attack would have been explicitly addressed.

The centrality of state variables in defining and understanding the meaning of the vulnerability and resilience of CP Complex SoS should not be underestimated. Both the vulnerability and the resilience of a system are multidimensional: a *vector* in mathematical terms. *Resilience of CP Complex SoS is defined as the ability of the CP Complex SoS to withstand a major disruption within acceptable degradation parameters and to recover within an acceptable composite cost and time.* Recall that the resilience of a system is a manifestation of the states of the system; perhaps most critically, it is a vector that is time dependent (Haimes, 2016). Moreover, resilience is similar to vulnerability in that it cannot simply be measured in a single unit metric; its importance lies in the ultimate and aggregated outputs (the consequences) of the system for any specific inputs (intrusion threats). Indeed, *vulnerability* is defined as the manifestation of the inherent states of the system (e.g. physical, technical, organizational, cultural) that can be exploited by an adversary to adversely affect (cause harm or damage to) that system. As state variables of the CP Complex SoS, both resilience and vulnerability are vectors of substate variables. Consider, for example, a supervisory control and data acquisition (SCADA) system that controls oil and gas pipelines through the Internet or through other unsecured telephone lines. The question "What is the resilience of this SCADA system?" is unanswerable for a variety of reasons. First, at least four major subsystems constitute the SCADA system – the *operator*, *master terminal unit* (MTU), *communications*, and *remote terminal unit* (RTU) – each of which can be modeled with a vector of state variables (Ezell et al., 2001). For example, the operator as a subsystem may be represented in terms of her knowledge, her experience and competence, her trustworthiness and loyalty to the organization, and her mental stability, among many other attributes. Similar substates can be identified for each of the three other SCADA subsystems; indeed, there can be a hierarchy of resilience attributes. It is worth noting that similar to quality, resilience is a composite of the attributes of the system, in this case the cyber–infrastructure system. For the SCADA system, these attributes may be related to its hardware, software, personnel, and organization. Furthermore, the resilience of the SCADA system is not only a function of the states of the system but also, and most importantly, a function of the specific type and level of the malevolent cyber intrusion (the input) and the corresponding consequences (outputs) in terms of time of recovery and the associated composite costs.

Note that the consequences that are considered as part of the risk function are in fact the outputs of the system model and that the input of the system's model is parallel to the concept of threat of cyber intrusion. Indeed, the risk associated with a cyber attack on a simple CP system will depend not only on the resilience of the system but also on the type and sophistication of the cyber attack. This is because the resilience of a system can be measured only in terms of the input (threat), and the system's recovery time and the associated composite costs and risks, which are functions of the states of the system. Furthermore, by their nature, malevolent cyber intrusions are probabilistic in nature; thus, the consequences (the outputs), which are also functions of the vector states of the CP Complex SoS, can be measured only in probabilistic terms.

One of the challenges facing managers and cyber security experts is determining the needed and desired business investment in cyber security and then providing that security. The implicit and explicit tradeoffs that organizations make in their quest to mitigate the potential impact of cyber intrusions (e.g. theft of intellectual property, disruption of operations, or other losses associated with viruses) are complex and are driven by various conflicting considerations. On one hand, for instance, investment in the current resilience of a CP Complex SoS is a real and deterministic problem; on the other hand, achieving sufficient protection against likely future intrusions is uncertain and probabilistic. Another challenge facing these decision makers is how to understand, evaluate, and determine the desired level of resilience of their cyber–infrastructure system and, most importantly, what metric or metrics to use to measure resilience, if any.

12.II.4 CRITICAL FACTORS AFFECTING THE RESILIENCE OF CYBER–PHYSICAL COMPLEX SYSTEMS OF SYSTEMS

It is constructive to revisit the following definition: CP Complex SoS connotes a generalized complex and large-scale CP SoS network that is comprised of hardware and software as well as organizational policies and procedures and that is connected to the Internet. The resilience (i.e. the states) of each element of the CP Complex SoS constitutes a vector of resilience of substates. For example, the states of the communications unit of the SCADA system are best modeled by a vector of substates, including the Internet (which itself is composed of myriad sub-substates) and human procedures, among others. In fault-tree analysis the system is ultimately simplified and presented as a composition of critical subsystems connected in series (known as minimal cut sets), such that if one of them fails, the entire system fails (i.e. the occurrence of the top event) (Haimes, 2016). Since in the aggregate, the resilience of the SCADA SoS is a function of all subresiliencies of its subsystems. The following question deserves an in-depth answer: What is the most dominant function or activity associated with CP Complex SoS, which renders them vulnerable to failure (including internal or external failure due to malevolent intrusion)? The answer is systems integration, *because systems integration not only connotes the integration of components but also involves an understanding of the functionality that emerges from that integration.*

Central to understanding systems integration is an awareness of the dynamic nature of the systems integrator. Note that the evolution of the systems integration from hardware to software and the profound impact of that change the (i) displacement of hardware by software is evident, and (ii) functioning of software as a systems integrator, i.e. as a coordinator of all a system's activities without which the system will not operate or will not function correctly or effectively, adds a new dimension to software risk management.

Thus, the emerging dominance of software as a systems integrator, whereby the various system components are organized, directed, and controlled to perform a specific set of functions, will necessarily bring with it new sources of risk of failure of the system. As a software-intensive system's complexity increases, with the ever increasing development of integrated hardware/software Complex SoS, it becomes more difficult to comprehend all positive (synergistic) and negative (adverse) interactions of its components. Furthermore, it also becomes more difficult to plan for the integration, testing, and evaluation of the system. The difficulties transcend (i) grouping and ordering of components for integration purposes; (ii) selecting the type, sequence, and methods of testing; and (iii) determining the best method and execution of regression testing as new components are added.

Comprehending the complexity of software-intensive systems is essential to the security of CP Complex SoS for the following sample of reasons:

1. Correctly integrating the myriad components and subsystems of a CP Complex SoS system is an imperative requisite for ensuring the proper operations and functionality of systems and information engineering.
2. Adhering to the six-sigma quality control, the systems integration process must ensure that all components and subsystems that constitute the ultimate CP Complex SoS must meet its design and quality specifications; otherwise, the system would operate on a different roadmap from the one for which it was designed. Indeed, effective systems integration necessitates the accountability of all functions, aspects, and components of the system.
3. Well-designed and executed malevolent cyber intrusions into CP Complex SoS would likely cause an intentional adverse change of the system's configuration. Therefore, performing proper systems integration that ensures a complete knowledge of the system's functionality and operations according to the roadmap design

would more likely enable the discovery of any malevolent-forced changes introduced by intruders (Chittister and Haimes 1993, 1994).

12.II.5 THE ART AND SCIENCE OF MODELING CYBER–PHYSICAL COMPLEX SYSTEMS OF SYSTEMS

To manage Complex SoS, one must at least understand the causal relationships among their major subsystems and the environment within which they operate. Without a doubt, the art and science of modeling constitutes one of the central enabler instruments of systems engineering, helping designers, builders, and managers alike to gain a better understanding of the intricacy of systems behavior. Building a good model of CP Complex SoS requires several elements: effective communication, careful data on which to base assumptions, valid metrics, and a clear idea of the outcome and the questions for which answers are sought. For example, when NASA first considered building a space station, the agency tried out a variety of models. Indeed, it is critical for models to be reviewed, updated, and scrutinized for accuracy, relevance, representativeness, and practicality.

If systems engineering did not exist as a discipline, it would have to be invented to address and model the large-scale integrated CP Complex SoS that affect cyber security. However, systems engineering cannot be practiced effectively, if at all, without models (analytical, conceptual, or simulated). Models, experiments, and simulations are conceived and built to answer specific questions. For example, when technology is used, we must update the processes that build on such technology and then update them again as the technology changes and evolves over time. One question that arises, among many others, is how to update, modify, respond, and manage these processes using new technology, given the reality of forced changes in technology and the adaptive resourcefulness of adversaries. In other words, how do we understand and respond to a new reality, given that the old design and functionality of the system might have been adversely changed? Systems modeling and systems integration have been the hallmark mechanism with which to overcome significant scientific and technological challenges, such as sending the first man to the moon or the incredible achievements in the use of technology in medicine.

In general, models assess what is optimal, given what we know, or what we think we know about EFCs. Other questions may include: What do we need to know? What value might appear from reducing risk and by producing more precise and updated knowledge about a given CP Complex SoS? What knowledge is needed for an effective decision-making process and where should that knowledge be applied? Furthermore, in considering these questions we must remember that mathematical models are the requisite mechanism with which to perform quantitative systems engineering analysis of complex problems and particularly those problems involving CP Complex SoS. These mathematical models are essential for systems engineers, enabling them to better understand and thus to manage a system by making explicit the relations among its subsystems and its relevant or critical interdependent systems.

Modeling is also an art because successful models must build on the artistic traits of experimentation, imagination, creativity, independent thinking, vision, and entrepreneurship. Systems engineers must possess and merge values and traits offered by both the arts and sciences. However, in contrast to scientific knowledge, whose validity can and must be proven, mathematical models cannot always be subjected to such metrics. In fact, the more complex is the system to be modeled, the lower is the modeler's ability to verify or validate the resulting models. Some scholars even argue that no complex model can be verified or validated, in part because of the dynamic and probabilistic nature of all natural and constructed systems. Heisenberg's principle is at work here, as well, that is, once a system's model is deployed, the essence of that system has been changed.

Thus, because CP Complex SoS security is essentially multidimensional and because its wholeness is characterized by multiple and varied paths of interdependence, a holistic approach is requisite in order to understand not only its myriad parts and functionality but also to address the forced changes imposed upon it both from within and without its domains. That is, due to the inherent intricacy of CP Complex SoS (including the causal relationships among their subsystems), modeling security for them poses an extremely challenging task, especially for those systems, which serve large public and private organizations. The enormity of this task is further magnified by the fact that models are built to answer specific questions, at the same time, models *ought to be as simple as possible but as complex as desired and needed.* Perhaps the greatest challenge in modeling deals with accounting for the almost countless entities that *reside outside the organization* of the CP Complex SoS and whose diverse (and not always fully known or understood) interests may be counter to the latter. Indeed, the main objective of these semi-phantom outside intruder entities is to promote *cyber insecurity* within the targeted organization (Chittister and Haimes, 2005, 2009). In these cases, the modeling challenge originates from the need to know and understand the interplay, the I-I among all subsystems not just of one's own, but of both CP entities: (i) the organization itself whose cyber security is the domain of interest and protection and (ii) the almost countless entities that reside outside this organization and whose interests are counter to the latter.

Another source of the intricacy of CP Complex SoS stems from their multiscale compositions – from molecular to macro dimensions. CP Complex SoS are envisioned, designed, and built to meet multiple users and prospective needs, and often their ultimate functionality frequently remains flexible and uncertain. For example, multiscale phenomena are encountered in climate models that model phenomena of vastly differing time and length scales, and multiple complex physical interactions, including transport, adsorption, and chemical reactions (Takahashi et al., 2006). Multiscale models can help answer limited questions about the behavior of these systems under both steady-state conditions and dynamic forced changes. The multiple perspectives that characterize each system – small or large – require developing a set of models to represent the essence of the multiplicity of perspectives, attributes, functions, and dimensions of the system. *Merriam-Webster Dictionary* defines *essence* as: "The most significant elements, attributes, quality, property, or aspect of a thing." Physical, chemical, biological, and other natural laws serve as the first principles and the foundations of such models. Although it is necessary, however, to take these natural laws into account, they are not in and of themselves sufficient for model construction, because of the above mentioned multiple organizational and procedural perspectives and dimensions inherent in CP Complex SoS. The essence of the multiple perspectives of any system, whether it is the human body, the environment, a bridge, a building, an airplane, or CP Complex SoS, cannot be adequately modeled in a single model. This fact presents a challenge to modelers. Thus, what is needed is a mechanism, a systemic framework, with which to augment the natural and physical laws with imagination, invention, innovation, entrepreneurship, out-of-the-box thinking, and boundless experimentation. The PSM introduced in Chapter 4 and briefly discussed in Part I can serve as one approach to represent a real and virtual laboratory for multiscale model building.

12.II.6 INTRUSION DETECTION

Probably the worst scenario for any chief information officer (CIO) of any organization is when malicious intrusions into its CP Complex SoS remain undetected. Indeed, no problem can be addressed, let alone solved, if it remains either unknown or invisible. Similarly, no repairs or preventions can be made unless the wide variety of viruses, worms, malicious codes, and other technological threats to the CP Complex SoS can be detected and understood. Models can play a central role in intrusion detection and thus enable information managers to develop protective and remedial courses of action. This can be done through well-designed, planned, and executed intelligence operations that follow, explore, and connect the many dots of intrusion into a CP Complex SoS. Such detection can also serve as a predictive mechanism as to what the would-be intruders might be planning to do in the future. CIOs must cope with and decipher a blizzard of diverse, often sparse, seemingly unrelated, and apparently or actually unconnected initiating events from multiple sources. "Connecting the dots" continues to challenge intelligence agencies worldwide. The Coordinated Hierarchical Bayesian Model (CHBM) (see Chapter 6) builds on the strengths of information available from indirect but related databases, borrowing from them in order to augment direct databases; hence it is most suitable for problems and applications with sparse data. The CHBM has been applied in evaluating safety solutions for mission-critical systems, including the analysis of runway incursion (RI) incidents, which the US Federal Aviation Administration has underscored in its list of "most wanted" areas for improvement. At present, the rates of RI are based primarily on point estimates derived from historical records; consequently, these rates are highly sensitive to changes in observed RI. Thus, extended analysis must consider not only those RI estimates directly derived from existing records, but also the uncertainties surrounding such estimates. Through coordination of multiple direct and indirect databases in a hierarchical structure, uncertainties relating to safety solutions with little or no documented historical precedence can be better described and assessed. Through analysis of real data and of simulations, results from the CHBM (see Chapter 6) have demonstrated that it offers great advantages compared with current statistical methods for understanding the impact of new solutions on levels of safety.

12.II.7 BUILDING DECEPTION INTO CYBER–PHYSICAL COMPLEX SYSTEMS OF SYSTEMS

The art and science of modeling can be used to answer a wide range of questions, including how to deceive the would-be intruders and lead them to believe that they are inflicting damage on the real intended system. Of course, the use of deception in warfare is an old trade; however, what makes cyber deception an effective instrument for CP Complex SoS security are the capabilities provided by the ingenuity and versatility of systems modeling and simulation. Two premises drive the following discussion on cyber deception: the (i) difficulty of detecting all cyber intrusions into CP Complex SoS (and to a large extent the relatively limited capability to detect consistently the signature sources of the intruders); and (ii) unlimited opportunities and capabilities available to the defenders of CP Complex SoS through modeling and simulation.

The PSM (see Chapter 4) enables research teams to effectively analyze major forced changes on the characteristics and performance of assured CP Complex SoS and other major socioeconomic systems. Forced changes are manifestations of the states of antagonistic systems, X_2, which have a direct impact on the states of the assured system, X_1. The PSM builds on and incorporates input from HHM (see Chapter 3) and by doing so seeks to develop causal relationships through various modeling and simulation tools; it imbues phantom ideas associated with emergent systems with a level of lifelikeness and realism that otherwise would have never been realized. In other words, PSM legitimizes the exploration and experimentation of out-of-the-box and seemingly "crazy" ideas with different modeling and simulation tools and ultimately discovers insightful implications that otherwise would have been completely missed and thus dismissed. In this sense, it allows for "nonconsensus" ideas or an "agree-to-disagree" process for further exploration and study. The output of the HHM is a taxonomy of identified risk scenarios, offering multiple perspectives of a system for modeling (see Chapter 3). Alternatively, the output of the PSM can provide justification for investment in preparedness, or learning activities necessary to protect against critical forced changes or emergent risks, or activities that might not otherwise have been justifiable. *Through logically organized and systemically executed models, the PSM is aimed at providing a reasoned experimental modeling framework with which to explore and thus understand the intricate relationships that characterize the nature of* CP Complex SoS. The PSM philosophy rejects a dogmatic approach to problem solving that relies on a single modeling approach structured on one school of thinking. Rather, the PSM attempts to draw on a modeling schema that builds on the multiple perspectives gained through generating multiple scenarios. This approach allows researchers to construct appropriate models with which to deduce tipping points and to determine information values for logical conclusions as well as for future actions. Currently, we assess what is optimal, given what we know, or what we think we know. We want to extend these models to answer two pressing questions: (i) What do we need to know and what value might appear from risk reduction results from having more precise and updated knowledge about Complex SoS? (ii) Where is that knowledge needed for acceptable risk management and decision making? The above capabilities of PSM to explore causalities among various components of CP Complex SoS, and to construct phantom subsystems within the real one to divert the would-be intruder to a fictitious path, constitute a valuable technology.

12.II.8 EPILOGUE

Cyber intrusion has multiple ramifications with multiple dimensions and scales due to the complex interrelated, multidimensional interdependencies within our critical civilian and military infrastructures. To appreciate the multiple dimensions of risk associated with the violation of CP Complex SoS security and integrity, it is important to understand and to configure those infrastructures against unwanted propagation of malicious cyber intrusions into the functionality and operations of these systems. In particular, the predominant reliance of the physical, economic, interregional, international, and workforce interdependencies on the CP Complex SoS places them at extreme risk for malicious cyber intrusion – a fact that has become increasingly evident as these infrastructures have grown increasingly dependent on SCADA systems. Exacerbating the potential dire consequences to the functionality and operations of physical infrastructures subjected to cyber intrusion is their common reliance on a single path of communication. Such reliance is typically due to a desire for economic efficiency; however, this efficiency comes at the expense of not achieving higher reliability nor improved security and system resilience.

The current economic, intellectual property, and other tangible and intangible losses accrued to countless individuals and organizations alike could be seen in the future as insignificant compared with likely concentrated, well-planned, and designed comprehensive cyber attack on a country's economy. Such events, accomplished by penetrating critical physical infrastructures controlled through SCADA systems, have become a reality for a variety of reasons:

1. Prevailing imbalance of tradeoffs practiced by the private and public sectors that favor economic efficiency at the expense of the security of CP Complex SoS.
2. Lack of comprehensive systems integration of the critical subsystems that constitute CP Complex SoS.
3. Prevailing tragedy-of-the-commons syndrome manifested by the reluctance of the private sector either to invest the required resources for information assurance and cyber protection, or to share information on the malevolent intrusions into the private sector enterprises.
4. Ever-increasing sophistication of the technology and methods practiced by individual- and state-supported intruders.
5. Increasing linkages and interconnected functionalities of several safety-critical CP Complex SoS, thus reducing the reliability of such systems by unduly increasing the number of single paths to failure.
6. Lack of sufficient focus on the social and behavioral sciences in the roadmap design and development of CP Complex SoS and,
7. Vulnerability of successful companies, which have more to lose due to the disruption technology syndrome than marginal ones and are thus more likely to be targeted.

In sum, the Internet is such a powerful tool that it has very rapidly become the foundation for all business interactions. Furthermore, using the Internet no longer just provides a competitive advantage; rather, it has become a requirement for doing business. Like any technology, it brings with it a set of processes to make it work. Originally, the Internet was created for an open use – unlimited point-to-point communication – not to be reliable, secure, or protected. And without being controlled or managed, it might change some part of the processes of CP Complex SoS. The emerging generation is doing business on the Internet, getting its news from the Internet, and even sharing personal lives on the Internet. Thus, the question is not *whether* to use the Internet, but *how* to use it advantageously. Organizations must be able to integrate the Internet into their policies, procedures, and processes. However, this must be done not only in light of the Internet's advantages but also by accounting for the risks using the Internet brings.

Organizations must learn how to protect their assets in light of the Internet vulnerability which may mean doing some part of the business separate from the Internet, or being isolated in some way to minimize or eliminate the associated risks.

REFERENCES

Arquilla, J. and Ronfeldt, D. (2001). *Networks and Netwars: The Future of Terror, Crime, and Militancy*. Santa Monica, CA: RAND Corporation https://www.rand.org/pubs/monograph_reports/MR1382.html (accessed 14 November 2017).

Ashkenas, R., Ulrich, D., Jick, T., and Kerr, S. (2002). *The Boundaryless Organization: Breaking the Chains of Organizational Structure*. New York: Jossey Bass.

Bronowski, J. (1978). *The Origin of Knowledge and Imagination*. New Haven, CT: Yale University Press.

Chittister, C.G. and Haimes, Y.Y. (1993). Risk associated with software development: a holistic framework for assessment and management. *IEEE Transactions on Systems, Man and Cybernetics* **23**(3): 710–723.

Chittister, C.G. and Haimes, Y.Y. (1994). Assessment and management of software technical risk. *IEEE Transactions of Systems, Man, and Cybernetics* **24**(2): 187–202.

Chittister, C.G. and Haimes, Y.Y. (2004). Risks of terrorism to information technology and to critical interdependent infrastructures. *Journal of Homeland Security and Emergency Management* **1**(4): 1–21.

Chittister, C.G. and Haimes, Y.Y. (2005). A roadmap for quantifying the efficacy of risk management of information security and interdependent SCADA systems. *Journal of Homeland Security and Emergency Management* **2**(2): 1117, Article 12.

Chittister, C.G. and Haimes, Y.Y. (2009). Harmonizing high performance computing (HPC) with large-scale complex systems in computational science and engineering. *Systems Engineering* **13**(1): 47–57.

Chittister, C.G. and Haimes, Y.Y. (2011). The role of modeling in the resilience of cyberinfrastructure systems and preparedness for cyber intrusions. *Journal of Homeland Security and Emergency Management* **8**(1): 1–19. http://www.bepress.co.

Chopra, D. (1994). *Ageless Body, Timeless Mind: The Quantum Alternative to Growing Old*. New York: Harmony.

Christensen, C.M. (2003). *The Innovator's Dilemma: The Revolutionary Book That Will Change the Way You Do Business*. New York: HarperCollins.

Christensen, C.M., Baumann, H., Ruggles, R., and Sadtler, T.M. (2006). Disruptive innovation for social change. *Harvard Business Review* **84**(12): 94–101.

Collins, J. (2011). *Good to Great: Why Some Companies Make the Leap and Others Don't*. NewYork: Harper Business.

Collins, J. and Porras, J. (2004). *Built to Last: Successful Habits of Visionary Companies*. New York: Harper Business Essentials.

Covey, S.M.R. (2008). *The SPEED of Trust: The One Thing That Changes Everything*. New York: Free Press.

Crowther, K. (2014). Understanding and overcoming information sharing failures. *Journal of Homeland Security and Emergency Management* **11**(1): 131–154. doi: 10.1515/jhsem-2013-0055.

Davenport, T. and Prusak, L. (2000). *Working Knowledge*, 2e. New York: Harvard Business Review Press.

Desourdis, R.I. (2012) Pearl Harbor, 9-11, Katrina, Virginia Tech shootings, Deepwater horizon planning deficiencies: a sense-respond information-sharing solution. *IEEE Conference on Homeland Security Technology*, Waltham, MA (13–15 November 2012), 220–231.

Desourdis, R.I. and Contestabile, J.M. (2011). Information sharing for situational understanding and command coordination in emergency management and disaster response. *2011 IEEE Conference on Homeland Security Technology*, Waltham, MA (15–17 November 2011), 26–32.

Desourdis, R.I., Rosamilia, P., Jacobson, C. et al. ed. (2009). *Achieving Interoperability in Critical IT and Communication Systems*. Boston, MA: Artech House.

deWeck, O., Roos, D., and Magee, C.L. (2016). *Engineering Systems: Meeting Human Needs in a Complex Technological World*. Cambridge, MA: MIT Press.

Drucker, P. (2001). *The Essential Drucker*. New York: Harper Business.

Ezell, B.C., Haimes, Y.Y., and Lambert, J.H. (2001). Risks of cyber attack to water utility supervisory control and data acquisition systems. *Military Operations Research* **6**(2): 23–33.

Friedman, T. (2005). *The World Is Flat*. New York: Farrar, Straus and Giroux.

Haimes, Y.Y. (1991). Total risk management. *Risk Analysis* **11**(2): 169–171.

Haimes, Y.Y. (2009). On the definition of resilience in systems. *Risk Analysis* **29**(4): 498–501.

Haimes, Y.Y. (2016). *Risk Modeling, Assessment, and Management*, 4e. Hoboken, NJ: Wiley.

Haimes, Y.Y. (2018). Risk modeling of interdependent complex systems of systems: Theory and practice. *Risk Analysis* **38**(1): 84–98.

Haimes, Y.Y. and Chittister, C. (1996). Systems integration via software risk management. *IEEE Transactions on Systems, Man, and Cybernetics* **26**(9): 521–532.

Haimes, Y.Y. and Chittister, C. (2012). Risk to cyberinfrastructure systems served by cloud computing technology as systems of systems. *Systems Engineering* **15**: 213–224. doi: 10.1002/sys.20204.

Kahneman, D. (2013). *Thinking Fast and Slow*. New York: Farrar, Straus and Giroux.

Katzenbach, J. and Smith, D. (2015). *The Wisdom of Teams: Creating the High-Performance Organization*. New York: Harvard Business Review Press.

Prange, G.W., Goldstein, D., and Dillon, K. (1991). *Pearl Harbor: The Verdict of History*. New York: McGraw-Hill.

Spirkin, A. (1983). *Dialectical Materialism*. Moscow: Progress Publishers.

Spirkin, A. (1985). Fundamentals of Philosophy. https://www.marxists.org/reference/archive/spirkin/works/dialectical-materialism/index.html (accessed 2 November 2017).

Takahashi, K., Peng, X., Odaira, M., and Onishi, R. (2006). Earth simulator multi-scale, multi-physics: simulations of weather/climate on the earth simulator. *Bulletin of the Japan Society for Industrial and Applied Mathematics* **16**(2): 112–124.

U.S. Nuclear Regulatory Commission (1981). *Fault Tree Handbook*, NUREG-81/0492. Washington, DC: Government Printing Office.

13

Software Engineering: The Driver of Cyber–Physical Complex Systems of Systems

OVERVIEW*

In Chapter 13 we extend our exploration of the role of cyber–physical (CP) complex systems of systems (SoS) (CP Complex SoS) to software engineering; and introducing the role of hardware/software in systems integration. In their essence CP Complex SoS are an amalgamation of (i) engineered physical systems, which undoubtedly also include cyber components (subsystems), and (ii) software engineering, which include million lines of code. As noted in Chapter 1, the terms complexity and Complex SoS have evolved and assumed different connotations during the last 50 years and thus the way to model and manage complexity. Decision makers and stakeholders by *necessity aspire to a reliable and trustworthy risk modeling and analysis, especially for safety-critical CP Complex SoS, not only for mundane single systems, but especially for interdependent and interconnected CP Complex SoS*. The ultimate efficacious contributions of CP Complex SoS are varied and multifarious by serving multiple needs, purposes, and constituencies at different time frames. Indeed, in their essence, most, if not all, CP Complex SoS are driven by shared organizational and government enterprises, shared resources, decisions, and decision makers. The cyber systems are primarily composed of an amalgamation of multiple software systems, guided by algorithms, with millions of lines of code, often designed, written, programmed, and integrated by multiple individuals and entities, responding to multiple requirements that are often competing and/or conflicting with others. CP Complex SoS are commonly characterized by multiple often conflicting and competing objectives at different time frames for different clients, and under varied conditions. The physical systems are commonly composed of primarily a large number of engineered systems, albeit not without their own software for some of them.

Two major fundamentals guide our reevaluation of current practices of risk analysis when addressing the risk to CP Complex SoS, and these are grounded on understanding and evaluating the interdependencies

*The author served for twenty years (1990–2010) as a visiting scholar at the Software Engineering Institute (SEI), Carnegie Mellon University. During those two decades he worked closely with, learned from, and published numerous technical papers with Clyde Chittister, Director of the SEI's Risk Management Program. This chapter draws from, and expands on the two decades of learning the art and science and management of software engineering.

Modeling and Managing Interdependent Complex Systems of Systems, First Edition. Yacov Y. Haimes.

and interconnectedness (I-I) within and among the CP Complex SoS manifested through shared/common states and other *essential entities* that make up CP Complex SoS. (We redefine the term "essential entities" in the context of CP Complex SoS to connote (i) millions of lines of code, (ii) computer algorithms, (iii) decision makers and stakeholders, (iv) resources, (v) organizational setups, (vi) emergent behavior, (vii) cultural identity, (viii) time frame, and (ix) being cognizant of and responsive to the emergent nature of Complex SoS. The Evolving Base, introduced in Chapter 9, is represented by the following dynamic shifting rules and realities for each subsystem and for the entire Complex SoS: (i) goals and objectives; (ii) stakeholders, decision makers, and interest groups; (iii) organizational, political, and budgetary baselines; (iv) reorganization and reallocation of key personnel; (v) emergent technology and its deployment; and (vi) requirements, specifications, delivery, users, and clients. *The above fundamental departures from performing risk analysis from a single system to CP Complex SoS constitute the core challenges facing all parties involved in them.*

In this chapter, we also address the understanding, modeling, composition, and management of CP Complex SoS in terms of their I-I and as CP Complex SoS. The challenges and associated difficulties in understanding and modeling CP Complex SoS have many origins – historical, cultural, organizational, and cognitive differences associated with the conception, design, development, and integration of cyber and physical systems that ultimately make up what we term CP Complex SoS. Furthermore, we often neither understand nor appreciate the two distinctive worlds in which both *software* and *hardware* are conceived, developed, and ultimately integrated. This integration commonly results in interdependent and interconnected CP Complex SoS. When the design, development, and ultimate integration of cyber and physical systems are executed separately and without a strict adherence to compatibility between the two CP Complex SoS, the results are likely to contain myriad predictable and invisible sources of risk. Some of this incompatibility between the two systems (cyber and physical) can be traced to differences in the background, training, culture, and orientation that commonly characterize, though not always, hardware and software engineers and related professionals. In particular, among the many differences between hardware and software engineering are the physical laws that constrain the skills and creativity of hardware engineers, but by and large do not constrain software engineers.

Clearly, there are distinctive skills, artistic creativity, and ingenuity in the development of software and hardware devices and systems. The gulf between the background and orientation of hardware and software engineers and other professionals must be accounted for as an important source of risk that renders CP Complex SoS vulnerable to malevolent intrusions with adverse consequences. Indeed, the variations in cognitive orientation and cultural perspective between hardware and software engineers, and consequently the differences in the products that each develops, cannot be overlooked. They are serious sources of risk to CP Complex SoS that serve safety-critical SoS.

When we consider that the intricacy of most Complex SoS stems primarily from the I-I that characterize them and that we create CP Complex SoS by merging and imbedding hardware and software, it becomes clear that we must not only be cognizant of, but also be proactive in, discovering, understanding, modeling, and managing the sources of failure due to the above characteristics of CP Complex SoS. Furthermore, when we also consider the diverse training, organizational, cultural, and technical orientations between hardware and software developers, it is not difficult to understand the disharmony that can evolve, when unattended, between hardware and software products that make up CP Complex SoS. The above disconnect between hardware and software development and integration characterizes the entire life cycle of conception, planning, design, development, testing, production, and ultimate integration resulting in functional CP Complex SoS.

PART I: SYSTEMS INTEGRATION VIA SOFTWARE RISK MANAGEMENT

13.I.1 INTRODUCTION

This part addresses the evolutionary process that has been taking place in software engineering: during the last two decades; namely, the shift from hardware to software, where the role of software engineering is increasing and is becoming more central in systems integration especially for interdependent and interconnected CP Complex SoS (Haimes and Chittister, 1996).

This shift also markedly affects the sources of risk that are introduced throughout the life cycle of a system's development – their requirements, specifications, architecture, process, testing, and end product.

Risk is commonly defined as a measure of the probability and severity of adverse effects (Lowrance, 1976). Software technical risk is defined as a measure of the probability and severity of adverse effects inherent in the development of software.

Consequently, risk assessment and management has become an overall cross-functional system integration agent, evaluating the changes that ought to take place in response to this shift in the overall two challenges. One is the need to reassess the role of a new breed of software systems engineers/systems integrators. The other is the need to develop new and appropriate metrics for measuring software technical risk.

Effective systems integration necessitates that all functions, aspects, and components of the system must be accounted for along with an assessment of most risks associated with the system. Furthermore, systems integration for software-intensive Complex SoS is not only the integration of components, but is also an understanding of the functionality that emerges from the integration. Indeed, when two or more software components are integrated, they often deliver more than the sum of what each was intended to deliver; this integration adds synergy and enhances functionality. In particular, the thesis advanced in this part of the chapter is that the process of risk assessment, management, and communication is an imperative requirement for successful systems integration; this is especially true for CP Complex SoS.

In addition, we advance the premise that the process of risk assessment, management, and communication is also an imperative requirement for ensuring against unwarranted time delay in a project's completion schedule, cost overrun, and failure to meet performance criteria. To achieve the aspired goal of systems integration, we deploy Hierarchical Holographic Modeling (HHM) introduced in Chapter 3. This HHM framework constitutes seven major considerations, perspectives, venues, or decompositions of CP Complex SoS, each of which identifies the sources of risk in systems integration from a different, albeit with some overlap, viewpoint: *(i) software development, (ii) temporal perspective, (iii), leadership, (iv) environment, (v) acquisition process, (vi) quality, and (vii) technology.*

13.I.2 ROLE OF RISK ASSESSMENT AND MANAGEMENT IN SYSTEMS INTEGRATION

We address the evolutionary process that has been taking place in engineering systems: the shift from hardware to software, to CP Complex SoS, where the role of software engineering is increasing and is becoming more central in systems integration. This shift also markedly affects the sources of risk that are introduced throughout the development of the CP Complex SoS: their requirements, specifications, architecture, process, integration and testing, and end product.

The life cycle developmental process of software encompasses several interconnected functions. The customer's interpretation of needs and desires as perceived by the developer is established through a set of *requirements*. To build the product, the developer projects and transforms these perceptions into step-by-step detailed instructions known as *specifications* (i.e. the "developer" does not build a product directly on the basis of the requirements per se.) The specifications are further translated into a cognitive roadmap, commonly known as architecture. The resulting *architecture* and specifications are then used by the developer as "blueprints" for the ultimate realization of the product; this is accomplished through the developmental process. The *developmental process* itself involves (i) appropriate use of technology and know-how, (ii) appropriate allocation of human and other resources, and (iii) availability of an organization that is at the right maturity level and is endowed by and has in place an appropriate organizational and managerial infrastructure. *The integration and testing* are an important function for any product and is particularly critical for CP Complex SoS. The *end product*, however, may or may not represent a true realization of the actual customer's needs and desires as were originally conceived by the developer.

The software engineering community accomplishes the software engineering life cycle via different models: two of the most common are the two-decade-old "waterfall" model consisting of system feasibility, software plans and requirements, product design, detailed design, coding, integration, implementation, and operations and maintenance and the spiral model (Boehm, 1988). Imbedded within these phases are *verification* (e.g. are we building the product right?), *validation* (e.g. are we building the right product?), and configuration management (e.g. are we properly managing change throughout the developmental process of the system?). The life cycle of software development from the systems engineering perspectives may be defined by the following seven stages or segments: (i) definition of requirements and specifications, (ii) technical and financial feasibility analysis, (iii) effectiveness of plan development of the project, (iv) effectiveness of the physical design and its logical enterprise, (v) effectiveness of implementation of the design and of the system tests, (vii) operational deployment of the entire enterprise, and (viii) final review, evaluation, and retirement.

From the perspectives of the temporal process of CP software development, these functions or phases do not exist in isolation from each other; rather, they are connected and exert influence on each other. This inter- and intrafunctional connectedness yields three sources of increased risk: (i) internal sources (originating from within a developmental function), (ii) interactive sources (originating from interactions among the various functions), and (iii) external sources (originating from outside the project). The assessment, management, and communication of these three sources of software technical risk should not only result in a better understanding of their I-I as CP Complex SoS, but can also provide a systemic framework for directing, streamlining, coordinating, and synergizing these functional relationships (Chittister and Haimes, 1994).

Risk is defined here as a measure of the probability and severity of adverse effects. Software technical risk is defined as a measure of the probability and severity of adverse effects inherent in the development of software (Haimes, 2016). In subsequent discussion, we will address in more detail the centrality of assessing and managing software-based defects and thus the risks thereto.

In this chapter *we define systems integration as the process of amalgamation and coordination among all the couplings and interactions of the system's components so that the entire system can perform its intended functions as a unit.* Indeed, we think of systems integration as a process that is not limited in its occurrence to an "after development," but to a continuous process from cradle to grave. Note that each subsystem of the CP Complex SoS can achieve its limited, compartmentalized function; however, only through system integration (i.e. through an orchestrated coordination) can the entire CP Complex SoS achieve its intended functions. This necessarily leads to the concept of a system within a system federated (coordinated) systems and Complex SoS. In previous chapters we emphasized that the key to evaluating the risk to Complex SoS lies in understanding the genesis characterizing the I-I manifested through shared states and other *essential entities* within and among the systems that constitute Complex SoS. The term "essential entities" includes *shared/common decisions, decision makers, stakeholders, resources, organizational behavior and norms, policies and procedures, management, culture*, and others. This undertaking can be accomplished by building on state-space theory, which is fundamental to systems engineering and process control. A special case for *systems federation, the subsystems may be loosely integrated, may act independently, and are coordinated primarily by humans* (e.g. the pilot changes the speed of an airplane or readjusts the position of the flaps). In integrated CP Complex SoS, on the other hand, the SoS act as a unit, and the integrating software performs all the functions without direct human interference. In other words, the more automated the system, the greater is its need for an effective systems integration. Since humans serve as integrating agents, more integration becomes necessary with less human involvement, and the systems integration agent thus becomes increasingly valuable. With automated control (with less human intervention), there is a risk that a problem might not be detected. The centrality of software continues to evolve in systems integration, and with this evolution there is an overriding need for the development of an effective process for assessing and managing the risks associated with systems integration for software-intensive systems (Chittister and Haimes, 2010).

Studies conducted on the reliability of software and on its role in overall system operation continuously point to the failure of such systems at the initial operating capability (IOC) phase. Failures are commonly caused by errors, which are caused by faults. A *fault* is a physical defect, imperfection, or flaw that occurs within some hardware or software component. A failure is the nonperformance of some action that is due or expected. Failures are commonly some manifestation of either poor systems integration, or situations where critical sources of risk have not been appropriately identified, assessed, and managed. In fact, all individual subcomponents and components can be perceived to be "correctly" designed and constructed as specified, and the system may still fail or not function as intended integration reveals the bugs that are otherwise not discovered in isolation. This phenomenon is also common in phased or incremental delivery, where functions are added to the product over time and where major design modifications take place between the IOC and the final operating capability (FOC) phase of the system development. These changes may be uncoordinated and thus constitute basic sources of risk. Furthermore, software engineering is an *intellectually intensive activity* and is subject to human errors and to personal interpretation by the software engineer of the architecture and of the overall system's requirements. These and other issues explored in this chapter are what make integration of software-intensive systems, such as CP Complex SoS, different from the integration of pure hardware systems.

The identification, assessment, management, and communication of all critical sources of risk in software-intensive Complex SoS constitute a critical requirement for successful systems integration. While assessing and managing the risks associated with software engineering during the entire life cycle of a system are an imperative requirement for successful systems integration, it is also a vital requirement for protecting against unwarranted time delay in project completion schedule, cost overrun, failure to meet performance criteria, and lack of quality. Of course all this is done in the spirit of team collaboration and close interpersonal communication. Figure 13.I.1 is a conceptual illustration of the life cycle project cost with and without risk mitigation strategies, given that, at time zero, sources of risk have been identified and assessed.

Although risk mitigation is not free, it pays off in later stages as suggested by Figure 13.I.1. Consider, for example, a subsystem deployment that is on the critical path. At time $t = 0$, it was assessed that if "business continues as usual," then the likely time delay may exceed 10 weeks (for a one-year total schedule). In this case management may consider any or all of the following options: add more personnel to the project, authorize more working hours as overtime, introduce new technology, stretch the schedule, delay the completion of select functions, reassign priorities, etc. Given that the cost of risk mitigation can be as high as one hundred times the cost of risk prevention (or even higher in some instances), prudence calls for the following precursors to failures and an early assessment of all major risks and for the initiation of appropriate mitigating actions in order to confront, control, and manage these risks. Clearly, the savings that may accrue as a result of risk prevention and early risk mitigation can be compounded, given the adverse domino effect that a delay in the completion of one subsystem may have on the entire system. Furthermore, the consequences of "doing nothing" may

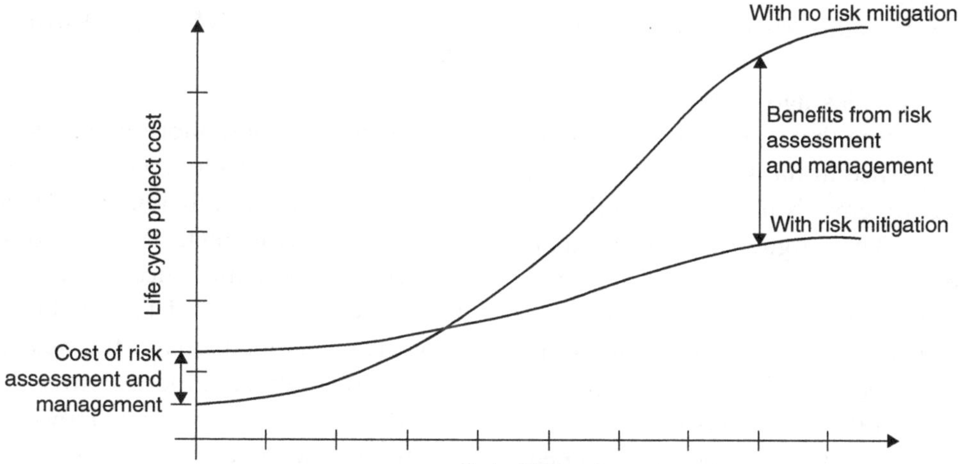

Figure 13.I.1 *The cost of no risk mitigation.*

manifest themselves not only in terms of dollars and time but also in a loss of good will, adverse effects on the employees, and culture of the organization as well as on the prospect of securing future contracts. In sum, the timely assessment of sources of risk associated with the life cycle of any project may be essential for successful systems integration.

The shift that favors software engineering should not be thought of as only from hardware to software; rather, it is a shift from a limited, piecemeal vision of software development to a holistic, system-wide vision. Such a visionary approach considers the entire life cycle path of software development through its six major milestones: requirements, specifications, architecture, process, testing, and product. *The risk assessment and management process serves as the overall cross-functional systems integration agent that connects and unites these milestones along the life cycle path.* The increasing role of software as an overall systems integrator, however, is becoming more evident as the shift from hardware to software continues. Furthermore, as the demarcation line between hardware and software becomes fuzzier and fuzzier with the improved technology of microprocessors and target machines, the dominance of software as the overall systems coordinator and integrator becomes more visible. *This necessarily leads to the conclusion that the process of risk assessment and management itself becomes central to effective overall systems integration.* For example, the use of higher-level language and the subsequent conversion to machine language and downloading to target machines (e.g. microprocessors) introduce critical sources of risks and uncertainties that must be addressed methodically and systematically.

More than two decades ago White et al. (1993) made a forceful argument that a new cadre of systems engineers (with expertise in the systems engineering of computer-based systems) is needed to do the systems integration job. They argue that "In too many cases, systems engineers do not understand what information should be provided to Computer-Based Systems (CBS) implementers. Systems engineers provide information in the wrong sequence and with insufficient detail" and that "most systems engineers do not have the skills to understand the consequences, perform the necessary analysis, or establish the required risk-avoidance measures." In addition, many upper level managers are hardware-oriented, and younger software-oriented engineers may face a challenging task in convincing hardware-oriented managers of the centrality or efficacy of software engineering. The thesis advanced in this part not only supports the notion that there is a need to "perform the necessary analysis, and establish the required risk-avoidance measures," but also further advocates that this need is so important that the process of risk assessment, management, and communication itself has become the quintessential instrument for systems integration. The two seemingly distinct groups mentioned earlier – *engineers as managers of risk and risk experts as managers of engineering systems* – are a clear manifestation of this emerging need for a new breed of engineers. There is a role of separation between the two and a new role for each one. These are engineers who understand how the various components of the system interact with each other and who also have the tools to investigate what can go wrong and what strategies to follow to mitigate potential risks. Today, computer-based systems can be so complex that the systems engineers cannot rely on their expertise alone and must seek the cooperation of other disciplines. This also means that systems engineers must empower other engineers in the team to do the risk analysis and the systems integration. *In particular, to be a good software engineer or a good systems engineer/integrator, one must also be a good systems modeler and risk analyst.*

13.1.3 SHIFT FROM HARDWARE TO SOFTWARE

As CP SoS become more complex, the risks associated with their development increase. At the lowest level (step 1), the individual (manager or engineer) can assess and manage the risks based on personal knowledge and experience. As complexity increases the individual's span of knowledge, and experience soon becomes inadequate, the next level (step 2) is to employ outside expertise to gain more specific knowledge and experience. However, as a system eventually becomes even more complex (step 3), interaction and communication among its engineers, managers, and other experts must be facilitated and managed in a systemic manner. At this step a systematic and a structured set of processes, methods, and tools for assessing and managing risks becomes imperative. Without such a framework professionals cannot effectively work as a team to manage the myriad sources of risk, and in fact they become part of the sources of the system's risk. In many cases this is a common situation in software-intensive systems. The systems engineer is often not intimately familiar with software and is lacking a process for effectively using his or her software expertise to assess and manage software risks. Since software is playing a more central role in providing the infrastructure for systems integration, software itself is increasingly becoming a source of risk for other components of the system.

This lack of process, methods, and tools and the lack of systems engineers with software knowledge have led some software engineers to assume some of the roles of systems integration and management.

The complexity of the supply chain (see Chapter 11) and the goods and services delivered around the world on a daily basis have led to extensive subspecializations in many fields, most notably in engineering and management. Although this subspecialization stems from our need for detailed expertise in narrow fields, it has brought with it a parochial and often limited vision of broader, overall system perspectives. Indeed, few software engineers have either the opportunity or the expertise to appreciate their own contribution (within a project) to the entire CP Complex SoS and to other Complex SoS. This evolution in software engineering – a shift from hardware-based decision making in engineering to software-based decision making – is further explored in this chapter. The increased influence of software in decision making has also introduced a new dimension to the way business is done in engineering quarters. Many of what used to be engineering decisions have been transferred and transformed, albeit in a limited and controlled manner, to the software function. This power shift in software functionality, the explicit responsibility and accountability of software engineers, and the expertise required of technical professionals on the job have interesting manifestations, implications, and challenges to adapt to new realities and to change – all of which affect the assessment and management of risk associated with software development.

The intricacy and complexity of the risk assessment and management process (when applied to CP Complex SoS) and the need for quantitative analysis that requires knowledge in probability and statistics have contributed to the emergence of the subspecialization of risk management in engineering. Thus, seeds for two seemingly distinct groups – *engineers as managers of risk and risk experts as managers of engineering systems* – have been sown. In a parallel way, one may trace the distinction between the (i) "engineer" as a technical expert, one who is primarily concerned with the technical aspects of a project and to a lesser degree with managerial issues, and the (ii) "manager," an individual who is primarily concerned with management (in the broader and more encompassing sense of the term) and to a lesser degree with technical aspects. Here again, on the one hand the *engineer (as a local manager)* and the *manager (as a more global manager with a broader vision and perspective)* share responsibilities, tools, and methodologies, yet at the same time each performs distinct functions, matures in different professional cultures, often uses a different jargon, and communicates with a different language. Understanding this emerging power shift surrounding the three entities – software engineering, management, and risk analysis – is at the heart of understanding the emergence of software technical risk management of CP-intensive Complex SoS.

Software engineers have been increasingly assuming the bridging role between the customer and the product. The degree of *closeness* between the customer's needs and desires and the attributes and quality of the product (that were supposed to meet these needs and desires) depend to a large extent on the correctness of the intermediate transformation stage from requirements to product and, especially, on the detailed requirements that are given to the software designer. And this intermediate transformation stage is now more and more being performed by software engineers. It is here where risk assessment and management can and should play a more central role of ensuring that this degree of closeness is acceptable to the *customer*. By its systemic approach in assessing all conceivable risks, the risk assessment process seeks answers to the following four questions (Kaplan and Garrick, 1981, Haimes 1991, 2016): What can go wrong? What is the likelihood that it would go wrong? What might be the consequences? What is the time frame?

Clearly, the complex cognitive path (that involves many individuals) toward understanding and then translating the customer's needs and desires through *requirements*, *specifications*, *architecture*, and *process* has an inherent likelihood of being diverted from its intended path into a chaotic one. This likelihood, however, can be markedly reduced if the progress along this path is continuously subjected to the scrutiny of the risk assessment process and ultimately to the prevention, abatement, and control of such risks through risk management.

13.I.4 SOFTWARE AS INTEGRATOR OF COMPLEX SYSTEMS OF SYSTEMS

The displacement of hardware by software is evident. Purely mechanical and electrical devices are continuously being replaced by software-based systems. The intricacy of CP Complex SoS and the basic operation of parking meters and public telephones with software-driven electronic cards are typical examples. The functioning of software as a systems integrator, i.e. as a coordinator of all a system's activities without which the system will not function correctly or effectively, adds a new dimension to software risk management. In fact, software can assume the control role in a system and provide a mechanism that drives all of the elements to work as a unit. Thus, the emerging dominance of software as a systems integrator, whereby the various system components are organized, directed, and controlled to perform a specific set of functions, will necessarily bring with it new sources of risk of failure of the emergent Complex SoS. As software assumes the role of systems integrator, risk assessment should provide a systemic approach to identifying, quantifying, and evaluating most, if not all, sources of software risk. Both technical and nontechnical risks and risk management will ensure the reliable performance of software and thus that of the overall integrated Complex SoS. Software technical risk was defined earlier: *Software nontechnical risk is defined as a measure of the probability and severity of adverse effects that are inherent in the development of software and are associated with the programmatic aspects in the development process.*

As the software-intensive system's complexity increases, it becomes more difficult to comprehend all the positive (synergistic) and negative (adverse) interactions of its components. Furthermore, it also becomes more difficult to plan for the integration, testing, and evaluation of the system. The difficulties transcend (i) grouping and ordering of components for integration purposes, (ii) selecting the type, sequence, and methods of testing, and (iii) determining the best method and execution of regression testing as new components are added. Thus, it is important that the systems integrator be able to answer not only the set of risk assessment questions posed earlier but also the set of questions in risk management; Namely, (i) What can go wrong? (ii) What is the likelihood that it would go wrong? (iii) What are the consequences? (iv) What is the time frame? (v) What can be done? (vi) What options are available and what are the costs, benefits, and risks associated with these tradeoffs? (vii) What are the impacts of current policy decisions on future options? (Haimes, 2016). For CP Complex SoS, we ought to extend the above set of questions to also include (beyond each system) the entire Complex SoS. Furthermore, these systems integrators should address these questions from the perspective of decision making under risk and uncertainty and with the appropriate tools, methodologies, and practices available in the field of risk-based decision making.

In software-intensive systems, as well as in many other systems, the systems integration process for delivery may be executed in a "big bang" or a single act, with no intention of adding further functionality in the future, or the integration may be performed in an incremental fashion. Each case has its own, albeit partially overlapping, sources of risk. The systems integrator must be able to understand these two sources of risk and select, in each case, the appropriate and most effective risk-mitigating integration strategy. Since the systems integration process is commonly performed by more than one person, it is imperative that the systems integration process itself facilitates an effective communication within the team through a common set of risk-based decision-making methods and tools.

13.I.5 THE INTERFACE BETWEEN USERS AND BUYERS

Software-intensive CP Complex SoS are commonly initiated, commissioned, designed, developed, deployed, and ultimately used and maintained by individuals and organizational entities. The opportunities for miscommunication, adversarial communication, or no communication are unfortunately numerous. It is not always that a win–win situation is advanced and promoted among the involved parties. The common parochial mentality of some individuals or organizations is to care only about their limited role, scope, and responsibility and, with this attitude, to sacrifice opportunities for synergism. This state of mind not only prevents the achievement of higher performance of the parties involved but also is the source of major risks – both technical and nontechnical risks. Understanding this vital interface among the involved parties in the life cycle development of the system (and the software that serves as its overall integrator) is a key to effective risk assessment and management.

Furthermore, in CP Complex SoS, the user and the buyer are often two separate entities, possibly SoS themselves. As a matter of fact, there is often a chain of individuals and organizations that separate the buyer (e.g. a manufacturing plant manager) and the ultimate user (e.g. a manufacturing product manager). The necessary involvement of these intermediate parties constitutes yet another source of risk and uncertainty that must be understood and addressed. *The fact that software is becoming more and more, the hub of the integration process of CP Complex SoS makes it imperative (in terms of sound management) for software risk to be assessed and managed.* This central role of software engineering in systems integration also implies and requires that the assessment and management of risk associated with each subsystem and its software must also be integrated within the overall system risk management.

13.I.6 SYSTEMS INTEGRATION: SOFTWARE ENGINEERING AND THE SOFTWARE ENGINEER INTEGRATOR

Since software engineering is an intellectually intensive activity, where both human intelligence and fallibility play a role, human errors are a major source of both software technical and nontechnical risks. One can identify multidimensional sources of risk within each of the functional perspectives (i.e. requirements, specifications, architecture) in the software life cycle developmental process. The customer's perception of product needs and of the utility and efficacy of the final product is one example. Other examples of human errors that constitute potential sources of risk include lack of common technical language, misinterpretation of requirements, wrong perception of what constitutes an acceptable response time, and level of training and education of the personnel involved. Therefore, the risk of human failure can be of paramount importance throughout the software development life cycle.

We have focused our discussion on the emerging role of software as an overall systems integrator. We pose the following two questions: (i) Who is the software engineer/integrator? (ii) Are the functions performed by the software engineer different from those performed by the software engineer/integrator? Our premise is that these roles are still *evolving* as the power shift from hardware to software system centrality continues to evolve. What is clear is that software is continually assuming a more *pivotal* role in systems integration, not because software must play this role, rather because software engineering is *de facto* playing this role. This rapid shift in the functionality, responsibility, and influence of the individuals involved in the development of software engineering has created a new paradigm. This is a paradigm where software engineers no longer limit their vision solely to the architectural design of the component, rather they incorporate within their vision a more holistic and systemic perspective of the overall system's model. This new paradigm is also necessarily forcing the introduction of a new cadre of software engineers who must be familiar with systems modeling and integration, and not limited to software architecture design, algorithms, or coding in higher languages. Indeed, the evolving role of software engineering is also forcing the creation of a software engineer/integrator, but the ultimate attributes and characteristics of the software engineer/integrator are increasingly being recognized and promoted. To gain an insight into this evolving role of the software engineer/integrator, we observe the following sample of activities that are commonly being performed by the software engineer/integrator:

1. The display format of information and functions of most systems, including all their decision-making ramifications, is being determined and orchestrated by the software engineer/integrator, because software is the implementation mechanism. Even though the applications engineer plays a role in coordinating these efforts, many technical decisions, such as performance, error detection, and conflict resolution, are left to the software engineer, who frequently does not have the information necessary to make the proper tradeoffs and decisions.
2. The control system, which is commonly modeled by the designer, is interpreted (and in some sense redesigned) and implemented by the software engineer/integrator. Although this interpretation, redesign, and implementation of the control system may not affect the functions of the control, they certainly do affect the *performance* of the overall system.
3. Increasingly, the systems modeler or designer has either to know the capability of the software pertaining to that design (which is rarely the case), or he or she must ask the software engineer whether the implementation of a specific design can be realized through software engineering. In either case, the systems designer is increasingly delegating an integrating role and responsibility to the software engineer. To further explore the evolving role of the software engineer/integrator, we will use as an example the implementation of the systems requirements of a combat airplane via software engineering through sensors; the software provides data on threats and informs a pilot through the display. This display was, of course, redesigned and coded by the software engineer/integrator. For example, data and other vital information must also be displayed and communicated, and once again the role of the systems integrator is indispensable in providing the partnership between the hardware and software engineers. For example, an airplane may be forced to be in various flight regimes, and during adverse weather, the priority of the functions performed by the airplane must have been modified from the previous flight regime, and the integration of the various systems components must be translated by the software engineer/integrator into command and control statements. These statements must be in congruence with the designated functions of all involved systems components. Furthermore, this task also requires that the pilot receive online feedback on the severe weather status – involving yet another essential systems integration function that should be deployed by the software engineer/integrator. Finally, there are also the default, online, and preprogrammed decisions that are automatically executed when the situation is too complex or too fast for the pilot to get involved. Although these critical controls are modeled and designed by others, it is the software engineer/integrator who ultimately translates and implements them into the system. For example, in commercial aircraft such as Boeing 777, critical functions can no longer be designed into the hardware because of the implementation of the system software. One might argue that this interfacing role of the software engineer has always existed; however, the latter is functioning more and more as the designer and deployer of vital systems decisions – decisions that not only transcend any single component's functionality but also involve the entire system.

Thus, the software engineer/integrator is now deciding not only on the implementation and deployment of specific functions of the system but also on the functional interface among the system's components and is often making decisions in support of the client/user. Although this is also true for hardware engineers, the shift from hardware to software predominance is resulting in an ever-increasing importance of software engineering within system integration.

13.I.7 HIERARCHICAL HOLOGRAPHIC MODELING AND THE COMPLEXITY OF SYSTEMS INTEGRATION

Effective systems integrators' team must (i) possess a broad and wide range of expertise in both hardware and software; (ii) understand the role and mission of the CP Complex SoS on which to perform the intricate integration; (iii) ask good questions, have a good framework with which to work, and design and follow a well-structured plan of work; and (iv) have assembled a technically capable staff with whom to perform the systems integration effectively and efficiently. Indeed, effective systems integration in general, and especially for CP Complex SoS, necessitate that all functions, aspects, and components of the SoS be accounted for along with an assessment of all the associated risks thereto. To be responsive to the above requirements and to accomplish this risk assessment task, HHM (Haimes, 1981) introduced in Chapter 3 and utilized throughout this book will be deployed here. Fundamentally, HHM is grounded on the premise that large-scale and Complex SoS, such as software development and integration, should be studied and modeled from more than a single perspective, decomposition, vision, or schema. And because such complexities cannot be adequately modeled, or represented through a planar or a single vision, overlapping among these visions not only is unavoidable but can also be helpful in a holistic appreciation of the I-I among the various components, aspects, objectives, and decision makers associated with such systems. This comprehensive and holistic approach is particularly essential in systems integration, where the couplings and interconnectedness among all parts and functions of the system as well as the influence of its environment on it must be accounted for (Hall, 1989). Indeed, if all sources of risk are not identified, successful systems integration cannot be expected. The HHM framework presented here (and in Chapter 3) ensures that most, if not all relevant, categories of risk from the multiple visions and perspectives are taken into account. In other words, when all sources of risk are identified and assessed systemically, and thus managed, then the process of systems integration is controllable; otherwise, the process is not controllable, and the success of systems integration is left to chance.

The HHM framework for the identification of all conceivable sources of risks associated with systems integration consists of seven visions, perspectives, points of view, or decompositions (see Figure 13.I.2).

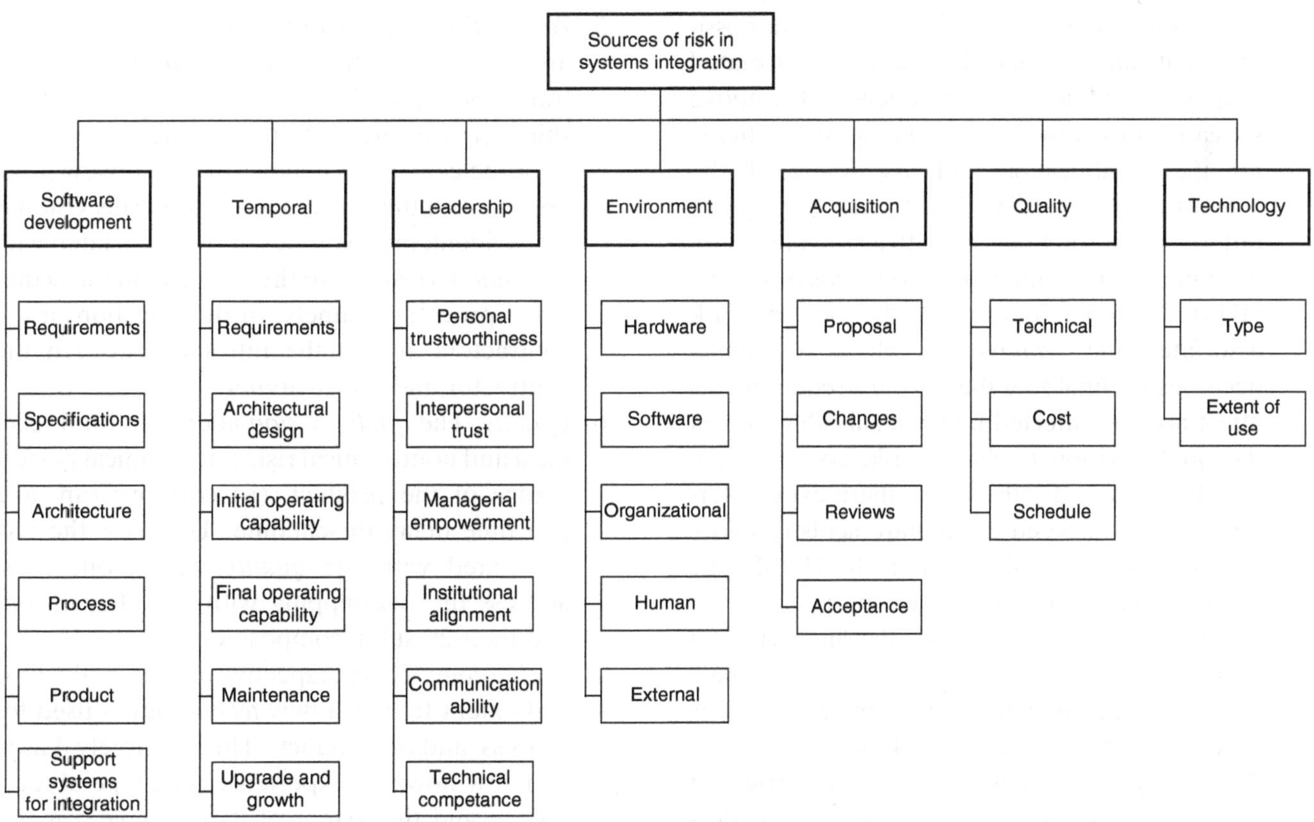

Figure 13.I.2 *Hierarchical Holographic Modeling framework for the identification of sources of risk in systems integration.*

In their totality, the seven disparate visions of sources of risk associated with systems integration constitute the mosaic building blocks of risk identification. Briefly:

1. *Software development*: Seven functions constitute the *software development* vision: (i) requirements, (ii) specifications, (iii) architecture, (iv) process, (v) testing, (vi) product, and (vii) support systems for integration. In other words, in this vision we attempt to identify and investigate most sources of risk associated with each function, stage, or phase of the software developmental life cycle. Furthermore, each of these functions is related (in the risk assessment process) not only to all other functions within the developmental life cycle, but also to all other visions of the HHM.
2. *Temporal*: Six categories of sources of risk are identified for the *temporal* vision: (i) requirements phase, (ii) architectural design, (iii) IOC phase, (iv) maintenance phase, (v) FOC phase, and (vi) upgrade and growth phase. At each stage of the system's life cycle – from cradle to grave – one should attempt to assess, to the extent possible, what can go wrong along with the associated frequency, severity, and consequences. The above six categories provide a comprehensive scheme for the identification and assessment of the sources of risk associated with each category at different points in time and with all the other subdecompositions (within the six other visions of the HHM). Consider, for example, the sources of risk associated the *requirements* phase. Certainly, decisions on finalizing the system's requirements affect and are affected by most subcategories of the *quality* vision (technical risk, cost overrun, schedule delay). Furthermore, there exists interdependence between the requirements category and the other six visions within the HHM. The *temporal* vision addresses the propagation of the development effort within the life cycle of a product. For example, the requirements are likely to change over time; this change in requirements is very prevalent from IOC to FOC.
3. *Leadership*: The lack of four categories of sources of risk identified by Covey for leadership (Covey, 1992): (i) personal trustworthiness, (ii) interpersonal trust, (iii) managerial empowerment, and (iv) institutional alignment. We add two more to make up for the *leadership* vision: (v) communication ability and (vi) technical competence. The *leadership vision* probably constitutes the best demonstration of the efficacious framework of the HHM. This vision, according to Covey (1992), "recognizes that people are the highest value because people are the programmers-they produce everything else at the personal, interpersonal, managerial, and organizational levels." One can easily see how each of the six categories of the *leadership* vision relates to all the subdecompositions of the other six visions within the HHM.
4. *Environment*: The environment perspective addresses five sources of possible risk of failures: (i) hardware failure, (ii) software failure, (iii) organizational failure, (iv) human failure, and (v) failure due to sources external to the system. The *environment vision* is the most general and generic of all other visions. Its mission is to provide coverage of sources of risk other visions may miss.
5. *Acquisition*: Four categories of sources of risk are associated with the *acquisition* process that transcends the customer–contractor relationships: (i) the proposal phase, (ii) the changes after the contract is issued, (iii) the review process, and (iv) the acceptance phase. This vision is self-evident; most sources of risk associated with the *quality vision* have their genesis in the acquisition process – namely, in the selection of the contractors and in the interface between the contractor and the customer.
6. *Quality*: The *quality vision* addresses both technical and nontechnical risks: (i) technical performance of the product, (ii) cost overrun, and (iii) time delay in schedule. To assess the risk associated with the *quality vision*, one must address the overlapping with all other visions and their 29 subdecompositions.
7. *Technology*: This perspective addresses the type and extent to which new *technology* is used for process and/or product. This is largely based on know-how and experience. The type of technology and the extent of its use have a major impact on the risk associated with the *quality*

vision and the *acquisition vision*, among others. Often the acquisition community (buyer and contractor) may underestimate the technical difficulty of a project and overestimate the availability of new technology.

In many ways, Figure 13.I.2 does not do justice to the task of communicating the contributions of the HHM concept to the reader; thus, the recommendation to consult Chapter 3. The most critical shortcoming of Figure 13.I.2 is that the HHM philosophy builds on a multidimensional representation of a system, in this discussion, the multidimensional representation of the sources of risk arise in systems integration. Yet, the two-dimensional depiction of the HHM in Figure 13.I.2 conceals the couplings, I-I, and the interactions among the various subsystems of the CP Complex SoS that contribute to and also constitute the sources of risk. Each of the seven visions (and as our knowledge of the role of software in systems integration in software-intensive systems) improves over time, the list is likely to extend further and incorporate new aspects or visions that are not foreseen today) can be viewed as *the primary vision* from the engineer's, manager's, or analyst's perspectives. For example, a project manager may want to focus on the *quality vision* (see Figure 13.I.3).

We envision that *technical risk, cost overrun risk, and schedule delay risk* are the manager's primary concerns. The HHM framework then enables this manager to trace, assess, and analyze all other factors affecting and affected by these primary sources of risk to the integrated CP Complex SoS.

The ultimate efficacy of the proposed HHM framework lies in at least two dimensions: (i) its capability to account for and display a complete a set of sources of risk associated with systems integration from their multidimensional perspectives as the analysts and experts can envision and (ii) its facility to provide in-depth varied interpretations of the multifarious dimensions of risks arising whether from the functional, temporal, managerial, developmental, quality, or from other perspectives.

A cardinal principle that transcends all risk management activities is the essential need for prior identification and assessment of all conceivable sources of risk (commonly achieved via HHM). The seven visions/decompositions of sources of risk and their corresponding 32 subdecompositions depicted in Figure 13.I.2 should provide an informative coverage of the sources of risk associated with systems integration. Otherwise, additional visions must be introduced. For the generation of mitigation strategies to control these risks, the reader is referred to the Fourth Edition textbook (Haimes, 2016). It is important to keep in mind, however, that since the HHM approach recognizes that no single vision or perspective of a system is adequate to represent a decomposed system by its components, only when all seven visions are considered in their overlapping nature would a comprehensive identification of all sources of risk be achieved.

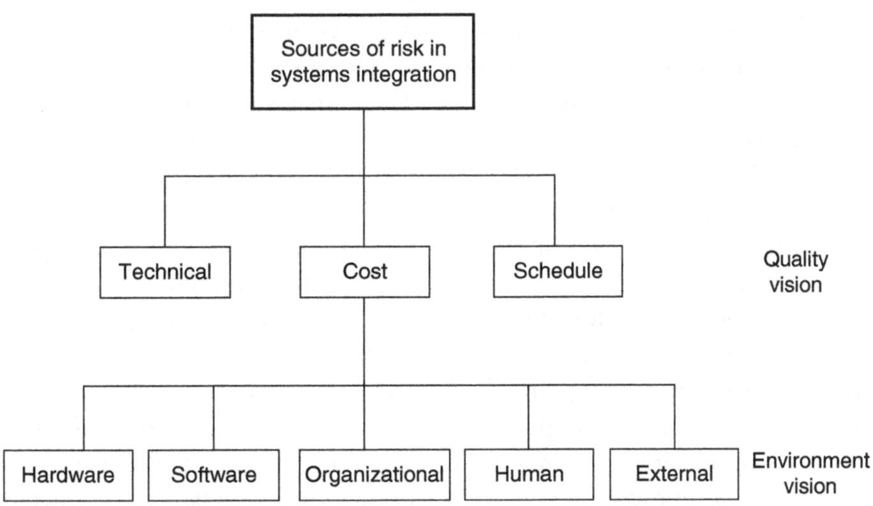

Figure 13.I.3 *Risk assessment quality-based HHM structure.*

13.I.8 ACQUISITION AS A PRECURSOR TO SUCCESSFUL SYSTEMS INTEGRATION

This section will focus on the *acquisition vision* to demonstrate the efficacious attributes of the HHM framework for risk identification. The fact that most acquisitions of software-intensive systems face chronic cost overrun and delay in their delivery has become a common affair. Many government and nongovernment software acquisitions have also had severe cost overruns and time delays in delivery, as well as cancellations, or the final product was never used.

Software acquisition and development, especially within government agencies, continue to be a significant management problem. With the ever-increasing importance of software as the *critical component* of today's economy as Complex SoS, a better understanding of the acquisition of software products by industry and their development is becoming a critical issue.

Clearly, one way is to focus on the acquisition process (i.e. consider the problem from the *acquisition vision* of the HHM as depicted in Figure 13.I.2) and delineate all other sources of risk and adverse consequences from that vision (while tracing the other 25 decompositions that are associated with the six other visions). In this case, selecting the contractor(s) and addressing the legal, organizational, technical, and financial issues associated with the selection process become the anchoring point. Furthermore, from a systems integration venue, an ad hoc acquisition process may prove to be the Achilles' heel for the entire project life cycle. One may trace, for example, changing the category of risk within the *acquisition vision* into the *temporal vision* and analyzing its impact on IOC or FOC. Such a systemic process of risk assessment markedly adds to the assurance that a successful systems integration can be achieved.

The need to revamp the software acquisition process to more effectively manage the risks of cost overrun and time delay associated with software development is widely discussed in the literature. In sum, the purpose of this discussion on the acquisition process is to demonstrate the efficacy of deploying the HHM in risk identification, and to highlight the dependency of successful systems integration on a comprehensive risk assessment framework.

13.I.9 THE NEED FOR METRICS

Risk, which was defined earlier as a measure of the probability and severity of adverse effects (Lowrance, 1976), is a quantitative measure. To manage risk, one must be able to quantify it to the extent possible (Haimes, 2016). This premise, which has been upheld in improving quality and effectiveness in the manufacturing of hardware – both for computer and noncomputer products – has merits that are applicable to software development. The almost revolutionary improvements in quality and productivity that have been realized in manufacturing during the last decade or two are attributed by many to new ethics that embody a commitment to quality management and to the improvement of human resources at all levels of the organizational structure. We can neither escape the principles for a better management advanced by quality improvement activists nor the several common denominators; the most important of which are attention to the centrality of human resources and the importance of quantification in achieving greater reliability and predictability and in the management of quality. Indeed various metrics and other statistical measurement tools have been developed and successfully applied as a holistic instrument of management in manufacturing.

The software engineering developmental process has almost reached that level of integrated management, in contrast, for example, with the high level of efficiency in manufacturing. The number of lines of code remains a dominant metric used to represent a variety of other attributes of the software complexity and architecture. Furthermore, the centrality of software as an overall systems integrator within larger scopes and domains has not been fully recognized. The shift away from hardware and toward increasing reliance on software in the systems integration scheme has not been matched by a related development of appropriate software metric/metrics with the capability to measure, predict, and control the software engineering development process. Such metrics can lead to a more cost- effective type of software engineering that promises quality, reliability, and adherence to prespecified time requirements and initially agreed-upon costs. Some argue for the emergence of another power shift in software engineering: one from emphasis on process to emphasis on product. The development of appropriate software engineering metrics would actually accelerate this shift, not necessarily from process to product but to a better systems integration and a holistic approach driven by systemic risk assessment and management. In this case, the interface between process and product would be better understood and cease to be viewed in dichotomous terms.

Furthermore, it is not sufficient to identify the universe of risk factors along the life cycle path of software development. To understand and appreciate the potential consequences of these risks and ultimately to take measures to mitigate their adverse effects or to prevent them, they must be measured and quantified to the extent possible. The introduction of appropriate metrics for software development thus becomes very important.

Observing the evolution of the field of multiple objectives, also termed as multiple criteria decision making (MCDM) over the last five decades, one cannot escape the lesson learned from the challenges that faced (and continue to face) modelers and systems analysts during that period. Note that Chapter 5 is devoted in entirety to multiple objectives and MCDM. In their quest to model, understand, and optimize systems with multiple, often conflicting and noncommensurate objectives, whether for hardware engineering-based systems or for intricate socioeconomic Complex SoS, experts in the MCDM field recognized the need for new measures (metrics) with which to quantify these objectives in nonmonetary units. Once the tools for generating Pareto-optimal solutions (policies) associated with multiple objectives were developed, a strong impetus emerged in the advancement in the art and science of systems modeling and in the measurement and quantification of these multiple objectives. One can safely project that as the field of software engineering continues to grow and as the importance of software risk management of CP Complex SoS continues to dominate management decisions, more effort will be devoted to the creation of metrics for software development.

The examples of old metrics in software development demonstrate their "primitive" nature: complexity in software engineering may be measured through the number of branches, number of decision blocks, number of lines of code, and/or number of functions to be performed. Size of software development is often measured by the number of lines of code, memory needed, and/or the time required for coding and design. And productivity is often measured through the number of lines or function points per unit time.

Clearly, significant concerted effort is needed for the development of more quantitative metrics so that an overall systems integration may be performed through the use of available quantitative risk-based decision-making methods and tools. The premise "to manage risk one must identify, assess and measure it" has multiple implications. First and foremost of those implications is the desire to evaluate the extent of the probability and severity of the associated risk. Other implications include the desire to evaluate the effectiveness of various policy options for risk mitigation or to assess the cost-effectiveness of such actions. Furthermore, the ability to measure risk with credible metrics provides an environment that is conducive to the acceptance by management of the cost–benefit–risk tradeoffs associated with mitigating actions. These and other reasons that justify the need for, and importance of, metrics in software engineering may not be new. The emergence of software systems engineers as systems integrators, however, accentuates the important role of metrics in systems integration, particularly for software-intensive systems.

13.I.10 EPILOGUE

The dominance of software in Complex SoS as a powerful instrument in today's society and technology and the subsequent power shift from hardware to software call for a new assessment of the role of software as a systems integrator and the corresponding new role of software engineers. This shift also represents an important change in functionality, where software assumes more of the systems integration and implementation role.

The previous sections examined this evolutionary process that has taken place in software engineering and in its role in software-intensive systems. Along with the need to manage this change, we also addressed the emergence of a new cadre of software engineers who at one time or another act as managers of risk, as systems integrators, or as risk experts who manage engineering systems. In all cases, however, the role of software engineering in software-intensive systems has become more critically important as systems are becoming more intricate as Complex SoS. In the quest for achieving a successful systems integration, we also addressed the developmental life cycle of software and highlighted six major functions or milestones – *requirements, specifications, architecture, process, testing, and end product.* To manage risk one must identify all sources of risk, assess them, and ultimately develop mitigating strategies for their management. The thesis advanced in the previous sections maintains that risk assessment, management, and communication as a process, serves as the overall cross-functional systems integration agent that connects and unites the above six functions/milestones along the life cycle path. This thesis is indeed grounded on the premise that the process of risk assessment, management, and communication is an imperative requirement for ensuring against unwarranted time delay in a project's completion schedule, cost overrun, and failure to meet performance criteria. Here, two sets of questions are advocated in the risk assessment and management process. First, in risk assessment, one should ascertain for each CP system and for the entire CP Complex SoS: What can go wrong; what is the likelihood that things can go wrong; what might be the consequences; and finally what is the time frame? Then, in risk management, one should also ascertain for each CP system and for the entire CP Complex SoS what risk mitigation options are available, what are their associated tradeoffs in terms of risk, costs, and benefits, and what is the impact of current policy decisions on future options. The previous sections relied on the HHM philosophy/methodology for the identification of most sources of risk associated with systems integration. The HHM, to which Chapter 3 is devoted in its entirety, builds (on the software engineering addressed here) on seven visions or decompositions of sources of risk and on 29 subdecompositions. These seven primary visions are *(i) software development, (ii) temporal, (iii) leadership, (iv) environment, (v) customer–contractor interface, (vi) quality, and (vii) technology.* Although each vision presents only a one-dimensional perspective, all, considered together, offer a multidimensional holographic representation of most, if not all, sources of risk. We devoted one section to the importance of developing metrics in software engineering in order to improve measuring these sources of risk.

PART II: HIGH PERFORMANCE COMPUTING TECHNOLOGY (HPC) WITH COMPLEX SYSTEMS OF SYSTEMS IN COMPUTATIONAL SCIENCE AND ENGINEERING

13.II.1 INTRODUCTION

Within the realm of computational science and engineering (CSE), two major engineering-based developments have been emerging through entirely different paths whose ultimate destination is to merge into one integrated system: one is software Complex SoS and the other, high performance computing (HPC) technology. Historically, CSE has been the principal driver of HPC technology. But within the past four decades, the development of Software Complex SoS and HPC technology has diverged somewhat and become multidisciplinary in nature. In many instances HPC vendors no longer provide turnkey systems to their customers, as they once did. Indeed, HPC software development and run-time environments now often come from third parties (open source, academic, or commercial sources), not the hardware vendors themselves. For our purposes the term "HPC technology" excludes hardware design and engineering but includes the software development and run-time environments that are the manifestations of HPC that CSE software developers, programmers, and users confront. Clearly CP Complex SoS benefit from and are an integral part of the above.

Lingering questions, which motivate this part, arise from the interaction among the three groups cited earlier, namely, the group that encompasses the application developers of large-scale scientific and engineering software systems, especially those requiring HPC, the HPC software development and run-time environments, and the integrators of the two groups: (i) To what extent does HPC meet the needs of Complex SoS in CSE and other domains that HPC is designed and programmed to solve? (ii) Is the complexity of the CP SoS understood well enough by software engineers in charge of developing HPC software to ensure that they can address Complex SoS challenges? (iii) To what extent can the development of software for HPC technology keep pace with the ever-staggering scientific and technological demand for performance and efficiency? (iv) To what extent can the CSE software engineering enterprise keep pace with HPC technology? (v) Given that HPC hardware architectures have become increasingly more complex (with hierarchical, distributed memories and multiple processors, tending toward tens or even hundreds of thousands) and include exotic processors such as cell processors. Thus, to what extent are engineers who specialize in software development environments addressing the complexity in a way that application developers of Complex SoS may not grasp? (vi) To what extent would a systems-driven, risk-based approach (that is embedded in the life cycle development of software engineering for HPC systems) manage the risks of failures to bring them to acceptable levels?

13.II.2 THE INTERPLAY BETWEEN COMPLEX SYSTEMS AND HPC SOFTWARE DEVELOPMENT ENVIRONMENT TECHNOLOGY

CP Complex SoS, which depend on HPC systems for their successful deployment, necessitate that software engineering teams fully understand the interdependencies and functionalities of the subsystems for which the software is designed and coded. The lack of such knowledge constitutes one of the major sources of risk of not meeting the system's performance criteria of time delay and of cost overrun in project completion. Conceptual mistakes are commonly traced to the educational and technical disconnect between two groups/schools of software engineers/developers. One group deals with Complex SoS and the other with HPC software development environments. The latter are not hardware engineers, but they must be close to them. The challenge is to bring together the groups of software engineers, systems engineers, and developers. This imperative need for understanding between HPC and Complex SoS practitioners should be extended beyond mere communications and exchange of information to bridging the gap in cultural, educational, and professional perspectives that apparently exists between them. A disparity exists in the way each group is educated and trained, and perceives and thinks about the problem, and how each envisions the solution and how to execute the mission.

13.II.3 RISKS ASSOCIATED WITH THE INTRA- AND INTERDEPENDENCIES BETWEEN HPC TECHNOLOGY AND COMPLEX SYSTEMS OF SYSTEMS

Effective software engineering and systems integration are by any measure critical for the ultimate performance of Complex SoS using HPC technology. In particular, HPC system development environments require software engineering processes that are more compatible with and responsive to the associated complex system integration. For example, electrical engineers may design a hardware system, but not necessarily build it themselves; however, software engineers and software developers are expected to do both. These conditions have cost US taxpayers billions of dollars in addition to major delays in project completion schedules and failures to meet project performance criteria. Sometimes a fatal inconsistency or omission is at fault, but more often, the overall design is vague and poorly thought out. Moreover, software engineers and developers often do not understand the interdependencies among the critical parts of Complex SoS. In other words, a holistic systems-based approach to the engineering problem seems to be lacking, namely, there is little or no understanding and response to the I-I, interplay, and overlapping among the myriad components and functions of the system. Exacerbating this reality, neither the objectives (e.g. mission, requirements) nor the stakeholders associated with these Complex SoS are fixed; rather, they often continue to evolve and change. On the other hand, most such systems are not adaptive, and their original designs do not accommodate adaptation and resiliency to changes. In sum, the intricacy of Complex SoS cannot be addressed with business-as-usual processes using fragmented and uncoordinated life cycle development that ignores their I-I.

Five major classes of risks are associated with the I-I that affect and, to a large extent, determine the successes and failures of Complex SoS that build on HPC technology:

1. The intricacy of Complex SoS and their myriad intertwined components, their almost-uncountable execution paths, and the centrality

of systems engineering to understand the explicit tradeoffs between software performance and hardware design.

2. The rapidly changing and evolving requirements, architectural design, performance optimization, coding, testing, and systems integration of both the software development environments and the hardware of HPC technology. (This constitutes the heart of the associated sources of risk and uncertainty.)
3. The omnipresence and centrality of human–machine interplay that remains under studied and less understood for its role in human and organizational failures.
4. The myriad fluid and ever-changing stakeholders and the allocated resources, which require high levels of modulation, adaptability, and the will and courage to manage change.
5. The numerous ways that Complex SoS and HPC technology can be decomposed to enhance our understanding of their intra- and interdependency and hardware–software configuration. A case in point is the strong coupling and overlapping between the development of an integrated software engineering for security and nonsecurity purposes.

As a prototype example, consider a simplified operation and control of only a small part of a modern complex steelmaking plant. The plant may be decomposed from *functional, temporal, and organizational* perspectives, where raw materials, energy, labor, quality control, and so on are transformed into finished products, often through very complex sequences of operations and production processes. Many control and decision-making functions are involved in determining, for example, the operating conditions, resource allocation, and scheduling of production units that affect plant performance. These control and decision-making functions are organized into a *temporal control hierarchy* that is based on the relative time scales of the associated actions. On the other hand, the same steelmaking plant can be viewed from the *hierarchical functional perspective*, namely, steelmaking furnaces, slab-casting machines, slab-heating furnaces, and rolling mills. Each subsystem has its own computer control that generates the control inputs to the subsystem in order to best satisfy one or more objectives associated with its subsystem and to the overall system as a whole. There is a synergy between software architecture and modeling the multiple perspectives of a system. For example, it is critical in the design of complex software systems to have architecture that describes the software from multiple points of view, such as *functionality, survivability, maintainability, security, and performance.* And, in the steel example, understanding and accounting for the overlapping computer control of the steelmaking plant from the multiple perspectives makes it likely to simplify the software engineering architecture and its integration with the complex hardware system. Similarly, architectures help in understanding and appreciating the integration of hardware and software through the multiple perspectives of the system. In particular, the centrality of systems integration to the ultimate performance of the system is highly dependent on the proper development of the system architecture and thus facilitates the system decomposition and hierarchical modeling of intricate HPC Complex SoS.

13.II.4 SYSTEMS INTEGRATION IN HIGH PERFORMANCE COMPUTING (HPC)

As the complexity of engineered systems increases, so do the methods and tools used to design and deploy these systems. Thus the need is greater for in-depth knowledge and specification for all parts of HPC Complex SoS and the HPC development environments that support them. The solution to this problem is now centered in the software (Complex SoS developers commonly see the hardware through the HPC software development environment). The issues we raised in risk analysis cannot be addressed through cosmetic modifications of legacy system components. It is conceivable that prospective system components have been developed under assumptions that are in conflict and cannot be rectified without major surgery. Rather, a paradigm shift in acknowledging, understanding, and managing the changes is required to streamline the roles and functionalities of the parties involved throughout the developmental life cycle of HPC Complex SoS. One example relates to the imperative needs to close the gaps in the education, training, skills, orientation, and culture that exist among the domain experts (e.g. physicists, chemists, and other scientists), systems engineers, and software engineers. In the past, hardware vendors provided "turnkey" development environments and systems engineers, who were trained in process control and understood the centrality of the state-space concept, which is central to this entire book, in system modeling, and control, and were responsible for systems integration. Both *technical risks* (not meeting a project's designed performance criteria) and programmatic risks (delay in meeting a project's completion schedule and cost overrun) were mostly within acceptable levels. It is worth noting that Complex SoS in CSE emerged after the hardware vendors stopped providing turnkey development environments. One critical question remains unaddressed at least in most academic programs, if not in the industry at large: To what extent should the developers of HPC Complex SoS software environments know about the issues that confront them? When the demarcation between hardware development environments and application software became fuzzier and fuzzier, then software engineers – not all of whom were formally educated and trained in holistic systems engineering – became the systems integrators, with all the technical ramifications thereto. The major technical and programmatic risks associated with legacy Complex SoS cannot be sustained and tolerated in HPC technology if the latter is to achieve the required agility, modularity, adaptability, robustness, resiliency, security, and seamless systems integration. The roles of systems engineers and software engineers are different. In particular, their success or failure may ultimately be determined by the requirements and tradeoffs between the demands made by the hardware architecture and software performance (and possibly even functionality) or by which function goes to the hardware development environment and which to the software of the HPC Complex SoS. These challenges include decisions on how to conceptualize and design the system, how to distribute the functions of the system, and how to implement the desired attributes. Among all challenges, the following questions deserve special consideration: (i) Who should decide, in addition to cost, whether to target one HPC-based system with 50 processors, with 50 distributed computers in a loosely coupled network (ii) or with a single processor with 50 cores? (iii) In what terms ought the tradeoffs be determined? (iv) How should we decompose the system for operation and implementation? (v) How should we highlight the common and different roles among systems and software engineers?

In sum, for software-intensive HPC Complex SoS systems (including many military systems), systems integration is not only the integration of components; it is also an understanding of the functionality that emerges from such integration and the appropriate distribution of roles, expertise, and responsibilities among the various and diverse parties. More on the centrality of systems integration in the performance of CP Complex SoS will be discussed later in this chapter.

13.II.5 THE ROLE(S) OF SYSTEMS ENGINEERS, SOFTWARE ENGINEERS, AND SCIENTISTS IN HPC COMPLEX SYSTEMS OF SYSTEMS TECHNOLOGY

An effective systems integration of HPC Complex SoS requires the utmost understanding, communication, collaboration, and coordination among the principals involved. Systems engineers, software engineers, computer scientists, and programmers usually create the development and run-time environments for HPC Complex SoS. For better or worse, scientists and engineers often subsume the roles of systems engineers and software engineers within Complex SoS software development projects because the latter professionals are perceived not to have adequate scientific backgrounds. The key role of software engineers is to architect/design/develop software using sound software engineering principles and processes. However, a core set of activities and functions centers around HPC Complex SoS where *the respective roles have no demarcation lines*. Should the tradeoffs among requirements, security, speed, and precision be performed by systems engineers, software engineers, computer scientists, programmers – or all four? Clearly, making the right or best decision on what and how to implement functions through HPC technology is critical; this process should be done by a team representing the strengths of all of these disciplines. (Most of the management and implementation functions of Complex SoS are in the hands of scientists, who have no backgrounds in either software or systems engineering.)

The challenge, and thus the remedy, is not only technological. Rather, it is grounded in understanding and acting on the historical evolution of systems engineering, software engineering, and risk analysis as recognized disciplines in academia and industry. Most dramatically, the emergence and growth of separate systems engineering and computer engineering societies within the Institute of Electrical and Electronics Engineers (IEEE), the International Council on Systems Engineering (INCOSE), and the Society for Risk Analysis (SRA) tend to highlight the differences, rather than the commonalities, among these most interdependent disciplines. At the same time, undergraduate and graduate programs at colleges and universities have not confronted the challenge of harmonizing the urgent needs for a new cadre of systems/computer engineers capable of functioning as systems integrators of HPC Complex SoS systems. Today, much of this needed harmony between the involved disciplines is achieved, albeit not uniformly, through imperative on-the-job training. Many of the problems identified here are likely to persist, unless the engineering community (industry, the professional societies, and academia) – including the rank and file of software and systems engineers – addresses them with a commitment to managing change.

The risks and opportunities associated with HPC Complex SoS systems are directly related to the engineering decisions made with respect to the roles and responsibilities, not only of systems and software engineers but also of the other players, especially scientists who develop code and manage the projects. Furthermore, uncertainty and confusion exist about the appropriate roles of the participants. Unfortunately, software and systems engineers do not play a large enough role. In particular, as systems become more complex, requirements for advanced design methods and tools are increasing, raising this question: What are the requirements for the key critical body of knowledge that engineers must possess (through education and/or experience) to enable them to expertly engineer HPC Complex SoS systems in a specific field? Two important organizations – INCOSE and the IEEE – have been engaged in such an important dialogue. For example, one cannot be a practicing physician just because he or she earned an MD degree; additional requirements include completing an internship and residency and commonly selecting a specific field of specialty. Should similar requirements be asked of systems engineers? An incompetent physician may kill one person at a time, whereas an incompetent engineer can kill 100 people at once due to a catastrophic system failure. A certified surgeon can perform only specific surgeries, but how about engineers and scientists? Can they design every system? For example, long ago the American Society of Civil Engineers (ASCE) established, through the American Academy of Water Resources Engineers (AAWRE), a thorough certification of water resources engineers with education and experience beyond the masters-degree level. What can engineers responsible for HPC Complex systems learn from the economic disaster and the human misery experienced following hurricane Katrina (see Chapter 14)? That disaster is attributed, at least in a significant part, to markedly deficient *knowledge management* – the lack of communication, sharing of critical and vital information, coordination, collaboration, and trust among the myriad local, state, and federal officials and the nongovernmental and public volunteer organizations. Indeed, organizational and human failures were at the heart of the misery experienced by thousands of Katrina victims. To what extent does this kind of deficient knowledge management still exist among the principal engineers cited above? In sum, effective knowledge management, which should dominate the communication and collaboration among the involved principals in charge of HPC Complex SoS and other SoS, is critical and imperative to avoid organizational failures such as those that took place in the wake of hurricane Katrina.

13.II.6 THE ROLE OF MODELS IN HPC COMPLEX SYSTEMS OF SYSTEMS

Modeling, an essential element of systems engineering, is imperative to build and bridge the functionality of HPC Complex SoS successfully and to operate them effectively. Invariably we predict the behavior of a system and its responses to various demands placed on it under different conditions through models – conceptual, heuristic, mathematical/analytical, simulation, and others. A model is said to represent a system properly if for every set of inputs, the model output closely coincides (according to some metric) with the system's true output. For example, the Boeing 777 jetliner was designed entirely via computer modeling. However, to be able to predict through modeling efforts the true response of a jetliner to various random perturbations, we must understand and represent the interactive causal relationship within its various components and subsystems as well as its interface with high altitude and wind environments. To the extent that the modeling of HPC Complex SoS is precise and is based only on theoretical principles and physical laws, it is not real; to the extent that it is real and is based on all relevant engineering, socioeconomic, political considerations, heuristics, and experimental data, it is not precise. To do justice to the multifarious aspects and dimensions of HPC Complex SoS, the modeler must make use of every possible avenue of knowledge. This includes information that draws from physical and other natural laws, as well as theoretical, empirical, computational, statistical, experimental, heuristic, and expert evidence. Clearly, the larger the number of components or subsystems that must be considered in a dynamic, nonlinear configurations, and nondeterministic HPC Complex SoS environment, the more difficult and challenging will be the modeling efforts.

To meet, at least partially, the challenge of risk modeling, assessment, and management associated with Complex SoS, given their unique attributes and characteristics, we discuss two complementary approaches: (i) decomposition and hierarchical modeling, when applied in conjunction with conventional aggregate modeling methods, offers several promising advantages (see Chapters 2 and 4), and (ii) Phantom System Models, used as a real-time and virtual modeling laboratory for systems integration and risk assessment and management (see Chapter 4).

13.II.7 CONCLUSIONS

The future and effectiveness of HPC technology as Complex SoS and the realization of their goals and objectives are clearly dependent upon better CSE, thus the need for more harmonized and integrated HPC Complex SoS. By this, we mean better definition of requirements; better communications and understanding among scientists, systems, and software engineers; and a more disciplined use of improved methodologies and tools for developing software. Defining the relationships and interdependencies among the domain experts in Complex SoS in (i) computational science (scientists acting in all of the typical software development roles) and in (ii) *HPC systems* (a more typical mixture of systems engineers, software engineers, software developers, and programmers) is an important step in harmonizing the integration of the two systems. For example, scientists need to appreciate the use of disciplined methods and tools in the development of software systems. Indeed, unless software engineering is developed from a holistic systems engineering perspective, it will be neither agile nor flexible. As the requirements for Complex HPC systems grow, the ability to engineer them becomes most challenging, maybe even not realizable. All of these goals are intimately dependent on the team of software engineers and systems engineers (also serving as systems integrators) who can communicate with each other and value each other's discipline.

PART III: ASSESSMENT AND MANAGEMENT OF SOFTWARE TECHNICAL RISK

13.III.1 INTRODUCTION

In the following part we address the modeling and management of the life cycle of software development within the framework of CP infrastructures as Complex SoS. We present an (i) overview of the emergence of software as a powerful instrument in today's society and technology, (ii) insight into the evolution of subspecializations in engineering fields, and (iii) outline of the conceptual framework for the modeling and management of software technical risk to Complex SoS. We subsequently establish the foundation for the conceptual framework, focusing on Complex SoS, and then discuss the differences between technical and nontechnical software risk. The power shift paradigm – the shift of importance from hardware to software – and its profound implications for Complex SoS are also discussed. The quintessential consequence of the power shift from hardware to software in

Complex SoS – in which hardware assumes the subsystems implementation role and software assumes the Complex SoS implementation role – is its influence on software technical risk. We conclude with the challenges and opportunities facing the professional community in communicating Complex SoS software technical risk. Four major internal and external forces that operate in the software development environment are discussed: the (i) culture of the organization, (ii) nature of the organization, (iii) software power shift paradigm, and (iv) resultant shift in functionality and decision making from hardware to software engineers, technological innovation and expertise, and other forces, such as the marketplace. The software development activity itself is characterized by five major traits: the (i) management of change, (ii) protection against organizational failure, (iii) maturity of the development process, (iv) design and technological capability, and (v) technological know-how. Influenced and driven by the internal and external environment, these five traits generate software risk of two types – technical risk and nontechnical risk. To assess and manage these risks, a holistic framework based on HHM introduced in Chapter 3 is used. The assessment and management of risk should ultimately enable any organization involved in software development to meet its product quality and performance goals while controlling costs and schedule.

The increased role of software CP, organizational, and other Complex SoS, as well as the increased dominance of computers in the design, manufacture, operation, maintenance, and management of most small and all Complex SoS, have made possible the resolution of many complex technologically-based Complex SoS. At the same time, the increased influence of software in decision making has introduced a new dimension to the way business is done in engineering quarters. Many of what used to be engineering decisions have been, and are being, transferred and transformed, albeit in a limited and controlled manner, to the software function. This power shift in software functionality, the explicit responsibility and accountability of software engineers, and the expertise required of technical professionals on the job have important challenges as they adapt to new realities, all of which affect the modeling and risk assessment and management of software development for Complex SoS.

Perhaps one of the most striking manifestations of this power shift relates to real-time control of CP Complex SoS. Quality control in the manufacture of an engineering subsystem, for example, is no longer primarily the responsibility of the operator. Instead, the software controlling the process also controls the quality. Thus, in many respects, the software, which is designed and developed by software engineers, actually controls the process, not the engineers who originally designed the product. *This implies that a shift has taken place from a strictly hardware engineering perspective to a hardware and software engineering perspective in the CP Complex SoS context.* Software now fundamentally influences the design of the system. Indeed, software is being used as a manufacturing or implementation mechanism and as a system design subsystem. For example, the decision to update or change operating parameters or entire algorithms based on real-time sensor data received from various sources may be embedded in the software system design. In another example, the data selected for display may be based on information received from other subsystems of the Complex SoS. That is, the types of changes or updates being implemented by software engineering today would have, in the past, required either system hardware modification, or a fundamental redesign of the system. Indeed, software risk assessment and management, as a specialized entity, with all its importance and implications for other engineering SoS, and for humans, will continue to serve as an emerging and evolving critical driver of all Complex SoS.

The evolving role of the software engineer in decision making has created and continues to create enormous new challenges. The risk of not meeting specified product quality has also shifted. What was once solely the responsibility of traditional engineers who had technical know-how, expertise, and experience, is now a responsibility shared with software engineers, who design and develop the controlling software.

In the broader sense, we must recognize the intersection of engineering and management, focusing on risk and uncertainty as important common denominators between the two disciplines. Figure 13.III.1 provides a roadmap to assist the reader through the evolution of the ideas and concepts presented in this part.

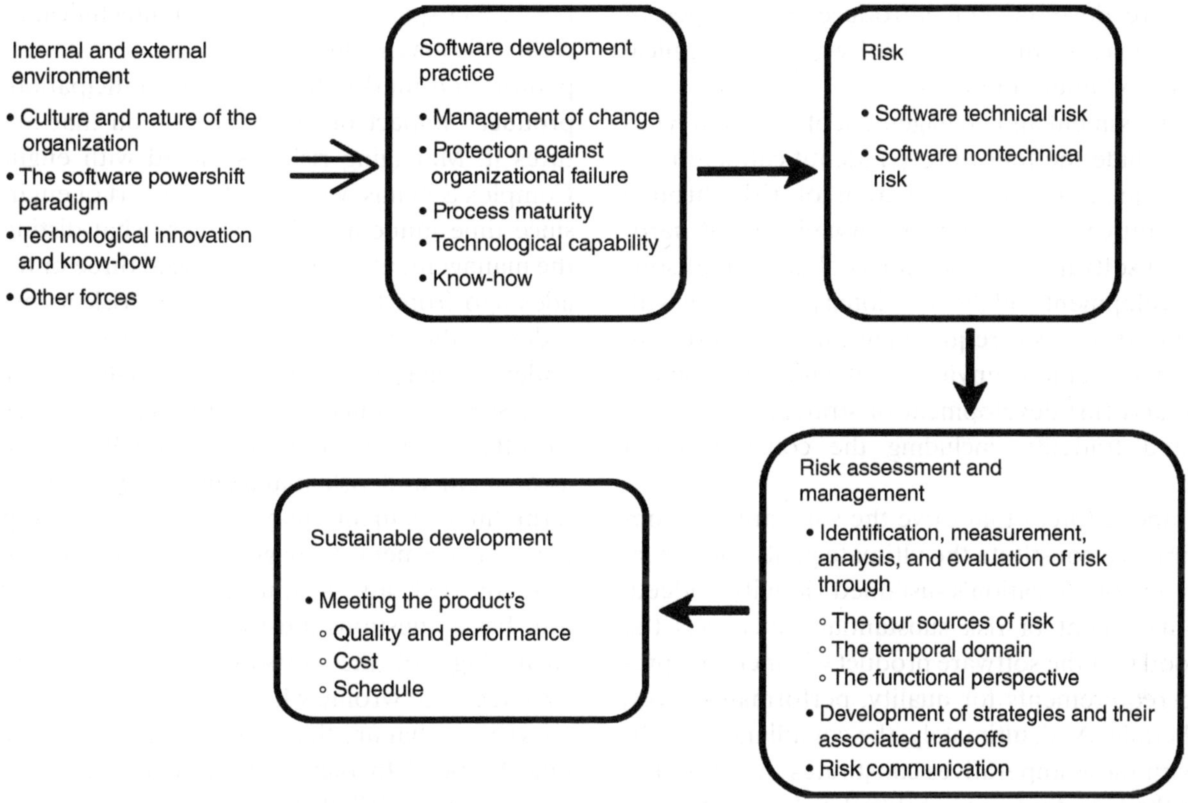

Figure 13.III.1 *A roadmap of this chapter's conceptual framework.*

13.III.2 A CONCEPTUAL FRAMEWORK

To realize the ultimate management of software technical risk, one must systematically follow the software development life cycle and understand the environment within which software is developed. Although Figure 13.III.1 provides a roadmap for the overall philosophy that has shaped the ideas presented here, for pedagogical and other conceptual reasons, the sections in this chapter do not necessarily follow the path outlined in Figure 13.III.1. The conceptual framework, composed of four major building blocks, is driven by internal organizational forces and the external marketplace and technologies that, together, shape and influence the development of software. The four building blocks of the framework for the assessment and management of software technical risk are (i) software development practice (this includes the process, methodology, and tools employed in software development), (ii) introduction of software technical and nontechnical risks, (iii) risk assessment and management, and (iv) sustainable development. Each of these is discussed in this section, along with the major components of each block and the internal and external forces that drive its development, its functionality, and its management.

Four major organizational and market forces drive the practice of software development: (i) culture and nature of the organization within which the software is being developed, (ii) software power shift paradigm (which will be discussed in subsequent sections), (iii) technological innovation and expertise used in the development of software, and (iv) other forces such as the marketplace, customer needs, competition, and politics.

The success or failure of the practice of software development for interdependent and interconnected Complex SoS depends on six major traits: (i) management of change, (ii) protection against organizational failure, (iii) process maturity, (iv) design and technological capability, (v) technological expertise, and (vi) complexity of the hardware–software CP SoS. These major traits, which essentially determine the effectiveness of the activities involved in the life cycle of software development, introduce three types of risks: software technical risk, software nontechnical risk, and SoS integration risk.

The assessment and management of the three types of risk include the following: (i) identification, measurement, analysis, and evaluation of risk through the four sources of risk – human, organizational, hardware, and software; (ii) the temporal domain of software development and the functional perspective with its seven attributes – requirement, product, process, people, management, environment, and development system; and (iii) development of strategies and their associated tradeoffs including the communication of risk.

The above forces that drive the software development practice constitute the ultimate goals and objectives of any organization's sustained viability. Indeed, the management of risk substantially increases the likelihood that the software product will meet the prescribed requirements for quality, performance, cost, and schedule. And, ultimately, the compliance of software with these imperatives constitutes the prerequisite for the long-term sustainable development of the organization and secures not only its current success, but also its future success and growth.

Note that there is a burgeoning tendency toward subspecialization in almost all fields that has caused a regrouping of professionals into seemingly distinct subdisciplines, each with its own subculture and vocabulary. Numerous examples of this tendency toward subspecialization can be found in medicine, law, accounting, engineering, and management. While this tendency to specialize may have the virtue of enhancing communication among members of a given subspecialization group, it obstructs communication across the disciplines.

13.III.3 ASSESSING SOFTWARE TECHNICAL RISK

Most, if not all, engineering systems as Complex SoS are conceived, modeled, designed, constructed, marketed, and maintained under great unknowns and immense uncertainties. This lack of knowledge is not limited to technological issues, such as strength of material, functionality, performance, accuracy, and quality of the components and the total product, but in fact spans a diversity of nontechnical areas as well. This may include predictions of customer, competitor, and market behaviors, or anticipation of the product's impact on the organization that manufactures it. Managing risk associated with engineering Complex SoS has been an integral part of engineering since time immemorial; however, what distinguishes the management of risk as practiced three or four decades ago from the management of risk as practiced today is the systemic and methodical approach to understanding, modeling, and management of Complex SoS. For conceptual and practical convenience, we often distinguish, as in Chapter 9, between risk assessment and risk management, even though the term "management" may connote the entire process of risk assessment, management, and communication.

In the risk assessment component, the following four basic questions must be posed and answered at each stage of the software development process: "What can go wrong; what is the likelihood that it will go wrong; what are the consequences; and what is the time frame?" In risk management and communication, we ask: "What can be done and what options are available; what are the tradeoffs in terms of all risks, costs, and benefits; what is the impact of current decisions on future options?; and what is the time frame?"; Determining what can be done includes developing alternative design options, evaluating tradeoffs, selecting one or more acceptable options (in terms of cost, reliability, performance, total quality, and safety), and evaluating the impact of current policies on future options. To answer the questions in the risk assessment process, however, one may benefit from knowledge of the four major sources of system failure, in general, as well as specifically in software development: (i) hardware failure, (ii) software failure (including software used in the development of software), (iii) organizational failure, and (iv) human failure.

Although all engineering managers practice risk management in one way or another, only a minority follow this systemic process by looking for sources of failure across the entire system and Complex SoS. The intricacy and complexity of the risk assessment and management process (especially when applied to complex engineering SoS) and the need for quantitative analysis (which requires knowledge in probability and statistics, and often more) have contributed to the emergence of the subspecialization of risk management in engineering. Thus, seeds for two seemingly distinct groups – engineers as managers of risk, and risk experts as managers of engineering systems and Complex SoS – have been sown. In a parallel way, one may trace the distinction between the "engineer" as a technical expert, one primarily concerned with the technical aspects of a project and to a lesser degree with managerial issues, and the "manager," one primarily concerned with management (in the broader and more encompassing sense of the term) and to a lesser degree with technical aspects. The "engineer" (as a local manager) and the "manager" (as a more global manager with a broader vision and perspective) share responsibilities, tools, and methodologies, yet at the same time each performs distinct functions, matures in different professional cultures, often uses different jargon, and communicates using different terminology. Understanding this emerging power shift surrounding the three entities – software engineering, management, and risk analysis – is at the heart of understanding the emergence of software technical risk management of Complex SoS.

Furthermore, it is important to appreciate and understand the following hierarchical managerial structure and the consequences of its divisions:

1. Upper management: This group views risk almost exclusively in terms of profitability, schedule, and quality. Risk is also viewed in terms of the organization as a whole and the effect on multiple projects or a product line.
2. Program management: Although this group is concerned with profitability, it concentrates more on cost, schedules, product specificity, quality, and performance, usually for a specific program or project.
3. Technical staff (software engineers, hardware engineers, etc.): This group of professionals concerns itself primarily with technical details of components, subassemblies, and products for one or more projects.

Clearly, differences among the risk managers at each level of this hierarchical decision-making structure are caused by numerous factors, including the scope and level of responsibility, time horizon, functionality, and requirements for skill, knowledge, experience, and expertise. Consequently, these differences determine, to a large extent, the tools and methodologies employed by risk managers at various levels. The management of risk associated with the development of software is governed by the same hierarchical decision-making structure and by the same interconnected engineering–management–risk subspecialization paradigm.

13.III.4 SOFTWARE TECHNICAL AND NONTECHNICAL RISKS

Software has dominated most engineering activities. Not surprisingly, the software's ever-increasing presence has followed the trend of subspecialization. Compiler and language developers, real-time embedded systems experts, management information systems, user interface management systems, and infrastructure and CP Complex SoS are all examples of increasing subspecialization. This influence and dominance of software on Complex SoS necessarily accompanies an increase in the elements of risk and uncertainty. Although no single classification of risk associated with software development has been developed, a dichotomous model of software technical risk versus software nontechnical risk is adopted for the purposes of this part.

Software technical risk is a measure of the probability and severity of adverse effects that are inherent in the development of software and associated with its intended functions and performance requirements. Thus, software technical risk connotes the risk associated with those aspects in the software development process that are concerned with the quality, precision, accuracy, and performance over time of the software. On one hand, software technical risk connotes risk associated with building a software product that meets the intended functions and performance expectations. On the other hand, *software nontechnical risk* connotes risk associated with the software development processes involving general management, i.e. with personnel, contractor selection, scheduling, budgeting, and marketing. *Software nontechnical risk* is a measure of the probability and severity of adverse effects that are inherent in the development of software and are associated with the programmatic aspects in the development process of software.

This dichotomy between software technical and nontechnical risks is introduced not for the purpose of distinguishing between two types of software products; rather, this classification distinguishes various functions in the developmental process of software and is concerned with the expertise required to deliver each function. Clearly, software technical and nontechnical risks are dependent on, and influence one another. For example, during a systems integration phase, the developed software may not meet some performance criteria or requirements. In this case, management has several options, including fixing the product and thus delaying the delivery time (and possibly exceeding the budgeted cost) or shipping the product as is on time. In either case, however, the sources of software technical risk have not changed; only the consequences have been altered.

REFERENCES

Boehm, B.W. (1988). A spiral model of software development and enhancement. *Computer* **21**(5): 61–72.

Chittister, C. and Haimes, Y.Y. (1994). Assessment and management of software technical risk. *IEEE Transactions on Systems, Man, and Cybernetics* **24**(2): 187–202.

Chittister, C. and Haimes, Y.Y. (2010). Harmonizing high performance computing (HPC) with large-scale complex systems in computational science and engineering. *Systems Engineering* **13**: 47–57. doi: 10.1002/sys.20131.

Covey, S.R. (1992). *Principle-Centered Leadership*. New York: Fireside.

Haimes, Y.Y. (1981). Hierarchical holographic modeling. *IEEE Transactions on Systems, Man, and Cybernetics* **11**: 606–617.

Haimes, Y.Y. (1991). Total risk management. *Risk Analysis* **11**: 169–171.

Haimes, Y.Y. (2016). *Risk Modeling, Assessment, and Management*, 4e. New York: Wiley.

Haimes, Y.Y. and Chittister, C.C. (1996). Systems integration via software risk management. *IEEE Transactions on Systems, Man, and Cybernetics* **26**(5): 521–532.

Hall, A.D. (1989). *Metasystems Methodology: A New Synthesis and Unification*. Elmsford, NY: Pergamon Press.

Kaplan, S. and Garrick, B.J. (1981). On the quantitative definition of risk. *Risk Analysis* **1**: 11–27.

Lowrance, W.W. (1976). *Of Acceptable Risk: Science and the Determination of Safety*. Los Altos, CA: William Kaufmann.

White, S., Alford, M., Holtzman, J. et al. (1993). Systems engineering of computer-based systems. *Computer* **22**(11): 54–65.

14

Infrastructure Preparedness for Communities as Complex Systems of Systems

PART I: INFRASTRUCTURE PREPAREDNESS: PRIMER

14.I.1 INTRODUCTION

Labels are helpful because they tend to deliver a major message in a few words. However, significant problems can arise when labels, such as *critical infrastructure protection*, evolving to a point where their use is inconsistent. Protection of complex systems of systems (SoS) (Complex SoS) may include a variety of risk-related countermeasures, such as detection, prevention, hardening, and containment. These are all important risk management policy options aimed at increasing safety and security. To appreciate the limitations of these security measures if they are not balanced with resiliency (introduced in Part II of Chapter 1 and to be subsequently defined), it is important to understand the epistemology of risks to infrastructure in terms of threats to Complex SoS. The following definitions apply *to each system and to the entire Complex SoS:*

- *Vulnerability* is manifestation of the inherent states of a system (e.g. physical, technical, organizational, cultural) that can be exploited or otherwise adversely affected by terrorism, natural hazards, and accidents, resulting in harm or damage to that system.
- *Intent* is the desire or motivation to attack a target and cause adverse effects (a definition only applicable for terrorism assessments).
- *Capability* is the ability and capacity of an adversary to attack a target and cause adverse effects (a definition only applicable for terrorism assessments).
- *Hazard forecasts* are predictions based on existing information about the likelihood of natural disasters.
- *Hazard consequences* are predictions about the consequences that could result from particular natural hazards.
- *Threat* is the *intent* and *capability* to adversely affect (cause harm or damage to) the system by adversely changing its states.
- *Risk* is the result of a threat or an accident with adverse effects to a vulnerable system. Risk is

Modeling and Managing Interdependent Complex Systems of Systems, First Edition. Yacov Y. Haimes.

also a measure of the probability and severity of adverse effects.

- *Preparedness* connotes policies, plans, and supporting infrastructures that are implemented in advance of a natural or man-made disaster – aimed at reducing disaster consequences (e.g. response/recovery time and cost) – and for the case of terrorism, through deterrence, reducing the likelihood to a level considered acceptable. Such acceptable levels of risk are obtained through the decision makers' implicit and explicit acceptance of various risks and tradeoffs. Preparedness is time dependent and must include effective and timely incorporation of intelligence and its analysis so that policies and plans are consistent with current events.
- *Readiness* connotes the deployment of tactical and strategic preparedness plans in a timely manner.
- *Resilience* is the ability of the system to prevent or to withstand a major disruption within acceptable degradation parameters and to recover within an acceptable cost range and time frame. Resilience is also a manifestation of the states of the system.
- *Prevention* is the ability of the system to halt or sufficiently reduce the consequences of a disruption and, in the case of terrorism, either to deter an attack by reducing its consequences to a level where the attack has little impact or to deter an attack by providing a response that is sufficient to more than counterbalance the potential impact of the attack.

Numerous studies have been conducted that explore various risks due to terrorism and natural hazards and the corresponding risk reduction measures. These activities have highlighted the point that the costs for the prevention of even a small fraction of possible risk cases are prohibitive, and the concept of investing to manage, reduce, or eliminate risks is neither always viable nor desirable. This leads to the need for developing new risk reduction options that can hopefully result in both economically viable and desirable risk management policies.

Natural disasters and catastrophes are commonly magnified by the actions or inactions of humans and organizations, whether advertently (e.g. acts of terrorism) or inadvertently (e.g. triggered by nature or by dysfunctional organizations). In this chapter, the composite of humans, organizations, and cyber–physical infrastructures is, in its essence, interdependent and interconnected Complex SoS. Components of this Complex SoS include human stakeholders as well as the multiple functions and operations of physical infrastructures, such as roads and bridges, telecommunications networks, electric power generation, oil and gas pipelines and installations, and water treatment and supply utilities, to cite a few examples. Each of these individual entities constitutes a *subsystem*, and their integration with the other subsystems makes up the Complex SoS.

The subsystems may function autonomously despite their interdependencies and interconnectedness (I-I); however, each is susceptible to experiencing unique adverse consequences resulting from an initiating event that affects a subset or all of them. In turn, such consequences commonly propagate and inflict other disastrous results. The initiating events of these disasters are generalized as *emergent forced changes (EFCs)* and connote *natural, external, or internal sources of risk that may adversely affect the states of the system and thereby affect the system as a whole.* Government agencies and other organizations often exacerbate the initial incidents by their inadequate preparation and response.

In his second seminal book, *The Next Catastrophe*, Perrow (2011) describes society's complex systems and posits that this complexity is leading to multiple and unexpected failures, i.e. catastrophes, that are waiting to happen. He offers valuable insights into how to make us safer and proposes innovative approaches for disaster preparedness. Perrow highlights the dismal performance of executives who are responsible for preparedness and response to catastrophes; he characterizes this situation as "executive malfeasance" and argues that the term "has a broader reference to illegality."

We offer the following definition of strategic preparedness: *A decision-making process and its associated actions implemented in advance of a natural or human-induced disaster and aimed at reducing consequences (e.g. recovery time, community suffering, and cost) and/or controlling their likelihood to a level considered acceptable (through the decision makers' implicit and explicit acceptance of various levels of risks and tradeoffs).*

Strategic preparedness, thus, refers to actions performed before a disaster as well as to the level of risk that results from such actions. Acceptable levels of risk are obtained through decision makers' implicit and explicit tolerance of various risks and tradeoffs (Haimes, 2016). Emergent properties of Complex SoS must guide the entire process of strategic preparedness, response, and recovery. We define these *emergent properties* as those system features that are not designed in advance, but evolve based on sequences of collected events that motivate the need for properties that ultimately evolve into system features. Government agencies, public and private sectors, and communities facing catastrophic risk must negotiate a host of conflicting and competing goals and objectives, which are advanced or constrained by stakeholders and special interest groups, as well as by organizational, political, and budgetary considerations.

In earlier chapters we introduced the centrality of state variables in all modeling efforts, especially their role in defining the I-I among the subsystems and Complex SoS as a whole. We also characterized I-I in terms of shared states and other essential entities within the subsystems that make up the Complex SoS. The term "essential entities" includes shared/common decisions, decision makers, stakeholders, resources, organizational behavior and norms, policies and procedures, management, culture, and others. For example, the shared states and other essential entities in the human body include the functioning heart and brain, along with the nervous and the circulatory subsystems, which together provide invaluable diagnostic information to the physician. Similarly, a farmer whose crops are growing poorly can readily assess the causes of the states of his crops by measuring the critical shared states and other essential entities, e.g. soil moisture and soil nutrients. Note that the states of the nutrients in the farm's soil also depend on both past and present crops planted on that parcel of land. This is due to the fact that some crops can deplete specific soil nutrients excessively. Here again, interdependency serves as a source of vital information that can be invaluable when properly understood and effectively exploited. These interdependencies, in the natural and constructed environment, have been continuously studied and exploited by humans as our knowledge of them improves.

However, the similarities of I-I that exist between the natural and constructed environment have not been fully exploited in our preparation for natural disasters. More problematically, the I-I within and among cyber–physical (CP) infrastructure Complex SoS have been perceived as detrimental to managing the myriad sources of risk to them. Indeed, there is a misguided notion that such I-I among the subsystems is a "curse" that complicates managing the innumerable sources of risk to which they are exposed rather than providing opportunities to be exploited. This is because the notion overlooks the positive value of the intricate I-I. The common/shared states and other essential entities among the subsystems of Complex SoS provide an invaluable avenue through which we can better understand the intricate dynamics that characterize Complex SoS, thus justifying and enabling effective investments in preparedness and responses to severe and possibly catastrophic EFCs, whether they originate from natural or malevolent sources.

Modeling a simple system, or a Complex SoS, necessarily implies (i) determining its properties, (ii) constructing the relationships among its inputs and outputs through its state variables and other building blocks of the model, (iii) quantifying intra- and interdependencies within and among its various components and subsystems, (iv) determining the appropriate model topology (structure) and the parameters that best represent its essence and functionality, and (v) determining the appropriate database to be collected. To achieve this, modelers rely extensively on data collection through testing, observation, experimentation, and measurement and through a meticulous process of learning about the system being modeled, including using dynamic simulation (Heal and Kunreuther, 2007).

There is a fundamental difference, however, between the complexity and challenges associated with modeling physical and natural systems – which are primarily controlled and driven by physical laws – and the challenges that characterize organizational and social systems – which are primarily controlled and driven by human and organizational behaviors. Complex SoS commonly represent interconnected physical, natural, and social subsystems. Among the fundamental needs of modelers of Complex SoS are to (i) find ways and means with which to enhance their knowledge of their complexity and (ii) discover the dynamics of their interdependency's behavior and identify the extent and nature of the I-I among their subsystems and the environment. In many ways, modeling is a trial-and-error interplay between the accumulated theory and knowledge about the Complex SoS and their environment, and continuous experimentation, observation, and estimation, in a learn-as-you-go inquisitive and exploratory process. Recall from previous chapters that models are built to answer specific questions; thus, they must be as simple as possible but as complex as required. This trade-off between complexity and simplicity is at the heart of the art and science of model building, given that over-specification or overcomplexity of a model tends to impair its usefulness. Furthermore, both natural and constructed environment Complex SoS are ultimately influenced by humans – individually, in groups, and in organizations. This necessarily implies that modelers must recognize and account for human and organizational impacts.

Recall that the states of a system are central to modeling Complex SoS and to risk analysis. These topics have been primarily addressed not only in Chapters 1 and 4 but also throughout this book. Recall Chen's (2012) following succinct definition of state variable: Under certain conditions: "The state $x(t_o)$ of a subsystem at time t_o is the information at time t_o that, together with the input $u(t)$, for $t \geq t_o$, determines uniquely the output $y(t)$ for all $t \geq t_o$." The states of a subsystem, commonly a multidimensional vector, characterize the system as a whole and play a major role in estimating its future states for any given input. Thus, theoretically, the knowledge of the states of a subsystem, as a function of time, enables modelers to determine, under particular conditions, its future behavior for any given input or EFC (e.g. an earthquake, hurricane, or tornado, a volcanic eruption, a national or regional epidemic, or a major fire). Note that such disasters affect communities and their inhabitants, as well as the services on which the population depends. Although people and organizations represent critical subsystems within these Complex SoS, there are challenges in accounting for them in the modeling of CP subsystems, and they remain mostly unaccounted for.

Identifying the shared states and other essential entities within and among the subsystems that make up Complex SoS constitutes the fundamental core for understanding and modeling their inherent I-I. Shared states among subsystems can assume many forms. For example, consider a community served by one major water supply system that provides water to the local hospital, fire station, electric power system, and wastewater system, as well as to all households in the community. Assume that the water supply system becomes dysfunctional. At the first stage of interdependency, all subsystems are affected, since they all share the *state* of water flow. At a subsequent stage, the electric power generation system would shut down for lack of available water for cooling of its power generation units, causing all other subsystems to experience blackouts, because they share the *state* and output of electric power, and so forth. Although this scenario is obvious, the reliance on shared states, and thus outputs, among interdependent subsystems provides a potent advantage to their modeling and system identification and thus to their ultimate management.

It is logical to assess, manage, and communicate the sources of risk associated with disasters by focusing on the shared states and other essential entities of the many subsystems that constitute the Complex SoS of communities and their inhabitants and the services upon which they depend. However, not all subsystems experience similar consequences in response to the same EFC. This fact is grounded on the inherent specific relationship among the states of the affected subsystems. Subsequent sections in this chapter will present the relationships among the vulnerability, resilience, risk, states of a system, and specific EFCs.

Modeling Complex SoS by focusing on the innate meaning of dependencies, or couplings that characterize them, must start with the fundamental building blocks of mathematical models, e.g. random, exogenous, and decision variables; and inputs and outputs. In reality, all state variables affected by these building blocks are under continuous natural, desired, or forced changes, both positive and negative. Thus, it is logical to concentrate on the behavior of the states of each subsystem in order to assess and learn the specific impacts of EFCs on them and thereby on the entire Complex SoS. Consider the following dynamic shifting of states and functionality of systems: (i) rules and realities; (ii) goals and objectives; (iii) stakeholders, decision makers, and interest groups; (iv) organizational, political, and budgetary baselines; and (v) reorganized and reallocated resources. Individually, and in combination, they constitute a significant challenge in modeling emergent Complex SoS. (Also consult Chapter 9.)

14.I.2 DEVELOPING A PREPAREDNESS ROADMAP USING THE ADAPTIVE MULTIPLAYER HIERARCHICAL HOLOGRAPHIC MODEL (AMP-HHM)

Hierarchical Holographic Modeling (HHM), introduced in Chapter 3, is a conceptual, functional, and systemic graphical representation of a system (e.g. a critical transportation infrastructure or a hospital) aimed at describing its essence and its inherent diverse characteristics and attributes. Recall that the term *holographic* refers to having a multiview image, as opposed to a single view, or planar image of a system. The term *hierarchical* refers to learning about the organizational structure and decision making process at the various levels of the Complex SoS hierarchy, structurally or organizationally. The HHM stipulates that a complete risk assessment must recognize and address the "macroscopic" and "microscopic" relationships that exist at the various management levels of an organization. The HHM enables the gathering of knowledge up and down the hierarchy. It also allows a system to be decomposed in order to describe its holographic and hierarchical nature. Within the HHM, holographic views are represented by "Head-Topics." These are further decomposed hierarchically into "Subtopics and Sub-Subtopics," and perhaps even further. Because it examines the system from multiple perspectives and decomposes those perspectives hierarchically into subsystems, the HHM can be a valuable tool for identifying the numerous sources of risks to Complex SoS.

Two useful features of HHM in the context of preparedness are its use of *multiple teams* and the concept of *flipping* perspectives. To capture the multifarious perspectives of a system, the HHM uses multiple teams with diverse knowledge and experiences, each group with its own view of the system, to share different perspectives on the systems, the Complex SoS, and the myriad sources of risk thereto. Therefore, different HHMs are created with different teams to more fully characterize Complex SoS. All sub-HHMs are then compared, consolidated, and discussed by all team members until a final comprehensive and overarching HHM is created, describing as many views as possible, as well as the generation of a taxonomy of risk-based scenarios. Another key feature of the HHM is the concept of "flipping the HHM" or changing perspectives to show coordination or reliance among entities. Flipping the HHM refers to the user's ability to rotate multiple model views depending on priorities and interest.

Effective preparedness requires planning for multiple decision-making perspectives, as depicted in the HHM of Figure 14.I.1. This planning includes factors such as (i) human resources, technology, and policies; (ii) interface arrangements among agencies at all levels (readiness must involve the public and the private sectors, not only government and nongovernment organizations (NGOs)); and (iii) interoperability and information sharing that transcend security (such as police, fire, and emergency management services), health and safety, transportation, critical utilities and infrastructures, and NGOs, among others.

One view of what might be in the US Department of Homeland Security's (DHS's) perspective on preparedness is presented in Figure 14.I.1 (perspective A). This can be decomposed into three levels: the Federal Emergency Management Administration (FEMA) and other federal agencies, state and local emergency response agencies, and other NGOs such as the Red Cross and volunteers.

Each agency can be decomposed according to a geographical region of interest. For example, a country can be subdivided into N regions that span several adjacent states. For each state, decision makers need to optimally, effectively, and efficiently allocate equipment, materials, commodities, and emergency responders. At the next level of Subtopics, the responders include the emergency management agency, public safety communication entities, law enforcement, and so on. For each responder, there are a number of critical infrastructures that need special attention during emergencies. These include information technology, telecommunications, chemical facilities, and other infrastructures like those depicted in Figure 14.I.1.

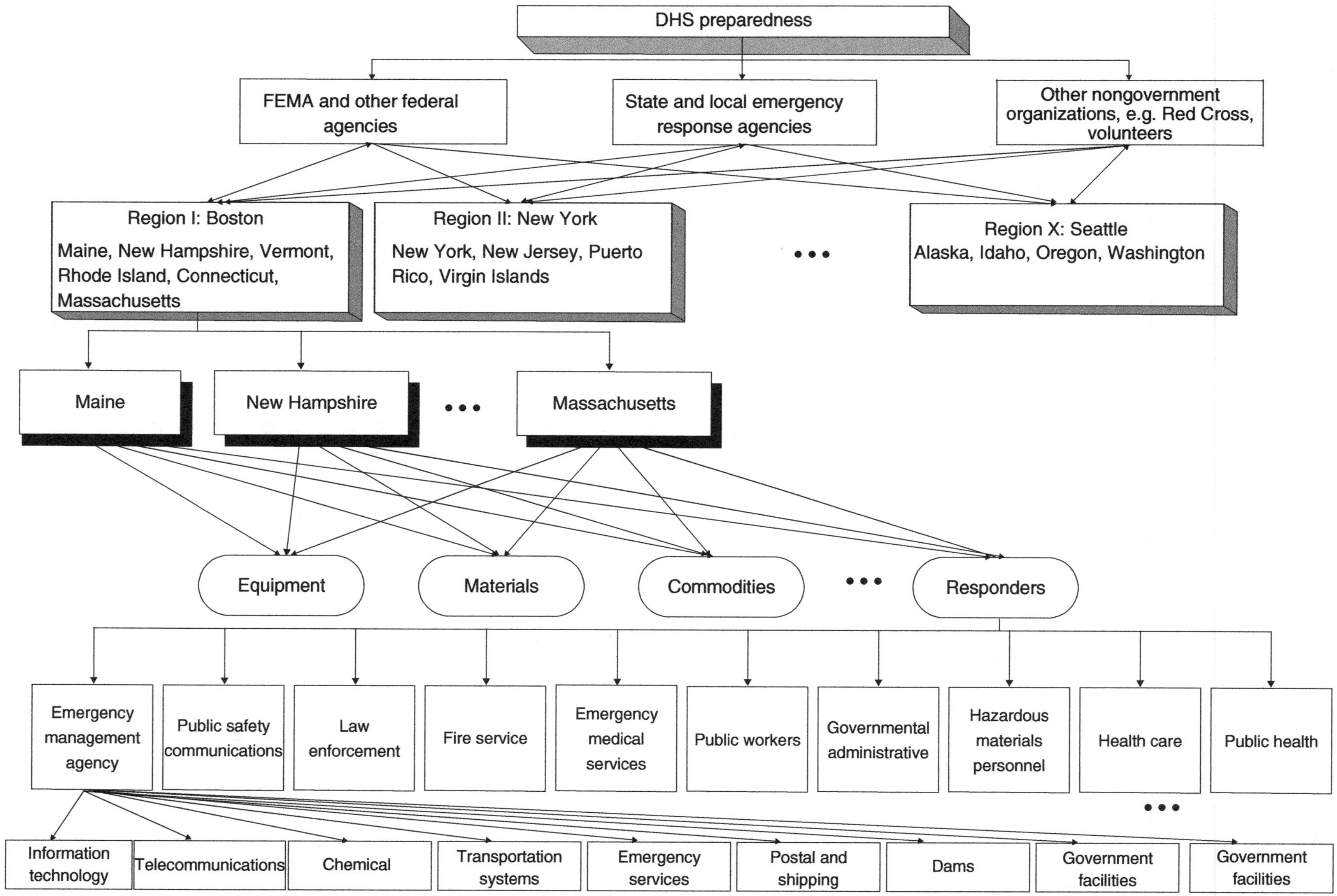

Figure 14.I.1 *An HHM for DHS preparedness – perspective A.*

Two alternative views of the HHM in Figure 14.I.1 are presented in Figure 14.I.2 (perspective A Part (i)) and Figure 14.I.3 (perspective A Part (ii)). Figure 14.I.2 shows what might be the DHS's perspective on a preparedness framework as viewed from the perspective of federal, state, and other emergency responders. These agencies need to coordinate their efforts based on their geographical and jurisdictional boundaries in order to generate efficient ways to allocate critical resources such as equipment, materials, commodities, and responders. In Figure 14.I.3, the HHM shows an alternative view of the preparedness roadmap, with the major focus being on critical resources. Often, emergency resources are limited, and hence they must be properly inventoried for efficient distribution to multiple agencies, taking into consideration geographical and jurisdictional policy-making factors.

Using the HHM's flipping feature, Figure 14.I.4 shows yet another view of what might be the government's perspective on preparedness (perspective B). In this view, the first layer shows the types of critical infrastructures as identified by the government. Each is located in one of the ten geographical regions. Underneath the geographical Subtopics are the three agencies that work to allocate resources, such as those shown in the next layer: equipment, material, commodities, and

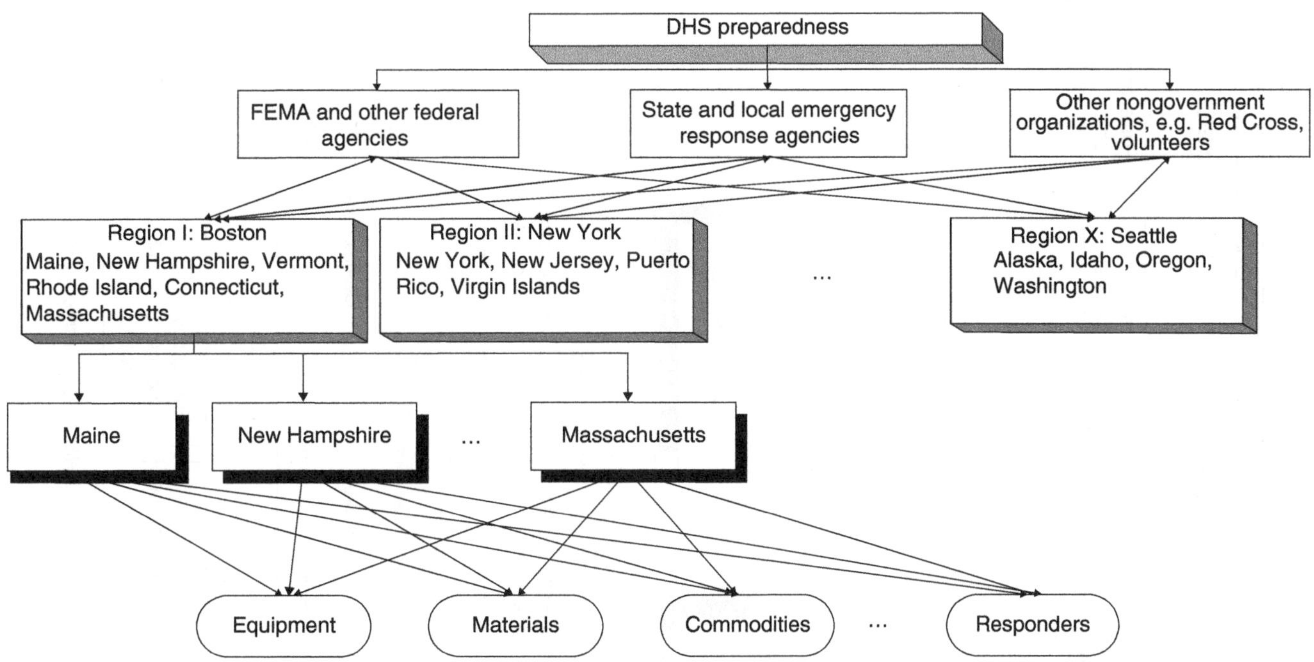

Figure 14.I.2 HHM for DHS preparedness – perspective A (Part (i)).

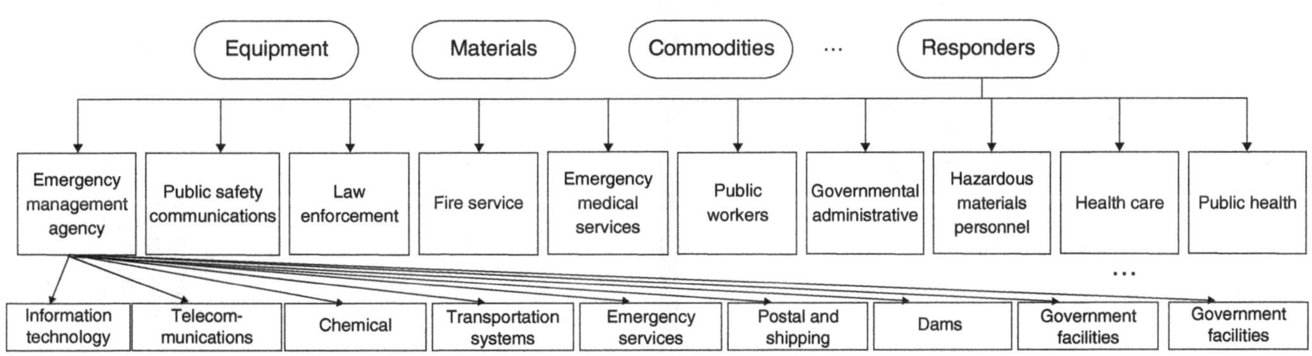

Figure 14.I.3 HHM for DHS preparedness – perspective A (Part (ii)).

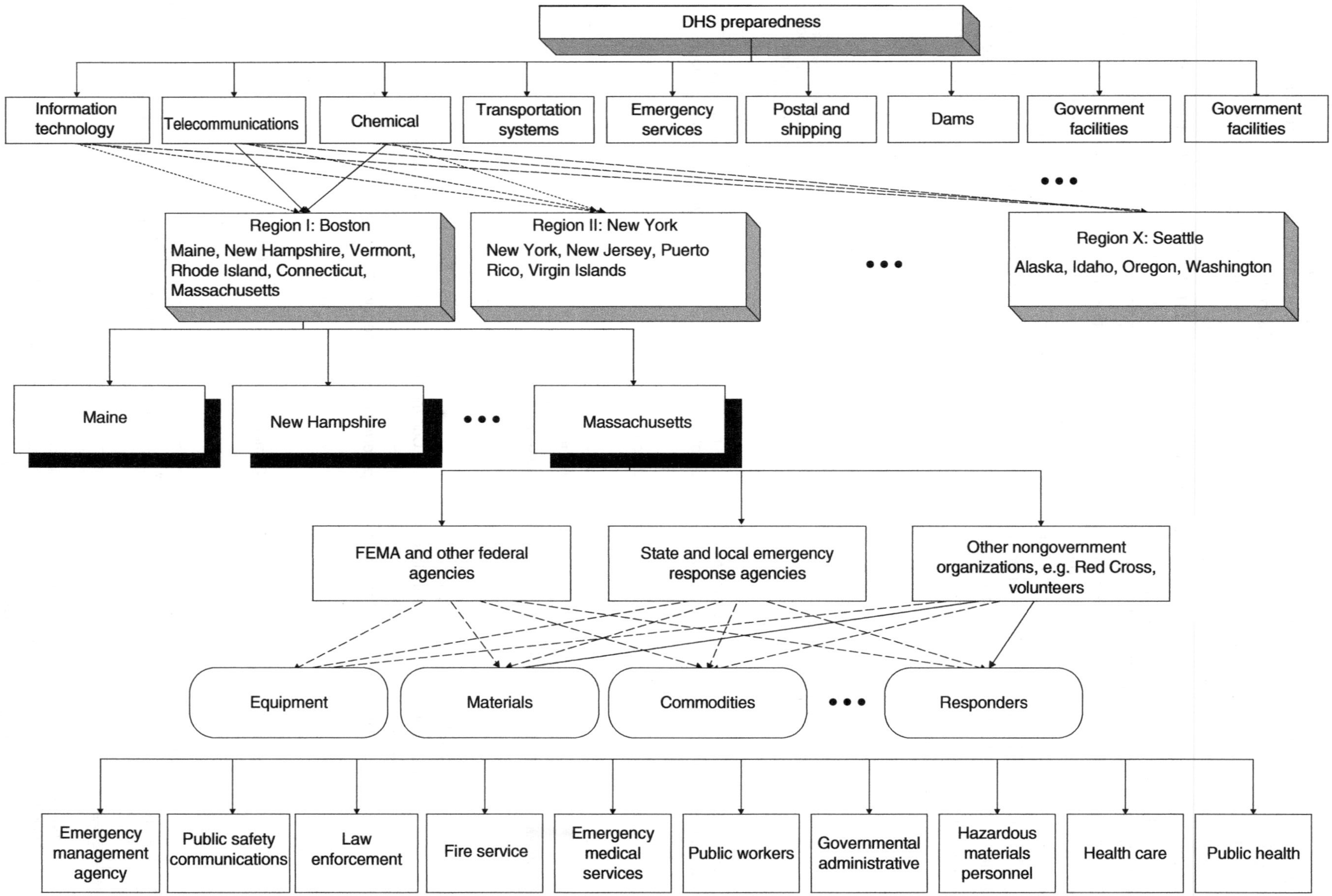

Figure 14.I.4 *HHM for government preparedness – perspective B.*

emergency responders. The final layer of Subtopics further categorizes the responders. Other alternative views of the HHM in Figure 14.I.4 may be imagined.

The adaptive multiplayer HHM (AMP-HHM), built on the HHM philosophy (Haimes, 1981, 2016; Haimes and Horowitz, 2004a; 2004b), is a computer-based, collaborative modeling paradigm aimed at the integration of multiple perspectives and the evidence of multiple experts for risk identification and analysis. For example, in a workshop conducted for the Institute for Information Infrastructure Protection (I3P), the AMP-HHM framework enabled over 100 participants from across several economic sectors to be divided into four teams (owners/operators, hackers, vendors, and government employees) and to construct a taxonomy of almost 150 classes of risk in 4 parallel 40-min sessions. In a second application, the Gubernatorial Inauguration Committee of the Commonwealth of Virginia distributed computing software implementing the AMP-HHM that enabled the involvement of 40 experts across 20 state agencies to be organized into three analysis teams. The AMP-HHM application was implemented in Microsoft's *Groove* collaborative computing environment. All expert participants contributed to, and interacted with, the application from their office desktops. This led to the identification of over 280 threat scenarios, which were reduced to 70, with the top 10 ranked and analyzed for risk management.

14.I.3 ON THE RELATIONSHIP BETWEEN PREPAREDNESS, RESILIENCE, AND RISK TO COMPLEX SYSTEMS OF SYSTEMS

The literature on risk analysis is replete with misleading definitions of vulnerability (Haimes, 2006, 2016). Of particular concern is the popular and misleading definition of risk as the product of "impact, vulnerability, and threat." Thus, in our quest to measure risk of terrorist attacks and natural disasters to critical infrastructures, we must account for the fundamental characteristics of the Complex SoS under consideration. More specifically, we must rely on the building blocks of mathematical models, focusing on the use of states of the Complex SoS, as discussed throughout this book.

In terms of vulnerability and criticality, infrastructures are represented (characterized) by a hierarchy of vectors of state variables. To relate the centrality of state variables in intelligence analysis to countering terrorism, it is important to redefine the following terms, which broadly apply to risk analysis as applied to single systems and to Complex SoS: (i) *Vulnerability* is the manifestation of the inherent states of the system (e.g. physical, technical, organizational, cultural) that can be exploited to adversely affect (cause harm or damage to) that system; (ii) *intent* is the desire or motivation to attack a target and cause adverse effects, (iii) *capability* is the ability and capacity to attack a target and cause adverse effects; (iv) *threat* is the *intent* and *capability* to adversely affect (cause harm or damage to) the system by adversely changing its states; and (v) *risk* is the probable result of a threat with adverse effects to a vulnerable system.

In sum, to assess the risks to a vulnerable system, or to Complex SoS, we need to (i) assess the likelihood of the threat (attack scenario), (ii) model the responses of the various interdependent state variables that characterize the system (i.e. its vulnerabilities) to the attack scenario (i.e. develop a "dose–response" function), (iii) assess the severities of consequences resulting from the dysfunctionality of the entire system or from a subset of its subsystems, and (iv) assess the likelihood of the severity of the projected consequences. *In the process of measuring risk, when the second imperative step – the critical modeling process that translates an attack scenario into consequences – is often masked or skipped by simply multiplying vulnerability directly into the risk measure, then the risk measure becomes detrimentally flawed.*

14.I.3.1 Defining Resilience in Measuring Risk to Complex Systems of Systems

In this book and in our earlier publications, we define the resilience of a system as a manifestation of the states of the system, and most importantly, it is a vector that is time dependent (Haimes, 2009, 2016). For the purposes of this chapter, resilience is also defined as the ability of each system, and Complex SoS as a whole, to withstand a major disruption within acceptable degradation parameters and to recover within an acceptable composite cost and time. Moreover, resilience, similar to vulnerability, is a manifestation of the states of the system, in that it cannot simply be measured in a single unit metric; its importance lies in the ultimate multidimensional outputs (the consequences) of the system for any specific inputs (threats). Note that the consequence, which is considered as part of the risk metric, is in fact the output of the systems model and that the input to the system's model is parallel to the concept of threat. Indeed, the risk associated with a cyber attack on a CP Complex SoS will depend not only on the resilience of the states of the system, but also on the specific type and sophistication of the cyber attack. This is because the resilience of a system can be measured in terms of the specific threat (input), the system's recovery time, and the associated composite consequences measured in terms of costs and risks. Thus, different attacks would generate different consequence (output) trajectories for the same resilient system.

Consider the immunization of a population against a major strain of a flu virus termed type B. Assume that the population develops resilience for multiple strains of viruses of type B, except for an evolving strain of type A. In this case, even though the population might have resilience (immunity) for type B, the appearance of strain A into this population would likely be infectious. Here again the risk to the population from a threat is dependent on the type of threat, the resilience of the system, and the ability of the system to withstand that specific threat.

Likewise, consider any large-scale physical infrastructure such as electric power, transportation, or telecommunication. In any such Complex SoS, the question "What is the resilience of infrastructure *x*?" is unanswerable, because the answer implicitly depends upon the I-I characterizing Complex SoS and knowing whether infrastructure *x* would recover following any attack *y* within an acceptable time and composite costs and risks. Thus, the only way such a question can be answerable is when the threat (or a set of threats) and its timing is specifically identified. Indeed, the system's resilience is not merely an abstract attribute of the system; rather, it is a state of the Complex SoS (composed of a vector of substates) for which any specific substate may respond differently to different inputs (threats). For example, a water distribution Complex SoS may have redundancy in its electric power subsystem, and thus it may be resilient to a major storm that would shut down one of the power lines to the water distribution SoS, leaving the other redundant line intact. On the other hand, suppose the water distribution SoS depends on only one main pipe to supply water to its customers, but is located in a region susceptible to earthquakes. The Complex SoS is resilient only to the extent that the main pipe is functioning and can withstand an earthquake up to level 4 on the Richter scale. However, the Complex SoS would likely fail during an earthquake of level 5 or 6. Here again, measuring the resilience of the water SoS is actually measuring the responses of the system to the specific threat (and time frame), in this case the scale of the earthquake.

Furthermore, one may associate a vector of resilience with each subsystem. Thus, there can be a hierarchy of resilience attributes for any Complex SoS natural or constructed environment. For example, the human body as a Complex SoS is made up of many subsystems (e.g. the digestive, pulmonary, and auditory systems, among others), each with a set of resilient organs and suborgans, where the level of such resilience depends on the input (physical or biological threats), the states of health, and the output (a temporary or long-term loss of functionality of specific organs or suborgans). This example reinforces the thesis that system resilience can be measured in terms of the outputs, which are functions of the states, for given inputs to the system. (Note that the inputs to the system, the states of the system, and the outputs are commonly time variant and probabilistic, as will be discussed subsequently.) To further appreciate the centrality of the system's input–output relationship to its resilience (states of the system), consider the fact that despite the resilience of the human body to various physical and biological attacks on it, its ultimate resilience depends upon the states of the body at the time as well as the type and strength of such attacks.

Complex SoS may also be characterized by the specific redundancy and robustness – both of which lead to a specific vector of resilience. Redundancy refers to the ability of certain subsystems of a system to assume the functions of failed subsystems without adversely affecting the performance of the system itself. Of course, redundancies constitute an integral and essential part of all safety-critical systems. Robustness refers to the degree of insensitivity of a system to perturbations, or to errors in the estimates of those parameters affecting the design choice.

14.I.3.2 On the Relationships among Preparedness, Vulnerability, and Resilience

Both vulnerability and resilience are manifestations of the states of each system and of the Complex SoS as a whole. In principle, they are two sides of the same coin; vulnerability addresses only a system's capacity of protection, whereas resilience focuses also on a system's recovery following an adverse event (Holling, 1973). As states of the same system, both represent the capability of the system to withstand threats. On the one hand, vulnerability represents those states of each system, and of the Complex SoS, that can be adversely affected by specific types and levels of magnitude of threats. On the other hand, resilience *also* represents the ability of the system to recover within an acceptable time and composite costs and risks, having been presented with a threat, that is, the vulnerability of a system does not provide information about the ability of the system to recover from a particular threat. What, then, is the relationship between preparedness and vulnerability, and what is the relationship between preparedness and resilience? If the primary objective of preparedness is reducing the vulnerability of a system to specific threats, it may (although not necessarily) also improve the resilience of the system to the same threats. For example, hardening a system against specific threat scenarios (e.g. adding more security by building fences or formulating policies and procedures that would limit access to infrastructures), but without addressing the recovery needs following a successful attack, would reduce the vulnerability to such threat scenarios; it might not, however, necessarily improve the resilience of the system in terms of its recovery time or composite costs. For example, an electric power generation unit might be hardened against terrorist attacks or major natural hazards, thus reducing its vulnerability to such events, but such hardening would not necessarily improve its resilience to an acceptable level of recovery. By the same token, improving the resilience of the electric power supply system by adding redundant power lines crossing different geographical sites could ensure an acceptable level of resilience to the same threat, but it would not lessen the vulnerability of the electric power generation unit to a physical threat.

Thus, improving a system's resilience offers significant advantages in managing risk. Namely, improving the resilience of a system constitutes an integral part of the risk management process. A fundamental benefit is that an acceptable level of residual risk to the system (i.e. an acceptable level of affordable safety) can be determined for each class of threat scenario. More specifically, because of the probabilistic nature of threats, given the occurrence of a class of threat scenarios, the outputs (consequences) are best represented with probability distribution functions. The resulting risks in terms of recovery time and composite costs can be calculated in a variety of ways, including the expected value of risks or the conditional expected value of risk of the extremes (Haimes, 2016). And ultimately, the tradeoffs among the various levels of risks and costs associated with each investment (e.g. through preparedness) in the system's resilience can be evaluated. Note that the severity levels of the consequences resulting from a threat to a system are used as the metric with which the system's vector of resilience is evaluated. Also, given that the inputs and thus the outputs to the system are probabilistic in nature: the above necessarily led again to the triplet of risk management questions posed in Chapter 9: What can be done, and what options are available? What are the associated tradeoffs in terms of all costs, benefits, and risks? What are the impacts of current management decisions on future options? (Haimes, 1991, 2016) The answers to these questions are specific to each system and to each scenario.

14.I.3.3 Preparedness

Preparedness includes the application of a mixture of preventive actions and measures to assure resilience. This section of the chapter focuses on the resilience resulting from of preparedness. There is a direct relationship between the resilience of infrastructures and the associated preparedness that can ensure effective utilization and performance of such infrastructures prior to, during, and following terrorist attacks or natural disasters. Preparedness, similar to resilience, is aimed at effectively coping with surprises and minimizing recovery time and cost. There is a variety of measures that can be taken to enhance preparedness. These include (i) hardening of critical systems by adding robustness and redundancy; (ii) providing an appropriate balance between logistics plans, including expanding and improving storage capacity and production capabilities, commodities, services, personnel, communications, and regional support; (iii) improving both the emergency response capabilities and the public's readiness to manage recoveries during a disaster situation through education and training; (iv) providing contingency resources (e.g. human resources, funds), technology, and policies for the entire hierarchy of the organizational infrastructure that operates and maintains the physical infrastructure; (v) coordinating an interface arrangement among agencies at all levels, thus strengthening the organizational/institutional resilience of the planning and response system; and (vi) assuring interoperability and information sharing that transcend security (e.g. police, fire), health and safety, transportation, critical utilities and infrastructures, and NGOs, among others.

Despite all of the above steps one might take for preparedness, it is not feasible to be prepared for all possible situations that can occur. As a result, the public and first responders must also react to situations for which they did not plan, that is, they must prepare for improvisation when plans do not work during catastrophic events. In this regard, preparedness should include planning for potential improvisation. This would include, for example, dynamically (i) adding redundancy to the communications system, (ii) providing remote access to up-to-date reliable and credible information, (iii) adjusting management and response responsibilities, (iv) establishing peer groups that communicate with each other, and (v) bringing the private sector into partnership with the overall preparedness efforts; and ideally working with the media.

Consider the following question: What are the overlapping fundamental commonalities and differences characterizing the development of risk-based approaches to preparedness for terrorism versus preparedness for natural hazards? Appreciating the similarity between terrorism and natural hazards can help in preparing for both events. A well-planned and well-executed preparedness plan can make the system resilient against both events. Also, while it is imperative to understand the difference in the public perception and psychological response to economic impacts and other devastations resulting from acts of terrorism vs. natural hazards, it is also constructive to distinguish between the two from the sociopolitical perspective. Indeed, although the consequences from the two events might be similar, the nature of the initiating events in the case of terrorism is critical in terms of public perception and acceptance, economic impacts (e.g. demand reduction), and impact on public policy and overall national security. In other words, it can be viewed as an unacceptable risk to be unprepared for certain types of terrorist attacks as compared with certain types of natural disasters, because the sentiment of public acceptance can be entirely different depending on the actual event.

In a nutshell, preparedness for a terrorist attack, or for a major natural hazard, is an essential requisite for reducing vulnerability and enhancing resilience. To better appreciate the interplay between preparedness and resilience, we define the following three terms: redundancy, robustness, and hardening.

Redundancy refers to the ability of certain components of a system to assume the functions of failed components without appreciably affecting the performance of the system itself. For a physical infrastructure, such as a transportation system, redundancy may manifest itself by adding alternative routings. For an information system, hardware redundancy may take the form of multiple backups of critical components, such as the central processing unit, storage, and power supplies. Similarly, information redundancy is achieved by backing up databases and data exchanges by way of, for example, disk mirroring. Software redundancy can be enhanced through replication, distribution of decision making, and so forth. There is usually a high overhead cost associated with enhancing a physical or an information system's redundancy. As a result, a completely redundant system is often too expensive, or operationally infeasible, to build and maintain within resource and budget limits. Redundancy can be modeled as a constrained optimization problem for which tradeoffs can be identified and Pareto-optimal policies formulated (in the context of a multiobjective tradeoff analysis, the subject of Chapter 5).

Robustness refers to the system's degree of insensitivity to errors in design parameter assumptions and to variations in the operational environment. Design errors caused by imprecise estimation of the model's parameters may arise from miscalculation or improper statistical sampling. *Hardening* a physical infrastructure, or an information system against terrorism, or natural disasters involves modifying or enhancing the system's design or, in effect, choosing a new optimal design. A system is "hardened" if the new or modified design is probabilistically more robust than the original system.

Since we define resilience as the ability of the system, or Complex SoS, to withstand a specific disruption within acceptable degradation parameters and to recover within an acceptable cost and time, resilience also builds on adding redundancy, robustness, and hardening. This is because redundancy, robustness, and hardening are all functions of the states of the system and may be viewed from multiple overlapping perspectives. For example, one may refer to the ability of a system to be operated close enough to its technical design and institutional performance objectives over the short run after an adverse event, such that the resulting economic or operational losses are held within manageable limits. Another perspective advanced in this chapter and book, recognizes that the *resilience* of critical infrastructures is a function of the states of the system and of Complex SoS and many related factors that can be impacted by the same adverse situation as the system itself (e.g. shortages of needed supplies to the systems and of logistics support, planning support, communications, and information assurance). This perspective builds on the premise that there will be a period where unavoidable and undesirable degradation will occur and defines resilience as achieving an *acceptable recovery time* of the systems at an *acceptable cost* following an attack or natural disaster.

For example, an information system is more resilient if it has the ability to restore its integrity after an adverse event. For many infrastructures, resilience is highly dependent on the ability of the workforce that operates them to quickly respond to disruptive events. A study performed for the commission on High-Altitude Electromagnetic Pulse (HEMP) regarding an attack against the United States (Haimes et al., 2005b, 2008) identified that rapidly reestablishing normal workforce operations after a HEMP attack is essential to reducing the most serious impacts on the nation's economy. The HEMP study revealed that significant amounts of economic loss can result from the lack of timely availability of skilled workers.

One approach to measuring the resilience of an infrastructure is to predict the trajectory of recovery time following a catastrophic event. In other words, how long would it take to achieve recovery from 10% to 90% of full capability, and at what level of resources? In some sense, cost and recovery time become synonymous with the resilience of the system and its interdependent systems (infrastructures). Consider, for example, the possibility of developing a nationally shared, very secure information infrastructure dedicated to supporting critical infrastructure Complex SoS. Such measures could add resilience to the country's critical infrastructures, particularly utilities and financial institutions that heavily rely on secure cyberspace to conduct their business. And, it could potentially be a cost-effective vehicle for reducing risks to critical interdependent infrastructures when compared with the alternative of hardening each of the individual infrastructures. Some of the ways that such a system could be used to enhance resilience are to support automation, distributed decision making, information sharing, remote human monitoring and control, automated sensing and control, machine-to-machine communication, and real-time network reconfiguration.

14.I.3.4 Infrastructure Interdependencies and the Tragedy of the Commons

The I-I among infrastructures, such as energy, telecommunications, banking and finance, transportation, and human services, have been widely acknowledged since the US President's Commission on Critical Infrastructure Protection report (PCCIP) (1997). Recognized less explicitly has been the central role that the resilience of one infrastructure plays in determining the resilience of other interdependent infrastructures. Consider, for example, the importance of a resilient water supply system (Haimes, 2016). To a varying degree, the failure of a water supply system (like any other interdependent and interconnected Complex SoS) affects the performance of other infrastructures. In particular, the operation of wastewater facilities may be hampered due to a shortage of finished (fresh) water, the emergency services may be strained, and the generation and distribution of electrical power may be disrupted. Furthermore, this Complex SoS is managed by multiple government agencies encompassing federal, state, regional, and local levels with multiple stakeholders and decision makers, who may have conflicting and often competing objectives. Also, these agencies have different missions, resources, agendas, and timetables. Finally, organizational and human errors and failures are common and may result in dire consequences. Thus, making a water supply infrastructure more resilient would affect the performance of other interdependent systems. Here is where "the tragedy of the commons" must be understood in a broad infrastructure resilience context.

Five decades ago, Garrett Hardin (1968) wrote a seminal paper on the *tragedy of the commons*, in which he described how the equal use of commonly owned property is jeopardized when individuals take unequal actions. In two books, three decades later, Bromley (1992) and Baden and Noonan (1998) provided extensive elaboration on the expansive implications of the tragedy of the commons. Currently, the "commons" connote the security of our infrastructures against risks of terrorism. In the United States today, a significant percentage of the physical and cyber infrastructures are owned and operated by the private sector. However, the uneven tactical and strategic investments the private sector has made in protecting these infrastructures against risks of terrorism are analogous to the tragedy of the commons. While it is recognized that different infrastructures face different risks and have different economic degrees of freedom, currently there is no shared view of appropriate accountability for the whole from each of the parts. Corporate leaders naturally focus on their companies' specific needs and do not view themselves as the guardians of the nation's critical infrastructures.

Over a decade ago, in a pertinent article in the 7 April 2006 issue of *Science*, titled "Cooperation, Punishment, and the Evolution of Human Institutions" Henrich (2006), an anthropologist, addressing the "commons" syndrome from a fresh perspective, wrote: "Explaining the scale, diversity, and historical dynamics of human cooperation is increasingly bringing together diverse empirical and theoretical approaches. For decades, this challenge has energized evolutionary and economic researchers to ask: Under what conditions will decisionmakers sacrifice their own narrow self-interest to help others?" (p. 60). About the same time, Gürerk et al. (2006) reinforced the above thesis; they argue that altruistic punishment can indeed induce corporations to cooperate. For example, they maintain that "Additionally, because the absolute number of punishers increases, the individual burden from effectively punishing free-riders becomes smaller over time." (p. 110). While the perspective of the tragedy of the commons is compelling, it must also be recognized that the contribution to spreading risks and to reducing risks is different for each company; thus, developing a "fair" rule for community action is no simple matter. Nonetheless, it is important to understand the human and corporate dynamics from the perspective of the tragedy of the commons and to do the work necessary to develop concepts that are understood by the various stakeholders as being fair and practical. In addition, the role of regulation is likely to be a critical part of the solution. Efforts to develop concepts for regulation that are compatible with the concepts for community action are a necessary part of treating resilience in a coherent manner.

An effective way to circumvent the tragedy of the commons is to highlight the socioeconomic and other dimensions of interdependencies and to demonstrate the efficacy of risk management, where the rate of return on investing funds for preparedness and added resilience to assets and infrastructures can be demonstrated.

14.I.3.5 Emergent Properties of Complex Systems of Systems

In this section we discuss the resilience of critical infrastructures as an emergent property of a nation's critical infrastructure Complex SoS. We define emergent properties of a Complex SoS as those system features that are not designed in advance, but rapidly evolve based on sequences of events that create the motivation and responses that ultimately develop into features characterizing the Complex SoS. Complex SoS that are more likely to have emergent properties are those that (i) have broad missions to fulfill, (ii) are created through the cooperation of many stakeholders that have overlapping, but not identical, objectives, (iii) have low capital–cost structures that reduce the financial obstacles related to emerging properties, and (iv) are subject to significant events that, should they occur, can stimulate the emergence of properties that otherwise might not be anticipated.

Systems less likely to have emergent properties are those that (i) are managed and controlled by a single organization, (ii) have relatively narrow missions to carry out. (iii) have high infrastructure capital costs that impede change due to excessive cost, and (iv) are less subject to single significant events, stimulating major changes to features of the system.

The Internet is recognized as emergent in nature. It is a system with many properties that have emerged due to the (i) low cost of entry for users, (ii) availability of technology from a multitude of companies that serve those users, (iii) broad mission of providing information to users, and (iv) initial driving forces of early information sources and corresponding demand for those sources, ranging from company websites to pornography. Nonetheless, parts of the Internet, such as the routing technology and corresponding protocols, are far less emergent since they require significant investment and support from technology companies. In this case, standard groups and sponsored research efforts must create new solutions and technologies in anticipation of stakeholder demands. This part of the Internet, as witnessed by the long lead time for the introduction and full-scale use of advanced routing protocols, is not nearly as emergent as new applications using existing technologies.

Consider the air traffic control system of the United States, which is far less emergent than the Internet. A single organization, such as the Federal Aviation Administration (FAA), (i) has a specifically defined mission and is principally responsible for the Complex SoS, (ii) is capital intensive, (iii) has important reliability and safety assurance features that require significant test and evaluation before replacing, and (iv) does not self-change at anything similar to the pace that the Internet changes, while single events such as midair collisions can cause large public responses. If one assesses resilience as an emergent system property in the context of terrorism and natural disasters, the question arises as to how can one prevent or influence negative emergence so as to avoid the tragedy of the commons. Some of the policies and actions for influencing emergence would be (i) punitive regulation or the threat of regulation, (ii) incentive-based regulation (e.g. tax cut incentives), (iii) technology that reduces the cost of particular aspects of resilience, (iv) analyses that influence value systems of stakeholders, and (v) using results of actual events, such as 9/11 or hurricane Katrina, as analogies that can influence behavior related to other possible scenarios.

Recognizing that the overall risk management for terrorism and natural disasters includes preventive measures and resilience measures, a framework for integrating these measures is provided in Figures 14.I.5 and 14.I.6, where Figure 14.I.5 presents the interdependencies between preparedness and resilience measures and Figure 14.I.6 presents the integrated risk management process.

Figures 14.I.5 and 14.I.6 illustrate the premise that preventive measures need to be adopted to account for the emergent risk management steps stakeholders are take to improve resilience and that the promotion of resilience measures through policy formulation and the creation of new solutions/technologies need to reflect the specific preventive measures that are going

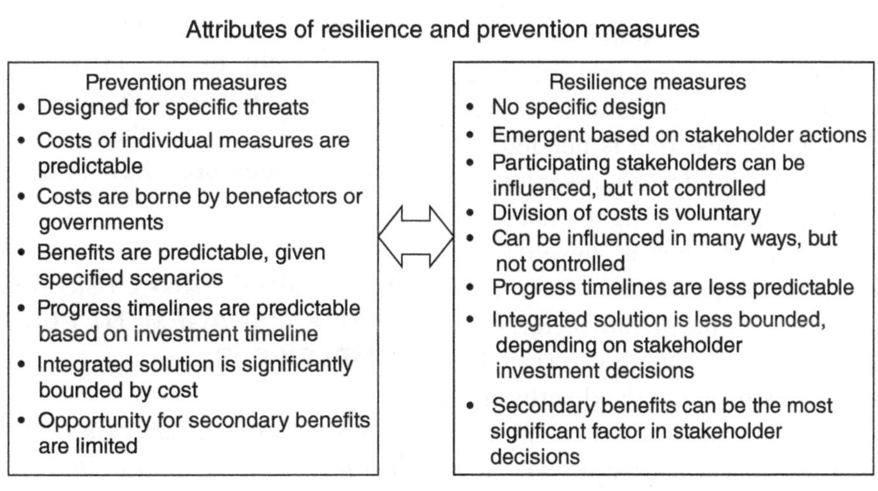

Figure 14.I.5 *Risk management of interdependencies: considering resilience measures and preventive measures.*

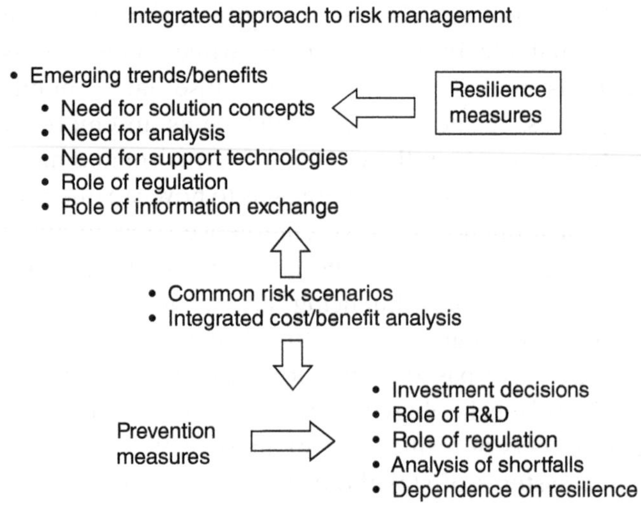

Figure 14.I.6 *An integrated approach to risk management.*

forward. The following discussion elaborates on these ideas by identifying approaches to stimulating resilience that have the potential to grow into important parts of the overall preparedness plans being undertaken.

One important factor in understanding the behavior of the private and public sectors relative to preparedness may be explained by the nonproductive and shortsighted nature of protection, which is a reactive expenditure to unpredictable threats with no associated product value. On the other hand, when balanced with protection, proactive investments add resilience to an infrastructure through better maintenance coupled with added robustness and appropriate redundancy. This may help reduce the effects of the tragedy of the commons syndrome, and bring corporate leaders to a more effective partnership with government in the quest for creating more secure critical infrastructures. This raises two interesting questions. What is an appropriate and acceptable balance between resilience and preparedness, and how can it be achieved? This is where research is needed to develop basic principles that guide risk assessment and management to help shed some light on these and other related questions. A start at developing needed principles is presented below.

In terms of risk assessment for the private sector, it is most likely that the executives in charge of privately owned critical infrastructures have obtained sufficient answers to the risk assessment triplet questions posed by Kaplan and Garrick (1981): What can go wrong? What is the likelihood? What are the consequences? The tragedy of the commons syndrome arises in the risk management process, namely, when answering the following triplet of questions (Haimes, 1991, 2016): What can be done and what options are available; what are the tradeoffs in terms of all relevant costs, benefits, and risks; and what is the impact of current decisions on future options? Of course, for Complex SoS we ought to appropriately modify the above set of questions.

Combining and paraphrasing the last two questions in terms of risk management might provide answers to the appropriate and acceptable balance between resilience and protection: *What are the tactical and strategic, short- and long-term tradeoffs associated with balancing protection with resilience, and what is the associated future impact on the enterprise?* This change in our perspective about the tradeoffs between protection and resilience invites the search for solutions that provide values in normal, everyday business situations and also provide added resilience in disaster situations.

14.I.3.6 The Efficacy of Preparedness and Resilience within Holistic Risk Management

The cost associated with risk management of Complex SoS can be reasonably assessed. However, the efficacy of their preparedness and resilience within holistic risk management remains elusive to homeland security analysts and policymakers. The inability to determine the exact level of future risk reduction associated with each risk management policy option, whether for preparedness or for added resilience, is probably a major reason that the private sector remains relatively aloof in this regard. Following are a sample of approaches that can potentially provide important contributions to influence emergent system properties related to resilience. Results would serve to identify which stakeholders ought to be persuaded to take their own actions, including consideration of what solutions are the most practical for them, as well as what definitive preventive measures they should pursue.

14.I.4 IMPACT ANALYSIS AND THE EFFICACY OF RISK MANAGEMENT PLANS FOR COMPLEX SoS

Data for Characterizing Risk Assessment and Risk Management Strategies

As risk management strategies are developed, a prioritized list of critical assets must be generated according to their effective impact on missions and mission-essential tasks of the enterprise (e.g. its CP infrastructure or other assets), and whether publicly or privately owned. To develop an assessment of the sources of risk to these assets, appropriate data must be collected, whether in a preexisting database or gathered with a support system. This is a costly undertaking whose efficacy cannot be accurately assessed. Similarly, appropriate data must be collected for the comparison of risk management strategies by their costs, risks, and benefits.

Metrics of Efficacy of Risk Management Strategies

Development of candidate risk management strategies for Complex SoS is needed to ensure the continuity of enterprise operations in the event of a disruptive risk scenario (e.g. a hurricane or an adversarial attack). A holistic analytical multiobjective cost–benefit–risk framework is needed for evaluating the efficacy of candidate strategies in terms of their risk reduction potentials, associated costs, and implementation challenges. It is essential that metrics can effectively characterize the costs, risks, and benefits of the strategies, including physical security, cyber security, integral hardening, and emergency protocols. Comparing the response and recovery times of several risk management strategies relative to the *status quo* is a challenging undertaking; however, it can provide a process for evaluating the net benefit or efficacy of implementing those strategies.

New Analysis Methods for Risk Management Strategies

There are three intertwined broad categories of analysis methods that can facilitate the evaluation of candidate risk management strategies in terms of their efficacy: (i) *multiobjective tradeoff analysis*, (ii) *hierarchical analysis*, and (iii) *dynamic analysis*. *Multiobjective tradeoff analysis* (the subject of Chapter 5) is a methodology that recognizes the existence of multiple, noncommensurate, and conflicting objectives, among others. *Hierarchical analysis* (the subject of Chapters 2 and 4 and most other chapters) enables the analysis of candidate risk management strategies as viewed from different levels of an organization: that is, risk scenarios are assessed at lower organizational levels, and hierarchical analysis is employed to generate an overall criticality rating that represents the multiple risk assessments, with the effectiveness of risk management strategies being evaluated based on their cost and perceived effectiveness. The challenge in implementing *dynamic analysis* is to model the recovery processes for multiple systems after a disruption. This is a challenge because an effective dynamic analysis continuously evaluates risk management strategies and their efficacy with respect to time.

14.I.4.1 Protection Strategies Complemented by Resilience and Recovery Strategies

Infrastructure protection strategies must address the interdependencies of preparedness plans; the dimensions of robustness, redundancy, and security; and the schedule and cost of recovery, as described below (Note that for completeness, minor repetition in the following one to three paragraphs is allowed for pedagogical purpose.):

1. *Interdependencies of preparedness plans*
   Preparedness is aimed at effectively coping with surprises and minimizing recovery time and cost. Hardening of critical systems by adding robustness and redundancy also improves preparedness. Preparedness planning addresses resources (human resources, funding, and others), technology, and policies affecting the entire hierarchy of the organizational infrastructure that operates and maintains the physical infrastructure and the interface arrangement among agencies at all levels, thus strengthening the organizational resilience of the system. Highlighting shortfalls can create an understandable demand for new solutions (technology, policy) and sensitize stakeholders to their role in providing resilience.
2. *Robustness, redundancy, and security*
   In risk management, it is imperative to evaluate critical assets in terms of their defensive properties for withstanding a threat scenario. The cost associated with defensive properties can be reasonably assessed but their efficacy cannot. Consider, for example, the challenge (and the need for more research and study) associated with assessing the return on investments in redundancy, robustness, and resilience. *Redundancy* refers to a strategy wherein extra components or backup systems are added to

substitute for failed components or a failed system. A strategy that increases *robustness* reduces the sensitivity of an infrastructure performance to internal or external stresses (an attack or added workload). Finally, a *resilient* risk management strategy is one that addresses the ability of an infrastructure to recover within a reasonable time at reasonable cost and with reasonable resources and/or to continue with an acceptable degraded state of operation during the recovery period.

3. *Schedule and cost of recovery*
   Realizing that a considerable investment in protection may still result in a degradation of infrastructure as the result of a threat scenario, there may be a period where unavoidable and undesirable degradation of infrastructure performance will occur. Therefore, decision makers are often challenged to determine acceptable recovery times and costs for restoring assets and operations to sufficient working order and to develop risk management strategies that achieve those acceptable recovery times and costs.

14.I.4.2 Impacts of Current Risk Mitigation Strategies on Future Options

To preserve and maximize future options, risk management of critical assets and infrastructures should incorporate adaptive analytical frameworks and quantitative impact analysis of risks and options in the near term relative to risks and options in the long term. This is done through:

1. *Adaptive frameworks for robust future actions, given the ever-changing threats, objectives, and stakeholders*
   The challenge to developing risk management strategies for publicly and privately owned Complex SoS with attributes and characteristics including agility, modularity, adaptability, robustness, and resiliency is due to the fact that changes in the objectives, functionality, and stakeholders of these systems are inevitable.
2. *Impact analysis of current risk management strategies on future options*
   Public and private organizations and their operating environments and risk concerns are ever changing. Thus, an essential role of risk management is to address the impact of current decisions on future options. Risk management analysts and decision makers associated with Complex SoS must assess and evaluate plausible future threat scenarios that may require changes and adaptation to respond to such risks.

14.I.4.3 National Resource Allocation for Catastrophic Scenarios

The allocation of resources to risk management of critical publicly and privately owned assets and infrastructures must consider the hierarchical and holographic (multiple stakeholder, multiple perspective) features of the problem. The approach to resource allocation needs to be repeatable and uniform but nevertheless able to be particularized to local needs in terms of time, geography, and organizational attributes. The technical analyses of risk management strategies must be cognizant of, and able to support, the broader organizational and political decision-making considerations. (Recall that Chapter 3 is devoted to Hierarchical Holographic Modeling.)

14.I.4.4 Classes of Risks Requiring Management

The categories and classes of risks that threaten public and private assets and infrastructures, and the likelihood that those risks will occur, must be considered in developing risk management strategies. The following three pairs of consequence and likelihood are discussed in further detail:

1. *Rare catastrophic*
   Threats that result in catastrophic consequences with rare occurrence must be addressed in the context of the broader enterprise security agenda; (regardless of the probability, in terms of resource allocation and the attendant priorities)
2. *Noncatastrophic, more likely*
   This class of threats is more likely to occur, but would have no catastrophic consequences, where the severity or combination of

consequence and likelihood is less than that of catastrophic events. However, risk management should consider threats whose consequence may not be as dire but may occur with a higher likelihood.

3. *Minor consequence, occurring with regularity*
   A final class of threats encompasses those whose consequences are minor, but occur more frequently. Such threats, whose severity is the lowest of the three classes, represent perhaps the "expected threat." However, developing strategies to mitigate only the most likely threat is an unsound risk management policy.

There is a need to quantify the value of investment in preparedness and added resilience to enterprises, whether assets, infrastructures, or sectors of the economy. The inoperability input–output model (IIM) and its extensions (see Chapter 8) can be used to quantify the economic losses for specific scenarios with and without risk management, that is, with and without preparedness and added resilience.

14.I.5 EPILOGUE

In Part I of this chapter, we formally introduced the concept for system planning efforts related to Complex SoS that consist of a rich mixture of designed subsystems and emergent subsystems. Planning methods and solution possibilities are presented for such systems so that emergence is, at a minimum, not prevented, and where possible, stimulated and supported by the formally designed portion of the overall system.

To make the proposed system planning approach tangible and to guide conceptual thinking, Part II is devoted to planning for a critical national problem and corresponding system, focusing on, *establishing a national preparedness system for terrorism and natural disasters.* The cost for creating a government-designed and government-implemented national preparedness system for significantly reducing risks resulting from the wide range of possible natural and terrorist threats is too high for most economies to absorb – a fact that has led to a period of relative stagnation in terms of selecting solutions that need to be implemented.

PART II: BALANCING HURRICANE PROTECTION AND RESILIENCE TO COMPLEX SYSTEMS OF SYSTEMS

OVERVIEW

Consider the following example of hurricane risk mitigation efforts associated with unavailable potable water supply following a hurricane. Risks associated with unavailable potable water are mitigated by contingent supply chains that are dynamically created from existing component systems in the region. This could include posthurricane transportation mobility, regional operating points of distribution, availability of contracts, and availability of private resources of corporations (e.g. supermarket chains) and of many region-specific system resources. *Protective options* to mitigate risks of water shortage include storage and maintenance of potable water inventories, hardening of transportation assets, and hardening of the potable water distribution infrastructure. *Resilience options* include integrating public and private resources to prestage emergency potable water in response to a forecast, and distributing the water through a coordinated strategy, educational strategies that would result in a personally owned and maintained inventory of potable water, and improved forecasting and warning methods, among other possibilities. We illustrate how an integrative modeling approach can provide a means of quantifying some of the tradeoffs between protective options and resilience options in an emergent region. Consider the following three metrics as measures of regional resilience: (i) cost of posthurricane emergency potable water distribution (in US dollars), (ii) quantity of potable water demand shortfall, 8 h after a hurricane strike (in US gallons), and (iii) time required after a hurricane strike to reduce potable water demand shortfall to 10% (in hours).

Resilience measures the capability of Complex SoS to recover at an acceptable time and degradation and reflects the capability of a region both to absorb the strike through hardened infrastructure and to recover from it through emergency potable water distribution strategies (such as the distribution of bottled water). The capability to perform potable water distribution is one aspect of the region's resilience with respect to potable water availability. Each of the resilience measures (i) through (iii) is a reflection of the way the emergent regional system commonly behaves following disruptions. Resilience measure (i) is in competition with the other two. The fact that (iii) can be derived from (ii) with some additional effort, leads us to focus for simplicity on (i) and (ii) only. In this example, the resilience of a region is the set of nondominated (Pareto-optimal) resilience measures (i, ii) that result from various available decision strategies. This set of nondominated measures reflects the system-level capability of a region, as explained below. (Note that Chapter 5 is devoted in its entirety to multiple objectives and tradeoff analysis.)

Evaluating resilience measures (i) through (iii) requires an understanding of many components in the regional system, including the process by which hurricanes exploit infrastructure vulnerabilities to result in the adverse loss of potable water supply. Furthermore, such an analysis requires the capability to predict potable water demand based on population, tourism, and population behaviors (such as voluntary evacuation). Potable water supply shortfall is the difference between the demand for and available supply of potable water. Once the a priori level of resilience has been established for a region, protective options (that change the vulnerabilities of assets) can be modeled by reevaluating the prior model with posterior parameters of an asset's vulnerability to hurricane wind, rain, and surge. However, evaluating resilience options requires integrating information, decision criteria, system understanding, and associated uncertainties. This section illustrates how this evaluation might be accomplished to compare the protective option of facility hardening against the resilience option of prestaging emergency water supplies. This summary is only for illustration; therefore actual facility and regional data have been removed.

14.II.1 CALCULATING FORECAST TRANSITION PROBABILITIES

The capability of a region to forecast storms is a component of the regional preparedness system. For example, the US National Hurricane Center (NHC) uses analytical tools to forecast the track and intensity of these storms and thus to warn local authorities of approaching threats. Prestaging decisions are made in response to these forecasts. Thus, the certainty in the hurricane forecasts contributes to the resilience of the region. To simplify the model for this illustration, consider all hurricanes having one of three strengths: stronger (more than 200-year frequency), medium (about 100-year frequency), and weak (<50-year frequency). To estimate the uncertainties from the forecasts, we analyzed data from all forecasts of all Atlantic storms hitting the East Coast for the past fifteen years. This amounted to more than six thousand forecasts.

Assuming that a storm forecast predicts a direct hit to the region of interest, we can use the data to predict approximately the expected probability with which the storm forecast will differ from later forecasts and actual storms. (In reality, the agency responsible for forecasting has knowledge of whether a forecast is either more or less certain than the expectation.) Based on the characteristics of hurricanes, it is important to analyze the error both in wind speed and in the position of the hurricane. As one moves away from the center of the hurricane, the wind dissipates. Using data from prior regional hurricanes, the wind speed decreases at a mean rate of 0.144 knots/nautical mile away from the center. (This number can be replaced if a particular storm is considered rather than the average of all regional storms.) Therefore, if a storm shifts 100 nautical miles from the region, the wind speed of the region can be changed by approximately 15 knots (0.144×100), thus changing the infrastructure impact to the region.

A probability distribution describing the joint probability of a forecast error resulting in either a wind speed or track error was partitioned according to the specific geography of the region and the average characteristics of hurricanes. The results produced probability estimates of a specific impact, given the 72-h forecast of the 200-year probabilistic storm. These data, coupled with the rate of decay of wind speed away from the center of the hurricane, indicated that the wind will cross a threshold of frequency classification at approximately 90 nautical miles. Figure 14.II.1 shows the uncertainty of 24-h forecasts for a 200-year storm and the partitions for probability counts based on hurricane and region characteristics.

The data indicate that over the past two decades, a 24-h forecast of a 200-year storm results in approximately a 40% chance that the storm will be weaker

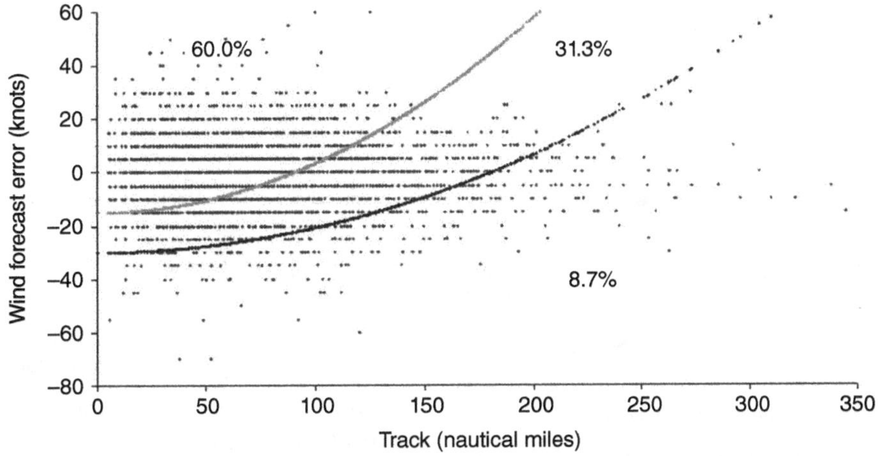

Figure 14.II.1 *Transition probabilities for 24-h forecast of a 200-year storm.*

TABLE 14.II.1 Transition Probabilities for 72-h Forecast of a 200-Year Hurricane

		24-h Forecast		
		50-year	100-year	200-year
72-h Forecast	200-year	0.054	0.645	0.301

TABLE 14.II.2 Transition Probabilities for Various 24-h Forecasts

		Actual storm		
		50-year	100-year	200-year
24-h Forecast	50-year	0.933	0.067	0.000
	100-year	0.288	0.672	0.040
	200-year	0.087	0.314	0.598

when it reaches the region of interest. Tables 14.II.1 and 14.II.2 show the estimates of transition probabilities, given the NHC data and the methodology described above.

Representing forecast uncertainty, as forecast transition probabilities, enables us to integrate information models with causal loss models to understand the efficacy of decisions. Decision efficacy will be represented as a frontier of available tradeoffs. Analysis-driven preparedness decisions that are implemented well in advance of hurricane forecasts can change the shape or position of the Pareto-optimal tradeoff frontier, as shown in the following section.

14.II.2 MODEL INTEGRATION FOR CALCULATING RESILIENCE MEASURES

To integrate the available data and decision options for a region, we build a multiobjective decision tree (MODT) (Haimes, 2016). (Recall that Chapter 5 is devoted in its entirety to multiple objectives.) In the following pedagogical application of MODT, forecasts are given at 72 h and 24 h prior to a storm. At each time period, the emergency planner decides how much water to order from outside corporations ("order" or "no action" in this illustrative model).

There exists a strong conditional relationship between the 72-h and the 24-h forecasts, and as the trend for better forecasts continues, the correlation will strengthen. At each node, the decision maker must weigh his/her decision based on two objectives: to minimize storage costs and to minimize the expected posthurricane shortage of water.

Illustrative MODT Results and Discussion

The results of this illustrative MODT analysis are shown in Figure 14.II.2. Sixteen courses of action (COA) are possible across the two time frames, approximately 50% of which were shown to be inferior because of information uncertainty or unacceptable costs. Each COA is represented by one point on the graph. The set of noninferior solutions, or the Pareto-optimal frontier, is highlighted by the line and represents the tradeoffs that exist among the noninferior forecast-responsive preparedness strategies and provides a measure of the region's resilience

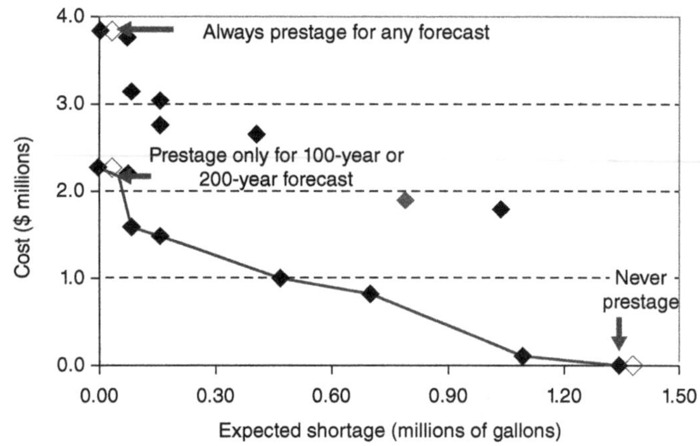

Figure 14.II.2 Result of MODT analysis with Pareto-optimal frontier.

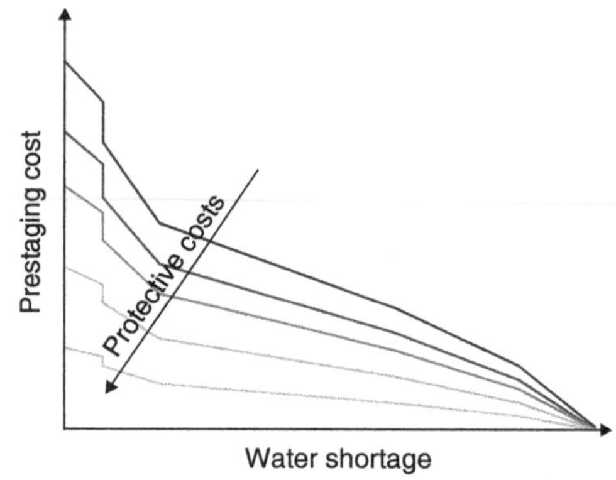

Figure 14.II.3 Pareto-optimal frontier shifts as a result of analysis-responsive preparedness activities, such as protective infrastructure hardening.

capability. Three decision strategies are labeled in Figure 14.II.2 for illustration. The strategy to always order prestage materials independent of hurricane forecast (labeled at the top left of the figure) is dominated by strategies to order only in the event of a 100-year storm (labeled at the middle left of the figure) and to never prestage emergency water supplies (labeled at the bottom right of the figure).

The results from the MODT highlight the noninferior COAs and provide the decision maker with quantifiable tradeoffs between two different resilience objectives. To formulate an operational strategy, a decision maker must decide what costs, or levels of potable water shortage, are acceptable. If no shortage in potable water can be tolerated, then the best strategy (according to the illustrative results in Figure 14. II.2) is to prestage emergency water at every forecast of a 100-year storm, or worse. The cost of this strategy can be compared against protective measures that would decrease the likelihood of potable water outage (e.g. strategies that would prevent flooding into potable water distribution facilities), decrease the costs of contingent distribution of potable water (e.g. better contingency contracts or available local inventories), or increase effectiveness of contingent delivery (e.g. hardened transportation roadways). *It is critical at this stage to underscore the fact that this tradeoff curve depicted in Figure 14.II.2 is a measure of the resilience of the region with respect to potable water distribution.* Figure 14.II.3 illustrates how the Pareto-optimal frontier of noninferior solutions shifts downward with increased investments in protective preparedness measures. One such measure is infrastructure hardening, where activities may include storing and maintaining potable water inventories, hardening transportation assets, or hardening the potable water distribution infrastructure. Recall that Chapter 5 is devoted in its entirety to multiobjective tradeoff analysis.

Alternative views of the tradeoffs in Figure 14.II.3 provide more insight into the ability of a region to make tradeoffs among forecast-responsive preparedness measures and analysis-responsive measures. Other results can be evaluated similar to those shown in Figure 14.II.3, wherein the noninferior decision strategies are graphed in multiple objectives.

Figure 14.II.4 is similar to Figure 14.II.3, but with more generic axis labels to emphasize its general applicability to understanding resilience. These graphs indicate the sets of Pareto-optimal (or noninferior) decision strategies that constrain response behavior, given the system characteristics, component system functionality, and loss mechanisms built into the model schema. However, the data in the graph may be "folded" or "flipped" to redraw the axis with a different focus. Similarly, the tradeoffs between responsive action costs and protective action costs can be made explicit and graphed with a constant level of lines of a particular resilience measure, as shown in Figure 14.II.5. Such analyses enable us to understand the constrained resource allocation decisions a region must make to achieve a level of acceptable resilience.

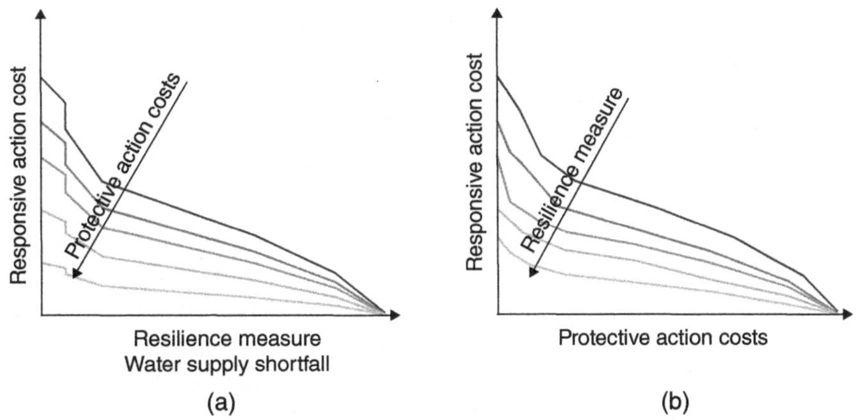

Figure 14.II.4 *Example charts representing model results with resilience measures graphed against responsive and protective action costs.*

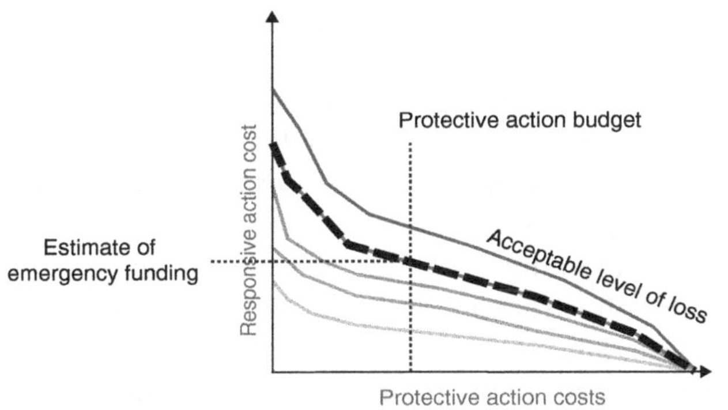

Figure 14.II.5 *Decision strategy given tradeoffs between protection and responsive costs.*

Figure 14.II.5 illustrates a potential method for using the calculated tradeoffs between protective and responsive action costs.

In Figure 14.II.5, a decision maker can choose an *acceptable level of loss*. In the case of emergency water, for example, a regional authority could decide to accept up to a 10% potable water shortfall for 72 h. This choice results from a set of tradeoffs that demonstrate the estimated effectiveness of various strategies combining investments in protective and responsive actions. Or, more generally, this decision could reflect the tradeoffs between investments in resilience and the protection of a region. Because protective actions are set early, they are constrained by a particular budget. The maximum expenditure of this budget then yields an estimate of emergency funding required for responsive actions when the hazardous event occurs. Understanding the tradeoffs will enable better-directed efforts at protective action costs that maximally reduce the responsive costs within the acceptable level of loss tolerated by the region.

This simple example has illustrated the complexity associated with evaluating system resilience. Such an analysis requires integrating characteristics including component system actions, decisions, constraints, information forces, and uncertainty. However, this complexity cannot be ignored if we were to achieve effective and efficient management of homeland security systems. Moreover, the results illustrated in this example are reflective of only a small set of decision strategies, over a limited time horizon, and for a limited set of hazards. As the number of factors increases, the complexity of the computations expands. Currently this specific methodology yields insights for only limited combinations of hazards, decision strategies, and time horizons. The principles illustrated here nonetheless provide insight into the effectiveness of considering broad tradeoffs between resilience and protection in a regional preparedness framework for homeland security.

This simple example is intended to be pedagogical and exploratory, not policy prescriptive. However, similar analyses might have identified the inferiority of certain decisions that led to the New Orleans disaster. They could have clearly illustrated a set of Pareto-optimal decisions and provided a quantitative method to evaluate levee maintenance and hardening compared with other regional investment choices. The greatest benefit to planners will be to develop quantitative and justifiable methods, as described in the previous section. This would enable establishing a clear, specific, and regionally customized concept of operation for public servants, elected officials, and component system owners and operators. Such methods would justify investments in protective actions and system-level actions for the efficient improvement of strategic preparedness.

14.II.3 EPILOGUE

This case study introduces a concept for the system planning efforts related to regional Complex SoS that consist of a rich mixture of formally designed subsystems and emergent systems and subsystems.

The cost is extremely high to create a government-designed and government-implemented national preparedness system that focuses on protection methods that would significantly reduce risks resulting from the wide range of possible natural and terrorist threats. We posit that this fact has led to a period of relative stagnation in terms of selecting solutions to be implemented. The focus on resilience has already increased in the last two decades, and an appropriate balance between protective actions and those that build system resilience will provide an efficient preparedness solution that our regional economies will capably absorb.

PART III: INSIGHTS INTO DECENTRALIZING RISK MANAGEMENT FOR STRATEGIC PREPAREDNESS THROUGH DECOMPOSITION AND THE INOPERABILITY INPUT–OUTPUT MODEL

14.III.1 INTRODUCTION

People implicitly sacrifice wealth and certain freedoms for a stronger guarantee of a desired society and culture (Crowther and Haimes, 2005). A social contract between citizens and their government evolves over the years as new officials are elected and as legislative and planning authorities debate and make decisions that impact society (Rousseau, 1762, 1997 translation). In particular, the historical social contract of the United States has held that the government guarantees an appropriate level of safety of critical infrastructures in return for taxes and political support. Also, in the United States, recent catastrophic hazards and protection against severe weather and terrorist events have added burdens to the government's traditional responsibilities. The additional burden of managing terrorism and natural hazard risks, combined with the fact that most of the US critical infrastructure is controlled by the private sector and States, has led to a change in the social contract, which is manifested as a more decentralized approach to critical infrastructure protection and risk management.

This decentralized paradigm has resulted in many changes. For example, the 20-year old Critical Infrastructure Information Act (2002) provides a means by which private organizations that control various aspects of critical infrastructures can share sensitive, operational information that will be protected by the government and would only be used to disseminate relevant intelligence and strategically allocate preparedness resources. The US government identified several critical infrastructures and key resources and assigned decentralized agencies to be responsible for collaborating with the owners and operators of these infrastructure systems. These sector-specific agencies have reorganized themselves in order to support collaboration among themselves and private owners of infrastructure systems.

This section formulates a strategic preparedness problem that is an abstraction of the overall risk problem of the United States. The formulation uses the inoperability IIM introduced in Chapter 8 to constrain the risk problem by economic sector interdependencies based on the commodity transactions that are recorded for each sector of the national economy (Haimes and Jiang, 2001; Santos and Haimes, 2004; Haimes et al., 2005a; Santos, 2006). We then apply various decomposition techniques to decompose the complexity of preparedness, thereby illustrating some of the challenges in developing a decentralized paradigm of strategic preparedness (Lasdon and Schoeffler, 1966).

Derived from a Nobel Prize-winning economic model by Leontief (1966) and built on data from the US Department of Commerce Bureau of Economic Analysis (BEA) (1998), the IIM is a quick, inexpensive, holistic method for estimating certain economic impacts and sector interdependencies (Miller and Blair, 1985). Note that the IIM was also introduced in Chapter 8.

$$\mathbf{q} = \mathbf{A}^*\mathbf{q} + \mathbf{f}^* \qquad (14.1)$$

In Eq. (14.1), the vector $\mathbf{q}$ represents the inoperability or system dysfunction that is measured by normalized production loss, $q_i = \frac{x_i - \widetilde{x}}{x_i}$, where x_i is the nominal predisaster production output of sector i and $\widetilde{x}_i$ is the postdisaster level of output. The perturbation vector $\mathbf{f}^*$ is a measure of the lost requirement or demand of sector output with respect to nominal output, $f_i^* = \frac{f_i - \widetilde{f}_i}{x_i}$, where f_i is the predisaster level of final (nonintermediate) consumption of sector i and $\widetilde{f}_i$ is the postdisaster level of final consumption. The industry interdependency matrix $\mathbf{A}^*$ is a linear transformation of the Leontief technical coefficient matrix $\mathbf{A}$, and contains the linkage information indicating the relative strength of propagation between sectors. The purpose of the transformation is to emphasize the metric of inoperability or percent of sector dysfunction.

Figure 14.III.1 depicts the underlying framework of the IIM for a simplified two-sector economy.

Intuitively, the operability of *sector 1* is determined by three inputs, namely, consumption of *sector 1* goods, availability of necessary products and operations from *sector 2*, and availability of necessary internal products and operations. Figure 14.III.1 (representing Eq. (14.1) for a two-sector system) focuses on inoperability or dysfunctionality, which is the result of a demand perturbation (e.g. downed power lines that impede demand) or a breakdown in the sector's own or another sector's production.

To introduce an example of decomposition, we define *pseudo-variables* as inputs to interconnected sectors. (Consult Chapters 2 and 4 for more information on *pseudo-variables.*) These pseudo-variables enable the decomposition at the interdependency points shown in Figure 14.III.2. Also, pseudo-variables are represented in Figures 14.II.1 and 14.II.2 as $q_{ij} = a_{ij}^* q_j$ for notational simplicity.

By decomposing the model at the coupling points (illustrated as decomposition points in Figure 14.II.2), we obtain separate and distinct decentralized models of sectors that can be coordinated in various ways. An example of decomposition is shown in Figure 14.III.3, where a higher-level "composing manager" coordinates the sectors using shadow prices (Lagrangian multipliers) of the overall dual problem. In other words, the composing manager monitors only the integration of sectors and establishes a pricing scheme that will result in coordination among sectors. The sectors then operate independently, yet the global solution remains optimal because a composing manager is setting coordination guidelines. In strategic preparedness these prices are manifested as laws, regulations, policies, incentives, etc.

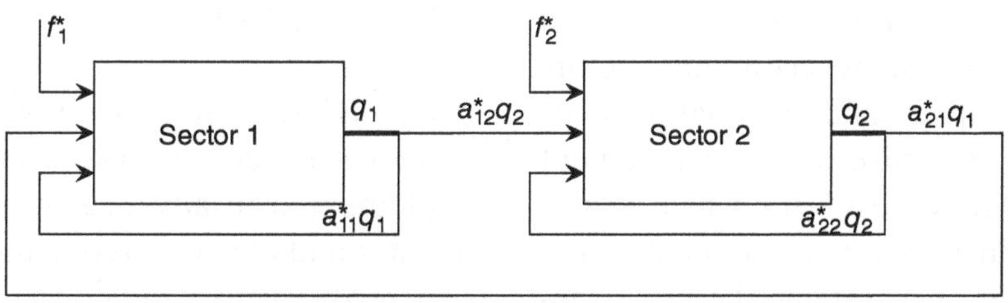

Figure 14.III.1 *Diagram of product flows in the two-sector IIM.*

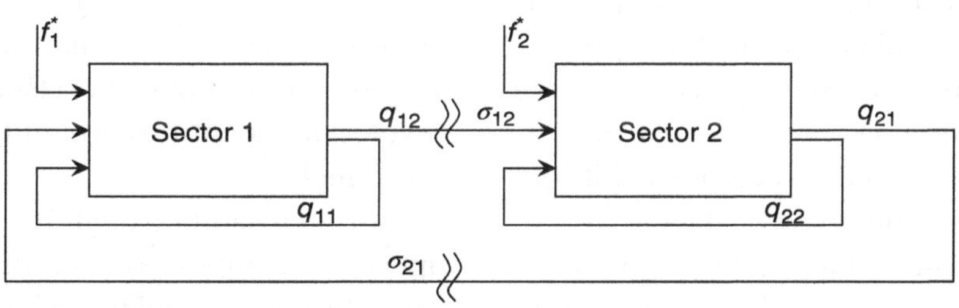

Figure 14.III.2 *Diagram of decomposition points in the two-sector IIM.*

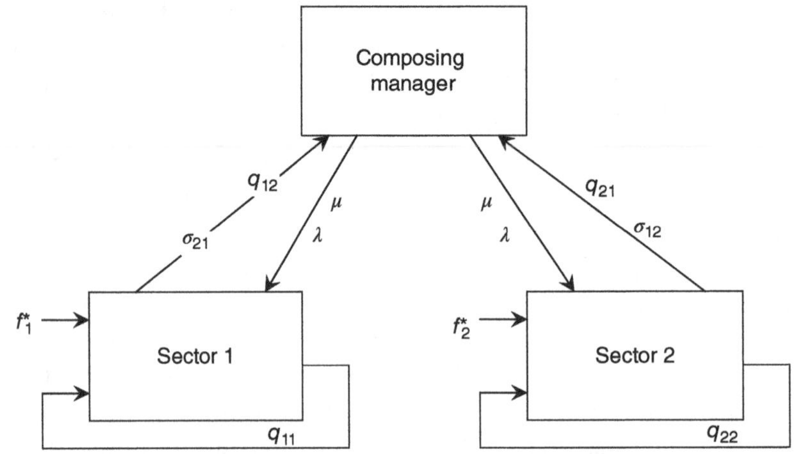

Figure 14.III.3 *Decomposed two-sector IIM coordinated by shadow prices (λ and μ).*

The decomposition shown in Figure 14.III.3 is often called price-directed or nonfeasible decomposition, because while partially iterated solutions are often globally optimal, they are infeasible. Another decomposition strategy, discussed subsequently, is a resource-directed or feasible decomposition, wherein the composing manager is an information sharing entity that passes information among sectors for coordination. This technique results in partially iterated solutions that are feasible but suboptimal.

The strategic preparedness problem has the requisite structure to make decomposition effective. Regional preparedness authorities (e.g. national or state government entities) desire to optimize preparedness investments across many economic sectors through legislation, incentives, and information sharing, but they desire to minimize direct involvement due to the greater effectiveness with which sector organizations and business leaders understand and are able to manage their own systems. The IIM captures the degree to which sectors are interconnected through economic transactions of goods and services. Decomposition theoretically enables regional preparedness authorities to coordinate lower-level preparedness across sectors by setting coordinating parameters and distributing preparedness investments to those who make sector decisions, without loss of overall system optimality. This part of Chapter 14 provides insight into strategic preparedness through the application of decomposition to a preparedness model that is built on the interdependency transaction data of the IIM. The following sections build the multiobjective strategic preparedness model, apply Lagrangian decomposition to the model, and illustrate the results with several simplified numerical examples with application to strategic preparedness. Finally, Part III offers some conclusions that can be drawn from the above discussion. Note that the Lagrange multipliers and the Lagrangian function were introduced in Chapters 2 and 4.

14.III.2 MULTIOBJECTIVE STRATEGIC PREPAREDNESS

As a reminder, strategic preparedness in this chapter connotes a decision process and its associated actions implemented in advance of a natural or man-made disaster and aimed at reducing disaster consequences (e.g. recovery time and cost) and/or their likelihood to an acceptable level. These acceptable levels of risk are obtained only through the decision makers' implicit and explicit acceptance of risks and tradeoffs resulting from the multiobjective nature of strategic preparedness decision problems. (Recall that Chapter 5 is devoted in its entirety to multiple objectives and tradeoff analysis.)

To gain insight through the application of the principles of decomposition, we must first construct a basic multiobjective problem of strategic preparedness. This will be used to frame the problem through the "optimization" of preparedness and resilience. Recall that resilience is the ability of the system to withstand a major disruption within acceptable degradation parameters and to recover within an acceptable cost and time (Rose and Liao, 2005; Woods, 2005, 2006; Hollnagel et al., 2006; Westrum, 2006; Haimes, 2016, 2018). Greater investments in preparedness either directly reduce the consequences of disaster impacts or reduce the likelihood with which certain consequences will occur, resulting in improved resilience. Recovery activities face their own needed tradeoffs that manifest the resilience of the system.

This chapter generalizes the strategic preparedness problem to three objectives, one related to investment in preparedness planning and execution and two objectives related to the ability of a system to absorb and recover from the impact of disasters. The resulting multiobjective system has the following form:

> Minimize (*Cost of Preparedness [$]; Cost of Recovery [$]; Ineffectiveness of Recovery* (time [days], or total economic losses [$] during recovery)) subject to *Limited Resources; Impact of Preparedness on Recovery/Resilience; Socio-economic Interdependencies.*

This model, while it oversimplifies the preparedness problem, is sufficient to illustrate the complexity of the problem and to gain some insights through the application of decomposition in the analysis of system tradeoffs of multiobjective systems. The cost of preparedness in dollars, $\mathbf{v}$, and the cost of recovery in dollars, $\mathbf{u}$, are vectors of investments made by each sector of the economy before and after the disaster, respectively. Premultiplying each vector by a row vector of ones produces a scalar total of investment dollars. Economic sectors are defined according to the North American Industry Classification System (NAICS), in order to utilize economic interdependency data from the inoperability IIM (see Chapter 8). The ineffectiveness of recovery is expressed as a measure of the total amount of time that sectors are experiencing dysfunction or not operating at their predisaster production levels. This measure is used (instead of resilience effectiveness) to streamline the multiobjective model by enabling us to minimize all three objectives. This measure is constructed using three vectors: $\mathbf{x}$ is a vector of nominal or predisaster production output scaled by producers' prices, $\widetilde{\mathbf{x}}$ is a vector of average degraded production level for some preset length of time after the disaster (in this case 1 year), and $\mathbf{c}$ is a vector of relative criticality of sectors per dollar output that could be used to adjust the value of sector recovery (e.g. electric power may be considered more critical per dollar of output than retail trade). It follows that the vector $(\mathbf{x}-\widetilde{\mathbf{x}})$ is the dollar value of production lost or economic losses for each sector of the economy. The measure of recovery ineffectiveness is represented as a weighted sum of differences squared or $\mathbf{C}^{\mathrm{T}}\mathrm{diag}(\mathbf{x}-\widetilde{\mathbf{x}})(\mathbf{x}-\widetilde{\mathbf{x}}) \Leftrightarrow \sum_i c_i(x_i-\widetilde{x}_i)^2$. This results in a dollar value representative of the slowness of recovery (higher values indicate longer recovery times) and penalizes any deviation from predisaster production levels for some preselected period of time following a disaster. Equation (14.2) shows simultaneous minimization of the three objectives contained in the preparedness problem described above. The notation diag(·) indicates a square matrix with the vector values on the matrix diagonal and the remaining matrix elements set to zero:

$$\min_{\mathbf{u},\mathbf{v}} \left| \begin{matrix} \mathbf{1}^{1\times n}\mathbf{u} \\ \mathbf{1}^{1\times n}\mathbf{v} \\ \mathbf{c}^{\mathrm{T}}\mathrm{diag}(\mathbf{x}-\widetilde{\mathbf{x}})(\mathbf{x}-\widetilde{\mathbf{x}}) \end{matrix} \right| \tag{14.2}$$

The solution to Eq. (14.2) is an efficient set of competing Pareto-optimal solution vectors that will be found using the ε-constraint method introduced in Chapter 5. This efficient set is also called the set of nondominated, noninferior, or Pareto-optimal solutions. The first objective in Eq. (14.2) represents the cost of preparedness, whereas the second and third objectives represent the resilience of the system. As decision makers gain a better system understanding of the tradeoffs inherent in the various solution sets, they will make better decisions in preparing against risk.

Two constraints reflect the complexity of the socioeconomic system. The first constraint is shown in Eq. (14.3), and it defines the relationship between recovery investment $\mathbf{u}$ and postdisaster perturbation level $\widetilde{\mathbf{f}}$. The vector function $\mathbf{g}$ is the relationship between recovery investments $\mathbf{u}$ and other system parameters that characterize the system during recovery. These system parameters are impacted by preparedness investments $\mathbf{v}$ to improve the overall resilience of the system:

$$\widetilde{\mathbf{f}}=\mathbf{g}\left(\mathbf{u},\mathbf{h}(\mathbf{v})\right)\Leftrightarrow\left\{\widetilde{f}_i=g_i(u_i,h_i(v_i))\right\},\forall i \tag{14.3}$$

Equation (14.3) represents the most difficult aspect of this problem because the efficacy of preparedness and recovery investment is grounded in public policy, political will, governmental resource allocation (also political), and other factors that are often difficult to quantify. This part of the chapter on system decomposition does not solve the unquantifiability problem, but it provides a framework where efficacy of preparedness investment need only be evaluated on a per sector basis. In other words, the evaluation of the efficacy of preparedness investment in oil refineries can be evaluated separately and independently from similar evaluations for pipelines or food processing plants. Decomposition provides a means by which to structure the strategic preparedness problem so that separate and independent evaluations or assumptions can be incorporated into a single framework for understanding system investment tradeoffs.

Equation (14.4) builds the interdependent relationships that exist among the various sectors of the economy into the multiobjective model through the use of the IIM, as described above, and results in the following multiobjective strategic preparedness problem:

$$\min_{\mathbf{u},\mathbf{v}}\left|\begin{array}{c}\mathbf{1}^{1\times n}\mathbf{u}\\ \mathbf{1}^{1\times n}\mathbf{v}\\ \mathbf{c}^{\mathrm{T}}\mathrm{diag}(\mathbf{x}-\widetilde{\mathbf{x}})(\mathbf{x}-\widetilde{\mathbf{x}})\end{array}\right| \tag{14.4}$$

$$\text{s.t.}:\widetilde{\mathbf{f}}=\mathbf{g}(\mathbf{u},\mathbf{h}(\mathbf{v})) \tag{14.5}$$

$$\mathbf{q}=\mathbf{A}^{*}\mathbf{q}+\mathbf{f}^{*} \tag{14.6}$$

Equations (14.4) through (14.6) are referred to as the multiobjective strategic preparedness problem throughout the remainder of this chapter. By embedding the preparedness assumptions from Eq. (14.5) and the economic interdependency model from Eq. (14.6) into a multiobjective preparedness framework, we are able to quantitatively explore the impact of preparedness alternatives and assumptions on the lack of system resilience. This framework is built on the concept first presented by Haimes and Chittester (2005) to study the efficacy of risk management for cyber vulnerabilities. Decomposition is used to reorganize the data in the multiobjective system in order to derive system tradeoffs. The result is that decomposition provides a means to construct a flexible modeling paradigm and to advance the ability to integrate multiple disparate data sources. Other advantages of decomposition are discussed in the following section.

14.III.3 APPLICATION OF DECOMPOSITION FOR STRATEGIC PREPAREDNESS

This section addresses the strategic preparedness problem through mathematical decomposition. At the conclusion of each aspect of the derivation, there is a brief summary of the insights that can be gleaned from the mathematical process.

Consider the interdependency matrix $\mathbf{A}^*$ in the IIM constraint of Eq. (14.6). A quick analysis of the national-level data shows that many industries consume products from relatively few industry sectors that operate up the supply chain (i.e. they are not very interconnected), while a few other sectors, such as power and wholesale trade, interact with many other sectors. For example, a value $a_{ij}=0.01$ means that for every dollar of output produced by industry j, industry j consumes one penny's worth of production from industry i. Of all 447 industries that have at least one technical coefficient more than 0.01 per year, 431 have transactions with fewer than 20 industries, and only 16 depend on more than 50. By filtering the technical coefficient matrix to strong interdependencies, we can focus on those with high degrees of interdependence, resulting in a separable structure that enables the decomposition of a highly complex interdependent socioeconomic system for strategic preparedness. Aggregating similar sectors to approximate the structure of the 17 Critical Infrastructures/Key Resources (CI/KR) sectors further improves the ability to decompose the strategic preparedness problem.

Decomposition, introduced in Chapters 2 and 4, is the process of breaking up large problems into a set of smaller subproblems of lower dimensionality so that the union forms a problem equivalent to the original.

The use of large-scale computers in the late 1960s led to the rapid expansion of decomposition techniques for optimization, process control, electrical network problems, and a wide variety of seemingly unconnected problems (Lasdon, 2002). Dantzig and Wolfe (1960) provided the first strong motivation for mathematical decompositions by publishing an algorithm that takes advantage of special structures in large matrices. The *process of decomposition* is the identification of composite problems and the equivalence transformation that generates them. It is critical that the equivalence transformation is known – otherwise it cannot be guaranteed that the union of the composite problems guarantees the same solution as the original problem. In other words, the strategic preparedness problem cannot be simply divided up and distributed among sector-specific agencies; there must be some sort of carefully designed decomposition architecture to guarantee an equivalent solution.

Recall from Chapter 2 and elsewhere that Lagrangian decomposition is a generalized multicomponent decomposition. In many respects it is a generalization of Dantzig–Wolfe decomposition for nonlinear models that still relies on a more general block-separable structure of a problem. The Lagrange multiplier acts as a price that coordinates the component optimizations of the problem (Lasdon, 2002). The solution algorithm must be adjusted to include coordination among components. There are two major coordination approaches to decomposed optimization problems, with each providing insight into the decomposition of a risk management problem. Both the feasible and nonfeasible decompositions are introduced in Chapters 2 and 4.

1. *Feasible decomposition* approaches preserve feasibility throughout the solution computation by directing the allocation of available constraints. They are often termed *resource-directed* or *primal* methods. They are typically characterized as problems where the sectors share information that relates to the variables of interdependence (e.g. the pseudo-variables from Figures 14.II.2 and 14.II.3).
2. *Nonfeasible decomposition* approaches coordinate the overall systems model by adjusting a theoretical price on the usage of constraints by component problems. These prices are adjusted until the solution is feasible, at which point it becomes optimal. Nonfeasible methods are frequently called *price-directed* or *dual* methods. These methods are a centralized coordinating body attempting to assess the marginal benefits to the overall system operation by forcing sectors to internalize interdependent risks through regulation, policy, or incentives. For example, the unreliability of the electric power infrastructure after a storm results in water suppliers purchasing and maintaining backup generators, gas-powered pumps, and other equipment to increase their own reliability. Thus, water suppliers have to internalize the interdependent risk transferred from the power sector, in order to guarantee an increased level of reliability. Alternatively, the power company could be forced to improve reliability through a regulation or policy. In both cases there are costs to the final consumer, but they manifest themselves differently. In the first case of the power sector's externalization of social risks caused by its unreliability, the increased water cost is passed onto the consumer directly from the water supplier. In the second case, the increased power cost is passed on to all the consumers, including the water supplier, who in turn passes on the cost to the final consumer. The question is which set of policies will guarantee the highest level of strategic preparedness at the lowest cost. With decomposition methods, we solve the problem by having each sector come to a solution independently, but with some set of influencing factors (e.g. policies, incentives, information sharing) that should provide for coordination and a socially optimal solution. Both methods are used in this section to estimate system tradeoff functions for strategic preparedness and illustrate insights into decentralized risk management.

14.III.4 CONCLUSIONS

Preparedness connotes decisions and actions implemented in advance of a disaster. Making decisions reflecting strategic preparedness is complicated by the phenomenon's multiobjective nature (minimizing the cost of preparedness and the impact of disasters). Thus, its decomposition is complicated by the interdependencies among the many industries within each economic sector. Information sharing entities are commonly sector specific, and sector-specific agencies do not have clear agendas to pass this interdependency information across information sharing entities, nor do they have a comprehensive mechanism in place for sharing it across critical infrastructures. Moreover, centralized bodies do not have the appropriate organizational structure to formulate policies that rely on knowledge from both the collective information of the subgroups (join through the information sharing organizations or the sector-specific agencies). Furthermore, information that measures or represents the interdependencies of the various critical infrastructures and key resource sectors is also scarce, or completely lacking. Currently, a multilayer decomposition of the preparedness problem seems to be emerging, where the preparedness problem is first decomposed by geography (e.g. state governments), after which specific geographical entities decompose the problem by sectors. This process seems to provide a means by which the individual regions will be more capable of understanding the specific interdependent structures that must be addressed to appropriately regulate or share information between critical infrastructure sectors.

Research on the principles of decomposition in system design is recommended to study the complex problem of strategic preparedness. By embedding preparedness assumptions from Eq. (14.5) and the economic interdependency model from Eq. (14.6) into a multiobjective preparedness framework, we are able to quantitatively explore how various assumptions of direct preparedness efforts impact the resilience of the socioeconomic systems of a region. The framework is built on the concept of studying the efficacy of risk management for cyber vulnerabilities (Haimes, 2016). Decomposition is commonly used to reorganize the data in the multiobjective system to derive system tradeoff functions. During the decomposition process, care is taken to preserve the composing structure, which in turn enables the integration of additional information into the framework. The result is that decomposition provides a means to construct a more flexible modeling paradigm and to advance the ability to integrate multiple disparate data sources while still preserving the overall optimality of the solution.

The basic multiobjective problem of strategic preparedness is optimizing (in the Pareto-optimal sense) preparedness and resilience, where resilience in this chapter connotes a system measure characterizing the time and cost required to return to a predisturbance level of operation. Greater investments in preparedness either directly reduce the consequences of disaster impacts or reduce the likelihood with which certain consequences will occur. Recovery activities face their own needed tradeoffs that reflect the resilience characteristics of the system.

We also proposed various methodologies to decompose the preparedness problem. Through the decomposition of these models, we have gained insight into the complicated structure of sector interdependencies and have noted methods by which understanding of structure simplifies interdependency analysis. Additionally, these insights have led to enhancing the capability of the preparedness-modeling framework to integrate analytically with other sector-specific models and databases. To illustrate these decomposition principles, we constructed two-sector examples that provide clear, illustrative numerical examples of feasible and nonfeasible decomposition. Such procedures will help to enhance the data richness and resolution of preparedness problems while simultaneously preserving the aggregated information needs of decision makers.

The chapter also formally introduced a concept for system planning efforts related to Complex SoS that consist of a rich mixture of formally designed subsystems and emergent subsystems. Emergent subsystems are those parts of a system that may or may not come into being based on events, and system stakeholder responses to those events that may or may not ultimately lead to a transformation, resulting in a new subsystem that becomes a formal and supported part of the overall system. To make the proposed system planning approach tangible and to guide conceptual thinking, the chapter addresses the planning for a critical national problem and corresponding system for *establishing a national preparedness system for terrorism and natural disasters*. The cost for creating a government-designed and government-implemented national preparedness system for significantly reducing risks resulting from the wide range of possible natural and terrorist threats is too high for any economy to absorb – a fact that can lead to a period of relative stagnation in terms of selecting solutions that need to be implemented.

The chapter also recognizes that parts of the preparedness system must emerge through the integrated outcomes of individual stakeholders' decisions and efforts, and other parts can be systemically created through organized and focused activities. Our analysis of a preparedness system divides the system properties into categories that are more or less likely to be addressed through either formally designed or, alternatively, emergent subsystems. The properties for resilience in the face of terrorist attacks or natural disasters are defined and discussed as principally dependent on emergent subsystems, while the properties for prevention or minimization of adverse consequences from attacks, or would-be disasters, are defined and discussed as being more dependent on formally designed subsystems. Research efforts are required to enrich the ideas presented in this chapter by applying the ideas to a national preparedness system.

REFERENCES

Baden, J.A. and Noonan, D. (1998). *Managing the Commons*, 2e. Bloomington, IN: Indiana University Press.

Bromley, D.W. ed. (1992). *Making the Commons Work*. San Francisco, CA: ICS Press, Institute for Contemporary Studies.

Chen, C. (2012). *Linear System Theory and Design*, 4e. New York: Oxford University Press.

Critical Infrastructure Information Act (2002). Section 214.

Crowther, K.G. and Haimes, Y.Y. (2005). Applications of the inoperability input-output model (IIM) for systemic risk assessment and management of interdependent infrastructures. *Systems Engineering* **8**(4): 323–341.

Dantzig, G.B. and Wolfe, P. (1960). Decomposition principle for linear programs. *Operations Research* **8**: 101–111.

Gourevitch, V. ed. (1997). Rousseau: *The Social Contract and other later political writings*. UK: Cambridge University Press.

Gürerk, O., Irlenbusch, B., and Rockenbach, B. (2006). The competitive advantage of sanctioning institutions. *Science* **312**: 108–111.

Haimes, Y.Y. (1991). Total risk management. *Risk Analysis* **11**(2): 169–171.

Haimes, Y.Y. (2006). On the definition of vulnerabilities in measuring risks to infrastructures. *Risk Analysis* **26**(2): 293–296.

Haimes, Y.Y. (2009). On the complex definition of risk: a systems-based approach. *Risk Analysis* **29**(12): 1647–1654.

Haimes, Y.Y. (2016). *Risk Modeling, Assessment, and Management*, 4e. Hoboken, NJ: Wiley.

Haimes, Y.Y. (2018). Risk modeling of interdependent complex systems of system: Theory and practice. *Risk Analysis* **38**(1): 84–98.

Haimes, Y.Y. and Chittister, C.G. (2005). A roadmap for quantifying the efficacy of risk management of information security and interdependent SCADA systems. *Journal of Homeland Security and Emergency Management* **2**(2): Article 12.

Haimes, Y.Y. and Horowitz, B. (2004a). Adaptive two-player hierarchical holographic modeling game for counterterrorism intelligence analysis. *Journal of Homeland Security and Emergency Management* **1**(3): 1038.

Haimes, Y.Y. and Horowitz, B. (2004b). Modeling interdependent infrastructures for sustainable counterterrorism. *Journal of Infrastructure Systems* **10**: 33–42.

Haimes, Y.Y. and Jiang, P. (2001). Leontief-based model of risk in complex interconnected infrastructures. *ASCE Journal of Infrastructure Systems* **7**(1): 1–12.

Haimes, Y.Y. (1981). Hierarchical holographic modeling. *IEEE Transactions on Systems, Man, and Cybernetics-SMC* **11**(9): 606–617.

Haimes, Y.Y., Horowitz, B., Lambert, J. et al. (2005a). Inoperability input-output model (IIM) for interdependent infrastructure sectors: theory and methodology. *ASCE Journal of Infrastructure Systems* **11**(2): 67–79.

Haimes, Y.Y., Horowitz, B., Lambert, J. et al. (2005b). Inoperability input-output model (IIM) for interdependent infrastructure sectors: case study. *ASCE Journal of Infrastructure Systems* **11**(2): 80–120.

Haimes, Y.Y., Crowther, K., and Horowitz, B.M. (2008). Homeland security preparedness: balancing protection with resilience in emergent systems. *Systems Engineering* **11**: 287–308. doi: 10.1002/sys.20101.

Hardin, G. (1968). The tragedy of the commons. *Science* **162**: 1343–1348.

Heal, G. and Kunreuther, H. (2007). Modeling interdependent risks. *Risk Analysis* **27**: 621–634. doi: 10.1111/j.1539-6924.2007.00904.x.

Henrich, J. (2006). Cooperation, punishment, and the evolution of human institutions. *Science* **312**: 60–62.

Holling, C.S. (1973). Resilience and stability of ecological systems. *Annual Review of Ecology and Systematics* **4**(1): 1–23.

Hollnagel, E., Woods, D.D., and Leveson, N. ed. (2006). *Resilience Engineering: Concepts and Precepts*. Aldershot: Ashgate Press.

Kaplan, S. and Garrick, B. (1981). On the quantitative definition of risk. *Risk Analysis* **1**(1): 11–27.

Lasdon, L.S. (2002). *Optimization Theory for Large Systems*. New York: Dover Publications.

Lasdon, L.S. and Schoeffler, J. (1966). Decentralized plant control. *ISA Transactions* **52**: 175–183.

Leontief, W.W. (1966). *Input-Output Economics*. New York: Oxford University Press.

Miller, R.E. and Blair, P.D. (1985). *Input-Output Analysis: Foundations and Extensions*. Englewood Cliffs, NJ: Prentice-Hall, Inc.

PCCIP (1997). Critical Foundations: Protecting America's Infrastructures, The Report of the President's Commission on Critical Infrastructure Protection, Washington, DC, October.

Perrow, C. (2011). *The Next Catastrophe: Reducing our Vulnerabilities to Natural, Industrial, and Terrorist Disasters*. Princeton, NJ: Princeton University Press.

Rose, A. and Liao, S. (2005). Modeling regional economic resilience to disasters: a computable general equilibrium analysis of water service disruptions. *Journal of Regional Science* **45**(1): 75–112.

Rousseau, J.J. (1762). *The Social Contract. Or Principles of Political Right*.

Santos, J.R. (2006). Inoperability input-output modeling of disruptions to interdependent economic systems. *Systems Engineering* **9**(1): 20–34.

Santos, J.R. and Haimes, Y.Y. (2004). Modeling the demand reduction input-output inoperability due to terrorism of interconnected infrastructures. *Risk Analysis* **24** (6): 1437–1451.

U.S. Department of Commerce, Bureau of Economic Analysis (1998). *Benchmark Input-Output Accounts of the United States, 1992*. Washington, DC: US Government Printing Office.

Westrum, R. (2006). A typology of resilience situations. In: *Resilience Engineering: Concepts and Precepts* (ed. E. Hollnagel, D.D. Woods and N. Leveson), 49–60. Aldershot: Ashgate Press.

Woods, D. (2005). Creating foresight: lessons for resilience from Columbia. In: *Organization at the Limit: NASA and the Columbia Disaster* (ed. M. Farjoun and W.H. Starbuck), 289–308. Malden, MA: Wiley-Blackwell.

Woods, D. (2006). Essential characteristics of resilience. In: *Resilience Engineering: Concepts and Precepts* (ed. E. Hollnagel, D.D. Woods and N. Leveson), 21–34. Aldershot: Ashgate Press.

15

Modeling Safety of Transportation Complex Systems of Systems via Fault Trees

PART I: MODELING, UNDERSTANDING, AND MANAGING THE RISK TO TRANSPORTATION INFRASTRUCTURE AS A COMPLEX SYSTEM OF SYSTEMS*

15.I.1 INTRODUCTION

> The *2017 Infrastructure Report Card* reveals that the U.S. has made some incremental progress toward restoring the U.S. infrastructure; however, the progress is not sufficient. As in 2013, America's cumulative Grade-Point Average (GPA) is once again a D+.
>
> (ASCE, 2017)

Current approach to civil infrastructure management is not sustainable. For example, the American Society of Civil Engineers (ASCE) in its most recent report card in 2017 assigned an overall grade of D+ to America's infrastructure and estimated that more than $4 trillion is needed just to maintain a state of good repair. This figure has increased from the $2.2 trillion estimated in the 2009 report card and does not include the investment needed to increase capacity or put new infrastructure into service. Clearly, the United States is losing ground, and we contend that the transportation Complex SoS of other countries might be in the same dire conditions as the United States. Thus, the expected efficacious contributions of the following analysis might be relevant to other countries around the world. It is not sufficient merely to recognize the problem, or even to establish national priorities and actions to deal with the backlog, as vital as these steps are. A new paradigm must be established for the management of civil transportation infrastructure, which must take into account the multiple and interrelated socio-economic factors associated with transportation infrastructure design, construction, maintenance, and replacement. Indeed, the major factor contributing to the increased current critical situation is that most infrastructures have not been designed and constructed to provide projected future level of service required. The current approach to transportation management is based upon condition and performance measures that are very subjective and heavily weighted toward ordinal assessments based upon nonquantitative criteria and methods. Past and current practice is to design and construct

*The author acknowledges the lessons learned on the life cycle of bridges from his colleague, Professor Steven B. Chase.

Modeling and Managing Interdependent Complex Systems of Systems, First Edition. Yacov Y. Haimes.

transportation complex systems of systems (Complex SoS) with almost no regard for adequate inspection, maintenance, preservation, and renewal. The current approach to assessing the condition and integrity of transportation infrastructure Complex SoS and their subsystems relies on visible indications of deterioration. This approach is by design, is reactive, and is limited to deterioration that has advanced to the point where substructures are damaged. Deterioration could, and probably does, exist that has not progressed to the point where visible indications are present. There have been many sudden failures of bypasses, bridges, and other transportation systems and subsystems that have been traced back to hidden or latent defects, many of which were directly related, for example, to bridges that were not constructed according to the detailed planned and with materials that did not meet specifications. In addition, there could be deterioration processes occurring that are not yet understood or recognized. Such challenges must be addressed by considering the transportation infrastructure as *interdependent and interconnected Complex SoS* and by applying risk assessment, management, and communication principles to identify those factors critical to their viability.

15.I.2 MODELING AND MANAGING TRANSPORTATION INFRASTRUCTURE COMPLEX SYSTEMS OF SYSTEMS

To study and manage transportation infrastructure as Complex SoS, we identify two key steps in this chapter:

1. Discover the primary factor(s) limiting the service life of highway transportation infrastructure.

The service life of highway transportation infrastructure, including bridges, is generally limited by factors that result from systemic failures, including (a) failure to deliver transportation infrastructure constructed as designed; (b) lack of knowledge about deterioration processes; (c) failure to detect damage before failure occurs; (d) failure to adequately maintain transportation infrastructure and bridges; (e) failure of current technical, management, and institutional systems to prevent these failures; and (f) failure to apply a holistic approach to modeling transportation infrastructure as interdependent and interconnected Complex SoS. Indeed, a holistic approach that studies and models the encompassing transportation infrastructure Complex SoS – their planning, design, construction, planning for operation, operation, and maintenance – can provide new insight and prove to be a more effective method for understanding the complexity and performance of such an enterprise than the current separate, disjointed approach.

2. Establish a method for assessing and managing multiple sources of risk to transportation infrastructure by building on state-space theory and methodology of Complex SoS.

Multiple sources of risk to the viability of a country's transportation infrastructure can be best modeled, assessed, managed, and communicated by studying and modeling the transportation infrastructure as Complex SoS. The multiple perspectives of the subsystems and systems that constitute the transportation infrastructure Complex SoS cannot be modeled within one single model. Thus, multiple models are needed to represent the essence of the respective subsystems through their shared states, resources, decisions, decision makers, and stakeholders and, most importantly, through their maintenance and the demands placed on them.

We present a paradigm shift for identifying, documenting, and quantifying the technical and institutional factors that limit the service life of existing transportation infrastructure as Complex SoS. Furthermore, there is a need for a body of knowledge and understanding of the factors that limit the useful service lives of a country's highways and bridges by utilizing data on the prevailing interdependencies and interconnectedness (I-I) within and among the subsystems that constitute the Infrastructure Complex SoS. Indeed, such knowledge is obtained by exploring and studying the vast database on transportation systems.

There is also a need to characterize, quantify, and model the (i) organizational, (ii) budgetary, (iii) maintenance, and (iv) entire life cycle of transportation infrastructure Complex SoS. The above would enable a more rational and systems-based approach to the planning, design, fabrication, construction, operation, maintenance, and preservation of the I-I transportation infrastructure Complex SoS. This knowledge can also be used to develop maintenance and preservation strategies to enable a country's transportation infrastructure Complex SoS to be kept in service and in a state of good repair for as long as it is needed. This knowledge can be used to influence engineering curricula at universities to help bring about a change of focus from newly designed and constructed transportation infrastructure to the challenges of managing and maintaining the existing I-I of transportation infrastructure Complex SoS.

Consider the following factors that have caused the dismal state of the transportation infrastructure in many countries: (i) most civil infrastructures have not been designed and constructed to provide the level of service required; (ii) bridge infrastructure has been designed without regard for adequate maintenance and renewal for the required service levels; (iii) current approach to transportation management is commonly founded on data obtained by subjective visual condition and nonquantitative performance measures; and (iv) current practice is to plan, design, construct, and maintain the transportation infrastructure Complex SoS with minimal regard for adequate funds with which to support inspection, maintenance, preservation, and rehabilitation.

15.I.2.1 Importance of the Transportation Infrastructure Complex Systems of Systems to the Public

Physical infrastructures, serving as the foundations of society's well-being and encompassing household and the entire private and public sectors, are the driving force of today's worldwide economic prosperity. They also serve the interest of diverse stakeholders, including individuals, families, households, companies, industries, and local, state, and federal governments. However, political and organizational forces and doctrines often overshadow the goal of understanding and responding to the critical societal need for the resources required to support the development, construction, maintenance, and replacement of transportation infrastructure. Drawing on different disciplines, there is a need to identify a broad-based knowledge with which to address the multifarious requirements, specifications, and appropriate technology and materials, which are requisite for the design, construction, operation, maintenance, and ultimate replacement of transportation infrastructure Complex SoS. Such a systemic process would enable countries to improve their understanding of the underpinnings characterizing every stage of the life cycle of the I-I characterizing the transportation infrastructure Complex SoS.

Important outcomes and by-products resulting from a systemic maintenance and management of transportation infrastructure Complex SoS include (i) defining the role of big data collection and metrics development in guiding future transportation requirements, specifications, and appropriate technology and materials, as well as transportation planning, construction, operation, and maintenance; (ii) understanding the dynamics that drive current public policy decisions on transportation financing, construction, operation, and maintenance; and (iii) understanding the role of systems-based risk assessment, management, and communication in guiding policies and procedures that ensure effective integration of all phases of the life cycle of the transportation sector as interdependent and interconnected Complex SoS.

15.I.2.2 The Role of State Space in the Modeling of Transportation Infrastructure Complex Systems of Systems

Transportation Complex SoS epitomize the criticality of physical infrastructures; they are visible and utilized by all of us as we travel from one location to another. It would be virtually impossible to effectively address the intricate life cycle of our roads and highways (or other physical infrastructures) without models that build on the indispensible domain knowledge of the complexities that stem from their multiscale structural composition. Indeed, structures which span the physical molecular and macro dimensions, as well as the myriad involved outside parties, e.g. stakeholders, organizations, the public, and political forces, and determine the fate of our transportation infrastructure as Complex SoS. In this sense, all transportation modes, as well as all physical infrastructures, constitute interdependent and interconnected Complex SoS. Models can help answer limited questions about the behavior of these systems under both steady-state conditions and dynamic forced changes. Because each transportation infrastructure is comprised of multiple subsystems, each of which, small or large, and is characterized by multiple perspectives, it is necessary to develop a set of models with which to represent this multiplicity. Such set of models must reflect the essence of the multifarious attributes, functions, and dimensions of the transportation Complex SoS. Physical, chemical, biological, and other natural laws affecting the structural health of a country's transportation system as Complex SoS serve as the first principles and the foundations of such models. Although they are most necessary, these natural laws are not sufficient for their model construction. This is because the multiple societal and political perspectives constitute critical driving forces for which all decision-making models must account. Furthermore, the above multiple perspectives cannot be adequately modeled in a single model, a fact that presents a challenge to modelers. Thus what is needed is a mechanism, a systemic framework, with which to augment (e.g. via Hierarchical Holographic Modeling (HHM), the subject of Chapter 3), the natural and physical laws with imagination, inventions, innovation, entrepreneurship, out-of-the-box thinking, and boundless database and experimentation. Thus, effective management of transportation infrastructure Complex SoS must benefit from, and build on, emergent theory and methodologies for their modeling.

The Phantom System Model (PSM) and the PSM laboratory (PSML), introduced in Chapter 4, represent a real and virtual laboratory for such model building (Haimes, 2007, 2012b, 2016). The essence of metamodel coordination and integration is to build on all relevant direct and indirect sources of information to gain insight into the I-I among the submodels and on the basis of which to develop representative models of the Complex SoS under consideration. The coordination and integration of the results of the multiple models are achieved at the metamodeling phase within the PSM, thereby yielding a better understanding of the Complex SoS as a whole (see Figure 15.I.1).

More specifically, we achieve the above by modeling the shared states and other essential entities within and among the subsystems of Complex SoS. The term "essential entities" includes *shared/common decisions*, *decision makers*, *stakeholders*, *resources*, *organizational behavior and norms*, *policies and procedures*, *management*, *culture*, and others. This process requires an understanding of the intricate relationships that characterize the dynamics within and among the states of the subsystems. This very important task is achieved at the metamodeling level of the PSM by observing, estimating, and assessing the outputs for given inputs and by building on the intrinsic common states and other essential entities within and among the subsystems. Note that although the intrinsic common/shared states and other essential entities within and among the systems and subsystems constitute a key element of the PSM, the extrinsic (input–output) relationships are also very important and augment the intrinsic one. Indeed, the selection of the trial inputs to the model, along with the inquisitive process of making sense of the corresponding outputs – all are at the heart of system identification and parameter estimation in model building. This is not a one-shot process; rather, it can be at best characterized by tireless testing, trial and error, experimentation with model topology, and parameter estimation and adjustments and by questioning whether the assumed model's topology is representative of the Complex SoS being modeled.

The following characteristics, or other relationships, must exist between and among any two systems within Complex SoS:

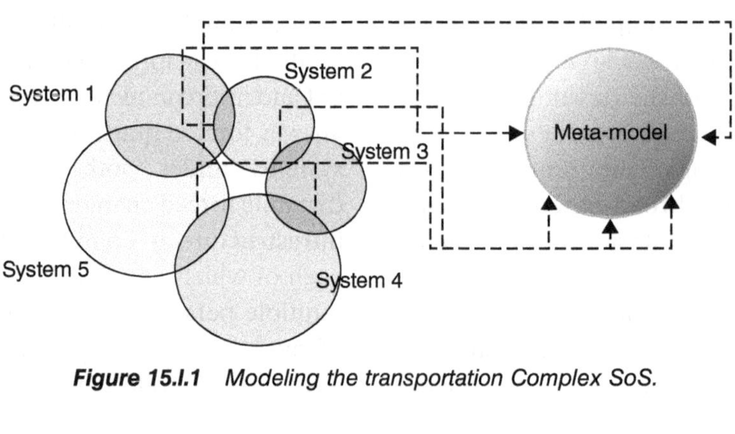

Figure 15.I.1 *Modeling the transportation Complex SoS.*

1. Recall from Chapter 4 that a metamodel represents the overall coordinated and integrated submodels of the Complex SoS. We redefine a metamodel as a family of submodels, each representing specific aspects of the subsystem for the purpose of gaining knowledge and understanding of the multiple I-I among the submodels and thus to comprehend the Complex SoS as a whole. For example, the physical structure of a bridge system is composed of multiple subsystems, e.g. a deck and wearing surface, which provides the riding surface and distributes load to the superstructure, the main structural support of the bridge, transfers load to the substructure, and distributes the load to the foundation soils. Each of these subsystems in turn contains and relies upon multiple subsystems, such as bearings and joints and various materials, such as concrete, structural steel, reinforcing steel, prestressing steel and coatings.
2. The essence of each subsystem can be represented by a finite number of variables. (The term *essence* of a system connotes the quintessence of the system, the heart of the system, namely, everything critical about the system.) Given that a system may have a large number of essential state variables, the term *essential* states of a system connotes the minimal number of state variables in a model with which to represent the system in a manner that permits the questions at hand to be effectively answered. Thus, these essential state variables become fundamental for an acceptable model representation.
3. For a properly defined Complex SoS, any interconnected subsystem will have at least one (typically more) shared state and other essential entities and objective(s) with at least one other subsystem. (Recall that the term "essential entities" includes shared/common decisions, decision makers, stakeholders, resources, organizational behavior and norms, policies and procedures, management, culture, and others.) This requirement constitutes a necessary condition for modeling interdependencies among the subsystems (and thus I-I across Complex SoS). This ensures an overlapping of state variables within the subsystems. Of course, the more we can identify and model joint (overlapping) shared states and other essential entities among the subsystems, the greater is the representativeness of the submodels and the metamodel of the Complex SoS.
4. The importance of the availability of multiple, albeit overlapping, databases can be effectively utilized by multiple submodels, each of which is commonly built to answer the specific questions for which it is built. Furthermore, each characterization of a submodel, whether modeled separately or in groups, is likely to share common state variables – a fact that facilitates the ultimate coordination and integration of the multiple submodels at the metamodeling level.
5. The fusion of multiple submodels via the intrinsic metamodeling coordination and integration enhances our understanding of the inherent behavior and interdependencies of existing and emergent Complex SoS.

15.I.2.3 Ten Systems-Based Guiding Principles Revisited for the Next Generation of Transportation Design, Construction, Operation, Maintenance, and Replacement

By its nature, risk analysis is an intricate, dynamic process – an amalgamation of the arts and sciences – tailoring an appropriate risk assessment, management, and communication to each specific set of sources of the risks to specific systems of the transportation Complex SoS. It follows, then, for any subsystem of the transportation infrastructure Complex SoS, the balance between quantitative and qualitative risk analysis will be system/problem and domain specific. The 10 principles we introduced in Chapter 9 are intended to serve as an efficacious, valuable risk analysis instrument for guiding the quantitative and qualitative decisions made by transportation infrastructure engineers and other professionals (Haimes, 2012a, 2016). Indeed, violations of these basic principles could lead to disastrous results to the transportation Complex SoS of a country. More specifically, the tenets behind these guiding principles are to provide engineers and other professionals with a checklist that will help them avoid the most common errors, while still providing them with the flexibility to make appropriate tradeoffs in constructing, deploying, and executing various functions of the transportation Complex SoS. (For further general discussion of the effectiveness of this approach, see *The Checklist Manifesto* by Atul Gawande, MD (2011).)

A transportation infrastructure engineer or a manager might ask: "Why do we need a set of principles to perform risk analysis?" Although no single answer can do justice to this broad question, one may posit that any decision-making process that addresses important probable, including dire, consequences resulting from changes emanating from within or without the transportation Complex SoS ought not be performed on an ad hoc basis. Thus, the life cycle of the transportation infrastructure, to be studied and modeled, must adhere to the following guiding principles and be cognizant of and responsive to the inevitable emergent forced changes (EFCs): trends in external or internal sources of risk to a system/project that may adversely affect specific states of that system/project. Examples of EFCs may include elements of the *Evolving Base*, (introduced in Chapter 9), which is characterized by dynamic shifting rules and realities, including (i) goals and objectives; (ii) stakeholders, decision makers, and interest groups; (iii) organizational, political, and budgetary baselines; (iv) reorganization and reallocation of key personnel; (v) emergent new technology; (vi) requirements, specifications, delivery, and users; and (vii) natural hazards.

The 10 principles were introduced in Chapter 9 and can be applied to the transportation sector of the economy.

15.I.3 MAINTENANCE OF THE BRIDGE COMPLEX SYSTEMS OF SYSTEMS

The most frequently cited reason by bridge maintenance engineers for bridges failing to provide adequate service is that the bridge Complex SoS was not constructed as designed. Perhaps the most common example is the lack of adding concrete cover for reinforcing steel during bridge construction. The primary protection for reinforcing steel against corrosion is the passive layer that forms on the steel in the alkaline environment within the concrete. The diffusion of chloride ions and the carbonization of the concrete due to adsorption of carbon dioxide from the atmosphere into the concrete are the two primary mechanisms by which this passivation is compromised. The most common way in which bridge Complex SoS can deal with this problem is by providing a thick layer of concrete between the surface and the reinforcing steel to reduce, or eliminate, the destructive I-I between the two. Typical minimum cover is 2 inches for normal concrete and up to 4 inches for concrete in severe exposure conditions. Indeed, many instances have been documented where the cover actually delivered on a bridge and put into service is less than 1 inch, resulting in severe corrosion of the reinforcing steel much earlier in the service life of the bridge than expected. It is conceivable that a detailed examination of bridges that are being taken out of service would reveal that in many cases the failure of the bridge was due to the above I-I, leading to a failure to deliver bridges that were appropriately constructed as designed (Andrijcic, 2016; Andrijcic and Haimes, 2017; Guo and Haimes, 2016; 2017).

A few examples of failure mechanisms that were not anticipated and not considered in the design of highway bridges that led to premature failures include (i) distortion induce cracking, (ii) brittle fracture of triaxially restrained welds, (iii) alkali–silica reactions, and (iv) excessive bleed water segregation leading to corrosion and failure of posttensioning tendons (Chase et al., 2013). The existence and importance of these failures in structural mechanisms of Complex SoS are not commonly known until premature failures occur. One hypothesis is that a detailed examination of bridges and other transportation Complex SoS that are being taken out of service would discover new and previously unappreciated deterioration mechanisms that exist and might be responsible for their failure. Also, a combination of factors – technical, professional, managerial, economic, political, and institutional – has brought about a situation in which transportation Complex SoS fail to provide adequate service life. These factors need to be understood and modeled within the I-I, characterizing the transportation Complex SoS. Their premature failure can result in widespread impacts, and often these impacts are not immediately obvious.

15.I.4 MODELING THE TRANSPORTATION COMPLEX SYSTEMS OF SYSTEMS

Transportation Complex SoS cannot be effectively modeled through a single model. Indeed, multidisciplinary submodels must be developed to represent the multiple players and systems including (i) stakeholder-level perspectives; (ii) interdependencies among the transportation subsystems (submodels), which we model primarily through shared states and other essential entities (e.g. physical condition of structural components, unrestricted load capacity, maintenance budget, traffic demand), and other shared/common essential entities; and (iii) a metamodel must be developed to coordinate and integrate the various interdependent transportation submodels (e.g. highways, tunnels, and bridges). Such metamodels, given their integrative capacity, should provide a deeper insight into the interdependencies among the systems and subsystems of transportation Complex SoS.

In addition, a metamodel must account for the wide range of factors that may affect the bridge subsystems and their interdependencies. An example includes (i) load, capacity, and foundations; (ii) the nature of infrastructure's design, in particular the fact that transportation infrastructures are built from bottom-up, whereas the load comes from top-down; (iii) heavy trucks, corrosion, combined forces, and changes in technology; and (iv) load distribution, change in capacity, extreme events, risk, economics, and stakeholder expectations based on today's standards, not on what people expected 100 years ago. Traditional transportation infrastructure management models, such as load rating (US AASHTO, 2017) using girder line, as well as more complex methods such as finite element analysis (FEA) and statistical deterioration, are well known to and applied by the transportation infrastructure engineering community. However, these models provide a very limited picture of the bridge maintenance and management problem, and they often ignore other factors (e.g. socioeconomic factors, regional development factors of historical importance) that could influence the choice of an optimal maintenance plan. A preferred maintenance approach would be more holistic and would consider the objectives and needs of all stakeholders, ranging from system owners to system users.

The bridge infrastructure is an integral system of the transportation Complex SoS that encompasses not only physical components but also human, social, and economic elements. The interplay of these elements is of critical importance, given the broad impacts of bridge failures. It is important to consider the concurrent use of a multidisciplinary set of models to explore the interdependencies among the various elements of the system, with the goal of discovering how current transportation Complex SoS maintenance practices affect, and are affected by, social and economic elements. The common use of multidisciplinary models allows us to examine the operation of the transportation infrastructure from various perspectives, and can help us gain insight into robust transportation infrastructure management.

15.I.5 SUMMARY

We ought to model and study transportation infrastructures as Complex SoS from their multiple perspectives by building on our combined experiences and expertise in their life cycle. In our quest to improve our understanding, modeling, and thus managing transportation Complex SoS, we also ought to rely on the theory and methodology drawn from our body of knowledge in (i) the planning, development, design, construction, assessment, management, communication, and modeling of transportation as Complex SoS and (ii) understanding, modeling, and testing transportation infrastructure as interdependent and interconnected Complex SoS. Note that the service life of transportation Complex SoS is limited by factors that result from systemic failures, including (i) failure to deliver roads and bridges that were constructed as designed, (ii) lack of knowledge about their deterioration processes, (iii) failure to detect damage before failure occurs, (iv) failure to adequately fund and maintain highways and bridges and other transportation Complex SoS, and (v) failure of applying current technical, management, and institutional resources and management systems to prevent these failures. New insight is provided by examining the transportation Complex SoS from their multiple perspectives, including the following:

1. *Engineering perspectives*: (i) Investigating and documenting the prevailing disharmonious coordination and not uncommon disregard for

the dependencies among (i) the requirements, specifications, and use of appropriate materials in construction and (ii) the operation and maintenance of the transportation Complex SoS.

2. *Organizational and budgetary perspectives*: (i) Identifying the genesis of the prevailing dysfunctional and disharmonious management of the engineering life cycle of the transportation infrastructure, with a focus on the interdependencies and shared states among the subsystems of the transportation Complex SoS; (ii) exploring effective risk management policy options, given limited resources and budgetary constraints; and (iii) expanding upon and demonstrating the efficacy of recently developed and published systems thinking advocating the exploitation of state-space theory as they apply to transportation Complex SoS.

PART II: MODELING TRANSPORTATION COMPLEX SYSTEMS OF SYSTEMS VIA FAULT TREES

OVERVIEW

Part II of this chapter builds on fault trees and fault-tree analyses for modeling transportation Complex SoS (Kuzminski et al., 1995; Joshua and Garber, 1992). One of the leading documents on fault-tree analysis is the *Fault Tree Handbook* written by the US Nuclear Regulatory Commission (1981). This handbook remains the primer for all students of fault trees.

We intend to achieve the following objectives with Part II of this chapter by:

1. Framing highway transportation as Complex SoS, involving myriad contributing factors to accidents, termed in this chapter as "system's failure."
2. Among the various decomposition methods introduced and developed throughout this book especially in Chapters 2 and 4, we highlight the modeling power of fault trees as another decomposition schema that effectively augments the multiple decomposition and higher-level coordination methods introduced in earlier chapters.
3. We relate the decomposition achieved via "AND" Gates (systems connected in parallel) and "OR" Gates (systems connected in series) by (a) highlighting the I-I that exist and characterize the modeling of Complex SoS; (b) quantifying the failure modes (e.g. accidents in transportation) resulting from the highly I-I with other vehicles and pedestrians characterizing highway Complex SoS; (c) developing measures with which to manage (not only to assess and quantify) the risk of accidents; and (d) using the transportation Complex SoS as a case study that can be applied to many other infrastructure Complex SoS.

Consider the following sample of failures of transportation Complex SoS to which this case study not only is applicable, but also provides a roadmap for the following sample: (i) the failure of hospitals to prevent, or at least reduce, discharged patients with infection, thus, require them to return to the hospital for emergency treatment; (ii) cascading failures, major disruptions, or inability to effectively perform intended functions of such sectors of the economy as supply chain, cyber–physical infrastructures, and smart manufacturing with an increased role of automation and robotics (e.g. slowing down or completely stopping production lines in the manufacturing sector); and (iii) failures from which Complex SoS suffer due to uncommon extreme and rare events that often escape management and operators. Indeed, events of very low probabilities, but with possible catastrophic consequences, are commonly unplanned for, yet the associated disruptions can be very costly in financial terms, delay in production time and schedule, or not meeting performance criteria in terms of schedule or quality.

The above is only a sample of the efficacious contributions of fault trees to our understanding, modeling, and managing transportation Complex SoS.

15.II.1 INTRODUCTION

Transportation is an emergent safety-critical interdependent and interconnected sector of the economy, which in its essence constitutes Complex SoS. Almost uncounted factors and sources of risk characterize and

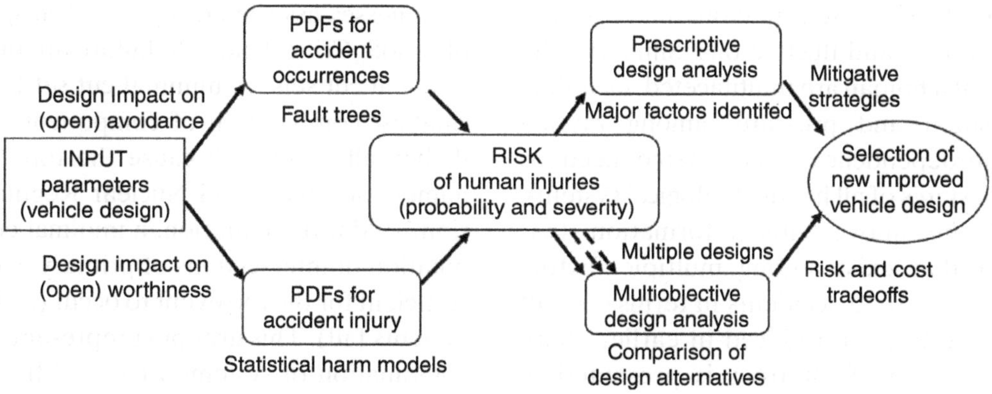

Figure 15.II.1 *Methodological framework.*

determine the ultimate safety of transportation on the highways. As a starting point, we investigate the modeling of automobile safety as a function of its design (assuming an average uniform level of safety characterizing all drivers). We focus primarily on how automobile accidents occur and address the broader challenge of quantifying and managing the risk inherent in particular automobile designs and automobiles and the environment within which they operate as Complex SoS. Figure 15.II.1 displays the structure developed to address accident safety, also depicting the accident causation in the fault tree.

Recall that we have added a primer on fault trees in the Appendix of this book for the uninitiated reader on the subject. An entire chapter on fault trees is dedicated in the Fourth Edition text book (Haimes, 2016). The modeling process of improving highway safety can be used as a prototype for modeling other emergent transportation Complex SoS. For example, over 210 million registered drivers were in the United States in 2016, and there were over 37 000 fatal motor vehicle crashes in the United States in that year (Federal Highway Administration, 2017). An average of 96 people died each day in motor vehicle crashes in 2015, 1 fatality every 15 min, that they comprise the sixth leading cause of fatality in the United States and the number one cause of death due to injury (Blincoe and Faigin, 1990; National Highway Transportation Safety Administration, 2017).

It is constructive to study and model the causes of vehicle accidents due to the I-I characterizing the transportation Complex SoS, by analyzing the interacting factors that contribute to automobile accident as a prelude to any risk analysis (Treat 1980). Figure 15.II.1 presents a sample of the I-I that characterize the transportation Complex SoS. We develop a framework for identifying significant contributory factors and their interdependencies and interconnections and thus contributions to highway accidents. Current studies reveal that researchers have made an effective use of accident databases, simulations, and crash testing to examine narrowly defined factors that contribute to automobile safety. An evaluation of the contributions of poor driver judgment, vehicle failure, poor weather conditions, and other causative factors offer insight into steering and other failures.

Recall from our earlier chapters that all outputs from a system are functions of the states of that system, and the latter are functions of all decisions, and exogenous and all inputs to the system. This notion is also of particular significance and central to modeling transportation Complex SoS. Recall Chen's (2012) succinct definition of *state variable*: "The state $x(t_o)$ of a system at time t_o is the information at time t_o that, together with the input $u(t)$, for $t \geq t_o$, determines uniquely the output $y(t)$ for all $t \geq t_o$." The states of a system, commonly a multidimensional vector, characterize the system as a whole and play a major role in estimating its future behavior for any given inputs. Thus, the behavior of the states of the system, as a function of time, enables modelers to determine, under certain conditions, its future behavior for any given *inputs*, or *initiating events*. For example, as noted in an earlier chapter, to determine the reliability and functionality of a car, one must know the *states* of the fuel, oil, tire pressure, and other mechanical and electrical components. All systems are characterized at any moment by their respective state and the conditions thereof, and these conditions are subject to continuous variations and fluctuations. Similarly, the states of health of a human are multifaceted, including blood composition and pressure, among myriad others and interdependencies and interconnections exist among the states of all human biological systems.

We model the above vital information that relates the I-I that exist among multiple factors affecting the rate of road accidents in terms of fault trees, which have been introduced in earlier chapters, and a primer on fault trees is presented in the Appendix of this book. Recall that fault trees will enable us to identify the consequences resulting from the I-I that characterize the transportation sector. This basic discovery of the latter enables the identification and thus the tracking of the essence of their complexity as manifested in the common/shared states and other among the systems that compose Complex SoS. (*A sample of shared states and other essential entities* includes states of vehicles and road condition, road congestion, weather, driver skill and situational awareness, road capacity and quality, visibility, among others.)

Fault trees have been deployed in almost all safety-critical systems from nuclear reactors to the space shuttles, and to numerous other cyber–physical systems cited in this book. In all of the above applications, the reliability of each system and the probability of failure of Complex SoS were not central to the efficacious contribution of fault-tree analyses. *The novelty is reinventing the use of fault trees (without neither the need for, nor the reliance on, the reliability of any component of the fault trees) in Chapter 10 of this book, and in our study of accidents in the transportation sector of the economy. Translating the I-I manifested via shared states and other essential entities (that characterize transportation Complex SoS) into AND Gates (systems connected in parallel) and into OR Gates (systems connected in series)*, and doing so without the need for having any access to the reliability nor of each system nor of the transportation Complex SoS. *More specifically, we are able to translate knowledge on shared states and other essential entities (within and among the systems that constitute the transportation Complex SoS) into quantitative metrics via "OR" and "AND" Gates.*

One other important advantage of building on fault-tree analysis is to identify potential critical weaknesses associated with the transportation Complex SoS. Critical paths to failure are discovered using minimal cut sets. A minimal cut set is defined as the smallest combination of component failures, which, if they all occur, will cause the top event to occur; namely, a failure (US Nuclear Regulatory Commission, 1981). By definition, a minimal cut set is a combination of intersections of primary events in parallel sufficient for the top event to occur (if all parallel components fail). This grouping represents the "smallest" combination of systems where all the failures in the minimal cut set are needed to occur for the top event (system failure) to occur. A fault tree consists of a finite number of minimal cut sets, all of which are in series, which are unique for the top event to occur. Since the combination of all minimal cut sets is in series, then the failure of any cut set will cause the failure of the entire system. *In other words, once the minimal cut sets are known, then any system can be written as the series arrangements of its cut sets, and the components of each minimal cut set (itself) are arranged in parallel.* The minimal cut sets are vital in projecting future unforeseen consequences from current decisions associated with any configuration of physical systems, especially of safety-critical transportation Complex SoS (Haimes, 2016).

This part provides an effective modeling schema of transportation Complex SoS that captures the I-I among human, vehicular, and environmental factors, all of which together contribute to the occurrence of accidents. Using these descriptive fault trees, a methodology is developed to aid automakers in their attempts to focus on and then adopt crash avoidance technologies that increase automobile safety.

In the following sections we develop a framework that would permit the (i) evaluation of design attributes on accident causation, (ii) prescription of possible design alternatives to improve safety, and (iii) subsequent analysis of competing design options. The automotive industry heavily relies on models, such as fault trees, for quantifying the I-I exemplifying cyber–physical infrastructure Complex SoS. The chief advantages of fault-tree analysis have been extensively discussed and used throughout this book. Fault trees provide a structured methodological approach to the study of failures in Complex SoS, thus, avenues with which to avoid or control them. Upon the completion of a fault-tree structure, quantitative analysis can readily provide a statistical depiction of the causal aspects of failure and a measure of the relative contribution of each element of fault tree to overall accident occurrence.

15.II.2 A FUNCTIONAL DECOMPOSITION OF THE HIGHWAY TRANSPORTATION COMPLEX SYSTEMS OF SYSTEMS

Consider the overall function of the driver-vehicle-roadway as Complex SoS, which is to transport passengers and goods and services from one location to another and to achieve the above in a manner that allows for flexibility of origin and destination points, efficiency of energy and time, ease of use, and comfort and safety of people and items in transit. This is only one of the many possible characterizations of the transportation Complex SoS.

Accident causes may be modeled in a variety of ways; for pedagogical illustration, we use a single driver in a single vehicle. Other vehicles are considered a subset of the roadway environment, whereas other drivers are not modeled. We begin with a functional description of the three subsystems: *environment*, *driver*, and *vehicle*, which comprise the greater transportation Complex SoS. Included in these functional descriptions is a brief explanation of the operating conditions of each subsystem and a discussion of the significant interfaces among them.

Consider the multiple competing and conflicting objectives characterizing the roadway transportation Complex SoS. They provide comfort, speed, and a safe and informative environment between the end points of transit that is amenable to the needs of a moving vehicle and its driver. For our purpose, the roadway system is comprised of two subsystems. One supports the motions of the vehicle, and the other provides information to the driver. The roadway system is expected to perform its intended functions in a wide variety of weather and traffic conditions. Exceptionally severe conditions, such as dense fog or gridlock, are the extremes at which the environment might be expected to fail and adversely affect a vehicle's safe operation. Unless otherwise noted, other vehicles, which are considered part of the roadway environment, shall be expected to follow the law and perform according to historical traffic motions.

The operating environment of a vehicle consists of an empty space over a traction surface in which physical forces support movement and maneuverability at a variety of speeds. Primary interaction between a vehicle and the roadway system is through the frictional contact of the tires with the road surface. Secondary interaction between the environment and vehicle takes the form of gravity, wind forces, and momentum. A third type of interaction between the environment and vehicle takes place in the space within which the vehicle occupies and must be modeled by the presence of other objects in the roadway.

Failure of the roadway system's surface, which provides proper traction, can occur due to a poor condition, or to natural or artificial influences. Poor condition of the surface includes holes and ruts or pavement with a poor coefficient of friction. This condition arises due to poor design, lack of maintenance, or wear and tear. Natural influences might include water, snow, dirt, rocks, or ice on the roadway; artificial influences include oil, other chemicals, and loose materials.

The transportation Complex SoS must also be designed to make allowances for gravitational and wind forces that affect vehicles in a number of operating conditions. Even if a vehicle is moving in a reasonable manner, any of these forces may adversely affect its safety. Extremely steep hills and failure of support structures can lead to gravitational failures. Similarly, wind failures are most apparent on bridge spans and in poor weather. Sharp curves or short acceleration and deceleration lanes are design faults that may not allow for the momentum of the vehicle. The above forces contribute to the intricacy of the transportation Complex SoS.

Another responsibility of the overall transportation Complex SoS to the vehicles is the provision of space in which vehicles can operate safely. Most typically, failures in this category are caused by other vehicles, pedestrians, or other animate objects on the road. We consider the following to be a fault of the roadway system when it inherently causes such an event, e.g. traffic congestion, and when the occurrence is accidental, e.g. deer crossing the road. Another type of failure to perform intended functions is a simple lack of space in the roadway design. This includes lack of shoulders, restrictive lane widths, and poor exit and entrance ramp geometries; failure of the traffic control system is included in this category.

The second subsystem of the roadway is the one that provides both natural and artificial information to the driver. Natural information is the raw, unenhanced data that is conveyed through direct observation of the roadway. Natural information that the roadway should provide includes the location of roadway components and users, e.g. edge of road or lane, other roads, and other vehicles. We define artificial information to be that which the roadway presents to the driver to facilitate navigation and driver awareness, such as vehicle routes and current roadway conditions. Interasction between the roadway system and the driver is nearly always filtered in some respect by the vehicle. Filtering is primarily through the tires, the passenger compartment, and the windows – although minor effects of the environment can also be felt through the steering wheel and pedals.

15.II.2.1 Drivers as Complex Systems of Systems

Much of the difficulty in properly modeling traffic accidents stems from human factors, e.g. defining the causes for and how the driver fails. Many studies and databases of accidents offer only vague descriptions of failure, but a proper delineation of possible driver failures requires a thorough understanding of the functions of the driver in the highway transportation system (Fell, 1974). Here, we submit that the function of the driver is to act on the steering and speed controls of the vehicle in order to travel under given environmental and vehicle conditions while maintaining operating conditions for themselves and the vehicle. As mentioned earlier, the interactions of the environment and the driver are nearly always filtered through the vehicle. In this sense, the driver's linkage with the environment will consist mostly of his positioning of the vehicle. The driver's influence over the vehicle takes place through a limited number of controls. These include steering and speed interfaces, passenger compartment controls, and extravehicular controls, e.g. blinkers, headlights, and horns.

The driver's perception system collects relevant natural and artificial information from the environment and vehicle. Three further subsystems are involved in this pursuit: the *visual*, *auditory*, *and tactile* systems. The visual system is expected to perform correctly in a given range of brightness, contrast, and distance. The auditory system is expected to relate sounds of loud enough volume, in a certain frequency, and of distinctive pitch. The tactile system operates through touch of the hands, feet, and lower torso of the driver and detects vibrations and other forces that affect the steering wheel and seat. Furthermore, the driver's comprehension system is expected to function properly when presented with correct information from the perception system. Its purpose is to convey and translate present and future situations for decision consideration. The comprehension system pieces together perceived information, knowledge, and experience.

The decision system functions to develop appropriate plans and to maintain operating conditions for the driver, vehicle, and environment. In other words, the decision system is responsible for choosing a suitable course of action. The operating conditions of the decision system include proper comprehension of the traffic situation and reasonable time to think and react. Part of the decision system is to generate alternative options, which formulate the different possible courses of action. A second part of the decision system is choice; its purpose is to select alternatives that match the situation and maintains operating conditions. All of the above human factors further contribute to the complexity of the transportation Complex SoS.

The action system relates its output entirely through the body of the driver. Controls are manipulated by the hands, fingers, and feet. Orientation of the perception system is governed by the movement of the driver's body and head. Maintenance of the driver's physical and mental capabilities is chiefly a responsibility of the driver himself and is governed by go/no-go actions.

15.II.2.2 The Vehicle Complex Systems of Systems

The function of the Vehicle Complex SoS is to physically transport passengers in safety and comfort according to the directional and speed instructions given to it by the driver. Furthermore, the vehicle is responsible for other tasks, such as passenger compartment conditions and enhancement of the roadway environment, e.g. headlights and horns. A vehicle plays a large role in supporting its own operating conditions and a smaller role in maintaining the operating conditions of the driver. Perhaps most importantly, the vehicle must operate in a manner that is safe and consistent with the movement of other objects or vehicles in the roadway system. This implies that the vehicle must share the responsibility of the roadway environmental system by keeping spaces, or paths for properly positioned vehicles and objects inviolate. Failure to uphold this last responsibility is the definition of an accident (Kuzminski et al., 1995).

We consider the following five significant subsystems of the transportation Complex SoS that contribute to the vehicle's responsibilities: the (i) directional system, (ii) speed governance system, (iii) environment–driver interface control system, (iv) passenger compartment control system, and (v) vehicle artificial information provision system.

Directional System

The directional system transforms the actions of the driver upon the steering controls into the directional movement of the vehicle. This interaction contributes to both the completion of the driver's plans and the avoidance of objects in the roadway environment. Acting as inputs to the system are the orientation of the steering wheel and position of the gearshift, and acting as output is the orientation of the tires. The steering system is expected to achieve or maintain certain changes in course within a corresponding range of momenta, frictional, gravitational, and wind forces on the vehicle and in the absence of other objects blocking the path of the vehicle.

Speed Governance System

The speed governance system of the vehicle transmutes the actions of the driver upon the speed controls into the actual speed of the vehicle. In special cases, such as cruise control, or antilock braking systems, it is also this system's function to regulate speed independent of the driver's actions. Similar to the directional system, the speed governance system also helps accomplish the driver's plans and aids in avoiding objects in the roadway environment. Actions of the driver upon the accelerator, brake, gearshift, and cruise control regulator comprise the inputs to this system. Output takes the form of distance covered in time. The speed governance system is expected to achieve and maintain certain speeds within a given time period and over a certain distance within a corresponding range of momenta, frictional, gravitational, and wind forces on the vehicle.

Environment–Driver Interface System

The environment–driver interface system is the collection of devices on the vehicle, which aid in the transfer of information from the environment to the driver. The vehicle supports the roadway environment's responsibilities to provide natural and artificial information to the driver. The windows and mirrors of the vehicle are the primary media for conveying visual information to the driver, and both regulate brightness to some extent. Vehicle headlights, wipers and wiper fluid, and defoggers enhance the information collection process. The steering column, seat, floor, and body of the vehicle all relay environmental information to the tactile perception system of the driver. The radio is available for auditory information if the roadway environment includes transmission of relevant signals, e.g. a travel advisory frequency.

Conversely, the driver often signals the surrounding roadway environment, usually to make other drivers aware of his intentions or presence. Signals are either light-or sound-based. Vehicle devices intended for this use include the parking, braking, and reverse lights, the horn, and the blinkers. Proper functioning of these devices contributes to the comprehension system of the driver, because the knowledge that other drivers are aware of his own actions and position aids in predicting future situations, which is part of the driver comprehension function. Figure 15.II.2 presents a top-level OR Gate fault tree.

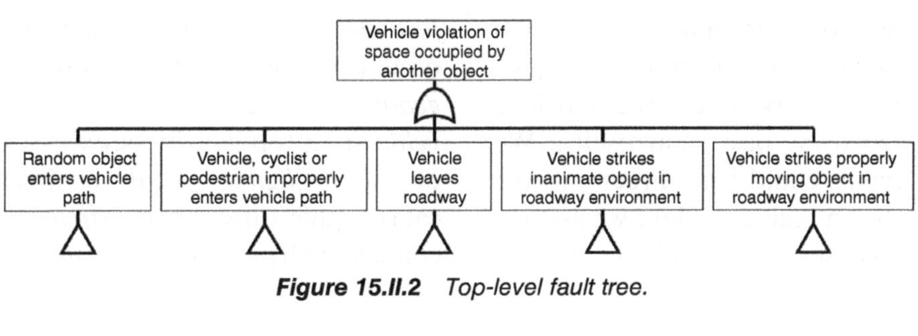

Figure 15.II.2 *Top-level fault tree.*

Passenger Compartment Control System

The passenger compartment control systems also support the operating conditions of the driver. It provides a reasonably stable environment in which the driver can be seated comfortably for periods of time. Simple access to controls and the absence of unreasonable distractions are necessary for the performance of his actions. Devices that support these conditions include the seat and its controls, the dashboard and steering column position controls, and the ventilation systems such as windows, vents, heating, and air conditioning.

Vehicle Artificial Information Provision System

The vehicle provides information about itself to the driver. The dashboard includes monitors of speed, engine temperature, fuel, oil, battery levels, RPM, mileage, and passenger compartment temperature. All of these information signals are usually light based. An autonomous car (also known as a driverless car, self-driving car, robotic car) and unmanned vehicles (a vehicle that is capable of sensing its environment and navigating without direct human input) are evolving, but as of 2018 no cars in the United States permitted on public roads were fully autonomous. They all require a human at the wheel, who must be ready to take control at any time.

15.II.3 ACCIDENT CAUSATION FAULT TREE

Having identified the functional responsibilities for the (i) environment, (ii) vehicle, and (iii) driver, we can now develop a structure for relating the I-I among the three systems to a failure model via fault-tree analysis. The resulting fault tree represents the intuitive and physical linkages, namely, the causal interactions that exist between system components due to the I-I characterizing the transportation Complex SoS.

Creation of a fault tree is a top-down process. We begin by defining the top-level event, *an accident*, as a vehicle's violation of space occupied by another object. In this sense, we consider an accident to be a collision with another vehicle, a pedestrian, an object, or a component of the roadway system (i.e. a bridge support or even the roadway pavement itself). We then consider what events could lead directly to the top-level event, namely, an accident. Here we assume the following five causes of accidents to be comprehensive: (i) *random object enters vehicle path (deer, falling rocks, etc.); (ii) vehicle, cyclist, or pedestrian improperly enters vehicle path; (iii) vehicle leaves roadway; (iv) vehicle strikes inanimate object in roadway environment (island, sign, stopped vehicle); and (v) vehicle strikes properly moving object in roadway environment.*

At least one of these events needs to happen for the vehicle to experience a collision accident, namely, the top event occurs. Thus, the top-level event of the tree occurs through the output of an OR Gate as shown in Figure 15.II.3.

Each of the five intermediate events contributing to the top-level event is similarly decomposed. Repetition of this process leads to a detailed hierarchy of contributing events. However, only the event vehicle leaves roadway (shown at the center of Figure 15.II.2) will be explicitly decomposed and quantified. The results of the complete fault-tree modeling and application can be found in Figure 15.II.3 (Haimes et al., 1994).

The intermediate event *vehicle leaves roadway* is a demonstrative case for the creation of the rest of the fault tree. We recognize that the vehicle's movement is governed by its momentum and four forces: (i) the friction of the road with the tires, the weight of the vehicle, (ii) the associated normal force of the road, and (iii) wind forces. Two systems that we have outlined in Figure 15.II.3 govern the momentum of these forces. They are the environmental support of vehicle's motion and the vehicle's directional system. Failure of one of these two systems to perform properly would lead to incorrect positioning of the vehicle, possibly off the roadway. In examining the *vehicle leaves roadway* subtree shown in Figure 15.II.3, we first consider the event *failure of environmental support system* – the system which is responsible for providing suitable frictional and gravitational support, negligible wind forces, and open space suitable for vehicle maneuvers. *Failure of the frictional support system* may occur in the event of *poor road surface condition, bad weather surface condition* (e.g. ice, snow, rain), or *poor road surface design. Failure of gravitational support system* occurs with significant frequency only when *poor road geometry* is coupled with *gravitational stress from the vehicle* (e.g. top-heaviness, excessive weight). The event of *nonnegligible wind conditions* tends to be catastrophic only when *extreme wind conditions* are coupled with *poor roadway support design* and *vehicles*

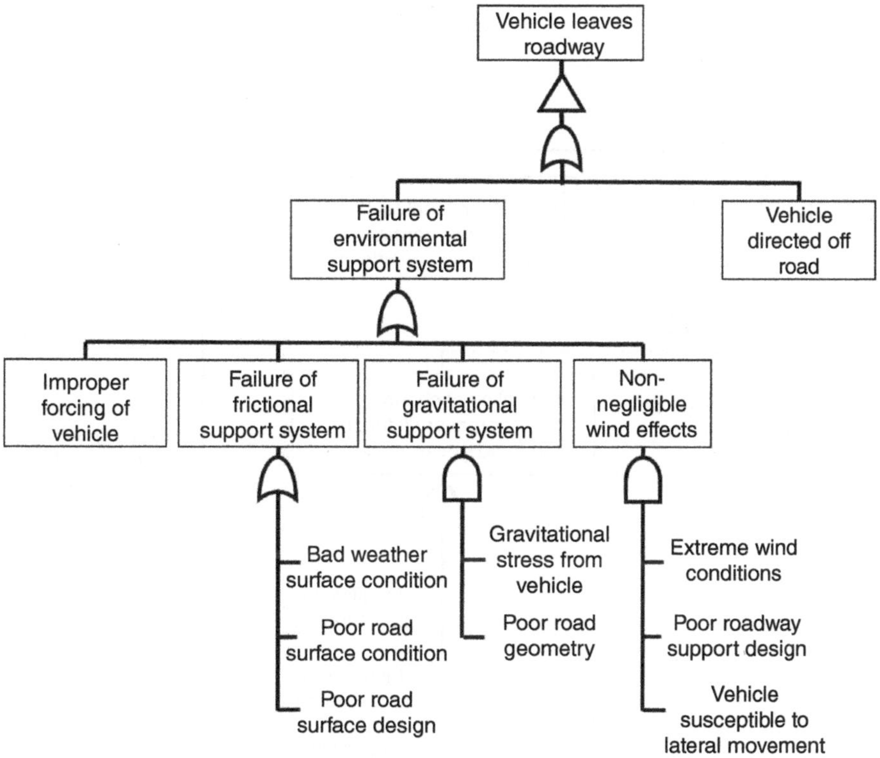

Figure 15.II.3 *Vehicle leaves roadway subtree.*

susceptible to lateral movement – extremely light vehicles.

A last liability of the roadway environment is the *improper "forcing" of the vehicle off the road.* For example, an improper lane change by a larger vehicle could well be considered a type of force in which the driver has no choice but to maneuver his vehicle off the road. We distinguish between this event and the event *vehicle, cyclist, or pedestrian improperly enters vehicle path* appearing higher in the tree. The latter case is that it would cause a collision, while the former case is a type of force equivalent to the environment not fulfilling its open space responsibilities.

If the environment is performing its intended functions properly, the vehicle would leave the roadway only if it is directed off the road by its directional system. The event of *vehicle directed off road* occurs if one of three events occurs: *driver-steering error, directional fault under normal operating conditions* (Figure 15.II.4).

We consider *directional fault under normal operating conditions* to be a basic vehicle-failure event. Further, we regard driver-steering error as a subset of the more general event driver action error, which is described in the next subsection. Finally, *directional fault beyond normal operating conditions* can take two forms: *(i) insufficient change of direction over distance*, or *(ii) excessive change of direction over distance*. While we consider the latter a basic event, the former is due to the condition of excessive momentum or inertia overtaking the grip of the tires.

We also consider excessive momentum to be the responsibility of the speed governance system (forces imposed by environmental conditions were considered above) and the mass of the vehicle to be the other component of momentum to be uncontrollable. Failure of the speed governance system is assumed to occur due to one of three events: *insufficient acceleration*, *insufficient deceleration*, and *excessive speed by driver*. The failure of driver action will be considered later. The normal operating conditions of the speed governance system consist of reasonable slope (for gravitational pull), reasonable traction (for frictional forces), reasonable time, and reasonable weather conditions. For the speed governance system to provide excessive momentum, it must attempt to increase speed too quickly, or be unable to decrease speed sufficiently. *Excessive speed by driver*, on the other hand, is a failure beyond normal operating

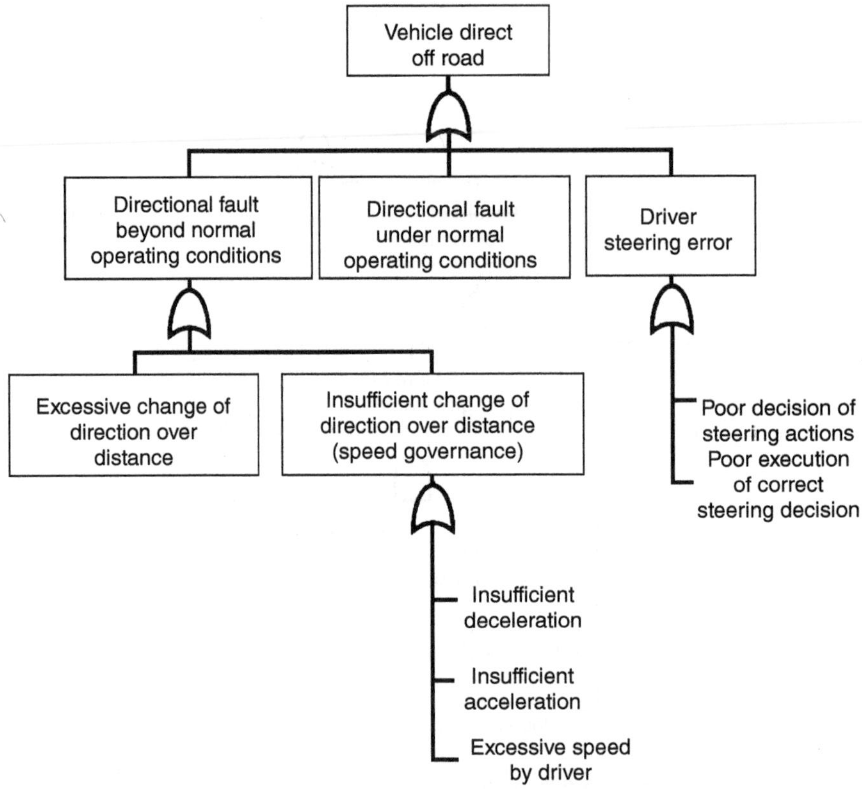

Figure 15.II.4 *Vehicle-directed off-road subtree.*

conditions, i.e. excessive speed (not an environmental/vehicular failure, which has inhibited lateral movement and prevents the vehicle from maintaining a proper position within the roadway). This part of the tree thus considers failures under normal conditions as well as those that occur while operating beyond normal conditions.

Failure of the driver to act in a correct manner is an event that can be decomposed, in general, with modifications for specific subtrees (e.g. steering action error, speed-governance action error). The top of this subtree, as shown in Figure 15.II.5, is the event *driver action error.*

This fault is due to one of two immediate events: *poor execution of correct decision* or *poor decision of course of action.* Poor execution with correct input is a failure of the action system. This is due to *execution error under normal operating conditions*, or to *execution error beyond operating condition.* The former is due to *inexperience* or *impairment*, and the latter is due to some *disruption of the driver's body* or *lack of suitable execution time.*

Poor decisions are failure of the decision system to process available input. This is due to a *poor decision-making process*, or a *poor understanding of the situation.* The former is caused by *inexperience, impairment, or lack of suitable decision time.* Poor understanding of the situation is caused by a *poor understanding of the driver's own vehicle* or a *poor understanding of environment* (Figure 15.II.6). Poor understanding of the situation leads to a poor decision when the decision system is given incorrect inputs.

To further decompose poor understanding of the situation, we divide *poor understanding of own vehicle into* poor understanding of three types of important information: the (i) position, (ii) velocity, and (iii) future position of the vehicle. Poor understanding of each will occur if *either correct artificial information is not provided or correct natural information is not provided.* Poor understanding of the position, velocities, and future position of objects is similarly used to detail the event *poor understanding of environment*, which is decomposed in Figure 15.II.7.

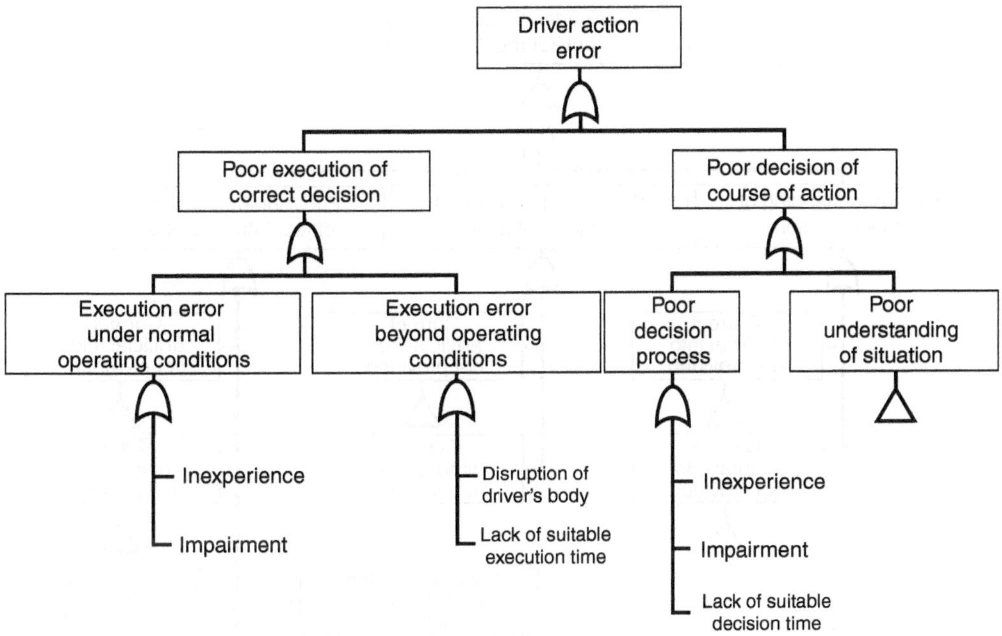

Figure 15.II.5 *Driver action error subtree.*

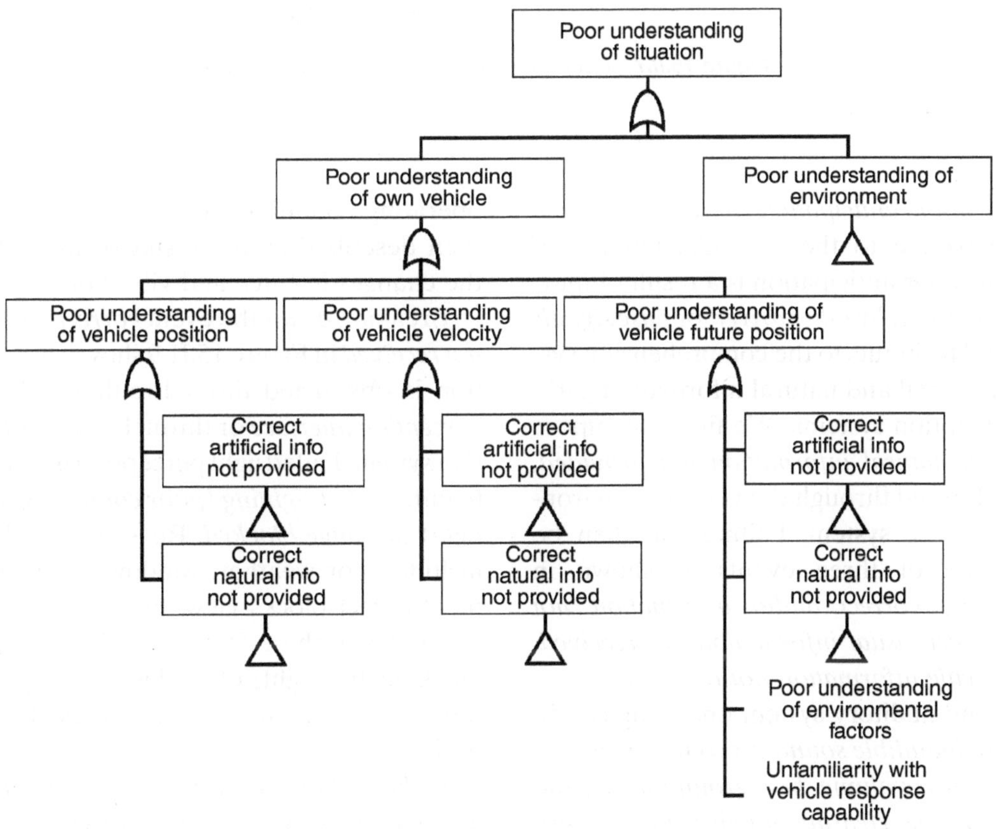

Figure 15.II.6 *Poor understanding of situation subtree.*

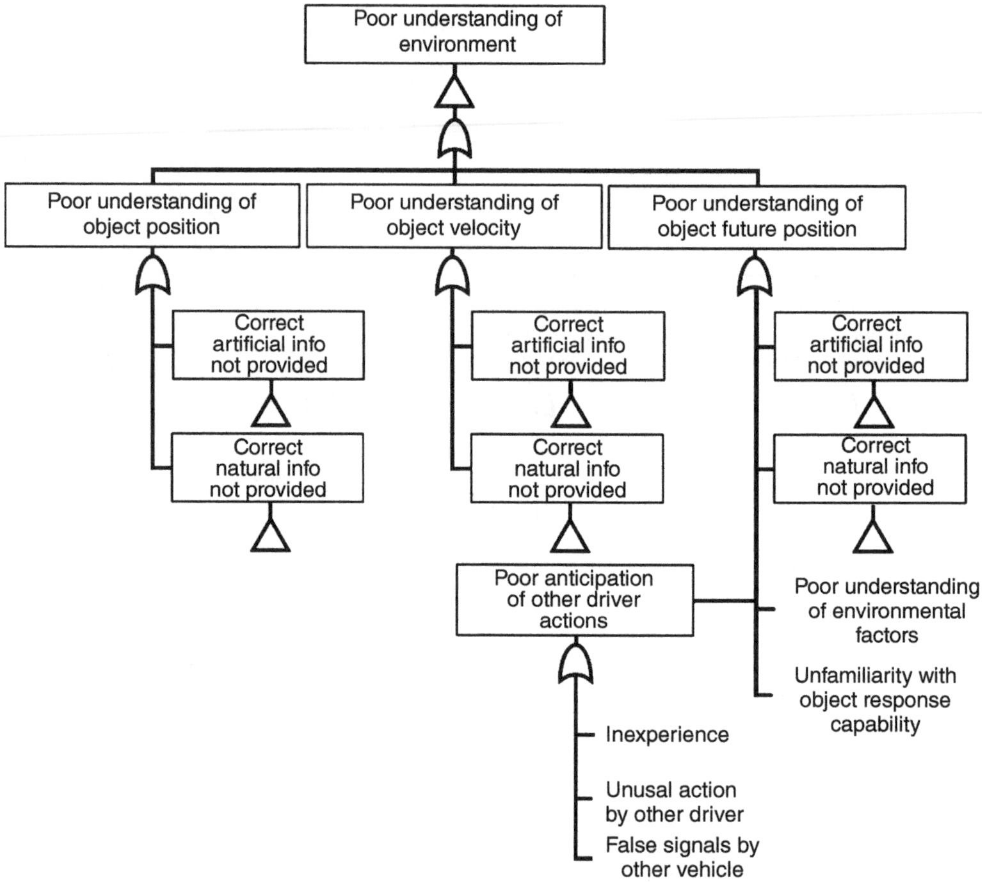

Figure 15.II.7 *Poor understanding of environment subtree.*

In addition, *poor anticipation of other drivers' actions* will contribute to the misunderstanding of future situations. Poor anticipation is a result of *inexperience, unusual action by other driver*, or *false signals by other vehicle.* The inputs to the comprehension system consist of artificial and natural information gathered by the perception systems. We first decompose the event *correct natural information not provided,* which is always filtered through the vehicle's environment–driver interface system. Failure can then be attributed to one of three events as shown in Figure 15.II.8: *(i) correct audio information not received, (ii) correct visual information not received,* or *(iii) correct tactile information not received.*

Failures beyond auditory system operating conditions in which an *inaudible sound* is produced include the causal events *low volume, background noise,* and *out of frequency range. Hearing impairment* is considered a failure of the auditory subsystem within operating conditions. Failure of the tactile system is then described as the result of the vehicle masking the changes in force and vibration.

Breaking down the event *correct visual information not received* in Figure 15.II.9 shows that visual information is obstructed through either *failure of the light perception interface* or through *failure of visual perception system.* Failed light perception interface occurs due to either *poor lighting* (*poor contrast, glare,* or *too little light*) or *image blocked.* Reasons for a blocked image include poor weather, window opaqueness, environment object blocks view, or *vehicle part blocks view.* Examples of these failures might include rain or dust blocking the light, obstacles blocking the light (e.g. other vehicles, overhanging trees, hills), or simple darkness.

Failure of the visual perception system may be due to *poor focus* or *wrong field of vision.* The former is caused by *temporary impairment, permanent*

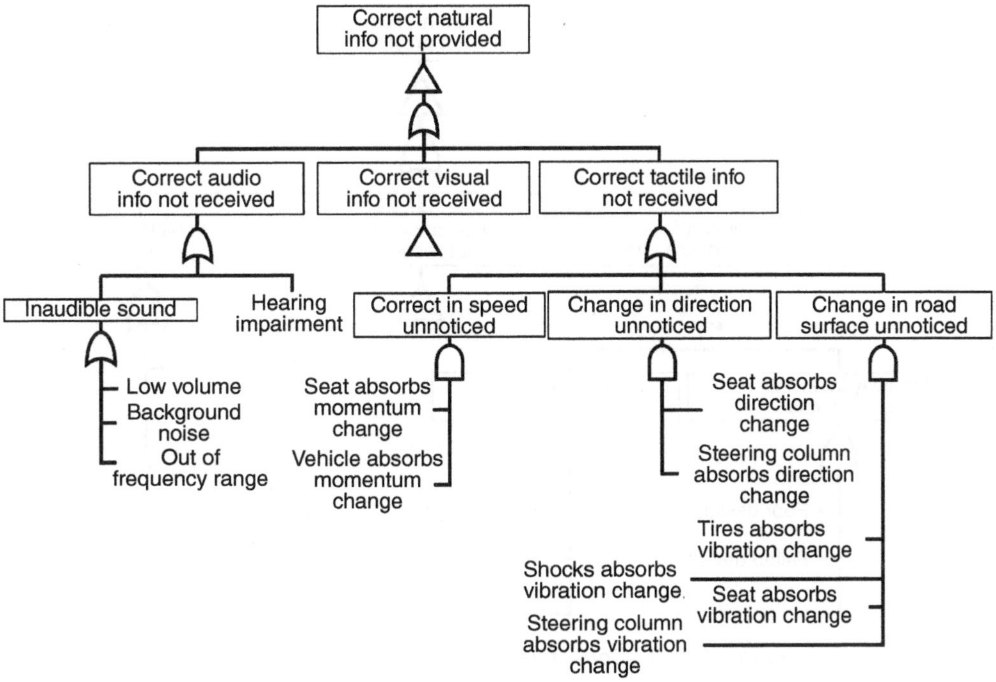

Figure 15.II.8 *Correct natural information not provided subtree.*

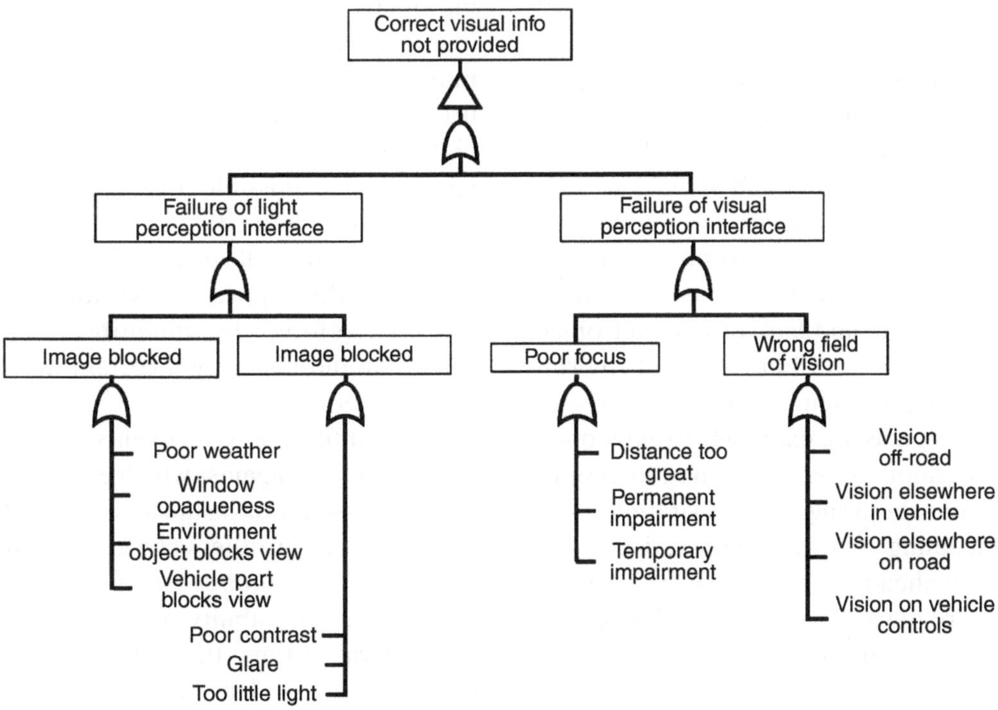

Figure 15.II.9 *Correct visual information not received subtree.*

impairment, or a *too great distance*, while the latter is caused by *vision on vehicle controls*, *vision elsewhere on road*, *vision elsewhere in vehicle*, or *vision off-road*.

From the poor understanding of the *environment subtree* of Figure 15.II.7, we next decompose the event "correct artificial information not provided." In

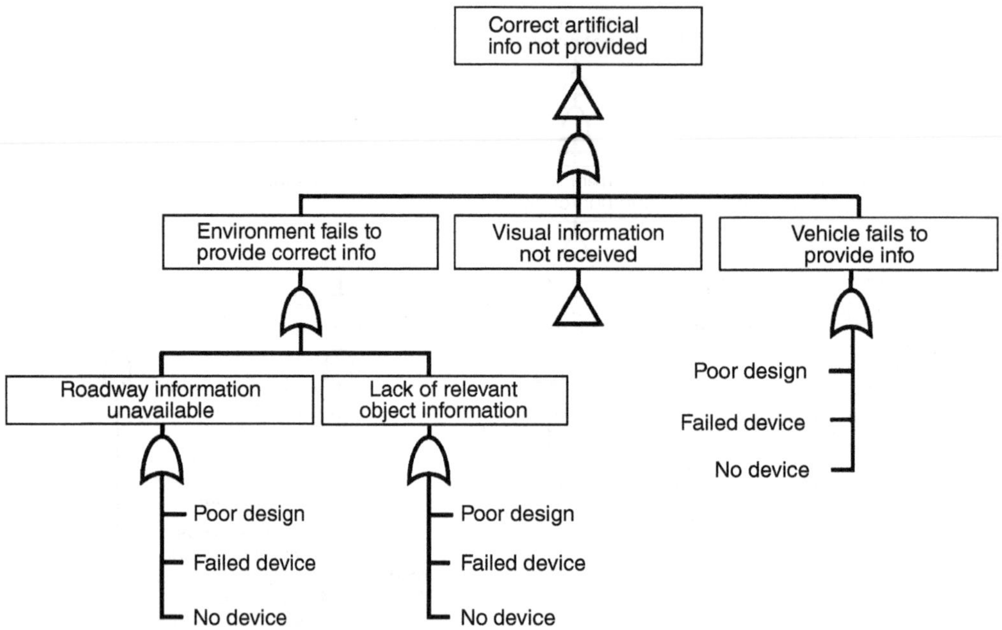

Figure 15.II.10 *Correct artificial information not provided subtree.*

Figure 15.II.10, *environment fails to provide correct information; visual information not received; vehicle fails to provide information.*

These failures are similar to failures in providing natural information, except that failures of the perception system beyond normal operating conditions are instead due to the environment and vehicle artificial information provisions. In the case of environmental information failure, these basic events specifically describe how the intermediate events – roadway information unavailable, and lack of relevant object information – occur and contribute to failed environmental information provision. Poorly located signs and lane direction signals are examples of inadequate information concerning imminent routing directions. Lack of information about suggested operating conditions due to upcoming roadway characteristics (e.g. speed limits, curve ahead signs, slippery bridge signs) are faults of the roadway system. In all cases, poor design, failed device, and no device are fundamental causes.

15.II.4 PARAMETERIZATION OF THE MODEL

A primary difficulty with quantifying inputs to the fault-tree model is the limited level of details that the model can meaningfully provide without becoming overly complex. Fault-tree analysis leads to an identification of the failure modes, which contribute most to certain types of traffic accidents. These failure modes can be further decomposed to obtain a high resolution examination of significant causative factors. The possible occurrence of basic causal events depends on the particular situation of concern. To include all such situations in the qualitative model would require an extraordinarily large fault tree. Therefore, the quantification of the given model should abide by a certain number of guiding parameters.

To quantify the basic events, we need to establish some standards against which each probability of failure is assessed. For instance, consider the basic event of false signals by other *vehicle in the poor understanding environment subtree* (see Figure 15.II.7). To make sense of the probability of this event, we must know the interval of time, the distance, or over the course of other event, such as the probability of a false signal. Furthermore, do we want the probability of a false signal in 10 s? Or when 50 yards is traveled? These standards must be established in order for the probability of the event to have meaning.

We assess the probability of failure events against the background of the vehicle's maneuver.

Table 15.II.1 lists one feasible set of movements. The analyst needs to assess the probability of failure occurring during one of the sample of 16 maneuvers. These standard events are judged to be significantly different from one another. Accordingly, the probabilities of the basic failure events occurring during each should differ.

Furthermore, there are conditions of the transportation Complex SoS that will affect the assessment of probabilities of basic (failure) event. We divide these conditions into four categories: (i) roadway characteristics, (ii) weather and light conditions, (iii) driver conditions, and (iv) vehicle conditions. Table 15.II.2 defines the parameters and possible values of roadway characteristics. Table 15.II.3 lists the parameters and values for weather and lighting conditions. Table 15.II.4 lists the parameters of the driver that might affect basic failure event probabilities.

In addition to the parameter set described above, we consider other criteria reflecting the design of the vehicle. Using different vehicle parameter sets would provide values of crash likelihood for specific vehicle designs and would establish a base for comparing various design alternatives. Indeed, the fault tree is applicable to a large number of driving situations. To make use of this capability, the analyst must quantify the likelihoods of the basic events. The quantification of basic events against all standard maneuvers and with all permutations of parameter values is ideal; nevertheless, impossible. Therefore, the analyst and vehicle design engineers must arrive at a set of useful permutations, which simulate and represent situations of concern. While this seems to be a daunting task, further inspection will indicate that only smaller subsets of the parameters influence any one basic event.

TABLE 15.II.1 Standard Vehicle Maneuver Events

Standard events
10 seconds straight forward (backward) movement
10 seconds of gradual forward curve to the right
10 seconds of gradual forward curve to the left
Right turn from forward (backward) movement
Left turn from forward (backward) movement
Straight forward (backward) movement from stop
Forward (backward) right turn from stop
Forward (backward) left turn from stop
Stopping from forward (backward) movement

TABLE 15.II.2 Roadway Parameters

Parameter name	Possible values
Roadway geometry	Straight, curved, percent gradient
Traffic direction	Oncoming, one way
Street designation	Rural, urban, secondary, primary, interstate
Functional location	Segment, acc/dec ramp, bridge, tunnel, controlled i-section, uncontrolled i-section, parking lot, driveway, circle
Traffic congestion	Heavy, light
Traffic speed (mph)	0–20, 21–50, 50+
Maintenance condition	Good, fair, poor

TABLE 15.II.3 Weather and Lighting Parameters

Parameter name	Possible values
Natural light	Dawn dusk, day, night
Artificial light	Lighted, unlit
Weather	Clear, drizzle, heavy rain, snow, fog
Weather effect on road surface	Normal, icy, wet slippery, flooded, snow
Wind conditions	Calm, heavy

TABLE 15.II.4 Driver Condition Parameters

Parameter name	Possible values
Experience	New, moderate, extensive
Age	15–24, 24–50, 50–65, 65+
Gender	Male, female
Familiarity with area	Familiar, comfortable, unfamiliar
Driver attitude	Aggressive, defensive
Permanent physical ability	Able, physically challenged
Temporary physical ability	Intoxicated, exhausted, ill, normal

15.II.5 UNCONDITIONAL PROBABILITY ASSESSMENT

In earlier chapters (most notably, we presented four case studies in Chapter 10) we attributed the following significant modeling virtue and contributions of fault trees to modeling the I-I characterizing Complex SoS by distinguishing between the following two basic interdependencies and interconnections: (i) systems connected in series (OR Gates) and (ii) systems connected in parallel (AND Gates). Furthermore, we did not build on, nor did we have knowledge of, the probabilities (more specifically, the reliability) associated with each connection.

One difficulty in conducting fault-tree analyses stems from the need for the quantification of the probabilities of a basic event – a significant challenge. Systems are not designed to fail, and therefore historical data on failure rate characterization is commonly sparse. Finding a method to assess the probability that a particular causative factor will occur continues to be challenging. For example, accident databases do not commonly provide information on the most causative factors to the level of detail necessary for useful model input. Some quantification, however, can take place at a less detailed level of a fault tree. Also, rather than describing the probability that a driver was inexperienced in a particular steering maneuver, it would be more practical to assess the probability of an intermediate event, such as an improper drivers steering action.

Furthermore, most studies and databases provide information based on situations involving traffic accidents. The probabilities of causative factors that can be obtained from databases are conditional on an accident occurring. In fault-tree analysis, however, the probabilities of occurrences of basic events are not conditional upon a top-level failure having occurred. Our studies have considered several approaches to the generation of input probability statistically. Among the methods that are suggested to augment information available from accident databases are (i) tagging studies, (ii) direct observation, (iii) police reports, and (iv) expert assessment. The example discussed in this part merges statistical information from the Virginia accident database with exposure information obtained therefrom (National Highway Traffic Safety Administration, 2016).

Fault-tree analysis is typically used for the failure analysis of systems that undergo a predictable, usually repetitive, process. In such cases, the occurrence of some combination of faults usually has a predictable outcome. In a traffic situation, however, the basic events, or the contributing faults of the many subsystems, do not necessarily have predictable consequences. For example, the failure of a driver to signal a turn could cause an accident in some cases and would have no consequence in others.

The reason for nondeterminism is that a multitude of other variables may offset the consequence of concern. For example, if an inexperienced driver fails to signal a left turn and then executes such a turn in the path of an oncoming vehicle, we have the ingredients for a head-on collision. However, a number of variables may prevent such an effect. The inexperienced driver may realize his error and abort the turn, or he may speed up to leave the path of the oncoming vehicle more quickly. The other driver may stop in time or alter the path of his own vehicle. To have our model reflect reality to a reasonable degree, we need to consider the random causal effects.

One possible approach is the creation of nondeterministic Gates. These would simply imply that the output event of a Gate occurs with some probability given that the input combination occurs. Alternatively, the connection between the output event and the nondeterministic Gate could be replaced by a deterministic AND Gate whose inputs include the output of the original nondeterministic Gate and a dummy event of an appropriate probability of occurrence.

15.II.6 AN APPLIED QUANTIFICATION OF THE FAULT TREE

The current limitations of the availability of specific data require the quantification of basic events of the fault tree. Thus, several simplifying steps were taken within this case study. The first simplification was the truncation of the fault trees at the level below which information was not practically obtainable. A second simplification was the use of general fleet statistics rather than individual design-based statistics for the quantification of basic event probabilities. The truncation of the fault trees is justifiable in many situations. For example, while it may be informative to decompose a poor driver decision into component causes, such as lack of experience, fatigue, or misunderstanding of situation, the analysis of events further up in the fault tree will be well served as long as the poor driver decision event can be quantified in terms of a probability relative to exposure. Meanwhile, the use of general fleet statistics is necessary at this time because the availability of information focusing on driver's behavior and fault exposure for specific vehicle designs is very limited.

One source of information with detailed pre-crash factors is provided in the Virginia accident database. This database is a compilation of statistics gathered from police reports and currently includes information concerning all automobile accidents occurring within the State of Virginia that involve at least \$1000 in damages or any degree of personal injury. Statistics are generally applicable to the United States because of the large sample size of approximately 100 000 accident reports for each year, which was obtained from a region whose driving patterns do not seem significantly different from those of the United States as a whole. Furthermore, by coupling the Virginia accident database statistics with information concerning the number and type of vehicle miles traveled within the state of Virginia, it is possible to obtain an approximation of the unconditional contribution of the basic events to accident causation. The latest available accident database in the state of Virginia is for the year 2015: There were 126 000 total crashes, 7500 of which were alcohol related, leading to close to 5000 injuries and 240 fatalities. This means that about 30% of all traffic fatalities resulted from alcohol-related collisions.

A truncated fault tree for vehicle leaves roadway is shown in Figure 15.II.11. The differences between this truncated fault tree and the general trees in Figures 15.II.3 and 15.II.4 are that in this case several events left undeveloped due to lack of statistical details.

The degree of truncation for the vehicle leaves roadway available from the Virginia accident database as shown in Table 15.II.5. The database was used to tally the number of accidents occurring in the presence of various causative factors and under specific sets of operating conditions to obtain a basis for the calculation of unconditional probabilities.

15.II.6.1 Quantifying the Fault Tree

Let B_i be the ith basic event and T the top event. Note that the Virginia accident database does not provide the value of Pr be a mile traveled and driving

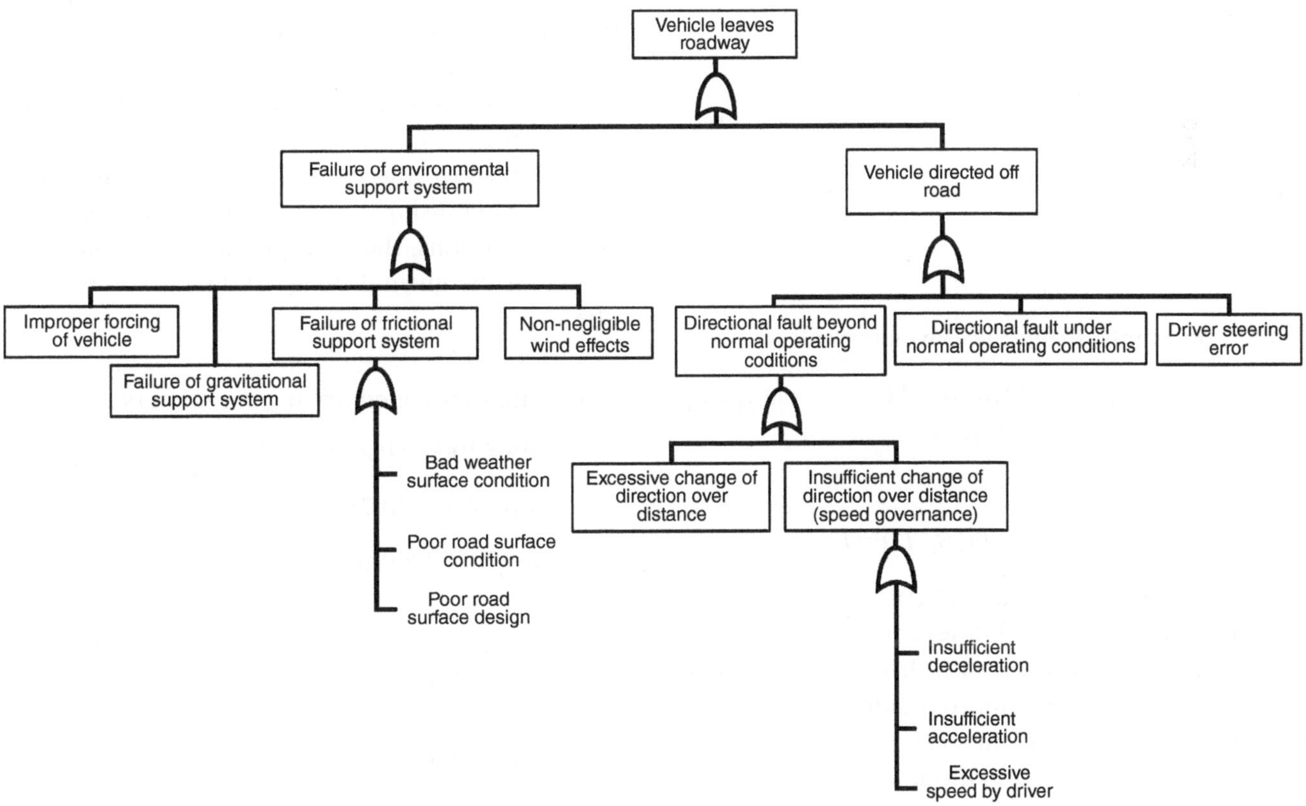

Figure 15.II.11 *Vehicle leaves roadway truncated subtree.*

TABLE 15.II.5 Basic Events and Relevant Virginia Database Fields

Basic event	Relevant Virginia database field
Failure of gravitational support	Alignment: hill crest or dip (col. 45)
Nonnegligible wind	N/A
Bad weather surface condition	Surface condition (col. 47)
Poor road surface condition	Surface condition (col. 47) Roadway defects (col. 48)
Poor road surface design	Surface condition (col. 47) Roadway defects (col. 48)
Driver steering error	Driver action failure (cols. 80–81)
Fault in directional system (normal operating conditions)	Vehicle failure: steering (col. 89)
Excessive change of direction (beyond normal operating conditions)	Crash avoidance: skidding before application of brakes (col. 90)
Excessive speed due to driver action	Driver action failure: exceeded speed limit/safe speed limit (col. 80–81)
Insufficient acceleration (beyond normal operating conditions)	Motor trouble (col. 89)
Insufficient deceleration (beyond normal operating conditions)	Crash avoidance: skidding after application of brakes (col. 90)

parameters for basic events. Fortunately, the fault trees developed earlier link basic events to the top event entirely through OR Gates, or AND Gates, making

$$\Pr[T \mid B_i] = 1 \tag{15.1}$$

for all basic events. From the Bayes formula

$$\Pr[B_i] = \frac{\Pr[B_i \text{ and } T]}{\Pr[T \mid B_i]} \tag{15.2}$$

we have

$$\Pr[B_i] = \Pr[B_i \text{ and } T] = \Pr[B_i \mid T]\Pr[T] \tag{15.3}$$

For simplicity, we stated explicitly in Eqs. (15.1)–(15.3) that all of the probabilities are conditioned on neither miles traveled nor the driving parameters under consideration. To approximate the exposure probability that an accident occurs, it is necessary to find two pieces of information, the (i) number of miles traveled by automobiles, conditioned on a chosen environment, driver, and vehicle parameter configuration (in this case general fleet design); and (ii) number of accidents that occur in those driving situations. It is possible to obtain basic estimates of mileage exposure through sources, which classify annual vehicle miles of travel by highway category and vehicle type. The number of accidents that occur in those driving situations can then be estimated from statistics gathered from the Virginia accident database. Using the parameter set suggested in the previous subsection (male or female driver, driving less than 55 mph, in clear, daylight conditions, on a straight, primary road) in conjunction with the accident type specification that a vehicle leaves the roadway, we can obtain the following relevant statistics to calculate the conditioned accident exposure.

Consider the following statistics for the calculation of Pr (vehicle leaves roadway | driving parameters): (i) number of vehicle roadway departures | driving parameters: 1562; (ii) primary road miles traveled in Virginia in 2 years 39 042 000 000 miles; (iii) % primary straight road miles traveled: 75.0%; (iv) % vehicle trips taken 6:00 am–7:00 p.m.: 80.7%; (v) % driving conducted in clear weather conditions: 75.0%; (vi) % vehicles traveling less than 55 mph on primary roads: 80.0%; and (vii) based on 59.1% less than 55 mph where speed limit = 55 mph.

Although it is unlikely that exposure to the above conditioning events is independent, to facilitate the calculation of overall exposure, that assumption will be made. Thus, using the assumption of independence among conditions, the total primary road miles traveled under the given driving conditions can be found to be

$$\begin{aligned}&(\text{No.miles traveled}|\text{driving parameters})\\&= 39042000000 \times (0.75)\\&\quad \times (0.807) \times (0.75) \times (0.8)\\&= 14178102300 \text{miles}\end{aligned} \tag{15.4}$$

We can now obtain the general equation for calculating the conditioned probability of an accident:

$$\begin{aligned}&\Pr(\text{Accident}\,|\text{driving parameters})\\&= \frac{(\text{Accidents} \mid \text{driving parameters})}{(\text{No.miles traveled} \mid \text{driving Parameters})}\end{aligned} \tag{15.5}$$

Applying this formula to the example, it is possible to arrive at the probability that a vehicle roadway departure accident will occur by using the roadway departure tally listed above and the exposure mileage calculated in Eq. (15.4).

$$\begin{aligned}&\Pr(\text{Vehicle leaves roadway}|\text{driving parameters})\\&=\frac{(\text{No.leaving roadway}|\text{driving parameters})}{(\text{No.miles traveled}|\text{driving parameters})}\\&=\frac{1562\,\text{accidents}}{14178102300\,\text{miles}}\\&=1.10^{*}10^{(-7)}\,\text{Off}-\text{road accidents/primary road mile}\end{aligned} \tag{15.6}$$

15.II.6.2 Obtaining Pr(Basic Event|Accident)

Using the set of standard driving conditions to obtain a sample space (male or female driver, driving less than 55 mph, in clear, daylight conditions, on a straight, primary road), it is possible to search the Virginia accident database for the number of times each basic event is marked as a factor associated with the cause of an accident type. The statistic Pr(Basic event|accident and driving parameters) can generally be found as

$$\begin{aligned}&\Pr(\text{Basic event}|\text{accident and driving parameters})\\&=\frac{(\text{No.basic event occurrence}|\text{accident and driving parameters})}{(\text{No.accident occurrence }|\text{driving parameters})}\end{aligned} \tag{15.7}$$

For example, the probability that a poor steering action was made in an accident involving a roadway departure is found to be

$$\begin{aligned}&\Pr\left(\begin{matrix}\text{Poor steering action}|\text{ vehicle leaves roadway}\\ \text{and driving parameters}\end{matrix}\right)\\&=\frac{\left(\begin{matrix}\text{No.poor steering action observations}|\\ \text{vehicle leaves roadway and driving parameters}\end{matrix}\right)}{(\text{No. roadway departures}|\text{driving parameters})}\\&=\frac{821\,\text{accidents}}{1562\text{ accidents}}\\&=0.525\end{aligned} \tag{15.8}$$

The probability shown in Eq. (15.8) is effectively the rate at which poor steering contributes to the type of accident under consideration.

*To obtain the Pr (*basic event), we use Eqs. (15.3), (15.5), and (15.7) to arrive at the probability for basic events. For example, combining Eqs. (15.6) and (15.8) yields

$$\begin{aligned}&\Pr(\text{Poor steering action})\\&=\Pr(\text{Poor steering action and a roadway departure accident})\\&=\Pr(\text{Poor steering action}|\text{roadway departure acc.})\\&\quad\times\Pr(\text{Roadway departure acc.})\\&=0.525\times1.10\times10^{-7}\\&\quad=5.5775\times10^{-8}\text{ per mile of parameter travel condition}\end{aligned} \tag{15.9}$$

The probability obtained in Eq. (15.9) implies that there are 5.78 steering-related accidents for approximately every 100 000 000 miles of travel under the parameter driving conditions. Repetition of these procedures will lead to estimates of the likelihoods for all basic events in the fault trees developed earlier. Table 15.II.6 displays the results of this analysis.

15.II.6.3 Representing Uncertainty

It could be argued that there is a significant degree of uncertainty about the obtained statistics. To address this problem of uncertainty, we use a distribution to represent the range of possible input values. The triangular distribution is an effective and easily applied distribution, requiring the assessment of only three values: a lowest likely, most likely, and highest likely probability. Using the basic event probabilities shown in Table 15.II.6 as a basis for distribution assessment (the most likely value), it is possible to formulate a procedure for creating a distribution about that single point. In our analysis, we use a distribution of basic event probabilities ranging from 95% to 110% of the calculated values to reflect the uncertainty in obtaining the vehicle exposure and the skew toward higher exposure probabilities (due to the use of accidents involving injury or more than \$1000 in damages). This would imply that the probability of occurrence of a poor steering action causing a roadway departure, which had previously been calculated to be 5.78×10^{-8} per mile, would be represented in terms of a distribution as shown in Figure 15.II.12.

TABLE 15.II.6 Roadway Parameters and Relevant Virginia Database Fields

Parameter name	Possible values	Relevant Virginia database field
Roadway geometry	Straight, curved, percent gradient	Alignment (col. 45)
Traffic direction	Oncoming, one-way	Kind of highway (col. 33)
Street designation	Frontage, secondary, primary, interstate	Route (cols. 18–21)
Functional location	Segment acc/dec ramp, bridge, tunnel, controlled i-section, uncontrolled i-section, parking lot, driveway, circle	Accident location (col. 42) traffic control (cols. 43–44)
Traffic congestion	Heavy, light	N/A
Traffic speed (mph)	0–20, 21–50, 50+	Speed (cols. 73–75, 93–95)
Maintenance cond.	Good, fair, poor	Roadway defects (col. 48)

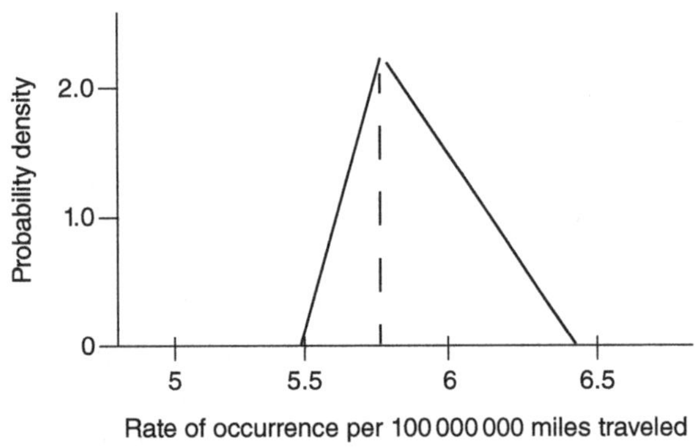

Figure 15.II.12 Unconditional probability of poor driver steering decision.

15.II.6.4 Solution of the Fault Tree

Once the basic event probabilities are obtained and the fault tree is described in terms of probability distributions for each of those basic events, a Monte Carlo simulation tool, DARE (Haimes, 2016), is used to solve the vehicle leaves roadway subtree in Figure 15.II.11. The basic events and the components of their distributions are shown in Table 15.II.7.

Table 15.II.8 displays the results of the simulations in terms of the 5th percentile, mean, and 95th percentile distribution values obtained for the failure rates of intermediate and top-level events.

TABLE 15.II.7 Basic Event Probability Quantifications

Event	Least likely failure rate	Most likely failure rate	Largest likely failure rate
Failure of gravitational support	0.004 49	0.004 73	0.005 20
Nonnegligible wind	0.000 4	0.000 4	0.000 4
Bad weather surface condition	0.005 56	0.005 85	0.006 43
Poor road surface condition	0.002 35	0.002 47	0.002 72
Poor road surface design	0.001 34	0.001 41	0.001 55
Poor steering action	0.055 01	0.057 15	0.063 69
Improper forcing of vehicle	0.001 34	0.001 41	0.001 55
Fault in directional system (normal operating conditions)	0.001 87	0.001 97	0.002 17
Excessive change of direction (beyond normal conditions)	0.012 53	0.013 19	0.014 51
Excessive speed due to driver action	0.008 24	0.008 67	0.009 54
Insufficient acceleration (beyond normal conditions)	0.000 40	0.000 42	0.000 46
Insufficient deceleration (beyond normal conditions)	0.033 90	0.035 69	0.039 26

TABLE 15.II.8 Intermediate and top-level event probability quantifications.

Event	5th percentile failure rate	Mean failure rate	95th percentile failure rate
Failure of frictional support system	0.0096	0.0099	0.0103
Insufficient change of direction over distance	0.0613	0.0635	0.0660
Directional fault beyond normal operating conditions	0.0748	0.0769	0.0795
Failure of environmental support system	0.0161	0.0165	0.0170
Vehicle-directed off roadway	0.1340	0.1377	0.1420
Vehicle leaves roadway	0.1504	0.1542	0.1585

The repeated application of the simulation process eventually results in the assessment of a distribution for the top-level events, where a top-level event describes accident causation for a single design and parameter set. Figure 15.II.13 exhibits a top-level distribution calculated for the entire parameter example.

15.II.6.5 Results of the Fault-Tree Analysis

In this case study a number of parameter subsets all based on a single automobile design (or group of designs, as in this example) are solved. The result is the generation of an accident probability profile as shown in Figure 15.II.14.

This profile presents probability distributions for the top-level event, namely, the vehicle's violation of space occupied by another object in relation to the driver/environment parameter set used for analysis. From these distributions it is then possible to obtain a single accident causation distribution by performing a Monte Carlo simulation to probabilistically merge the mutually exclusive accident probability distributions. Representing this information in an integrated manner, it is possible to bring the final information derived from various fault trees for a single design option together as is shown in Figure 15.II.14.

By pinpointing an accident type of parameter set of particular concern, or a specific causative factor that is significant across parameter subsets, it is possible to facilitate the generation of crash avoidance design alternatives. The following are examples of design alternatives that could possibly minimize important factors contributing to accident causation.

Consider the following factors associated with "crash avoidance alternatives": (i) self-aligning steering wheel; (ii) high-traction tire designs; (iii) center high-mounted stop lamps; (iv) antilock braking system (ABS); (v) head-up displays (HUD); (vi) intelligent

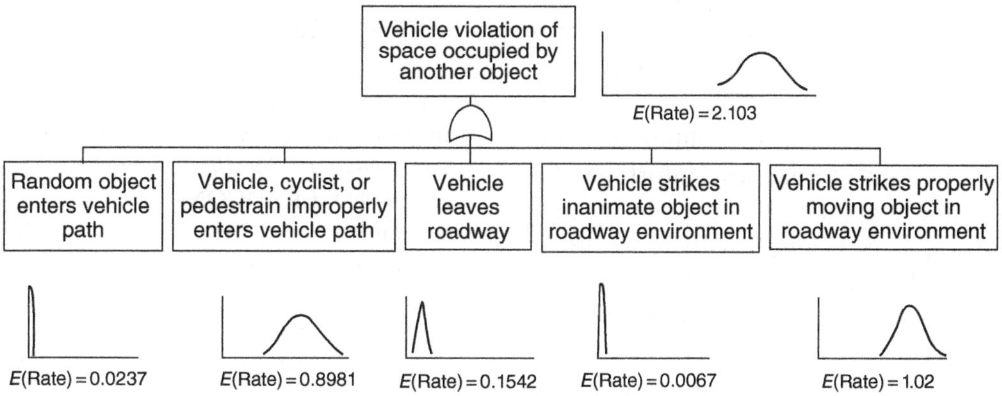

Figure 15.II.13 *Overall probability of accident causation.*

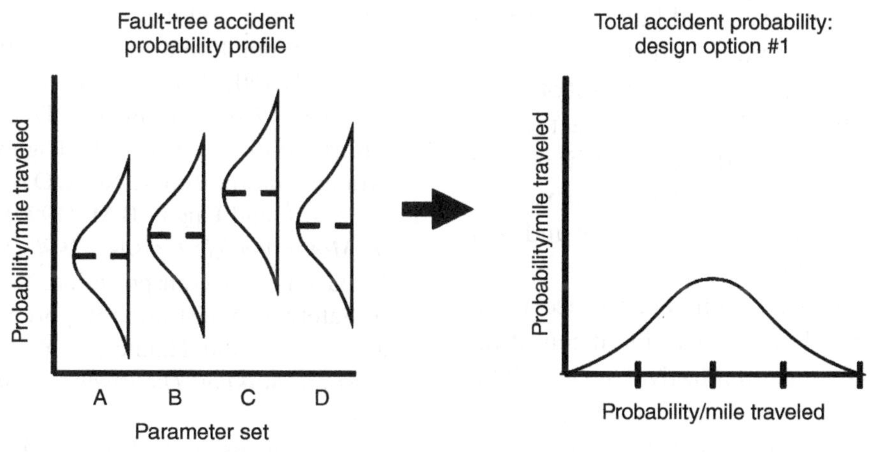

Figure 15.II.14 *Accident probability profile.*

vehicle/highway components; (vii) improved vehicle visibility; (viii) adaptive cruise control; (ix) improved lighting; (x) collision-warning radar; (xi) lane-sensing system; (xii) navigation aids; (xiii) driver monitoring; (xiv) near obstacle detection system (e.g. infrared sensors; short-range radar); and (xv) night vision enhancement system. For example, in the calculation of the vehicle leaves roadway subtree, the event poor steering action contributed to about 50% of these accidents. If the decomposition of subtrees for parameters of other accident types reveals that poor steering action is also very significant in those cases, then it would be worthwhile to examine the impact that a crash avoidance technology such as driver monitoring might have in mitigating this cause.

15.II.7 CONCLUSIONS

In the foregoing case study on highway Complex SoS, we have demonstrated how an analysis of the automobile fleet could be conducted in general; however, to allow for the evaluation of specific design options, it was necessary to obtain event probabilities that are based on individual design alternatives. Unfortunately, this requirement adds a great deal of complexity that we have only been able to manage by focusing on an "average" automobile as represented within the state/region accident database. Similar to most databases, the accident database does not provide design-specific information. Therefore, it is suggested that the needed statistics for specific automobile designs be extrapolated from the fleet-wide statistics available from accident databases. Furthermore, statistics should then be refined through the use of additional information obtained from tagging studies, direct observation, police reports, or expert assessment. Once a set of alternative designs has been evaluated with respect to the fault-tree analysis described in this chapter, it should be possible to conduct a comparative analysis of vehicle design options in terms of accident avoidance and cost (Chankong and Haimes, 2008).

Fault-tree modeling of accident occurrence is an effective framework for describing accident causation in both quantitative and qualitative terms. The decomposition of accident types using a set of parameter conditions provides a logical framework for approaching automobile safety research. At the same time, the statistical representation of the interacting faults provides insight into how various causative factors contribute to accident occurrence. One drawback that hampers the present utilization of this, or any model that closely examines accident occurrence, is that information about exposure of the vehicle–driver system to the identified basic causes is limited. Nonetheless, this well-defined information deficiency acts as a guide to future research. By examining an applied fault-tree analysis in light of the ideal availability of information for quantifying the general framework, it is possible to gain an understanding of where future information collection should be focused and what gains may be accomplished by obtaining that information. The fault-tree model developed within this part holds the potential of creating safer driving conditions by identifying the effectiveness of prospective crash-avoidance technologies.

REFERENCES

American Association of State Highway and Transportation Officials (AASHTO) (2017). *2017 LRFD Bridge Construction Specifications*, 4e. Washington, DC: AASHTO, LRFDCONS-4.

American Society of Civil Engineers (ASCE) (2017). Report Card for America's Infrastructure (2017 Report Card). www.infrastructurereportcard.org (accessed 14 November 2017).

Andrijcic, E. (2016). Intrinsic meta-modeling of systems of systems: a study in bridge infrastructure management. Ph.D. dissertation. University of Virginia.

Andrijcic, E. and Haimes, Y.Y. (2017). Metamodeling of interdependent systems: application to bridge infrastructure management. *Journal of Infrastructure Systems* **23**(2): 04016028. doi: 10.1061/(ASCE)IS.1943-555X.0000322.

Andrijcic, E. (2016). *The Economic Cost of Motor Vehicle Crashes, 1990*. Washington, DC: US Department of Transportation, National Highway Traffic Safety Administration Report No. DOT-HS-807-876.

Blincoe, L.J. and Faigin, B.M. (1990). *The Economic Cost of Motor Vehicle Crashes, 1990*. Washington, DC: US Department of Transportation, National Highway Traffic Safety Administration Report No. DOT-HS-807-876.

Chankong, V. and Haimes, Y.Y. (2008). *Multiobjective Decision Making: Theory and Methodology*. New York: Dover.

Chase, S. Haimes, Y.Y. Andrijcic, E. and Guo, Z. (2013). Modeling, understanding, and managing the risk and the life cycle of bridge infrastructure as a complex systems of systems. Technical Report No. 12. Center for Risk Management of Engineering Systems, University of Virginia. December 10, 2013.

Chen, C. (2012). *Linear System Theory and Design*, 3e. New York: Oxford University Press.

Fell, J.C. (1974). *A Motor Vehicle Accident Causal System: The Human Element*. Washington, D.C.: U.S. Department of Transportation, Research, and Development, Office of Statistics and Analysis, National Traffic Safety Administration, U.S. Government Printing Office.

Gawande, A. (2011). *The Checklist Manifesto: How to Get Things Right*. New York: Picador.

Guo, Z. and Haimes, Y.Y. (2016). Risk assessment of infrastructure system of systems with precursor analysis. *Risk Analysis* **36**(8): 1630–1643. doi: 10.1111/risa.12559.

Guo, Z. and Haimes, Y.Y. (2017). Exploring systemic risks in systems-of-systems within a multiobjective decision framework. *IEEE-SMC Systems* **47**(6): 906–915.

Haimes, Y.Y. (2007). Phantom system models for emergent multiscale systems. *Journal of Infrastructure System* **13**(2): 81–87.

Haimes, Y.Y. (2012a). Systems-based guiding principles for risk modeling, planning, assessment, management, and communication. *Risk Analysis* **32** (9): 1451–1467.

Haimes, Y.Y. (2012b). Modeling complex systems of systems with phantom system models. *Systems Engineering* **15**(3): 333–346.

Haimes, Y.Y. (2016). *Risk Modeling, Assessment, and Management*, 4e. Hoboken, NJ: Wiley.

Haimes, Y.Y., Li, D., Garber, N. et al. (1994). Improvement of Highway Safety Through Optimal Vehicle Design: Fault-Tree and Multiobjective Analysis. *NSF Project SES-9308247* Report.

Joshua, S.C. and Garber, N.J. (1992). A causal analysis of large vehicle accidents through fault-tree analysis. *Risk Analysis* **12**: 173–188. doi: 10.1111/j.1539-6924.1992.tb00666.x.

Kuzminski, P., Eisele, J.S., Garber, N. et al. (1995). Improvement of highway safety I: identification of causal factors through fault-tree modeling. *Risk Analysis* **15**: 293–312. doi: 10.1111/j.1539-6924.1995.tb00323.x.

Treat, J.R. (1980). A study of precrash factors involved in traffic accidents. *The HSRI Research Review* **10**(6): 1–35.

U. S. National Highway Transportation Safety Administration (2017). *Traffic Safety Facts:2015 Data-DOT HS 812-376*. Washington, DC: U.S. Government Printing Office.

U.S. Federal Highway Administration (2017). *Highway Statistics 1991*. Washington, DC: U.S. Government Printing Office https://www.fhwa.dot.gov/policyinformation/statistics/2016 (accessed 15 March 2018.

U.S. National Highway Transportation Safety Administration (2016). NCSA Data Resource. https://www-fars.nhtsa.dot.gov/Main/index.aspx. (accessed 16 March 2018).

U.S. Nuclear Regulatory Commission (1981). *Fault Tree Handbook*, NUREG-81/0492. Washington, DC: U.S. Nuclear Regulatory Commission.

Appendix

A.1 INTRODUCTION TO MODELING AND OPTIMIZATION

Systems engineering provides systematic methodologies for studying and analyzing the various structural and nonstructural aspects of a system and its environment by using mathematical and/or physical models.[1] It also assists in the decision-making process by selecting the best alternative policies subject to all pertinent constraints by using simulation and optimization techniques.

In general, to obtain a way to control or manage a physical system, we introduce a mathematical model that closely represents the physical system. A mathematical model is a set of equations that describes and represents the real system. This set of equations uncovers the various aspects of the problem, identifies the functional relationships between all of the system's components and elements and its environment, establishes measures of effectiveness and constraints, and thus indicates what data should be collected to deal with the problem quantitatively. These equations could be algebraic, differential, or other, depending on the nature of the system being modeled. The mathematical model is solved, and its solution is applied to the physical system.

Optimization is the procedure of selecting that set of decision variables (also known as manipulated variables) that maximizes the objective function (also known as performance function or index of performance) subject to the system's constraints.

The following is a general optimization problem.

Select the set of decision variables, $x_1^*, x_2^*, \ldots, x_n^*$, that maximize (minimize) the objective function $f(x_1, x_2, \ldots, x_n)$:

$$\max_{x_1,\ldots,x_n} f(x_1, x_2, \ldots, x_n)$$

subject to the constraints

$$\begin{aligned} g_1(x_1, x_2, \ldots, x_n) &\le b_1 \\ g_2(x_1, x_2, \ldots, x_n) &\le b_2 \\ &\vdots \\ g_m(x_1, x_2, \ldots, x_n) &\le b_m \end{aligned} \tag{A.1a}$$

where $b_1, \ldots, b_m$ are known values.

Let $\mathbf{x}^T = [x_1, x_2, \ldots, x_n]$ denote an n-dimensional row vector. The superscript T denotes the transpose

[1] Appendix A.1 is based on the Appendix to the Fourth Edition of Risk Modeling, Assessment, and Management (Haimes, 2016).

Modeling and Managing Interdependent Complex Systems of Systems, First Edition. Yacov Y. Haimes.

operation. Thus the system in Eq. (A.1a) can be rewritten as

$$\left.\begin{array}{l}\max\limits_{\mathbf{x}} f(\mathbf{x}) \\ g_j(\mathbf{x}) \le b_j,\ j=1,2,\ldots,m\end{array}\right\} \quad \text{Subject to the constraints} \tag{A.1b}$$

Depending on the nature of the objective function and the constraints, the general optimization problem posed by Eq. (A.1b) can be classified accordingly. The following are four possible ways that may be considered to classify mathematical models:

1. Linear versus nonlinear.
2. Deterministic versus probabilistic (stochastic).
3. Static versus dynamic.
4. Lumped parameters versus distributed parameters.

1. *Linear Versus Nonlinear*. A *linear* model is one that is represented by linear equations; that is, all constraints and the objective function(s) are linear. A *nonlinear* model is represented by nonlinear equations; that is, part or all of the constraints and/or the objective function are nonlinear.

   Examples

   Linear equations: $y = 5x_1 + 6x_2 + 7x_3$

   Nonlinear equations: $y = 5x_1^2 + 6x_2x_3$

   $y = \log x_1$

   $y = \sin x_1 + \log x_2$

2. *Deterministic Versus Probabilistic*. *Deterministic* models or elements of models are those in which each variable and parameter can be assigned a definite fixed number or a series of fixed numbers for any given set of conditions.

   In *probabilistic* (stochastic) models, the principle of uncertainty is introduced. Neither the variables nor the parameters used to describe the input–output relationships and the structure of the elements (and the constraints) may be precisely known.

   Example

   "The value of x is in $(a-b,\ a+b)$ with 90% probability," meaning that in the long run, the value of x will be less than $(a-b)$ or greater than $(a+b)$ in 10% of the cases.

3. *Static Versus Dynamic*. *Static* models are those that do not explicitly take the variable time into account. In general, static models are of the form given by Eq. (A.1a).

   Dynamic models are those involving difference or differential equations. An example is given in Eq. (A.2):

$$\min_{u_1,\ldots,u_M} \int_{t_0}^{t} F(x_1,\ldots,x_N,u_1,\ldots,u_M,t)dt \tag{A.2}$$

   subject to the constraints

$$\frac{d}{dt}(x_i) = G_i(x_i,\ldots,x_N,u_1,\ldots,u_M,t) \quad i=1,2,\ldots,N$$

$$x_t(t_0) = x_i^0,\quad i=1,2,\ldots,N$$

   Static optimization problems are often referred to as mathematical programming, while dynamic optimization problems are often referred to as optimal control problems.

4. *Distributed Parameters Versus Lumped Parameters*. A *lumped parameter* model ignores variations, and the various parameters and dependent variables can be considered homogeneous throughout the entire system.

   A *distributed parameter* model takes into account detailed variations in behavior from point to point throughout the system.

   Most physical systems are distributed parameters systems. For example, the diffusion equation

$$T\left[\frac{1}{r}\frac{\partial}{\partial r}\left(r\frac{\partial P}{\partial r}\right)\right] = S\frac{\partial P}{\partial t} + Q \tag{A.3}$$

   represents a distributed parameter system.

A.1.1 The Lagrangian Function

For nonlinear equality constraints, the Lagrangian formulation can be utilized. Consider the following simple optimization problem with two decision variables, x_1 and x_2:

$$\min_{x_1x_2} f(x_1,x_2)$$

subject to the equality constraint

$$g(x_1,x_2)=b$$

where

$$f(x_1,x_2)\in C^1$$
$$g(x_1,x_2)\in C^2$$

writing the constraint as

$$g(x_1,x_2)-b=0$$

We define a function L, called the Lagrangian, as follows:

$$L(x_1,x_2,\lambda)=f(x_1,x_2)+\lambda[g(x_1,x_2)-b]$$

where λ is a Lagrange multiplier. Note that if the constraints are satisfied at (x_1^*,x_2^*), then $g(x_1^*,x_2^*)=b$ and $g(x_1^*,x_2^*)-b=0$; therefore $L(x_1^*,x_2^*,\lambda^*)=f(x_1^*,x_2^*)$, where λ^* is the optimal Lagrange multiplier; that is, the value of the Lagrangian is the same as the optimal value of the objective function.

Necessary conditions for a stationary point of L are

$$\frac{\partial L}{\partial x_1}=\frac{\partial L}{\partial x_2}=\frac{\partial L}{\partial \lambda}=0$$

or, by expanding,

$$\frac{\partial L}{\partial x_1}=\frac{\partial f}{\partial x_1}+\lambda\frac{\partial g}{\partial x_1}=0$$

$$\frac{\partial L}{\partial x_2}=\frac{\partial f}{\partial x_2}+\lambda\frac{\partial g}{\partial x_2}=0$$

$$\frac{\partial L}{\partial \lambda}=g-b=0$$

These are three equations with three unknowns x_1, x_2, λ; their solution yields the stationary points x_1^*, x_2^*, and λ^*. If the sufficiency conditions for minimum are satisfied, then (x_1^*,x_2^*,λ^*) yields the minimum to $f(x_1, x_2)$.

A.1.1.1 General Formulation of the Lagrangian Function

Consider the following general, equality-constrained, nonlinear optimization problem:

$$\min_{\mathbf{x}} f(\mathbf{x})$$

subject to the equality constraints

$$g_j(\mathbf{x})=b_j,\quad j=1,2,\ldots,m$$

where

$$f(\mathbf{x})\in C^1$$
$$g_j(\mathbf{x})\in C^2,\quad j=1,2,\ldots,m$$

Form the Lagrangian L:

$$L(\mathbf{x},\boldsymbol{\lambda})=f(\mathbf{x})+\sum_{j=1}^{m}\lambda_j\left[g_j(\mathbf{x})-b_j\right]$$

where

$$\boldsymbol{\lambda}^{\mathbf{T}}\triangleq(\lambda_1,\lambda_2,\ldots,\lambda_m)$$

Necessary conditions for a stationary point of L are

$$\frac{\partial L}{\partial x_i}=0 \text{ yields} \frac{\partial f(x)}{\partial x_i}+\sum_{j=1}^{m}\lambda_j\frac{\partial g_j}{\partial x_i}=0,\quad i=1,2,\ldots,n$$

$$\frac{\partial L}{\partial \lambda_k}=0 \text{ yields} \frac{\partial f(x)}{\partial \lambda_k}+\sum_{j=1}^{m}\left[g_j(x)-b_j\right]\frac{\partial \lambda_j}{\partial \lambda_k}=0,$$

$$k=1,2,\ldots,m$$

Note that

$$\frac{\partial \lambda_j}{\partial \lambda_k}=\delta_{jk}$$

where

$$\delta_{jk}=\begin{cases}1 & \text{for } j=k\\ 0 & \text{for } j\neq k\end{cases}$$

A.1.1.2 Example Problem

Find the radius, r, and the length, h, of a water reservoir of a closed cylinder shape and of a given volume V_0, having the minimum surface area.

We know that $V_0=\pi r^2 h$. Let S = surface area = $2\pi rh$. The objective function is

$$\min_{r,h}\left\{f(r,h)=2\pi rh+2\pi r^2\right\}$$

subject to the constraint

$$g(r,h)=\pi r^2 h=V_0$$

Form the Lagrangian, L:

$$\begin{aligned}L(r,h,\lambda)&=f(r,h)+\lambda[g(r,h)-V_0]\\&=2\pi rh+2\pi r^2+\lambda\left(\pi r^2 h-V_0\right)\end{aligned}$$

Necessary conditions for optimum are

$$\frac{\partial L}{\partial r} = \frac{\partial f}{\partial r} + \frac{\partial g}{\partial r} = 0$$
$$\frac{\partial L}{\partial h} = \frac{\partial f}{\partial h} + \lambda\frac{\partial g}{\partial h} = 0$$
$$\frac{\partial L}{\partial \lambda} = \pi r^2 h - V_0 = 0$$

Substituting the following derivatives

$$\frac{\partial f}{\partial r} = 2\pi(h+2r),\ \frac{\partial f}{\partial h} = 2\pi r,\ \frac{\partial g}{\partial r} = 2\pi rh,\ \frac{\partial g}{\partial h} = \pi r^2$$

yields the following three equations:

$$2\pi(h+2r) + 2\lambda\pi rh = 0$$
$$2\pi r + \lambda\pi r^2 = 0$$
$$\pi r^2 h - V_0 = 0$$

or

$$h + 2r + \lambda rh = 0$$
$$2r + \lambda r^2 = 0$$
$$\pi r^2 h - V_0 = 0$$

Solving the last three equations simultaneously yields

$$\lambda^* = -\frac{2}{r}$$
$$r^* = \sqrt[3]{\frac{V_0}{2\pi}}$$
$$h^* = \sqrt[3]{\frac{4}{\pi}V_0}$$

and

$$f(r^*,h^*) = 2\pi\sqrt[3]{\frac{V_0}{2\pi}}\sqrt[3]{\frac{4}{\pi}V_0} + 2\pi\left(\sqrt[3]{\frac{V_0}{2\pi}}\right)^2$$

A.1.1.3 Lagrange Multipliers and Inequality Constraints

The Lagrangian approach given above is suitable for handling a nonlinear programming problem with equality constraints. Generalized Lagrange multipliers (also called the Kuhn–Tucker multipliers) can be introduced to handle nonlinear problems with inequality constraints. In order to facilitate the discussion on the Kuhn–Tucker theory in nonlinear programming, the classical method for inequality constraints will be discussed first. Consider the problem

$$\min_{x_1,x_2} f(x_1,x_2)$$

subject to the constraints

$$g(x_1,x_2) = b$$

and

$$x_1 \geq a$$

We convert this to equality by introducing the variable θ, which is defined by

$$\theta^2 = x_1 - a$$

We require θ to be real; and if $x_1 \geq a$, then $\theta^2 > 0$. (If $x_1 < a$, then $\theta^2 < 0$, θ is imaginary.) Hence, the two constraints can be rewritten as

$$g(x_1,x_2) - b = 0$$
$$\theta^2 - x_1 + a = 0$$

We form the Lagrangian L:

$$L(x_1,x_2,\theta,\lambda_1,\lambda_2) = f(x_1,x_2) + \lambda_1[g(x_1,x_2) - b] + \lambda_2[\theta^2 - x_1 + a]$$

The necessary conditions for stationary points are

$$\frac{\partial L}{\partial x_1} = \frac{\partial f}{\partial x_1} + \lambda_1\frac{\partial g}{\partial x_1} - \lambda_2 = 0$$
$$\frac{\partial L}{\partial x_2} = \frac{\partial f}{\partial x_2} + \lambda_1\frac{\partial g}{\partial x_2} = 0$$
$$\frac{\partial L}{\partial L_1} = g - b = 0$$
$$\frac{\partial L}{\partial \lambda_2} = \theta^2 - x_1 + a = 0$$
$$\frac{\partial L}{\partial \theta} = 2\lambda_2\theta = 0$$

In analyzing condition 5, two cases can be distinguished for $2\lambda_2\theta = 0$:

Case 1: $\theta = 0$, then $x_1 = a$. The solution in this case is on the boundary; that is, the constraint is binding. Often a binding constraint is referred to as an active constraint; then λ_2 is not necessarily equal to zero.

Case 2: $\lambda_2 = 0$, then $\theta \neq 0$. The solution in this case is not on the boundary; that is, the constraint is not binding. Often a nonbinding constraint is referred to as an inactive one.

Example Problem. A desalination plant produces freshwater in each of three successive periods. The requirements for freshwater are at least 5 units (acre-ft) at the end of the first period, 10 units at the end of the second period, and 15 units at the end of the third period for a total of 30. The cost of producing x units in any period is $f(x) = x^2$.

Additional water may be produced in one period and carried over to a subsequent one. A holding cost of \$2 per unit is charged for any freshwater carried over from one period to the next. Assuming no initial inventory, how many units should be produced each period?

Formulation: Let x_1, x_2, and x_3 represent production in periods 1, 2, and 3, respectively. Total cost = production cost plus holding cost:

$$f(x_1,x_2,x_3) = x_1^2 + x_2^2 + x_3^2 + 2(x_1 - 5)) + 2(x_1 + x_2 - 15)$$

The constraints are

1. $x_1 \geq 5$
2. $x_1 + x_2 \geq 15$
3. $x_1 + x_2 + x_3 = 30$
4. $x_2 \geq 0$
5. $x_3 \geq 0$

The optimization problem is

$$\min_{x_1,x_2,x_3} f(x_1,x_2,x_3)$$

subject to constraints 1 through 5.

One possible approach is to ignore inequality constraints and form the Lagrangian L and then check the solution for feasibility:

$$L(x_1,x_2,x_3,\lambda) = x_1^2 + x_2^2 + x_3^2 + 2(x_1 - 5) + 2(x_1 + x_2 - 15)$$
$$\lambda(x_1 + x_2 + x_3 - 30)$$

The necessary conditions for minimum are

$$\frac{\partial L}{\partial x_1} = 2x_1 + 2 + 2 + \lambda = 0 \quad \text{or} \quad 2x_1 = -4 - \lambda$$

$$\frac{\partial L}{\partial x_2} = 2x_2 + 2 + \lambda = 0 \quad \text{or} \quad 2x_2 = -2 - \lambda$$

$$\frac{\partial L}{\partial x_3} = 2x_3 + \lambda = 0 \quad \text{or} \quad 2x_3 = -\lambda$$

$$\frac{\partial L}{\partial \lambda} = x_1 + x_2 + x_3 - 30 = 0 \quad \text{or} \quad x_1 + x_2 + x_3 = 30$$

Solving the above simultaneously yields

$$-4 - \lambda - 2 - \lambda - \lambda = 60; \text{ thus } \lambda^* = -\frac{66}{3} = -22$$

and

$$x_1^* = 9, \; x_2^* = 10, \; x_3^* = 11, \; f(x_1^*,x_2^*,x_3^*) = 318$$

The above result should be tested for feasibility. Substituting the values of $x_1 = 9$, $x_2 = 10$, and $x_3 = 11$ into constraints 1 and 2 does not violate them. Thus, the optimal solution is feasible and constraints 1 and 2 are not binding [not active]; that is,

1. $x_1 > 5$
2. $x_1 + x_2 > 15$

A.1.2 Dynamic Programming

Dynamic programming is the most used nonlinear optimization technique in water resources systems. This is because the Markovian and sequential nature of the decisions that arise in water resource problems nicely fits into Bellman's principle of optimality on which dynamic programming is based (Bellman and Dreyfus, 1962).

A.1.2.1 Principle of Optimality and Recursive Equation

Bellman's principle of optimality, states that:

> An optimal policy has the property that whatever the initial state and initial decisions are, the remaining decisions must constitute an optimal policy with regard to the state resulting from the first decision.

This principle will be used in the derivation of a recursive equation that relates the state of the system and the decisions from one stage to another. In order to avoid deriving a purely theoretical formula in this section, an allocation problem will be utilized as a vehicle for constructing the general recursive dynamic programming equation.

Given a resource (water in a reservoir) of capacity Q that can be supplied to N consumers (for example, N cities), let x_i denote the amount of water supplied to the ith city with a return of $g_i(x_i)$. The problem is how many units of water, x_i, to allocate to the ith city in order to maximize the total net return subject to certain constraints. Mathematically, the problem can be formulated as follows.

Given:

$$N \text{ decision}: x_1, x_2, \ldots, x_N$$

$$N \text{ return functions}: g_1(x_1), g_2(x_2), \ldots, g_N(x_N)$$

Let $f_N(Q)$ represent the maximum return from the allocation of the resource Q to the N consumers. The overall optimization problem is

$$\max_{x_i} \{g_1(x_1) + g_2(x_2) + L + g_N(x_N)\}, \quad x_i \geq 0, \quad i = 1, 2, \ldots, N$$

subject to a limited quantity Q

$$\sum_{i=1}^{N} x_i \leq Q \tag{A.4}$$

It is assumed that the functions $g_i(x_i)$, $i = 1, 2, \ldots, N$, possess the following properties:

1. They are bounded for $0 \leq x_i \leq Q$.
2. They need not be continuous (often these return or cost functions are given in a tabulated or a graphical form).
3. Each $g_i(x_i)$ is a function of only one decision variable.

Note that in this problem there is one state variable only; namely, the water to be allocated to the various consumers. The state variable will be represented by q, indicating the amount of resource available for allocation. The number of decision variables is N, and so is the number of stages. Note that the number of decisions and stages is not always the same.

At the first stage, we assume that there is only one potential user of the resource, which will be designated by the subscript 1. Then, since we would still wish to make maximum use of the resource, we define

$$f_1(q) = \max_{\substack{0 \leq x_1 \leq q \\ 0 \leq q \leq Q}} \{g_1(x_1)\} \tag{A.5}$$

At the second stage, we assume that there are two potential users of the resource. The new user is designated by the subscript 2. If we allocate to this user an amount x_2, $0 \leq x_2 \leq q$, there will be a return $g_2(x_2)$, and a remaining quantity of the resource $(q - x_2)$ can be allocated to user 1. Applying the principle of optimality, the optimal return of the resource for two potential users is

$$f_2(q) = \max_{\substack{0 \leq x_2 \leq q \\ 0 \leq q \leq Q}} \{g_2(x_2) + f_1(q - x_2)\} \tag{A.6}$$

The recursive calculation is now established and f_3, f_4, …, f_N can be written and solved in succession for all possible values of q. When this process is completed, $f_N(q)$ represents the return from allocating the resource optimally to N users as a function of the quantity of the resources, whatever it may be.

The general recursive relationship for N stages is

$$f_N(q) = \max_{\substack{0 \leq x_N \leq q \\ 0 \leq q \leq Q}} \{g_N(x_N) + f_{N-1}(q - x_N)\} \tag{A.7}$$

A direct derivation of the above dynamic programming recursive equation is given below, following Bellman and Dreyfus (1962):

$$\max_{\substack{x_1 + x_2 + \cdots + x_N = q \\ x_j \geq 0}} = \max_{0 \leq x_N \leq q} \left[\max_{\substack{x_1 + x_2 + \cdots + x_{N-1} = q - x_N \\ x_1 \geq 0}} \right] \tag{A.8}$$

We can write

$$f_N(q) = \max_{\substack{x_1 + x_2 + L + x_N = q \\ x_1 \geq 0}} [g_N(x_N) + g_{N-1}(x_{N-1}) + \cdots + g_1(x_1)]$$

$$= \max_{0 \leq x_N \leq q} \left[\max_{\substack{x_1 + x_2 + L + x_{N-1} = q - x_N \\ x_1 \geq 0}} [g_N(x_N) + g_{N-1}(x_{N-1}) + \cdots + g_1(x_1)] \right]$$

$$= \max_{0 \leq x_N \leq q} \left[g_N(x_N) + \max_{\substack{x_1 + x_2 + L + x_{N-1} = q - x_N \\ x_1 \geq 0}} [g_{N-1}(x_{N-1}) + \cdots + g_1(x_1)] \right]$$

$$= \max_{0 \leq x_N \leq q} [g_N(x_N) + f_{N-1}(q - x_N)]$$

A.2 FAULT TREES

A.2.1 Introduction

One common denominator – unreliability – unifies all of the following undesired events: Two cars collide due to the malfunction of one car's brakes, pollutants are discharged from a wastewater treatment plant due to the failure of a pump, and a nuclear reactor power plant is shut down due to the failure of a relay. Safe operation in these three examples depends on the proper functioning, namely, reliability, of all critical components that constitute these systems. Dual brakes installed in parallel in the disabled car, two pumps installed in parallel in the wastewater treatment plant, and two relays installed in parallel in the nuclear reactor could have prevented the above failures.

Myriad components constitute a technologically based system, each with a given reliability and configuration. We define reliability as the conditional probability that the system (or a component thereof) will perform its intended function(s) throughout an interval, given that it was functioning correctly at time t_0. Evaluating the composite reliability of the overall system without a systematic process is a daunting task. Fault-tree analysis is a systematic and quantitative process that takes into account the unreliability contributions of the various components to the overall system (see Apostolakis (1991) and Henley and Kumamoto (1992)). Recent additions to the literature on fault trees include works from Ebeling (2005), Limnios (2007), and Birolini (2007).

Fault-tree analysis was first conceived in 1961 by H. A. Watson of Bell Telephone Laboratories in connection with a US Air Force contract to study the Minuteman launch control system. At a safety symposium held in 1965 at the University of Washington, cosponsored by the Boeing Company, several papers expounded the virtues of fault-tree analysis. These presentations marked the beginning of a widespread interest in using fault-tree analysis as a safety and reliability tool for complex dynamic systems such as nuclear reactors. Since then, fault-tree analysis has been widely used for evaluating the safety and reliability of complex engineering systems. Thus far, the most widespread use of fault trees has been in the nuclear industry beginning with the *Reactor Safety Study* (U.S. Nuclear Regulatory Commission, 1975) conducted over a two-year period.

One of the leading documents on fault-tree analysis is the *Fault Tree Handbook* written by the US Nuclear Regulatory Commission (1981). This handbook remains the primer for all students of fault trees. The following is a succinct description of the fault-tree model (U.S. Nuclear Regulatory Commission, 1981):

> A fault tree analysis can be simply described as an analytical technique, whereby an undesired state of the system is specified (usually a state that is critical from a safety standpoint), and the system is then analyzed in the context of its environment and operation to find all credible ways in which the undesired event can occur. The fault tree itself is a graphic model of the various parallel and sequential combinations of faults that will result in the occurrence of the predefined undesired event. The faults can be events that are associated with component hardware failures, human errors, or any other pertinent events which can lead to the undesired event. A fault tree thus depicts the logical interrelationships of basic events that lead to the undesired event – which is the top event of the fault tree. It is important to understand that a fault tree is not a model of all possible system failures or all possible causes for system failure. A fault tree is tailored to its top event which corresponds to some particular system failure mode, and the fault tree thus includes only those faults that contribute to this top event. Moreover, these faults are not exhaustive – they cover only the most credible faults as assessed by the analyst.

Fault-tree analysis, which is one of the principal methods for analyzing systems safety, can be used to identify potential weaknesses in a system or the most likely causes of a system's failure. The method is a detailed deductive analysis that requires considerable system information and can also be a valuable design or diagnostic tool.

In fault-tree analysis, the sequence of events leading to the probable occurrence of a predetermined event is systematically divided into primary events whose failure probabilities can be estimated. Several methods have been suggested for handling uncertainty in the failure probabilities of the primary event of interest; however, most of them develop merely an interval of uncertainty. Also, most available research results in fault-tree analysis are applicable only to cases with "point probability distributions."

Most current methods for fault-tree analysis do not provide the means to use probability distributions for the primary components. When these methods do use probability distributions, at best they develop an interval of uncertainty for the probability of the undesired event of interest. Also, most current methods use the unconditional expected value as a measure of risk. This chapter introduces a relatively new method that incorporates conditional expectations and multiobjective analysis with fault-tree analysis. It provides managers and decision makers with more information about the system rather than merely providing a single point probability for the undesired event.

The conventional approach to fault-tree analysis has been the use of point probabilities for the analysis of the system. The approach is valid when we have accurate data on the component failure rate along with a point distribution. This is, however, practically never the case in most applications. In most cases, the database available for component failure rate is sketchy or has a wide uncertainty interval associated with it. Also, since fault trees deal with rare events, often the failure of some components of the system may not have occurred in the past and thus would not be included in the database.

To overcome the limitations imposed by the unavailability of data, it is common practice to approximate the available data and/or the subjective estimates of the failure rates by a probability distribution.

When different probability distributions are used for basic component failure, existing analytical methods are not very useful because there are no closed-form solutions available for the products and the sums of these distributions. Methods based on analytical techniques (e.g. variance decomposition, variance partitioning, and system moments) develop, at best, an interval of uncertainty or confidence intervals for the overall system failure rate. This is accomplished by approximating the basic component distributions to normal or lognormal distributions and then using known relationships for adding normal or multiplying lognormal distributions. These methods tend to be computationally complex and are difficult to adopt for large systems.

In such cases, methods based on a combination of random variables through numerical simulation are very useful. Numerical methods, when used for fault-tree analysis, can handle most well-known probability distributions, such as normal, lognormal, exponential, and Weibull. System components having these failure rate distributions may be connected in series or in parallel.

Numerical methods are based on the generation of pseudorandom numbers to approximate known or assumed probability distributions for system components. Random numbers generated to approximate a probability distribution can be augmented to obtain the required information about the top event of the system.

The use of simulation methods has grown rapidly with the increased use of high-speed digital computers, since they overcome the one major limitation of numerical methods – the requirement of a large amount of computer time. Personal computers, which can run dedicated programs without having to share processor time with other applications, have also contributed to the widespread use of numerical methods based on simulation.

The main limitation of many current methods stems from the fact that it is not possible to obtain the complete probability distribution for the top event. Among the exception is the Integrated Reliability and Risk Analysis System (IRRAS), which is an integrated computer software for performing probabilistic risk assessment using fault trees (Russell et al., 1987). These methods can develop the moments of the distribution only for the top event, which can then be used to approximate the top event distribution using empirical distributions. Alternatively, these methods can develop measures for the components at the basic level. Cox (1982) uses variance as the primary measure in his approach and ranks the input variables according to their contributions to the output uncertainty. Also, all analytical methods use approximations at one stage or another in order to simplify the analytical expressions obtained. This naturally affects the results. The moments of the distributions are represented instead of the distributions themselves. This factor is an approximation in itself, in that two distributions that may have the same moments are treated in the same way even though they may be completely different. Most analytical methods cannot be used to model dependencies among components.

A.2.2 Basic Fault-Tree Analysis

A.2.2.1 Fault Trees and Extreme Events

The theory, methodology, and utilization of fault trees have become so extensive over the last two decades that no one chapter can do justice to the subject. This section is intended to provide introductory material on fault trees to readers who are interested in the broader subject of risk analysis. They can then consult any of several references on fault tree analysis, such as Apostolakis (1991), Henley and Kumamoto (1992), Hoyland and Rausand (1994), Johnson (1989), Martensen and Butler (1987), NASA (1996), Rao (1992), Storey (1996), and US Nuclear Regulatory Commission (1981).

A.2.2.2 Procedure for Fault-Tree Analysis

To analyze a system using fault trees, we first specify the undesired state of the system whose occurrence probability we are interested in determining. This state may be the failure of the system or of a subsystem. Once this undesired state has been specified, a list is made of all the possible ways in which this event can occur. Each of the possible ways is then examined independently to find out how it can occur, until it is no longer feasible or cost-effective to carry out the analysis further.

The lowest-level events are called primary events. All the events are laid out in a "tree" form connected by "gates" that show the relationships between successive levels of the tree. A few of the most common symbols used for fault-tree construction and analysis are shown in Figure A.1.

A fault tree is a graphic model of the various sequential and parallel combinations of faults (see Figures A.2 and A.6) that will result in the occurrence of the predefined undesired event. The faults can be associated with component hardware failures, human errors, or any other pertinent events that can lead to the undesired outcome. A fault tree thus depicts the logical interrelationships of the basic events that lead to the undesired top event.

A.2.2.3 Limitations of Fault-Tree Analysis

One major limitation of fault-tree analysis concerns the qualitative aspects of fault-tree construction. It is possible that significant failure modes may be overlooked during the analysis. It is thus very important that the analyst thoroughly understands the system before the fault tree is constructed.

Another limitation is the difficulty in applying Boolean logic to describe the failure modes of some components when their operation can be partially successful. Techniques exist to address this problem, but they increase the complexity of the analysis.

Also, there is the lack of appropriate data on failure modes; even though data might be available, they may

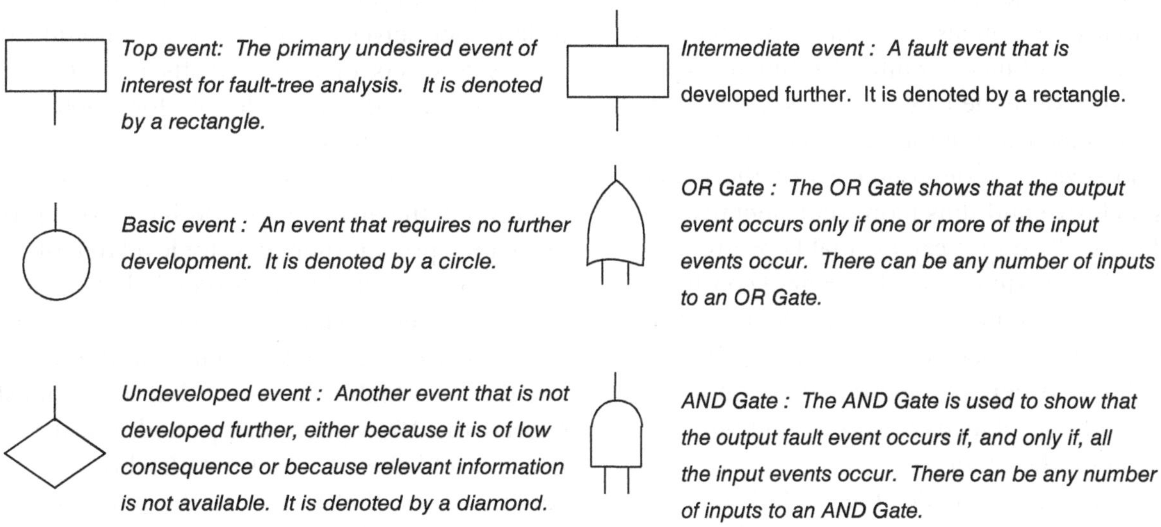

Figure A.1 *Basic components of a fault tree.*

Figure A.2 Components in series.

not be applicable to the system under consideration. Data on human reliability are very sketchy if at all available.

A.2.3 Reliability and Fault-Tree Analysis

A.2.3.1 Risk Versus Reliability Analysis

The distinction between reliability and risk is not merely a semantic issue; rather, it is a major element in resource allocation throughout the life cycle of a product (whether in design, construction, operation, maintenance, or replacement). The distinction between risk and safety, well articulated over two decades ago by Lowrance (1976), is vital when addressing the design, construction, and maintenance of physical systems, since by their nature such systems are built of materials that are susceptible to failure. The probability of such a failure and its associated consequences constitutes the measure of risk. Safety manifests itself in the level of risk that is acceptable to those in charge of the system. For instance, the selected strength of chosen materials, and their resistance to the loads and demands placed on them, is a manifestation of the level of acceptable safety. The ability of materials to sustain loads and avoid failures is best viewed as a random process – a process characterized by two random variables: (i) the load (demand) and (ii) the resistance (supply or capacity).

Unreliability, as a measure of the probability that the system does not meet its intended functions, does not include the consequences of failures. On the other hand, risk as a measure of the probability (i.e. unreliability) and severity (consequences) of the adverse effects is inclusive and thus more representative.

Clearly, not all failures can justifiably be prevented at all costs. Thus, system reliability cannot constitute a viable metric for resource allocation unless an a priori level of reliability has been determined. This brings us to the duality between risk and reliability on the one hand and multiple objectives and a single-objective optimization on the other.

In the multiobjective model, the level of acceptable reliability is associated with the corresponding consequences (i.e. constituting a risk measure) and is thus traded off with the associated cost that would reduce the risk (i.e. improve the reliability). In the single-objective model, on the other hand, the level of acceptable reliability is not explicitly associated with the corresponding consequences; rather it is predetermined (or parametrically evaluated) and thus is considered as a constraint in the model.

There are, of course, both historical and evolutionary reasons for the more common use of reliability analysis rather than risk analysis as well as substantive and functional justifications. Historically, engineers have always been concerned with strength of materials, durability of product, safety, surety, and operability of various systems. The concept of risk as a quantitative measure of both the probability and consequences (or an adverse effect) of a failure has evolved relatively recently. From the substantive–functional perspective, however, many engineers or decision makers cannot relate to the amalgamation of two diverse concepts with different units – probabilities and consequences – into one concept termed risk. Nor do they accept the metric with which risk is commonly measured. The common metric for risk – the expected value of adverse outcome – essentially commensurates events of low probability and high consequences with those of high probability and low consequences. In this sense, one may find basic philosophical justifications for engineers to avoid using the risk metric and instead work with reliability. Furthermore and most important, dealing with reliability does not require the engineer to make explicit tradeoffs between cost and the outcome resulting from product failure. Thus, design engineers isolate themselves from the social consequences that are by-products of the tradeoffs between reliability and cost. The design of levees for flood protection may clarify this point.

Designating a "one-hundred-year return period" means that the engineer will design a flood protection levee for a predetermined water level that on the average is not expected to be exceeded more than once every 100 years. Here, ignoring the socioeconomic consequences, such as loss of lives and property damage due to a high water level that would most likely exceed the 100-year return period, the design engineers shield themselves from the broader issues of consequences, that is, risk to the population's social well-being. On the other hand, addressing the multiobjective dimension that the risk metric brings requires much closer interaction and coordination between the design engineers and the decision makers. In this case, an interactive process is required to reach acceptable levels of risks, costs, and benefits. In a nutshell, complex issues, especially those involving public policy with health and socioeconomic dimensions, should not be addressed through overly simplified models and tools. As the demarcation line between hardware and software slowly but surely fades away, and with the ever-evolving and increasing role of design engineers and systems analysts in technology-based decision making, a new paradigm shift is emerging. This shift is characterized by a strong overlapping of the responsibilities of engineers, executives, and less technically trained managers.

The likelihood of multiple or compound failure modes in infrastructure systems (as well as in other physical systems) adds another dimension to the limitations of a single reliability metric for such infrastructures (Park et al., 1998; Schneiter et al., 1996). Indeed, because one must address multiple reliabilities of a system, the need for explicit tradeoffs among risks and costs becomes more critical. Compound failure modes are defined as two or more paths to failure with consequences that depend on the occurrence of combinations of failure paths. Consider the following examples: (i) a water distribution system, which can fail to provide adequate pressure, flow volume, water quality, and other needs; (ii) the navigation channel of an inland waterway, which can fail by exceeding the dredge capacity and by closing to barge traffic; and (iii) highway bridges, where failure can occur from deterioration of the bridge deck, corrosion or fatigue of structural elements, or an external loading such as flood. Water quality could be used as another basis for the reliability of the water distribution system. None of these failure modes is independent of the others in probability or consequence. For example, deck cracking can contribute to structural corrosion. Structural deterioration in turn can increase the vulnerability of the bridge to floods; nevertheless, the individual failure modes of bridges are typically analyzed in isolation of one another. Acknowledging the need for multiple metrics of reliability of capacity, pressure, hydraulic capacity (joint requirements for flow volume and pressure in the system), or quality could markedly improve decisions regarding maintenance and rehabilitation, especially when these multiple reliabilities are augmented with risk metrics.

Over time, most, if not all, man-made products and structures ultimately fail. Reliability is commonly used to quantify this time-dependent failure of a system. Indeed, the concept of reliability plays a major role in engineering planning, design, development, construction, operation, maintenance, and replacement.

To streamline our discussion on fault-tree analysis, we define the following terms associated with reliability and its modeling:

- Reliability $R(t)$: The probability that the system operates correctly (or performs its intended function) throughout the interval $(0, t)$ given that it was operating correctly at $t = 0$.
- Unreliability $Q(t)$: The probability that the system fails during interval $(0, t)$, given that it was operating correctly at $t = 0$.
- Failure density $f(t)$: The term $f(t)dt$ is the probability that the system fails in time dt about t.
- Failure rate $\lambda(t)$: The term $\lambda(t)dt$ is the conditional probability of system failure in time dt about t, given that no failure occurs up to time t:

$$Q(t) = 1 - R(t) \tag{A.9}$$

$$f(t) = \frac{dQ(t)}{dt} = -\frac{dR(t)}{dt} \tag{A.10}$$

$$\lambda(t) = \frac{f(t)}{R(t)} = -\frac{1}{R(t)}\frac{dR(t)}{dt} \tag{A.11}$$

$$R(t) = \exp\left[-\int_0^t \lambda(\tau)d\tau\right] \tag{A.12}$$

A.2.3.2 Series System

When subsystems are connected in series (see Figure A.2), the system fails when at least one of its components fails:

$$R(t) = R_A(t)R_B(t) \tag{A.13}$$

$$\begin{aligned} Q(t) &= 1 - [1 - Q_A(t)][1 - Q_B(t)] \\ &= Q_A(t) + Q_B(t) - Q_A(t)Q_B(t) \end{aligned} \tag{A.14}$$

To generalize Eq. (A.13), let $R_i(t)$ represent the reliability of the ith subsystem and let $R_s(t)$ represent the reliability of the entire system:

$$R_s(t) = \prod_{i=1}^{n} R_i(t) \tag{A.15}$$

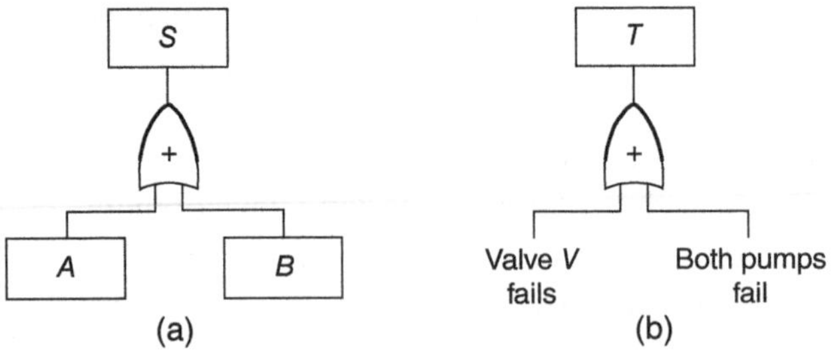

Figure A.3 *(a) OR Gate. (b) OR Gate for pumping system. (components in series).*

$$Q_s(t) = 1 - R_s(t) = 1 - \prod_i R_i(t) = 1 - \prod_i (1 - Q_i(t)) \tag{A.16}$$

$$R_s(t) < \min_i \{R_i(t)\} \tag{A.17}$$

Note that Eq. (A.17) is correct for subsystems in series, unless all components have $R_i(t) = 1$; then the inequality sign should be modified.

Quantitative fault-tree analysis is based on Boolean algebra, where the events either occur or do not occur. The two basic gates used in fault-tree analysis are the OR gate and the AND gate.

The OR Gate. The OR gate represents the union of the events attached to the gate. Any one or more of the input events must occur to cause the event above the gate to occur. The OR gate is equivalent to the Boolean symbol +. For example, the OR gate with two input events (as shown in Figure A.3a) is equivalent to the Boolean expression

$$S = A + B = A \cup B \tag{A.18}$$

In terms of probability,

$$\begin{aligned} P(S) &= P(A) + P(B) - P(AB) \\ &= P(A) + P(B) - P(A)P(B|A) \\ &= P(A) + P(B) - P(B)P(A|B) \end{aligned} \tag{A.19}$$

If A and B are independent events, then $P(B|A) = P(B)$ or $P(A|B) = P(A)$; therefore

$$P(S) = P(A) + P(B) - P(A)P(B) \tag{A.20}$$

The Nuclear Regulatory Commission uses rare-event approximation in its *Fault Tree Handbook* (U.S. Nuclear Regulatory Commission, 1981). In this case, we have

$$P(S) \approx P(A) + P(B) \tag{A.21}$$

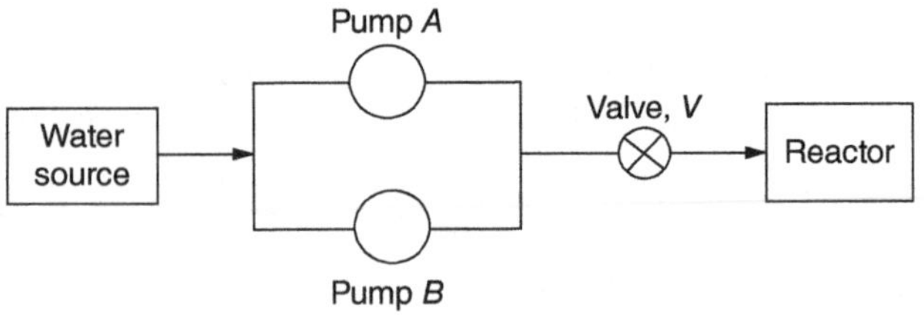

Figure A.4 *Water pumping system. (After US Nuclear Regulatory Commission (1981)).*

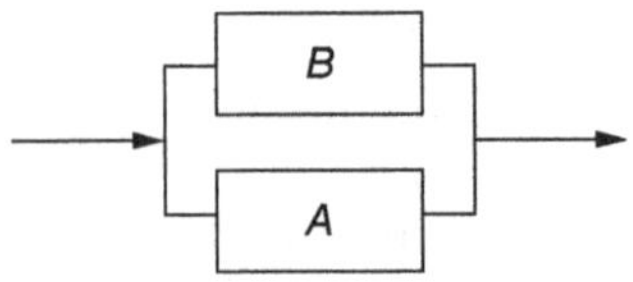

Figure A.5 *Schematic diagram for the two pumps in parallel.*

Consider a simple water pumping system (U.S. Nuclear Regulatory Commission, 1981) consisting of a water source, two pumps in parallel, a valve, and a reactor (see Figure A.4). A no flow of water to the reactor constitutes the undesired event – that is, a failure of the system.

Denote the failure of the system as the top event, T. Then we can represent this simple water pumping system as shown in Figure A.3b.

If either valve V or both pumps fail, the top event will occur – failure of the system. The two pumps are designed in parallel as discussed next.

A.2.3.3 Parallel System

When subsystems are connected in parallel (see Figure A.5), the system fails only when all of its components fail.

For the system in Figure A.5, the unreliability of the pumps in parallel is

$$\begin{aligned} Q(t) &= Q_A(t) + Q_B(t) \\ R(t) &= 1 - Q(t) = 1 - [1 - R_A(t)][1 - R_B(t)] \\ &= R_A(t) + R_B(t) - R_A(t)R_B(t) \end{aligned} \tag{A.22}$$

In general,

$$Q_s(t) \cong \prod_i Q_i(t) \tag{A.23}$$

$$R_s(t) = 1 - \prod_i Q_i(t) = 1 - \prod_i (1 - R_i(t)) \tag{A.24}$$

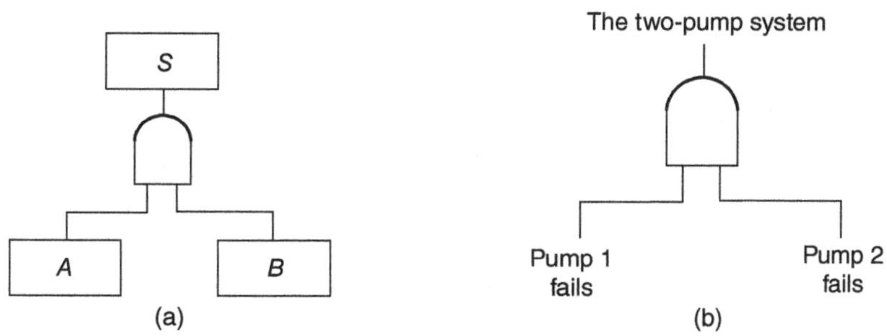

Figure A.6 *(a) AND Gate (b) AND Gate for water pumping system. (components in parallel).*

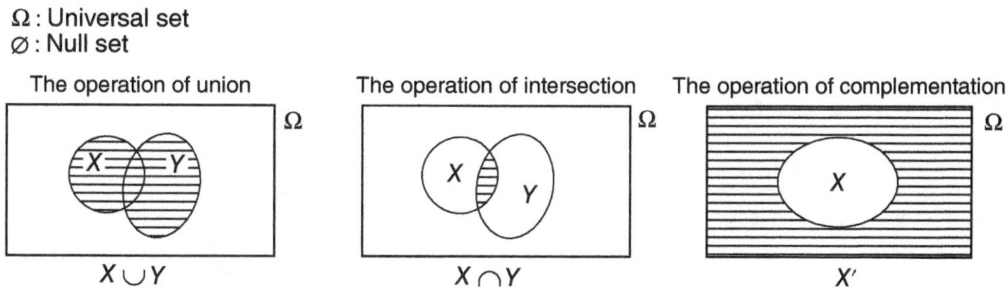

Figure A.7 *Venn diagram representation.*

$$R_s(t) > \max_i \{R_i(t)\} \qquad \text{(A.25)}$$

Note that Eq. (A.25) is correct for parallel subsystems only.

The AND Gate. The AND Gate represents the intersection of the events attached to the gate, where the components are in parallel. All of the input events must occur to cause the event above the gate to occur.

The AND Gate is equivalent to the Boolean symbol •. For example, the AND Gate with two input events (as shown in Figure A.6a) is equivalent to the Boolean expression

$$S = A \bullet B \qquad \text{(A.26)}$$

If A and B are independent events, then $P(B|A) = P(B)$ or $P(A|B) = P(A)$; therefore,

$$P(S) = P(A)P(B|A) = P(B)P(A|B) = P(AB) \qquad \text{(A.27)}$$

$$P(S) = P(AB) = P(A)P(B) \qquad \text{(A.28)}$$

The AND Gate is used to demonstrate that the output fault occurs only if all the input faults occur, as Figure A.6b illustrates.

A.2.3.4 Venn Diagram Representation of Sets

The operational rules of set theory and their graphical representation through Venn diagrams markedly simplify the complexity of fault trees. As will be demonstrated in a subsequent discussion, a system with a large number of components (subsystems) that are connected in series and parallel (through OR Gates and AND Gates) can be reduced to a simple connection through the use of operational rules of set theory. A brief review of the notation and laws of the algebra of sets is presented in Figure A.7 and Table A.1.

A.2.3.5 Boolean Algebra

Boolean algebra is the algebra of events; it is especially important in situations where events either occur or do not occur. Understanding the rules of Boolean algebra contributes toward the construction and simplification of fault trees.

Operation	Probability	Mathematics	Engineering	Symbol	Structure
Union of A and B	A or B	$A \cup B$	$A + B$		Series
Intersection of A and B	A and B	$A \cap B$	$A \bullet B$ or AB		Parallel
Complement of A	Not A	A' or $\bar{A}$	A' or $\bar{A}$		

TABLE A.1 Laws of the Algebra of Sets

Absorption law	
1a. $A \cup A = A$	1b. $A \cap A = A$
Associative law	
2a. $(A \cup B) \cup C = A \cup (B \cup C)$	2b. $(A \cap B) \cap C = A \cap (B \cap C)$
Commutative law	
3a. $A \cup B = B \cup A$	3b. $A \cap B = B \cap A$
Distributive law	
4a. $A \cup (B \cap C) = (A \cup B) \cap (A \cup C)$	4b. $A \cap (B \cup C) = (A \cap B) \cup (A \cap C)$
Operations with $\emptyset$ and Ω	
5a. $A \cup \emptyset = A$	5b. $A \cap \Omega = A$
6a. $A \cup \Omega = \Omega$	6b. $A \cap \emptyset = \emptyset$
Complementation law	
7a. $A \cup A' = \Omega$	7b. $A \cap A' = \emptyset$
8a. $(A')' = A$	8b. $\Omega' = \emptyset, \emptyset' = \Omega$
DeMorgan's theorem	
9a. $(A \cup B)' = A' \cap B'$	9b. $(A \cap B)' = A' \cup B'$

Source: From US Nuclear Regulatory Commission (1981).

Example

Show that $[(A \bullet B) + (A \bullet B') + (A' \bullet B')]' = A' \bullet B$

$= (A \bullet B)' + (A \bullet B')' + (A' \bullet B')'$ deMorgan's theorem

$= (A' + B') \bullet (A' + B) \bullet (A + B)$ deMorgan's theorem

$= (A' + (B' \bullet B)) \bullet (A + B)$ Distributive law

$= (A' + \emptyset) \bullet (A + B)$ Complementation law

$= A' \bullet (A + B)$

$= (A' \bullet A) + (A' \bullet B)$ Distributive law

$= \emptyset + (A' \bullet B)$

$= A' \bullet B$

A.2.4 Minimal Cut Sets

A minimal cut set is defined as the smallest combination of component failures, which, if they all occur, will cause the top event to occur (U.S. Nuclear Regulatory Commission, 1981).

By definition, a minimal cut set is a combination of intersections of primary events in parallel sufficient for the top event to occur (if all parallel components fail). This combination is the "smallest" combination in that all the failures in the minimal cut set are needed to occur for the top event (system failure) to occur. If any one component in the parallel combination does not occur, then the top event will not occur (by this combination).

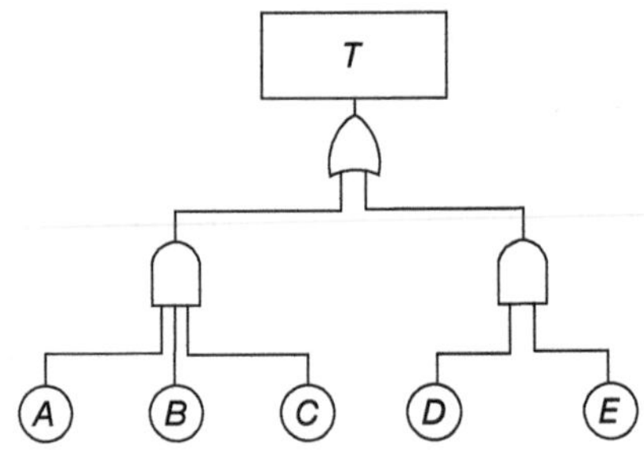

Figure A.8 *A five-component fault tree.*

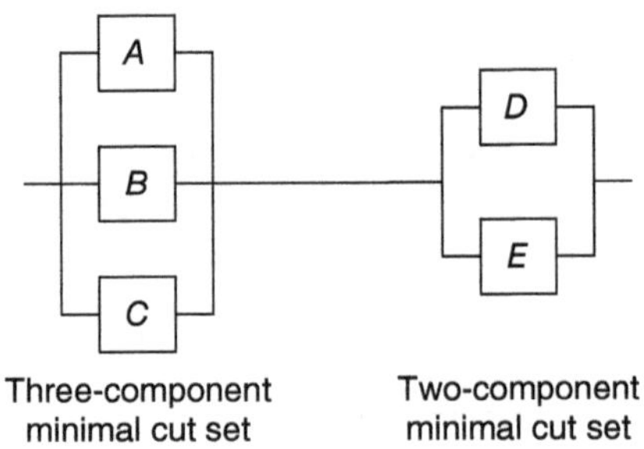

Figure A.9 *Minimal cut sets.*

A fault tree will consist of a finite number of minimal cut sets, all of which are in series, which are unique for the top event to occur. Since the combination of all minimal cut sets are in series, then the failure of any cut set will cause the failure of the entire system. *In other words, once the minimal cut sets are known, then any system can be written as the series arrangements of its cut sets, and the components of each minimal cut set are arranged in parallel.* Figures A.8 and A.9 are a representation of a two-component minimal cut set.

In sum, the one-component minimal cut set represents a single failure that will cause the top event to occur. The two-component minimal cut set represents double failures that together will cause the top event to occur. For an n-component minimal cut set, all n components in the cut set must fail in order for the top event to occur.

The general expression of the minimal cut set for the top event can be written as a combination of OR Gates (elements in series):

$$T = M_1 + M_2 + \cdots + M_k \quad (A.29)$$

where T is the top event and each M_i, $i = 1, 2, \ldots, k$, is a minimal cut set and where

$$M_i = X_i \bullet X_2 \bullet \cdots \bullet X_{n_i} \qquad (A.30)$$

and X_i are basic events that can be written as a combination of AND Gates (elements in parallel). For the fault tree in Figure A.3 (OR Gate), the minimal cut set expression is

$$T = A + B \qquad (A.31)$$

with A and B as the two minimal cut sets. Similarly, for the fault tree in Figure A.6 (AND Gate), the minimal cut set expression is

$$T = A \bullet B \qquad (A.32)$$

with $A \bullet B$ as the only minimal cut set.

Fault-tree Evaluation. Denote the unreliability of the basic event (component) by $q_j(t)$. Then the unreliability of the minimal cut set i, $Q_i(t)$, with n_i components, is given by Eq. (A.33):

$$Q_i(t) = q_1(t) q_2(t) \cdots q_{n_i}(t) \qquad (A.33)$$

The unreliability of the system (top event), $Q_s(t)$, is given by Eq. (A.34):

$$Q_s(t) \cong \sum_{i=1}^{n} Q_i(t) \qquad (A.34)$$

The fraction of system unreliability contributed by minimal cut set i, $E_i(t)$, is given by Eq. (A.35):

$$E_i(t) = \frac{Q_i(t)}{Q_s(t)} \qquad (A.35)$$

The fraction of system unreliability that is contributed by the failure of component k, $e_k(t)$, which represents the importance of component k at time t, is given by Eq. (A.36):

$$e_k(t) = \frac{\sum_{k \text{ in } i} Q_i(t)}{Q_s(t)} \qquad (A.36)$$

The importance of the minimal cut sets and of Eqs. (A.33)–(A.36) will become more evident in the specific example problems.

Example (U.S. Nuclear Regulatory Commission, 1981) Consider the fault tree given in Figure A.10. The fault tree can be constructed by following either the top-down or bottom-up approaches:

$$T = E_1 \cdot E_2;\ E_1 = A + E_3;\ E_3 = A + C;\ E_2 = C + E_4;$$
$$\text{and } E_4 = A \cdot B$$

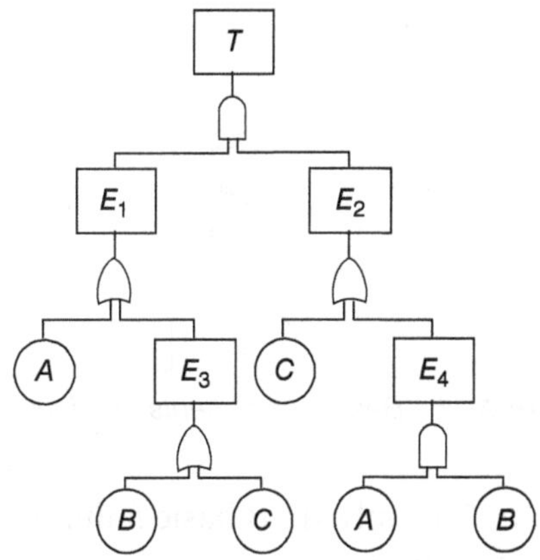

Figure A.10 *Example fault tree.*

Top-Down Approach

$$T = E_1 \cdot E_2 = (A + E_3) \cdot (C + E_4) = A \cdot C + (E_3 \cdot C) + (E_4 \cdot A) + (E_3 \cdot E_4)$$

Substituting for E_3,

$$T = (A \cdot C) + (B + C) \cdot C + E_4 \cdot A + (B + C) \cdot E_4$$
$$= A \cdot C + B \cdot C + C \cdot C + E_4 \cdot A + E_4 \cdot B + E_4 \cdot C$$

By the idempotent law $C \cdot C = C$

$$\therefore\ T = A \cdot C + B \cdot C + C + E_4 \cdot A + E_4 \cdot B + E_4 \cdot C$$

But $A \cdot C + B \cdot C + C + E_4 \cdot C = C$ by the law of absorption

$$\therefore\ T = C + E_4 \cdot A + E_4 \cdot B.$$

By substitution for E_4 and applying the law of absorption twice,

$$T = C + (A \cdot B) \cdot A + (A \cdot B) \cdot B$$
$$= C + A \cdot B + A \cdot B, \text{ note that } A \cdot B + A \cdot B = A \cdot B$$
$$= C + A \cdot B$$

The minimal cut sets of the top event are thus

$$C \text{ and } A \cdot B$$

that is, one simple-component minimal cut set and one double-component minimal cut set. The equivalent final tree is shown in Figure A.11.

Bottom-Up Approach

$$T = E_1 \cdot E_2;\ E_1 = A + E_3;\ E_3 = B + C;\ E_2 = C + E_4; \text{ and } E_4 = A \cdot B$$

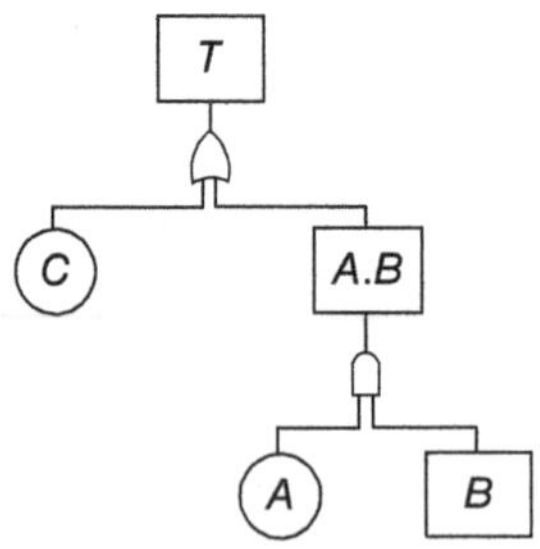

Figure A.11 *Basic components of a fault tree.*

Because E_4 has only $A \cdot B$ basic failures, we substitute into E_2 to obtain

$$E_2 = C + E_4 = C + A \cdot B$$

$$E_1 = A + E_3 = A + B + C$$

$$\begin{aligned} T &= E_1 \cdot E_2 = (A + B + C) \cdot (C + A \cdot B) \\ &= A \cdot C + A \cdot A \cdot B + B \cdot C + B \cdot A \cdot B + C \cdot C + C \cdot A \cdot B \\ &= A \cdot C + A \cdot B + B \cdot C + A \cdot B + C + A \cdot B \cdot C, \end{aligned}$$

note that $A \cdot C + B \cdot C + C + A \cdot B \cdot C = C$

by law of absorption

Thus

$$T = C + A \cdot B$$

The minimal cut sets are of two components: (i) C one-component cut set and (ii) $A \cdot B$ two-component cut set. Indeed, both the top-down and bottom-up approaches led to the same cut sets.

A.3 THE PARTITIONED MULTIOBJECTIVE RISK METHOD

Before the Partitioned Multiobjective Risk Method (PMRM) was developed, problems with at least one random variable were solved by computing and minimizing the unconditional expectation of the random variable representing damage. In contrast, the PMRM isolates a number of damage ranges (by specifying so-called partitioning probabilities) and generates conditional expectations of damage, given that the damage falls within a particular range. A *conditional expectation* is defined as the expected value of a random variable, given that this value lies within some prespecified probability range. Clearly, the values of conditional expectations depend on where the probability axis is partitioned. The analyst subjectively chooses where to partition in response to the extremal characteristics of the decision making problem. For example, if the decision maker is concerned about the once-in-a-million-years catastrophe, the partitioning should be such that the expected catastrophic risk is emphasized.

The ultimate aim of good risk assessment and management is to suggest some theoretically sound and defensible foundations for regulatory agency guidelines for the selection of probability distributions. Guidelines for the selection of probability distributions should help incorporate meaningful decision criteria, accurate assessments of risk in regulatory problems, and reproducible and persuasive analyses. Since these risk evaluations are often tied to highly infrequent or low-probability catastrophic events, it is imperative that the guidelines consider and build on the statistics of extreme events in the selection of probability distributions. Selecting probability distributions to characterize the risk of extreme events is a subject of several studies in risk management [Haimes et al., 1992, Lambert et al., 1994, Leemis, 1995, and Bier et al., 2004].

There is abundant literature that reviews the methods of approximating probability distributions from empirical data. Goodness-of-fit tests determine whether hypothesized distributions should be rejected as representations of empirical data. Approaches such as the method of moments and maximum likelihood are used to estimate distribution parameters. The caveat in directly applying accepted methods to natural hazards and environmental scenarios is that most deal with selecting the best matches for the "entire" distribution. The problem is that natural hazards and environmental assessments and decisions typically address worst-case scenarios on the tails of distributions. The differences in distribution tails can be very significant even if the parameters that characterize the central tendency of the distribution are similar. A normal and a uniform distribution that have similar expected values can markedly differ on the tails. The possibility of significantly misrepresenting potentially the most relevant portion of the distribution, the tails, highlights the importance of bringing the consideration of extreme events into the selection of probability distributions.

More time and effort should be spent to characterize the tails of distributions along with modeling the entire distribution. Improved matching between extreme events and distribution tails provides policy-makers with more accurate and relevant information. Major factors to consider when developing distributions that account for tail behaviors include (1) availability of data, (2) characteristics of the distribution tail, such as shape and rate of decay, and (3) value of additional information in assessment.

The PMRM is a risk analysis method developed for solving multiobjective problems of a probabilistic nature [Asbeck and Haimes, 1984]. Instead of using the traditional expected value of risk, the PMRM generates a number of conditional expected-value functions, termed "risk functions," that represent the risk given that the damage falls within specific ranges of the probability of exceedance. Before the PMRM was developed, problems with at least one random variable were solved by computing and minimizing the unconditional expectation of the random variable representing damage. In contrast, the PMRM isolates a number of damage ranges (by specifying so-called partitioning probabilities) and generates conditional expectations of damage, given that the damage falls within a particular range. In this manner, the PMRM generates a number of risk functions, one for each range, which are then augmented with the original optimization problem as new objective functions.

The conditional expectations of a problem are found by partitioning the problem's probability axis and mapping these partitions onto the damage axis. Consequently, the damage axis is partitioned into corresponding ranges. *A conditional expectation is defined as the expected value of a random variable given that this value lies within some prespecified probability range.* Clearly, the values of conditional expectations are dependent on where the probability axis is partitioned. The choice of where to partition is made subjectively by the analyst in response to the extreme characteristics of the problem. If, for example, the analyst is concerned about the once-in-a-million-years catastrophe, the partitioning should be such that the expected catastrophic risk is emphasized. Although no general rule exists to guide the partitioning, Asbeck and Haimes [1984] suggest that if three damage ranges are considered for a normal distribution, then the $+1s$ and $+4s$ partitioning values provide an effective rule of thumb. These values correspond to partitioning the probability axis at 0.84 and 0.99968; that is, the low-damage range would contain 84% of the damage events, the intermediate range would contain just under 16%, and the catastrophic range would contain about 0.032% (probability of 0.00032). In the literature, catastrophic events are generally said to be events with a probability of exceedance of 10^{-5} (see, for instance, the NRC Report on dam safety [National Research Council, 1985]). This probability corresponds to events exceeding $+4s$.

A continuous random variable X of damages has a cumulative distribution function (cdf) $P(x)$ and a probability density function (pdf) $p(x)$, which are defined by the relationships

$$P(x) = \text{Prob}[X \le x] \tag{A.37}$$

and

$$p(x) = \frac{dP(x)}{dx} \tag{A.38}$$

The cdf represents the *nonexceedance probability* of x. The *exceedance probability* of x is defined as the probability that X is observed to be greater than x and is equal to one minus the cdf evaluated at x.

The expected value, average, or mean value of the random variable X is defined as

$$E[X] = \int_0^\infty xp(x)\,dx \tag{A.39}$$

For the discrete case, where the universe of events (sample space) of the random variable X is discretized into I segments, the expected value of damage, $E[X]$ can be written as

$$E[X] = \sum_{i=1}^{I} p_i x_i \tag{A.40}$$

$$p_i \ge 0 \tag{A.41}$$

$$\sum p_i = 1 \tag{A.42}$$

where x_i is the ith segment of the damage.

In the PMRM, the concept of the expected value of damage is extended to generate multiple *conditional expected-value functions*, each associated with a particular range of exceedance probabilities or their corresponding range of damage severities. The resulting conditional expected-value functions, in conjunction with the traditional expected value, provide a family of risk measures associated with a particular policy.

Let $1-\alpha_1$ and $1-\alpha_2$, where $0<\alpha_1<\alpha_2<1$, denote exceedance probabilities that partition the domain of X into three ranges, as follows. On a plot of exceedance probability, there is a unique damage β_1 on the damage axis that corresponds to the exceedance probability $1-\alpha_1$ on the probability axis. Similarly, there is a unique damage β_2 that corresponds to the exceedance probability $1-\alpha_2$. Damages less than β_1 are considered to be of low severity, and damages greater than β_2 are of high severity. Similarly, damages of a magnitude between β_1 and β_2 are considered to be of moderate severity. The partitioning of risk into three severity ranges is illustrated in Figure A.12. If the partitioning probability α_1 is specified, for example, to be 0.05, then β_1 is the 5th exceedance percentile. Similarly, if α_2 is 0.95 (i.e., $1-\alpha_2$ is equal to 0.05), then β_2 is the 95th exceedance percentile.

For each of the three ranges, the conditional expected damage (given that the damage is within that particular range) provides a measure of the risk associated with the range. These measures are obtained through the definition of the *conditional expected value.* Consequently, the new measures of risk are $f_2(\cdot)$, of high exceedance probability and low severity; $f_3(\cdot)$, of medium exceedance probability and moderate severity; and $f_4(\cdot)$, of low exceedance probability and high severity. The function $f_2(\cdot)$ is the conditional expected value of X, given that x is less than or equal to β_1:

$$f_2(\cdot)=E[X\,|\,X\le\beta_1] = \frac{\int_0^{\beta_1} xp(x)\,dx}{\int_0^{\beta_1} p(x)\,dx} \tag{A.43}$$

Similarly, for the other two risk functions, $f_3(\cdot)$ and $f_4(\cdot)$

$$f_3(\cdot)=E[X\,|\,\beta_1\le X\le\beta_2]$$

$$f_3(\cdot)=\frac{\int_{\beta_1}^{\beta_2} xp(x)\,dx}{\int_{\beta_1}^{\beta_2} p(x)\,dx} \tag{A.44}$$

and

$$f_4(\cdot)=E[X\,|\,X>\beta_2]$$

$$f_4(\cdot)=\frac{\int_{\beta_2}^{\infty} xp(x)dx}{\int_{\beta_2}^{\infty} p(x)dx} \tag{A.45}$$

Thus, for a particular policy option, there are three measures of risk, $f_2(\cdot)$, $f_3(\cdot)$, and $f_4(\cdot)$, in addition to the traditional expected value denoted by $f_5(\cdot)$. The function $f_1(\cdot)$ is reserved for the cost associated with the management of risk. Note that

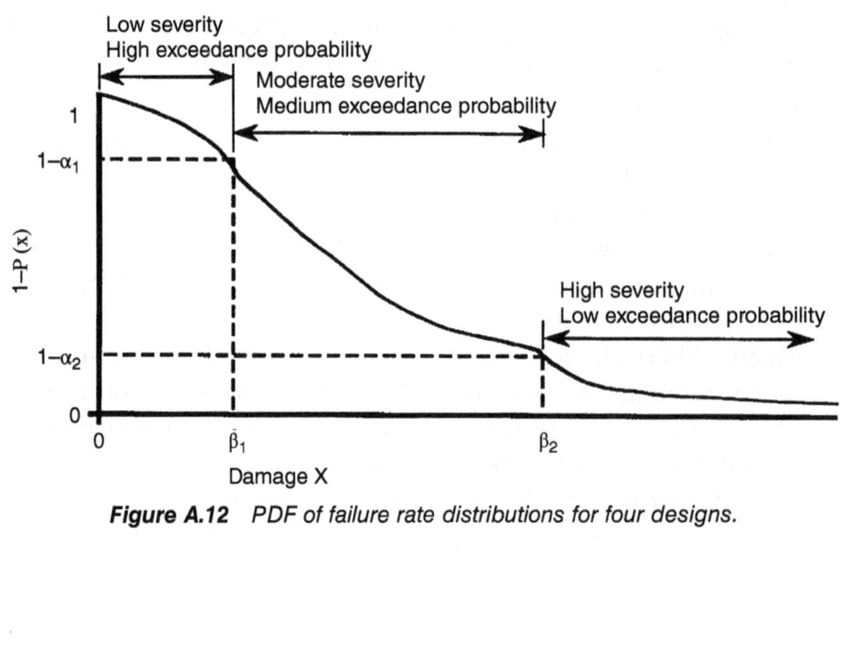

Figure A.12 *PDF of failure rate distributions for four designs.*

$$f_5(\cdot) = \frac{\int_0^\infty xp(x)\,dx}{\int_0^\infty p(x)\,dx} = \int_0^\infty xp(x)\,dx \tag{A.46}$$

since the total probability of the sample space of X is necessarily equal to one. In the PMRM, all or some subset of these five measures are balanced in a multiobjective formulation.

REFERENCES

Apostolakis, G. (1991). *Probabilistic Safety Assessment and Management*, vol. **1 and 2**. New York: Elsevier.

Asbeck, E., Haimes, Y.Y. (1984). The partitioned multiobjective risk method, *Large Scale Systems* **6**(1): 13–38.

Bellman, R.E. and Dreyfus, S.E. (1962). *Applied Dynamic Programming*. Princeton, NJ: Princeton University Press.

Bier, V.M., S. Ferson, Y.Y. Haimes, J. H. Lambert, M. J. Small, 2004, *Risk of Extreme and Rare Events: Lessons from a Selection of Approaches, Risk Analysis and Society*, Timothy McDaniels and Mitchell Small (Eds.) Cambridge, pp. 74–118.

Birolini, A. (2007). *Reliability Engineering: Theory and Practice*, 5e. New York: Springer-Verlag.

Cox, D.C. (1982). An analytic method for uncertainty analysis of nonlinear output functions, with applications to fault-tree analysis. *IEEE Transactions of Reliability* **R-31**(5): 465–462.

Ebeling, C.E. (2005). *Introduction to Reliability and Maintainability Engineering*. Long Grove, IL: Waveland Press.

Haimes, Y.Y. (1977). *Hierarchical Analyses of Water Resource Systems: Modeling and Optimization of Large-Scale Systems*. New York: McGraw-Hill.

Haimes, Y.Y., J.H. Lambert, and D. Li, (1992). Risk of extreme events in a multiobjective framework, *Water Resources Bulletin* **28**(1): 201–209.

Henley, E. and Kumamoto, H. (1992). *Probabilistic Risk Assessment: Reliability Engineering, Design, and Analysis*. New York: IEEE Press.

Hoyland, A. and Rausand, M. (1994). *System Reliability Theory, Models and Statistical Methods*. New York: Wiley.

Intrilligator, M.D. (1971). *Mathematical Optimization and Economic Theory*. Englewood Cliffs, NJ: Prentice Hall.

Johnson, B.W. (1989). *Design and Analysis of Fault-Tolerance Digital Systems*. Reading, MA: Addison-Wesley.

Kuhn, H.W. and Tucker, A.W. (1951). Nonlinear programming. In: *Proceedings of Second Berkeley Symposium on Mathematical Statistics and Probability, 1950* (ed. J. Neyman), 481–492. Berkeley, CA: University of California Press.

Lambert, J.H., N.C. Matalas, C.W. Ling, Y.Y. Haimes, and D. Li, (1994). Selection of probability distributions in characterizing risk of extreme events, *Risk Analysis*, **149**(5): 731–742.

Lasdon, L.S. (1968). Duality and decomposition in mathematical programming. *IEEE Transactions on Systems Science and Cybernetics* **SSC-4**(2): 86–100.

Lasdon, L.S. (1970). *Optimization Theory for Large Systems*. New York: Macmillan.

Leemis, M.L. (1995). *Reliability; Probabilistic Models and Statistical Methods*. Englewood Cliffs, NJ: PrenticeHall.

Limnios, N. (2007). *Fault Trees*. London: ISTE Publishing Company.

Lowrance, W.W. (1976). *Of Acceptable Risk*. Los Altos, CA: William Kaufmann.

Martensen, A.L. and Butler, R.W. (1987). *The Fault-Tree Compiler*, NASA Technical Memorandum, vol. **89098**. Hampton, VA: Langley Research Center.

NASA (1996). NASA Guidelines for Critical Software Analysis and Development. *NASA-GB-1740*. 13–96.

National Research Council (NRC) (1985). *Committee on Safety Criteria for Dams, Safety of Dams—Flood and Earthquake Criteria*, National Academy Press, Washington, DC.

Park, J.I., Lambert, J.H., and Haimes, Y.Y. (1998). Hydraulic power capacity of water distribution networks in uncertain conditions of deterioration. *Water Resources Research* **34**(2): 3605–3614.

Rao, S.S. (1992). *Reliability-Based Design*. New York: McGraw-Hill.

Russell, K.D., Snider, D.M., Sattison, M.B. et al. (1987). Integrated reliability and risk analysis system (IRRAS), User's Guide—Version 1.0.

Schneiter, C., Haimes, Y.Y., Li, D., and Lambert, J.H. (1996). Capacity reliability of water distribution networks and optimum rehabilitation decision making. *Water Resources Research* **32**(7): 2271–2278.

Storey, N. (1996). *Safety Critical Systems*. New York: Addison Wesley.

U.S. Nuclear Regulatory Commission (1981). Fault Tree Handbook. *NUREG-81/0492*.

U.S. Nuclear Regulatory Commission (1975). Reactor Safety Study: An Assessment of Accident Risks in U.S. Commercial Nuclear Power Plants, Appendix I, Accident and Use of Event Trees. *Report No. WASH-1400 (NUREG 75/014)*, October.

Author Index

Modeling and Managing Interdependent Complex Systems of Systems, First Edition. Yacov Y. Haimes.

Subject Index

Modeling and Managing Interdependent Complex Systems of Systems, First Edition. Yacov Y. Haimes.